The Impact of Environmental Factors on the Quality of Horticultural Commodities

The Impact of Environmental Factors on the Quality of Horticultural Commodities

Guest Editor

Grzegorz P. Łysiak

MDPI

Basel • Beijing • Wuhan • Barcelona • Belgrade • Novi Sad • Cluj • Manchester

Guest Editor
Grzegorz P. Łysiak
Department of Ornamental
Plants, Dendrology and
Pomology
Poznan University of Life
Science
Poznań
Poland

Editorial Office
MDPI AG
Grosspeteranlage 5
4052 Basel, Switzerland

This is a reprint of the Special Issue, published open access by the journal *Agriculture* (ISSN 2077-0472), freely accessible at: www.mdpi.com/journal/agriculture/special_issues/IH1N3SGNIZ.

For citation purposes, cite each article independently as indicated on the article page online and as indicated below:

Lastname, A.A.; Lastname, B.B. Article Title. *Journal Name* **Year**, *Volume Number*, Page Range.

ISBN 978-3-7258-5315-1 (Hbk)
ISBN 978-3-7258-5316-8 (PDF)
https://doi.org/10.3390/books978-3-7258-5316-8

Contents

About the Editor

Grzegorz P. Łysiak

Grzegorz P. Łysiak is a Polish scientist, fruit grower, pomologist, professor of agricultural and horticultural sciences, and employee of the Poznan University of Life Sciences. He has authored of numerous scientific and popular works, books, and chapters in books on the subject of growing, storing, and processing fruit. Professor Łysiak also worked at the University of Georgia in the United States and cooperated with several European universities. He developed innovative methods for determining the harvest date, using mathematical models based on continuous temperature measurement. In cooperation with domestic and foreign institutions, he conducts research on the content of biologically active compounds in fruit plant fruits. He conducts research on the impact of climate on the quality of fruit and its storability, as well as suitability for processing. He is the author and co-author of more than 220 publications, about 50 of which are listed in the Web of Science database.

The publications focus mainly on topics related to the cultivation of fruit trees, but also on the storage and processing of fruit. In addition, it publishes research on the impact of the consumption of fruits rich in biologically active substances, other parts of fruit plants, and edible flowers on the human health. In recent years, he has been studying the impact of changing climatic conditions on the chemical composition and quality of horticultural products. He is a member of several editorial boards of scientific journals, scientific councils of institutions in Poland, and several international scientific associations.

Preface

Horticultural production is one of the most intensive sectors of agriculture, which develops very dynamically. As yields increase and quality criteria increase, greater emphasis is placed on the biological balance of ecosystems. Therefore, new environmentally friendly production systems are being developed, in which growing methods and techniques are strongly based on environmental conditions. This involves breeding new cultivars with improved resistance to adverse factors, using crop rotation to limit exposure to pathogens in soil, controlling fertilization and organic mulching, stimulating biological processes in the rhizosphere, and applying biological preparations for plant protection and nutrition to improve product quality.

Just like other food products, fruit, vegetables, edible flowers, and other horticultural commodities products offered to consumers should be of good quality and, above all, safe to eat. Each step from field to table must be subject to procedures that ensure that relevant quality requirements are met. Aside from the growing method, pruning, pollination, protection and fertilization, product quality is affected by environmental conditions, the most notable of which include all macro- and microclimatic elements and soil environment. Among these conditions are air temperature, precipitation, sun radiation, and wind conditions. These factors play a significant role in shaping the quality of products intended for fresh consumption, processing, or therapeutic purposes.

Grzegorz P. Łysiak
Guest Editor

agriculture

MDPI

Editorial

The Impact of Environmental Factors on the Quality of Horticultural Commodities

Grzegorz P. Łysiak

Department of Ornamental Plants, Dendrology and Pomology, University of Life Sciences in Poznan, Dąbrowskiego 159, 60-594 Poznan, Poland; glysiak@up.poznan.pl

The Sustainable Development Goals popularised by the United Nations include achieving food security, improving nutrition, eliminating hunger and poverty, promoting sustainable agriculture, and ensuring healthy lifestyles [1]. Limited access to safe and nutrient-rich foods increases the risk of malnutrition and hunger, and it can even cause mental imbalances. Horticultural production, as a branch of agriculture, is placing increasing emphasis on the sustainable development and biological balance of ecosystems. Without losing sight of this overarching goal, measures are being taken to ensure that crops are highly productive, high-quality, and environmentally friendly. All stages of horticultural activities, including cultivation, agrotechnical activities, distribution, storage, and preparation for consumption, have a direct impact on the quality, safety, and sustainability of crops, as well as their effects on human health. So far, the main goals of horticultural activity have been to increase the volume of production and improve commercial quality. The change in approach is also due to the fact that the intensive use of chemicals in agriculture has a negative impact on the environment, biodiversity, natural resources, and human health [1]. Research currently focuses on friendly production systems in which cultivation methods and techniques are strongly based on environmental conditions.

Another element with an increasingly disruptive impact on the growing ecosystem is climate change, which can largely limit the cultivation of many horticultural species or at least necessitate measures to support plants through changes in agricultural technology, protection, fertilisation, and cultivation systems [2,3]. Currently, the cultivation systems used often do not tolerate or poorly tolerate large fluctuations in temperature, rainfall, solar radiation, and violent winds. This is compounded by the enormous pressure of pathogens, which, due to their short development period and high multiplication rate, adapt more quickly to changing conditions. Horticultural production, which is an intensive form of production, is therefore under very strong pressure from climate change. In this field, researchers are currently focusing on reducing the impact of climate change through adapting plant cultivars or even introducing new species, making changes and improvements to the cultivation technologies used [4,5], enhancing protection, and introducing new ways of storing and processing horticultural products. Natural substances are increasingly being used to biostimulate transformations in plants and promote greater accumulation of health-promoting substances [6,7].

The popularisation of organic farming and agroecological approaches is conducive to reducing environmental changes [8]. This reduces dependence on chemicals in production and the harmful effects of chemical products on humans, animals, and the environment. Assessing the sensitivity or tolerance of plants to harmful substances found in the soil or air due to human activity is also a research topic that facilitates sustainable development [9]. Such a change in approach is also expected by consumers. Desirable methods include the

Agriculture **2025**, *15*, 1829 https://doi.org/10.3390/agriculture15171829

breeding and selection of varieties with increased resistance to adverse factors, as well as activities that support the preservation of biodiversity, including the improvement of soil fertility with the use of natural resources and biofertilizers [10–12] and the stimulation of biological processes in the rhizosphere through crop rotation and organic [13,14] and living mulching [15,16]. With the aim of maintaining biodiversity and pro-environmental activities, natural biological preparations are being increasingly used for plant protection and controlled plant nutrition is being used in fertilisation. Research is also focusing on developing strategies to monitor pests and diseases, which minimises excessive pesticide use. Additionally, innovations in integrated technology increase environmental safety and balance horticultural production. Newly introduced species are of particular interest to breeders. Self-pollinating cultivars are sought after in order to obtain high yields with high-quality fruit [17]. One of the most important features of new cultivars is their qualitative assessment and storability [18]. For the storage of horticultural products, in order to maintain high quality, a controlled atmosphere (KA) in cold stores is increasingly used. Modified Atmosphere Packaging (MAP) is also used to store products, along with additional substances to delay ripening [19,20].

Monitoring and adapting to changes are also possible through the use of increasingly sophisticated risk assessment methods, growth rate assessment, and monitoring the types of changes in a plant [5,12,21]. Factors affecting the accumulation and stability of health-promoting substances in fruits and vegetables, as well as edible flowers, have been studied [22,23]. Mathematical patterns are increasingly being developed to assess growth rates and simulate possible processes or plant behaviour, increasingly supported by the use of artificial intelligence [21].

Like other food products, horticultural products offered to consumers should be high-quality and, above all, safe for consumption. Each stage from farm to fork must be subject to procedures that ensure that the relevant quality requirements are met. Fruit, vegetables, edible flowers, and other horticultural goods must meet the highest quality standards, which, according to the literature, comprise sensory quality, health-promoting properties, and their availability, as well as symbolic quality.

Horticultural production is one of the fastest changing branches of agriculture. Regardless of where it is conducted, new challenges are being faced. The main factors necessitating changes, apart from environmental considerations, are the growing requirements of consumers, who not only expect an aesthetically appealing product but also high nutritional value, an abundance of health-promoting substances, and environmentally friendly production methods. Due to the complexity of this problem, the following question arises: how can results be achieved that satisfy the producer, the consumer, and the trade? The papers published in this Special Issue attempt to address some relevant questions by modifying cultivation technology and horticultural product handling procedures at every stage, from production to their preparation for sale and consumption, ensuring that the growing demands of consumers are met to the greatest extent possible.

Conflicts of Interest: The author declares no conflict of interest.

References

1. What are the Sustainable Development Goals? United Nations Development Programme. Available online: https://www.undp.org/sustainable-development-goals (accessed on 10 August 2025).
2. Raju, C.; Pazhanivelan, S.; Perianadar, I.V.; Kaliaperumal, R.; Sathyamoorthy, N.K.; Sendhilvel, V. Climate Change as an Existential Threat to Tropical Fruit Crop Production—A Review. *Agriculture* **2024**, *14*, 2018. [CrossRef]
3. Danko, R.; Pavloušek, P.; Kapłan, M.; Klimek, K.E. Conception, Consequences and Design of Cool Climate Viticulture Training Systems. *Agriculture* **2024**, *14*, 1966. [CrossRef]

4. Małachowska, M.; Majak, T.; Krupa, T.; Tomala, K. Increasing Productivity and Fruit Quality of 'Mutsu' Apple Orchard by Dwarfing Treatments. *Agriculture* **2024**, *14*, 1838. [CrossRef]
5. Kapłan, M.; Klimek, K.E.; Buczyński, K. Analysis of the Impact of Treatments Stimulating Branching on the Quality of Maiden Apple Trees. *Agriculture* **2024**, *14*, 1757. [CrossRef]
6. Wojciechowicz-Żytko, E.; Kunicki, E.; Nawrocki, J. Influence of Biostimulants and Microbiological Preparations on the Yield and the Occurrence of Diseases and the European Corn Borer (*Ostrinia nubilalis* Hbn, Lepidoptera, Crambidae) on Sweet Corn (*Zea mays* L. Var. saccharata). *Agriculture* **2024**, *14*, 1754. [CrossRef]
7. Kunicki, E.; Capecka, E.; Wojciechowicz-Żytko, E. The Effect of Chosen Biostimulants on the Yield of White Cabbage. *Agriculture* **2024**, *14*, 1700. [CrossRef]
8. Mikiciuk, G.; Miller, T.; Kisiel, A.; Cembrowska-Lech, D.; Mikiciuk, M.; Łobodzińska, A.; Bokszczanin, K. Harnessing Beneficial Microbes for Drought Tolerance: A Review of Ecological and Agricultural Innovations. *Agriculture* **2024**, *14*, 2228. [CrossRef]
9. Antkowiak, W.; Bosiacki, M.; Sowelo, M. Potential of Selected Species of *Spiraea* L. for Phytoremediation of Heavy Metals from Soils of Urban Areas. *Agriculture* **2024**, *14*, 1916. [CrossRef]
10. Przybyłko, S.; Marszał, J.; Kowalczyk, W.; Szpadzik, E. Effect of Kaolin Clay on Post-Bloom Thinning Efficacy, Cropping, and Fruit Quality in 'Gala Vill' Apple (*Malus* × *domestica*) Cultivation. *Agriculture* **2025**, *15*, 440. [CrossRef]
11. Kozioł, A.; Paliwoda, D.; Mikiciuk, G.; Benhadji, N. Biochar as a Multi-Action Substance Used to Improve Soil Properties in Horticultural and Agricultural Crops—A Review. *Agriculture* **2024**, *14*, 2165. [CrossRef]
12. Rokosa, M.; Mikiciuk, M.; Ptak, P. Assessment of Physiological Traits of *Fragaria vesca* Genotypes Under Water Deficit Conditions. *Agriculture* **2025**, *15*, 70. [CrossRef]
13. Adamczewska-Sowińska, K.; Bykowy, J.; Jaworska, J. Effect of Biodegradable Mulch and Different Synthetic Mulches on Growth and Yield of Field-Grown Small-Fruited Tomato (*Lycopersicon esculentum* Mill.). *Agriculture* **2025**, *15*, 212. [CrossRef]
14. Adamczewska-Sowińska, K. The Use of Soil Surface Mulching on Melon (*Cucumis melo* L.) Production under Temperate Climate Conditions. *Agriculture* **2024**, *14*, 1398. [CrossRef]
15. Licznar-Małańczuk, M. Impact of High Nitrogen Doses and Living Mulch on Growth, Yield and Fruit Quality of Young Apple Trees (cv. 'Sampion'). *Agriculture* **2025**, *15*, 724. [CrossRef]
16. Licznar-Malanczuk, M.; Baluszynska, U.B. Do Living Mulches or Environmental Conditions Have a Greater Impact on the External Quality of the Apple Fruit 'Chopin' Cultivar? *Agriculture* **2024**, *14*, 610. [CrossRef]
17. Parveze, M.U.; Mir, M.M.; Rehman, M.U.; Iqbal, U.; Khan, S.Q.; Khan, F.A.; Khan, I.; Qayoom, S.; Mushtaq, I.; Shah, H.K.; et al. Regulation of crop load and quality in sweet cherry cv. 'Sweet Heart' using blossom thinning. *Folia Hortic.* **2024**, *36*, 311–321. [CrossRef]
18. Małachowska, M.; Tomala, K. Apple Quality during Shelf-Life after Long-Term Storage and Simulated Transport. *Agriculture* **2023**, *13*, 2045. [CrossRef]
19. Kowalczyk, B.A.; Bieniasz, M.; Błaszczyk, J. Effect of Short-Term Storage in Modified Atmosphere Packaging (MAP) and Controlled Atmosphere (CA) on Total Polyphenol Content and Antioxidant Activity in Juices from Haskap Berry (*Lonicera caerulea* L.). *Agriculture* **2025**, *15*, 253. [CrossRef]
20. Małachowska, M.; Tomala, K. Effect of Preharvest and Postharvest Application of 1-MCP on the Quality of Gala Schniga® SchniCo Red(s) Apples during Long-Term Storage. *Agriculture* **2022**, *12*, 2073. [CrossRef]
21. Łysiak, G.P.; Szot, I. The Use of Temperature Based Indices for Estimation of Fruit Production Conditions and Risks in Temperate Climates. *Agriculture* **2023**, *13*, 960. [CrossRef]
22. Hęś, M.; Antkowiak, W.; Stuper-Szablewska, K.; Dziedzic, K.; Jessa, M.; Ratajczak, P. Assessment of the Antioxidative Properties of Extracts from the Fruits of *Pyrus pyraster* (L.) Burgsd and *Pyrus* × *myloslavensis* Czarna & Antkowiak Grown under Natural Environmental Conditions. *Agriculture* **2024**, *14*, 1790. [CrossRef]
23. Aninbon, C.; Surach, N.; Phungam, N.; Janket, A.; Bunphan, D.; Charirak, P.; Ruttanaprasert, R. Stability of anthocyanin, carotenoid and root yield of 10 sweet potato genotypes. *Folia Hortic.* **2025**, *36*, 475–490. [CrossRef]

agriculture

MDPI

Article

Impact of High Nitrogen Doses and Living Mulch on Growth, Yield and Fruit Quality of Young Apple Trees (cv. 'Sampion')

Maria Licznar-Małańczuk

Department of Horticulture, Faculty of Live Sciences and Technology, Wroclaw University of Environmental and Life Sciences, Grunwaldzki 24a, 50-363 Wroclaw, Poland; maria.licznar-malanczuk@upwr.edu.pl

Abstract: The experiment aimed to evaluate the effectiveness of using high doses of nitrogen (N) and delaying the sowing of living mulch in the apple tree (*Malus domestica* Borkh.) orchard as factors stimulating tree growth and influencing fruit quality. Blue fescue (*Festuca ovina* L.), as a cover crop, was sown in the tree rows in the second and fourth year after tree planting. It was compared to herbicide fallow control treatment. In each floor management, four N doses (50, 80, 110, and 140 kg ha^{-1}) were used every year. The study indicated the dominance of living mulch, especially sowing in the second year after tree planting as a factor influencing the reduction in growth and yield but improving the red blush surface on the fruit skin. The growth of trees was also strongly influenced by biotic and abiotic stress factors, which did not allow for a clear assessment of N fertilization. With objectively low cropping levels and poor growth of all trees tested in the experiment, their N needs were satisfied by a dose of 50 kg ha^{-1}. The application of the highest dose of 140 kg N ha^{-1} had a significant and negative impact on the red color of the fruit skin.

Keywords: cover crop; fescue; nitrogen; apple tree; red blush; size of fruit; orchard management; sustainable agriculture

1. Introduction

Fruit tree cultivation with cover crops is recommended for organic production [1–3] whose importance in European large-scale apple production is still small, with the exception of Italy [4]. This ecological weed management [5,6], depending on the selection of cover crop, limits the presence of weeds to a varying degree [7,8]. In the orchard, living mulch often replaces herbicide fallow [9–11]. At the same time, it has a positive effect on the biodiversity of flora [12], affects soil properties [13–16], and determines its microbial communities [17,18]. Living mulch primarily plays the role of a 'catch crop', preventing the leaching of nutrients, including N [19,20]. This may affect the N supply to trees because greater availability of this nutrient in orchard ensures floor management with free vegetation soil surface in the rows, e.g., herbicide fallow [21].

N has a clear stimulating effect on the shoot growth of apple trees and the surface of leaves [22]. Increasing the N dose did not have an effect on other tree growth parameters, such as trunk cross sectional area [23,24]. Stronger shoot growth is negatively correlated with the concentration of phenolic compounds in the apple leaves [22] and, in case of the apple skin, phenolics belonging to the group of anthocyanins. They are responsible for the red blushing of fruit [25]. On the one hand, agrotechnical methods ensuring a greater availability of N for trees affect the decrease in the red color of apple fruits [26]. On the other hand, the proper choice of production methods may have a positive impact on the phenolic synthesis by plants [27–30]. Such technologies include replacing chemical weed control with living mulch because competitive N uptake by living mulch restricts its availability to trees [31].

The presence of green mulch has a positive impact on the red skin color of apples [32,33] or other fruits [34,35]. Apples from a young orchard with living mulch had a red blush area of about 70–80% [36,37], compared to the herbicide fallow where it reached less than 30% of the skin surface [37]. The positive effect of living mulch on fruit coloration unfortunately does not apply to other fruit quality characteristics. The cover crop may contribute to a reduction in the size [37,38] or other parameters of fruit [39]. Even with low tree yields, in young orchard with living mulch, the share of apples with a diameter below 7.5 cm increased, while the 7.0–8.0 cm or above classes decreased [37,38].

Different floor managements, including living mulch, affect the availability of N to trees to varying degrees [19,20,40]. Compared to herbicide fallow, the overall N balance in grass living mulch can be positive. This may be due to greater N recycling of groundcover biomass and soil mineralization [20]. Application of legume living mulch can guarantee good N nutrition of trees [37,41]. Nevertheless, living mulch in the orchard is perceived as a competitive cover plant for fruit trees, which is associated with decrease in the tree growth or N leaf status, or both [37,42–44]. Its presence often significantly reduces yield compared to the crop in a herbicide or mechanical fallow [37,40,44].

A good supply of N to trees is ensured more by the use of annual fertilization than by proper orchard floor management [40]. The problem of N fertilization was revealed by older studies conducted on vigorous apple trees in a less intensive orchard with 550 trees ha^{-1} and living mulch. Compensation for competition between the trees and cover crop sod required higher N rates than 166 kg N ha^{-1} [45]. Increased foliar N concentration was possible at a very high rate of 180 kg N ha^{-1}. This dose of N, compared to a three-times lower dose, ensured a high N concentration (over 24.0 g kg^{-1}) in leaves when white clover or mixed grasses mulch replaced other floor managements [46]. Direct use of these recommendation is not possible in a modern apple orchard, with several times more dwarf trees ha^{-1}. Methodology of integrated apple production currently recommends an N dose not exceeding 80 kg N ha^{-1} of fertilizing soil surface [47,48]. These recommendations apply to trees maintained in herbicide fallow or mechanical cultivation in the tree rows. To what extent the N dose can be increased remains an open question in an orchard where living mulch replaces chemical weed control.

The aim of the present study was the assessment of high doses of N as a factor mitigating the competition between living mulch and apple tree 'Sampion' cv. and the impact of evaluated factors on the increase red blush on the fruit area and counteracting the decrease in the size and weight of apple.

2. Materials and Methods

2.1. Experimental Site and Design

The research work was carried out in 2015–2020 at the Fruit Experimental Station in Samotwor (51°06′12″ N, 16°49′52″ E), which belongs to the Wrocław University of Environmental and Life Sciences, in Poland. Unbranched one-year-old apple trees with an average trunk diameter 1.15 cm were planted on haplic Luvisol, in spring 2015. 'Sampion'—an important apple cultivar in middle Europe [4]—was used, grafted onto a semi-dwarf M.26 rootstock. The distance between rows in the orchard was 3.5 m and between trees in a single row of 1.2 m, which allowed 2380 trees ha^{-1} to be obtained.

An experiment was established following a two-way randomized block design, with four replications. In each replication, a single treatment was represented by a plot (6 m^2) with 5 apple trees. The first experimental factor was a orchard floor management system: herbicide fallow and living mulch. Additionally, two terms of sowing seeds of living mulch were used, which were performed in the second (2016) and in the fourth (2018) year after trees planting. Herbicide fallow, as a control was treated with a mix of glyphosate (1.44–1.96 kg ha^{-1}) and MCPA (2-methyl-4-chlorophenoxyacetic acid, 0.60–1.00 kg ha^{-1}). Their application was in spring (April or May), summer (July), and at the end of vegetation period in some years. Herbicide fallow was also applied until both living mulches were sown. The perennial living mulch was the blue fescue. The cultivars 'Noni' and 'Ridu' were sown (ratio 1:1, 50 kg seeds ha^{-1}). The grass sod was mowed with a string trimmer (two or three times per year). Treatment with later sowing of living mulch, in the year 2018, was introduced to assess the potential for delaying the development of competition between cover crop sod and the apple trees.

The second factor in the experiment was a dose of N fertilization: 50, 80, 110, and 140 kg N ha^{-1}. The recommendations from the then applicable methodology of integrated production [47] for apple orchards with herbicide fallow were taken into account when determining the N dose. On this basis, two doses of 50 and 80 kg N ha^{-1} were recommended. Two over-doses of 110 and 140 kg N ha^{-1} also were chosen. Compared to existing studies on N management in the orchard, the dose of 140 kg N ha^{-1} exceeds the N doses used in the experiments in which herbicide fallow was applied in the tree rows [23,49]. It was introduced due to the presence of an additional plant in the orchard—cover plant. It is expected that N uptake by cover crop will prevent the negative impact of a very high dose on the agricultural environment. Similarly, the use of only living mulch will eliminate the negative effects of agrochemicals in weed control, such as groundwater pollution and others that were reviewed by Mia et al. [50].

2.2. *Agrotechnical Methods in Experiment*

Before planting the trees, a standard fertilization dose of 100 kg P_2O_5 ha^{-1} and 200 kg K_2O ha^{-1} was applied. The soil was not limed. In the year of planting the trees (2015), in the spring all trees were fertilized with ammonium nitrate equally (15 g N tree^{-1}). The apple tree canopies were trained into the slender spindle. The pruning and plants protection were in accordance with the current recommendations [48]. Hand thinning of fruit sets was performed manually in the years 2016 and 2019. Irrigation interventions were applied using a relocated overhead irrigation system. Irrigation was carried out twice in the summer of 2019 and once in 2020, using a rate of water approximately 5 L tree^{-1}.

2.3. *Evaluation of Tree Growth, Yield and Fruit Quality*

Number of all annual shoots were counted, and their length were measured on two selected trees in each plot, in the autumn of 2015 and 2016. The measurements of two trunk diameters, north–south and east–west direction, were made 30 cm above the grafting point of each tree, every year. Based on these, the trunk cross-sectional area (TCSA) and its annual increment were calculated, from autumn 2015 until 2020.

Three interior trees, out of five trees in each plot, were assessed for the number of flower clusters and fruit sets (2016–2018) and the number of fruits during the harvest and yield quantity (2016–2020). The crop efficiency coefficient (CEC) was computed as a ratio of the total yield of five years (2016–2020) and TCSA in autumn 2020. All collected apples were sorted manually into three groups: >75%, 25–75%, and <25% of red blush on the skin surface. The same apples were classified according to fruit diameter: <6.5; 6.5–7.5; 7.5–8.5; and >8.5 cm.

2.4. *Estimation of Nutritional Status of Trees*

Samples of 25 leaves from the midparts of extension shoots of three apple trees, on each plot were collected in September in the 2018 and 2020 years. Due to the fact that the sowing date of the living mulch in the fourth year after the establishment of the orchard was mid-May 2018, treatments covered by the early developing sod were omitted in the assessment of macronutrients in the 2018 year. Leaf samples were weighed, dried, once more weighed, and then grounded. On this basis, the mean fresh and dry weight of one leaf was determined. The analyses were performed at the Environmental Quality Analysis Center of the Wrocław University of Environmental and Life Sciences in the 2020 year. Leaf N concentration was determined with the automated method using an elemental CHNS analyzer [51]. Dried apple leaves were thoroughly mixed with a spoon in a paper bag and then weighed in the range of 6.1–8.7 mg using a Mettler Toledo AT21 analytical balance (Mettler-Toledo GmbH, Greifensee, Switzerland) with an accuracy of 0.01 mg. The sample was placed in a tin capsule, and total N was directly determined using a CHNS analyzer (CHNS EA 1110, CE Instruments, Wigan, UK). The sample was heated in the presence of oxygen in a quartz tube at 1030 °C, followed by combustion in the presence of tungsten trioxide as a catalyst. The combustion products were separated on a chromatographic column and measured using a thermal conductivity detector. The peak area was converted to a concentration based on a calibration curve, using EAGER 200 software, version 1:00.

Concentrations of magnesium (Mg), potassium (K), and phosphorus (P) were determined by Inductively Coupled Plasma Optical Emission Spectrometry—ICP-OES [52]. Dried and grounded apple leaves were thoroughly mixed with a spoon in a paper bag, and then weighed in the range of 0.20–0.26 g using a Sartorius AG Göttingen AC 120 S analytical balance (Sartorius, Göttingen, Germany), with an accuracy of 0.1 mg. The samples were transferred to an XP1500 Teflon vessel for mineralization. To the vessel, 2 mL of water and a mixture of aqua regia (6 mL of concentrated hydrochloric acid, 37%, and 2 mL of concentrated nitric acid, 65%) were added, and the samples were left under a lid for 24 h. Subsequently, the vessels were placed in a MARS 5 microwave mineralizer (CEM Corporation, Matthews, NC, USA) on the DRY 1.0 program, which consisted of two steps: step 1—600 W, 60% power for 20 min, and step 2—600 W, 90% power for 10 min. After cooling, the mineralizate was filtered into a 25 mL Class A volumetric flask.

The prepared sample was then analyzed using an ICP-OES inductively coupled plasma spectrometer (ICAP 7400, Thermo Fisher Scientific, Waltham, MA, USA). The sample solution was drawn using a peristaltic pump and delivered to the nebulizer. The sample, in aerosol form, was transported by argon to the plasma core, where the atoms were excited. Upon returning to their ground state, the atoms emitted electromagnetic radiation, which was directed to the spectrometer, where it was dispersed in the optical system. The optical signal was then redirected to the detector, where it was converted into a digital signal. The concentrations of the selected elements were determined based on the following spectral lines: Mg—285.213 nm (in the axial position), P—185.942 nm (in the axial position), and K—766.490 nm (in the axial position).

The interpretation of the macronutrient concentration in apple leaves was made according to Sobiczewski [48].

2.5. Weather Conditions

The average annual temperature was 9.9 °C and the sum of rainfall was 534.7 mm during last two decades (2001–2020) at the Fruit Experimental Station near by Wrocław city. Heavy hail damaged the leaves and bark on the trunks and branches in 2015. This has caused the slower growth and development of young trees in the first years after the orchard establishment. In 2017, due to average hail damage to the fruit, apple quality assessment was omitted. Detailed information on average monthly air temperatures and precipitation totals is provided in Table S1.

2.6. Statistical Processing of Data

Statistical analysis was performed using the two-way ANOVA for randomized block design, in R Studio software (version 4.3.2). Significant differences were marked * at $p \leq 0.05$, ** at $p \leq 0.01$, *** at $p \leq 0.001$. For the data describing the share of red blush color and fruit size classes (expressed as a percentage) that do not fulfill the requirement, a normal distribution was transformed angularly by the function of Bliss. The exponential transformations were also used in the case of non-normal distribution of remaining data. Due to the death of some trees, the measurement set was balanced using the missing plot technique procedure. Both in cases when a significant ANOVA result was obtained, as well as when it was not obtained, the variables were subjected to further post hoc testing: Tukey procedure. Significant differences between means were calculated with a Tukey test at $p = 0.05$.

3. Results and Discussion

3.1. Growth of Trees

Total number and length of annual shoots were not significantly different on the young trees in herbicide fallow and living mulch (Table 1). However, it is worth noting that the tested population of trees in herbicide fallow was smaller than in living mulches, due to dead trees. This was most probably caused by the presence of more grubs in the soil under herbicide. This could have influenced the final result of the experiment regarding the growth of trees. Shoot length (2015–2016) was about 70 cm shorter in the herbicide fallow compared to living mulch. This is in contradiction with other experiments in which the significant reduction in total shoot growth [38] or shoot extension [44] were noted in the apple tree orchard with grass living mulch. In the case of trunk cross sectional area assessment, our results were more transparent (Table 1, Figure S1). They are also consistent with predictions, based on previous reports [37,40,44], which showed a reduction in the trunk increase or of the cross-sectional area of the trunk due to the presence of cover crops in the orchard [32,40].

In our experiment, the number and total length of shoots tended to grow when the N dose also increased (Table 1). However, there were no significant differences between treatments, which is similar to the increase in trunk cross sectional area. The application of 140 kg N ha^{-1}, compared to 50 kg significantly affected only the weight of fresh and dry leaf, in the last year (2020) of the study (Table S2). In the pot experiment, the leaf area of the apple tree increased with the increase of the N fertilization dose [22]. In field experiments in an apple orchard with herbicide fallow, varying the N dose usually did not result in a significant effect on the tree growth [23,24]. Additionally, it is worth noting that in our experiment mean leaf weight from trees in the herbicide fallow was always significantly heavier compared to two living mulch treatments.

Table 1. Growth of the apple tree 'Sampion' cv. depending on the floor management and dose of N in the tree row, in the years 2015–2020 (n = 4).

Floor Management and Dose of N (kg ha^{-1})		Total Number of Annual Shoots			Total Length of Annual Shoots 2015–2016 (cm·Tree^{-1})	Trunk Cross Sectional Area (cm^2)			% of Dead Trees Up to the End of 2020 °
		2015	2016	Total 2015–2016		2015	2020	Incerase 2015–2020	
Means for interaction: N dose (D) × floor management (FM)									
HF	50	9 a	18 a	27 a	242 a	1.20 a	7.91 a	6.71 a	58
	80	9 a	26 a	35 a	406 a	1.37 a	5.57 a	4.20 a	25
	110	10 a	25 a	35 a	466 a	1.39 a	7.28 a	5.90 a	8
	140	10 a	26 a	36 a	379 a	1.27 a	5.59 a	4.32 a	8
LM 2	50	11 a	26 a	37 a	424 a	1.31 a	4.60 a	3.29 a	25
	80	10 a	25 a	35 a	388 a	1.33 a	4.95 a	3.61 a	–
	110	11 a	29 a	40 a	433 a	1.35 a	5.17 a	3.82 a	8
	140	9 a	29 a	38 a	513 a	1.50 a	5.99 a	4.49 a	–
LM 4	50	–	–	–	–	1.36 a	5.02 a	3.66 a	25
	80	–	–	–	–	1.18 a	4.56 a	3.38 a	33
	110	–	–	–	–	1.35 a	4.59 a	3.24 a	–
	140	–	–	–	–	1.34 a	5.65 a	4.31 a	–
Statistical significance D × FM		NS	NS	NS	NS	NS	NS	NS	–
Means for N doses (D)									
50		10 a	22 a	32 a	333 a	1.29 a	5.84 a	4.55 a	36
80		10 a	26 a	36 a	397 a	1.29 a	5.03 a	3.73 a	20
110		11 a	27 a	38 a	450 a	1.36 a	5.68 a	4.32 a	5
140		10 a	28 a	38 a	446 a	1.37 a	5.74 a	4.37 a	3
Statistical significance		NS	NS	NS	NS	NS	NS	NS	–
Means for floor managements (FM)									
HF		10 a	24 a	34 a	373 a	1.31 a	6.59 b	5.28 b	25
LM 2		10 a	27 a	37 a	440 a	1.37 a	5.18 a	3.80 a	8
LM 4		–	–	–	–	1.31 a	4.96 a	3.65 a	15
Statistical significance		NS	NS	NS	NS	NS	*	**	–

°—without statistical evaluation; HF—herbicide fallow; LM 2—living mulch sowing in the second year after trees planting (2016); LM 4—living mulch sowing in the fourth year after trees planting (2018); *, **, ***—significantly different (ANOVA) at $p \leq 0.05$, $p \leq 0.01$ and $p \leq 0.001$, respectively; NS—not significantly different. Means marked with different letters in each separate columns represent statistical differences among eight treatments (D × FM) or four N doses (D), or three floor managements (FM) (Tukey test, $p = 0.05$).

3.2. N Status of Tree

Leaf N concentration in 2018 was high in all treatments (>24.0 g kg^{-1}) [48] and was significantly affected only by floor management (Table 2). On the contrary, two years later, it showed a significant effect only on the N dose. In 2020, the N concentration remained at a high (>24.0 g kg^{-1}) [48] or optimal (21.0–24.0 g kg^{-1}) [48] level. It decreased to a low (18.0–20.9 g kg^{-1}) [48] level only in the herbicide fallow treatment with fertilization of 50 kg N ha^{-1}.

Table 2. Concentration of macronutrients in leaves of the apple tree 'Sampion' cv. depending on the floor management and dose of N in the tree row in the years 2018 and 2020 (n = 4).

Floor Management and Dose of N (kg ha^{-1})		Macronutrient (g kg^{-1})							
		N		Mg		K		P	
		2018	2020	2018	2020	2018	2020	2018	2020
Means for interaction: N dose (D) × floor management (FM)									
HF	50	28.0 a	20.5 a	2.0 a	1.0 a	7.3 ab	4.6 a	1.8 ab	1.1 a
	80	28.0 a	24.0 ab	2.0 a	1.6 bc	6.4 a	8.8 b	1.4 a	1.7 ab
	110	29.0 a	24.8 ab	2.1 a	1.8 c	6.9 a	8.6 b	1.9 ab	1.6 ab
	140	30.5 a	24.5 ab	2.0 a	1.6 bc	6.8 a	8.9 b	1.4 a	1.5 ab
LM 2	50	24.3 a	24.3 ab	1.9 a	1.9 c	9.6 c	11.4 b	4.1 c	4.0 d
	80	24.8 a	24.3 ab	2.0 a	1.8 c	9.2 bc	10.9 b	3.4 bc	3.0 cd
	110	26.3 a	27.0 b	2.0 a	1.9 c	8.4 abc	10.3 b	2.3 abc	2.6 bc
	140	27.5 a	23.5 ab	2.0 a	1.7 bc	7.7 abc	10.2 b	3.1 abc	2.6 bc
LM 4	50	–	21.0 ab	–	1.2 ab	–	8.8 b	–	3.5 cd
	80	–	22.5 ab	–	1.6 abc	–	10.3 b	–	3.6 cd
	110	–	24.8 ab	–	1.6 abc	–	11.0 b	–	3.4 cd
	140	–	24.5 ab	–	1.6 bc	–	10.7 b	–	2.6 bc
Statistical significance D × FM		NS	NS	NS	*	NS	**	NS	**
Means for N doses (D)									
50		26.1 a	21.9 a	1.9 a	1.4 a	8.4 a	8.3 a	2.9 a	2.9 b
80		26.4 a	23.6 ab	2.0 a	1.7 b	7.8 a	10.0 b	2.4 a	2.8 b
110		27.6 a	25.5 b	2.0 a	1.8 b	7.6 a	10.0 b	2.1 a	2.5 ab
140		29.0 a	24.2 ab	2.0 a	1.7 b	7.2 a	10.0 b	2.2 a	2.2 a
Statistical significance		NS	**	NS	**	NS	**	NS	*
Means for floor managements (FM)									
HF		28.9 b	23.4 a	2.0 a	1.5 a	6.8 a	7.7 a	1.6 a	1.5 a
LM 2		25.7 a	24.8 a	2.0 a	1.8 b	8.7 b	10.7 b	3.2 b	3.0 b
LM 4		–	23.2 a	–	1.5 a	–	10.2 b	–	3.3 b
Statistical significance		**	NS	NS	***	***	***	***	***

HF—herbicide fallow; LM 2—living mulch sowing in the second year after tree planting (2016); LM 4—living mulch sowing in the fourth year after tree planting (2018); *, **, ***—significantly different (ANOVA) at $p \leq 0.05$, $p \leq 0.01$ and $p \leq 0.001$, respectively; NS—not significantly different. Means marked with different letters in each separate columns represent statistical differences among eight treatments (D × FM) or four N doses (D), or three floor managements (FM) (Tukey test, $p = 0.05$).

Compared to herbicide fallow, the N concentration in leaves (2018) was significantly lower in living mulch, sown already in the second year after the orchard establishment (Table 2). Despite this, it still remained at a high level as in the experiment with young apple trees of the 'Elshof' cv. with grass living mulch [38]. It did not decrease to a low or deficit level like in the young orchard of 'Ligol' cv. [37]. Two years later (2020), apple trees grown in living mulch, sown in the second year after their planting, still had a high N concentration in leaves. The optimal level was observed in the trees in herbicide fallow and in living mulch sown in the fourth year. However, the N concentration differences among these floor managements were not significant.

In 2020, the N concentration in leaves was primarily significantly influenced by the applied fertilizer dose (Table 2). It increased with the use of larger doses. A significantly higher concentration was shown only with the use 110 kg ha^{-1}, compared to 50 kg. This trend was transparent on trees grown in herbicide fallow, which is consistent with other studies on fertilization in apple orchards [26,49,53]. Hogue et al. [40] suggested that N tree status is mostly determined by annual N fertigation instead of by floor management. Other experiments confirmed that, because of a high organic fertilization rate corresponding to about 110 kg N ha^{-1}, foliar N concentration remained at the same level regardless of whether mechanical tillage or legume living mulch was used in the tree rows [19]. These results are consistent with our studies, which emphasize the importance of N dose selection rather than floor management as a factor regulating N supply to trees. It is worth emphasizing, however, that too small an N supply may be insufficient to overcome the competition of cover crops in the apple orchard [44]. The relatively weak growth of all trees in our experiment (Table 1) resulted in a low N demand of trees. This is probably why the interaction of the studied factors was often insignificant (Table 2).

3.3. Mg, K, and P Status of Tree

The two studied factors and their interaction had no effect on Mg concentration in the leaf in 2018, probably because it remained on low level (1.8–2.1 g kg^{-1}) [48] in all studied treatments (Table 2). Two years later, the significance of statistical analysis was shown; however, its meaning was reduced by the fact that the Mg concentration was already deficient (<1.8 g kg^{-1}) [48]. Such a low level of Mg was not recorded in apple leaves in other experiments that evaluated living mulch [37,40,54]. The leaves of the apple trees grown in herbicide fallow and fertilized with various doses of N, also in other studies, did not contain such little concentration of this macronutrient [49,53]. The deficiency of this element has serious consequences for the tree growth and yield [55].

In 2018, the concentration of K in the leaves was often low (7.0–9.9 g kg^{-1}) [48] and in the case of some treatments in herbicide fallow deficit (<7.0 g kg^{-1}) [48] (Table 2). Two years later (2020), apple trees in the herbicide fallow often had low K levels. Significantly higher levels, at an optimal level (10.0–15.0 g kg^{-1}) [48], were recorded when living mulch was present.

Leaf K concentration was significantly affected by floor management and, only in 2020, the N dose and the interaction of both factors (Table 2). It was in contrary to the study of Hogue et al. [40], in which the use of different living mulches and varying N supply had no effect on the K concentration in apple leaves. Similarly, a double N dose for apple trees maintained in the herbicide fallow had no impact on K leaf status [49,53].

The concentration of P significantly varied depending on estimated factors (Table 2). Similar to K, it was significantly affected by floor management, N dose, and the interaction two factors, but only in 2020. The concentration of this macronutrient was optimal (1.5–2.6 g kg^{-1}) [48] in herbicide fallow and high (>2.6 g kg^{-1}) [48] in living mulches, in 2018 and 2020. It confirmed other study reports that the presence of cover plants in the orchard stabilizes P status of the tree [37]. In 2020, its concentration was significantly higher in all living mulch with different N dose treatments, compared to the herbicide fallow with 50 kg N ha^{-1}.

3.4. Yield of Tree and Crop Efficiency Coefficient

Total yield was significantly influenced by floor management (Table 3). The expected effect of the living mulch presence was a significant reduction in yield in the living mulch sown in the second year after trees planting. Our studies confirmed earlier observations in young apple orchards with grass living mulches [37,38,40,56]. Delaying grass living mulch seeds sowing until the fourth year after tree planting also significantly reduced yield compared to herbicide fallow. The competition of grass sod with apple trees was particularly evident after two years of the cover crop presence (Figure S2). The year 2018 concerned trees cultivated in living mulch, which was sown in the second year after the orchard establishment. Analogously, in 2020, it was observed on trees that have been grown in living mulch since the fourth year after their planting.

The presence of grass sod in the orchard was probably the direct cause of the high yield variability between the study years. The especially drastic decrease in the number of fruit sets and fruit observed on trees in 2018 was due to the competition of sod already sown in the second year (Table S3). Abiotic factors, e. g. high mean temperature in May 2018 (Table S1), which increased evapotranspiration and reduced water availability for the tree, also influenced the formation of very few fruit sets. This affected the lowest yield of the tree (Table 3). A low Mg and K level in leaves (Table 2) also contributed to the significant decrease in total yield of the tree in both living mulches. However, the yield level in herbicide fallow was also relatively low for a highly valued 'Sampion' cv. in Central Europe [4].

The unexpectedly low yield of all experimental trees was the result of biotic and abiotic factors shaping the growth and cropping of apple trees. Unbranched one-year old trees were planted in the experiment. The use of such nursery plant material delays the entry of trees into the fruiting period [57]. In the year of planting (2015) as well as in 2018–2019, very low rainfall was recorded (Table S1). Unfavorable water conditions at high air temperatures caused spring or summer drought periods, noticeable in cultivation without irrigation. In addition, the trees suffered severe damage to the leaves and bark of the trees in the year of their planting (2015), as a result of strong hail in July. Grubs appeared in the experiment causing damage to the root system or even the death of trees (Table 1). Trees with damaged root systems were weaker. Additionally, once more it is important to mention that leaf analyses showed a proper N supply, but the trees already had low Mg concentrations in 2018, and at the end of the study (2020) they were most often deficient. This nutrient, present in chlorophyll, serves important biochemical functions in the plant, and its serious deficiency can cause a reduction in canopy photosynthesis and fruit size [55].

On the one hand, it can be assumed that the low annual cropping level of all trees in the experiment depended on the complex of the above-mentioned biotic and abiotic factors. Multi-interaction impact could also contribute to the insignificant effect of different N doses on total yield and crop efficiency coefficient (Table 3). On the other hand, in reports focused on apple trees in herbicide fallow, increasing the N dose of fertilization [23,24] or fertigation [26] also did not have a significant effect on total yield.

The presence of grass sod under trees, irrespective of four tested N doses, significantly reduced the total yield compared to herbicide fallow with 110 kg N ha^{-1}. Such difficulties in eliminating the competition of living mulch for fruit trees are confirmed by the study of Tahir et al. [44]. In our experiment, the potential of grass sod to uptake N was probably higher than its importance in N cycling. Unfortunately, our study did not include soil testing or water availability after several years of cultivation. In addition, equal N fertilization of the entire plot surface favored good N uptake by the grass population. Not only the fruit tree but the cover crop was privileged. This also could have determined the insignificant importance of N doses for apple trees and the failure of different N doses as a factor mitigating living mulch competition in the orchard.

3.5. Red Blush, Mean Weight, and the Size of Fruit

Good fruit coloration was noted every year (Figure 1). The share of yield with sporadic red blush that did not exceed 25% of the skin surface was marginal. It appeared only in the first and last year of the study. The largest share of yield with the best colored apples was recorded in 2018. In this year, in September the precipitation was low (Table S1). During the apple ripening period, more days with minimum air temperature below 5 °C were recorded compared to the following years of the study [56]. Such thermal and water conditions favor the red blushing of apples [58].

Table 3. Yield, fruit number, and the crop efficiency coefficient of the apple tree 'Sampion' cv. depending on the floor management and the dose of N in the tree row, in the years 2016–2018 (n = 4).

Floor Management and Dose of N (kg ha^{-1})		Total Yield 2016–2020 (kg·Tree^{-1})	Total Number of Fruit 2016–2020 (no·Tree^{-1})	Crop Efficiency Coefficient 2020 (kg·cm^{-2})
Means for interaction: N dose (D) × floor management (FM)				
HF	50	11.64 ab	131 ab	1.48 abc
	80	13.92 ab	161 ab	2.56 de
	110	20.77 b	219 b	2.77 e
	140	14.92 ab	174 ab	2.54 cde
LM 2	50	6.26 a	75 a	1.38 ab
	80	6.42 a	66 a	1.24 ab
	110	8.49 a	89 ab	1.63 abcd
	140	5.32 a	76 a	0.89 a
LM 4	50	10.58 ab	166 ab	2.14 bcde
	80	6.27 a	86 ab	1.40 ab
	110	9.54 ab	136 ab	2.00 bcde
	140	12.83 ab	163 ab	2.24 bcde
Statistical significance D × FM		NS	NS	**
Means for N doses (D)				
50		9.49 a	124 a	1.66 a
80		8.87 a	104 a	1.73 a
110		12.93 a	148 a	2.13 a
140		11.02 a	138 a	1.89 a
Statistical significance		NS	NS	NS
Means for floor managements (FM)				
HF		15.31 b	171 b	2.34 c
LM 2		6.62 a	77 a	1.28 a
LM 4		9.81 a	138 b	1.94 b
Statistical significance		***	***	***

HF—herbicide fallow, LM 2—living mulch sowing in the second year after trees planting (2016), LM 4—living mulch sowing in the fourth year after trees planting (2018). *, **, ***—significantly different (ANOVA) at $p \leq 0.05$, $p \leq 0.01$, and $p \leq 0.001$, respectively; NS—not significantly different. Means marked with different letters in each separate columns represent statistical differences among eight treatments (D × FM) or four N doses (D), or three floor managements (FM) (Tukey test, $p = 0.05$).

Two studied agrotechnical factors—floor management and the dose of N fertilization significantly affected the red blushing of apples, but their interaction did not (Table 4). The average share of yield with fruit blush exceeding 75% of the skin surface was significantly higher on trees grown in live mulch sown in the second year after trees planting, compared to the fourth year. This phenomenon could result from a significantly lower total number of fruits on these trees, especially in 2018 (Table S3). In both living mulches, compared to herbicide fallow, the share of apples with red blush over 75% and 25–75% of the skin surface was not significantly different. Experiments in other young orchards showed the opposite results, and the presence of the cover crop was considered as a factor guaranteeing improvement in apple red coloration [37,38]. However, in the case of older trees, differences in red color on the fruit surface were not significant [37], nor was the difference in anthocyanin synthesis in the skin of apples [33].

Table 4. Fruit quality of the apple 'Sampion' cv. depending on the floor management and dose of N in the tree row (mean for years 2016–2020 (n = 4)).

Floor Management and Dose of N (kg ha^{-1})		% of Fruit with Blush on the Skin Surface			% of Fruit with Diameter (cm)				Mean Fruit Weight (g)
		>¾	¼–¾	<¼ °	>8.5 °	7.5–8.5	6.5–7.5	<6.5	
Means for interaction: N dose (D) × floor management (FM)									
HF	50	77 a	21 a	2	2	18 a	31 a	50 a	108 a
	80	69 a	30 a	1	6	17 a	28 a	49 a	110 a
	110	63 a	36 a	1	7	21 a	32 a	41 a	124 a
	140	59 a	40 a	–	4	13 a	37 a	45 a	107 a
LM 2	50	83 a	16 a	1	–	18 a	23 a	60 a	98 a
	80	73 a	27 a	–	–	14 a	36 a	49 a	96 a
	110	76 a	23 a	1	3	17 a	31 a	49 a	105 a
	140	68 a	31 a	–	–	30 a	22 a	48 a	98 a
LM 4	50	70 a	30 a	–	8	6 a	19 a	66 a	99 a
	80	73 a	26 a	1	–	11 a	30 a	59 a	94 a
	110	53 a	46 a	1	3	14 a	29 a	54 a	93 a
	140	56 a	43 a	1	4	24 a	29 a	43 a	111 a
Statistical significance D × FM		NS	NS	–	–	NS	NS	NS	NS
Means for N doses (D)									
50		77 b	22 a	1	3	14 a	24 a	58 a	102 a
80		72 ab	28 ab	1	2	14 a	31 a	52 a	100 a
110		64 ab	35 ab	1	4	17 a	31 a	48 a	107 a
140		61 a	38 b	1	3	22 a	30 a	46 a	105 a
Statistical significance		*	*	–	–	NS	NS	NS	NS
Means for floor managements (FM)									
HF		67 ab	32 ab	1	5	17 a	32 a	46 a	112 a
LM 2		75 b	24 a	1	1	20 a	28 a	52 a	99 a
LM 4		63 a	36 b	1	4	14 a	27 a	56 a	99 a
Statistical significance		*	*	–	–	NS	NS	NS	NS

°—without stastistical evaluation; HF—herbicide fallow; LM 2—living mulch sowing in the second year after trees planting (2016); LM 4—living mulch sowing in the fourth year after trees planting (2018); *, **, ***—significantly different (ANOVA) at $p \leq 0.05$, $p \leq 0.01$ and $p \leq 0.001$, respectively; NS—not significantly different. Means marked with different letters in each separate columns represent statistical differences among eight treatments (D × FM) or four N doses (D), or three floor managements (FM) (Tukey test, $p = 0.05$).

Granatstein and Mullinix [41] suggested that floor management providing higher N in the soil tends to affect a smaller share of red color surface on the apple skin. Our studies confirmed such trends. A significantly higher share of apples with blush exceeding 75% of the skin surface was observed when the dose of 50 kg N ha^{-1} was used, compared to 140 kg. The limited red blush of apples is related to the good N supply of trees. High availability of this macronutrient affects excessive shoot growth, which is in conflict with the quantitatively satisfactory synthesis of anthocyanins, which conditions the red skin color [25]. The total length of annual shoots of 'Sampion' trees increased with the change in N dose from 50 to 140 kg N ha^{-1}, but it was not significantly different (Table 1). However, it can be assumed that the trend of stronger growth of apple canopies continued in the subsequent years of the trees' lives. The application of 110 and 140 kg N ha^{-1} probably impacted the synthesis of red pigment in the fruit. It determined the limitation of the share of yield with red skin surface over 75%, observed in 2019 and especially in 2020 (Figure 1).

In all tested treatments, apples of the 'Sampion' cv. were small, often with a diameter below 6.5 cm (Table 4). This was particularly visible in 2019, and to a lesser extent in

2018 (Figure 2). These two years, compared to the years 2016 and 2020 were characterized by very low total precipitation and high unevenness of its distribution in the summer (Table S1). Additionally, very high air temperatures were recorded from July to August. According to Musacchi and Serra [58], the final quality of the fruit is strongly determined by environmental and agronomic factors, which were observed in our experiment. Environmental conditions dominated over the tested factors: floor management and N doses. Probably for this reason, both the fruit size and weight were not significantly affected by both studied factors (Table 4). In studies by other authors [37,38,59], replacing herbicide fallow with living mulch affected the diminutive of fruit in the first years of fruiting trees. In the light of these studies, the expected loss of weight or increase share of yield with small size apples in living mulch was considered marginal and insignificant. However, it should be taken into account that the fruit number was significantly lower in the case of trees grown in living mulch sown in the second year after planting (Table 3). This limited mutual competition of fruit sets allowed them to grow to the level achieved by apples on significantly better yielding trees in herbicide fallow (Table 3, Figure S2). Due to the very small number of fruits, the negative effect of living mulch on the apple size was not revealed in the presented experiment.

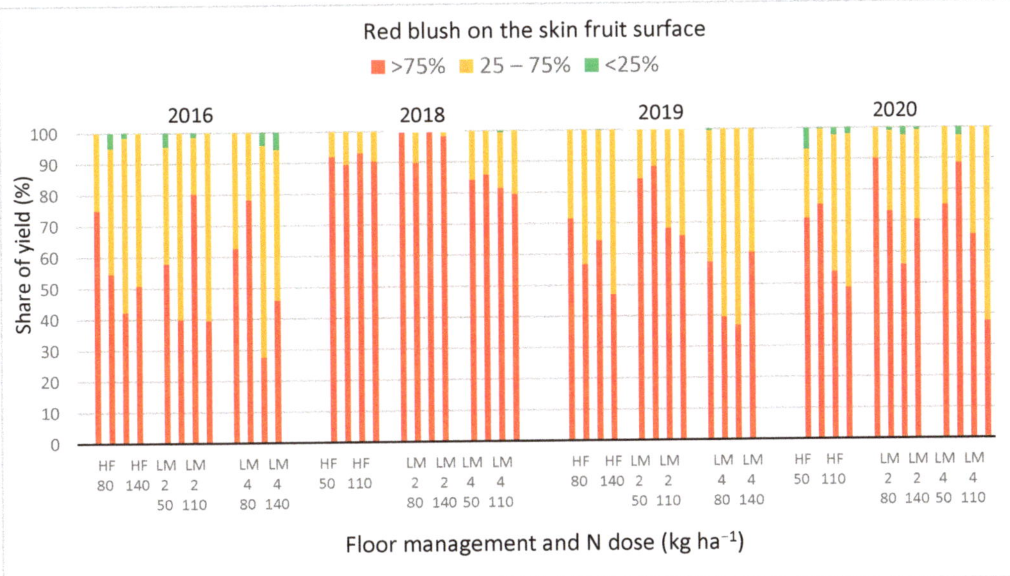

Figure 1. Red blush on the apple skin surface of 'Sampion' cv. depending on dose of N, from trees grown in herbicide fallow (HL) or cultivated in living mulch sowing in the second (LM 2) or in the fourth (LM 4) year after trees planting, in the years 2016–2020.

The second experimental factor—N doses had no significant effect on size, weight, and the yield of apple trees (Tables 3 and 4). Similarly to our experiment, the cultivar 'Sampion' [23] and also 'Jonagored' [24] did not respond with a significant change in the fruit weight, under the influence of different N treatments in the orchard with herbicide fallow. It is worth noting, however, that the tendency to increase the total fruit number, when doses above 100 kg N ha^{-1} were used (Table 3), was accompanied by a tendency to increase the share of large size apples, with a diameter exceeding 7.5 cm (Table 4).

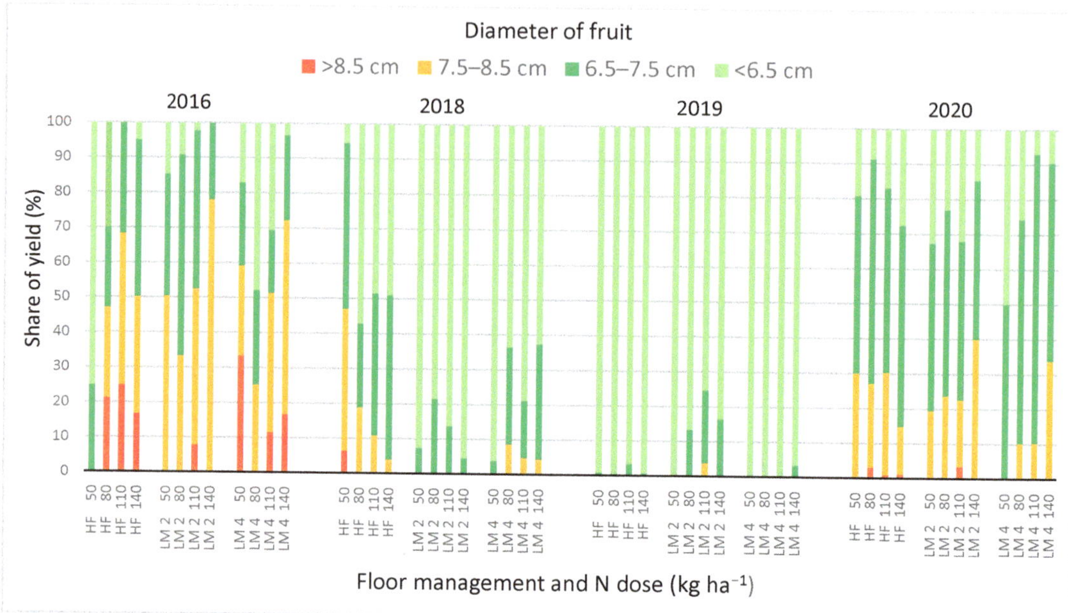

Figure 2. Sizes of the apple fruit 'Sampion' cv. depending on dose of N, from trees grown in herbicide fallow (HL) or cultivated in living mulch sowing in the second (LM 2) or in the fourth (LM 4) year after tree planting, in the years 2016–2020.

4. Conclusions

The growth and development of 'Sampion' cv. under the condition at the Fruit Experimental Station near by Wrocław (southwest part of Poland) were strongly influenced by both biotic and abiotic stress factors as well as the two experimental factors. The result of their action was the impact on low yield and a reduction in growth of all the tested trees. Despite these challenges, the conducted studies have been shown to have an effect of living mulch as a factor significantly limiting the level of apple tree production, but they improve the red color on the fruit skin surface. In contrast to floor management, the assessment of the second factor—dose of N is not that clear. Different doses of fertilization from 50 to 110 kg N ha^{-1} had no significant effect on annual shoot growth, increase in trunk cross sectional area, total yield, fruit weight and size on trees in herbicide fallow and living mulches. Application of the highest dose of 140 kg N ha^{-1} had a significant negative effect on the red blushing of apples. Different N doses had a minor effect on the tree compared to the impact of biotic and abiotic stress factors, including Mg deficiency. As a final outcome, with objectively low yield and poor growth, of all trees tested in the experiment, their N needs were satisfied by a dose of 50 kg ha^{-1}.

The cultivation of apple trees of the 'Sampion' cv. in the presence of cover crops should be more associated with the elimination of abiotic stress factors, which can promote the increase in importance of agrotechnical factors. It can be assumed, that the growth and yield of apple trees in living mulch would be better with the addition of more than 50 kg N ha^{-1}. Considering the development of red blush on the fruit skin surface, this dose should not exceed about 110 kg N ha^{-1}. It also seems important to pay attention to giving up the even distribution of N over the entire surface in the tree rows. Precise local fertilization under the tree would be more conducive to the uptake of fertilizer by apple trees than by the grass sod. Thus, it would enable more effective use of this macronutrient. The development of proper N fertilization technology for apple trees with living mulch requires further research on varied apple tree cultivars under different climate and soil conditions.

Supplementary Materials: The following supporting information can be downloaded at: https://www.mdpi.com/article/10.3390/agriculture15070724/s1, Table S1. Total precipitation and mean temperatures at the Wrocław-Strachowice Station (51°12′ N. 16°87′ E) in the years 2015–2020; Table S2. Mean weight of the fresh and dry leaf of the apple tree 'Sampion' cv. depending on the floor management and dose of nitrogen in the tree row, in the year 2018 and 2020 (n = 4); Table S3. Number of flower clusters, fruit sets and fruits on the apple tree 'Sampion' cv. depending on the floor management and dose of N in the tree row, in the years 2016–2018 (n = 4); Figure S1. Trunk cross sectional area of the apple tree 'Sampion' cv. depending on the floor management and dose of N in the tree row, in the years 2015–2020 (cm^2); Figure S2. Yield per tree of the apple tree 'Sampion' cv. depending on the floor management and dose of N in the tree row, in the years 2015–2020 (kg $tree^{-1}$).

Funding: This work was supported by the Wrocław University of Environmental and Life Sciences (Poland) as a part of the research program 'MISTRZ', no. N090/0012/22.

Institutional Review Board Statement: Not applicable.

Data Availability Statement: All data are present in the manuscript and Supplementary Materials.

Acknowledgments: I would like to thank each person who help me in setting up the experiment and valuable suggestions on how to evaluate the trees.

Conflicts of Interest: The author declares no conflicts of interest.

References

1. Mia, M.J.; Furmanczyk, E.M.; Golian, J.; Kwiatkowska, J.; Malusá, E.; Neri, D. Living Mulch with Selected Herbs for Soil Management in Organic Apple Orchards. *Horticulturae* **2021**, *7*, 59. [CrossRef]
2. Furmanczyk, E.M.; Malusà, E.; Kozacki, D.; Tartanus, M. Insights into the Belowground Biodiversity and Soil Nutrient Status of an Organic Apple Orchard as Affected by Living Mulches. *Agriculture* **2024**, *14*, 293. [CrossRef]
3. Diyanat, M. Weed Management in Organic Horticulture by Cover Crop in Iran. *Int. J. Adv. Biol. Biomed. Res.* **2015**, *3*, 153–162.
4. Muder, A.; Garming, H.; Dreisiebner-Lanz, S.; Kerngast, K.; Rosner, F.; Klickova, K.; Kurthy, G.; Cimer, K.; Bertazzoli, A.; Altamura, V.; et al. Apple Production and Apple Value Chains in Europe. *Eur. J. Hortic. Sci.* **2022**, *87*, 1–22.
5. Lemessa, F.; Wakjira, M. Cover Crops as a Means of Ecological Weed Management in Agroecosystems. *J. Crop Sci. Biotechnol.* **2015**, *18*, 133–145. [CrossRef]
6. Haring, S.; Gaudin, A.C.M.; Hanson, B.D. Functionally Diverse Cover Crops Support Ecological Weed Management in Orchard Cropping Systems. *Renew. Agric. Food Syst.* **2024**, *38*, e54. [CrossRef]
7. Schappert, A.; Linn, A.I.; Sturm, D.J.; Gerhards, R. Weed Suppressive Ability of Cover Crops Under Water-Limited Conditions. *Plant Soil Environ.* **2019**, *65*, 541–548. [CrossRef]
8. Smith, R.G.; Warren, N.D.; Cordeau, S. Are Cover Crop Mixtures Better at Suppressing Weeds than Cover Crop Monocultures? *Weed Sci.* **2020**, *68*, 186–194. [CrossRef]
9. Żelazny, W.R.; Licznar-Małańczuk, M. Living Mulch Persistence in an Apple Orchard and Its Effect on the Weed Flora in Temperate Climatic Conditions. *Weed Res.* **2022**, *62*, 85–99. [CrossRef]
10. Bałuszyńska, U.B.; Chaploutskyi, A.; Polunina, O.; Slobodianyk, L.; Licznar-Małańczuk, M. Grass Species as a Living Mulch—Impact of Sod and Its Weediness on Apple Trees. *Acta Agrobot.* **2023**, *76*, 172257. [CrossRef]
11. Golian, J.; Anyszka, Z.; Kwiatkowska, J. Multifunctional Living Mulches for Weeds Control in Organic Apple Orchards. *Acta Sci. Pol. Hortorum Cultus* **2023**, *22*, 73–84. [CrossRef]
12. Bałuszyńska, U.B.; Rowińska, M.; Licznar-Małańczuk, M. Grass Species as Living Mulches—Comparison of Weed Populations and Their Biodiversity in Apple Tree Rows and Tractor Alleys. *Acta Agrobot.* **2022**, *75*, 758. [CrossRef]
13. Mulumba, L.N.; Lal, R. Mulching Effects on Selected Soil Physical Properties. *Soil. Tillage Res.* **2008**, *98*, 106–111. [CrossRef]
14. Qian, X.; Gu, J.; Pan, H.J.; Zhang, K.Y.; Sun, W.; Wang, X.J.; Gao, H. Effects of Living Mulches on the Soil Nutrient Contents, Enzyme Activities, and Bacterial Community Diversities of Apple Orchard Soils. *Eur. J. Soil. Biol.* **2015**, *70*, 23–30. [CrossRef]
15. Ramos, M.E.; Robles, A.B.; Sánchez-Navarro, A.; González-Rebollar, J.L. Soil Responses to Different Management Practices in Rainfed Orchards in Semiarid Environments. *Soil Tillage Res.* **2011**, *112*, 85–91. [CrossRef]
16. Brewer, M.; Kanissery, R.G.; Strauss, S.L.; Kadyampakeni, D.M. Impact of Cover Cropping on Temporal Nutrient Distribution and Availability in the Soil. *Horticulturae* **2023**, *9*, 1160. [CrossRef]

17. St. Laurent, A.; Merwin, I.A.; Thies, J.E. Long-Term Orchard Groundcover Management Systems Affect Soil Microbial Communities and Apple Replant Disease Severity. *Plant Soil* **2008**, *304*, 209–225. [CrossRef]
18. Yang, J.; Duan, Y.; Zhang, R.; Liu, C.; Wang, Y.; Li, M.; Ding, Y.; Awasthi, M.K.; Li, H. Connecting Soil Dissolved Organic Matter to Soil Bacterial Community Structure in a Long-Term Grass-Mulching Apple Orchard. *Ind. Crops Prod.* **2020**, *149*, 112344. [CrossRef]
19. Hoagland, L.; Carpenter-Boggs, L.; Granatstein, D.; Mazzola, M.; Smith, J.; Peryea, F.; Reganold, J.P. Orchard Floor Management Effects on Nitrogen Fertility and Soil Biological Activity in a Newly Established Organic Apple Orchard. *Biol. Fertil Soils* **2008**, *45*, 11–18. [CrossRef]
20. Atucha, A.; Merwin, I.A.; Purohit, C.K.; Brown, M.G. Nitrogen Dynamics and Nutrient Budgets in Four Orchard Groundcover Management Systems. *HortScience* **2011**, *46*, 1184–1193. [CrossRef]
21. Brunetto, G.; Oliveira, B.S.; Ambrosini, V.G.; Couto, R.d.R.; Sete, P.B.; dos Santos Junior, E.; Loss, A.; Stefanello da Silva, L.O.; Gatiboni, L.C. Nitrogen Availability in an Apple Orchard with Weed Management. *Cienc. Rural* **2018**, *48*, e20160895. [CrossRef]
22. Rühmann, S.; Leser, C.; Bannert, M.; Treutter, D. Relationship Between Growth, Secondary Metabolism, and Resistance of Apple. *Plant Biol.* **2002**, *4*, 137–143. [CrossRef]
23. Wrona, D. Effect of Nitrogen Fertilization on Growth, Cropping and Fruit Quality of "Šampion" Apple Trees During 9 Years After Planting. *Folia Hortic.* **2004**, *16*, 55–60.
24. Wrona, D. The Influence of Nitrogen Fertilization on Growth, Yield and Fruit Size of "Jonagored" Apple Trees. *Acta Sci. Pol. Hortorum Cultus* **2011**, *10*, 3–10.
25. Treutter, D. Biosynthesis of Phenolic Compounds and Its Regulation in Apple. *Plant Growth Regul.* **2001**, *34*, 71–89.
26. Neilsen, G.H.; Neilsen, D.; Herbert, L. Nitrogen Fertigation Concentration and Timing of Application Affect Nitrogen Nutrition, Yield, Firmness, and Color of Apples Grown at High Density. *HortScience* **2009**, *44*, 1425–1431.
27. Veberic, R. The Impact of Production Technology on Plant Phenolics. *Horticulturae* **2016**, *2*, 8. [CrossRef]
28. Mikulic-Petkovsek, M.; Slatnar, A.; Stampar, F.; Veberic, R. The Influence of Organic/Integrated Production on the Content of Phenolic Compounds in Apple Leaves and Fruits in Four Different Varieties over a 2-Year Period. *J. Sci. Food Agric.* **2010**, *90*, 2366–2378. [CrossRef]
29. Roussos, P.A.; Gasparatos, D. Apple Tree Growth and Overall Fruit Quality under Organic and Conventional Orchard Management. *Sci. Hortic.* **2009**, *123*, 247–252. [CrossRef]
30. Le Bourvellec, C.; Bureau, S.; Renard, C.M.G.C.; Plenet, D.; Gautier, H.; Touloumet, L.; Girard, T.; Simon, S. Cultivar and Year Rather than Agricultural Practices Affect Primary and Secondary Metabolites in Apple Fruit. *PLoS ONE* **2015**, *10*, e0141916. [CrossRef]
31. TerAvest, D.; Smith, J.L.; Carpenter-Boggs, L.; Hoagland, L.; Granatstein, D.; Reganold, J.P. Influence of Orchard Floor Management and Compost Application Timing on Nitrogen Partitioning in Apple Trees. *HortScience* **2010**, *45*, 637–642. [CrossRef]
32. Slatnar, A.; Licznar-Malanczuk, M.; Mikulic-Petkovsek, M.; Stampar, F.; Veberic, R. Long-Term Experiment with Orchard Floor Management Systems: Influence on Apple Yield and Chemical Composition. *J. Agric. Food Chem.* **2014**, *62*, 4095–4103. [CrossRef] [PubMed]
33. Slatnar, A.; Kwiecinska, I.; Licznar-Malanczuk, M.; Veberic, R. The Effect of Green Cover within Rows on the Qualitative and Quantitative Fruit Parameters of Full-Cropping Apple Trees. *Hortic. Environ. Biotechnol.* **2020**, *61*, 41–49. [CrossRef]
34. Lee, J.; Steenwerth, K.L. "Cabernet Sauvignon" Grape Anthocyanin Increased by Soil Conservation Practices. *Sci. Hortic.* **2013**, *159*, 128–133. [CrossRef]
35. Sosna, I.; Fudali, E. Usefulness of Living Mulch in Rows in a Dwarf Pear, *Pyrus communis* L., Orchard. *Agriculture* **2023**, *13*, 2145. [CrossRef]
36. Kühn, B.F.; Lindhard Pedersen, H. Cover Crop and Mulching Effects on Yield and Fruit Quality in Unsprayed Organic Apple. *Eur. J. Hortic. Sci.* **2009**, *74*, 247–253. [CrossRef]
37. Licznar-Małańczuk, M.; Kwiecińska, I. Application of Living Mulch in Rows of the Apple Trees on Several Rootstock—Long-Term Evaluation. *Acta Sci. Pol. Hortorum Cultus* **2023**, *22*, 79–90. [CrossRef]
38. Andersen, L.; Kühn, B.F.; Bertelsen, M.; Bruus, M.; Larsen, S.E.; Strandberg, M. Alternatives to Herbicides in an Apple Orchard, Effects on Yield, Earthworms and Plant Diversity. *Agric. Ecosyst. Environ.* **2013**, *172*, 1–5. [CrossRef]
39. Kopta, T.; Ragasová, L.N.; Sotolář, R.; Sedláček, J.; Ferby, V.; Hurajová, E.; Winkler, J. The Influence of Different Methods of Under-Vine Management on the Structure of Vegetation and the Qualitative Parameters of the Grapes in the Moravian Wine Region. *Folia Hortic.* **2024**, *36*, 235–257. [CrossRef]
40. Hogue, E.J.; Cline, J.A.; Neilsen, G.; Neilsen, D. Growth and Yield Responses to Mulches and Cover Crops under Low Potassium Conditions in Drip-Irrigated Apple Orchards on Coarse Soils. *HortScience* **2010**, *45*, 1866–1871. [CrossRef]
41. Granatstein, D.; Mullinix, K. Mulching Options for Northwest Organic and Conventional Orchards. *HortScience* **2008**, *43*, 45–50. [CrossRef]
42. Tworkoski, T. Response of Potted Peach Trees to Pruning and Grass Competition. *HortScience* **2000**, *35*, 1209–1212. [CrossRef]

43. Hoagland, L.; Carpenter-Boggs, L.; Granatstein, D.; Mazzola, M.; Peryea, F.; Smith, J.; Reganold, J. Nitrogen Cycling and Partitioning under Alternative Organic Orchard Floor Management Strategies. In Proceedings of the Western Nutrient Management Conference, Salt Lake City, UT, USA, 4–5 March 2007; pp. 117–123.
44. Tahir, I.I.; Svensson, S.-E.; Hansson, D. Floor Management Systems in an Organic Apple Orchard Affect Fruit Quality and Storage Life. *HortScience* **2015**, *50*, 434–441. [CrossRef]
45. Merwin, I.A.; Stiles, W.C. Orchard Groundcover Management Impacts on Apple Tree Growth and Yield, and Nutrient Availability and Uptake. *J. Am. Soc. Hort. Sci.* **1994**, *119*, 209–215. [CrossRef]
46. Neilson, G.H.; Hogue, E.J. Comparison of White Clover and Mixed Sodgrass as Orchard Floor Vegetation. *Can. J. Plant Sci.* **2000**, *80*, 617–622. [CrossRef]
47. *Metodyka Integrowanej Produkcji Jabłek (Methodology of Integrated Apple Production)*; Sobiczewski, P., Ed.; Institute of Horticulture: Skierniewice, Poland, 2013.
48. *Metodyka Integrowanej Produkcji Jabłek (Methodology of Integrated Apple Production)*; Sobiczewski, P., Ed.; Institute of Horticulture: Skierniewice, Poland, 2020.
49. Kowalczyk, W.; Wrona, D.; Przybyłko, S. Effect of Nitrogen Fertilization of Apple Orchard on Soil Mineral Nitrogen Content, Yielding of the Apple Trees and Nutritional Status of Leaves and Fruits. *Agriculture* **2022**, *12*, 169. [CrossRef]
50. Mia, M.J.; Massetani, F.; Murri, G.; Neri, D. Sustainable Alternatives to Chemicals for Weed Control in the Orchard—A Review. *Hortic. Sci.* **2020**, *47*, 1–12. [CrossRef]
51. Gautam, V.P.; Mishra, S.; Ahmed, H. Comparison of Total Nitrogen Estimation by Kjeldahl Method and CHNS Analyzer in Dry Tropical Grassland. *Int. J. Plant Environ.* **2023**, *9*, 180–182. [CrossRef]
52. Sharma, I. ICP-OES: An Advance Tool in Biological Research. *Open J. Environ. Biol.* **2020**, *5*, 027–033. [CrossRef]
53. Kowalczyk, W.; Wrona, D.; Przybyłko, S. Content of Minerals in Soil, Apple Tree Leaves and Fruits Depending on Nitrogen Fertilization. *J. Elem.* **2017**, *22*, 67–77. [CrossRef]
54. Sánchez, E.E.; Giayetto, A.; Cichón, L.; Fernández, D.; Aruani, M.C.; Curetti, M. Cover Crops Influence Soil Properties and Tree Performance in an Organic Apple (*Malus domestica* Borkh) Orchard in Northern Patagonia. *Plant Soil* **2007**, *292*, 193–203. [CrossRef]
55. Neilsen, G.H.; Neilsen, D. Nutritional Requirements of Apple. In *Apples: Botany, Production, and Uses*; Ferree, D.C., Warrington, I.J., Eds.; CABI Publishing: Wallingford, UK, 2015; pp. 267–302.
56. Licznar-Malanczuk, M.; Baluszynska, U.B. Do Living Mulches or Environmental Conditions Have a Greater Impact on the External Quality of the Apple Fruit 'Chopin' Cultivar? *Agriculture* **2024**, *14*, 610. [CrossRef]
57. Gudarowska, E.; Szewczuk, A. The Influence of Agro-Technical Methods Used in the Nursery on Quality of Planting Material and Precocity of Bearing in Young Apple Trees in the Orchard. *J. Fruit Ornam. Plant Res.* **2004**, *12*, 91–96.
58. Musacchi, S.; Serra, S. Apple Fruit Quality: Overview on Pre-Harvest Factors. *Sci. Hortic.* **2017**, *234*, 409–430. [CrossRef]
59. Atucha, A.; Merwin, I.A.; Brown, M.G. Long-Term Effects of Four Groundcover Management Systems in an Apple Orchard. *HortScience* **2011**, *46*, 1176–1183. [CrossRef]

agriculture

MDPI

Article

Do Living Mulches or Environmental Conditions Have a Greater Impact on the External Quality of the Apple Fruit 'Chopin' Cultivar?

Maria Licznar-Malanczuk and Urszula Barbara Baluszynska *

Department of Horticulture, Faculty of Live Sciences and Technology, Wroclaw University of Environmental and Life Sciences, Grunwaldzki 24a, 50-363 Wroclaw, Poland; maria.licznar-malanczuk@upwr.edu.pl
* Correspondence: urszula.baluszynska@upwr.edu.pl

Abstract: Research was carried out to assess the yield and quality of fruits from the new Polish apple tree 'Chopin'—a 'green peel', scab-resistant cultivar under grass living mulch management. Blue fescue and red fescue were tested in experiment no. 1. In experiment no. 2, meadow grass and perennial ryegrass were used. Every species of grass was sown in two doses of 50 and 150 kg per ha. Herbicide fallow was introduced as a control in both experiments. Strongly variable temperatures and precipitations in the years of evaluation made it possible to estimate year—a function of variable environmental conditions—as an additional experimental factor. An unexpected effect of the presence of living mulch was its significant impact on the appearance of blush on the 'green peel' fruit. However, living mulches had little effect on weight and fruit size. Cool days during apple ripening enhanced the process of fruit skin red coloration. The effect of both agrotechnical and environmental factors on fruit quality was more visible in the case of less vigorous trees, which were more susceptible to experimental, stressful conditions. However, increasing the sowing dose of each grass seed did not influence red blushing, weight, or fruit size. An additional difficulty for the trees was the competition caused by the early germination of these living grass mulches, reinforced by the presence of *Trifolium repens* L.

Keywords: blushing; fruit size; cover crop; thermal conditions; water conditions

1. Introduction

A modern view of agricultural production emphasizes the need for the sustainable use of agrochemicals due to the potential, undesirable impact of active substances on the safety of produced food [1,2] and the agricultural environment [3]. Therefore, the use of glyphosate-based herbicides in integrated agricultural systems should be eliminated [4,5]. Fruit production faces the challenges of providing a valuable, quality product for the consumer [6,7] and transforming the orchard into a place that serves as an element of a sustainable environment [8]. Chemical weed control can be replaced by alternative methods of weed control, for example, using living mulch [9–11]. In addition to weed control [11–14], living mulch offers other benefits for the orchard agroecosystem, including an increase in flora biodiversity [8] and organic matter in the soil [15–17], and it also affects the chemical [15,17,18] and biological [19,20] properties of soil. Species of the Poaceae family are often used as cover plants in perennial fruit crops [21]. Among grasses, species of the *Festuca* genus deserve special attention, especially blue fescue (*F. ovina* L.) [7,22] and red fescue (*F. rubra* L.) [23–26]. Other grasses used in fruit production include meadow grass (*Poa pratensis* L.) [23,26,27] and perennial ryegrass (*Lolium perenne* L.) [15,25,27].

The presence of living mulches causes discomfort for fruit trees, which may result in reduced tree yield [22,24,26,27]. With the appropriate choice of both cover crops and agrotechnical methods, the competition of living mulch relative to the cultivated plant can be mitigated [7,22,28]. However, there is an impact on fruit quality [7,22]. The external features of apples like size or the color of the skin are determined by both agrotechnical and environmental growing conditions [29] and are important for the individual choice of fruit of the consumer [30,31]. Cover crops in apple orchards significantly increased the area of 'red skin' [7,27], but the presence of additional vegetation, even weeds, did not determine the intensity of the basic color of the fruit in the case of 'yellow peel' apple [32]. On the other hand, mean apple fruit weight and size may be reduced compared to fruits from trees grown in herbicide fallow [22,25]. Environmental factors can also play an important role in fruit quality. External quality factors such as color, size, shape, and russeting are determined by the following factors: temperature, light, and precipitation, which vary in individual years and areas of fruit cultivation [29,33,34]. Warmer temperatures inhibit anthocyanin accumulation, which is responsible for the red blushing of apple skin [35], although they are necessary for the proper development of fruit in terms of size [36].

The aim of the study was to investigate the effect of living grass mulches on the external parameters of fruit from the new Polish apple tree 'Chopin'—a green peel, scab-resistant cultivar. Strongly variable temperatures and precipitations in the years of evaluation made it possible to estimate yearly conditions as an additional experimental factor and their impact on apple quality as a direct effect or result of the interaction with the presence of a cover crop.

2. Materials and Methods

2.1. Plant Material and Two Field Experiments

The influence of grass living mulches on the external parameters of fruit from the Polish apple tree 'Chopin'—a new, 'green peel', scab-resistant cultivar—was estimated. This study was conducted at the Fruit Experimental Station of the Wrocław University of Environmental and Life Sciences in Wrocław (Poland), Samotwór (51°06′12″ N, 16°49′52″ E) in the years of 2016–2021. Two-year-old apple trees (*Malus domestica* Borkh.) grafted on a semi-dwarf MM.106 rootstock were planted in the spring of 2016 on haplic luvisol. The chosen spacing was 3.6 m between tree rows and 1.2 m as a distance between trees in a row. In the spring of 2016, soil samples were taken from the entire area of the experimental field before planting the trees. Initial soil analysis showed an average level of available phosphorus and potassium and a high level of magnesium. The K:Mg ratio was proper (1.00). On this basis, the fertilizer doses of phosphorus (P_2O_5 100 kg per ha), potassium (K_2O 100 kg per ha), and liming (CaO 500 kg per ha) were determined.

Two separate experiments were established, each following a one-way randomized block design in four repetitions. Each repetition was represented by an experimental plot with four trees growing on it. In experiment no. 1, blue fescue (*Festuca ovina* L.) of the 'Noni' cv. and red fescue (*F. rubra* L.) of the 'Adio' cv. as living mulches were sown. The mean cross-sectional area of the tree trunk was 2.44 cm^2, in this experiment, when grass cover crops were sown in the early summer of 2017. The living mulches in experiment no. 2 were represented by meadow grass (Kentucky bluegrass) (*Poa pratensis* L.) of the 'Niweta' cv. and perennial ryegrass (*Lolium perenne* L.) of the 'Info' cv. The apple trees in this experiment were weaker than in experiment no. 1 and their mean trunk cross-sectional area was only 1.72 cm^2. Compared to trees in experiment no. 1, the growth of 'Chopin' cv. was lower than with all treatments in experiment no. 2. This could have been related to the presence of grubs causing damage to the root systems that, in some cases, even killed several trees in the year 2016–2017. This affected the growth and development of young trees in this year and most likely in the subsequent years. On the contrary, in experiment no. 1, the presence of grubs was sporadic and only caused the death of two trees.

Each estimated grass was sown separately at a dose of 50 kg and a dose of three times more—150 kg seeds per ha—in the 1.0 m wide tree rows as well as in the 2.6 m wide adjacent tractor alleys on the western side of each row. At the beginning of June 2017, the grass seeds were scattered by hand and raked into the soil in the tree rows, after which the floor was rolled with a hand roll. Like in the tree rows, the grass seeds in the alleys were scattered by hand, but the following cultivation was limited to rolling with a roll pulled by a tractor. The grass sod in the tree rows was mowed manually, twice or three times per year, with a string trimmer every year. But, the grass sod in the tractor alleys was mowed with a lawn mower several times per year. In both cases, the mowed biomass of grass was left.

A control treatment using herbicide fallow was created separately for each experiment. It was treated with a mix of glyphosate (1.44–1.96 kg ha^{-1}) and MCPA (2-methyl-4-chlorophenoxyacetic acid, 0.60–1.00 kg ha^{-1}) two or three times a year: in spring (April or May), summer (July), and, in some years, at the end of the vegetation period. A grass mixture was sown in the alleys adjacent to the herbicide fallow plots. The trees were trained into a slender spindle canopy form. After the second year of orchard establishment, each tree was fertilized with ammonium nitrate (15 g N in the years 2017−2018 and 20 g N in the years 2019−2021) separately. The cutting of the trees as well as plant protection were managed in accordance with current commercial orchard recommendations. Manual hand thinning of fruit sets was performed only at the beginning of June in the year 2020.

2.2. Living Mulch Sod and Weediness Estimation

The separate plot area for each individually estimated grass was 17.28 m^2 (3.6 m × 4.8 m) and contained four apple trees. In the middle part of the plot area, two sub-plots were designated—the first for the assessment of living mulch sod and weed infestation in the tree row (1.80 m^2) and the second for only sod estimation in the tractor alley (4.68 m^2). Whole-plot evaluation was abandoned to avoid the non-representative weed communities that occurred at the edges of the investigated living grass mulches. The degree of soil coverage by living mulches was assessed as the percentage of the total area of sub-plots occupied by the cover crop sod. Using the same plots, but only in the tree rows, the percentage of the soil surface occupied by weed taxa was determined. For this purpose, a noninvasive method of weed population assessment, conforming to the modified methodology of Lipecki and Janisz [37], was employed. The original share of the scale was modified by splitting the range over 0% up to 20% into two separate groups and, as a final result, each taxon was expressed using a discrete percentage scale: 0%, 1%, 20%, 40%, 60%, 80%, and 100%. The assessments were performed separately for each species and, in some cases, genus, of the weed. As the share of each species was assessed independently, it was impossible to express the relationship between the total share of all weed populations and the sod of the living mulch at a scale of 100%. The living mulch sod cover estimation was performed in the autumn of every year (2017–2021). The weed taxa assessment took place in the last year of the research, in the spring of 2021. The nomenclature of vascular plants was based on POWO [38].

2.3. Soil Analysis

In the spring of 2022, 'A horozon' topsoil samples were collected from the tree rows in each sub-plot in experiment no. 2. Soil analyses were performed at the Chemical and Agricultural Station in Wrocław (Poland). The available elements (K, Mg, Ca) were determined using the Mehlich-3 method [39,40]. Phosphorus was determined colorimetrically using the vanadium method [41]. In addition, pH was determined in 1 mol KCl L^{-1} by the potentiometric method [41] and soil organic carbon (SOC) was determined by the Tiurin method. The assessment of soil nutrient content was carried out in relation to the limit numbers developed by Sadowski et al. [42].

2.4. Yield, Fruit Quality, and the Growth of Apple Trees

In both experiments, each tree was individually assessed for the abundance of flowering, number of fruits, and yield in the years 2017–2021. Flowering was assessed on each tree according to a scale of 0–5, where 0 meant no flowers and 5 meant a very abundant cover of the tree with flowers. During harvest, the number of fruits on each tree was counted and its yield was weighed. The fruit was harvested in the last days of September or the first days of October. Fruit quality assessments were carried out in 2018–2021; the year 2017 was omitted due to hail damage to the apples. The fruit weight was calculated as a ratio of yield mass to the number of fruits and was computed separately for each tree every year. All harvested apples per repetition were sorted into classes—over 75%, 25–75%, or less than 25% of fruit skin red blush surface. It should be noted that the color and intensity of the blush appearing on the 'green-peel' 'Chopin' cv. did not reach the typical red color but rather oscillated towards light red-pink blushing (Scheme S1). Apples were also divided into classes of fruit diameter: less than 6.5, 6.5−7.5, 7.5−8.5, and over 8.5 cm. As a measure of each tree's growth, trunk cross-sectional area (TCSA) and its increment were calculated as an average of the diameter measured in two directions (north–south and east–west directions), measured 30 cm above the grafting point. The measurements were made in autumn 2016 and autumn 2021. The crop efficiency coefficient (CEC) was computed as a ratio of the total yield of five years (2017−2021) and TCSA in the autumn of 2021.

2.5. Statistical Evaluation

All annual and mean data connected with the growth and yield of the trees were evaluated statistically using one-way analysis of variance (ANOVA) for a randomized block design in RStudio software (version 2023.12.1). Prior to the analysis, in order to—at least approximately—fulfill the assumptions of the analysis of variance, angular (by the function of Bliss), logarithmic, or exponential transformations were applied to some of the data. Due to the death of some trees, the measurement system (data presented: yield, growth, and fruit quality) was balanced using the missing plot technique procedure. Significant differences between treatment means were calculated with the Tukey test ($p \leq 0.05$). Means were presented with standard deviations (mean ± SD).

In addition, a two-way analysis of variance (ANOVA) was also performed when data from individual years was taken into account as a second research factor to identify the effects of yearly conditions (YEAR), living grass mulch species with either their seed sowing dose or herbicide fallow (GLMHL), and the interaction of these two factors (YEAR × GLMHL) on the variables connected with fruit quality parameters only. Significant differences in interactions were noted at $p \leq 0.05$, $p \leq 0.01$, and $p \leq 0.001$. This statistical analysis of data was supplemented with a linear correlation analysis ($p = 0.05$), in which the selected thermal condition and precipitation impacts on the quality of individual fruits were assessed.

3. Results and Discussion

3.1. Annual Temperature and Total Precipitation with a Focus on the Fruit Ripening Period

The year of tree planting (2016) and the year of sowing living mulches (2017) were characterized by moderate mean monthly and annual air temperatures (Tables 1 and S1). The distribution of monthly total precipitation was even, except for the peak in both years in July. In the next two years (2018–2019), very high mean monthly summer temperatures were recorded, and the annual rainfall showed a deficit (less than 450 mm) compared to the first two years of the experiment. On the contrary, in the next study year (2020), rainfall from May to October was very high, especially in the relatively cold month of June. In the last year of research, the mean annual air temperature was low, despite the records for June and July when mean monthly air temperatures exceeded 20 °C. Precipitation was less than in the first year of research.

In the first five years of cropping, which is associated with the end of September and the beginning of October in Poland, the fruits ripened in very diverse weather conditions (Tables 1 and S1). In the summer of 2017, hail damaged the skin of the apples, which disrupted their growth. In subsequent years of research, September 2018 was exceptionally warm (Figure S1). The apple harvest was carried out on 2 October. In the 30 days preceding harvest, the maximum air temperature reached above 25.0 °C for as many as 10 days. In this year (2018), the number of days with a minimum temperature below 5.0 °C reached 5 days. This was 2 days higher compared to the remaining years of the study. Noticeable cooling was noted mainly before fruit harvest (Figure S1). It can be assumed that such thermal conditions in 2018 were conducive to fruit skin red coloring. Anthocyanin synthesis, responsible for the red blushing of apple skins, increases with decreasing temperatures during fruit ripening [33,35].

Table 1. Mean annual temperatures and total precipitation at the Wrocław-Strachowice Station (51°12′ N, 16°87′ E) and characteristics of thermal conditions and rainfall during fruit ripening a month before fruit harvest in the years 2017–2021.

Specification	Year				
	2017 *	2018	2019	2020	2021
Mean annual temperature (°C)	9.8	10.8	11.0	10.6	9.4
No. of days during fruit ripening with minimum temperature:					
0.0–5.0 °C	3	5	3	3	3
5.1–10.0 °C	17	9	11	13	12
10.1–15.0 °C	10	15	15	14	14
15.1–20.0 °C	–	1	1	–	1
No. of days during fruit ripening with maximum temperature:					
10.1–15.0 °C	4	4	5	5	4
15.1–20.0 °C	24	8	14	9	8
20.1–25.0 °C	2	8	10	8	14
25.1–35.0 °C	–	10	1	8	4
Total precipitation (mm):					
January–December	619.3	441.4	430.8	808.4	540.3
January–September	478.6	343.1	348.3	667.5	457.2
September	65.1	45.9	53.0	94.2	24.2
No. of days in September with precipitation => 0.1 mm	15	8	15	9	11

*—hail shower.

In the following years (2019–2021), the distribution of the number of days with individual ranges for minimum temperatures was similar. It is worth noting that in 2019, as many as 10 days with a minimum temperature below 10 °C were recorded in the second half of September (Figure S1), just before the fruit harvest, which was on the last day of the month. In 2020 and 2021, the number of such days in the second half of September was lower. At the same time, they were accompanied by relatively high maximum temperatures, which additionally did not favor the red coloring of the apple skins compared to more favorable thermal conditions in 2018 and 2019.

In 2018–2019, from January to the end of September, the total rainfall was very low (Tables 1 and S1). Trees with ripening fruits had better conditions in 2017 and 2021, and the best conditions were in 2020, which was exceptionally abundant in precipitation. This was the only year in the study in which the amount of rainfall could be classified as ensuring 'good fruit quality' in apple production [29]. The highest rainfall in September was also recorded in 2020; it was twice as low in 2018 and four times lower in the last year of tree fruiting (2021). The number of days with precipitation ranged in September from 8—in the exceptionally warm fruit ripening period in 2018—to 1 day more, which was recorded in the exceptionally rainy year of 2020, and, in other years, it even reached 15 days.

3.2. Characteristics of Living Mulch Sod and Weed Infestation

The choice of the beginning of June 2017 as the seed sowing date for the tested grasses turned out to be beneficial because, during establishment and in summer, further development of the living mulch sod was accompanied by adequate moisture (Table S1). Proper growth conditions were provided that stimulated seed germination and the abundant development of cover crop biomass [43,44]. In both experiments, however, the percentage of the soil surface covered with four living grass mulches varied greatly in the year of sowing (Table 2). In experiment no. 1, the development of blue fescue sod was very slow and the soil cover in the autumn of 2017 was unsatisfactory compared to that of red fescue. This could have been due to the differential ability of fescue cultivars to rapidly establish [45]. In experiment no. 2, perennial ryegrass, regardless of the seed sowing dose used, filled 100% of the orchard area in the year of sowing. Meadow grass gained by such a year later.

Table 2. Percentage of the soil surface under the covering of different living grass mulch sods in the tree rows and alleyways in the succeeding years 2017−2021.

Year	Living Grass Mulch Species and Seed Sowing Dose (kg·ha^{-1})							
	Experiment 1				Experiment 2			
	Blue Fescue		Red Fescue		Meadow Grass		Perennial Ryegrass	
	50	150	50	150	50	150	50	150
	Tree rows							
2017	10	5	65	85	60	80	100	100
2018	70	85	90	95	95	100	95	100
2019	95	100	100	100	100	100	100	100
2020	95	100	100	100	100	100	95	100
2021	100	85	100	100	100	100	100	100
	Alleyways							
2017	20	50	90	95	80	85	100	100
2018	60	85	95	95	100	100	90	100
2019	80	100	95	100	100	100	100	100
2020	75	85	100	100	100	100	100	100
2021	70	75	75	85	100	100	95	100

For the perennial ryegrass, the very early germination of this species contributed to good sodding. It had already appeared between 10 and 20 days from the date of sowing the seeds compared to meadow grass and red or blue fescue, which germinated after 20 days at the earliest [46]. The soil cover of perennial ryegrass and meadow grass in 2018–2021 was stable and satisfactory. Despite this, weeds were spreading in the grass sod (Table 3).

Table 3. Diversity of the structure of identified weed taxa number and soil covering within four different living grass mulch sods in the tree rows in spring 2021.

Specification		Living Grass Mulch Species and Seed Sowing Dose (kg·ha^{-1})							
		Experiment No. 1				Experiment No. 2			
		Blue Fescue		Red Fescue		Meadow Grass		Perennial Ryegrass	
		50	150	50	150	50	150	50	150
Percentage of soil surface under individual weed taxa covering within sod:									
Achillea millefolium L.	ACHMI	0	5.0	0	0	0	5.0	0	0
Arabidopsis thaliana (L.) Heynh.	ARBTH	0	0	0	0	<1.0	<1.0	5.0	0
Artemisia vulgaris L.	ARTVU	0	0	5.0	0	<1.0	<1.0	0	<1.0
Calamagrostis epigejos (L.) Roth	CLMEP	5.0	0	0	0	0	0	0	0
Capsella bursa-pastoris (L.) Medik.	CAPBP	0	0	0	0	0	0	5.3	<1.0
Cerastium holosteoides Fr.	CERVU	5.0	0	0	0	15.3	10.3	20.0	5.5
Crepis spp.	CVPG	<1.0	0	0	0	0	0	0	<1.0
Draba verna L.	ERPVE	5.0	0	0	0	5.3	10.0	15.0	10.0
Elymus repens (L.) Gould	AGRRE	<1.0	0	0	0	15.0	<1.0	0	0
Hieracium spp.	HIEG	0	5.0	0	<1.0	5.0	<1.0	<1.0	0
Hypericum perforatum L.	HYPPE	<1.0	0	<1.0	0	0	<1.0	<1.0	0
Lamium purpureum L.	LAMPU	<1.0	0	0	0	5.8	5.5	15.3	10.3
Leontodon hispidus L.	LEBHI	<1.0	0	0	0	0	<1.0	<1.0	0
Malva sylvestris L.	MALSI	<1.0	0	0	0	0	<1.0	0	<1.0
Plantago major L.	PLAMA	0	0	0	<1.0	0	<1.0	0	0
Plantago lanceolata L.	PLALA	10.0	<1.0	0	0	0	0	<1.0	5.0
Poa annua L.	POAAN	0	0	0	0	0	0	0	<1.0
Poaceae—other species	GRAF	25.0	15.3	0	<1.0	15.3	5.5	10.0	<1.0
Ranunculus repens L.	RANRE	0	5.0	0	0	0	0	0	0

Table 3. *Cont.*

Specification		Living Grass Mulch Species and Seed Sowing Dose (kg·ha^{-1})							
		Experiment No. 1				Experiment No. 2			
		Blue Fescue		Red Fescue		Meadow Grass		Perennial Ryegrass	
		50	150	50	150	50	150	50	150
Rumex acetosa L.	RUMAC	<1.0	0	0	0	0	0	0	0
Stellaria media (L.) Vill.	STEME	<1.0	0	0	0	5.5	5.5	15.3	20.0
Taraxacum sp.	TARG	10.5	5.8	<1.0	<1.0	20.0	15.3	20.3	20.0
Trifolium repens L.	TRFRE	10.5	10.3	25.0	10.5	65.0	75.0	75.0	55.0
Veronica arvensis L.	VERAR	0	0	0	0	0	0	0	<1.0
Veronica serpyllifolia L.	VERSE	0	0	0	0	5.3	<1.0	0	0
Veronica sp.	VERG	0	0	0	0	<1.0	0	<1.0	<1.0
No. of weed taxa covering soil surface within sod:									
<1.0%		8	1	2	4	3	9	5	8
1.1–5.0%		3	3	1	–	1	1	1	1
5.1–10.0%		1	1	–	–	4	4	2	2
10.1–20.0%		2	2	–	1	4	2	4	3
20.1–40.0%		1	–	1	–	–	–	1	–
40.1–100.0%		–	–	–	–	1	1	1	1
Total number of weed taxa		15	7	4	5	13	17	14	15

In the spring of 2021, a large share of covering by *Trifolium repens* L. and *Taraxacum* sp. was already observed in the tree rows. These were two perennial weeds that were exceptionally bothering for orchard production using living mulch [16,25]. The use of different seed sowing doses did not have a significant impact on the coverage of these weed species. The share of typical monocotyledonous weeds in cover crops, such as *Elymus repens* (L.) Gould [27,47] or other species from the Poaceae family [16,47], was noted clearly in the present study when living meadow grass mulch was sown at the basic standard of only 50 kg of seeds per ha. In both grasses tested in experiment no. 2, the number of identified weed taxa increased from 13 to 17 and was similar to the blue fescue sod, which was sown at the standard dose of 50 kg of seeds per ha. The coverage of several of these weed taxa did not exceed 1% of the soil surface. They can therefore be considered as species contributing to the biodiversity of the agricultural landscape [48].

3.3. Soil Properties under Living Mulch Cover

The presence of living grass mulches and their weediness compared to the herbicide fallow did not result in significant differences in soil conditions in the first five years of apple tree cultivation (Table 4).

The soil reaction in all treatments was similar, and the organic C content did not exceed 11.0 g per kg. A similar content of organic C was determined by Ramos et al. [49] under native grass cover. In this experiment, only the frequent tillage of soil contributed to a significant decline in organic matter after several months. A significant effect of various living mulches, including perennial ryegrass, on the increase of organic C compared to mechanical soil cultivation was also shown after three years of research by Qian et al. [15]. In our five-year study, the control treatment, herbicide fallow, did not have such a strong effect on reducing soil organic C. But, the twelve-year impact of two living grass mulches in comparison with herbicide fallow and black woven polypropylene fabric improved the humus content of the soil [16]. Also, the three-year cultivation of various grass mixtures with dicotyledons or only dicotyledons contributed to an increase in the organic matter content of the soil [17].

Table 4. Selected chemical and physical–chemical properties of soil under living meadow grass or perennial ryegrass mulch sods and herbicide fallow in the tree rows—experiment no. 2, spring 2022 (mean ± SD, n = 4).

Specification	Herbicide Fallow	Living Grass Mulch Species and Seed Sowing Dose (kg·ha^{-1})			
		Meadow Grass		Perennial Ryegrass	
		50	150	50	150
pH $_{KCl}$	5.5 ± 0.8 a	5.6 ± 0.2 a	5.4 ± 0.5 a	5.4 ± 0.6 a	5.6 ± 0.9 a
$C_{org.}$ (g·kg^{-1})	10.8 ± 0.9 a	10.6 ± 1.1 a	10.5 ± 1.0 a	11.0 ± 1.2 a	10.6 ± 1.5 a
P (mg·kg^{-1})	36 ± 17 a	39 ± 9 a	42 ± 14 a	42 ± 13 a	42 ± 18 a
K (mg·kg^{-1})	103 ± 46 a	113 ± 22 a	105 ± 36 a	130 ± 68 a	127 ± 41 a
Mg (mg·kg^{-1})	88 ± 16 a	85 ± 11 a	80 ± 8 a	86 ± 15 a	87 ± 13 a
K:Mg	1.13 ± 0.32 a	1.31 ± 0.16 a	1.31 ± 0.36 a	1.45 ± 0.54 a	1.43 ± 0.25 a

Means marked with different letters in rows represent statistical differences among treatments (one-way ANOVA, Tukey test, $p \leq 0.05$).

The content of available phosphorus in the soil remained at an average level with herbicide fallow treatment. It was high in meadow grass sown at a dose of 150 kg per ha and in both treatments with perennial ryegrass. Regardless of the treatment, potassium and magnesium remained at medium and high levels, respectively, and the K:Mg ratio was always proper. A significant increase in soil phosphorus and potassium content was noted by Qian et al. [15] after three years of research in which mechanical cultivation was replaced by the use of several cover crops. Among them, living perennial ryegrass mulch provided significantly higher potassium contents. In other studies, compared to long-term herbicide fallow treatment, the presence of grass cover crops had a positive impact on the total soil nutrient contents [16].

3.4. Growth and Yield of Apple Trees

In the autumn of 2016, before the sowing of living mulches, stronger growth of the 'Chopin' apple tree on the strong semi-dwarf rootstock MM.106 was recorded in experiment no. 1 compared to experiment no. 2 (Tables 5 and 6).

Table 5. Yield and tree growth of 'Chopin' cv. depending on living blue fescue or red fescue mulch sods and herbicide fallow—experiment no. 1 (mean ± SD, n = 4).

Specification	Herbicide Fallow	Living Grass Mulch Species and Seed Sowing Dose (kg·ha^{-1})			
		Blue Fescue		Red Fescue	
		50	150	50	150
TCSA autumn 2016 (cm^2)	2.46 ± 0.88 a	2.69 ± 0.72 a	2.69 ± 0.41 a	2.16 ± 0.48 a	2.20 ± 0.56 a
TCSA increment spring 2016–autumn 2021 (cm^2)	24.62 ± 9.22 a	21.93 ± 4.25 a	23.96 ± 5.46 a	20.64 ± 5.71 a	18.51 ± 8.87 a
TCSA autumn 2021 (cm^2)	27.08 ± 10.09 a	24.63 ± 4.97 a	26.65 ± 5.86 a	22.80 ± 6.17 a	20.70 ± 9.40 a
Mean fruit number 2017–2021 (no·tree^{-1})	49 ± 8 b	30 ± 10 a	27 ± 5 a	24 ± 3 a	22 ± 3 a
Total yield 2017–2021 (kg·tree^{-1})	26.74 ± 3.30 b	18.82 ± 6.24 ab	16.66 ± 3.31 a	14.27 ± 0.90 a	12.34 ± 3.71 a
Crop efficiency coefficient 2017–2021 (kg·cm^{-2})	1.13 ± 0.55 a	0.76 ± 0.17 a	0.65 ± 0.20 a	0.67 ± 0.24 a	0.64 ± 0.14 a

TCSA—trunk cross-sectional area. Means marked with different letters in rows represent statistical differences among treatments (one-way ANOVA, Tukey test, $p \leq 0.05$).

Table 6. Yield and tree growth of 'Chopin' cv. with living meadow grass or perennial ryegrass mulch sods or herbicide fallow—experiment no. 2 (mean ± SD, n = 4).

Specification	Herbicide Fallow	Living Grass Mulch Species and Seed Sowing Dose (kg·ha^{-1})			
		Meadow Grass		Perennial Ryegrass	
		50	150	50	150
TCSA autumn 2016 (cm^2)	1.79 ± 0.11 a	1.58 ± 0.21 a	1.82 ± 0.14 a	1.70 ± 0.32 a	1.73 ± 0.20 a
TCSA increment spring 2016–autumn 2021 (cm^2)	16.64 ± 1.73 b	10.16 ± 1.61 a	11.64 ± 3.28 ab	10.44 ± 2.72 a	8.22 ± 0.80 a
TCSA autumn 2021 (cm^2)	18.42 ± 1.69 b	11.74 ± 1.80 a	13.46 ± 3.23 ab	12.13 ± 3.01 a	9.95 ± 0.98 a
Mean fruit number 2017–2021 (no·tree^{-1})	47 ± 14 b	23 ± 3 a	26 ± 3 a	16 ± 6 a	19 ± 2 a
Total yield 2017–2021 (kg·tree^{-1})	25.87 ± 5.72 b	11.99 ± 2.96 a	13.29 ± 1.47 a	8.50 ± 4.07 a	9.48 ± 2.05 a
Crop efficiency coefficient 2017–2021 (kg·cm^{-2})	1.39 ± 0.19 b	1.03 ± 0.25 ab	1.04 ± 0.31 ab	0.68 ± 0.22 a	0.96 ± 0.24 ab

TCSA—trunk cross-sectional area. Means marked with different letters in rows represent statistical differences among treatments (one-way ANOVA, Tukey test, $p \leq 0.05$).

These differences determined the growth of trees in subsequent years of research. Five years later, regardless of the sowing dose for the living grass mulches, all trees grown in fescue treatments had similar growth and trunk cross-sectional areas. Trees in herbicide fallow did not grow significantly stronger. This was similar in the experiment with good-quality nursery material, i.e., apple trees of the 'Ligol' cv. on semi-dwarf rootstocks with living blue fescue mulch [22]. Smaller trees whose trunk diameter did not exceed 2 cm^2 at the time of sowing living mulches in experiment no. 2 reacted more strongly to their presence (Table 6). Their increment and trunk cross-sectional area were often significantly lower compared to those noticed in cases of trees cultivated with herbicide fallow. Such a negative impact of living grass mulches was observed in other studies, not only in young apple orchards on dwarf rootstocks [22,26] but also when trees were grafted on the strongly growing MM.111 rootstock with the use of the M.9 interstock [24]. With the same soil nutrient content in all treatments in experiment no. 2 (Table 3), it can be assumed that the rapid development of living mulch sod in the orchard contributed to the limitation of tree growth. This was also due to the more abundant weed cover for meadow grass and perennial ryegrass than in experiment no. 1 for both fescue sods (Tables 2 and 3). Unfortunately, a negative impact of the presence of soil grubs on the condition and growth of the tree root system in experiment no. 2 cannot be excluded. This could have weakened the growth of young trees in the first years after establishing the orchard and additionally strengthened the negative impact of living mulch on the level of tree yield (Table 6).

The influence of both fescue species in experiment no. 1 sown at a dose of 150 kg per ha and of red fescue sown at a dose of 50 kg per ha were also factors that significantly decreased the number of fruits and yield of trees cultivated with living mulches compared to herbicide fallow (Table 5). This reaction of trees to the presence of a cover crop was expected because other previous studies showed the influence of the competition of an additional plant in the orchard on the level of apple tree fruiting [22,26]. However, the very low yield of all trees in both experiments is worthy of attention. It increased in subsequent years of the study, but even the most cropping trees in both control treatments with herbicide fallow gained only over 10 kg per tree in the fifth year (Tables S2 and S3). Such a low yield could have been the result of an inappropriately selected MM.106 rootstock for the new scab-resistant variety 'Chopin'. Although it is a semi-dwarf rootstock, it grows relatively strongly [50,51], which unfortunately delays trees' entry into full fruiting [50]. In the present study, this was also confirmed by the assessment of tree blooming, which most often remained below a value of 2, which indicated blooming below the average level (Figures S2 and S3). In the present research, the low yield per tree cultivated with fescue resulted in the crop efficiency coefficient being below unity, as in another experiment with red fescue [24]. This was similar in both treatments with perennial ryegrass.

3.5. Blushing, Mean Weight, and Size of Fruit

An unexpected effect of using living mulch in the orchard that was significant in experiment no. 2 was an increase in the share of apples with a red-pink color exceeding 25% of the skin surface compared to the herbicide fallow control (Table 7). However, the share of 'green peel' 'Chopin' cv. with blushing greater than 75% was occasional. The influence of cover plants on the increase in the red blush color area has been shown in various research experiments, but it has been connected with 'red skin' apples [7,22]. However, Atay et al. [32] did not observe differences in the basic yellow color of the fruit 'yellow skin' 'Golden Delicious' cv. when assessing the different intensities of weed infestation in the orchard. The competition of weeds against the tree can, to some extent, be compared to the competition created by our cover plants. Additionally, in the experiment of Atay et al. [32], the intensity of only the deep yellow color changed under the influence of yearly weather conditions. In the case of our research, different thermal conditions in September, in the period preceding fruit harvest, contributed to a significant increase in the share of red blush apples of the 'green-peel' variety (Tables 1 and S1). This was observed in both research experiments, but the interaction effect (living mulch × year) was significant only in the meadow grass and perennial ryegrass experiments.

Sawicka et al. [52] reported that the mean fruit weight of 'Chopin' cv. was 183 g. In our experiment, the fruit was smaller in all years of the study (Table 7). In 2019–2020, in experiment no. 1, the mean fruit weight was significantly lighter compared to 2018. Similar trends were noted in experiment no. 2. In both experiments, in the first two years of evaluation, a high share of apples with a diameter of 6.5–7.5 cm was recorded. In the next two years, the share of larger apples (7.5–8.5 cm) increased, even though the yield per tree most often increased compared to 2018–2019 (Tables S1 and S2). This could have been due to a much better supply of water to trees in 2020 and 2021 compared to 2018–2019 (Tables 1 and S1). The use of living mulch had no effect on fruit weight and diameter in experiment no. 1, similar to the findings of Bałuszyńska et al. [7]. However, other authors mentioned the negative impact of cover crops on fruit weight and size [22,25]. This tendency was present in experiment no. 2. Significant differences were noted when the herbicide fallow was replaced with meadow grass sown at a rate of 50 kg of seeds per ha.

Table 7. Mean fruit weight, blushing, and fruit size of 'Chopin' cv. with living grass mulch sods or herbicide fallow by year (mean ± SD, n = 4).

Specification	Mean Fruit Weight (g)	% of Fruit with Blush on Skin Surface Area			% of Fruit with Diameter (cm)			
		>75%	25–75%	<25%	>8.5	7.5–8.5	6.5–7.5	<6.5
Experiment No. 1								
GLMHL:								
Herbicide fallow	121 ± 9 a	–	12 ± 4 a	88 ± 4 a	1 ± 1	17 ± 5 a	57 ± 5 a	26 ± 10 a
Blue fescue 50	134 ± 8 a	–	20 ± 8 a	80 ± 8 a	2 ± 3	19 ± 8 a	57 ± 8 a	21 ± 8 a
Blue fescue 150	133 ± 6 a	–	20 ± 4 a	80 ± 4 a	1 ± 1	21 ± 4 a	59 ± 6 a	19 ± 6 a
Red fescue 50	138 ± 5 a	–	26 ± 1 a	74 ± 1 a	2 ± 2	23 ± 5 a	53 ± 4 a	22 ± 7 a
Red fescue 150	135 ± 13 a	–	22 ± 13 a	78 ± 14 a	3 ± 2	24 ± 7 a	53 ± 6 a	21 ± 12 a
YEAR:								
2018	142 ± 10 c	–	42 ± 2 c	58 ± 2 a	–	14 ± 9 a	71 ± 2 b	15 ± 9 a
2019	126 ± 6 ab	–	23 ± 7 b	77 ± 8 b	1 ± 1	14 ± 6 a	65 ± 2 b	20 ± 6 a
2020	120 ± 8 a	–	9 ± 12 a	91 ± 12 c	2 ± 1	23 ± 12 ab	46 ± 5 a	29 ± 9 a
2021	140 ± 15 bc	–	6 ± 3 a	94 ± 3 c	4 ± 3	32 ±12 b	41 ± 4 a	23 ± 18 a
GLMHL × YEAR	NS	–	NS	NS	–	NS	NS	NS

Table 7. *Cont.*

Specification	Mean Fruit Weight (g)	% of Fruit with Blush on Skin Surface Area			% of Fruit with Diameter (cm)			
		>75%	25–75%	<25%	>8.5	7.5–8.5	6.5–7.5	<6.5
				Experiment No. 2				
GLMHL:								
Herbicide fallow	127 ± 3 b	–	14 ± 4 a	86 ± 4 c	3 ± 3	19 ± 4 b	51 ± 7 a	27 ± 8 a
Meadow grass 50	108 ± 11 a	4 ± 6	17 ± 6 ab	77 ± 3 bc	–	4 ± 5 a	39 ± 9 a	57 ± 13 b
Meadow grass 150	120 ± 5 ab	–	25 ± 6 bc	75 ± 6 abc	1 ± 1	9 ± 3 ab	51 ± 9 a	39 ± 8 ab
Perennial ryegrass 50	126 ± 12 b	–	26 ± 3 bc	73 ± 3 ab	1 ± 1	15 ± 8 b	50 ± 5 a	34 ± 13 a
Perennial ryegrass 150	119 ± 9 ab	1 ± 2	34 ± 6 c	64 ± 4 a	5 ± 5	11 ± 2 ab	43 ± 10 a	40 ± 13 ab
YEAR:								
2018	128 ± 6 b	–	50 ± 5 c	50 ± 5 a	–	10 ± 5 ab	50 ± 10 b	40 ± 13 a
2019	111 ± 7 a	–	25 ± 3 b	75 ± 3 b	–	6 ± 3 a	50 ± 5 b	44 ± 8 a
2020	119 ± 13 ab	5 ± 4	11 ± 5 a	84 ± 4 bc	6 ± 3	16 ± 6 b	35 ±7 a	43 ± 8 a
2021	123 ± 11 ab	–	7 ± 2 a	93 ± 2 c	2 ± 1	15 ± 5 b	51 ± 5 b	31 ± 9 a
GLMHL × YEAR	***	–	***	**	–	***	*	NS

GLMHL— living grass mulch species and seed sowing dose (kg·ha^{-1}) or herbicide fallow. NS—not significant. * Statistically significant differences at p value ≤ 0.05. ** Statistically significant differences at p value ≤ 0.01. *** Statistically significant differences at p value ≤ 0.001. Means marked with different letters in separate columns represent statistical differences among treatments (two-way ANOVA, Tukey test, $p \leq 0.05$).

3.6. Influence of Agrotechnical Methods versus Weather Conditions on Fruit Quality

Living mulch is one of the floor management options that eliminates the use of glyphosate in rows of fruit trees. It has a positive influence on the agricultural environment. On the other hand, living mulch competes with fruit trees, which was also visible in the present study (Tables 5 and 6). In these studies, there was also a contrasting milder effect of the tested fescues on the growth and yield of apple trees compared to the stronger effect of perennial ryegrass and meadow grass. However, it should be taken into account that at the time of sowing mulch seeds, the trees in experiment no. 2 were weaker compared to those grown in experiment no. 1 with both fescues. Access to available forms of potassium, phosphorus, and magnesium in the soil can be considered correct (Table 3).

The development of trees in experiment no. 2 was probably limited by the nitrogen availability, which also benefited the cover crops, as in the studies of TerAvest et al. [53]. Another limiting factor was probably water. Its availability is closely related to the biomass production of cover crops [43] and is also used by perennial weeds, the appearance of which in experiment no. 2 was more abundant than in experiment no. 1. Such conditions limiting the growth and development of trees in perennial ryegrass and meadow grass also determined the quality of the fruit, especially its blush color. The synthesis of anthocyanins, responsible for the red appearance of apples, is not favored by excessive tree vigor, especially excessive shoot growth [54]. As shown by Andersen et al. [26], in the presence of cover plants, the annual increment of shoots is significantly reduced. Hence, also in the present research in experiment no. 2, in which grass cover crops significantly weakened the growth of trees (Table 6), an increase in the share of apples was observed where the red blush color reached 25–75% of the fruit skin surface. Limiting the conditions of tree growth and development had a weaker impact on the average weight and size of apples. But such a situation occurred only with meadow grass cultivation with sowing of 50 kg of seeds per ha, when both the mean fruit weight and the share of fruit in the 7.5–8.5 cm apple class were significantly lower compared to the herbicide fallow control.

Fruit quality is determined by agronomic factors, such as orchard design and the training system or pruning of trees, and environmental factors, such as temperature and light distribution. It is also regulated by crop load and fruit set thinning [29]. In our research, the unification of agrotechnical methods in all treatments with living mulches and herbicide fallow excluded the influence on the weight, size, or blushing of the fruit of factors other than the presence of a cover crop. However, it seems that the relatively low level of yield, especially in the case of weaker-growing and low-yielding trees in experiment no. 2, also did not have a negative impact on fruit quality in the young orchard of 'Chopin' cv. (Tables S2 and S3). However, weather conditions were variable (Tables 1 and S1). In both experiments, the share of apples with red blush on the skin surface from 25 to 75% was significantly higher in 2018 compared to 2019 (Table 7). The share of red color apples in the first two years of fruiting differed significantly from those obtained in 2020 and 2021. This phenomen was related to low temperatures during fruit ripening, which is a factor that stimulates the red color of apples [33,35]. In our studies, the number of days with a minimum temperature ranging from 0 to 5.0 °C was positively correlated with an increase in the share of fruits with a red color covering 25–75% of the skin surface in each orchard floor management (Table 8).

Table 8. Linear correlation coefficient between selected thermal climatic conditions during fruit ripening or total annual precipitation in the years 2018–2021 and two fruit quality features of 'Chopin' cv. in relation to living grass mulch sods and herbicide fallow.

Specification	No. of Days with Low Temperature (°C)		No. of Days with High Temperature (°C)		No. of Days with Daily Temperature Amplitude (°C)		Total January–September Precipitation (mm)
	0.0–5.0	0.0–10.0	>20.0	>25.0	10.0–20.0	15.0–20.0	
Experiment No. 1							
% of fruit with blush on skin surface area 25–75%:							
Herbicide fallow	0.71	−0.71	NS	NS	NS	NS	−0.67
Blue fescue 50	0.65	−0.65	NS	NS	NS	NS	−0.62
Blue fescue 150	0.80	−0.73	NS	NS	0.57	NS	−0.67
Red fescue 50	0.78	−0.75	NS	NS	0.57	NS	−0.67
Red fescue 150	0.56	NS	NS	NS	0.59	NS	NS
% of fruit with diameter 7.5–8.5 cm:							
Herbicide fallow	NS	NS	NS	NS	NS	NS	NS
Blue fescue 50	NS	NS	NS	NS	NS	NS	NS
Blue fescue 150	NS	NS	0.56	NS	NS	NS	NS
Red fescue 50	NS	NS	NS	NS	NS	NS	NS
Red fescue 150	NS	0.52	NS	NS	NS	NS	NS
Experiment No. 2							
% of fruit with blush on skin surface area 25–75%:							
Herbicide fallow	0.50	−0.81	NS	NS	NS	NS	−0.77
Meadow grass 50	0.68	−0.63	NS	NS	NS	NS	−0.58
Meadow grass 150	0.64	NS	NS	NS	0.59	NS	NS
Perennial ryegrass 50	0.86	−0.79	NS	NS	0.58	NS	−0.73
Perennial ryegrass 150	0.93	−0.53	NS	0.64	0.80	0.59	NS
% of fruit with diameter 7.5–8.5 cm:							
Herbicide fallow	NS	NS	0.50	NS	NS	NS	NS
Meadow grass 50	NS	NS	NS	NS	NS	NS	NS
Meadow grass 150	−0.56	0.71	NS	NS	NS	NS	0.65
Perennial ryegrass 50	NS	NS	NS	NS	NS	NS	NS
Perennial ryegrass 150	NS	0.92	NS	NS	NS	NS	0.96

NS—not significant (linear correlation $p = 0.05$).

There were the most such days in 2018 (Table 1). In the following year, 2019, which also stimulated the red coloring of apples, there were fewer such days. However, in the last decade before harvest, the number of days with a daily temperature amplitude range from 10.0 to 20.0 °C was high, which was also significantly correlated with the increase in the share of red-skinned apples cultivated in living mulch. The influence of weather conditions on the weight and size of the fruit was small. However, it is worth noting that in experiment no. 2, with the increase in precipitation, the share of large fruits with a diameter of 7.5–8.5 cm increased significantly in two treatments with the sowing of meadow bluegrass and grass ryegrass at a dose of 150 kg of seeds per ha (Table 8). This proves that the improvement of water conditions mitigated the competitive impact of dense sod against fruit trees.

In our experience, the external quality of the fruit of new, scab-resistant 'green peel' 'Chopin' cv. was primarily influenced by weather conditions. Agricultural practice, similar to the experiment of Le Bourvellec et al. [6], was of secondary importance. Other comprehensive assessments of new 'Chopin' cv. trees have been carried out in recent years and they showed significant changes in the quality of the internal fruit as a result of changing weather conditions in subsequent growing seasons [52,55]. Additionally, attention was paid to the high acidity of apples, which seems to be an advantage of this cultivar compared to commonly produced 'red skin' apples in Poland, e.g., 'Idared' or 'Sampion' [55]. Such properties of the fruit also broaden the spectrum of its use compared to other sweeter 'green peel' apples like 'Golden Delicious' or 'Mutsu' [52]. As we have shown, in the example of a young orchard with 'Chopin' cv., the first effect of the presence of living mulch on the group of 'green apple' cultivars was the appearance of a certain amount of fruit with slightly red-pink colored skin. In favorable growing seasons, the share of such fruits will increase significantly. The producer's care for the agricultural environment and an increase in consumer awareness should favor the acceptance of the presence of red blushing on 'green peel' apples (Scheme S1). This is also justified by the fact that as the area of red skin color increases, the share of anthocyanins in the apple peel also increases [54], which are phenolic compounds that are important for human health [56,57].

4. Conclusions

Several years of research on the new scab-resistant apple tree cultivar 'Chopin' shows that the yield and growth of the trees were influenced by cover crop presence as well as grafting trees on one of the strongest semi-dwarf rootstocks—MM.106. The quality of fruit was determined by environmental rather than agrotechnical factors. Nevertheless, an unexpected effect of the presence of living mulch was its significant influence on fruit appearance, showing a red-pink blush on the 'green peel' fruit of 'Chopin' cv. Low temperatures and moderate rainfall during the apple ripening period enhanced the red coloration of the fruit skin. Modifications of fruit quality by the presence of a cover crop and weather conditions were stronger in the case of weaker trees cultivated with meadow grass and perennial ryegrass compared to more vigorous trees grown in fescue grasses. This was influenced by the early germination and faster development of these living grass mulches and the presence of weeds, mainly *Trifolium repens* L., in their sod. Increasing the sowing rate of grass seeds had no direct impact on the quality of the fruit. The presence of living mulch and weather conditions had little effect on fruit mean weight and fruit size.

Supplementary Materials: The following supporting information can be downloaded at https://www.mdpi.com/article/10.3390/agriculture14040610/s1. Figure S1: Daily minimum and maximum temperatures at the Wrocław-Strachowice Station (51°12′ N, 16°87′ E) in September in the years 2018–2021; Figure S2: Blooming of 'Chopin' trees with living blue fescue or red fescue mulch sods or herbicide fallow—experiment no. 1 (scale 1–6); Figure S3: Blooming of 'Chopin' trees with living meadow grass or perennial ryegrass mulch sods or herbicide fallow—experiment no. 2 (scale 1–6); Table S1: Mean temperatures and total precipitation at the Wrocław-Strachowice Station (51°12′ N, 16°87′ E) in the years 2016–2021; Table S2: Fruit number and yield of 'Chopin' cv. with living blue fescue or red fescue mulch sods or herbicide fallow—experiment no. 1, in the individual years since 2017 up to 2021 (mean ± SD, n = 4); Table S3: Fruit number and yield of 'Chopin' cv. with living meadow grass or perennial ryegrass mulch sod or herbicide fallow—experiment no. 2, in the individual years since 2017 up to 2021 (mean ± SD, n = 4); Scheme S1: Red-pink blushing on the fruit skin surface of the green–peel 'Chopin' cv. in the autumn of 2018.

Author Contributions: Conceptualization, M.L.-M.; methodology, M.L.-M.; software, U.B.B.; validation, M.L.-M. and U.B.B.; formal analysis, M.L.-M. and U.B.B.; investigation, M.L.-M. and U.B.B.; resources, M.L.-M. and U.B.B.; data curation, M.L.-M.; writing—original draft preparation, M.L.-M. and U.B.B.; writing—review and editing, M.L.-M. and U.B.B.; visualization, M.L.-M. and U.B.B.; supervision, M.L.-M.; project administration, M.L.-M.; funding acquisition, M.L.-M. All authors have read and agreed to the published version of the manuscript.

Funding: This work was supported by the Wrocław University of Environmental and Life Sciences (Poland) as a part of the research program "MISTRZ", no. N090/0012/22.

Institutional Review Board Statement: Not applicable.

Data Availability Statement: All data are present in the manuscript and Supplementary Materials.

Conflicts of Interest: The authors declare no conflicts of interest.

References

1. Sivaperumal, P.; Thasale, R.; Kumar, D.; Mehta, T.G.; Limbachiya, R. Human Health Risk Assessment of Pesticide Residues in Vegetable and Fruit Samples in Gujarat State, India. *Heliyon* **2022**, *8*, e10876. [CrossRef]
2. Jankowska, M.; Hrynko, I.; Łozowicka, B. Human Health Risk Assessment of Pesticide Residues in Fruit, Vegetable and Cereal Samples from Poland–a 5-Year Survey. *J. Plant Prot. Res.* **2023**, *62*, 385–392. [CrossRef]
3. Qi, Y.; Li, J.; Guan, X.; Yan, B.; Fu, G.; He, J.; Du, L.; Zhao, C.; Zhang, D. Effects of Herbicides on Non-Target Plant Species Diversity and the Community Composition of Fallow Fields in Northern China. *Sci. Rep.* **2020**, *10*, 9967. [CrossRef] [PubMed]
4. Fogliatto, S.; Ferrero, A.; Vidotto, F. Current and Future Scenarios of Glyphosate Use in Europe: Are There Alternatives? *Adv. Agron.* **2020**, *163*, 219–278. [CrossRef]
5. Paušič, A.; Tojnko, S.; Lešnik, M. Permanent, Undisturbed, in-Row Living Mulch: A Realistic Option to Replace Glyphosate-Dominated Chemical Weed Control in Intensive Pear Orchards. *Agric. Ecosyst. Environ.* **2021**, *318*, 107502. [CrossRef]
6. Le Bourvellec, C.; Bureau, S.; Renard, C.M.G.C.; Plenet, D.; Gautier, H.; Touloumet, L.; Girard, T.; Simon, S. Cultivar and Year Rather than Agricultural Practices Affect Primary and Secondary Metabolites in Apple Fruit. *PLoS ONE* **2015**, *10*, e0141916. [CrossRef]
7. Baluszynska, U.B.; Licznar-Malanczuk, M.; Medic, A.; Veberic, R.; Grohar, M.C. Influence of Living Mulch and Nitrogen Dose on Yield and Fruit Quality Parameters of *Malus domestica* Borkh. cv. 'Sampion'. *Agriculture* **2023**, *13*, 921. [CrossRef]
8. Bałuszyńska, U.B.; Rowińska, M.; Licznar-Małańczuk, M. Grass Species as Living Mulches–Comparison of Weed Populations and Their Biodiversity in Apple Tree Rows and Tractor Alleys. *Acta Agrobot.* **2022**, *75*, 758. [CrossRef]
9. Demir, Z.; Tursun, N.; Dogan, I. Effects of Different Cover Crops on Soil Quality Parameters and Yield in an Apricot Orchard. *Int. J. Agric. Biol.* **2019**, *21*, 25–33.
10. Mia, M.J.; Massetani, F.; Murri, G.; Neri, D. Sustainable Alternatives to Chemicals for Weed Control in the Orchard–A Review. *Hortic. Sci.* **2020**, *47*, 1–12. [CrossRef]
11. Golian, J.; Anyszka, Z.; Kwiatkowska, J. Multifunctional Living Mulches for Weeds Control in Organic Apple Orchards. *Acta Sci. Pol. Hortorum Cultus* **2023**, *22*, 73–84. [CrossRef]
12. Hammermeister, A.M. Organic Weed Management in Perennial Fruits. *Sci. Hortic.* **2016**, *208*, 28–42. [CrossRef]
13. Hussain, S.; Sharma, M.K.; War, A.R.; Hussain, B. Weed Management in Apple Cv. Royal Delicious by Using Different Orchard Floor Management Practices. *Int. J. Fruit. Sci.* **2020**, *20*, 891–921. [CrossRef]
14. Haring, S.; Gaudin, A.C.M.; Hanson, B.D. Functionally Diverse Cover Crops Support Ecological Weed Management in Orchard Cropping Systems. *Renew. Agric. Food Syst.* **2024**, *38*, e54. [CrossRef]
15. Qian, X.; Gu, J.; Pan, H.J.; Zhang, K.Y.; Sun, W.; Wang, X.J.; Gao, H. Effects of Living Mulches on the Soil Nutrient Contents, Enzyme Activities, and Bacterial Community Diversities of Apple Orchard Soils. *Eur. J. Soil Biol.* **2015**, *70*, 23–30. [CrossRef]
16. Żelazny, W.R.; Licznar-Małańczuk, M. Soil Quality and Tree Status in a Twelve-Year-Old Apple Orchard under Three Mulch-Based Floor Management Systems. *Soil Tillage Res.* **2018**, *180*, 250–258. [CrossRef]
17. Brewer, M.; Kanissery, R.G.; Strauss, S.L.; Kadyampakeni, D.M. Impact of Cover Cropping on Temporal Nutrient Distribution and Availability in the Soil. *Horticulturae* **2023**, *9*, 1160. [CrossRef]
18. Chen, G.; Liu, S.; Xiang, Y.; Tang, X.; Liu, H.; Yao, B.; Luo, X. Impact of Living Mulch on Soil C:N:P Stoichiometry in Orchards across China: A Meta-Analysis Examining Climatic, Edaphic, and Biotic Dependency. *Pedosphere* **2020**, *30*, 181–189. [CrossRef]
19. Yang, J.; Duan, Y.; Zhang, R.; Liu, C.; Wang, Y.; Li, M.; Ding, Y.; Awasthi, M.K.; Li, H. Connecting Soil Dissolved Organic Matter to Soil Bacterial Community Structure in a Long-Term Grass-Mulching Apple Orchard. *Ind. Crops Prod.* **2020**, *149*, 112344. [CrossRef]
20. Furmanczyk, E.M.; Malusà, E.; Kozacki, D.; Tartanus, M. Insights into the Belowground Biodiversity and Soil Nutrient Status of an Organic Apple Orchard as Affected by Living Mulches. *Agriculture* **2024**, *14*, 293. [CrossRef]
21. Capri, C.; Gatti, M.; Fiorini, A.; Ardenti, F.; Tabaglio, V.; Poni, S. A Comparative Study of Fifteen Cover Crop Species for Orchard Soil Management: Water Uptake, Root Density Traits and Soil Aggregate Stability. *Sci. Rep.* **2023**, *13*, 721. [CrossRef] [PubMed]
22. Licznar-Małańczuk, M.; Kwiecińska, I. Application of Living Mulch in Rows of the Apple Trees on Several Rootstock-Long-Term Evaluation. *Acta Sci. Pol. Hortorum Cultus* **2023**, *22*, 79–90. [CrossRef]
23. Kühn, B.F.; Lindhard Pedersen, H. Cover Crop and Mulching Effects on Yield and Fruit Quality in Unsprayed Organic Apple. *Eur. J. Hortic. Sci.* **2009**, *74*, 247–253.
24. Atucha, A.; Merwin, I.A.; Brown, M.G. Long-Term Effects of Four Groundcover Management Systems in an Apple Orchard. *HortScience* **2011**, *46*, 1176–1183. [CrossRef]
25. Tworkoski, T.J.; Michael Glenn, D. Weed Suppression by Grasses for Orchard Floor Management. *Weed Technol.* **2012**, *26*, 559–565. [CrossRef]

26. Andersen, L.; Kühn, B.F.; Bertelsen, M.; Bruus, M.; Larsen, S.E.; Strandberg, M. Alternatives to Herbicides in an Apple Orchard, Effects on Yield, Earthworms and Plant Diversity. *Agric. Ecosyst. Environ.* **2013**, *172*, 1–5. [CrossRef]
27. Tahir, I.I.; Svensson, S.-E.; Hansson, D. Floor Management Systems in an Organic Apple Orchard Affect Fruit Quality and Storage Life. *HortScience* **2015**, *50*, 434–441. [CrossRef]
28. Sosna, I.; Fudali, E. Usefulness of Living Mulch in Rows in a Dwarf Pear, *Pyrus communis* L., Orchard. *Agriculture* **2023**, *13*, 2145. [CrossRef]
29. Musacchi, S.; Serra, S. Apple Fruit Quality: Overview on Pre-Harvest Factors. *Sci. Hortic.* **2017**, *234*, 409–430. [CrossRef]
30. Seppä, L.; Railio, J.; Vehkalahti, K.; Tahvonen, R.; Tuorila, H. Hedonic Responses and Individual Definitions of an Ideal Apple as Predictors of Choice. *J. Sens. Stud.* **2013**, *28*, 346–357. [CrossRef]
31. Drkenda, P.; Ćulah, A.; Spaho, N.; Akagić, A.; Hudina, M. How Do Consumers Perceive Sensory Attributes of Apple? *Foods* **2021**, *10*, 2667. [CrossRef] [PubMed]
32. Atay, E.; Gargin, S.; Esitken, A.; Guzel, N.P.; Atay, A.N.; Altindal, M.; Senyurt, H.; Emre, M. The Effect of Weed Competition on Apple Fruit Quality. *Not. Bot. Horti Agrobot. Cluj-Napoca* **2017**, *45*, 120–125. [CrossRef]
33. Saure, M.C. External Control of Anthocyanin Formation in Apple. *Sci. Hortic.* **1990**, *42*, 181–218. [CrossRef]
34. Singh, A.; Kishore, K.; Kumar, P.; Khapte, P.S.; Mishra, D.S.; Singh, D.; Kothyari, H.S. Phenological Growth and Development Stages of Common Fig (*Ficus Carica* L.) under Arid Climate of India. *Folia Hortic.* **2023**, *35*, 395–402. [CrossRef]
35. Honda, C.; Moriya, S. Anthocyanin Biosynthesis in Apple Fruit. *Hort. J.* **2018**, *87*, 305–314. [CrossRef]
36. Warrington, I.J.; Fulton, T.A.; Halligan, E.A.; De Silva, H.N. Apple Fruit Growth and Maturity Are Affected by Early Season Temperatures. *J. Am. Soc. Hortic. Sci.* **1999**, *124*, 468–477. [CrossRef]
37. Lipecki, J.; Janisz, A. Rozmieszczenie Przestrzenne Oraz Zmiany w Występowaniu Niektórych Gatunków Chwastów w Sadzie Jabłoniowym Gospodarstwa Doświadczalnego Felin Koło Lublina [Spatial Distribution and Changes in Occurrence of Some Weeds Species in the Orchard in AES Felin near Lublin]. *Acta Agrobot.* **2000**, *53*, 85–104.
38. POWO Plants of the World Online. Facilitated by the Royal Botanic Gardens, Kew. Published on the Internet. Available online: http://www.Plantsoftheworldonline.Org/ (accessed on 18 March 2024).
39. Ziadi, N.; Sen Tran, T. Mehlich-3 Extractable Elements. In *Soil Sampling and Methods of Analysis*; Carter, M.R., Gregorich, E.G., Eds.; CRC Press: Boca Raton, FL, USA, 2008; pp. 81–87.
40. Korzeniowska, J.; Stanisławska-Glubiak, E.; Lipiński, W. Development of the Limit Values of Micronutrient Deficiency in Soil Determined Using Mehlich 3 Extractant for Polish Soil Conditions. Part I. Wheat. *Soil Sci. Annu.* **2019**, *70*, 314–323. [CrossRef]
41. Buurman, P.; van Lagen, B.; Velthorst, E.J. *Manual for Soil and Water Analysis*; Bakhuys Publishers Leiden: Amsterdam, The Netherlands, 1996.
42. Sadowski, A.; Nurzyński, J.; Pacholak, E.; Smolarz, K. *Określanie Potrzeb Nawożenia Roślin Sadowniczych II. Zasady, Liczby Graniczne i Dawki Nawożenia. [Determining the Fertilization Needs of Fruit Plants II. Rules, Limit Numbers and Fertilization Doses]*; Warsaw University of Life Sciences: Warszawa, Poland, 1990.
43. Nielsen, D.C.; Lyon, D.J.; Hergert, G.W.; Higgins, R.K.; Holman, J.D. Cover Crop Biomass Production and Water Use in the Central Great Plains. *Agron. J.* **2015**, *107*, 2047–2058. [CrossRef]
44. Haring, S.C.; Hanson, B.D. Agronomic Cover Crop Management Supports Weed Suppression and Competition in California Orchards. *Weed Sci.* **2022**, *70*, 595–602. [CrossRef]
45. Bertin, C.; Senesac, A.F.; Rossi, F.S.; Ditommaso, A.; Weston, L.A. Evaluation of Selected Fine-Leaf Fescue Cultivars for Their Turfgrass Quality and Weed Suppressive Ability in Field Settings. *Horttechnology* **2009**, *19*, 660–668. [CrossRef]
46. Gawryluk, A. The Influence of Sowing Date on Initial Growth and Development of Selected Lawn Varieties of *Festuca Arundinacea*, *Festuca Rubra*, *Festuca Ovina*, *Lolium Perenne* and *Poa Pratensis* on a Roadside Bank. *Agron. Sci.* **2019**, *74*, 57–66. [CrossRef]
47. Licznar-Małańczuk, M. Occurrence of Weeds in an Orchard Due to Cultivation of Long-Term Perennial Living Mulches. *Acta Agrobot.* **2020**, *73*, 1–11. [CrossRef]
48. Mézière, D.; Petit, S.; Granger, S.; Biju-Duval, L.; Colbach, N. Developing a Set of Simulation-Based Indicators to Assess Harmfulness and Contribution to Biodiversity of Weed Communities in Cropping Systems. *Ecol. Indic.* **2014**, *48*, 157–170. [CrossRef]
49. Ramos, M.E.; Robles, A.B.; Sánchez-Navarro, A.; González-Rebollar, J.L. Soil Responses to Different Management Practices in Rainfed Orchards in Semiarid Environments. *Soil Tillage Res.* **2011**, *112*, 85–91. [CrossRef]
50. Kosina, J. Effect of Dwarfing and Semi Dwarfing Apple Rootstocks on Growth and Productivity of Selected Apple Cultivars. *Hort. Sci.* **2010**, *37*, 121–126. [CrossRef]
51. Zenginbal, H.; Öztürk, T.; Faizi, Z.A. Bacterial Strains Effect on the Nursery Plants Growth of 'Granny' Apple Grafted on M-9, MM-106, and MM-111 Rootstocks. *Biocatal. Agric. Biotechnol.* **2023**, *50*, 102747. [CrossRef]
52. Sawicka, M.; Latocha, P.; Łata, B. Peel to Flesh Bioactive Compounds Ratio Affect Apple Antioxidant Potential and Cultivar Functional Properties. *Agriculture* **2023**, *13*, 478. [CrossRef]
53. TerAvest, D.; Smith, J.L.; Carpenter-Boggs, L.; Hoagland, L.; Granatstein, D.; Reganold, J.P. Influence of Orchard Floor Management and Compost Application Timing on Nitrogen Partitioning in Apple Trees. *HortScience* **2010**, *45*, 637–642. [CrossRef]
54. Treutter, D. Biosynthesis of Phenolic Compounds and Its Regulation in Apple. *Plant Growth Regul.* **2001**, *34*, 71–89. [CrossRef]
55. Kistechok, A.; Wrona, D.; Krupa, T. Quality and Nutritional Value of 'Chopin' and Clone 'JB' in Relation to Popular Apples Growing in Poland. *Agriculture* **2022**, *12*, 1876. [CrossRef]

56. Nezbedova, L.; McGhie, T.; Christensen, M.; Heyes, J.; Nasef, N.A.; Mehta, S. Onco-Preventive and Chemo-Protective Effects of Apple Bioactive Compounds. *Nutrients* **2021**, *13*, 4025. [CrossRef] [PubMed]
57. Hyson, D.A. A Comprehensive Review of Apples and Apple Components and Their Relationship to Human Health. *Adv. Nutr.* **2011**, *2*, 408–420. [CrossRef]

agriculture

MDPI

Article

The Use of Soil Surface Mulching on Melon (*Cucumis melo* L.) Production under Temperate Climate Conditions

Katarzyna Adamczewska-Sowińska

Department of Horticulture, Wroclaw University of Environmental and Life Sciences, 50-375 Wroclaw, Poland; katarzyna.adamczewska-sowinska@upwr.edu.pl; Tel.: +48-71-320-17-25

Abstract: Despite its significant thermal requirements, melon is a vegetable species that holds the potential for expanding the crop range in temperate climate regions. The selection of appropriate varieties and agronomic practices facilitates its cultivation in these regions. This experiment, employing a randomized block design, was conducted from 2019 to 2021, and this study evaluated the response of three melon varieties—'Seledyn F1', 'Melba', and 'Malaga F1' (factor I)—to various mulching materials (factor II): black polyethylene film (PE), black polypropylene nonwoven (PP), biodegradable film (Fbio), and giant miscanthus straw. Control plots were left unmulched. Melon seeds were sown on 15 April, and seedlings were transplanted on 31 May at a spacing of 100 × 80 cm. This study assessed yield, fruit number, individual fruit weight, and vertical and horizontal fruit diameters. Under optimal conditions, the varieties Seledyn F1 and Malaga F1 produced fruits with the highest individual mass. The application of synthetic mulches led to a two-fold increase in fruit yield compared to unmulched plots, with a 23.7% increase in fruit number. On average, the largest fruits were obtained with PE mulch. Melons grown on Fbio mulch yielded, on average, 40% less and produced 18.8% fewer fruits compared to those grown with PE mulch.

Keywords: synthetic mulch; organic mulch; biodegradable film; yield; fruit parameters

1. Introduction

Climate change adversely affects agricultural production and contributes to the decline in productivity of traditional major crops [1]. Stress factors such as a very high temperature and a lack of rainfall will appear more and more often, regularly, so that crops are permanently exposed to them [2].

A classification of thermal conditions in Poland indicates that since the beginning of the second half of the 20th century, the conditions have changed a lot [3]. Until the mid-1980s, they were classified as cold and cool conditions every year. Since the second half of the 1980s, conditions described as normal or warmer occurred more often. In the last decade, however, conditions from very warm to extremely warm prevailed. The average air temperature in Poland in 2022 reached 9.5 °C and was 0.8 °C higher than the annual multi-annual average (1991–2020). In the study area (Lower Silesia) was the warmest region, with an average annual air temperature of 10.5 °C. The average total rainfall in Poland in 2022 was 534.4 mm, which is 87.4% of the normal average value determined on the basis of 30-year data (1991–2020). Rainfall was characterized by strong spatial variability (350 to 950 mm) and unevenness during the growing season. The climatic water balance, i.e., the difference between total precipitation and evaporation, in the period from May to October, is often negative in Lower Silesia.

One of the outcomes of global warming is the expansion or limitation of the range of cultivated plants and the introduction of new species and varieties that require higher thermal conditions [4,5]. This direction can positively impact the economy by introducing these species to the local market, reducing imports, and decreasing energy consumption for transportation, thereby creating a shorter supply chain [6–8].

One species that could diversify the vegetable offerings for cultivation in Central Europe is the melon (*Cucumis melo* L.). This species likely originates from Northeast Africa [9]. Presently, it is cultivated in 88 countries worldwide, covering over 1 million hectares, with a yield of 28.6 million tons [10]. In Europe, it is grown on 77,000 hectares, predominantly in Italy, Spain, and Ukraine. In Central Europe, due to unfavorable climatic conditions, melon cultivation is mainly pursued by amateur gardeners and noncommercial production. However, there is increasing interest in its field cultivation. This thermophilic species is highly sensitive to low temperatures and spring and autumn frosts [11], is light-dependent, and has a significant water demand, especially during the fruit-setting period [12]. For the cultivation of thermophilic species, it is recommended to delay the sowing or planting date to improve growth conditions [13]. Another approach is the use of covers made of perforated polyethylene film or polypropylene nonwoven fabric [14–16]. Covering the soil surface with synthetic materials also ensures optimal conditions post-planting and during growth, by increasing soil temperature, reducing evapotranspiration, storing water, and limiting weed growth [17]. The application of synthetic mulches in melon cultivation not only raises soil temperature but also enhances sunlight availability, improves water utilization, and protects the plantation from insects and diseases [18]. Similarly, Parmar et al. [19] highlight the positive effects of mulches on the growth and yield of *Cucurbitaceae* vegetables, such as watermelon.

The enhancement of habitat conditions has a positive impact on plant development, as well as on the yield and quality of melon fruits. High-quality fruits are characterized by appropriate taste, good texture, aroma, and optimal sugar content. The sugar content, which can be positively correlated with the content of soluble substances, is the primary determinant of fruit quality [18]. Nevertheless, a fundamental step in commencing melon cultivation is the selection of a variety suitable for the prevailing climatic conditions. In Central Europe, it is advisable to cultivate varieties adapted to temperate climate conditions, tolerant to cold, and belonging to early and medium-early maturity groups. Singh et al. [20] demonstrated that, in addition to the cultivation location, the variety is a critical factor in determining fruit quality and sugar content. According to these authors, it is crucial in melon breeding programs to precisely identify the factors influencing the phytochemical composition of the fruits. This allows for the recommendation of specific varieties for particular environments, maximizing yield and quality, and enhancing the sensory and functional properties of the fruits.

The objective of this research was to determine the influence of variety and cultivation conditions on the yield and quality of melon fruits. It was hypothesized that the selection of an appropriate variety, adapted to climatic conditions, and the type of mulching material used for the soil surface could have a decisive impact on the yield and fruit quality of melons cultivated under the climatic conditions of Central Europe.

2. Materials and Methods

2.1. Experiment Localization and Design

The experiment was conducted from 2019 to 2021 at the Research and Didactic Station in Psary (51°19′08″ N, 17°03′37″ E), affiliated with the Department of Horticulture at the Wroclaw University of Environmental and Life Sciences. A two-factor experiment was established using a randomized block design with three replications. The study evaluated the response of three Polish-bred melon varieties from W. Legutko Sp. z o.o. [21], 'Seledyn F1', 'Melba', and 'Malaga F1' (factor I), to the application of different mulching materials (factor II). The mulches included black polypropylene nonwoven (PP) with a mass of 60 g m^{-2}, black polyethylene film (PE), 0.05 mm thick, and biodegradable film BioAgri, 0.025 mm thick. BioAgri is produced from Mater-Bi®, a complex bioplastic raw material made from starch with polyesters, certified as biodegradable and compostable (European Standard EN 13432 and the US Standard ASTM D6400). For organic mulch, cut plants of giant miscanthus (*Miscanthus* × *giganteus* Greef et Den.) were used. Control plots were left without mulch.

2.2. Experiment Methodology

Melon seeds were sown in a greenhouse on April 15, point-wise into multi-cell trays (with a cell volume of 0.273 dm^3) filled with peat substrate. Care procedures included watering the seedlings and pinching the shoot tips before transplanting to stimulate branching. Pruning the shoots was performed when the plants had developed 4–5 leaves. The prepared seedlings were acclimate to harsher conditions for a week before planting in the field. Planting took place on 31 May, with a spacing of 100 × 80 cm. The size of each plot was 2.4 × 2.0 m. Before planting, the soil surface was leveled, and mulches were laid in strips 2.0 m wide. The gaps between the mulched areas were 0.5 m.

The field experiment was conducted on degraded calcic black earth, characterized by a light, moderately loamy texture derived from medium clay, classified as class IIIa according to the FAO-WRB Gleyic Calcic Chernozems soil classification and Polish soil classification system [16,22]. In the soil profile, there was a humus content of 1.8%, a pH level of 7.0, and a salinity of 103.1 $\mu S\ cm^{-1}$.

Prior to planting, deep plowing was conducted in autumn following the application of phosphorus and potassium fertilizers. Fertilizer application rates were adjusted based on current soil nutrient levels determined through chemical analysis. Melons were cultivated with optimal nutrient concentrations: 80 mg P dm^{-3} and 200 mg K dm^{-3}. In spring, soil leveling and harrowing operations were performed. Immediately before transplanting, ammonium nitrate was applied at a rate of 150 kg N ha^{-1}, which was thoroughly incorporated into the soil using a rotary tiller.

During the vegetative growth period in the field, plants were irrigated as required, with each irrigation event delivering a 30 mm water dose. Harvesting occurred in 2019 on four occasions spanning from 29 July to 26 August, in 2020 on seven occasions from 3 August to 24 August, and in 2021 on five occasions from 5 August to 26 August. Fruits were harvested at full maturity, when ripe and easily detachable from the peduncle. Marketable yield and fruit count were assessed, accompanied by biometric measurements to determine individual fruit weight (in kg). Vertical and horizontal diameters measure a wider fruit diameter (in cm) and shape coefficient as a ratio of the vertical to horizontal diameter performed. The above-listed parameters depended to a large extent on the mulch type.

Throughout the growing season, soil temperature was monitored at a depth of 10 cm in each experimental plot using a DT-34 digital thermometer (Termoprodukt, Bielawa, Poland). Concurrently, air temperature at the experimental site was continuously recorded using an electronic Temp Logger AZ 8828, and precipitation was measured with a Hellmann rain gauge. The collected data were compared with meteorological records spanning from 1980 to 2010 obtained from the Institute of Meteorology and Water Management.

2.3. Experimental Statistical Analysis

Statistical analyses of the experimental results regarding plant yield and biometric measurement were conducted using an analysis of variance (ANOVA/MANOVA) in Statistica 13 software. Tukey's test was applied to calculate confidence intervals at a significance level of $\alpha = 0.05$.

3. Results and Discussion

3.1. Weather Conditions

The weather conditions prevailing during the study period are presented in Tables 1 and 2. In 2019, the average air temperature exceeded the long-term average. Late May precipitation facilitated sufficient soil moisture, promoting rapid root growth and seedling establishment post-transplanting. Showers in late July and early August intensified melon flowering and fruit set.

Table 1. Temperature within melon growing period in years 2019–2021 [°C].

Months	Period (Month Days)	2019	2020	2021	Average from Years 1991–2020
June	01.–10.	23.2	17.0	19.3	16.8
	11.–20.	25.0	21.5	20.9	
	21.–30.	25.6	20.2	20.6	
	average	24.6	19.6	20.3	
July	01.–10.	19.7	22.0	19.5	18.8
	11.–20.	20.1	19.6	21.2	
	21.–31.	24.7	22.4	20.7	
	average	21.6	21.4	20.5	
August	01.–10.	22.2	23.3	17.7	18.4
	11.–20.	21.3	24.1	19.2	
	21.–31.	24.0	20.4	15.5	
	average	22.5	22.5	17.4	
Average for growing season		22.7	21.2	19.4	18.0

Table 2. Rainfall within melon growing period in years 2019–2021 [mm].

Months	Period (Month Days)	2019	2020	2021	Average from Years 1991–2020
June	01.–10.	0.9	38.4	1.2	73.0
	11.–20.	19.4	68.8	18.2	
	21.–31.	6.6	64.0	28.6	
	sum	26.9	171.2	48.0	
July	01.–10.	2.9	8.2	60.8	88.9
	11.–20.	6.4	18.7	10.4	
	21.–31.	32.2	2.9	10.8	
	sum	41.5	29.8	82.0	
August	01.–10.	14.3	41.5	18.4	64.0
	11.–20.	5.5	6.6	16.4	
	21.–31.	30.5	30.1	84.6	
	sum	50.3	78.2	119.8	
Sum for growing season		118.7	279.2	249.8	203.8

In 2020, excessive rainfall adversely affected melon growth and subsequent yield. These rains occurred post-transplanting and persisted throughout June, surpassing twice the long-term average precipitation. July, however, experienced notably low precipitation (60% below the long-term average). Air temperatures during the vegetative phase were above the long-term average. High temperatures in July, combined with water deficits and during the first two decades of August, were unfavorable for flowering and fruit set.

In 2021, June and July air temperatures averaged 3.5 °C and 1.7 °C higher, respectively, than the long-term average for these months. Low rainfall was recorded at the end of May and June, and in the second and third decades of July, while precipitation at the end of June and the beginning of July supported vegetative growth and flowering of the plants.

During the three-year study period, initial measurements conducted at the onset of melon growth revealed that soil temperatures at a depth of 10 cm remained consistent across plots with no mulch and those covered with PE and Fbio mulches (Figure 1). A slightly lower temperature was observed in plots mulched with PP nonwoven fabric approximately three weeks after planting. Throughout the measurement period, the lowest soil temperatures were consistently recorded under the miscanthus mulch. This temperature disparity deepened over time.

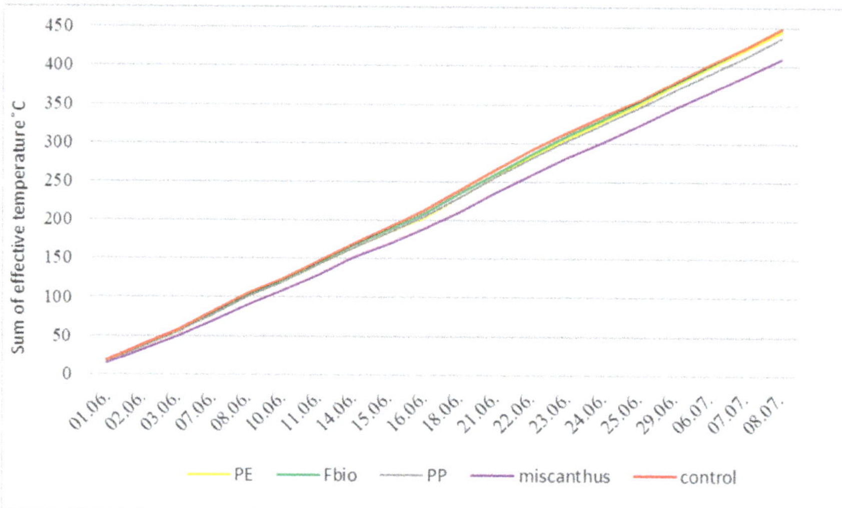

Figure 1. Sum of effective temperature under soil cover. Average from years 2019–2021.

3.2. *Melon Fruit Characteristics*

Melon is categorized among vegetables with the most stringent thermal requirements. Optimal temperatures, both in soil and air, facilitate robust vegetative growth, flowering, and fruit ripening. The species is intolerant to prolonged periods of cold and frost throughout its growth stages, with higher average temperatures accelerating plant development and hastening fruit maturity [23]. According to Kurtar [24], cucurbitaceous plants thrive under temperatures ranging between 22 and 32 °C, while melon can withstand maximum temperatures between 39 and 45 °C. Moreover, melon necessitates consistent water supply during vegetative growth, with peak water demand occurring during flowering and fruit set. As pointed out by Kumlay et al. [2], drought stress, at various stages of plant development, changes the physiological and biochemical processes, which results in reduced crop productivity.

Ekinci and Dursun [25] demonstrated that even in warm climates, melon yields vary between growing seasons. A statistical analysis of the data gathered during this study underscored the significant impact of weather conditions, as well as the choice of variety and mulching material, on the successful cultivation of melon in a moderate climate (Tables 3–6). The year 2019 emerged as the most thermally favorable, yet fruit quality benefited from the conditions observed in 2021, characterized by elevated average air temperatures and moderately distributed rainfall. Similarly, Zaniewicz-Bajkowska et al. [26] emphasized better melon yields in years characterized by high air temperatures and low rainfall during the fruit ripening period. In the present research, all varieties exhibited significantly smaller vertical fruit diameters in 2020 (Table 7). While horizontal fruit diameter did not exhibit statistical differences across study years, a trend towards smaller fruit sizes was observed in 2020 compared to 2019 and 2021, by 20.0% and 13.8%, respectively. The Melba variety consistently yielded fruits of uniform mass across all study years. Furthermore, fruits of the Malaga F1 and Seledyn F1 varieties demonstrated significantly higher weight in 2021 compared to 2019 and 2020, with average increases of 36.8% and 57.9%, respectively.

Table 3. Variances analysis of melon fruit horizontal diameter.

Factors	Sum of Squares	df	Mean Square	F	Significance
Year	1729.64	2	864.82	51.851	<0.001 ***
Variety	1096.16	2	548.08	32.861	<0.001 ***
Cover	1756.42	4	439.11	26.327	<0.001 ***
Year × variety	211.34	4	52.83	3.168	0.013 *
Year × cover	1443.25	8	180.41	10.816	<0.001 ***
Variety × cover	574.32	8	71.79	4.304	<0.001 ***
Year × variety × cover	476.45	16	29.78	1.785	0.028 *

df—degrees of freedom. The F test to determine whether group means are equal is calculated using the variation between sample means/variation within the samples. *** significant at $p \leq 0.001$, * significant at $p \leq 0.05$.

Table 4. Variances analysis of melon fruit vertical diameter.

Factors	Sum of Squares	df	Mean Square	F	Significance
Year	3452.1	2	1726.1	75.067	<0.001 ***
Variety	54.6	2	27.3	1.188	0.305
Cover	2606.9	4	651.7	28.344	<0.001 ***
Year × variety	470.7	4	117.7	5.117	<0.001 ***
Year × cover	1687.6	8	211.0	9.175	<0.001 ***
Variety × cover	927.4	8	115.9	5.041	<0.001 ***
Year × variety × cover	584.7	16	36.5	1.589	0.065

df—degrees of freedom. The F test to determine whether group means are equal is calculated using the variation between sample means/variation within the samples. *** significant at $p \leq 0.001$.

Table 5. Variances analysis of melon fruit weight.

Factors	Sum of Squares	df	Mean Square	F	Significance
Year	27.7	2	13.9	60.6	<0.001 ***
Variety	42.0	2	21.0	91.8	<0.001 ***
Cover	30.0	4	7.5	32.8	<0.001 ***
Year × variety	9.5	4	2.4	10.4	<0.001 ***
Year × cover	21.7	8	2.7	11.8	<0.001 ***
Variety × cover	18.7	8	2.3	10.2	<0.001 ***
Year × variety × cover	8.8	16	0.5	2.4	<0.01 **

df—degrees of freedom. The F test to determine whether group means are equal is calculated using the variation between sample means/variation within the samples. *** significant at $p \leq 0.001$, ** significant at $p \leq 0.01$.

Table 6. Variances analysis of melon fruit vertical to horizontal diameter ratio.

Factors	Sum of Squares	df	Mean Square	F	Significance
Year	23.2	2	11.6	71.5	<0.001 ***
Variety	14.0	2	7.0	43.2	<0.001 ***
Cover	7.3	4	1.8	11.3	<0.001 ***
Year × variety	3.1	4	0.8	4.7	<0.001 ***
Year × cover	9.0	8	1.1	6.9	<0.001 ***
Variety × cover	1.5	8	0.2	1.1	0.342
Year × variety × cover	3.5	16	0.2	1.3	0.167

df—degrees of freedom. The F test to determine whether group means are equal is calculated using the variation between sample means/variation within the samples. *** significant at $p \leq 0.001$.

Table 7. Melon fruit biometric characteristics in following research years.

Variety \ Year	2019	2020	2021
	Vertical diameter (cm)		
Melba	14.1 *b	11.5 a	13.6 b
Malaga F1	15.8 c	11.4 a	13.5 b
Seledyn F1	14.9 bc	11.0 a	15.7 c
	Horizontal diameter (cm)		
Melba	10.8	8.4	9.2
Malaga F1	13.8	11.2	12.8
Seledyn F1	12.9	10.4	12.8
	Fruit weight (kg)		
Melba	0.6 a	0.5 a	0.6 a
Malaga F1	0.9 b	1.0 b	1.3 c
Seledyn F1	0.9 b	1.0 b	1.5 d

* The same lowercase letters mark values belonging to the same homogeneous groups, determined on the basis of statistical analysis for $\alpha = 0.05$.

Fruit size and shape constitute fundamental varietal traits of melon. The study findings have demonstrated significant varietal differences in these characteristics, as well as in fruit mass, as detailed in Tables 8–11. The largest fruits were consistently observed in the Seledyn F1 and Malaga F1 varieties. Specifically, the horizontal diameter and unit mass of Seledyn F1 fruits remained statistically comparable to those of Malaga F1. Conversely, the Melba variety exhibited markedly smaller, elliptical fruits that weighed half as much as the others. Notably, Melba fruits were characterized by their elongated shape, while Malaga F1 fruits displayed a predominantly spherical form (Table 11).

Table 8. Melon fruit vertical diameter (cm). Interaction of cover type and variety. Average from 2019–2021 years.

Cover \ Variety	Melba	Malaga F1	Seledyn F1	Average for Cover
PE	14.5 **c	13.9 c	17.9 d	15.4 *C
Fbio	14.3 c	12.6 abc	13.6 bc	13.6 B
PP	13.3 bc	14.8 c	14.0 c	14.0 BC
miscanthus	9.8 a	11.5 abc	10.4 ab	10.5 A
control	12.6 abc	14.2 c	14.0 c	13.6 B
Average for variety	13.1	13.6	14.3	-

The same letters mark values belonging to the same homogeneous groups, determined on the basis of statistical analysis for $\alpha = 0.05$. * Capital letters refer to the cover; ** lower letters refer to the interaction of variety × cover.

Table 9. Melon fruit horizontal diameter (cm). Interaction of cover type and variety. Average from 2019–2021 years.

Cover \ Variety	Melba	Malaga F1	Seledyn F1	Average for Cover
PE	10.3 **bc	13 def	15.1 f	12.8 *C
Fbio	10.2 bc	11.9 bcde	11.8 bcde	11.2 B
PP	9.5 ab	13.9 ef	12.4 cde	11.7 BC
miscanthus	7.1 a	10.3 bc	8.9 ab	8.7 A
control	9.3 ab	13.3 ef	11.7 bcde	11.4 B
Average for variety	9.4 *A	12.7 B	12.2 B	-

The same letters mark values belonging to the same homogeneous groups, determined on the basis of statistical analysis for $\alpha = 0.05$. * Capital letters refer to the cover and variety; ** lower letters refer to the interaction of variety × cover.

Table 10. Melon fruit weight (kg). Interaction of cover type and variety. Average from 2019–2021 years.

Cover \ Variety	Melba	Malaga F1	Seledyn F1	Average for Cover
PE	0.69 **abc	1.18 ef	1.74 g	1.20 *C
Fbio	0.70 abc	0.79 bcd	1.05 def	0.84 B
PP	0.55 abc	1.27 f	1.16 ef	0.97 B
miscanthus	0.39 a	0.87 cde	0.66 abc	0.63 A
control	0.47 ab	1.10 def	1.04 def	0.87 B
Average for variety	0.57 *A	1.07 B	1.17 B	-

The same letters mark values belonging to the same homogeneous groups, determined on the basis of statistical analysis for $\alpha = 0.05$. * Capital letters refer to the cover and variety; ** lower letters refer to the interaction of variety × cover.

Table 11. Melon fruit ratio of vertical to horizontal diameter. Interaction of cover type and variety. Average from 2019–2021 years.

Cover \ Variety	Melba	Malaga F1	Seledyn F1	Average for Cover
PE	1.34 **a	1.03 a	1.19 a	1.17 *B
Fbio	1.30 a	0.92 a	1.15 a	1.15 B
PP	1.32 a	1.00 a	1.05 a	1.13 B
miscanthus	1.09 a	0.87 a	0.93 a	0.97 A
control	1.24 a	1.00 a	1.06 a	1.10 AB
Average for variety	1.27 *C	0.98 A	1.09 B	-

The same letters mark values belonging to the same homogeneous groups, determined on the basis of statistical analysis for $\alpha = 0.05$. * Capital letters refer to the cover and variety; ** lower letters refer to the interaction of variety × cover.

Moreover, this study highlighted the substantial influence of mulching materials on melon fruit size (Tables 8–11). Some sources attest to the positive impact of soil mulching on vegetable growth [25], primarily due to enhanced soil moisture conditions (elevated moisture), increased soil temperature, and reduced fluctuations of these parameters during development. On average, the largest fruits, measuring 15.4 × 12.8 cm with a mass of 1.20 kg, were harvested from plants mulched with PE film (Tables 8–10). Conversely, fruits from plots mulched with Fbio, PP, and unmulched controls exhibited an average 10.8% reduction in both vertical and horizontal diameters and a 25.6% decrease in mass. Similar findings were reported by Parmar et al. [19] regarding the positive effects of plastic mulches on watermelon fruit size, while Ekinci and Dursun [25] demonstrated significantly increased fruit mass in melon plants grown under plastic mulch compared to those grown without mulch. The disparities in fruit size between plants mulched with miscanthus straw and those mulched with PE (yielding the largest fruits) were substantial, with differences of 31.8% for vertical diameter, 32% for horizontal diameter, and 47.5% for mass. Furthermore, the shape coefficient of fruits mulched with miscanthus straw was significantly lower compared to those mulched with PE, PP, and Fbio (Table 11).

On average, across the three-year study period, the Seledyn F1 variety cultivated with PE mulch produced the heaviest fruits, weighing 1.74 kg. Slightly lighter fruits were observed from the Malaga variety on PP and PE mulches, as well as from Seledyn F1 on PP mulch.

The beneficial effects of mulching were consistently evident throughout the study years (Figures 2–4). In 2019, the smallest fruits were harvested from plants mulched with miscanthus, while significantly larger vertical and horizontal diameters were observed in other treatments (Figures 2 and 3). The heaviest fruits were obtained from PE, PP, and Fbio mulches, whereas fruits from the unmulched and miscanthus mulch treatments had, on average, 24.7% and 46.2% lower unit masses, respectively (Figure 4). Similarly, in 2021, melons grown with miscanthus mulch exhibited the smallest fruits. The vertical and horizontal diameters, as well as the unit mass, were on average 18.7%, 21.9%, and 23.8% smaller compared to fruits from plants mulched with other materials. In the less favorable growing conditions of 2020, the largest fruits were obtained from plants mulched with PE. These fruits displayed significantly larger vertical and horizontal diameters, averaging 40% and 37.8% larger, respectively, and had a unit mass 1.7 times greater than fruits from plants grown on PP and Fbio mulches. They were also 2.5 times and 1.7 times longer and, on average, twice as wide as fruits from miscanthus mulched or control plots. Their mass was on average 3.7 times greater than that from the aforementioned plots.

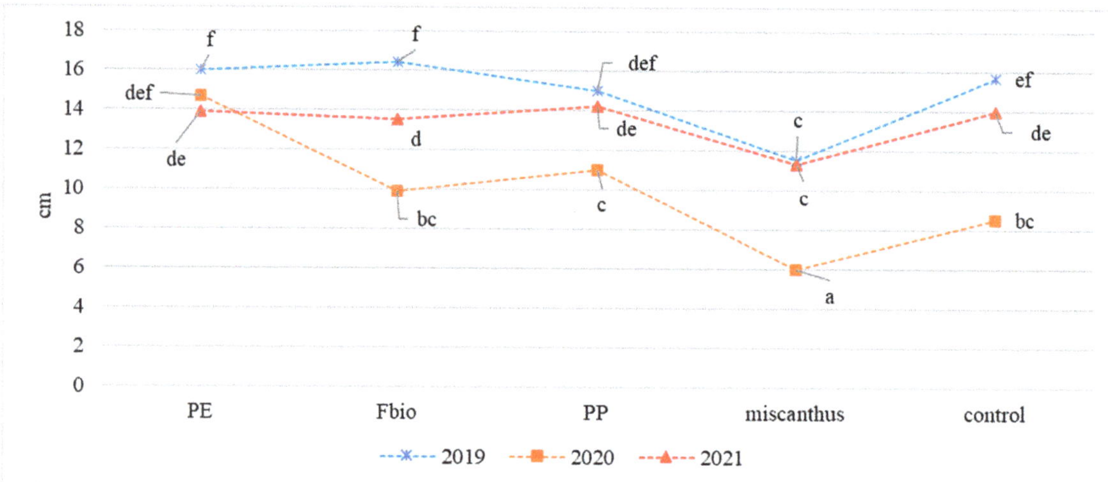

Figure 2. Fruit vertical diameter (cm). Interaction of the weather conditions in the years of study and the type of mulch. The same lowercase letters mark values belonging to the same homogeneous groups, determined on the basis of statistical analysis for α = 0.05.

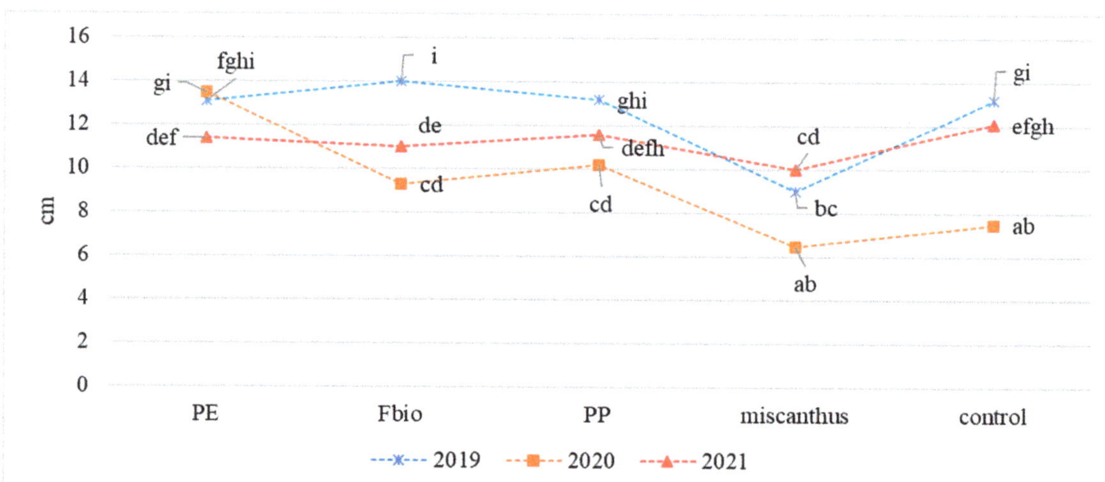

Figure 3. Fruit horizontal diameter (cm). Interaction of the weather conditions in the years of study and the type of mulch. The same lowercase letters mark values belonging to the same homogeneous groups, determined on the basis of statistical analysis for α = 0.05.

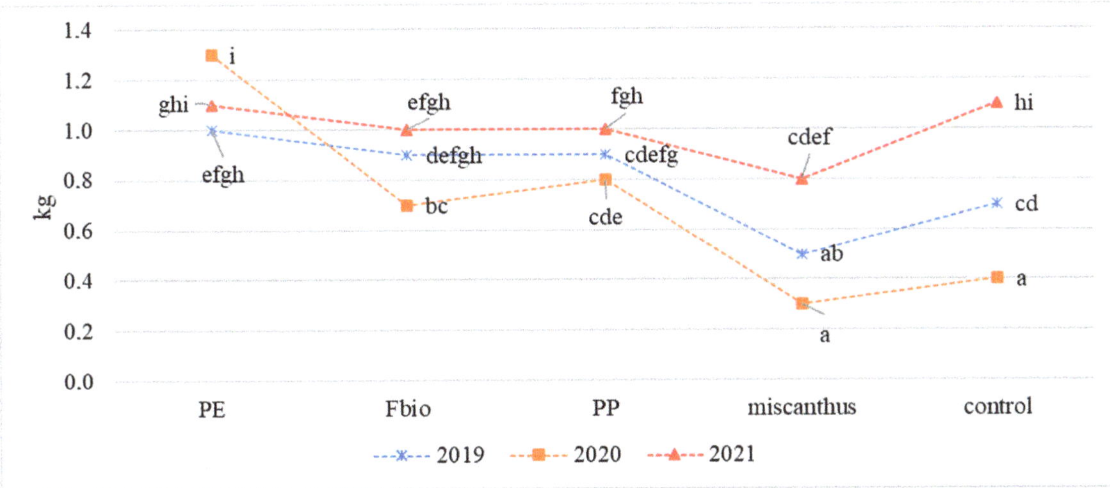

Figure 4. Fruit weight (kg). Interaction of the weather conditions in the years of study and the type of mulch. The same lowercase letters mark values belonging to the same homogeneous groups, determined on the basis of statistical analysis for $\alpha = 0.05$.

3.3. Melon Fruit Yield

The study results reveal pronounced variability in the yield of melon fruits under the influence of cultivation conditions. Yield outcomes were shaped not only by meteorological factors but also by the specific variety and type of mulching employed (Tables 12–14). In 2021, irrespective of the experimental factor, the average yield was twice as high as in 2019 and 2020, while in the control plots, it exceeded threefold. Across the study period, the average melon yield remained notably lower compared to yields typically achieved in warmer climates.

Table 12. Variance analysis of melon fruit yield.

Factors	Sum of Squares	df	Mean Square	F	Significance
Year	3241.81	2	1620.90	48.4437	<0.001 ***
Variety	1431.75	2	715.88	21.3953	<0.001 ***
Cover	2703.76	4	675.94	20.2017	<0.001 ***
Year × variety	521.66	4	130.41	3.8977	<0.01 **
Year × cover	2131.38	8	266.42	7.9625	<0.001 ***
Variety × cover	1031.54	8	128.94	1.4788	0.172
Year × variety × cover	1556.88	16	97.30	2.9081	<0.001 ***

df—degrees of freedom; The F test to determine whether group means are equal is calculated using the variation between sample means/variation within the samples. *** significant at $p \leq 0.001$, ** significant at $p \leq 0.01$.

Table 13. Variance analysis of melon fruit number per square meter.

Factors	Sum of Squares	df	Mean Square	F	Significance
Year	16.66	2	8.33	33.26	<0.001 ***
Variety	0.51	2	0.26	1.02	0.3641
Cover	9.93	4	2.48	9.91	<0.001 ***
Year × variety	1.33	4	0.33	1.33	0.2664
Year × cover	14.40	8	1.80	7.19	<0.001 ***
Variety × cover	2.48	8	0.31	1.24	0.2853
Year × variety × cover	10.41	16	0.65	2.60	<0.01 **

df—degrees of freedom; The F test to determine whether group means are equal is calculated using the variation between sample means/variation within the samples. *** significant at $p \leq 0.001$, ** significant at $p \leq 0.01$.

Table 14. Melon variety fruit yielding in following research years (t ha^{-1}).

Cover \ Variety	Melba	Malaga F1	Seledyn F1	Mean
		2019		
PE	10.03	8.62	16.56	11.74 *ABCD
Fbio	9.06	15.00	15.07	13.04 ABCDE
PP	5.89	10.73	14.04	10.22 ABCD
miscanthus	4.47	6.68	10.14	7.09 AB
control	7.20	10.99	9.24	9.14 AB
mean	7.33	10.40	13.01	-
		2020		
PE	12.32 **ab	30.24 d	32.91 d	25.16 EF
Fbio	8.95 ab	1.94 a	5.23 ab	5.38 AB
PP	6.99 ab	17.07 bc	8.67 ab	10.91 ABCD
miscanthus	1.60 a	1.70 a	2.21 a	1.83 A
control	5.99 ab	4.53 a	3.26 a	4.59 AB
	7.17	11.10	10.46	-
		2021+		
PE	10.39 ab	25.07 def	31.94 fg	22.47 DEF
Fbio	17.99 abcde	10.02 ab	22.82 bcdef	16.94 BCDE
PP	18.06 abcde	27.57 ef	41.11 g	28.91 F
miscantus	7.63 a	18.7 abcde	8.20 a	11.51 ABCD
control	6.32 a	33.52 fg	25.00 def	21.61 CDEF
	12.08	22.58	25.81	-

The same letters mark values belonging to the same homogeneous groups, determined on the basis of statistical analysis for $\alpha = 0.05$. * Capital letters refer to the interaction of year × cover; ** lower letters refer to the interaction of year × variety × cover.

Under favorable meteorological conditions in 2020 and 2021 (Table 14), cultivation of the Seledyn F1 or Malaga F1 varieties with PE and PP mulches resulted in fruit yields similar to those reported by Buczkowska et al. [27] using mycorrhizal applications and irrigation regimes. The Seledyn F1 variety also showed high yields in studies by Kosterna et al. [28].

The yield of Seledyn F1 and Malaga F1 varieties surpassed that of the Melba variety by an average of 76.4% (Table 15). Similar fruit counts per square meter were observed across these varieties (Table 16).

Table 15. Melon fruit yield t per ha. Interaction of cover type and variety. Average from 2019–2021 years.

Cover \ Variety	Melba	Malaga F1	Seledyn F1	Average for Cover
PE	10.9	21.3	27.1	19.8 *C
Fbio	12.0	9.0	14.4	11.8 AB
PP	10.3	18.5	21.3	16.7 BC
miscanthus	4.6	9.0	6.8	6.8 A
control	6.5	16.3	12.5	11.8 AB
Average for variety	8.9 **a	14.8 b	16.4 b	-

The same letters mark values belonging to the same homogeneous groups, determined on the basis of statistical analysis for $\alpha = 0.05$. * Capital letters refer to the cover; ** lower letters refer to the variety.

Table 16. Number of melon fruit per square meter. Interaction of cover type and variety. Average from 2019–2021 years.

Cover \ Variety	Melba	Malaga F1	Seledyn F1	Average for Cover
PE	1.5 a	1.7 a	1.6 a	1.6 *B
Fbio	1.6 a	1.0 a	1.2 a	1.3 AB
PP	1.7 a	1.4 a	1.6 a	1.6 B
miscanthus	0.9 a	0.8 a	0.8 a	0.8 A
control	1.3 a	1.4 a	1.0 a	1.2 AB
Average for variety	1.4 a	1.3 a	1.3 a	-

The same letters mark values belonging to the same homogeneous groups, determined on the basis of statistical analysis for $\alpha = 0.05$. * Capital letters refer to the cover, lower letters refer to the variety.

Utilization of synthetic mulches led to a twofold increase in fruit yield compared to plots without mulching, accompanied by a 23.7% higher fruit count (Tables 15 and 16). Optimal yields were achieved notably with PE film mulch and comparably with PP nonwoven mulch. The observed impact of mulching on melon cultivation aligns with findings from the previous literature. Ekinci and Dursun [25] reported melon yield increases of 15% with black plastic mulch and 25–28% with clear plastic mulch under Pakistan's climatic conditions compared to control plots. Parmar et al. [19] emphasized enhanced water retention and improved nutrient availability on silver/black and black/white PE mulches in India, facilitating water conservation and nutrient uptake. In temperate climates during periods of reduced rainfall, plastic mulches play an equally pivotal role, contributing to elevated soil temperatures, particularly beneficial during cooler growing seasons. Elevated soil temperatures promote robust root system development in warm-season vegetables, thereby fostering vigorous above-ground growth. Bucki et al. [29] noted minimal microclimate effects of PP nonwoven mulch in their zucchini experiments, highlighting the predominant influence of meteorological conditions on zucchini yield outcomes.

After the application of plastic films, it takes approximately 100 years for them to degrade [30]. Retrieving these materials post-use is challenging, often leading to their residual accumulation in the soil, which adversely affects the absorption of nutrients and water by plants. Consequently, this diminishes crop yields and contributes to environmental contamination [31,32]. Moreover, the deployment of plastic mulch entails considerable expenses related to procurement, field application, and disposal, prompting the exploration of alternatives such as biodegradable films [29,33,34].

Research by Wang et al. [35] revealed that during initial growth stages, soil temperatures under biodegradable films exceeded those under polyethylene films. However, in subsequent stages, biodegradable films degraded, resulting in reduced thermal insulation compared to polyethylene counterparts [36]. Despite the higher initial cost of biodegradable films relative to plastic films, comprehensive cost analysis is essential, considering their disposal cost advantage [37].

In the conducted research, melon plants mulched with Fbio exhibited a yield that was, on average, 40% lower compared to those mulched with PE (Table 15). This yield was statistically equivalent to the unmulched control. In the Fbio-treated plots, there was an average reduction of 18.8% in fruit yield compared to PE, though this difference did not achieve statistical significance (Table 16). The study revealed that the impact of Fbio on melon yield varied significantly across different years. In 2019, the commercial yield was statistically similar to that of PE and PP (Table 14). Conversely, in 2020, it was 4.7 times lower than in PE plots and twice as low as in PP plots, while in 2021, it decreased by 1.3 and 1.7 times, respectively. Cozzolino et al. [38] compared the effects of using 50 μm thick black PE film and 15 μm thick black biodegradable Mater-Bi® film and found no significant differences in commercial yield, average fruit mass, or melon pulp hardness. Numerous other studies have reported the positive effects of biodegradable films on the cultivation of various vegetables, such as pumpkins [39], zucchinis [29], potatoes [40], and maize [35].

Cultivation under organic mulch from miscanthus resulted in a yield reduction of 58.9% (Table 15), a decrease in individual fruit mass by 40.3% (Figure 4), and a reduction in fruit quantity by 40.7% compared to plots mulched with synthetic materials (Table 16). Parmar et al. [19] similarly observed that mulching watermelons with wheat straw led to an average yield reduction of 18.1% compared to plastic mulches, with a corresponding decrease in individual fruit mass by 14.8%. It is plausible that the smaller fruit dimensions and lower yield observed under miscanthus straw mulch in the present study stem from its light color, which reflects solar radiation. Consequently, the soil temperature under this mulch is lower compared to other treatments (Figure 1). Furthermore, the decomposition of organic matter releases nitrogen, making it available for competing microorganisms and potentially limiting its availability for plant uptake.

In 2019, an average of 0.9 to 1.67 fruits per square meter were harvested. The Seledyn F1 variety yielded the highest production, while yields of Malaga F1 and Melba were lower by 20% and 43.7%, respectively (Table 14, Figure 5) In 2020, the yields of varieties remained statistically consistent but showed variation which influenced by the type of mulch used. The highest fruit quantity and yield were achieved in plots mulched with PE (Table 14, Figure 6). Plots mulched with PP yielded half the number of fruits and had a yield of 2.3 times lower. Similar yields were observed in plots treated with Fbio and the control group. Mulching with miscanthus resulted in a 6-fold decrease in fruit quantity and a 13.7-fold decrease in fruit yield. In the third year of the study, the Seledyn F1 and Malaga F1 varieties produced the highest yields. Despite a similar number of fruits per square meter, the yield of the Melba variety was on average 50% lower (Table 14, Figure 5).

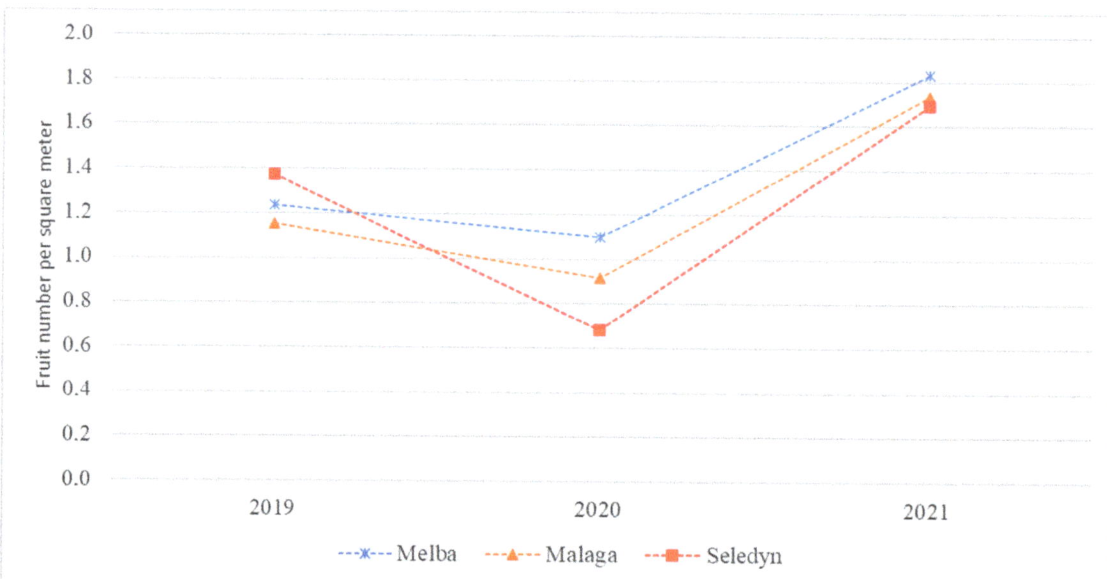

Figure 5. Fruit number per square meter. Interaction of the weather conditions in the years of study and varieties.

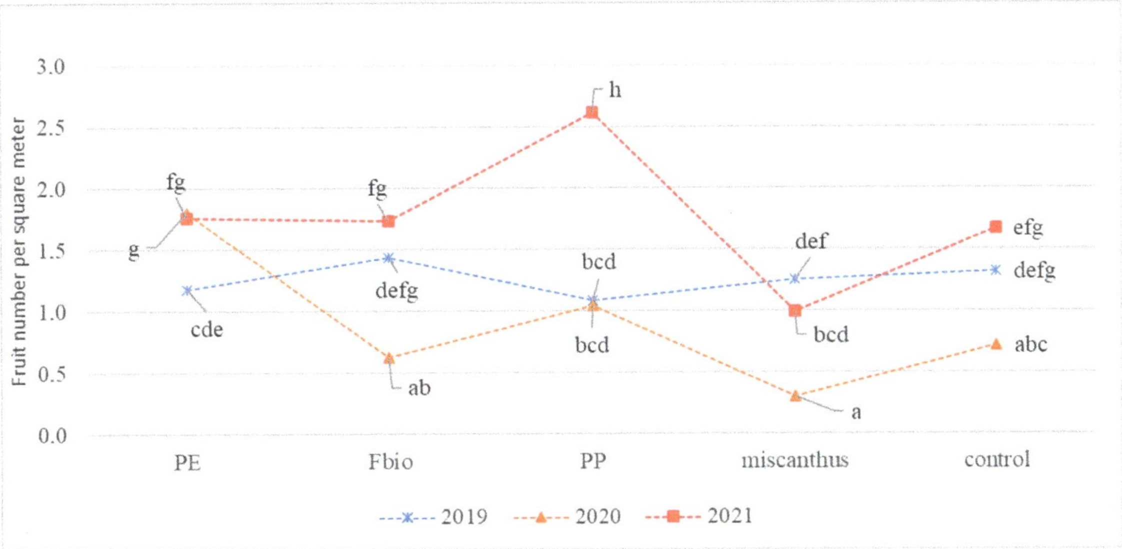

Figure 6. Fruit number per square meter. Interaction of the weather conditions in the years of study and the type of mulch. The same lowercase letters mark values belonging to the same homogeneous groups, determined on the basis of statistical analysis for $\alpha = 0.05$.

4. Conclusions

In temperate climates, amidst the global trend of rising average temperatures, there is an increasing inclination towards scaling up the cultivation of plants with higher thermal demands. In moderate climates, risks such as frost, cold spells, and frequent periods of precipitation deficit pose significant challenges. An exemplary vegetable species with stringent thermal and water requirements is the melon.

The findings from the conducted research underscore profound variability in the commercial yield of melon fruits influenced by weather conditions. Despite notably higher average temperatures during June–August 2019 compared to 2020 and 2021, superior yields with increased unit masses were achieved in 2021. Adequate, evenly distributed rainfall was recorded in 2021, ensuring sustained water availability for the plants. Periods of water deficit lasting 10–20 days result in diminished flowering and fruit set, emphasizing the critical role of water accessibility during plant growth, particularly during crucial stages like flowering and fruit development. Optimal coinciding of these factors is crucial for maximizing melon yield.

Addressing these challenges involves the deliberate selection of suitable varieties adapted to climatic conditions and the implementation of agronomic practices aimed at enhancing soil moisture and thermal conditions. Mulching represents one such pivotal practice.

For cultivation in the temperate climate of Central Europe, varieties with short to medium growing seasons, resilient to cold and adverse moisture conditions, are recommended based on the study results. Specifically, the Seledyn F1 and Malaga F1 varieties are deemed suitable for field cultivation. These varieties yield substantial fruits, each weighing over 1 kg. Under favorable weather conditions in 2021, yields of these varieties without mulching reached 25 t ha^{-1} and 33.52 t ha^{-1}, respectively.

Mulching with black polyethylene film or polypropylene nonwoven fabric exerts a positive influence on melon yield. The use of these materials resulted in an average increase in commercial fruit yield of 67.8% and 41.5%, respectively. Their efficacy varied depending on annual weather patterns; conversely, employing biodegradable film as mulch yielded inferior outcomes compared to synthetic counterparts. Organic mulching, meanwhile, is not recommended for melon cultivation.

Funding: The research is co-financed from the subsidy increased by the minister responsible for higher education and science for the period 2020–2026 in the amount of 2% of the subsidy referred to Art. 387 (3) of the Act of 20 July 2018—Law on Higher Education and Science, obtained in 2019.

Institutional Review Board Statement: Not applicable

Data Availability Statement: Data is contained within the article.

Conflicts of Interest: The authors declare no conflict of interest.

References

1. Kang, Y.; Khan, S.; Ma, X. Climate change impacts on crop yield, crop water productivity and food security—A review. *Prog. Nat. Sci.* **2009**, *19*, 1665–1674. [CrossRef]
2. Kumlay, A.M.; Kocak, M.Z.; Gohari, G.; Nouraein, M.; Celikcan, F.; Kaysim, M.G.; Kulak, M. Agronomic traits, secondary metabolites and element concentrations of *Lavandula angustifolia* leaves as a response to single or reiterated drought stress: How effective is the previously experienced stress? *Folia Hort.* **2022**, *34*, 1–16. [CrossRef]
3. Climate of Poland 2022, Polish Climate Monitoring Bulletin, IMGW-PIB Own Study 2023. Available online: https://www.imgw.pl/sites/default/files/2023-07/climate-of-poland-2023_report.pdf (accessed on 10 August 2024).
4. Ali, A.; Erenstein, O. Assessing farmer use of climate change adaptation practices and impacts on food security and poverty in Pakistan. *Clim. Risk Manag.* **2017**, *16*, 183–194. [CrossRef]
5. Raza, A.; Razzaq, A.; Mehmood, S.S.; Zou, X.; Zhang, X.; Lv, Y.; Xu, J. Impact of Climate Change on Crops Adaptation and Strategies to Tackle Its Outcome: A Review. *Plants* **2019**, *8*, 34. [CrossRef] [PubMed]
6. European Commission. *A Farm to Fork Strategy for a Fair, Healthy and Environmentally-Friendly Food System COM (2020) 381 Final*; European Commission: Brussels, Belgium, 2020.
7. Kiss, K.; Ruszkai, C.; Szucs, A.; Koncz, G. Examining the Role of Local Products in Rural Development in the Light of Consumer Preferences—Results of a Consumer Survey from Hungary. *Sustainability* **2020**, *12*, 5473. [CrossRef]
8. Pató, G.S.B.; Csiszárik-Kocsir, Á.; Herczeg, M.; Dominek, Á.; Varga, I.; Pató, B.; Kiss, F. Short Supply Chains from an Intermediary's Point of View. *Online J. Model. New Eur.* **2020**, 168–183. [CrossRef]
9. Swamy, K.R.M. Origin, distribution and systematics of culinary cucumber (*Cucumis melo* subsp. *agrestis var. conomon). J. Hort. Sci.* **2017**, *12*, 1–22. [CrossRef]
10. FAOSTAT 2022. Available online: https://www.fao.org/faostat/en/#data/QCL (accessed on 10 August 2024).
11. Rolbiecki, R.; Rolbiecki, S.; Figas, A.; Jagosz, B.; Wichrowska, D.; Ptach, W.; Prus, P.; Sadan, H.A.; Ferenc, P.-F.; Stachowski, P.; et al. Effect of Drip Fertigation with Nitrogen on Yield and Nutritive Value of Melon Cultivated on a Very Light Soil. *Agronomy* **2021**, *11*, 934. [CrossRef]
12. Fabeiro, C.; Martin de Santa Olalla, F.; De Juan, J.A. Production of muskmelon (*Cucumis melo* L.) under controlled deficit irrigation in a semi-arid climate. *Agric. Water Manag.* **2002**, *54*, 93–105. [CrossRef]
13. Franczuk, J.; Rosa, R.; Kosterna-Kelle, E.; Zaniewicz-Bajkowska, A.; Panasz, M. The effect of transplanting date and covering on the growth and development of melon (*Cucumis melo* L.). *Acta Agrobot.* **2017**, *70*, 1699. [CrossRef]
14. Jifon, J.L.; Lester, G.E. Foliar fertilization of muskmelon: Effects of potassium source on market quality and phytochemical content of field-grown fruit. In Proceedings of the 2007 Fluid Forum, Scottsdale, AZ, USA, 18–20 February 2007; pp. 154–161.
15. Majkowska-Gadomska, J. Mineral content of melon fruit (*Cucumis melo* L.). *J. Elem.* **2009**, *14*, 717–727. [CrossRef]
16. Adamczewska-Sowińska, K.; Sowiński, J. Reaction of Sweet Maize to the Use of Polyethylene Film and Polypropylene Non-Woven Fabric in the Initial Growth Phase. *Agronomy* **2020**, *10*, 141. [CrossRef]
17. Lopez, M.; Quezada, R.M.; De Larosa-Ibarra, M.; Cedeno-Ruvalcaba, B. Effect of plastic mulch on growth of melon (*Cucumis melo* L.) "Lacuna" hybrid. *Phyton* **2000**, *69*, 37–44.
18. Vescera, M.; Brown, R.N. Effects of three production systems on muskmelon yield and quality in New England. *HortScience* **2016**, *51*, 510–517. [CrossRef]
19. Parmar, H.N.; Polara, N.D.; Viradiya, R.R. Effect of Mulching Material on Growth, Yield and Quality of Watermelon (*Citrullus Lanatus* Thunb) Cv. Kiran. *Universal J. Agric. Res.* **2013**, *1*, 30–37. [CrossRef]
20. Singh, J.; Metrani, R.; Jayaprakasha, G.K.; Crosby, K.M.; Jifon, J.L.; Ravishankar, S.; Brierley, P.; Leskovar, D.L.; Turini, T.A.; Schultheis, J.; et al. Profiling carotenoid and sugar contents in unique *Cucumis melo* L. cultigens harvested from different climatic regions of the United States. *J. Food Comp. Anal.* **2022**, *106*, 104306. [CrossRef]
21. COBORU. *Polish National List of Vegetable Plant Varieties*; COBORU: Słupia Wielka, Poland, 2023; ISSN 1231-8280.
22. Kabała, C.; Charzyński, P.; Chodorowski, J.; Drewnik, M.; Glina, B.; Greinert, A.; Hulisz, P.; Jankowski, M.; Jonczak, J.; Łabaz, B.; et al. Polish Soil Classification, 6th edition—Principles, classification scheme and correlations. *Soil Sci. Annu.* **2019**, *70*, 71–97. [CrossRef]
23. Bouzo, C.A.; Küchen, M.G. Effect of temperature on melon development rate. *Agron. Res.* **2012**, *10*, 283–294.

24. Kurtar, E.S. Modelling the effect of temperature on seed germination in some cucurbits. *Afr. J. Biotechnol.* **2010**, *9*, 1343–1353.
25. Ekinci, M.; Dursun, A. Effects of different mulch materials on plant growth, some quality parameters and yield in melon (*Cucumis melo* L.) cultivars in high altitude environmental condition. *Pak. J. Bot.* **2009**, *41*, 1891–1901.
26. Zaniewicz-Bajkowska, A.; Franczuk, J.; Kosterna, E. Effect of foliar feeding on yield and fruit quality of three melon (*Cucumis melo* L.) cultivars. *Folia Hort.* **2009**, *21*, 65–75. [CrossRef]
27. Buczkowska, H.; Sałata, A.; Nurzyńska-Wierdak, R. Melon (*Cucumis melo* L.) Fruit Yield under Irrigation and Mycorrhiza Conditions. *Agronomy* **2023**, *13*, 1559. [CrossRef]
28. Kosterna, E.; Zaniewicz-Bajkowska, A.; Franczuk, J.; Rosa, R.; Chromińska, K.; Borysiak-Marciniak, I.; Panasz, M. Effect of synthetic mulches on melon (*Cucumis melo* L.) yielding. *Folia Hort.* **2011**, *23*, 151–156. [CrossRef]
29. Bucki, P.; Siwek, P.; Domagała-Świątkiewicz, I.; Puchalski, M. Effect of Agri-Environmental Conditions on the Degradation of Spunbonded Polypropylene Nonwoven with a Photoactivator in Mulched Organically Managed Zucchini. *Fibres Text. East. Eur.* **2018**, *2*, 55–60. [CrossRef]
30. Briassoulis, D.; Babou, E.; Hiskakis, M.; Kyrikou, I. Analysis of long-term degradation behaviour of polyethylene mulching films with pro-oxidants under real cultivation and soil burial conditions. *Environ. Sci. Pollut. Res. Int.* **2015**, *22*, 2584–2598. [CrossRef]
31. Zhang, D.; Liu, H.; Hu, W.L.; Qin, X.; Ma, X.; Yan, C.; Wang, H. The status and distribution characteristics of residual mulching film in Xinjiang. *China. J. Integr. Agr.* **2016**, *15*, 2639–2646. [CrossRef]
32. Borrowman, C.K.; Johnston, P.; Adhikari, R.; Saito, K.; Patti, A.F. 2020. Environmental degradation and efficacy of a sprayable, biodegradable polymeric mulch. *Polym. Degrad. Stab.* **2020**, *175*, 109126. [CrossRef]
33. Waterer, D. Evaluation of biodegradable mulches for production of warm-season vegetable crops. *Can. J. Plant Sci.* **2010**, *90*, 737–743. [CrossRef]
34. Wang, Y.; Jia, X.; Olasupo, I.O.; Feng, Q.; Wang, L.; Lu, L.; Xu, J.; Sun, M.; Yu, X.; Han, D.; et al. Effects of biodegradable films on melon quality and substrate environment in solar greenhouse. *Sci. Total Environ.* **2022**, *829*, 154527. [CrossRef]
35. Wang, Z.; Li, M.; Flury, M.; Schaeffer, S.M.; Chang, Y.; Tao, Z.; Jia, Z.; Li, S.; Ding, F.; Wang, J. Agronomic performance of polyethylene and biodegradable plastic film mulches in a maize cropping system in a humid continental climate. *Sci. Total Environ.* **2021**, *786*, 147460. [CrossRef]
36. Mei, N.; Liu, B.; Zheng, J.; Lv, K.; Tang, D.; Zhang, Z. A novel magnetic palladium catalyst for the mild aerobic oxidation of 5-hydroxymethylfurfural into 2,5-furandicarboxylic acid in water. *Catal. Sci. Technol.* **2015**, *5*, 3194–3202. [CrossRef]
37. Cozzolino, E.; Di Mola, I.; Ottaiano, L.; Bilotto, M.; Petriccione, M.; Ferrara, E.; Mori, M.; Morra, L. Assessing Yield and Quality of Melon (*Cucumis melo* L.) Improved by Biodegradable Mulching Film. *Plants* **2023**, *12*, 219. [CrossRef] [PubMed]
38. Cozzolino, E.; Bilotto, M.; Leone, V.; Zampella, L.; Petriccione, M.; Cerrato, D.; Morra, L. Produzione e qualità di melone retato su pacciamatura in mater-Bi. *Colt. Protette* **2015**, *6*, 66–71.
39. Ghimire, S.; Wszelaki, A.L.; Moore, J.C.; Inglis, D.A.; Miles, C. The Use of Biodegradable Mulches in Pie Pumpkin Crop Production in Two Diverse Climates. *HortScience* **2018**, *53*, 288–294. [CrossRef]
40. Zhao, H.; Xiong, Y.C.; Li, F.M.; Wang, R.Y.; Qiang, S.C.; Yao, T.F.; Mo, F. Plastic Film Mulch for Half Growing-Season Maximized WUE and Yield of Potato via Moisture-Temperature Improvement in a Semi-Arid Agroecosystem. *Agric. Water Manag.* **2012**, *104*, 68–78. [CrossRef]

agriculture

MDPI

Article

Effect of Biodegradable Mulch and Different Synthetic Mulches on Growth and Yield of Field-Grown Small-Fruited Tomato (*Lycopersicon esculentum* Mill.)

Katarzyna Adamczewska-Sowińska *, Joanna Bykowy and Janina Jaworska

Horticulture Department, Wroclaw University of Environmental and Life Sciences, 50-375 Wroclaw, Poland; joannabykowy@gmail.com (J.B.); janina.jaworska@upwr.edu.pl (J.J.)
* Correspondence: katarzyna.adamczewska-sowinska@upwr.edu.pl

Abstract: Mulching is a widely adopted practice in vegetable cultivation globally. This technique employs various plastic materials, such as polyethylene (PE) film or polypropylene (PP) nonwoven fabric, with an increasing trend toward the use of biodegradable materials. Between 2014 and 2016, field experiments were conducted to evaluate the performance of the small-fruited tomato Intrigo F1 cultivated using synthetic mulches. The trials, designed as single-factor experiments employing a randomized block layout with three replicates, assessed plant morphological traits, yield, and the biological value of the tomato fruits. Weather conditions and the type of mulch applied had a pronounced influence on the quality of tomato plants and yield. Compared to the control, the use of black, red, and aluminum PE films and brown PP resulted in a 7.2% increase in plant height. All mulching treatments, except white film, increased the lateral spread of the plants by an average of 24.2%. Plants cultivated on red PE film exhibited a 26.4% increase in leaf count with respect to the control. Mulched treatments achieved an average increase of 19.6% in marketable yield. The highest marketable fruit yield was recorded with black nonwoven fabric mulch. Mulching had a significant effect on the chemical composition of tomato fruits. Fruits on biodegradable foil had the most potassium, lycopene, and polyphenols.

Keywords: black polyethylene film; white polyethylene film; red polyethylene film; aluminum-coated polyethylene film; polypropylene nonwoven fabric; marketable yield; lycopene; phenolic; carotenoids; vitamin C

1. Introduction

The tomato (*Solanum lycopersicum* Mill.), a member of the nightshade family, stands as a paramount horticultural species globally. Its cultivation is notably facile, and the plethora of thousands of registered varieties facilitates its adaptability for commercial cultivation by both smallholder and large-scale producers alike. World tomato production is 192.3 million tons. This species is cultivated on over 5.4 million ha globally and the average yield is 35.5 t·ha^{-1}. In Europe, it occupies 395.5 thousand ha, the production is about 21.5 million tons, and the average yield is 54.3 t·ha^{-1}. In Poland, it is cultivated on 8.5 thousand ha, the harvest is around 0.9 million tons, and the yield is 103.9 t·ha^{-1} [1]. The tomato's diverse applications have led to a classification system that distinguishes between varieties destined for fresh market consumption and those intended for processing purposes. Varieties earmarked for direct consumption are cultivated in both protected environments, such as greenhouses, and in open-field settings.

Mulching the soil is a cultivation practice widely used for various reasons, especially on a large scale in vegetable production. Synthetic mulching has been popular since the 1950s and has proven to be very effective [2]. At the beginning of the 21st century, synthetic mulches were used on nearly 13 million hectares worldwide. In 2017, Mormile et al. [3] estimated that the total global area of agricultural land mulched was about 18 million hectares. The benefits of mulching include increased soil temperature and moisture, reduced nitrogen leaching losses, faster fruit ripening, increased yields, prevention of soil erosion, and weed suppression, which in turn reduces herbicide use [4–6].

In temperate climates, synthetic mulches play a significant role in enhancing growth conditions for thermophilic species belonging to the *Solanaceae* and *Cucurbitaceae* families. The effectiveness of such mulches is influenced by the color of the mulch. Mulches also contribute to optimizing the radiation balance, which depends on the type of material used. They can transmit, absorb, or reflect portions of solar radiation, directly influencing above-ground plant growth. Kader et al. [7] confirmed that mulching is an effective water conservation technique, reducing soil evaporation. Furthermore, Amare and Desta [8] emphasized that the color of the mulch significantly affects soil moisture levels and its water retention capacity. The color of the (mulch) mulching film determines its photoselective properties, thereby influencing the microclimate surrounding the cultivated plants. Additionally, it impacts soil temperature both at and below the surface. Bucki and Siwek [9] reported that in temperate climates, black polyethylene PE film raises daytime soil temperature by an average of 3–4 °C, while polypropylene (PP) fabric achieves an increase of 1–2 °C. Amare and Desta [8] observed that black and blue polyethylene (PE) films tend to increase soil temperature, whereas light-colored and white films reduce it. The factors mentioned above directly influence the above-ground growth of plants. Experimental studies have demonstrated that tomato plants cultivated on black or white mulching films exhibit an increased number of leaves and longer stems compared to those grown on red or silver mulching films [10]. Mutoro [11] found that using white mulching film for tomatoes enhanced plant height and the number of stems. Research conducted by Bhujbal et al. [12] revealed that specific mulch colors, such as black on silver film, silver on black film, and transparent film, promote flowering, fruiting, and yield while simultaneously reducing the incidence of pest-induced plant diseases. The accelerated development of the above-ground parts of vegetables grown with mulching directly translates into improved yields, a trend observed across multiple species and climate zones. Adamczewska-Sowińska et al. [13] demonstrated that the use of black PE film mulch significantly increased eggplant yield in temperate climates.

The growing issue of environmental pollution caused by synthetic materials has necessitated increased interest in biodegradable alternatives, such as those derived from corn starch and biodegradable polymers. Under aerobic conditions, these mulches are decomposed by soil microorganisms into carbon dioxide and water. Sękara et al. [14] demonstrated that tomatoes cultivated using biodegradable mulch exhibited superior quality and health compared to those grown with black PE film. Additionally, they noted that biodegradable mulches do not require manual removal from the field after the cultivation cycle. Research conducted by Gabryś et al. [15] established that the use of biodegradable viscose fabric mulch in tomato cultivation provided optimal thermal and water conditions while effectively suppressing weed growth.

The effect of mulching on the chemical composition of the edible portions of vegetables remains inconclusive. Nevertheless, certain studies suggest that the type and color of mulch may enhance the biological value of the harvested yield.

The objective of the study was to assess the response of small-fruited tomato plants to mulching with synthetic materials and biodegradable mulch. It was hypothesized that, under temperate climate conditions, the applied mulches would have a positive impact on the growth, yield, and nutritional value of the tomato fruits. The study aimed to evaluate the impact of various synthetic mulching materials on the growth, yield, and biological value of the tall-growing small-fruited tomato variety Intrigo F1.

2. Materials and Methods

The field experiment was conducted over the period of 2014–2016 at the Vegetable and Ornamental Plant Research and Teaching Station Psary of the Department of Horticulture, Wroclaw University of Environmental and Life Sciences (51°19′08″ N, 17°03′37″ E). The experimental design followed a randomized block layout with three replications. The following synthetic mulches were evaluated: PE films, 0.05 mm thick, in black (PE black) and white (PE white), as well as red PE film (PE red) and aluminum-coated PE film (PE alu) with a thickness of 0.025 mm. Additionally, PP nonwoven fabric with a weight of 50 g m^{-2} was tested in black (PP black) and brown (PP brown). A black biodegradable film, BioAgri (Fbio (I), Italy), 0.025 mm thick, was included. This film is manufactured from Mater-Bi®, a bioplastic material derived from complexed starch and biodegradable polyesters, and it is certified as biodegradable and compostable in compliance with European Standard EN 13,432 and American Standard ASTM D6400 [16]. Control plots (Control) consisted of unmulched soil. The films and nonwoven fabrics were laid out in 2.5-m-wide strips four days before transplanting the tomato plants into the field. Each experimental plot measured 3.75 m^2, and the plants were cultivated at a spacing of 80 × 50 cm.

The experiments were conducted on degraded black soil made from light clay with weak sandy on medium clay subsoil material with a humus content of 1.8% and classified as soil quality class IIIa (Polish Soil Classification System) [17]. In the autumn, prior to the experiments, deep pre-winter plowing was performed, followed in spring by cultivator tillage and soil loosening with a rotary tiller. Before establishing each experiment, chemical soil analysis was conducted in autumn.

The soil pH ranged from 7.2 to 7.9, with salinity between 96 and 240 $\mu S \cdot cm^{-1}$. The macronutrient levels in the soil were adjusted to optimal values: 70 mg $P \cdot dm^{-3}$, 200 mg $K \cdot dm^{-3}$, and 65 mg $Mg \cdot dm^{-3}$. Fertilization included the application of granular triple superphosphate and potassium sulfate. In spring, prior to transplanting the tomato plants, nitrogen was applied at a rate of 150 $kg \cdot ha^{-1}$ in the form of ammonium nitrate.

Tomato seeds were sown in a greenhouse on April 10th into seed trays filled with a peat substrate. After the development of cotyledons and the appearance of the first true leaf, the seedlings were transplanted into pots with a 10 cm diameter between 27th and 29th April. To harden the plants, they were moved to an unheated plastic tunnel 10 days prior to planting. The tomato seedlings were transplanted into the field on the following dates: 26 May 2014, 25 May 2015, and 29 May 2016. The plants were cultivated with 2-m-high wooden stakes and conducted using a two-stem system.

Throughout the vegetative period, tomato plants were irrigated as required, with a single water application rate of 30 mm per irrigation. Toward the conclusion of the seedling production phase and during field cultivation, chemical treatments were systematically applied to protect the plants from bacterial and fungal diseases. The plant protection procedures adhered to the guidelines outlined in the current Vegetable Plant Protection Program. Lateral shoots and lower leaves were routinely pruned to maintain plant structure and health. At the end of August, plant topping was performed by removing the apical section of the main stem. This operation ensured that two to three leaves were retained above the final, sixth fully developed inflorescence on each of the two conducted stems.

In 2014, the harvests were carried out on 1, 6, 18, and 25 August and 8 September. In 2015, harvesting took place at weekly intervals from 3 August to 5 October, while in 2016, harvesting occurred from 2 August to 12 September. Harvesting was performed when nearly all fruits within a cluster (except for the last 3–5 fruits) had reached full ripeness. The yields were categorized as total yield, marketable yield, and early marketable yield. The marketable yield consisted of ripe, uniformly colored, healthy fruits, while the early marketable yield included fruits from the first three harvests. The average number of fruits and the average fruit weight (g) from each cluster weighted for each harvest were assessed.

Biometric measurements of the plants were conducted during the first ten days of July (1–10 July) and August (1–10 August). Two plants were randomly selected from each plot for measurement. The following parameters were recorded: plant height from the base of the stem to the apex (cm), stem diameter at 1 cm above the soil surface (mm), the lateral spread of the plant at mid-height (cm), and the number of leaves.

At the end of August, during the full vegetative period, tomato fruit samples were collected for chemical analysis. A total of 30 mature fruits from each plot were selected for this purpose. The samples were analyzed for their content of dry matter, mineral nutrients, and organic compound composition.

2.1. Chemical Analysis Methodology

The dry matter content in tomato fruits was determined using the oven-drying method, with samples dried at 105 °C. Phosphorus and magnesium were quantified using the colorimetric method in dry matter (spectrophotometer type 102). The phosphorus content was measured using a spectrophotometer set to a factor of 1.36 and a wavelength of 470 nm, while magnesium was measured at a wavelength of 555 nm with a factor of 0.33. Potassium and calcium were analyzed by flame photometry in dry matter (flame photometer Carl Zeiss, Jena, Germany), and nitrate content was determined using potentiometric methods in fresh matter (ionometer Thermo Orion 5 Star, USA). Vitamin C content was analyzed using Tillmann's method in fresh matter (PN-90/A-75101/11), while total and reducing sugars were measured using the Lane–Eynon method in fresh matter. Polyphenols were determined using the Folin–Ciocalteu method in fresh matter. Total carotenoid content was assessed using the colorimetric method in fresh matter. Absorbance was measured using a spectrophotometer against a blank (80% acetone) at wavelengths of 663, 645, and 470 nm for a single sample. Lycopene content was determined using the Fish method, with absorbance measured at a wavelength of 503 nm against the blank (hexane). Polyphenols and lycopene were measured using a Spectroquant Pharo 100 Merck spectrophotometer (Switzerland). Antioxidant activity was evaluated using the DPPH method in fresh matter.

2.2. Weather Conditions

The tomato is a thermophilic vegetable, and as such, the success of its cultivation is largely dependent on temperature fluctuations as well as the amount of precipitation during the growing season. While it can be grown in a variety of climatic conditions, the optimal temperature for most varieties ranges from 21 to 24 °C. Boote et al. [18] indicated that the optimal lower temperature is 22 °C and the optimal upper temperature is 28 °C. Tomato tissues are susceptible to damage when temperatures fall below 10 °C or exceed 38 °C [19,20]. Ayankojo and Morgan [21] highlighted that mitigating heat stress caused by high temperatures in warm climates contributes to improved tomato yields. Throughout the course of the study in a temperate climate, the weather conditions varied across the years.

At the time of tomato seedling planting, the average temperatures were 17.3 °C, 14.0 °C, and 20.9 °C. During this period, precipitation was highly variable. In 2014, the total rainfall amounted to 42.8 mm; in 2015, it amounted to 0.5 mm; and in 2016, no precipitation occurred. In 2014, the conditions during planting were favorable for the rapid establishment of the plants. The average temperature in July was 2.7 °C higher than the long-term mean for the month. Rainfall during June and July was significantly low, constituting only 34.2% and 55.3%, respectively, of the long-term average for these months. Water deficits were compensated through irrigation. The low temperatures observed from 11 August to 20 August (average 17.4 °C) and from 21 August to 31 August (average 15.9 °C), along with substantial rainfall during this period, contributed to the onset of *Phytophthora infestans* (potato blight) on the tomato plants. The climatic conditions in 2014 were the least conducive to tomato yield.

In 2015, low air temperatures (average 14.0 °C) immediately following transplanting inhibited tomato growth. However, during the subsequent two months, temperatures were optimal for flowering, fruit set, and subsequent fruit development. The precipitation in July and August was insufficient, resulting in water deficits that were addressed through irrigation. The average temperature in August was 6.7 °C higher than the long-term mean for the month. Elevated temperatures and low humidity in August and September promoted plant health, and no symptoms of potato blight were observed. The final fruit harvest occurred in the first 10 days of October. This year represented the most favorable environmental conditions for tomato development and yield.

Between 21 May and 31 May and from 1 June to 10 June 2016, the air temperature was higher by 6.7 °C and 4.1 °C, respectively, compared to the long-term averages for these months, while rainfall was either absent or minimal. Following transplanting, the seedlings required intensive irrigation. In the subsequent two months, temperatures remained above the long-term averages (averaging 21.5 and 20.5 °C). Intense rainfall during the second 10 days of July (70.5 mm) favored vegetative growth. However, the precipitation in August and the first 10 days of September was low, while the average temperature at the beginning of September was 21.2 °C. These conditions were optimal for the maturation of tomato fruits.

2.3. Experimental Design and Statistical Analyses

The results obtained in the experiment were subjected to statistical analysis using analysis of variance (ANOVA/MANOVA) in the Statistica 13 software package. The experiment was two-factor. Years of study were considered as the first factor and the type of mulches was the second factor. The statistical analysis also included the interaction of years and type of mulch. Confidence intervals were calculated using Tukey's test at a significance level of $\alpha = 0.05$.

3. Results and Discussion

3.1. Morphological Characteristics

The analysis of the study results revealed that both the climatic conditions and the type of synthetic mulch applied significantly influenced the morphological characteristics of the tomato plants of the Intrigo F1 variety (Figures 1–4). During the measurements conducted in July, significant differences were observed in plant height and lateral spread across the study years (Table 1), as well as in plant height in relation to the type of mulch used (Figure 5). By contrast, in August, only stem diameter did not show significant variation between the years. The tallest plants with the greatest lateral spread were recorded in 2015, both during the first and second measurement periods (Table 1, Figures 1 and 3). In the August measurement, the height of the plants grown in 2016 was comparable to that of the 2015 plants, while the plants from 2014 exhibited the greatest leaf development.

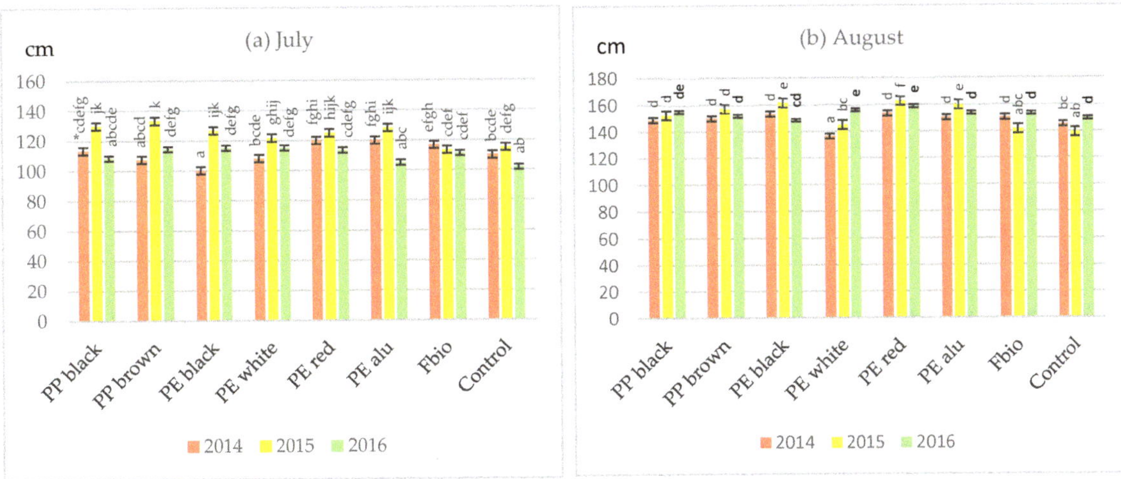

Figure 1. Tomato plant height depending on the type of mulch in 2014–2016 (cm). * The same letters mark values belonging to the same homogeneous groups, determined on the basis of statistical analysis for α = 0.05. Type of mulch: PP nonwoven: black and brown; PE film: black, white, red; PE alu—aluminum film; Fbio—biodegradable film.

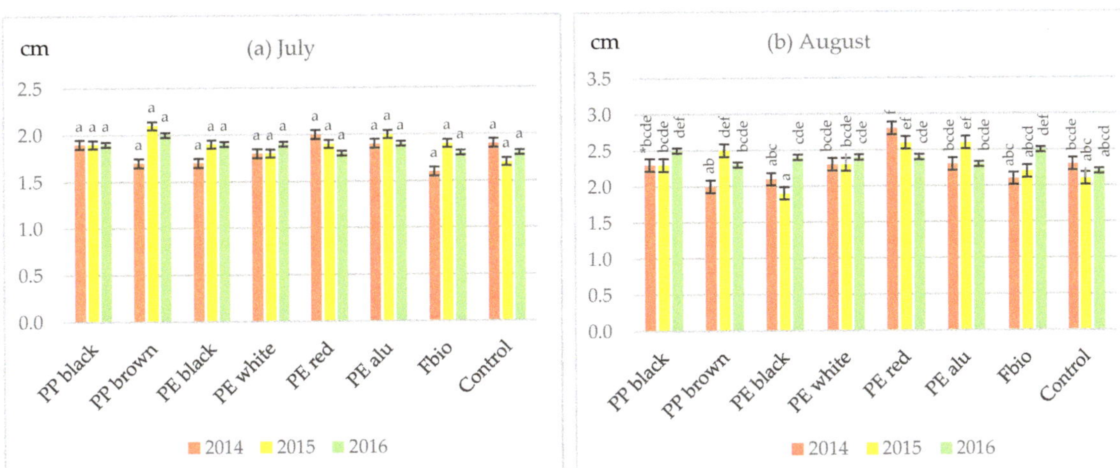

Figure 2. Tomato plant stem diameter depending on the type of mulch in 2014–2016 (cm). * The same letters mark values belonging to the same homogeneous groups, determined on the basis of statistical analysis for α = 0.05. Type of mulch: PP nonwoven: black and brown; PE film: black, white, red; PE alu—aluminum film; Fbio—biodegradable.

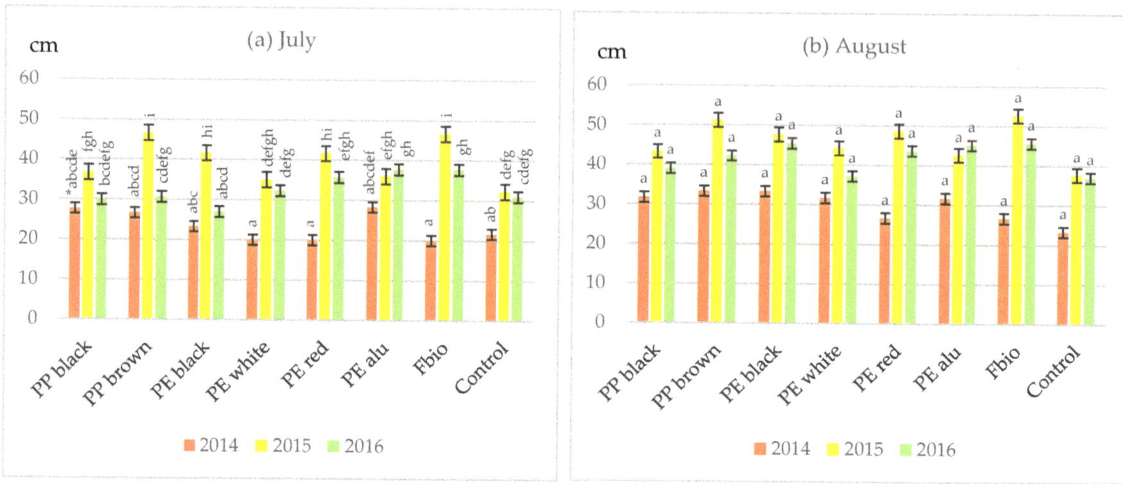

Figure 3. Tomato lateral spread depending on the type of mulch in 2014–2016 (cm). * The same letters mark values belonging to the same homogeneous groups, determined on the basis of statistical analysis for α = 0.05. Type of mulch: PP nonwoven: black and brown; PE film: black, white, red; PE alu—aluminum film; Fbio—biodegradable.

Figure 4. Tomato leaf number depending on the type of mulch in 2014–2016. * The same letters mark values belonging to the same homogeneous groups, determined on the basis of statistical analysis for α = 0.05. Type of mulch: PP nonwoven: black and brown; PE film: black, white, red; PE alu—aluminum film; Fbio—biodegradable.

Table 1. Tomato plant morphology parameters average for years.

Parameters	2014		2015		2016	
	July	August	July	August	July	August
Plant height (cm)	112.0 *a	148.5 A	124.2 b	152.5 B	110.5 a	153.4 B
Plant stem diameter (cm)	1.8 a	2.3 A	1.9 a	2.3 A	1.9 a	2.4 A
Lateral spread (cm)	23.5 a	29.8 A	39.6 c	46.0 C	32.8 b	41.9 B
Number of leaves (szt.)	10.7	46.4 B	21.3	34.2 A	20.9	34.7 A

* The same letters mark values belonging to the same homogeneous groups, determined on the basis of statistical analysis for α = 0.05. Lower letters mark significance differences in July; capital letters mark significance differences in August.

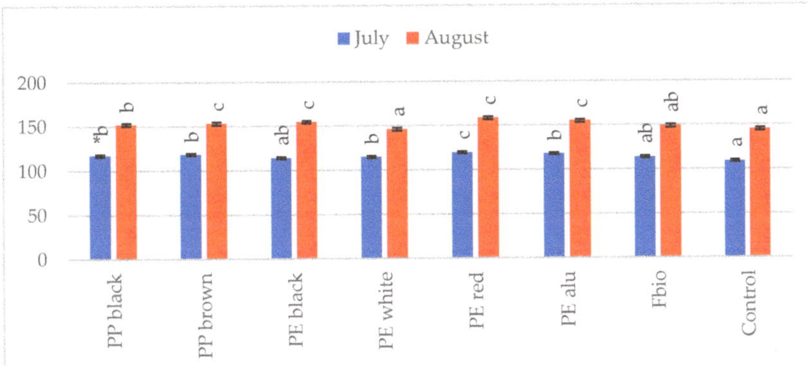

Figure 5. Tomato plant height in July and August depending on the type of mulch. Average for 2014–2016 (cm). * The same letters mark values belonging to the same homogeneous groups, determined on the basis of statistical analysis for α = 0.05. Letters refer to type of mulch. Type of mulch: PP nonwoven: black and brown; PE film: black, white, red; PE alu—aluminum film; Fbio—biodegradable.

In the present study, it was observed that field-grown tomato plants mulched with synthetic materials exhibited significantly greater height, stem diameter at the base, lateral spread, and leaf number compared to the control plot (Figure 6). Among the mulches applied, both black and red E films had a pronounced effect on tomato development [22].

Pinder et al. [22] reported that mulching with black PE film resulted in a significant increase in the growth of small-fruited tomato plants. Similarly, Islam [23] noted a positive impact of black PE film mulching on the vegetative development of tomatoes, with plants exhibiting increased height, more stems, and a higher fruit count, ultimately leading to significantly higher yields compared to plants grown without mulch. Onunva et al. [24] also observed enhanced plant height and stem and leaf numbers as well as improved flowering and fruiting in tomato plants grown with synthetic mulches. Furthermore, the beneficial effects of red PE film on plant growth were also confirmed by Agrawal et al. [25]. In July, plants grown with mulch were, on average, 7% taller than those in the control treatment. The stem diameter at the base ranged from 1.8 to 2.0 cm, while the lateral spread of the plants (28.3–34.8 cm) and the number of leaves (19.8–22.7) were not significantly influenced by the type of mulch used. It was observed that, during the second measurement period, mulched plants were, on average, 5.3% taller than those in the control group, with a 22.9% greater lateral spread and 5.4% more leaves (Figures 5 and 6). Plants mulched with red PE film exhibited the greatest height (158.3 cm), stem diameter (2.6 cm), and leaf number (46.4 leaves). Plants mulched with black PE and aluminum films showed similar heights, while those mulched with brown PP nonwoven fabric and aluminum film had comparable leaf numbers. All mulches applied resulted in an average 22.9% increase in lateral spread compared to the control. Tomatoes grown on white PE film mulch exhibited the lowest height. The use of Fbio mulch ensured a similar condition of the plants as in the other mulches. The exception was the height of the plants, which was among the smallest in both measurement terms, and the number of leaves in August.

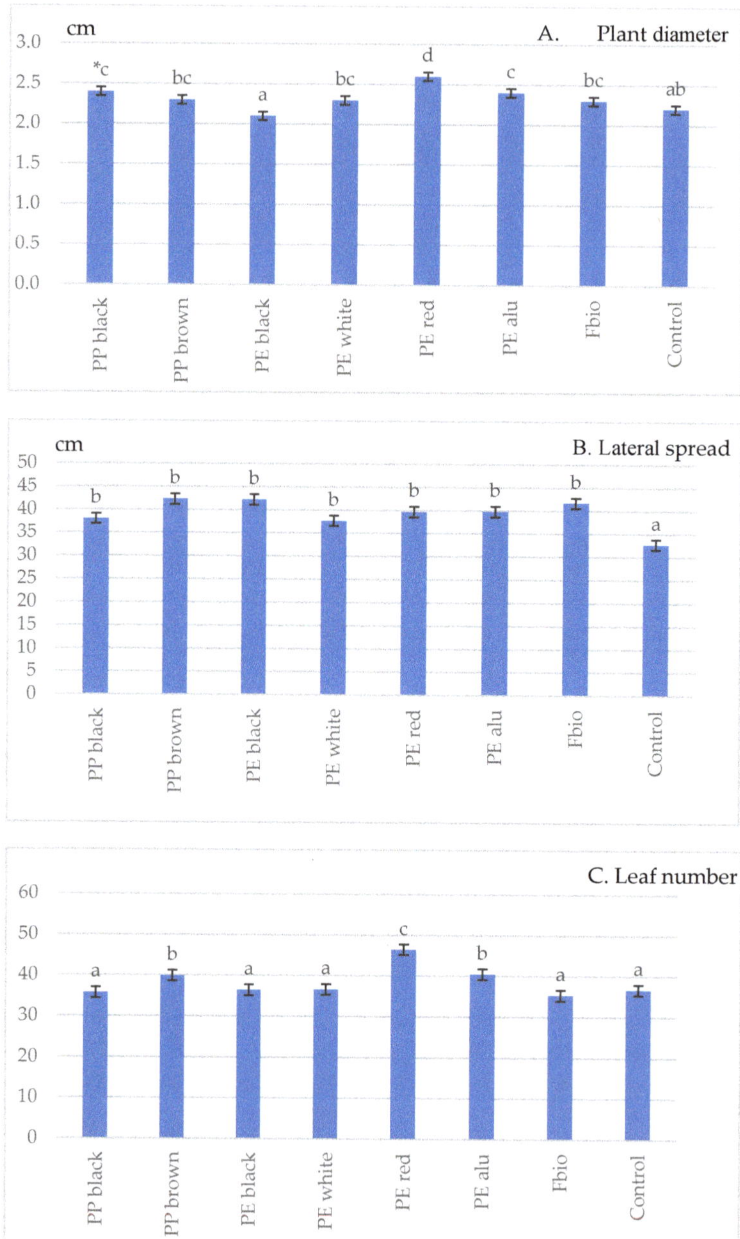

Figure 6. Tomato plant diameter, span, and leaf number in August depending on the type of mulch. Average for 2014–2016. * The same letters mark values belonging to the same homogeneous groups, determined on the basis of statistical analysis for $\alpha = 0.05$.

3.2. Yield Performance

The study results, subjected to statistical analysis, demonstrated that weather conditions and the type of synthetic mulch applied significantly influenced the yield of the Intrigo F1 tomato cultivar (Table 2). The lowest average total fruit yield was recorded in 2014 (8.04 t·ha^{-1}), attributable to adverse weather conditions. In 2016, the yield increased 2.3-fold, while in 2015, it was approximately three times higher than in 2014.

Table 2. Total, marketable, and early yield of Intrigo F1 tomato fruit depending on the type of mulch in 2014–2016 [t·ha^{-1}].

Type of Mulch	Total Yield				Marketable Yield				Early Yield			
	2014	2015	2016	Mean	2014	2015	2016	Mean	2014	2015	2016	Mean
PP black	11.36	26.96	19.30	19.71 **c	9.95	19.59	11.33	13.62 d	1.44	3.53	3.22	2.73 ab
PP brown	7.86	28.45	15.13	17.15 ab	6.38	20.65	8.93	11.99 c	1.16	3.85	2.32	2.44 a
PE black	4.35	27.28	17.82	16.48 ab	3.77	19.58	10.38	11.24 abc	0.79	4.03	3.85	2.89 b
PE white	8.23	25.31	19.08	17.54 ab	6.49	17.87	10.87	11.74 bc	1.71	4.56	4.08	3.45 cd
PE red	6.06	23.39	17.65	15.70 a	5.45	16.21	9.47	10.38 ab	2.00	4.28	3.43	3.24
PE alu	6.73	27.09	20.88	18.23 bc	5.58	18.20	12.33	11.97 c	1.83	4.41	5.02	3.75 d
Fbio	8.59	23.49	17.38	16.49 ab	6.72	17.34	9.12	11.06 abc	2.17	3.88	2.00	2.68 ab
Mean	**7.58**	**26.00**	**18.18**	**17.25**	**6.33**	**18.49**	**10.25**	**11.72**	**1.59**	**4.08**	**3.42**	**3.03**
Control	11.21	15.51	21.55	16.09 a	9.14	10.15	10.10	9.80 a	2.54	3.34	3.27	3.05 bc
Mean	**8.04 *A**	**24.69 C**	**18.60 B**	**17.10**	**6.67 A**	**17.45 C**	**10.29 B**	**11.48**	**1.70 A**	**3.99 B**	**3.40 B**	**3.03**

The same letters mark values belonging to the same homogeneous groups, determined on the basis of statistical analysis for α = 0.05. * Capital letters refer to years, ** lower letters refer to type of mulch. Type of mulch: PP nonwoven: black and brown; PE film: black, white, red; PE alu—aluminum film; Fbio—biodegradable film.

Extensive research has indicated that soil mulching fosters vegetable yield improvement [13,26,27]. The effectiveness of this practice, however, may vary depending on the type of mulching material utilized [28]. Based on the averages across the experimental years, the highest total fruit yield was achieved in cultivation using black PP nonwoven fabric (19.17 t·ha^{-1}), with a comparable yield observed when aluminum foil mulch was employed. In other treatments, the yield ranged from 15.70 to 17.54 t·ha^{-1}. Over the course of the study, the highest total yield was recorded in 2015 with brown PP mulch (28.45 t·ha^{-1}), whereas the lowest yield was observed in 2014 with black PE film mulch (4.35 t·ha^{-1}). During the hot and dry conditions of 2015, mulching significantly enhanced the total yield, resulting in an average increase of 67.6% in mulched treatments compared to the control.

The marketable yield of Intrigo F1 tomato fruits accounted for an average of 67% of the total yield, with values ranging from 6.67 t·ha^{-1} in 2014 to 17.45 t·ha^{-1} in 2015. Statistical analysis revealed a significant positive effect of synthetic mulching materials on marketable yield; however, the magnitude of this effect varied across the study years. In 2015, a yield increase of 82.2% in marketable yield was observed in mulched treatments compared to the control. On average, across all experimental years, mulching resulted in a 19.6% increase in marketable yield relative to the control. Di Mola et al. [29] demonstrated that cultivating large-fruited tomatoes using black PE and biodegradable mulches produced a comparable yield increase of 25%. Similarly, Agrawal et al. [25] reported that black and red PE mulches increased tomato yields by 45.5% and 40.1%, respectively, compared to the control. Rajablariani et al. [30] observed marketable yield increases of 65% with black-and-aluminum mulch, 50% with black PE mulch, and 26% with red PE mulch. In the present study, the highest marketable yield (13.62 t·ha^{-1}) was recorded in treatments with black nonwoven fabric, representing a 39% increase compared to the control. Marketable yields from brown PP, black, white, biodegradable, and aluminum PE mulches averaged 11.60 t·ha^{-1}, which exceeded the control yield by 18.4%. Over the experimental period, the maximum marketable yield was observed in 2015 on brown nonwoven fabric, reaching 20.65 t·ha^{-1}. For tomato cultivation on Fbio, the marketable yield of Intrigo fruit constituted an average of 67.1% of the total yield. The total and marketable yield was lower by 16.4% and 18.8%, respectively, compared to the highest yield obtained from black PP.

The early yield of Intrigo F1 tomato fruits exhibited variability across the study years, ranging from 1.70 t·ha^{-1} in 2014 to an average of 3.70 t·ha^{-1} in subsequent years. Statistical analysis confirmed that the type of synthetic mulch employed had a significant effect on early yield. The highest early yield was obtained in treatments with aluminum foil mulch and white PE film mulch, averaging 3.60 t·ha^{-1}. By contrast, the lowest early yields were recorded in treatments with black and brown PP nonwoven fabric and biodegradable mulches, averaging 2.62 t·ha^{-1}. The most pronounced effect of mulching was observed in 2015, where it led to an average yield increase of 22% compared to the control (non-mulched) treatments. Over the study period, the maximum early yield was recorded in 2016 on aluminum foil mulch (5.02 t·ha^{-1}), with a comparable yield observed in 2015 on white PE film and aluminum foil mulches.

The highest number of fruits per cluster was recorded in 2015 (16.6 fruits), whereas the number was approximately 31.4% lower in 2014 and 2016. The largest mean number of fruits per cluster was observed in plants grown on black PP nonwoven fabric (15.1 fruits), with comparable values recorded for brown PP nonwoven fabric and black PE film mulches (Figure 7). These numbers exceeded those recorded in the control treatments by 24.8% and an average of 14.5%, respectively. On the Fbio mulch, the number of fruits per cluster was at the same level as in the control. Di Mola et al. [29] observed a 31.3% increase in the number of fruits on black PE and biodegradable mulches compared to non-mulched cultivation. The mean fruit mass per cluster ranged from 266.3 g in 2014 to 230 g in 2016. Statistical analysis revealed no significant effect of synthetic mulch type on the average fruit mass per cluster (Figure 8). However, slightly higher fruit masses were observed in plants cultivated on black PP nonwoven fabric (266.7 g).

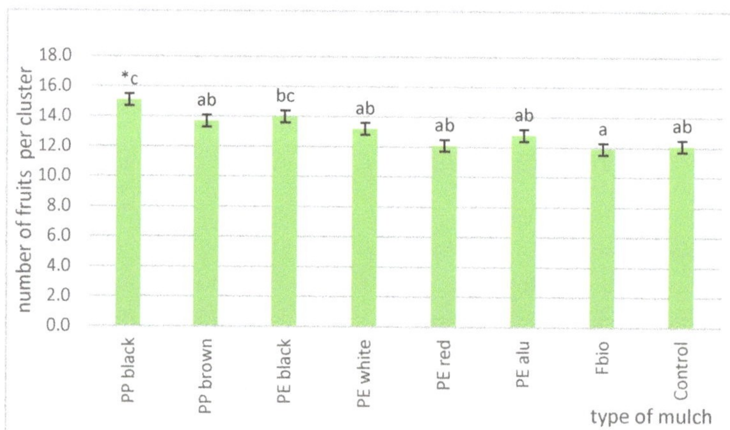

Figure 7. Number of fruits per cluster depending on the type of mulch. Average for 2014–2016. * The same letters mark values belonging to the same homogeneous groups, determined on the basis of statistical analysis for $\alpha = 0.05$. Type of mulch: PP nonwoven: black and brown; PE film: black, white, red; PE alu—aluminum film; Fbio—biodegradable film.

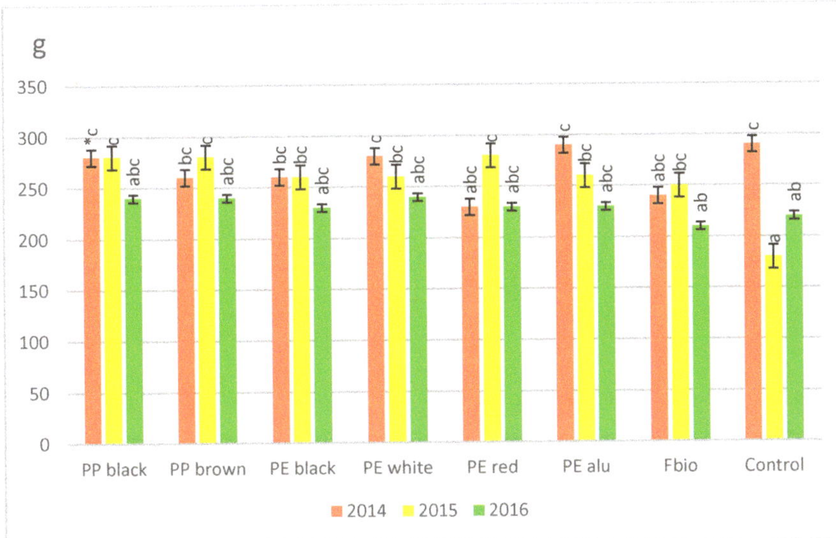

Figure 8. Fruit mass from one cluster depending on the type of mulch in 2014–2016 (g). * The same letters mark values belonging to the same homogeneous groups, determined on the basis of statistical analysis for $\alpha = 0.05$. Type of mulch: PP nonwoven: black and brown; PE film: black, white, red; PE alu—aluminum film; Fbio—biodegradable film.

3.3. Chemical Composition of Tomato Fruits

Tomatoes are among the most widely consumed vegetables worldwide due to their unique flavor and high nutritional profile. In addition to essential macro- and micronutrients, tomato fruits are rich in secondary metabolites, such as ascorbic acid, sucrose, hexoses, citrates, and malates. They also accumulate bioactive compounds beneficial to human health, including carotenoids, phenylpropanoids, and terpenoids [31–34]. Raffo et al. [35] reported that cherry tomatoes typically contain higher levels of antioxidants compared to standard-sized varieties. They demonstrated that fully ripe cherry tomatoes exhibit the highest concentrations of carotenoids and antioxidant activity, while ascorbic acid levels remain relatively stable throughout the ripening process.

The chemical composition of vegetables is primarily determined by genetic factors but can be significantly influenced by external variables, such as climatic conditions and agronomic practices, during the plant's growth cycle [35,36]. In this study, the application of synthetic mulches significantly affected the chemical composition of the Intrigo F1 tomato cultivar. The impact of mulching varied by the type of material and its color (Tables 3 and 4). On average, higher concentrations of organic compounds, phosphorus (P), and potassium (K), as well as elevated DPPH radical scavenging activity, were observed in fruits grown with PE film mulches compared to PP nonwoven fabric and control treatments. Tomatoes grown on biodegradable mulches demonstrated lower levels of dry matter, total sugars, phosphorus, vitamin C, and carotenoids while accumulating higher concentrations of polyphenols, lycopene, and potassium compared to fruits cultivated on PE film mulches.

Table 3. The content of dry matter, total sugars, and macronutrients in Intrigo F1 tomato fruit depending on the type of mulch (%). Average for 2014–2016.

Type of Mulch	Dry Matter	Total Sugars	P	K	Mg	Ca
			%			
PP black	9.02 *cd	4.11 a	0.25 a	3.43 cd	0.16 a	0.26
PP brown	8.71 ab	4.32 ab	0.31 bc	3.00 a	0.22 bc	0.27
PE black	8.48 a	4.09 a	0.35 d	3.48 d	0.20 abc	0.25
PE white	9.22 de	5.41 d	0.27 a	3.31 bc	0.22 bc	0.28
PE red	9.38 e	4.85 c	0.32 cd	3.31 bc	0.19 ab	0.28
PE alu	8.51 a	4.50 bc	0.32 cd	3.10 a	0.24 c	0.26
Fbio	8.81 bc	4.06 a	0.28 ab	3.82 e	0.20 abc	0.23
Control	8.57 a	4.71 c	0.33 c	3.25 b	0.20 abc	0.29

* The same letters mark values belonging to the same homogeneous groups, determined on the basis of statistical analysis for $\alpha = 0.05$. Type of mulch: PP nonwoven: black and brown; PE film: black, white, red; PE alu—aluminum film; Fbio—biodegradable film.

Table 4. The content of vitamin C, carotenoids, polyphenols, lycopene, and N-NO3⁻ and the value of DPPH in Intrigo F1 fruits depending on the type of mulch. Average for 2014–2016.

Type of Mulch	Vitamin C	Polyphenols	Carotenoids [μg·100 g^{-1} f.m.]	Lycopene [mg·kg^{-1} f.m.]	DPPH [%]	$N\text{-}NO_3^-$ [mg·kg^{-1} f.m.]
	[mg·100 g^{-1} f.m.]					
PP black	36.81 *a	16.19 a	38.22 c	19.10 abc	58.73 b	120.33 d
PP brown	39.87 cd	24.32 c	37.27 c	16.36 a	68.96 d	118.18 cd
PE black	41.79 e	19.57 b	41.44 d	17.67 ab	56.42 a	102.35 ab
PE white	41.88 e	21.11 b	39.26 cd	18.65 abc	71.52 e	98.86 a
PE red	39.74 c	26.97 d	32.27 ab	20.91 cd	65.22 c	118.54 cd
PE alu	40.59 d	36.46 f	30.10 a	20.33 bcd	70.45 de	106.69 abc
Fbio	37.84 b	30.69 e	31.11 ab	22.31 d	59.61 b	107.57 abcd
Control	36.57 a	27.41 d	33.64 b	20.35 bcd	56.18 a	111.55 abcd

* The same letters mark values belonging to the same homogeneous groups, determined on the basis of statistical analysis for $\alpha = 0.05$. Type of mulch: PP nonwoven: black and brown; PE film: black, white, red; PE alu—aluminum film; Fbio—biodegradable film.

Fruits harvested from red and white PE mulch exhibited the highest dry matter content, averaging 9.30%, with comparable levels observed in fruits grown on black PP nonwoven fabric (9.02%). Significantly lower dry matter levels were detected in fruits from treatments using black PE mulch, aluminum foil mulch (PE alu), and control plots. These findings align with those of Hallmann and Rembiałkowska [37], who reported an average dry matter content of 7.36% in the small-fruited tomato cultivar Koralik. Sugars, the primary component of tomato pulp extract, play a critical role in determining fruit flavor and palatability, with higher sugar concentrations enhancing fruit quality. In the current study, fruits grown on white PE mulch demonstrated the highest total sugar content (5.41%), followed by those grown on red PE film (4.85%) and aluminum foil mulch (4.50%). These results are consistent with findings by Shahzad et al. [38], who confirmed the significant contribution of sugars to the sweet taste and nutritional value of small-fruited tomatoes (*Solanum lycopersicum* var. *cerasiforme*). Hallmann and Rembiałkowska [37] reported a total sugar content of 2.62% in Koralik small-fruited tomatoes. By contrast, Kowalczyk et al. [39] observed a higher sugar concentration in the greenhouse-grown small-fruited cultivar Dasher F1, which exhibited 3.66 g of total sugar per 100 g of fresh weight, surpassing medium-fruited varieties. Similarly, Gharezi et al. [40], in their analysis of six small-fruited tomato cultivars, documented sugar concentrations ranging from 3.14 to 4.81 mg per 100 g.

Key bioactive compounds in tomatoes include carotenoids, with lycopene being the primary carotenoid responsible for the red color of tomatoes, along with β-carotene. According to Li et al. [33], β-carotene is considered the most efficient source of provitamin A among carotenoids. The present study revealed a significant influence of the type of synthetic mulch on the carotenoid content in Intrigo F1 tomato fruits. The highest carotenoid concentrations were recorded in fruits from plants grown under black PE mulch (41.44 μg·100 g^{-1} fresh weight), which was approximately 23% higher compared to the control. Slightly lower levels were observed in fruits from plants mulched with white PE (5.3% lower) and black and brown PP (on average 9.0% lower). The lowest carotenoid content (30.10 $\mu g \cdot 100\ g^{-1}$ fresh weight) was found in fruits from plants grown with aluminum foil mulch (PE alu). Di Mola et al. [29] reported that, under Southern European conditions, the average increase in carotenoid content in fruits of traditional tomato varieties grown on black mulches (PE and biodegradable) was 57%. Dobromilska et al. [41] observed carotenoid concentrations in the small-fruited Conchita F1 cultivar ranging from 0.43 to 0.62 $mg \cdot g^{-1}$ fresh weight.

Lycopene exhibits strong antioxidant and anticancer properties. According to Kuti and Konuru [32], both the cultivar and the growing environment have a significant influence on the lycopene content in tomatoes. The authors observed that compared to other varietal groups, cocktail tomato fruits accumulate higher levels of lycopene. In field cultivation, the lycopene content in these fruits was estimated to be an average of 91.9 mg kg^{-1} fresh weight (f.w.), while in greenhouse cultivation, it averaged 56.1 mg kg^{-1} f.w. Bilalis et al. [42] reported lycopene contents of 88.5 and 80.5 mg kg^{-1} f.w. for tomatoes fertilized organically and with inorganic fertilizers, respectively, while Jędrszczyk and Ambroszczyk [43] reported values of 4.29–6.82 mg 100 g^{-1} f.w. in their studies. The present research found a significant effect of the type of mulching material on the lycopene content in cocktail tomato fruits. Biodegradable mulch resulted in the highest lycopene concentration (22.31 mg kg^{-1} fresh weight). Fruits from plants mulched with red PE film, PE alu, and the control treatment exhibited similar lycopene levels, ranging from 20.91 to 20.33 mg kg^{-1} fresh weight.

Tomato fruits contain a variety of phenolic compounds, such as flavonoids, caffeoylquinic acids, and other hydroxycinnamates. Cruz-Carrion et al. [34] identified 57 different polyphenolic compounds in tomato fruits. Their concentration varied depending on the cultivation site. In the present experiment, the highest polyphenol content was recorded in tomatoes grown under aluminum foil (36.46 $mg \cdot 100\ g^{-1}$ fresh weight). Significantly lower levels, 15.8% less, were found in fruits from biodegradable film, and 25.4% less on average in fruits from red film and the control group. Cruz-Carrion et al. [34] reported an average total polyphenol content of 3554 $\mu g \cdot g^{-1}$ fresh weight in tomato fruits. Morra et al. [44] found that fruits from plants mulched with biodegradable black Mater-Bi foils had more polyphenols and lycopene and higher antioxidant activity compared to LDPE and unmulched soil.

The vitamin C content in Intrigo F1 tomato fruits ranged from 36.57 to 41.88 $mg \cdot 100\ g^{-1}$ fresh weight. It was demonstrated that mulching (except for black PP) significantly increased the content of this component compared to the control. The greatest increase, averaging 14.4%, was observed in the treatments with white and black PE film. In crops grown under red foil, aluminum foil, and brown PP fabric, the increase averaged 9.6%, and it averaged 3.5% on biodegradable film. Gharezi et al. [40] reported vitamin C levels in small-fruited tomato varieties ranging from 23.6 to 28.1 $mg \cdot 100\ g^{-1}$. In research by Jędrszczyk and Ambroszczyk [43], the amount of this component in tomato fruits grown in the field ranged from 10.77 to 28.9 mg 100 g^{-1} f.w.

Minoggio et al. [45] highlighted the high biological value of tomato fruits, particularly due to the significant antioxidant activity of the polyphenols they contain. Through studies on various tomato lines and cultivars examining the relationship between polyphenol, lycopene, and beta-carotene content and total antioxidant activity (TAA), it has been established that nearly all lines with low carotenoid content produce high levels of polyphenols and consequently exhibit the strongest antioxidant potential. In our own research, the highest DPPH index (71.52%) was observed in tomatoes grown on white film, with comparable levels found in fruits grown on aluminum film (70.45%) and brown PP film (68.96%). The lowest antioxidant activity was recorded in tomatoes grown on black PE film and in the control group (56.30% DPPH).

The lowest nitrate nitrogen content in fruits (98.86 mg·kg^{-1} fresh weight) was observed in fruits grown under white film mulch. At the same level of significance, nitrate levels were similar in fruits from treatments with black PE film, aluminum film, and biodegradable film and the control. The highest nitrate content was recorded in fruits from the black fabric mulch treatment (120.33 mg·kg^{-1} fresh weight). Dobromilska et al. [41] reported nitrate contents in cocktail tomato fruits grown in a plastic tunnel ranging from 200 to 245 mg·kg^{-1} fresh weight.

The macronutrient content in tomato fruits varied depending on the type of mulch used. The phosphorus content ranged from an average of 0.27% (black PP, white PE, biodegradable fabric) to 0.33% (red PE, aluminum fabric, black PE) in fresh weight. Potassium content ranged from 3.05% (brown PP, aluminum fabric) to 3.46% (black PE, black PP) and 3.82% (biodegradable fabric). Magnesium content ranged from 0.16% (black PP) to 0.24% (aluminum fabric). Dobromilska et al. [41] reported phosphorus contents in Conchita F1 fruits grown in a plastic tunnel of 4.86–6.40 g·kg^{-1} fresh weight, potassium contents of 24.90–32.7 g·kg^{-1}, calcium contents of 1.55–2.10 g·kg^{-1}, and magnesium contents of 1.00–1.40 g·kg^{-1}.

4. Conclusions

The growth of plants, as well as the quantity and quality of the yield of the small-fruited tomato variety Intrigo F1, were significantly influenced by the weather conditions during the growing season. According to the observations, it can be concluded that the primary yield-determining factor for tomatoes is the temperature regime at transplanting. The second crucial factor is water availability. During the first part of the growing season, adequate water supply promotes vegetative development, flower formation, and fruit set.

Plants grown with synthetic mulches exhibited greater height, stem diameter at the base, lateral spread, and leaf number compared to those grown without mulch. A somewhat weaker effect on tomato plants was observed when using biodegradable film. Tomatoes grown on biodegradable film yielded slightly more than those in the control. A broader use of this mulch is encouraged due to the fact that it has a similar effect on the soil environment and plant surroundings as other mulches, without the issue of removal and disposal from the field. Soil mulching improved tomato growth conditions and fruit quality due to higher soil temperatures and consistent moisture, especially compared to bare ground.

Author Contributions: Conceptualization K.A.-S. and J.B.; methodology, K.A.-S. and J.B. software, K.A.-S. and J.B.; validation, K.A.-S. and J.B.; formal analysis, K.A.-S. and J.J.; investigation, J.B.; resources, K.A.-S.; data curation, J.B.; writing—original draft preparation, K.A.-S.; writing—review and editing, K.A.-S.; visualization, K.A.-S.; supervision, K.A.-S.; project administration, K.A.-S.; funding acquisition, K.A.-S. All authors have read and agreed to the published version of the manuscript.

Funding: The research is co-financed from the subsidy increased by the minister responsible for higher education and science for the period 2020–2026 in the amount of 2% of the subsidy referred to Art. 387 (3) of the Act of 20 July 2018—Law on Higher Education and Science, obtained in 2019.

Institutional Review Board Statement: Not applicable.

Data Availability Statement: Data is contained within the article.

Conflicts of Interest: The authors declare no conflicts of interest.

References

1. FAOSTAT. Available online: https://www.fao.org/faostat/en/#data (accessed on 4 December 2024).
2. Mansoor, Z.; Tchuenbou-Magaia, F.; Kowalczuk, M.; Adamus, G.; Manning, G.; Parati, M.; Radecka, I.; Khan, H. Polymers Use as Mulch Films in Agriculture—A Review of History, Problems and Current Trends. *Polymers* **2022**, *14*, 5062. [CrossRef]
3. Mormile, P.; Stahl, N.; Malinconico, M. The world in plasticulture. In *Soil Degradable Bioplastics for a Sustainable Modern Agriculture, Green Chemistry and Sustainable Technology*; Malinconico, M., Ed.; Springer-Verlag GmbH: Berlin/Heidelberg, Germany, 2017; pp. 1–21. [CrossRef]
4. Steinmetz, Z.; Wollmann, C.; Schaefer, M.; Buchmann, C.; David, J.; Tröger, J.; Muñoz, K.; Frör, O.; Schaumann, G.E. Plastic mulching in agriculture. Trading short-term agronomic benefits for long-term soil degradation? *Sci. Total. Environ.* **2016**, *550*, 690–705. [CrossRef] [PubMed]
5. Gao, H.; Yan, C.; Liu, Q.; Ding, W.; Chen, B.; Li, Z. Effects of plastic mulching and plastic residue on agricultural production: A meta-analysis. *Sci. Total Environ.* **2019**, *651*, 484–492. [CrossRef] [PubMed]
6. Li, S.; Ding, F.; Flury, M.; Wang, Z.; Xu, L.; Li, S.; Jones, D.L.; Wang, J. Macro- and microplastic accumulation in soil after 32 years of plastic film mulching. *Environ. Pollut.* **2022**, *300*, 2–8. [CrossRef] [PubMed]
7. Kader, M.A.; Singha, A.; Begum, M.A.; Jewel, A.; Khan, F.H.; Khan, N.I. Mulching as water-saving technique in dryland agriculture: Review article. *Bull. Natl. Res. Cent.* **2019**, *43*, 1–6. [CrossRef]
8. Amare, G.; Desta, B. Coloured plastic mulches: Impact on soil properties and crop productivity. *Chem. Biol. Technol. Agric.* **2021**, *8*, 4. [CrossRef]
9. Bucki, P.; Siwek, P. Organic and non-organic mulches—Impact on environmental conditions, yield, and quality of Cucurbitaceae. *Folia Hort.* **2019**, *31*, 129–145. [CrossRef]
10. Decoteau, D.R. The emergence and early development of colored reflective plastic mulch technology in agriculture. In *Recent Advances in Agriculture*; Research Signpost: Thiruananthapuram, India, 2008; pp. 1–17.
11. Mutoro, K. Effect of organic and inorganic mulching materials on tomato growth and development in Western Kenya. *Acad. Lett.* **2021**, *1131*, 2–7. [CrossRef]
12. Bhujbal, P.D.; Tambe, T.B.; Ulemale, P.H. Effect of mulches on flowering, fruiting, yield and pest-disease incidence of tomato (Lycopersicon esculentum Mill). *Bioscan* **2015**, *10*, 465–468.
13. Adamczewska-Sowińska, K.; Krygier, M.; Turczuk, J. The yield of eggplant depending on climate conditions and mulching. *Folia Hort.* **2016**, *28*, 19–24. [CrossRef]
14. Sękara, A.; Pokluda, R.; Cozzolino, E.; Del Piano, L.; Cuciniello, A.; Caruso, G. Plant growth, yield, and fruit quality of tomato affected by biodegradable and non-degradable mulches. *Hort. Sci.* **2019**, *46*, 138–145. [CrossRef]
15. Gabryś, T.; Fryczkowska, B.; Grzybowska-Pietraś, J.; Biniaś, D. Modification and Properties of Cellulose Nonwoven Fabric-Multifunctioanl Mulching Material for Agricultural Applications. *Materials* **2021**, *14*, 4335. [CrossRef] [PubMed]
16. Available online: https://biobagworld.com/products/agriculture/ (accessed on 4 December 2024).
17. Adamczewska-Sowińska, K.; Sowiński, J. Reaction of Sweet Maize to the Use of Polyethylene Film and Polypropylene Non-Woven Fabric in the Initial Growth Phase. *Agronomy* **2020**, *10*, 141. [CrossRef]
18. Boote, K.J.; Rybak, M.R.; Scholberg, J.M.; Jones, J.W. Improving the CROPGRO-tomato model for predicting growth and yield response to temperature. *Hort. Sci.* **2012**, *47*, 1038–1049. [CrossRef]
19. Heisenberg, C.; Stewart, K. Field crop management. In *The Tomato Crop*; Springer: Berlin/Heidelberg, Germany, 1986; pp. 511–557.
20. Van Dam, B.; Goffau, M.; van Lidt, J.; Naika, S. La culture de la tomate: Production, transformation et commercialisation. *Ser. Agrodok* **2005**, *17*, 6–104.
21. Ayankojo, I.T.; Morgan, K.T. Increasing air temperatures and its effects on growth and productivity of tomato in south Florida. *Plants* **2020**, *9*, 2–16. [CrossRef]
22. Pinder, R.; Rana, R.; Maan, D.; Kumar, K. Impact of different mulching materials on the growth and yield of tomato (Solanum lycopersicum) in Dehradun region of Uttarakhand. *Int. J. Environ. Agric. Biotechnol. IJEAB* **2016**, *1*, 631–636. [CrossRef]
23. Islam, S. Effects of raised bed furrow irrigation and various mulching techniques on the growth, yield and water use efficiency of tomato cultivation. *Int. J. Hort. Sci. Technol.* **2023**, *10*, 33–40. [CrossRef]
24. Onunva, A.O.; Nwaiwu, C.J.; Madueke, C.O.; Nnabuihe, E.C.; Nwosu, T.V.; Iwuchukwu, T. Effect of different mulch materials on soil properties, growth and yield of tomato (*Lycopersicon esculentum* mill) at Awka. In Proceedings of the First Faculty of Agriculture International Conference, Nnamdi Azikiwe University, Awka, Nigeria, 22–24 March 2023; pp. 108–114.
25. Agrawal, N.; Panigrahi, H.K.; Sharma, D.; Agrawal, R. Effect of different colour mulches on the growth and yield of tomato under Chhattisgarh region. *Indian J. Hort.* **2010**, *67*, 295–300.
26. Kosterna, E. The effect of covering and mulching on the soil temperature, growth and yield of tomato. *Folia Hort.* **2014**, *26*, 91–101. [CrossRef]
27. Mutetwa, M.; Mtaita, T. Effects of mulching and fertilizer sources on growth and yield of onion. *J. Glob. Innov. Agric. Soc. Sci.* **2014**, *2*, 102–106. [CrossRef]

28. Haque, M.A.; Jahiruddin, M.; Clarke, D. Effect of plastic mulch on crop yield and land degradation in south coastal saline soils of Bangladesh. *Int. Soil Water Conserv. Res.* **2018**, *6*, 317–324. [CrossRef]
29. Di Mola, I.; Cozzolino, E.; Ottaiano, L.; Riccardi, R.; Spigno, P.; Petriccione, M.; Fiorentino, N.; Fagnano, M.; Mori, M. Biodegradable Mulching Film vs. Traditional Polyethylene: Effects on Yield and Quality of San Marzano Tomato Fruits. *Plants* **2023**, *12*, 3203. [CrossRef] [PubMed]
30. Rajablariani, H.R.; Hassankhan, F.; Rafezi, R. Effect of colored plastic mulches on yield of tomato and weed biomass. *Int. J. Environ. Sci. Dev.* **2012**, *3*, 590–593. [CrossRef]
31. Ghosh, P.K.; Dayal, D.; Bandyopadhyay, K.K.; Mohanty, M. Evaluation of straw and polythene mulch for enhancing productivity of irrigated summer groundnut. *Field Crops Res.* **2006**, *99*, 76–86. [CrossRef]
32. Kuti, J.O.; Konuru, H.B. Effects of genotype and cultivation environment on lycopene content in red-ripe tomatoes. *J. Sci. Food Agric.* **2005**, *85*, 2021–2026. [CrossRef]
33. Li, Y.; Wang, H.; Zhang, Y.; Martin, C. Can the world's favorite fruit, tomato, provide an effective biosynthetic chassis for high-value metabolites? *Plant Cell Rep.* **2018**, *37*, 1443–1450. [CrossRef]
34. Cruz-Carrion, A.; Calani, L.; Ruiz de Azua Ma, J.; Mena, P.; Del Rio, D.; Suarez, M.; Arola-Arnal, A. (Poly) phenolic composition of tomatoes from different growing locations and their absorption in rats: A comparative study. *Food Chem.* **2022**, *388*, 132984. [CrossRef]
35. Raffo, A.; Leonardi, C.; Fogliano, V.; Ambrosino, P.; Salucci, M.; Gennaro, L.; Bugianesi, R.; Giuffrida, F.; Quaglia, G. Nutritional value of cherry tomatoes (Lycopersicon esculentum Cv. Naomi F1) harvested at different ripening stages. *J. Agric. Food Chem.* **2002**, *50*, 6550–6556. [CrossRef]
36. Lee, S.K.; Kader, A.A. Preharvest and postharvest factors influencing vitamin C content of horticultural crops. *Postharvest Biol. Technol.* **2000**, *20*, 207–220. [CrossRef]
37. Hallmann, E.; Rembiałowska, E. Estimation of fruits quality of selected tomato cultivars from organic and conventional cultivation with special consideration of bioactive compounds content. *J. Res. Appl. Agric. Eng.* **2007**, *52*, 55–60. (In Polish)
38. Shahzad, T.; Ahmad, I.; Choudhry, S.; Saeed, M.K.; Khan, M.N. DPPH free radical scavenging activity of tomato, cherry tomato and watermelon: Lycopene extraction, purification and quantification. *Int. J. Pharm. Pharm. Sci.* **2014**, *6*, 224–228.
39. Kowalczyk, K.; Gajc-Wolska, J.; Radzanowska, J.; Marcinkowska, M. Assesment of chemical composition and sensory quality of tomato fruit depending on cultivar and growing conditions. *Acta Sci. Pol. Hortorum Cultus* **2011**, *10*, 133–140.
40. Gharezi, M.; Joshi, N.; Indiresh, K.M. Physico—Chemical and sensory characteristics of different cultivars of cherry tomato. *Mysore J. Agric. Sci.* **2012**, *46*, 610–613.
41. Dobromilska, R.; Mikiciuk, M.; Gubarewicz, K. Evaluation of cherry tomato yielding and fruit mineral composition after using of BIO-ALGEEN S-90 preparation. *J. Elem.* **2008**, *13*, 491–499. [CrossRef]
42. Bilalis, D.; Krokida, M.; Roussis, I.; Papastylianou, P.; Travlos, I.; Cheimona, N.; Dede, A. Effects of organic and inorganic fertilization on yield and quality of processing tomato (Lycopersicon esculentum Mill). *Folia Hort.* **2018**, *30*, 321–332. [CrossRef]
43. Jędrszczyk, E.; Ambroszczyk, A.M. The influence of NANO-GRO® organic stimulator on the yielding and fruit quality of field tomato (Lycopersicon esculentum Mill). *Folia Hort.* **2016**, *28*, 87–94. [CrossRef]
44. Morra, L.; Cozzolino, E.; Salluzzo, A.; Modestia, F.; Bilotto, M.; Baiano, S.; del Piano, L. Plant Growth, Yields and Fruit Quality of Processing Tomato (*Solanum lycopersicon* L.) as Affected by the Combination of Biodegradable Mulching and Digestate. *Agronomy* **2021**, *11*, 100. [CrossRef]
45. Minoggio, M.; Bramati, L.; Simonetti, P.; Gardana, C.; Iemoli, L.; Santangelo, E.; Mauri, P.L.; Spigno, P.; Soressi, G.P.; Pietta, P.G. Polyphenol Pattern and Antioxidant Activity of Different Tomato Lines and Cultivars. *Ann. Nutr. Metab.* **2003**, *47*, 64–69. [CrossRef]

agriculture

MDPI

Article

Effect of Kaolin Clay on Post-Bloom Thinning Efficacy, Cropping, and Fruit Quality in 'Gala Vill' Apple (*Malus* × *domestica*) Cultivation

Sebastian Przybyłko *, Jacek Marszał, Wojciech Kowalczyk and Ewa Szpadzik

Department of Pomology and Horticulture Economics, Institute of Horticultural Sciences, Warsaw University of Life Sciences (SGGW-WULS), Nowoursynowska 159 C Str., 02-787 Warsaw, Poland; jacek_marszal@sggw.edu.pl (J.M.); wojciech_kowalczyk@sggw.edu.pl (W.K.); ewa_szpadzik@sggw.edu.pl (E.S.)

* Correspondence: sebastian_przybylko@sggw.edu.pl

Abstract: Effective thinning methods that balance yield, fruit quality, and ripening dynamics are essential to ensure efficient and sustainable apple production. This study examined the effects of various thinning treatments on 'Gala Vill' apples (*Malus* × *domestica* Borkh.) to assess their impacts on fruit set, yield, and quality parameters. The experiment was conducted in 2020 at the experimental orchard of WULS located in Wilanów, Warsaw, Poland. The treatments included chemical thinning using different doses of metamitron (Brevis 150 SG) alone and in combination with kaolin clay at two concentrations (50 and 100 kg·ha^{-1}), and, as alternatives to the chemical method, using kaolin clay alone (50, 100, and 200 kg·ha^{-1}) and artificial shading. The results highlight the effectiveness of thinning treatments in modulating key agronomic traits. Artificial shading significantly reduced the number of fruitlets, demonstrating its utility as a non-chemical thinning option. Metamitron application effectively reduced the number of fruitlets in a dose-dependent manner. Combining metamitron with kaolin clay did not enhance the thinning effect compared with metamitron alone. However, kaolin clay applied independently, particularly at higher concentrations, was associated with improved fruit setting and yield. For instance, kaolin clay at 200 kg·ha^{-1} (KC200) resulted in the highest fruit set (85.2) and yield (10.1 kg·tree^{-1}), suggesting its adverse effect on thinning to promote fruit retention under certain conditions.

Keywords: apple; thinning methods; kaolin clay; metamitron; artificial shading; fruit retention

1. Introduction

Apple fruitlet thinning is a key agronomic practice in apple cultivation, significantly influencing both the yield quality and quantity [1]. This process involves reducing the number of flowers or fruitlets on trees to ensure optimal conditions for the growth of the remaining fruits and to maintain a balance between fruiting and the trees' health. The primary aim of thinning is to improve fruit quality parameters, such as size, color, and sugar content, while preventing undesirable alternate bearing [2–4]. Excess flowers or fruitlets can be removed by hand, chemically, or mechanically. Chemical thinning remains the primary approach for early thinning due to the high costs of manual thinning, the limited availability of mechanical equipment, and the lack of selectivity in mechanical methods [5–8].

The effectiveness of chemical thinning depends on various factors. The level of thinning and its impact on flowering in the subsequent seasons are influenced by the apple

https://doi.org/10.3390/agriculture15040440

variety, rootstock, tree age, growth vigor, health status, flowering intensity, pollination success, and weather conditions before, during, and after the treatment [9–11]. In apple cultivation, thinning during the flowering stage is achieved mainly by preventing fertilization by blocking pollen germination. Desiccants are commonly used, targeting the highly hydrated pistil and its stigma and causing them to dry out. This treatment limits the deposition, germination, and fertilization of pollen [12–15]. Research indicates that desiccants, apart from their direct physical effects, may also influence physiological processes, such as increasing ethylene production in damaged flowers, which intensifies plant stress and enhances thinning efficiency [16,17].

The use of ATS (ammonium thiosulphate) or S-Ca (calcium polysulfide) often reduces photosynthesis efficiency, with calcium polysulfide potentially lowering photosynthesis by up to 50% for several weeks [18]. A similar effect on trees' energy balance is observed with metamitron, a triazine-based compound used in horticulture for weed control. Metamitron disrupts photosynthesis by blocking electron transport in PSII, reducing the production of ATP and NADPH, which are essential for carbohydrate synthesis. This leads to plant stress caused by a negative energy balance, which results in the abscission of excessive fruitlets. Low light interception enhances its thinning effect by further decreasing carbohydrate availability, which promotes fruitlet abscission. High temperatures, on the other hand, can accelerate the plant's metabolic processes, potentially intensifying metamitron's impact, but may also reduce its persistence due to faster degradation [11]. The effects of metamitron are comparable to those of plant shading, whose beneficial impact on thinning and treatment efficiency has been confirmed in numerous studies [19–22]. Some researchers suggest that artificial shading may be an intriguing alternative to chemical thinning [23]. However, the practical application of this method is challenging due to its high material and labor costs [24].

For this reason, methods that achieve similar effects to shading while minimizing its drawbacks are worth exploring. One potentially effective method is covering trees with kaolin clay, a natural material belonging to the group of clay minerals that is widely used in apple and other fruit production [25–27]. Thanks to its light-reflecting properties, kaolin clay spraying helps mitigate stress caused by excessive sunlight and decreases losses from sunburn on developing fruitlets [28,29]. Additionally, kaolin has been shown to improve fruit quality in terms of better coloration and solid soluble content and promote better growth [25]. Typical doses range from 2 to 5% in solution, depending on environmental conditions and tree size. Applications are usually made 3–5 times during the growing season at 7–14-day intervals, starting before the onset of high temperatures [25,28,29]. Meanwhile, several studies have shown that kaolin clay application can reduce the leaf photosynthetic rate by decreasing the amount of light reaching the photosynthetic apparatus due to the improved reflective properties of the leaf surface and decreased energy absorption [30,31].

In our study, we hypothesized that applying high doses of kaolin clay by spraying trees after flowering will result in the formation of a light-reflecting layer on leaves, producing a similar effect to artificial shading or metamitron application, which reduces photosynthetic efficiency and leads to fruitlet thinning. The potential additive effect of combining kaolin clay with metamitron was also tested. The expected result was enhancing fruitlet abscission via dual mechanisms: kaolin clay reduces light interception by reflecting solar radiation as a physical barrier, while metamitron acts as a physiological photosynthetic inhibitor. This combined approach could optimize thinning efficacy and reduce the use of synthetic agents in fruit production. Providing new potential approaches for apple thinning strategies is of a great importance for improving fruit quality, ensuring consistent yields, and meeting the growing demand for more sustainable orchard management practices. Exploring alternative methods such as kaolin clay application offers a promising tool to reduce dependence on synthetic chemical agents while maintaining thinning efficiency and minimizing environmental impact.

2. Materials and Methods

2.1. Location, Plant Material, and Experimental Design

The experiment was conducted in the experimental orchard of Warsaw University of Life Sciences (WULS) in Wilanów, Warsaw, Poland (N 52°9′36.1″, E 21°5′58.2) in the growing season of 2020. The weather data on the plot were gathered using the Davis Vantage Pro weather station installed in the experimental plot and are presented in Figure 1.

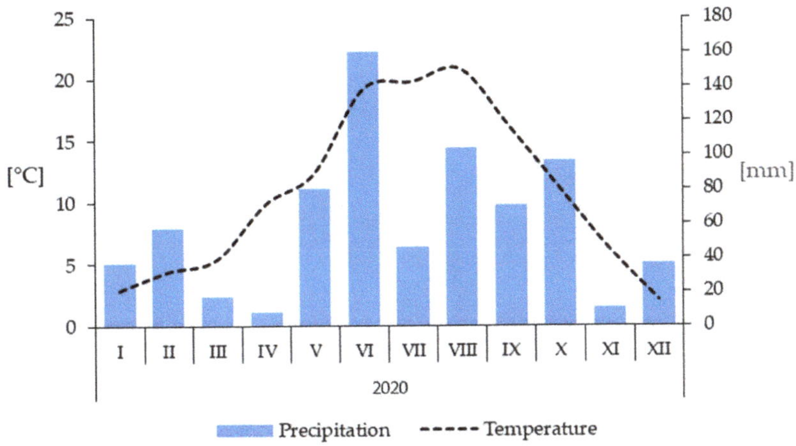

Figure 1. Weather data for the experimental plot in 2020.

The plant material consisted of apple (*Malus* × *domestica* Borkh.) trees of the 'Gala Vill' cultivar that were planted in the spring of 2016 on silty loam alluvial soil with 3.2 × 0.8 m spacing between them. The trees were trained as a standard spindle bush canopy. The floor management system in the experimental block maintained herbicide fallow within the rows of trees and turf grass in the inter-rows. Anti-hail nets were used to protect the fruits and trees against damage during hail storms. No irrigation was used on the experimental plots during the experiment.

Other agrotechnical treatments were applied following the local standards for the integrated production strategy and were the same for all trees in the experimental plot, excluding the thinning strategy.

During the trial, 10 different thinning treatments were tested as follows:

(1) The control (control), where no thinning treatments were applied;

(2) Artificial shading (AS), where a black agro-textile (density: 50 $g \cdot m^{-2}$), commonly used for mulching the soil to avoid weed competition, was applied over the trees for 5 days at a 6–8 mm fruitlet size (Figure 2a);

(3) Brevis 2.2 (B2), where single applications of Brevis (containing metamitron as an active ingredient) at a dose of 2.2 $kg \cdot ha^{-1}$ were applied at a 6–8 mm fruitlet size;

(4) Kaolin Clay 50 (KC50), where a single application of kaolin clay was applied at a 50 $kg \cdot ha^{-1}$ dose at a 6–8 mm fruitlet size (Figure 2b,c);

(5) Kaolin Clay 100 (KC100), where a single application of kaolin clay was applied at a 100 $kg \cdot ha^{-1}$ dose at a 6–8 mm fruitlet size (Figure 2d);

(6) Kaolin Clay 200 (KC200), where a single application of kaolin clay was applied at a 200 $kg \cdot ha^{-1}$ dose at a 6–8 mm fruitlet size (Figure 2e);

(7) Brevis 1.1 + Kaolin Clay 50 (B1 + KC50), where a single application of Brevis at a 1.1 kg·ha^{-1} dose was applied, followed by kaolin clay application at a 50 kg·ha^{-1} dose at a 6–8 mm fruitlet size;
(8) Brevis 1.1 + Kaolin Clay 100 (B1 + KC100), where a single application of Brevis at a 1.1 kg·ha^{-1} dose was applied, followed by kaolin clay application at a 100 kg·ha^{-1} dose at a 6–8 mm fruitlet size;
(9) Brevis 2.2 + Kaolin Clay 50 (B2 + KC50), where a single application of Brevis at a 2.2 kg·ha^{-1} dose was applied, followed by kaolin clay application at a 50 kg·ha^{-1} dose at a 6–8 mm fruitlet size;
(10) Brevis 2.2 + Kaolin Clay 100 (B2 + KC100), where a single application of Brevis at a 2.2 kg·ha^{-1} dose was applied, followed by kaolin clay application at a 100 kg·ha^{-1} dose at a 6–8 mm fruitlet size.

Figure 2. Trees in the experimental plot under the chosen treatments tested in the trial: (**a**) black agro-textile over an experimental plot; (**b**) trees after the application of kaolin clay at a 200 kg·ha^{-1} dose; (c) an apple leaf after the application of kaolin clay at a 50 kg·ha^{-1} dose; (**d**) an apple leaf after the application of kaolin clay at a 100 kg·ha^{-1} dose; (**e**) an apple leaf after the application of kaolin clay at a 200 kg·ha^{-1} dose.

The spraying of trees was carried out using a Stihl 430 motor sprayer (Andreas Stihl AG & Co., Waiblingen, Germany), applying the working liquid at a rate of 1000 L·ha^{-1}. All treatments were performed in three replications, consisting of 8 experimental trees in each (24 trees per treatment). Four buffer trees separated the plots. A complete randomized block design was used to perform the trial. Its layout is presented in Figure 3.

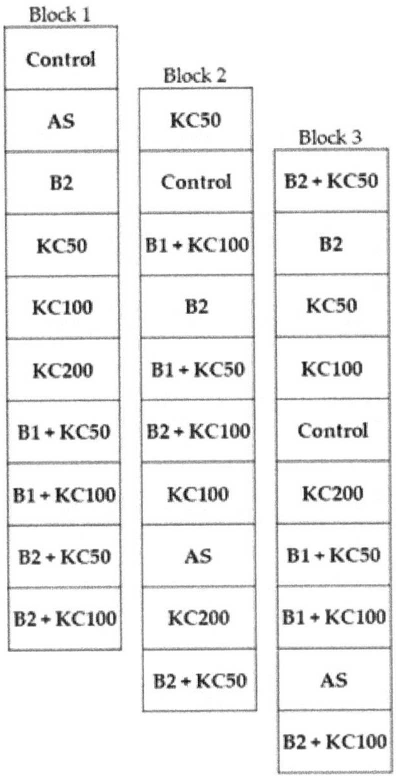

Figure 3. The layout of the trial designed as complete randomized blocks with three replications in each.

2.2. *Data Collection and Analysis*

2.2.1. Flowering, Fruit Setting, and Yield

The number of inflorescences per tree was recorded pre-flowering and is given in pieces per tree. The fruitlets post June drop were counted, and the setting rate was expressed as fruitlets per 100 inflorescences. The tree size was assessed as a trunk cross-sectional area (TCSA) for each plot. This parameter was calculated as a function of the tree trunk diameter measured with a hand caliper at a height of 30 cm above the soil level in a permanently marked spot. Trunk diameter measurements were performed twice during the trial in the beginning and at the end of the vegetation season. The results are presented in cm^2.

Fruits were harvested at the beginning of September according to a starch index test made for apples grown in a control plot. The yield was assessed during the harvest by weighing apples picked from each separate tree in a plot. The blooming efficiency index (BEI) was calculated and given in $pcs \cdot cm^{-2}$ based on the number of inflorescences per tree and the TCSA measured at the beginning of the season. The cropping efficiency index (CEI) was computed as a quotient of the fruit mass per tree, and the TCSA was measured at the end of the season. These values are presented in $kg \cdot cm^{-2}$.

2.2.2. Physiological Status and Inner and Outer Fruit Quality Determined Directly After Harvest

All physiological and quality fruit parameters, except for the mean fruit mass, were determined in a representative sample of ten apples from each replication.

The starch index was measured in the cross-sections of apples by dipping their surface in an iodine solution in potassium iodide. The color reaction of the fruit flesh was assessed from 1 to 10 according to standard tables of starch degradation.

The internal ethylene concentration (IEC) was measured using a gas chromatograph (Clarus 690, PerkinElmer, Inc., Waltham, MA, USA). A sample of air containing 1 cm^3 was taken from the seed socket for analysis using a syringe. The results are expressed in $\mu L \cdot L^{-1}$.

The fruit firmness was measured using an Instron 5540 penetrometer (Instron, Norwood, MA, USA). A narrow strip of skin was removed from both the blush side and the opposite side of the fruit to expose the flesh. The measurement was performed in the peeled areas directly in the flesh using a probe with a diameter of 11 mm, penetrating the flesh to a depth of 10 mm. The measurement results are expressed in Newtons (N).

The fruits' soluble solid content and acidity were determined in the juice extracted from the tested fruit using a juicer for each replication. The soluble solid content (SSC) was measured directly in the juice using an ATAGO PAL-1 (Atago Co. Ltd., Tokyo, Japan) refractometer and expressed in °Brix.

For titratable acidity (TA) measurements, the apple juice was diluted with distilled water in a 1:10 ratio, and the resulting solution was titrated with 0.1 M NaOH to a pH of 8.1 using a semi-automated titrator (TitroLine 5000, Xylem Analytics Germany GmbH, Weilheim, Germany). The TA was calculated and expressed as a percentage of malic acid based on the amount of NaOH used for each sample. Additionally, the SSC–TA ratio was calculated.

Considering the flesh firmness, SSC, and starch index value, the Streif index was calculated using the following formula:

$$\mathrm{SI} = \mathrm{F}/(\mathrm{R} \times \mathrm{S})$$

where SI denotes the Streif index, F denotes the firmness (expressed in kG), R denotes the soluble solid content (°Brix), and S denotes the starch test result (scale: 1–10).

The apple skin coloration was measured with a Minolta CR-400 colorimeter (Konica Minolta Inc., Tokyo, Japan) and presented in a CIE Lab system as the values of three parameters: 'L' (lightness), 'a' (red–green axis), and 'b' (yellow–blue). Additionally, the hue and chroma values were computed.

The mean fruit mass was measured in a sample of 50 apples per replication using a simple scale and was given in grams.

2.3. *Statistical Analysis of Data*

One-way ANOVA was used to analyze the data gathered during the trial. The normality of the data distribution of results was verified using the Shapiro–Wilk test. The means were separated according to the Newman–Keuls post hoc test at a significance level of $p \leq 0.05$. All tests were performed with the Statistica 13.3 software package (Statsoft Inc., Tulsa, OK, USA).

3. Results

The results presented in Table 1 demonstrate that the treatment applied to the trees had a significant impact on their flowering, fruiting, and yield characteristics.

Table 1. Blooming, fruit setting, and cropping of the tested apple trees depending on the treatment applied during the trial.

Treatment	No. of Flower Clusters [pcs·Tree^{-1}]	Blooming Efficiency Index (BEI) [pcs·cm^{-2}]	No. of Fruits [pcs·Tree^{-1}]	Fruit Set [pcs·100^{-1} Flower Buds]	Yield [kg·Tree^{-1}]	Cropping Efficiency Index (CEI) [kg·cm^{-2}]
Control	89.6 a [1]	11.9 a	66.3 cd	74.6 cd	6.35 d	0.59 de
AS	88.2 a	8.80 a	7.06 a	7.92 a	1.17 a	0.10 a
B2	90.0 a	12.0 a	10.7 a	11.9 a	2.71 ab	0.29 abc
KC50	92.5 a	11.4 a	57.4 c	63.1 c	6.67 d	0.60 de
KC100	87.2 a	9.72 a	58.7 c	67.6 c	6.93 d	0.65 e
KC200	88.1 a	9.20 a	75.4 cd	85.2 cd	10.1 e	0.85 f
B1 + KC50	91.1 a	9.51 a	28.5 b	31.2 b	4.79 cd	0.40 bcde
B1 + KC100	98.0 a	10.0 a	33.8 b	34.6 b	6.21 d	0.52 cde
B2 + KC50	94.8 a	12.1 a	18.5 ab	19.9 ab	2.62 ab	0.26 ab
B2 + KC100	88.9 a	10.1 a	18.3 ab	20.7 ab	3.93 bc	0.35 bcd

[1] Note: Means followed by the same letter in a column do not differ statistically according to the Newman–Keuls post hoc test at $p \leq 0.05$.

The number of flower clusters per tree did not significantly vary across treatments, with all values falling in a similar range. The highest number was observed for the B1 + KC100 treatment (98.0 pcs·tree^{-1}), followed closely by B2 + KC50 (94.8 pcs·tree^{-1}) and KC50 (92.5 pcs·tree^{-1}). The control group recorded 89.6 pcs·tree^{-1}, comparable to the other treatments. The blooming efficiency index (BEI) values also showed no statistically significant variation among treatments. The highest BEI value was recorded for B2 + KC50 (12.1 pcs·cm^{-2}), which was similar to the values observed for B2 (12.0 pcs·cm^{-2}) and the control group (11.9 pcs·cm^{-2}). The KC100, KC200, and B1 + KC50 treatments exhibited slightly lower BEI values, but the differences were not significant.

The number of fruits per tree significantly varied between treatments. The KC200 treatment resulted in the highest number of fruits per tree (75.4 pcs·tree^{-1}), followed by the control group (66.3 pcs·tree^{-1}) and KC50 (57.4 pcs·tree^{-1}). Lower values were noted for B1 + KC50 (28.5 pcs·tree^{-1}) and B1 + KC100 (33.8 pcs·tree^{-1}). The lowest numbers were recorded for AS (7.06 pcs·tree^{-1}) and B2 + KC100 (18.3 pcs·tree^{-1}). Based on the results of the calculation and analysis of fruit setting, fruit set significantly varied between treatments, with KC200 achieving the highest value (85.2 pcs·100^{-1} flower buds). The control group and KC100 showed moderate fruit set values (74.6 and 67.6 pcs·100^{-1} flower buds, respectively). The AS (7.92 pcs·100^{-1} flower buds) and B2 + KC100 (20.7 pcs·100^{-1} flower buds) treatments exhibited the lowest fruit set values.

The yield per tree displayed notable variations across treatments. The KC200 treatment achieved the highest yield (10.1 kg·tree^{-1}), while the control group recorded a significantly lower yield (6.35 kg·tree^{-1}), similar to KC50 (6.67 kg·tree^{-1}). AS produced the lowest yield (1.17 kg·tree^{-1}). Combined treatments such as B1 + KC50 and B2 + KC50 recorded intermediate yields of 4.79 and 2.62 kg·tree^{-1}, respectively. This also had a significant impact on the cropping efficiency index (CEI), which also significantly varied across treatments. KC200 exhibited the highest CEI value (0.85 kg·cm^{-2}), followed by KC100 (0.65 kg·cm^{-2}). The control group and KC50 had similar CEI values (0.59 and 0.60 kg·cm^{-2}, respectively). The lowest CEI was recorded for AS (0.10 kg·cm^{-2}), while intermediate values were noted for treatments such as B1 + KC50 (0.40 kg·cm^{-2}) and B2 + KC50 (0.26 kg·cm^{-2}).

According to the results presented in Table 2, the physiological status of fruits depended on the treatments tested during the trial, while there was no variation in fruit coloration.

Table 2. Physiological parameters and skin coloration of fruits measured directly after harvest depending on the treatment applied during the trial.

Treatment	Internal Ethylene Concentration [$\mu L \cdot L^{-1}$]	Starch Index [-]	Streif Index [-]	Apple Skin Coloration				
				L [-]	a [-]	b [-]	Hue (h°)	Chroma (C)
Control	0.143 ab [1]	4.60 a	0.143 c	42.1 a	23.8 a	7.93 a	1.25 a	25.1 a
AS	0.094 bcd	6.73 ab	0.094 abc	42.2 a	21.8 a	6.73 a	1.27 a	22.9 a
B2	0.096 abc	6.30 ab	0.096 abc	39.6 a	20.4 a	6.53 a	1.26 a	21.4 a
KC50	0.135 a	4.96 a	0.135 bc	41.6 a	23.4 a	7.86 a	1.25 a	24.7 a
KC100	0.140 a	4.70 a	0.140 c	42.7 a	22.5 a	7.70 a	1.24 a	23.8 a
KC200	0.130 a	5.60 ab	0.130 abc	41.2 a	23.0 a	7.16 a	1.27 a	24.1 a
B1 + KC50	0.101 ab	6.23 ab	0.101 abc	41.1 a	22.0 a	7.13 a	1.26 a	23.1 a
B1 + KC100	0.088 abc	6.56 ab	0.088 ab	40.5 a	20.7 a	6.33 a	1.28 a	21.7 a
B2 + KC50	0.082 cd	7.10 b	0.082 a	41.0 a	20.5 a	6.50 a	1.27 a	21.6 a
B2 + KC100	0.096 d	6.73 ab	0.096 abc	40.1 a	20.1 a	6.33 a	1.27 a	21.1 a

[1] Note: Means followed by the same letter in a column do not differ statistically according to the Newman–Keuls post hoc test at $p \leq 0.05$.

The internal ethylene concentration significantly varied across treatments. The highest ethylene concentration was recorded for the KC50 treatment (0.135 $\mu L \cdot L^{-1}$), followed closely by the KC100 treatment (0.140 $\mu L \cdot L^{-1}$) and the control group (0.143 $\mu L \cdot L^{-1}$). Conversely, the lowest concentration was observed in the B2 + KC50 treatment (0.082 $\mu L \cdot L^{-1}$), with similarly low values for B1 + KC100 (0.088 $\mu L \cdot L^{-1}$). Treatments such as AS and B2 recorded intermediate values (0.094 and 0.096 $\mu L \cdot L^{-1}$, respectively).

The analysis of the starch index values showed noticeable variations across treatments, indicating differences in fruit ripening. The highest starch index was observed in the B2 + KC50 treatment (7.10), followed by B2 + KC100 (6.73) and AS (6.73). Lower starch index values were recorded for KC50 (4.96), KC100 (4.70), and the control group (4.60).

The Streif index values, which are inversely related to fruit maturity, displayed a range of results. The control group and KC100 treatment had the highest Streif index values (0.143 and 0.140, respectively), indicating lower fruit maturity. The B2 + KC50 treatment recorded the lowest Streif index value (0.082), reflecting advanced fruit ripening. Intermediate values were observed for treatments such as AS (0.094) and B2 (0.096).

The parameters related to apple skin coloration, including L (lightness), a (red–green axis), b (yellow–blue axis), hue, and chroma, showed no significant differences across treatments.

The inner fruit quality and fruits were also affected by treatments used in the trial (Table 3). Flesh firmness varied across treatments but to a lesser extent. Relatively high fruit firmness was observed in the B2 + KC100 treatment (79.5 N), followed by B2 + KC50 (75.9 N). These values were significantly higher than those recorded for the control plots, KC50. Intermediate firmness values were noted for KC100 (75.4 N) and AS (74.7 N). The soluble solid content (SSC) was higher for B1 + KC100 and B2 + KC50 (13.3 °Brix and 13.2 °Brix, respectively) compared with the control, KC50, and KC200 treatments, while the AS, B2, and B1 + KC50 combinations showed intermediate levels of the tested parameter. Similar variations between treatments were observed for TA, with B2 + KC50 and B2 + KC100 exhibiting higher values (0.51% malic acid) than the control, KC50, and KC100 (0.39%). Intermediate acidity levels were observed for the AS and B2 treatments.

Table 3. Fruit quality parameters measured directly after harvest depending on the treatment applied during the trial.

Treatment	Flesh Firmness [N]	Soluble Solid Content [°Brix]	Titratable Acidity [% Malic Acid]	SSC–Acidity Ratio [-]	Fruit Mass [g]
Control	72.4 a [1]	11.2 a	0.39 a	28.6 ab	171 a
AS	74.7 ab	12.1 ab	0.44 ab	27.2 ab	197 b
B2	73.9 ab	12.6 ab	0.47 bc	26.8 ab	192 b
KC50	73.1 a	11.3 a	0.39 a	28.8 ab	196 b
KC100	75.4 ab	11.9 ab	0.39 a	29.9 b	188 b
KC200	73.8 ab	11.4 a	0.41 a	27.6 ab	169 a
B1 + KC50	73.6 ab	12.0 ab	0.40 a	29.5 b	186 b
B1 + KC100	73.7 ab	13.3 b	0.38 a	34.6 c	189 b
B2 + KC50	75.9 ab	13.2 b	0.51 c	26.0 ab	186 b
B2 + KC100	79.5 b	12.7 ab	0.51 c	24.7 a	183 b

[1] Note: Means followed by the same letter in a column do not differ statistically according to the Newman–Keuls post hoc test at $p \leq 0.05$.

The SSC–TA ratio showed distinct differentiation. B1 + KC100 showed the highest ratio (34.6), indicating a sweeter taste profile, while B2 + KC100 had the lowest one (24.7) due to its high acidity. Most of the other treatments, such as KC50, KC100, and the control, recorded intermediate values of the described parameter (27.2–29.9).

The fruit mass was the highest in the AS and KC50 treatments (197 g and 196 g, respectively). The smallest fruits were observed in the control and KC200 treatments (171 g and 169 g, respectively), while intermediate mass values were recorded for the B2, KC100, and B1 + KC50 treatments.

4. Discussion

The experiment conducted to complete this study on 'Gala Vill' aimed to identify an alternative method to using chemical thinning agents or enable a reduction in metamitron dosages during the post-bloom thinning of apple fruitlets. The tested combinations in the experiment were designed to limit, or potentially limit, the photosynthetic process and act as a factor increasing competition among fruitlets and causing weaker fruitlets to drop, which is a well-known phenomenon contributing to crop load management techniques [32].

One of the thinning methods examined in this study was the use of a black agro-textile for artificial shading. Shading trees for 5 days in combination with AS resulted in excessive thinning, with an almost 80% reduction in fruitlet retention compared with the control, leading to a large decrease in yield in our experiment. Previous studies reported good thinning effects when shading was applied for up to 5 days, with shorter periods being insufficient and longer periods, such as 7 days, causing overthinning [33–35]. The use of kaolin clay, in theory, was expected to produce similar results to shading by limiting the production of assimilates during the photosynthetic process. However, this study showed that kaolin clay had no significant effect on fruitlet thinning. Higher doses of kaolin clay positively influenced fruit set and led to a significantly increased yield. This may be attributed to several reasons related to kaolin clay's impact on environmental conditions and its indirect effect on tree physiology.

The kaolin layer that reflects sunlight can reduce the temperature on the plant's surface. This can be particularly beneficial during heat stress, which is responsible for improper fruit development. The light reflection by kaolin clay also reduces transpiration, enabling plants to cope better with water deficits. As a result, trees can allocate more resources to the development of fruitlets. Reducing heat and water stress can enhance photosynthetic activity and tree metabolism, allowing more energy to be directed toward fruit development. While this is probably not the case in the present experiment, because of the mild temperatures which are commonly present in Polish conditions during flowering and in the post bloom period (average temperature for May on the experimental site was 12.4 °C), other factors may have played a role. According to Wünsche [31], while a thin film of kaolin clay on a leaf can reduce the individual leaf's photosynthesis, it does not affect the global canopy due to better light distribution. Redistribution of light reflected from leaves covered with kaolin helps illuminate shaded areas within the canopy, as a result enhances the overall photosynthetic efficiency of the plant. This improved light distribution could support more balanced photosynthesis across the canopy [28], and might explain kaolin clay's improvement of fruit setting and cropping observed in our experiment.

Spraying trees with metamitron significantly reduced fruit set, confirming its strong thinning efficacy, which is well documented in several authors' reports [35–39]. The effect of metamitron sprayings followed by kaolin clay application did not enhance its thinning effect. In cases where lower doses of metamitron were used with kaolin clay, the results were comparable to those using the same lower doses of Brevis (metamitron) alone and were less effective than the standard dose of 2.2 kg·ha^{-1} used in the presented experiment.

The lack of an additive effect of metamitron and kaolin clay may be attributed to the different mode of action of both agents but mostly to the failure of kaolin clay as a thinning agent at all. In light of the revealed potential of kaolin clay to improve fruit setting, this seems to be a reasonable outcome.

The thinning of apple fruitlets is crucial for increasing the fruit size because it reduces competition between the fruit for limited tree resources. According to Robinson and Lakso's research [40], properly performed thinning increases the size of apples by up to 20–30% compared with unthinned trees, which makes this procedure an essential element of modern fruit-growing technology. Other authors have confirmed similar relationships. Our study demonstrated comparable findings, where an increase in fruit size was observed for all thinning treatments. Basak reported similar findings [41], where thinning resulted in a larger proportion of fruits exceeding 70 mm in diameter compared with control treatments.

Thinning treatments positively affected fruit firmness, which has been proven by the results provided by several authors [2,6,42,43]. While significantly higher firmness values were observed only with the B2 + KC100 combination, the tendency was observed among the other thinning treatments in our experiment. This can be attributed to the better conditions for growing young fruitlets and more dynamic cell division in the early stage of development, which improved their size.

The thinning of apple fruitlets has a positive effect on the sugar content and titratable acidity of the fruits. Several authors' studies [4,44,45] showed that fruits from thinned trees had a significantly higher soluble solid content compared with fruits from unthinned trees. Similar results were confirmed by Robinson's observations [40], indicating a clear relationship between the thinning intensity and the increase in fruit quality, including sweetness. Some combinations in the presented trial also exhibited such effects. Fruits from trees treated with B1 + KC100 and B2 + KC50 exhibited higher soluble solid contents. These findings may be linked to the lower number of fruits on the trees, allowing the remaining fruits to accumulate more assimilates. It is also widely known that well-developed fruits typically have higher sugar content, which contributes to better flavor and overall quality [2,3,24,44,45]. The highest levels of titratable acidity were observed in fruits from combinations involving higher metamitron doses, namely, B2, B2 + KC50, and B2 + KC100. This suggests that these combinations may enhance the fruits' acidity levels, contributing to a better flavor balance, which is an essential quality trait for apples.

One of the limitations of this study is the lack of assessment of the long-term effects of kaolin clay applications in terms of overall tree health and orchard ecosystems. While kaolin clay has not shown effectiveness in thinning, other benefits appeared such as improved fruit setting. Taking into account fruit crops is essential to investigate potential cumulative effects that may include changes in soil composition due to runoff and accumulation and the effect on pests and disease development. Such observations should be performed over multiple growing seasons.

5. Conclusions

The findings indicate that the application of kaolin clay does not show a thinning effect or enhance the thinning efficacy of metamitron, suggesting that its use as an additive for chemical thinning agents may not bring additional benefits. However, high doses of kaolin clay sprays positively affected fruit set and yield in trees where chemical thinning was not applied. This demonstrates the promising potential of kaolin clay usage to improve orchard productivity.

Kaolin clay application in the early stages of fruitlet development (6–8 mm diameter) does not exhibit a clear impact on internal fruit quality or skin coloration. This highlights its potential as a neutral treatment option that can improve yields without unwanted side effects.

Since the present study was an initial one-year investigation, future research will focus on proving the observed effects, revealing the exact mode of action, and evaluating its long-term effects on fruit yield, quality, and tree health under different management conditions. Special interest should be focused on kaolin clay's potential impact on non-target organisms, particularly beneficial insects. While the effect of kaolin clay is based more on its physical properties and influencing physiological processes in plants, it may also affect the activity or foraging efficiency of beneficial insects on coated surfaces. Taking this into account, it is crucial to investigate these effects more comprehensively to develop best practices that help mitigating any potential unwanted impact.

Author Contributions: Conceptualization, S.P. and E.S.; methodology, S.P., E.S. and W.K.; formal analysis, S.P. and J.M.; investigation, S.P. and J.M.; data curation, J.M.; writing—original draft preparation, S.P.; writing—review and editing, S.P., J.M. and W.K.; supervision, S.P. All authors have read and agreed to the published version of the manuscript.

Funding: This research received no external funding.

Institutional Review Board Statement: Not applicable.

Data Availability Statement: The data are contained within this article.

Conflicts of Interest: The authors declare no conflicts of interest.

References

1. Netsawang, P.; Damerow, L.; Lammers, P.S.; Kunz, A.; Blanke, M. Alternative Approaches to Chemical Thinning for Regulating Crop Load and Alternate Bearing in Apple. *Agronomy* **2023**, *13*, 112. [CrossRef]
2. Bound, S.A. Determination of Target Crop Loads for Maximising Fruit Quality and Return Bloom in Several Apple Cultivars. *Appl. Biosci.* **2023**, *2*, 586–606. [CrossRef]
3. Peifer, L.; Ottnad, S.; Kunz, A.; Damerow, L.; Blanke, M. Effect of Non-Chemical Crop Load Regulation on Apple Fruit Quality, Assessed by the DA-Meter. *Sci. Hortic.* **2018**, *233*, 526–531. [CrossRef]
4. Kurlus, R.; Rutkowski, K.; Łysiak, G.P. Improving of Cherry Fruit Quality and Bearing Regularity by Chemical Thinning with Fertilizer. *Agronomy* **2020**, *10*, 1281. [CrossRef]
5. Sidhu, R.S.; Bound, S.A.; Hunt, I. Crop Load and Thinning Methods Impact Yield, Nutrient Content, Fruit Quality, and Physiological Disorders in 'Scilate' Apples. *Agronomy* **2022**, *12*, 1989. [CrossRef]

6. Win, N.M.; Song, Y.-Y.; Nam, J.-C.; Yoo, J.; Kang, I.-K.; Cho, Y.S.; Yang, S.-J.; Park, J. Influence of Mechanical Flower Thinning on Fruit Set and Quality of 'Arisoo' and 'Fuji' Apples. *Int. J. Plant Biol.* **2023**, *14*, 503–511. [CrossRef]
7. Wang, S.; Wang, Q.; Jiang, W.; Wang, Y.; Yan, J.; Li, X.; Wang, J.; Guan, Q.; Ma, F.; Zhang, J.; et al. Evaluating the Sustainable Cultivation of 'Fuji' Apples: Suitable Crop Load and the Impact of Chemical Thinning Agents on Fruit Quality and Transcription. *Fruit Res.* **2024**, *4*, e0024. [CrossRef]
8. Rutkowski, K.; Łysiak, G.P. Thinning Methods to Regulate Sweet Cherry Crops—A Review. *Appl. Sci.* **2022**, *12*, 1280. [CrossRef]
9. Radivojevic, D.; Oparnica, C.; Milivojevic, J. Modeling of Apple Chemical Fruit Thinning. *Ann. Univ. Craiova Agric. Mont. Cadast. Ser.* **2023**, *53*, 257–264. [CrossRef]
10. Gonzalez, L.; Torres, E.; Àvila, G.; Bonany, J.; Alegre, S.; Carbó, J.; Martín, B.; Recasens, I.; Asin, L. Evaluation of Chemical Fruit Thinning Efficiency Using Brevis® (Metamitron) on Apple Trees ('Gala') under Spanish Conditions. *Sci. Hortic.* **2020**, *261*, 109003. [CrossRef]
11. Gonzalez Nieto, L.; Ottnad, S.; Kunz, A.; Damerow, L.; Blanke, M. Metamitron Thinning Efficacy of Apple Fruitlets Is Affected by Different Rates, Timings, and Weather Factors in New York State. *Horticulturae* **2023**, *9*, 1179. [CrossRef]
12. Kon, T.M.; Schupp, J.R.; Yoder, K.S.; Combs, L.D.; Schupp, M.A. Comparison of Chemical Blossom Thinners Using 'Golden Delicious' and 'Gala' Pollen Tube Growth Models as Timing Aids. *HortScience* **2018**, *53*, 1143–1151. [CrossRef]
13. De Ross Marchioretto, L.; De Rossi, A.; Michelon, M.F.; Orlandi, J.C.; Amaral, L.O.D. Ammonium Thiosulfate as Blossom Thinner in 'Maxi Gala' Apple Trees. *Pesqui. Agropecuária Bras.* **2018**, *53*, 1132–1139. [CrossRef]
14. Kaçal, E.; Öztürk, G.; Gür, İ.; Aydınlı, M.; Koçal, H.; Altındal, M.; Yıldırım, A.N. Crop Load Management with Blossom Thinners in 'Redchief' Apple and Their Effects on Fruit Mineral Composition. *Erwerbs-Obstbau* **2019**, *61*, 231–236. [CrossRef]
15. Szot, I.; Basak, A.; Lipa, T.; Krawiec, P. Thinning of Apple Flowers with Potassium Bicarbonate (Armicarb®) in Organic Orchard. *Acta Sci. Pol. Hortorum Cultus* **2016**, *15*, 29–43.
16. Wertheim, S.J. Chemical control of flower and fruit abscission in apple and pear. *Acta Hortic.* **1973**, *34*, 321–331. [CrossRef]
17. McArtney, S. Effective use of apple blossom thinners. In Proceedings of the Great Lakes Fruit, Vegetable & Farm Market EXPO, Grand Rapids, MI, USA, 5–7 December 2006.
18. Pavanello, A.P.; Zoth, M.; Ayub, R.A.; de Souza Los, K.K. Different Methods of Thinning Influenced by Variety and Hail Nets in Apple Orchards. *Agric. Res. Technol.* **2019**, *3*, 69–79. [CrossRef]
19. Goulart, G.; de Andrade, S.B.; Schiavon, A.V.; Aguiar, G.A.; Malgarim, M.B. Metamitron and Different Plant Growth Regulators Combinations in the Chemical Thinning of 'Eva' Apple Trees. *J. Exp. Agric. Int.* **2017**, *18*, 1–6. [CrossRef]
20. Gonzalez, L.; Ottnad, S.; Kunz, A.; Damerow, L.; Blanke, M. Hail Nets Do Not Affect the Efficacy of Metamitron for Chemical Thinning of Apple Trees. *J. Hortic. Sci. Biotechnol.* **2020**, *95*, 128–135. [CrossRef]
21. Gabardo, G.C.; Petri, J.L.; Hawerroth, F.J.; Couto, M.; Argenta, L.C.; Kretzschmar, A.A. Use of Metamitron as an Apple Thinner. *Rev. Bras. Frutic.* **2017**, *39*, e514. [CrossRef]
22. Gonzalez, L.; Ottnad, S.; Kunz, A.; Damerow, L.; Blanke, M. Effect of Different Application Rates of Metamitron as Fruitlet Chemical Thinner on Thinning Efficacy and Fluorescence Inhibition in 'Gala' and 'Fuji' Apple. *Plant Growth Regul.* **2019**, *89*, 259–271. [CrossRef]
23. Greene, D.W.; Schupp, J.R.; Winzler, H.W. Effect of abscisic acid and benzyladenine on fruit set and fruit quality of apples. *HortScience* **2011**, *46*, 1–6. [CrossRef]
24. Bound, S.A. Managing Crop Load in European Pear (*Pyrus communis* L.)—A Review. *Agriculture* **2021**, *11*, 637. [CrossRef]
25. Ergun, M. Postharvest quality of 'Galaxy' apple fruit in response to kaolin-based particle film application. *J. Agric. Sci. Technol.* **2012**, *14*, 599–607.
26. Shellie, K.C.; King, B.A. Kaolin-based foliar reflectant and water deficit influence malbec leaf and berry temperature, pigments, and photosynthesis. *Am. J. Enol. Vitic.* **2013**, *2012*, 12115.
27. Conde, A.; Pimentel, D.; Neves, A.; Dinis, L.-T.; Bernardo, S.; Correia, C.M.; Gerós, H.; Moutinho-Pereira, J. Kaolin Foliar Application Has a Stimulatory Effect on Phenylpropanoid and Flavonoid Pathways in Grape Berries. *Front. Plant Sci.* **2016**, *7*, 1150. [CrossRef] [PubMed]
28. Glenn, D.M.; Puterka, G.J. Particle films: A new technology for Agriculture. *Hortic. Rev.* **2010**, *31*, 1–44.
29. Schrader, L.E. Scientific basis of a unique formulation for reducing sunburn of fruits. *HortScience* **2011**, *46*, 6–11. [CrossRef]
30. Rosati, A.; Metcalf, S.G.; Buchner, R.P.; Fulton, A.E.; Lampinen, B.D. Physiological Effects of Kaolin Applications in Well-Irrigated and Water-Stressed Walnut and Almond Trees. *Ann. Bot.* **2006**, *98*, 267–275. [CrossRef] [PubMed]
31. Wünsche, J.N.; Lombardini, L.; Greer, D.H. Surround Particle Film Applications—Effects on Whole Canopy Physiology of Apple. *Acta Hortic.* **2004**, *636*, 565–571. [CrossRef]
32. Zakalik, D.; Brown, M.G.; Peck, G.M. Fruitlet Thinning Reduces Biennial Bearing in Seven High-tannin Cider Apple Cultivars. *HortScience* **2024**, *59*, 26–35. [CrossRef]
33. Bertschinger, L.; Stadler, W.; Stadler, P.; Weibel, F.; Schumacer, R. New methods of environmentally safe regulation of flower and fruit set and of alternate bearing of the apple crop. *Acta Hortic.* **1998**, *466*, 65–70. [CrossRef]

34. Byers, R.E.; Barden, J.A.; Polomski, R.F.; Young, R.W.; Carbaugh, D.H. Apple thinning by photosynthetic inhibition. *J. Am. Soc. Hortic. Sci.* **1990**, *115*, 14–19. [CrossRef]
35. Byers, R.E.; Carbaugh, D.H.; Presley, C.N.; Wolf, T.K. The influence of low light on apple fruit abscission. *J. Am. Soc. Hortic. Sci.* **1991**, *66*, 7–17. [CrossRef]
36. Petri, J.L.; Couto, M.; Gabardo, G.C.; Francescatto, P.; Hawerroth, F.J. Metamitron replacing carbaryl in post bloom thinning of apple trees. *Rev. Bras. Frutic.* **2016**, *38*, 14. [CrossRef]
37. McArtney, S.J.; Obermiller, J.D. Use of 1-Aminocyclopropane Carboxylic Acid and Metamitron for Delayed Thinning of Apple Fruit. *HortScience* **2012**, *47*, 1612–1616. [CrossRef]
38. Gonzalez, L.; Bonany, J.; Alegre, S.; Àvila, G.; Carbó, J.; Torres, E.; Recasens, I.; Martin, B.; Asin, L. Brevis thinning efficacy at different fruit size and fluorescence on 'Gala' and 'Fuji' apples. *Sci. Hortic.* **2019**, *256*, 108526. [CrossRef]
39. Cline, J.A.; Bakker, C.J.; Beneff, A. Multi-year investigation on the rate, timing, and use of surfactant for thinning apples with post-bloom applications of metamitron. *Can. J. Plant Sci.* **2022**, *102*, 628–655. [CrossRef]
40. Robinson, T.L. Crop Load Management of Apple to Optimize Fruit Quality and Economic Return. *Acta Hortic.* **2006**, *772*, 367–378.
41. Basak, A. Efficiency of Fruitlet Thinning in Apple 'Gala Must' by Use of Metamitron and Artificial Shading. *J. Fruit Ornam. Plant Res.* **2011**, *19*, 51–62.
42. Racskó, J. Crop Load, Fruit Thinning and Their Effects on Fruit Quality of Apple (*Malus domestica* Borkh.). *J. Agric. Sci.* **2006**, *24*, 29–35. [CrossRef]
43. Anthony, B.; Serra, S.; Musacchi, S. Optimizing Crop Load for New Apple Cultivar: "WA38". *Agronomy* **2019**, *9*, 107. [CrossRef]
44. Stopar, M.; Bolcina, U.; Vanzo, A.; Vecchione, A. Sugar and Organic Acid Content in Apple Juice. *Acta Aliment.* **2002**, *31*, 229–239. [CrossRef]
45. Szot, I.; Lipa, T. Apple Trees Yielding and Fruit Quality Depending on the Crop Load, Branch Type and Position in the Crown. *Acta Sci. Pol. Hortorum Cultus* **2019**, *18*, 205–215. [CrossRef]

agriculture

MDPI

Article

Potential of Selected Species of *Spiraea* L. for Phytoremediation of Heavy Metals from Soils of Urban Areas

Wojciech Antkowiak [1,*], Maciej Bosiacki [2] and Mateusz Sowelo [3]

[1] Department of Botany, Faculty of Agronomy Horticulture and Biotechnology, Poznań University of Life Sciences, Wojska Polskiego 28, 60-637 Poznań, Poland
[2] Department of Plant Physiology, Faculty of Agronomy Horticulture and Biotechnology, Poznań University of Life Sciences, Wołyńska 35, 60-637 Poznań, Poland; maciej.bosiacki@up.poznan.pl
[3] The Botanical Garden, Adam Mickiewicz University Poznań, Wieniawskiego 1, 61-712 Poznań, Poland; matsow@amu.edu.pl
* Correspondence: wojciech.antkowiak@up.poznan.pl

Abstract: Four taxa of *Spiraea* were selected for this study: *S.* × *cinerea* Zabel 'Grefsheim', *S. nipponica* Maxim. 'Snowmound', *S. splendens* É. N. Baumann ex K. Koch and *S.* × *vanhouttei* (Briot) Carrière growing for a minimum of 5 years along heavily trafficked traffic routes. This study included the genus *Spiraea* due to its popularity in horticultural practice (commercial availability, widespread in urban environments). In addition, the use of ornamental shrubs for phytoremediation in urban green spaces effectively combines the aesthetic needs of residents with those of caring for the urban environment. This study was conducted in Poznań (population 550,000, the fifth largest city in Poland). Soils and foliage were examined in spring and autumn. Soil pH and specific electrolytic conductivity (EC) were determined. The content of micronutrients (Cu, Fe, Mn, Ni, Zn) and toxic heavy metals (Cd, Cr, Pb) in soil dry matter and leaves was determined. The uptake capacity of bioavailable forms of heavy metals by *Spiraea* from the soil was analyzed by determining the bioconcentration factor (BCF). It was found that the studied taxa meet the basic requirements for plants used for soil phytoremediation processes, especially for chromium phytoextraction. The degree of salinity of the tested soils did not pose a threat to the shrubs growing there, and most of the sites, despite the alkaline reaction, are suitable for their cultivation. *S.* × *cinerea* and *S.* × *vanhouttei* have BCFs for lead <1. The remaining taxa are characterized by strong concentrations of all analyzed elements. A particularly high BCF, above 10, was recorded for chromium and high for manganese and nickel.

Keywords: *Spiraea* × *cinerea* 'Grefsheim'; *S.* × *vanhouttei; S. splendens; S. nipponica* 'Snowmound'

1. Introduction

The genus *Spiraea* L. represents deciduous shrubs of the family Rosaceae Juss. There are 80 to 100 species of this genus throughout the world. *Spiraea* is widespread in the temperate and subtropical zones of the northern hemisphere. The genus has the richest species diversity in the territory of China, with about 70 *Spiraea* species [1]. The genus *Spiraea* was included in this study due to its popularity in horticultural practice (commercial availability, prevalence in urban environments). The use of ornamental shrubs for phytoremediation purposes in urban green spaces can effectively combine the aesthetic needs of residents with the needs of caring for the urban environment.

Poznań (52°24′30.0″ N 16°56′03.0″ E; 52.408333, 16.934167) is located in west-central Poland, in the central part of the Greater Poland. It is the fifth largest city in Poland in terms of population (541,316 inhabitants in 2023) and ninth in terms of area (261.9 km^2). In total, public roads in Poznań in 2021 were 1073 km long. The city's green areas cover an area of 70 km^2, i.e., 27% of the entire city area. The city is an important road and railway junction, and the Poznań-Ławica international airport also operates here. Poznań is a center of industry, trade, logistics and tourism. The food industry, production of machines and electrical equipment, motor vehicles, chemicals and rubber products dominate. Poznań is also a large production, service, transport and trade center. Poznań is dominated by sandy and clay-sandy soils. They also encountered Tertiary clays of the Poznań facies (Poznań clays). Urban soils in which the tested plants grew are exposed to heavy metal contamination. Their main source is substances produced during the abrasion of car tires, abrasion of road surfaces and fuel combustion.

Heavy metal contamination of urban soils is an issue that has been studied in research centers around the world [2–8]. The term heavy metals is usually defined as elements with a minimum density of 4.5 g cm^{-3}, although some consider values in the range of 3.5 to 7 g cm^{-3} as the lower limit [9]. This group includes both micronutrients (Cu 8.95 g cm^{-3}, Fe 7.87 g cm^{-3}, Mn 7.47 g cm^{-3}, Ni 8.91 g cm^{-3}, Zn 7.14 g cm^{-3}) and toxic heavy metals such as Cd (8.65 g cm^{-3}), Cr (7.17 g cm^{-3}) and Pb (11.34 g cm^{-3}). High concentrations of micronutrients in the soil are harmful—they may even be toxic to plants.

The average content of selected heavy metals in the 0–20 cm layer of agricultural soils in Poland is Pb—13.7 mg kg^{-1}, Cd—0.22 mg kg^{-1}, Ni—6.3 mg kg^{-1}, Zn—32.5 mg kg^{-1}, Cu—6.5 mg kg^{-1} [10]. In turn, the average content of selected elements for the humus layer of European soils is Pb—15 mg kg^{-1}, Cd—0.145 mg kg^{-1}, Ni—14 mg kg^{-1}, Zn—48 mg kg^{-1} and Cu—12 mg kg^{-1} [11]. These values for urbanized areas may present themselves differently, depending on the type and degree of industrialization in present or historical terms. Typically, the presence of heavy metals on urban sites is mainly of anthropogenic rather than lithogenic origin [12,13]. This is related to, among other things, exhaust emissions, widespread industrialization and low-level emissions [14–16]. According to Bosiacki [17], the permissible standard for the content of Cd, Cr, Ni and Pb in soils in green areas located near routes in Poznań was not exceeded. The authors found that the content was Cd 0.16–0.42, Cr 0.26–0.67, Ni 0.43–1.25 and Pb 0.79–42.96 mg kg^{-1} d.m. The authors also found that in the vicinity of selected communication routes in the city of Poznań, most of the tested soils had an alkaline reaction pH > 7.4, which may significantly affect the habitat conditions of plants.

Some heavy metals, although counted as nutrients, can be excessively taken up by plants from overly abundant growing sites and have a detrimental effect on them. An example is iron, which is 80% present in photosynthesizing tissues, but an excess of it leads to the overproduction of reactive oxygen species, affecting the formation of oxidative stress and resulting in the disruption of basic plant metabolism [18,19]. Elements such as mercury, cadmium and chromium are characterized by toxic effects affecting reduced growth and biomass accumulation, disruption of gas exchange, photosynthesis and the mechanism of water and nutrient uptake [20–23]. In addition, heavy metals also have toxic effects on soil microorganisms, with the following degrees of toxicity being the most common: Cd > Cu > Zn > Pb [24]. This is an important aspect, as microorganisms determine the fertility and biological productivity of the soil, which consequently translates into the yield of crops grown [25]. Although the soils of roadside areas and road separation strips are not used for food production, it is beneficial from the point of view of growing ornamental plants to have as much microbial activity as possible. Ensuring optimal plant growth conditions determines the success of growing both ornamental [26] and utility crops. In the case of land intended for landscaping characterized by excessive heavy metal content, phytoremediation can be a process preceding the target crop [27]. For this purpose, plants that store large amounts of elements in their aboveground parts (hyperaccumulators) or plants that deposit smaller amounts of elements but produce a large biomass of aboveground parts are used [28]. In addition to heavy metals, high salinity is also a problem, which also contributes to inhibiting the growth of cultivated plants [29].

The bioconcentration factor (BCF) is used to assess the degree of heavy metal accumulation in plants. It is the quotient of the concentration of metals in plant tissues (mg kg^{-1}) to the concentration of metals in soil (mg kg^{-1}) [30–32]. Depending on the nature of the study, this method is used to determine the concentration of heavy metals in roots, in shoots or only in selected organs [33–36]. In addition, the BCF coefficient, based on the plant's ability to extract heavy metals from the soil, makes it possible to compare individual taxa against each other in terms of phytoremediation potential [35]. With few exceptions, most plants have a bioconcentration factor for heavy metals and semi-metals of less than 1 [37]. Plant-growing sites in the city may not only be characterized by a significant content of heavy metals in the soil but also by high salinity. Particularly vulnerable to this factor are plants growing on roadside sites, into which sodium chloride and, in 2% of cases, calcium chloride or magnesium chloride used for de-icing pavement penetrate [38], with NaCl showing greater toxicity to plants than $CaCl_2$ [39]. Studies have proven a correlation between sodium and chlorine content in the soil, and increased sodium and chlorine accumulation in plants affects the reduction of biomass, the occurrence of chlorosis and necrosis on leaves, and, in extreme cases, also leads to the death of entire plants [29,39–43]. According to Equiza et al. [44], significant post-winter soil salinity levels in roadside sites can persist for almost the entire growing season. Soil alkalinization due to the use of de-icing agents is also an important aspect [45,46]. To date, only a few spirea taxa (*S.* × *cinerea* 'Grefsheim', *S. betulifolia*, *S. japonica*, *S. media*, *S. nipponica* and *S. thunbergii*) have been included in studies to test their resistance to salinity [47,48].

The aim of the research was to determine the potential of the species *Spiraea* growing in urbanized areas for phytoremediation of heavy metals based on the content of micronutrients (Cu, Fe, Mn, Ni, Zn) and toxic metals (Cd, Cr, Pb) in the leaves and soil of their cultivation sites, as well as the bioconcentration factor. The soils of the cultivation sites in Poznań, Lublin, and Gdańsk were also analyzed for pH and specific electrical conductivity to expand knowledge about the cultivation conditions of these plants in major urban centers in Poland.

2. Materials and Methods

For this study, four taxa of *Spiraea* L. were selected: *S.* × *cinerea* 'Grefsheim', *S. nipponica* 'Snowmound', *S. splendens* (=*S. densiflora*) and *S.* × *vanhouttei* growing for a minimum of 5 years on three sites located directly on traffic routes with heavy traffic. In most cases, these plants were at least a dozen years old. Two medium samples were taken from each location of Poznań. Each consisted of 25 individual soil samples taken with an Egner stick from a depth of 0–20 cm. Two samples of fresh leaves were also collected from each site. Each sample consisted of leaves collected from 15 drawn shrubs. Leaves and soil were collected in spring (second half of May) and autumn (second half of September). The criteria for selecting sites from Gdańsk and Lublin and the sampling methods were the same as those for Poznań. GPS coordinates of the sampling sites are included in Tables 4 and 5.

All soil samples were dried, sieved and cleaned of mechanical impurities. Soil pH (pH in H_2O) was analyzed using the potentiometric method, and specific electrical conductivity (EC; mS cm^{-1}) was analyzed using the conductometric method. Both tests were carried out on an aqueous extract basis with a soil dry weight to water volume ratio of 1:2 (v/v). Available heavy metals, including micronutrients (Cu, Fe, Mn, Ni, Zn) and toxic heavy metals (Cd, Cr, Pb), were extracted from soil dry weight using a modified Lindsay solution [49], in which pentetic acid (DTPA) was replaced with ethylene-descorbic acid (EDTA). The composition of the solution—5 g of EDTA, 9 cm^3 of 25% NH_4OH solution, 4 g of citric acid and 2 g of $Ca(CH_3COO)_2$ $2H_2O$ in 1 dm^3—was adopted from Bosiacki et al. [27]. The content of micronutrients and heavy metals was determined by flame atomic absorption spectroscopy (FAAS) with an AAS-5 spectrophotometer (Carl Zeiss, Jena, Germany).

The leaves were dried in an extractor dryer at 105 °C for 48 h and mineralized in a mixture of HNO_3 and $HClO_4$ acids (3:1 v/v). The contents of Cu, Fe, Mn, Ni, Zn, Cd, Cr and Pb were determined using flame atomic absorption spectroscopy (FAAS) with an AAS-5 spectrophotometer (Carl Zeiss, Jena, Germany).

The studied spirea were analyzed for their ability to take up bioavailable forms of heavy metals from the soil. For this purpose, the bioconcentration factor BCF was determined. In order to interpret the results obtained, it was assumed, following Lugwisha and Othman [28,29], that a BCF above 1 indicates strong uptake of a given metal from the soil and its concentration in the organs studied. On the other hand, a result below 1 indicates weak or, in the case of a value close to 0, negligible uptake and concentration. The coefficient was calculated based on average values. BCF was calculated as follows:

$$\text{Bioconcentration Factor[BCF]} = \frac{\text{Metal concentration in plant leaves (mg kg}^{-1}\text{d.m.)}}{\text{Metal concentration in soil (mg kg}^{-1}\text{d.m.)}}$$

3. Results

The copper content of the soils in which *S. splendens* was grown ranged from 0.63 to 5.53 mg kg^{-1} d.m. and had strong variability (57.14% CV) (Table 1). In soils in which *S. nipponica* 'Snowmound' was grown, there was strong variability in copper content (62.76% CV), and the content of this metal ranged from 0.56 to 4.43 mg kg^{-1} d.m. (Table 1). The copper content in soils in which *S.* × *vanhouttei* was grown ranged from 2.54 to 6.08 mg kg^{-1} d.m. and had low variability (23.83% CV) (Table 1). In the soils in which *S.* × *cinerea* 'Grefsheim' was grown, there was strong variability in copper content (46.28% CV), and the content of this metal ranged from 1.82 to 6.21 mg kg^{-1} d.m. (Table 1). Regardless of the date of soil sampling, the highest average copper content was found in soils in which *S.* × *vanhouttei* was grown, while the lowest was found in soils in which *S. nipponica* 'Snowmound' was grown (Table 2).

Copper content in leaves of *S. splendens* ranged from 3.04 to 11.45 mg kg^{-1} d.m. and had an average variability of 44.18% CV (Table 1). There was a low variability of copper content of 21.57% CV in the leaves of *S. nipponica* 'Snowmound', and the content of this metal in the leaves ranged from 3.95 to 7.04 mg kg^{-1} d.m. (Table 1). The copper content in the leaves of *S.* × *vanhouttei* ranged from 3.34 to 7.93 mg kg^{-1} d.m. and had an average variability (29.48% CV) (Table 1). Leaves of *S.* × *cinerea* 'Grefsheim' also showed average variability in copper content (39.78% CV), and the content of this metal ranged from 4.67 to 14.42 mg kg^{-1} d.m. (Table 1). The highest copper content in leaves was found in *S.* × *cinerea* 'Grefsheim' (Table 3).

Iron content in the soils in which *S. splendens* was grown ranged from 25.45 to 88.51 mg kg^{-1} d.m. and had an average variability (41.30% CV) (Table 1). In soils in which *S. nipponica* 'Snowmound' was grown, there was low variability in iron content (15.90% CV), and the content of this metal ranged from 43.40 to 70.11 mg kg^{-1} d.m. (Table 1). Iron content in soils in which *S.* × *vanhouttei* was grown ranged from 38.70 to 90.10 mg kg^{-1} d.m. and had average variability (30.87% CV) (Table 1). In soils in which *S.* × *cinerea* 'Grefsheim' was grown, there was low variability in iron content (21.41% CV), and the content of this metal ranged from 39.26 to 79.10 mg kg^{-1} d.m. (Table 1). Regardless of the date of collection of soil samples, there were no significant differences in the iron content of the soils in which the studied spirea was grown (Table 2).

Iron content in leaves of *S. splendens* ranged from 218.27 to 507.75 mg kg^{-1} d.m. and had an average variability of 27.77% CV (Table 1). There was a low variability of 21.61% CV in the iron content of *S. nipponica* 'Snowmound' leaves, and the content of this metal in leaves ranged from 194.36 to 378.46 mg kg^{-1} d.m. (Table 1). Iron content in leaves of *S.* × *vanhouttei* ranged from 195.99 to 398.18 mg kg^{-1} d.m. and had a low variability (22.77% CV) (Table 1). The leaves of *S.* × *cinerea* 'Grefsheim' also showed low variability in iron content (22.33% CV), and the content of this metal ranged from 284.07 to 541.13 mg kg^{-1} d.m. (Table 1). The lowest iron content was obtained in the leaves of *S. nipponica* 'Snowmound' and *S.* × *vanhouttei* (Table 3), while the highest iron content was found in the leaves of *S.* × *cinerea* 'Grefsheim'.

The content of manganese in soils in which *S. splendens* was grown ranged from 2.97 to 11.11 mg kg^{-1} d.m. and was characterized by strong variability (49.13% CV) (Table 1). In soils in which *S. nipponica* 'Snowmound' was grown, there was low variability in manganese content (19.82% CV), and the content of this metal ranged from 3.83 to

7.85 mg kg^{-1} d.m. (Table 1). The content of manganese in soils in *which S.* × *vanhouttei* was grown ranged from 2.78 to 12.43 mg kg^{-1} d.m. and had a strong variability (56.67% CV) (Table 1). The soils in which *S.* × *cinerea* 'Grefsheim' was grown showed average variability in manganese content (33.49% CV), and the content of this metal ranged from 3.46 to 9.32 mg kg^{-1} d.m. (Table 1). There were no significant differences in the average manganese content of the soils in which the studied spirea was grown (Table 2).

The manganese content of *S. splendens* leaves ranged from 35.90 to 109.32 mg kg^{-1} d.m. and had an average variability of 38.35% CV (Table 1). There was a low variability of manganese content of 14.03% CV in the leaves of *S. nipponica* 'Snowmound', and the content of this metal in the leaves ranged from 31.43 to 48.61 mg kg^{-1} d.m. (Table 1). The manganese content in the leaves of *S.* × *vanhouttei* ranged from 47.48 to 111.00 mg kg^{-1} d.m. and had an average variability (27.35% CV) (Table 1). Leaves of *S.* × *cinerea* 'Grefsheim' also showed average variability in manganese content (38.99% CV), and the content of this metal ranged from 31.00 to 98.64 mg kg^{-1} d.m. (Table 1). The lowest manganese content was obtained in leaves of *S. nipponica* 'Snowmound' (Table 3), while the highest content was found in leaves of *S.* × *vanhouttei*.

The nickel content of the soils in which *S. splendens* was grown ranged from 0.03 to 0.24 mg kg^{-1} d.m. and was characterized by strong variability (50.00% CV) (Table 1). The soils in which *S. nipponica* 'Snowmound' was grown also showed strong variability in nickel content (66.67% CV), and the content of this metal ranged from 0.01 to 0.21 mg kg^{-1} d.m. (Table 1). The nickel content in soils in which *S.* × *vanhouttei* was grown ranged from 0.07 to 0.33 mg kg^{-1} d.m. and had strong variability (47.37% CV) (Table 1). The soils in which *S.* × *cinerea* 'Grefsheim' was grown showed average variability in nickel content (40.00% CV), and the content of this metal ranged from 0.09 to 0.28 mg kg^{-1} d.m. (Table 1). The highest nickel content was found in soils in which *S.* × *vanhouttei* was grown, while the lowest content was found in soils in which *S. nipponica* 'Snowmound' was grown (Table 2).

The nickel content of *S. splendens* leaves ranged from 0.74 to 1.92 mg kg^{-1} d.m. and had an average variability of 31.06% CV (Table 1). There was a low variability of 19.15% CV in the nickel content *of S. nipponica* 'Snowmound' leaves, and the content of this metal in leaves ranged from 0.64 to 1.15 mg kg^{-1} d.m. (Table 1). The nickel content in *S.* × *vanhouttei* leaves ranged from 0.94 to 1.58 mg kg^{-1} d.m. and had a low variability (16.41% CV) (Table 1). The leaves of *S.* × *cinerea* 'Grefsheim' showed low variability in nickel content (20.12% CV), and the content of this metal ranged from 1.18 to 2.13 mg kg^{-1} d.m. (Table 1). The lowest nickel content was obtained in the leaves of *S. nipponica* 'Snowmound' (Table 3), while the highest content was found in the leaves of *S.* × *cinerea* 'Grefsheim'.

Zinc content in soils in which *S. splendens* was grown ranged from 7.10 to 16.02 mg kg^{-1} d.m. and had average variability (30.68% CV) (Table 1). The soils in which *S. nipponica* 'Snowmound' was grown also showed average variability in zinc content (33.66% CV), and the content of this metal ranged from 3.56 to 13.65 mg kg^{-1} d.m. (Table 1). Zinc content in soils in which *S.* × *vanhouttei* was grown ranged from 8.13 to 23.45 mg kg^{-1} d.m. and had an average variability (33.47% Cv) (Table 1). The soils in which *S.* × *cinerea* 'Grefsheim' was grown showed average variability in nickel content (34.65% CV), and the content of this metal ranged from 4.78 to 18.23 mg kg^{-1} d.m. (Table 1). The highest zinc content was found in soils where *S.* × *vanhouttei* was grown (Table 2).

The zinc content in the leaves of *S. splendens* ranged from 19.63 to 30.11 mg kg^{-1} d.m. and had a low variability of 13.36% CV (Table 1). There was a low variability of 5.35% CV in the zinc content of the leaves of *S. nipponica* 'Snowmound', and the content of this metal in the leaves ranged from 16.46 to 19.56 mg kg^{-1} d.m. (Table 1). The zinc content in the leaves of *S.* × *vanhouttei* ranged from 31.29 to 50.10 mg kg^{-1} d.m. and had a low variability (16.41% CV) (Table 1). Leaves of *S.* × *cinerea* 'Grefsheim' showed average variability in zinc content (34.90% CV), and the content of this metal ranged from 33.70 to 92.04 mg kg^{-1} d.m. (Table 1). The lowest zinc content was obtained in the leaves of *S. nipponica* 'Snowmound' and *S. splendens* (Table 3), while its highest content was found in the leaves of *S.* × *cinerea* 'Grefsheim'.

The cadmium content in the soils in which *S. splendens* was grown ranged from 0.11 to 0.30 mg kg^{-1} d.m. and had average variability (38.89% CV) (Table 1). In soils in which *S. nipponica* 'Snowmound' was grown, there was low variability in cadmium content (21.74% CV), and the content of this metal ranged from 0.16 to 0.32 mg kg^{-1} d.m. (Table 1). Cadmium content in soils in which *S.* × *vanhouttei* was grown ranged from 0.10 to 0.34 mg kg^{-1} d.m. and had average variability (36.36% CV) (Table 1). In soils in which *S.* × *cinerea* 'Grefsheim' was grown, there was strong variability in cadmium content (95.24% CV), and the content of this metal ranged from 0.07 to 0.80 mg kg^{-1} d.m. (Table 1). There were no statistically significant differences in the cadmium content of the soils between the sites where the spirea was grown, and there was no effect of the date of soil sampling on the density of this metal in the soils (Table 2).

Cadmium content in *S. splendens* leaves ranged from 0.83 to 1.82 mg kg-1 d.m. and had an average variability of 27.41% CV (Table 1). There was a low variability of 23.70% CV in the cadmium content of *S. nipponica* 'Snowmound' leaves, and the content of this metal in the leaves ranged from 0.94 to 1.75 mg kg^{-1} d.m. (Table 1). The cadmium content in *S.* × *vanhouttei* leaves ranged from 0.75 to 1.61 mg kg^{-1} d.m. and had an average variability (36.36% CV) (Table 1). Leaves of *S.* × *cinerea* 'Grefsheim' showed average variability in cadmium content (28.91% CV), and the content of this metal ranged from 0.83 to 1.98 mg kg^{-1} d.m. (Table 1). There were no statistically significant differences in the cadmium content of soils between sites where spirea was grown, and there was no effect of the date of soil sampling on the density of this metal in soils. The lowest cadmium content was found in the leaves of *S.* × *vanhouttei* (Table 3). The other spirea were characterized by significantly higher amounts of cadmium in their leaves.

The chromium content in the soils in which *S. splendens* was grown ranged from 0.18 to 1.64 mg kg^{-1} d.m. and had strong variability (65.33% CV) (Table 1). There was also strong variability in chromium content (92.86% CV) in soils in which *S. nipponica* 'Snowmound' was grown, and the content of this metal ranged from 0.01 to 2.67 mg kg^{-1} d.m. (Table 1). The chromium content in soils in which *S.* × *vanhouttei* was grown ranged from 0.28 to 0.97 mg kg^{-1} d.m. and had an average variability (41.27% CV) (Table 1). In soils in which *S.* × *cinerea* 'Grefsheim' was grown, there was a strong variability in chromium content (68.13% CV), and the content of this metal ranged from 0.15 to 1.96 mg kg^{-1} d.m. (Table 1). The highest chromium content was found in soils in which *S. nipponica* 'Snowmound' was grown, while the lowest was found in soils in which *S.* × *vanhouttei* was grown (Table 2).

The chromium content of *S. splendens* leaves ranged from 5.80 to 16.70 mg kg^{-1} d.m. and had an average variability of 37.73% CV (Table 1). There was an average variation of 35.06% CV in the chromium content of *S. nipponica* 'Snowmound' leaves, and the content of this metal in the leaves ranged from 6.43 to 18.81 mg kg^{-1} d.m. (Table 1). The chromium content in *S.* × *vanhouttei* leaves ranged from 4.29 to 12.79 mg kg-1 d.m. and had an average variation (36.69% CV) (Table 1). Leaves of *S.* × *cinerea* 'Grefsheim' showed average variability in cadmium content (35.29% CV), and the content of this metal ranged from 7.02 to 19.61 mg kg^{-1} d.m. (Table 1). The lowest chromium content was obtained in the leaves of *S.* × *vanhouttei* (Table 3), while significantly higher contents of this metal were found in the leaves of the other spirea.

Soils in which *S.* × *cinerea* 'Grefsheim' was grown showed strong variability in lead content (71.52% CV), and the content of this metal ranged from 1.25 to 9.45 mg kg^{-1} d.m. (Table 1). The content of lead in soils in which *S. splendens* was grown ranged from 1.11 to 4.57 mg kg^{-1} d.m. and was characterized by strong variability (52.42% CV). Soils in which *S. nipponica* 'Snowmound' was grown showed average variability in lead content (42.15% CV), and the content of this metal ranged from 1.03 to 3.90 mg kg^{-1} d.m. Lead content in soils in which *S.* × *vanhouttei* was grown ranged from 0.64 to 12.85 mg kg^{-1} d.m. and was characterized by strong variability (62.54% CV). Regardless of the date of collection of soil samples, the lowest average lead content was found in soils in which *S. nipponica* 'Snowmound' (2.23 mg kg^{-1} d.m.) and *S. splendens* (2.27 mg kg^{-1} d.m.) was grown, while the highest was found in soils in which *S.* × *vanhouttei* was grown (6.94 mg kg^{-1} d.m.)

(Table 1). There was no effect of the harvest date of the soil samples on the content of this metal in the soils in which the spirea was grown.

Lead content in the leaves of *S.* × *cinerea* 'Grefsheim' ranged from 2.60 to 4.73 mg kg^{-1} d.m. (Table 1). As in the case of the other spirea, there was low variability in lead content (21.39% CV). In the leaves of *S. splendens*, the lead content ranged from 2.71 to 4.86 mg kg^{-1} d.m., in *S. nipponica* 'Snowmound' from 3.00 to 4.90 mg kg^{-1} d.m., in the leaves of *S.* × *vanhouttei* it ranged from 2.69 to 3.83 mg kg^{-1} d.m. Regardless of the date of harvesting the soil samples, the lowest average lead content was found in the leaves of *S.* × *vanhouttei* (3.16 mg kg^{-1} d.m.). The other spirea were characterized by significantly higher leaf lead content (Table 1).

Table 1. Heavy metal content in the soil of the cultivation sites of *S.* × *cinerea* 'Grefsheim', *S. nipponica* 'Snowmound', *S. splendens* and *S.* × *vanhouttei* in their leaves.

		Cu	Fe	Mn	Ni	Zn	Cd	Cr	Pb
S. × *cinerea* 'Grefsheim'	soils (mg kg^{-1} d.m.)								
	Min.	1.82	39.26	3.46	0.09	4.78	0.07	0.15	1.25
	Max.	6.21	79.10	9.32	0.28	18.23	0.80	1.96	9.45
	Mean	3.63	65.57	6.48	0.15	11.89	0.21	0.91	4.53
	SD *	1.68	14.04	2.17	0.06	4.12	0.20	0.62	3.24
	CV **	46.28	21.41	33.49	40.00	34.65	95.24	68.13	71.52
	leaves (mg kg^{-1} d.m.)								
	Min.	4.67	284.07	31.00	1.18	33.70	0.83	7.02	2.60
	Max.	14.42	541.13	98.64	2.13	92.04	1.98	19.61	4.73
	Mean	8.90	405.51	56.29	1.64	55.93	1.28	12.41	3.88
	SD	3.54	90.55	21.95	0.33	19.52	0.37	4.38	0.83
	CV	39.78	22.33	38.99	20.12	34.90	28.91	35.29	21.39
S. nipponica 'Snowmound'	soils (mg kg^{-1} d.m.)								
	Min.	0.56	43.40	3.83	0.00	3.56	0.16	0.00	1.03
	Max.	4.43	70.11	7.85	0.21	13.65	0.32	2.67	3.90
	Mean	2.39	53.89	6.66	0.12	9.24	0.23	1.26	2.23
	SD	1.50	8.57	1.32	0.08	3.11	0.05	1.17	0.94
	CV	62.76	15.90	19.82	66.67	33.66	21.74	92.86	42.15
	leaves (mg kg^{-1} d.m.)								
	Min.	3.95	194.36	31.43	0.64	16.46	0.94	6.43	3.00
	Max.	7.04	378.46	48.61	1.15	19.56	1.75	18.81	4.90
	Mean	5.10	279.55	39.84	0.94	17.93	1.35	12.95	3.90
	SD	1.10	60.42	5.59	0.18	0.96	0.32	4.54	0.60
	CV	21.57	21.61	14.03	19.15	5.35	23.70	35.06	15.38
S. splendens	soils (mg kg^{-1} d.m.)								
	Min.	0.63	25.45	2.97	0.03	7.10	0.11	0.18	1.11
	Max.	5.53	88.51	11.11	0.24	16.02	0.30	1.64	4.57
	Mean	2.66	57.33	5.76	0.14	10.30	0.18	0.75	2.27
	SD	1.52	23.68	2.83	0.07	3.16	0.07	0.49	1.19
	CV	57.14	41.30	49.13	50.00	30.68	38.89	65.33	52.42
	leaves (mg kg^{-1} d.m.)								
	Min.	3.04	218.27	35.90	0.74	19.63	0.83	5.80	2.71
	Max.	11.45	507.75	109.32	1.92	30.11	1.82	16.70	4.86
	Mean	7.04	354.32	61.39	1.32	25.67	1.35	10.76	3.97
	SD	3.11	98.39	23.54	0.41	3.43	0.37	4.06	0.74
	CV	44.18	27.77	38.35	31.06	13.36	27.41	37.73	18.64

Table 1. *Cont.*

		Cu	Fe	Mn	Ni	Zn	Cd	Cr	Pb
S. × *vanhouttei*					soils (mg kg^{-1} d.m.)				
	Min.	2.54	38.70	2.78	0.07	8.13	0.10	0.28	0.67
	Max.	6.08	90.10	12.43	0.33	23.45	0.34	0.97	12.85
	Mean	4.28	59.67	7.50	0.19	15.09	0.22	0.63	6.94
	SD	1.02	18.42	4.25	0.09	5.05	0.08	0.26	4.34
	CV	23.83	30.87	56.67	47.37	33.47	36.36	41.27	62.54
					leaves (mg kg^{-1} d.m.)				
	Min.	3.34	195.99	47.48	0.94	31.29	0.75	4.29	2.69
	Max.	7.93	398.18	111.00	1.58	50.10	1.61	12.79	3.83
	Mean	5.53	279.88	78.50	1.28	39.11	1.15	7.85	3.16
	SD	1.63	63.73	21.47	0.21	6.39	0.28	2.88	0.36
	CV	29.48	22.77	27.35	16.41	16.34	24.35	36.69	11.39

* SD—standard deviation; ** CV—coefficient of variation.

Table 2. Average content of heavy metals (mg kg^{-1} d.m.) in the soils of the cultivation sites of *S.* × *cinerea* 'Grefsheim', *S. nipponica* 'Snowmound', *S. splendens* and *S.* × *vanhouttei*, depending on the date of soil sampling.

Metal	Date of Soil Sampling	*S.* × *cinerea* 'Grefsheim'	*S. nipponica* 'Snowmound'	*S. splendens*	*S.* × *vanhouttei*	Średnia
Cu	spring	3.65^{b*}	3.10^{ab}	3.67^{b}	4.67^{b}	3.77^{b}
	autumn	3.60^{b}	1.69^{a}	1.65^{a}	3.88^{b}	2.71^{a}
	mean	3.63^{bc}	2.39^{a}	2.66^{ab}	4.28^{c}	
Fe	spring	57.42^{bc}	54.59^{ab}	38.17^{a}	53.29^{ab}	50.87^{a}
	autumn	73.72^{cd}	53.18^{ab}	76.50^{d}	66.06^{bcd}	67.36^{b}
	mean	65.57^{a}	53.89^{a}	57.33^{a}	59.67^{a}	
Mn	spring	7.26^{a}	6.13^{a}	5.14^{a}	6.16^{a}	6.17^{a}
	autumn	5.71^{a}	7.20^{a}	6.38^{a}	8.84^{a}	7.03^{a}
	mean	6.48^{a}	6.66^{a}	5.76^{a}	7.50^{a}	
Ni	spring	0.19^{bc}	0.18^{bc}	0.20^{c}	0.18^{bc}	0.19^{b}
	autumn	0.12^{ab}	0.06^{a}	0.08^{a}	0.21^{c}	0.12^{a}
	mean	0.15^{ab}	0.12^{a}	0.14^{ab}	0.19 b	
Zn	spring	11.77^{abc}	9.43^{ab}	11.07^{abc}	15.65^{c}	11.98^{a}
	autumn	12.02^{abc}	9.05^{a}	9.53^{ab}	14.53^{bc}	11.28^{a}
	mean	11.89^{ab}	9.24^{a}	10.30^{a}	15.09^{b}	
Cd	spring	0.20^{a}	0.24^{a}	0.23^{a}	0.24^{a}	0.23^{a}
	autumn	0.22^{a}	0.21^{a}	0.14^{a}	0.21^{a}	0.19^{a}
	mean	0.21^{a}	0.23^{a}	0.18^{a}	0.22^{a}	
Cr	spring	1.42^{c}	2.37^{d}	1.13^{bc}	0.85^{b}	1.44^{b}
	autumn	0.40^{a}	0.16^{a}	0.37^{a}	0.42^{a}	0.34^{a}
	mean	0.91^{b}	1.26^{c}	0.75^{ab}	0.63^{a}	
Pb	spring	4.20^{ab}	1.95^{a}	2.18^{a}	6.50^{b}	3.71^{a}
	autumn	4.86^{ab}	2.52^{a}	2.36^{a}	7.38^{b}	4.28^{a}
	mean	4.53^{ab}	2.23^{a}	2.27^{a}	6.94^{b}	

* Followed by the same letters do not differ significantly at ($\alpha = 0.05$).

The timing of the collection of leaves for analysis had a significant effect on their heavy metal content. Significantly higher contents of lead, cadmium, chromium, nickel, zinc, iron and manganese, regardless of the taxon of the spirea, were found in leaves harvested in spring compared with the content found in leaves harvested in autumn. In the case of copper, no significant differences were found in the content of this metal in leaves depending on their harvest date (Table 3).

Table 3. Average heavy metal content (mg kg^{-1} d.m.) in leaves of *S.* × *cinerea* 'Grefsheim', *S. nipponica*, 'Snowmound', *S. splendens* and *S.* × *vanhouttei* depending on the date of leaf sampling.

Metal	Date of Leaf Sampling	*S.* × *cinerea* 'Grefsheim'	*S. nipponica* 'Snowmound'	*S. splendens*	*S.* × *vanhouttei*	Średnia
Cu	spring	9.40 ^{c*}	5.49 ^{ab}	7.57 ^{abc}	5.79 ^{ab}	7.06 ^{a}
	autumn	8.40 ^{bc}	4.72 ^{a}	6.52 ^{abc}	5.27 ^{ab}	6.23 ^{a}
	mean	8.90 ^{b}	5.10 ^{a}	7.04 ^{ab}	5.53 ^{a}	
Fe	spring	474.74 ^{c}	326.96 ^{b}	424.34 ^{c}	323.56 ^{b}	387.40 ^{b}
	autumn	336.28 ^{b}	232.15 ^{a}	284.30 ^{ab}	236.20 ^{a}	272.23 ^{a}
	mean	405.51 ^{c}	279.55 ^{a}	354.32 ^{b}	279.88 ^{a}	
Mn	spring	69.77 ^{c}	44.52 ^{ab}	76.74 ^{c}	95.15 ^{d}	71.54 ^{b}
	autumn	42.81 ^{a}	35.17 ^{a}	46.04 ^{ab}	61.85 ^{bc}	46.47 ^{a}
	mean	56.29 ^{b}	39.84 ^{a}	61.39 ^{b}	78.50 ^{c}	
Ni	spring	1.78 ^{d}	1.00 ^{ab}	1.44 ^{cd}	1.35 ^{bc}	1.39 ^{b}
	autumn	1.50 ^{cd}	0.88 ^{a}	1.20 ^{abc}	1.21 ^{abc}	1.20 ^{a}
	mean	1.64 ^{c}	0.94 ^{a}	1.32 ^{b}	1.28 ^{b}	
Zn	spring	61.52 ^{e}	18.64 ^{a}	26.74 ^{ab}	43.94 ^{cd}	37.71 ^{b}
	autumn	50.34 ^{de}	17.21 ^{a}	24.60 ^{ab}	34.36 ^{bc}	31.63 ^{a}
	mean	55.93 ^{c}	17.93 ^{a}	25.67 ^{a}	39.15 ^{b}	
Cd	spring	1.53 ^{bc}	1.64 ^{c}	1.63 ^{c}	1.37 ^{b}	1.54 ^{b}
	autumn	1.04 ^{a}	1.07 ^{a}	1.06 ^{a}	0.93 ^{a}	1.02 ^{a}
	mean	1.28 ^{ab}	1.35 ^{b}	1.35 ^{b}	1.15 ^{a}	
Cr	spring	16.22 ^{c}	15.79 ^{c}	13.84 ^{c}	10.26 ^{b}	14.03 ^{b}
	autumn	8.60 ^{b}	10.11 ^{b}	7.68 ^{ab}	5.44 ^{a}	7.96 ^{a}
	mean	12.41 ^{b}	12.95 ^{b}	10.76 ^{b}	7.85 ^{a}	
Pb	spring	4.09 ^{bc}	4.17 ^{c}	4.22 ^{c}	3.32 ^{ab}	3.95 ^{b}
	autumn	3.68 ^{abc}	3.63 ^{abc}	3.72 ^{abc}	2.99 ^{a}	3.50 ^{a}
	mean	3.88 ^{b}	3.90 ^{b}	3.97 ^{b}	3.16 ^{a}	

* Followed by the same letters do not differ significantly at ($\alpha = 0.05$).

The values of bioconcentration factors for micronutrients and toxic heavy metals are shown in Figure 1. Only *S.* × *cinerea* 'Grefsheim' and *S.* × *vanhouttei* have a BCF for lead < 1. The remaining spirea are characterized by strong concentrations of all analyzed elements. A particularly high bioconcentration factor, above 10, was recorded for chromium and high for manganese and nickel.

The results of soil reaction and specific electrical conductivity for Poznań are summarized in Table 4; those obtained for Lublin and Gdańsk are shown in Table 5.

Both the smallest and the largest soil reaction in Poznań are characterized by the sites of *S.* × *vanhouttei* (pH 6.9–8.5); *S. nipponica* 'Snowmound' was planted in sites with the highest soil reaction (pH 7.70–7.95). The specific electrical conductivity of soil samples analyzed in spring is noticeably higher than that of those tested in autumn. Particularly noticeably higher total salt concentrations can be seen in the soils of Garbary Street (with *S. splendens* growing there) and Żeromskiego Street (*S.* × *cinerea* 'Grefsheim'). In Lublin, *S. splendens* has particularly unfavorable living conditions, growing at Nadbystrzycka in soil with a pH of 7.90, and *S.* × *cinerea* 'Grefsheim' from aleja Wincentego Witosa, in soil with an electrical conductivity of 0.258 mS cm^{-1}. In Gdańsk, the highest soil reaction was found at the Aleja Rzeczypospolitej traffic circle (pH H_2O—8.19). On the Wały Jagiellońskie, *S.* × *vanhouttei* grew in soil that showed an electrical conductivity of 0.471 mS cm^{-1} during the summer.

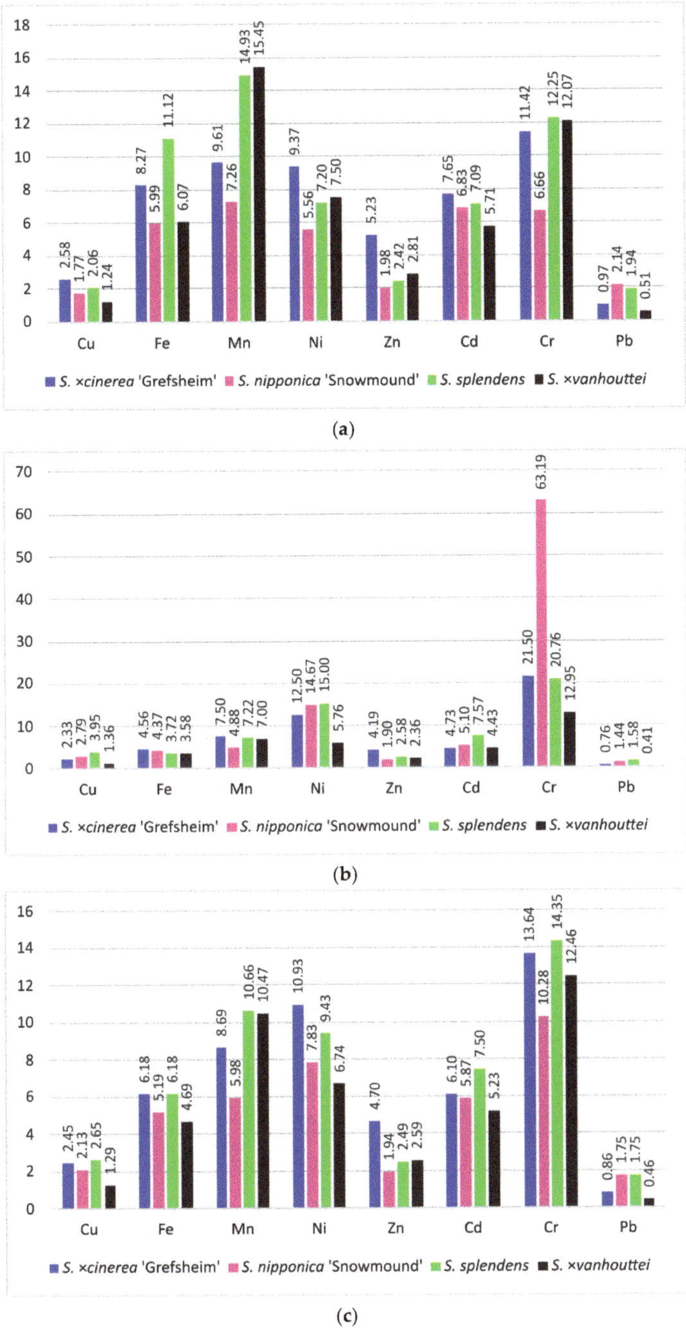

Figure 1. Bioconcentration coefficient of selected heavy metals in the leaves of spirea—(**a**) spring, (**b**) autumn, (**c**) mean.

Table 4. Soil reaction and salinity of spirea growing sites in Poznań.

Taxon/Street, Location		Sample No.	Spring Term		Autumn Term	
			pH (H_2O)	EC (mS cm^{-1})	pH (H_2O)	EC (mS cm^{-1})
S. × *cinerea* 'Grefsheim'	Karola Libelta	2A-1	7.38	0.233	7.93	0.260
	52.4124, 16.9189	2A-2	7.40	0.245	7.89	0.256
	Polska	2B-1	7.32	0.167	7.85	0.169
	52.4197, 16.8727	2B-2	7.35	0.178	7.85	0.163
	Żeromskiego	2C-1	7.88	0.378	7.77	0.249
	52.4165, 16.8923	2C-2	7.77	0.401	7.80	0.241
S. splendens	Rondo Śródka	3A-1	7.59	0.287	7.67	0.255
	52.4102, 16.9561	3A-2	7.61	0.293	7.63	0.254
	Garbary	3B-1	7.81	0.352	7.74	0.230
	52.4101, 16.9381	3B-2	7.89	0.347	7.71	0.228
	Strzeszyńska	3C-1	8.11	0.269	7.50	0.250
	52.4513, 16.8891	3C-2	8.06	0.297	7.49	0.251
S. nipponica 'Snowmound'	Winogrady	4A-1	7.84	0.209	7.72	0.260
	52.4249, 16.9271	4A-2	7.91	0.197	7.70	0.263
	Lechicka	4B-1	7.76	0.263	7.95	0.235
	52.4381, 16.9529	4B-2	7.89	0.267	7.89	0.230
	Polska	4C-1	7.79	0.282	7.70	0.222
	52.4184, 16.8717	4C-2	7.84	0.291	7.72	0.227
S. × *vanhouttei*	Słowiańska	1A-1	7.36	0.185	6.96	0.171
	52.4302, 16.9287	1A-2	7.28	0.210	6.98	0.168
	Serbska	1B-1	7.55	0.284	7.37	0.180
	52.4332, 16.9430	1B-2	7.51	0.278	7.30	0.182
	Lechicka	1C-1	7.66	0.246	8.51	0.227
	52.4396, 16.9506	1C-2	7.56	0.256	8.48	0.229

Table 5. Soil reaction and salinity of spirea growing sites in Lublin and Gdańsk (summer term).

Taxon	Street, Location	pH (H_2O)	EC (mS cm^{-1})
	Lublin		
S. × *cinerea* 'Grefsheim'	Aleja Wincentego Witosa 51.2304, 22.6128	7.85	0.258
S. splendens	Nadbystrzycka 51.2358, 22.5467	7.90	0.220
S. × *vanhouttei*	Wyżynna 51.2239, 22.5256	7.69	0.216
S. × *vanhouttei*	Pogodna 51.2335, 22.5950	7.67	0.190
	Gdańsk		
S. × *cinerea* 'Grefsheim'	Aleja Grunwaldzka 54.3779, 18.6083	7.89	0.258
S. splendens	Aleja Rzeczypospolitej 54.4029, 18.5917	8.15	0.148
S. × *vanhouttei*	Wały Jagiellońskie 54.3561, 18.6456	7.28	0.471
S. × *vanhouttei*	Aleja Rzeczypospolitej 54.4032, 18.5917	8.19	0.178

4. Discussion

From the point of view of caring for the urban environment, street greenery can provide a remedy for the lingering toxic metals in the soil. One of the methods used in the process of soil phytoremediation is the phytoextraction of elements with the help of aboveground parts of plants [17]. The effectiveness of phytoextraction is determined by two key factors: high biomass production and a bioconcentration factor higher than 1 [37]. It is difficult to estimate the biomass of the spirea differing in growth vigor and habit, but it is worth noting that many of the recognized hyperaccumulators are herbaceous plants reaching small sizes, such as *Thlaspi caerulescens*, *Arabidopsis halleri*, *Elsholtzia splendens* and *Viola baoshanensis* [50]. Against their background, the biomass production of the studied taxa of the spirea does not fare badly, and their bioconcentration coefficient is high, especially for chromium; however, such a state of bioaccumulation was found at a low content of heavy metals in the soil. The extraction of elements is strong, and the studied shrubs are promising for use in soil remediation processes, but there is not enough data to show the spirea as hyperaccumulators, especially since a taxon, in order to be classified as a hyperaccumulator and be included in the Global Hyperaccumulator Database, should contain in dry leaf mass >100 mg kg^{-1} of cadmium, thallium or selenium, >300 mg kg^{-1} of cobalt, copper or chromium, >1000 mg kg^{-1} of nickel, arsenic, lead or rare earth elements, >3000 mg kg^{-1} of zinc or >10,000 mg kg^{-1} of manganese under growth conditions in its natural environment [51]. The phrase "natural environment" may seem problematic for plants that are cultivars, such as *S.* × *vanhouttei* and *S.* × *cinerea* 'Grefsheim'. In addition, plants that are facultative hyperaccumulators may take up small amounts of heavy metals at their natural sites, as they are poor in them, while relocated to heavily contaminated areas may show significant remediation potential.

According to Ciążela and Siepak [52], soils adjacent to Poznań exit roads, within 5 m from the roadway in its near-surface layer, have the highest content of Pb, Cd, Cr, Ni, Zn and Cu. The factor that increases the migration of heavy metals is the wind; hence, this distance can increase with its direction. It is also worth mentioning that the metals that most often contaminate the soils of Poznań are zinc, lead and copper [53]. In the case of lead, only *S. splendens* and *S. nipponica* 'Snowmound' show strong bioconcentration with a BCF value of 1.75 each. In contrast, *S.* × *cinerea* 'Grefsheim' and *S.* × *vanhouttei* accumulate lead weakly in their leaves, with BCFs of 0.86 and 0.46, respectively. Meanwhile, the bioconcentration factor for zinc and copper of all spirea tested is above 1, with the highest BCF = 4.7 for zinc in *S.* × *cinerea* 'Grefsheim' and BCF = 2.65 for copper in S. splendens.

Using ornamental shrubs to remediate urban soils can be both practical and aesthetically pleasing. The fact is that shrub pruning treatments are commonly used in green spaces regardless of whether or not we use them to extract heavy metals from the soil. Thus, it is possible to combine attention to the decorative aspect of cultivated plants, which is associated with their maintenance pruning, with a deliberate process of phytoremediation. For example, *S. splendens* blooms at the top of this year's long shoots, and spring pruning of its shoots is carried out annually. As a result of the treatment, the plant develops numerous shoots of offshoots, which translates into more abundant flowering. It is possible to deliberately change the timing of pruning from spring to autumn before the onset of leaf fall. Phenological observations have determined the onset of leaf fall in *S. splendens* in late October or early November. However, a considerable part of the dead leaves remain on the stems until about mid-February. Such developmental peculiarities of the spirea allow it to be pruned in mid-October, during the period when the plant gradually enters dormancy, resulting in a negligible risk of activating dormant buds. Autumn pruning, on the other hand, makes it possible to remove shoots with extracted heavy metals from the growing site, and it is worth noting that such treatment will translate into more abundant flowering in the following growing season. In the case of *S.* × *vanhouttei* and *S. nipponica* 'Snowmound', inflorescences develop on short lateral shoots growing along last year's long shoots, while in *S.* × *cinerea* 'Grefsheim', they develop directly on last year's long shoots. These spirea are pruned after flowering in the first half of June. Then, the plant will have time in the same growing season to produce shoots on which it will set inflorescence buds for next year's spring. In these taxa, pruning cannot be postponed to the autumn date, as it would result in the complete loss of next year's flowers. However, matching the date of harvesting shoots with accumulated metals with the date of cutting shrubs consistent with horticultural practice is also not without a phytoremediation assumption. The results show that the concentration of heavy metals in leaves in autumn is not always higher than in spring. For example, the average content of lead in the leaves of *S. nipponica* 'Snowmound' was 4.17 mg kg^{-1} d.m. in spring (with the content in soil at 1.95 mg kg^{-1} d.m.; BCF 2.14), and in autumn the content in leaves was 3.63 mg kg^{-1} d.m. (in soil 2.52 mg kg^{-1} d.m.; BCF 1.44). The average cadmium content in the leaves of *S.* × *cinerea* 'Grefsheim' in spring was 1.52 mg kg^{-1} d.m. (0.20 mg kg^{-1} d.m. in soil; BCF 7.82), and in autumn was 1.04 mg kg^{-1} d.m. and 0.22 mg kg^{-1} d.m., respectively (BCF 4.76). Both examples indicate that for some heavy metals and taxa of spirea, spring harvesting of leafy shoots may be more advantageous than autumn harvesting. Nonetheless, the developmental peculiarities of individual spirea species make it possible to use both dates in a targeted soil treatment measure.

The use of ornamental shrubs for remediation purposes in green spaces can effectively combine the aesthetic needs of residents with those of caring for the urban environment. At the same time, these shrubs do not necessarily have to have the status of hyperaccumulators, as it is possible in their case to implement a strategy based on less accumulation of metals but high biomass production [28]. It is worth noting that the genus *Spiraea* has many representatives in Polish and foreign dendrological collections, and it is not excluded that some species may prove to be extremely effective in soil phytoremediation processes. The dual function that the plants could perform would undoubtedly be an added value for urban green areas.

Based on the results of the soil analysis, it can be concluded that the degree of salinity of the selected sites for the cultivation of spirea in Poznań, Gdańsk and Lublin does not translate into excessive stress for the shrubs growing there. No damage to the cultivated spirea caused by growth in a saline environment was recorded at any of the studied sites. Komosa [46] includes blue-green and then browning and drying of leaves, soaring habit, limited growth of aboveground organs, or brittleness and ease of falling of blades. The chlorosis of leaves and dying shoots mentioned by Łuczak et al. [43] mentions salt damage to plants. Of the 48 mixed samples taken from locations in Poznań, the highest salinity was 0.401 mS cm^{-1}, and the lowest was 0.163 mS cm^{-1}, while the average value was 0.246 mS cm^{-1}. For Lublin, on the other hand, the highest value of EC was 0.530 mS cm^{-1}, the lowest was 0.190 mS cm^{-1}, and the average was 0.257 mS cm^{-1}. Meanwhile, for Gdansk, it was 0.471 mS cm^{-1}, 0.124 mS cm^{-1}, and 0.236 mS cm^{-1}, respectively (Tables 2 and 3). These results are several times lower than the literature values [47,48] at which damage occurred in spirea. In Marosz [47] experiment, *S.* × *cinerea* 'Grefsheim' responded strongly with a reduction in growth and development only at an EC of 6 mS cm^{-1}, and at 12 mS cm^{-1} 1/3 of the analyzed specimens died—the taxon was classified as a plant with strong resistance to substrate salinity, which is suitable for street plantings.

Sun et al. [48] presented the results of an 8-week experiment during which *S. betulifolia*, *S. japonica*, *S. media*, *S. nipponica*, and *S. thunbergii* were watered with a solution of NaCl and $CaCl_2$ in two combinations—an EC of 3 mS cm^{-1} and an EC of 6 mS cm^{-1}. Except for *S. thunbergii*, which had slightly damaged leaves, all specimens of the other species survived the two-month application period of the 3 mS cm^{-1} solution and maintained good health. In the case of shrubs treated with the 6 mS cm^{-1} solution, all specimens of *S. thunbergii* died and *S. media* suffered severe leaf damage, and *S. betulifolia*, *S. japonica* and *S. nipponica* showed slight to moderate blade damage and a significant reduction in shoot dry weight.

The results of our study from the three cities are similar, both in terms of salinity and soil reaction, with a marked alkalinization of the substrate. Of the eight sites for the cultivation of spirea in Lublin, eight in Gdansk, and 12 in Poznań, as many as 22 had an alkaline soil reaction, and the rest were neutral. Such soil properties were also noted in studies conducted in Kielce, Cracow and Warsaw [54–56]. This is undoubtedly a characteristic feature of the land of urban agglomerations, which (as explained by Czarnowska [57] and Breś [58]) is due to the precipitation of alkaline dust from low emissions, salinity and admixture of calcium debris.

The lack of specific recommendations in the literature regarding the cultivation requirements of spirea in terms of soil pH limits the interpretation of the results. To date, the result of only one experiment documenting the growth of two American species (*S. alba* and *S. tomentosa*) in soils with different pH has been published [59]. The authors checked growth, leaf nutrient content and degree of greening for three combinations of pH (in H_2O)—6, 6.5 and 7—while reporting that the soil at their natural sites had a pH in the range of 4 to 6. The cited study showed that at the highest pH analyzed, the plants were lower and narrower than the shrubs from the other variants, although their leaf mass and degree of greening were the same as in the other combinations. As pH increased, the content of some nutrients in the blades (N, K, Zn, Mn) also decreased. Both spirea were evaluated as species that perform well outside their optimum, i.e., the soil in the area of their natural range. It is noteworthy that the lowest recorded soil pH at spirea growing sites in Poznań was 6.96, the highest was 8.51, and the average value was 7.69. In contrast, for Lublin, it was 7.36, 8.09 and 7.76, and for Gdańsk, it was 7.28, 8.40 and 7.96, respectively. The pH (in H_2O) in the range of 6 to 6.5 is considered the optimum for most crops, including useful and ornamental plants [46]. It would follow that urban soils are not the best environment for plant growth, and creating more favorable conditions would require deliberate acidification. The situation could be improved by applying sulfur fertilizers annually, which would lower the soil reaction in the long run [60]. A possible solution is also the selection of species that naturally occur on land with neutral and alkaline pH. Analyzing the pH of China's soils, it can be seen that the provinces of Shandong, Henan, Shanxi, Shaanxi, Gansu, Inner Mongolia Autonomous Region, Qinghai, Xinjiang and Tibet, together accounting for about half of the country's area, have soils with pH (in H_2O) ranging from 7.2 to 8.5 [61]. In such a vast area, there are natural stands of many spirea [62].

5. Conclusions

1. The taxa studied meet the basic requirements for plants used for soil phytoremediation processes, especially for chromium phytoextraction. Further laboratory studies are advisable (environment intentionally contaminated with heavy metals, varying increasing concentrations of individual metals), as well as on heavily contaminated sites, to confirm the usefulness of *Spiraea* taxa in heavy metal phytoextraction.

2. The degree of soil salinity of the studied sites of cultivation of spirea does not pose a threat to the shrubs growing there.

3. Most of the sites for the cultivation of spirea are characterized by an alkaline soil reaction, which, although not optimal for the growth of many plant species, is suitable for the cultivation of a number of spirea especially.

4. For the determination of the higher contents of Fe, Mn, Ni, Zn, Cd, Cr and Pb, the spring sampling date of spirea leaves is more appropriate.

Author Contributions: Conceptualization, W.A., M.B. and M.S.; methodology, M.B. and M.S.; material preparation, data collection and analysis, M.S. and M.B.; chemical analysis, M.B.; data curation, M.S.; writing—original draft preparation, M.S., W.A. and M.B.; writing—review and editing, W.A. and M.B. All authors have read and agreed to the published version of the manuscript.

Funding: This research received no external funding.

Institutional Review Board Statement: Not applicable.

Data Availability Statement: The original contributions presented in this study are included in the article; further inquiries can be directed to the corresponding author.

Conflicts of Interest: The authors declare no conflicts of interest.

References

1. Wu, Z.Y.; Raven, P.H.; Hong, D.Y. *Flora of China, Volume 9: Pittosporaceae Through Connaraceae*; Science Press: Beijing, China; Missouri Botanical Garden Press: St. Louis, MO, USA, 2003.
2. Maas, S.; Scheifler, R.; Benslama, M.; Crini, N.; Lucot, E.; Brahmia, Z.; Benyacoub, S.; Giraudoux, P. Spatial distribution of heavy metal concentrations in urban, suburban and agricultural soils in a Mediterranean city of Algeria. *Environ. Pollut.* **2010**, *158*, 2294–2301. [CrossRef] [PubMed]
3. Chabukdhara, M.; Nema, A.K. Heavy metals assessment in urban soil around industrial clusters in Ghaziabad, India: Probabilistic health risk approach. *Ecotoxicol. Environ. Saf.* **2013**, *87*, 57–64. [CrossRef]
4. Li, X.; Liu, L.; Wang, Y.; Luo, G.; Chen, X.; Yang, X.; Hall, M.H.P.; Guo, R.; Wang, H.; Cui, J.; et al. Heavy metal contamination of urban soil in an old industrial city (Shenyang) in Northeast China. *Geoderma* **2013**, *192*, 50–58. [CrossRef]
5. Luo, X.S.; Xue, Y.; Wang, Y.L.; Cang, L.; Xu, B.; Ding, J. Source identification and apportionment of heavy metals in urban soil profiles. *Chemosphere* **2015**, *127*, 152–157. [CrossRef]
6. Milenkovic, B.; Stajic, J.M.; Gulan, L.; Zeremski, T.; Nikezic, D. Radioactivity levels and heavy metals in the urban soil of Central Serbia. *Environ. Sci. Pollut. Res.* **2015**, *22*, 16732–16741. [CrossRef] [PubMed]

7. Mehr, M.R.; Keshavarzi, B.; Moore, F.; Sharifi, R.; Lahijanzadeh, A.; Kermani, M. Distribution, source identification and health risk assessment of soil heavy metals in urban areas of Isfahan province, Iran. *J. Afr. Earth Sci.* **2017**, *132*, 16–26. [CrossRef]
8. Tepanosyan, G.; Sahakyan, L.; Belyaeva, O.; Maghakyan, N.; Saghatelyan, A. Human health risk assessment and riskiest heavy metal origin identification in urban soils of Yerevan, Armenia. *Chemosphere* **2017**, *184*, 1230–1240. [CrossRef]
9. Duffus, J.H. "Heavy metals" a meaningless term? (IUPAC Technical Report). *Pure Appl. Chem.* **2002**, *74*, 793–807. [CrossRef]
10. Terelak, H.; Motowicka-Terelak, T. The heavy metals and sulphur status of agricultural soils in Poland. In *Soil Quality, Sustainable Agriculture and Environmental Security in Central and Eastern Europe*; Wilson, M.J., Maliszewska-Kordybach, B., Eds.; NATO Science Series 69; Springer: Dordrecht, The Nederlands, 2000. [CrossRef]
11. Salminen, R. Foregs Geochemical Atlas of Europe. EuroGeoSurveys. 2005. Available online: http://weppi.gtk.fi/publ/foregsatlas/ (accessed on 17 July 2024).
12. Manta, D.S.; Angelon, M.; Bellanca, A.; Neri, R.; Sprovieri, M. Heavy metals in urban soils: A case study from the city of Palermo (Sicily), Italy. *Sci. Total Environ.* **2002**, *300*, 229–243. [CrossRef]
13. Alloway, B.J. Sources of heavy metals and metalloids in soils. In *Heavy Metals in Soils. Environmental Pollution*; Alloway, B.J., Ed.; Springer: Dordrecht, The Nederlands, 2013; Volume 22. [CrossRef]
14. Wie, B.; Yang, L. A review of heavy metal contaminations in urban soils, urban road dusts and agricultural soils from China. *Microchem. J.* **2010**, *94*, 99–107. [CrossRef]
15. Steindor, K.A.; Franiel, I.J.; Bierza, W.M.; Pawlak, B.; Palowski, B.F. Assessment of heavy metal pollution in surface soils and plant material in the post-industrial city of Katowice, Poland. *J. Environ. Sci. Health* **2016**, *51*, 371–379. [CrossRef] [PubMed]
16. Silva, H.F.; Silva, N.F.; Oliveira, C.M.; Matos, M.J. Heavy metals contamination of urban soils—A decade study in the city of Lisbon, Portugal. *Soil Syst.* **2021**, *5*, 27. [CrossRef]
17. Bosiacki, M.; Kleiber, T.; Markiewicz, B. Continuous and Induced Phytoextraction-Plant-Based Methods of Remove Heavy Metals from Contaminated Soil. In *Environmental Risk Assessment of Soil Contamination*; Hernandez-Soriano, M.C., Ed.; InTech: Rijeka, Croatia, 2014; Volume 20, pp. 575–612. ISBN 978-953-51-1235-8.
18. Sharma, P.; Jha, A.B.; Dubey, R.S.; Pessarakli, M. Reactive oxygen species, oxidative damage, and antioxidative defense mechanism in plants under stressful conditions. *J. Bot.* **2012**, *2012*, 217037. [CrossRef]
19. Zaid, A.; Ahmad, B.; Jaleel, H.; Wani, S.H.; Hasanuzzaman, M. A critical review on iron toxicity and tolerance in plants: Role of exogenous phytoprotectants. In *Plant Micronutrients*; Aftab, T., Hakeem, K.R., Eds.; Springer: Berlin/Heidelberg, Germany, 2020. [CrossRef]
20. Nagajyoti, P.C.; Lee, K.D.; Sreekanth, T.V.M. Heavy metals, occurrence and toxicity for plants: A review. *Environ. Chem. Lett.* **2010**, *8*, 199–216. [CrossRef]
21. Chen, J.; Yang, Z.M. Mercury toxicity, molecular response and tolerance in higher plants. *Biometals* **2012**, *25*, 847–857. [CrossRef]
22. Nazar, R.; Iqbal, N.; Masood, A.; Khan, M.; Syeed, S.; Khan, N. Cadmium toxicity in plants and role of mineral nutrients in its alleviation. *Am. J. Plant Sci.* **2012**, *3*, 1476–1489. [CrossRef]
23. Singh, H.P.; Mahajan, P.; Kaur, S.; Batish, D.R.; Kohli, R.K. Chromium toxicity and tolerance in plants. *Environ. Chem. Lett.* **2013**, *11*, 229–254. [CrossRef]
24. Bååth, E. Effects of heavy metals in soil on microbial processes and populations. *Water Air Soil Pollut.* **1989**, *47*, 335–379. [CrossRef]
25. Kucharski, J.; Barabasz, W.; Bielińska, E.J.; Wyszkowska, J. *Biological and Biochemical Properties of Soils*; Mocek, G., Wydanie, I., Eds.; Wydawnictwo Naukowe PWN: Warszawa, Poland, 2016; ISBN 978-83-01-17944-6.
26. Shelton, J.S.; Florkowski, W.J.; Pennisi, S.V. Establishing native wildflower habitats in urban settings on a low budget. *Folia Hortic.* **2024**, *36*, 3. [CrossRef]
27. Bosiacki, M.; Bednorz, L.; Fedeńczak, K.; Górecki, T.; Mizgajski, A.; Poniży, L.; Spiżewski, T. Soil quality as a key factor in producing vegetables for home consumption—A case study of urban allotments in Gorzów Wielkopolski (Poland). *Agronomy* **2021**, *11*, 1836. [CrossRef]
28. Yan, A.; Wang, Y.; Tan, S.N.; Yusof, M.L.M.; Ghosh, S.; Chen, Z. Phytoremediation: A promising approach for revegetation of heavy metal-polluted land. *Front. Plant Sci.* **2020**, *11*, 359. [CrossRef] [PubMed]
29. EL-Bauome, H.; Doklega, S.; Saleh, S.; Mohamed, A.; Suliman, A.; Abd El-Hady, M. Effects of melatonin on lettuce plant growth, antioxidant enzymes and photosynthetic pigments under salinity stress conditions. *Folia Hortic.* **2024**, *36*, 1–17. [CrossRef]
30. Boim, A.G.F.; Melo, L.C.A.; Moreno, F.N.; Alleoni, L.R.F. Bioconcentration factors and the risk concentrations of potentially toxic elements in garden soils. *J. Environ. Manag.* **2016**, *170*, 21–27. [CrossRef] [PubMed]
31. Lugwisha, E.H.; Othman, C.O. Heavy metal levels in soil, tomatoes and selected vegetables from Morogoro Region, Tanzania. *Int. J. Environ. Monit. Anal.* **2016**, *4*, 82–88. [CrossRef]
32. Siyar, R.; Ardejani, F.D.; Norouzim, P.; Maghsoudy, S.; Yavarzadeh, M.; Taherdangkoo, R.; Butscher, C. Phytoremediation potential of native hyperaccumulator plants growing on heavy metal-contaminated soil of Khatunabad copper smelter and refinery, Iran. *Water* **2022**, *14*, 3597. [CrossRef]
33. Garg, V.K.; Yadav, P.; Mor, S.; Singh, B.; Pulhani, V. Heavy metals bioconcentration from soil to vegetables and assessment of health risk caused by their ingestion. *Biol. Trace Elem. Res.* **2014**, *157*, 256–265. [CrossRef]
34. Lee, C.S.; Nzioka, A.M.; Kim, Y.J.; Sakong, J. Evaluation of bioconcentration factors of metals and non-metals in crops and soil from abandoned mines in Korea. *Contemp. Probl. Ecol.* **2017**, *10*, 583–590. [CrossRef]

35. Amin, H.; Arain, B.A.; Abbasi, M.S.; Amin, F.; Jahangir, T.M.; Soomro, N.U. Evaluation of chromium phyto-toxicity, phyto-tolerance, and phyto-accumulation using biofuel plants for effective phytoremediation. *Int. J. Phytoremediat.* **2019**, *21*, 352–363. [CrossRef]
36. Usman, K.; Hareb, J.A.; Abu-Dieyeh, M.H.; Alsafran, M.H.S.A. Comparative assessment of toxic metals bioaccumulation and the mechanisms of chromium (Cr) tolerance and uptake in *Calotropis procera*. *Front. Plant Sci.* **2020**, *11*, 883. [CrossRef]
37. McGrath, S.P.; Zhao, F. Phytoextraction of metals and metalloids from contaminated soils. *Curr. Opin. Biotechnol.* **2003**, *14*, 277–282. [CrossRef]
38. Mazur, R. Impacts of Road Deicing Salt on Natural Environment. *Inżynieria I Ochr. Sr.* **2015**, *15*, 449–458.
39. Czerniawska-Kusza, I.; Kusza, G.; Duzyński, M. Effect of deicing salts on urban soils and health status of roadside trees in the Opole region. *Environ. Toxicol.* **2004**, *19*, 296–301. [CrossRef] [PubMed]
40. Munck, I.A.; Bennett, C.M.; Camilli, K.S.; Nowak, R.S. Long-term impact of de-icing salts on tree health in the Lake Tahoe Basin: Environmental influences and interactions with insects and diseases. *For. Ecol. Manag.* **2010**, *260*, 1218–1229. [CrossRef]
41. Zhang, Y.; Li, F.; Sun, T.; Wang, J. Effect of deicing salts on urban soils and the health of roadside pines (*Pinus tabulaeformis*) in Northeast China. *Appl. Mech. Mater.* **2012**, *178–181*, 353–356. [CrossRef]
42. Dmuchowski, W.; Baczewska-Dąbrowska, A.; Gozdowski, D.; Brągoszewska, P. Effect of salt stress on the chemical composition of leaves of different tree species in urban environment. *Fresenius Environ. Bull.* **2013**, *22*, 987–994.
43. Łuczak, K.; Czerniawska-Kusza, I.; Rosik-Dulewska, C.; Kusza, G. Effect of NaCl road salt on the ionic composition of soils and *Aesculus hippocastanum* L. foliage and leaf damage intensity. *Sci. Rep.* **2021**, *11*, 5309. [CrossRef]
44. Equiza, M.A.; Calvo-Polanco, M.; Cirelli, D.; Señorans, J.; Wartenbe, M.; Saunders, C.J.J.; Zwiazek, J.J. Long-term impact of road salt (NaCl) on soil and urban trees in Edmonton, Canada. *Urban For. Urban Green.* **2017**, *21*, 16–28. [CrossRef]
45. Gałuszka, A.; Migaszewski, Z.M.; Podlaski, R.; Dołęgowska, S.; Michalik, A. The influence of chloride deicers on mineral nutrition and the health status of roadside trees in the city of Kielce, Poland. *Environ. Monit. Assess.* **2011**, *176*, 451–464. [CrossRef]
46. Komosa, A. *Żywienie Roślin Ogrodniczych. Podstawy i Perspektywy [Horticultural Plant Nutrition. Fundamentals and Perspectives]*; Powszechne Wydawnictwo Rolnicze i Leśne: Poznań, Poland, 2012; ISBN 978-83-09-01141-5.
47. Marosz, A. Effect of soil salinity on nutrient uptake, growth, and decorative value of four ground cover shrubs. *J. Plant Nutr.* **2004**, *27*, 977–989. [CrossRef]
48. Sun, Y.; Li, L.; Wang, Y.; Dai, X. Morphological and physiological responses of *Spirea* species to saline water irrigation. *HortScience* **2020**, *55*, 55–888. [CrossRef]
49. Lindsay, W.L.; Norvell, W.A. Development of a DTPA soil test for zinc, iron, manganese, and copper. *Soil Sci. Soc. Am. J.* **1978**, *42*, 421–428. [CrossRef]
50. Ent, A.; Baker, A.J.M.; Reeves, R.D.; Pollard, A.J.; Schat, H. Hyperaccumulators of metal and metalloid trace elements: Facts and fiction. *Plant Soil* **2013**, *362*, 319–334. [CrossRef]
51. Reeves, R.D.; Baker, A.J.M.; Jaffré, T.; Erskine, P.D.; Echevarria, G.; Ent, A. A global database for plants that hyperaccumulate metal and metalloid trace elements. *New Phytol.* **2018**, *218*, 407–411. [CrossRef] [PubMed]
52. Ciążela, J.; Siepak, M. Environmental factors affecting soil metals near outlet roads in Poznań, Poland: Impact of grain size, soil depth, and wind dispersal. *Environ. Monit. Assess.* **2016**, *188*, 323. [CrossRef]
53. Program of Environmental Protection for the City of Poznań for 2017–2020, with an Outlook Until 2024. Resolution No. 4214/2017 of the Board of Directors of the Wielkopolska Region of September 6, 2017. Available online: https://bip.umww.pl/artykuly/2823361/pliki/20171010081435_4214.pdf (accessed on 21 July 2024).
54. Bach, A.; Pawłowska, B.; Kraus, D.; Malinowska, Z.; Pniak, M.; Bartyńska, M. Urban ornamental trees reaction to the soil sodium chloride salinity and pH factor in Krakow. *Zesz. Probl. Postępów Nauk Rol.* **2006**, *510*, 39–48.
55. Dusza-Dobek, A. Geochemikal studies of soils in some city parks of Warsaw. *Biul. Państwowego Inst. Geol.* **2012**, *450*, 35–46.
56. Widłak, M.; Stoińska, R.; Kowalik, R. Assessment of physical and chemical pollution of urban agglomeration soils. *Desalination Water Treat.* **2020**, *199*, 137–143. [CrossRef]
57. Czarnowska, K. Soils and plants in urban environment. *Zesz. Probl. Postępów Nauk Rol.* **1995**, *418*, 111–115.
58. Breś, W. Anthropopressure factors causing trees to die off in urban landscape. *Nauka Przyr. Technol.* **2008**, *2*, 31.
59. Mickelbart, M.V.; Gosney, M.J.; Camberato, J.; Stanton, K.M. Soil pH effects on growth and foliar nutrient concentrations of *Spiraea alba* Du Roi and *Spiraea tomentosa* L. *HortScience* **2012**, *47*, 902–906. [CrossRef]
60. Filipek-Mazur, B.; Tabak, M.; Gorczyca, O.; Bobowiec, A. Influence of mineral fertilizers containing sulfur on soil chemical properties. *Fragm. Agron.* **2018**, *35*, 55–65. [CrossRef]
61. Shangguan, W.; Dai, Y.; Liu, B.; Zhu, A.; Duan, Q.; Wu, L.; Ji, D.; Ye, A.; Yuan, H.; Zhang, Q.; et al. A China data set of soil properties for land surface modeling. *J. Adv. Model. Earth Syst.* **2013**, *5*, 212–224. [CrossRef]
62. Lu, L.; Crinan, A. Spiraea. In *Flora of China 2003*; Pittosporaceae Through Connaraceae; Science Press: Beijing, China; Missouri Botanical Garden Press: St. Louis, MO, USA, 2003; Volume 9, pp. 47–73.

agriculture

MDPI

Article

Assessment of the Antioxidative Properties of Extracts from the Fruits of *Pyrus pyraster* (L.) Burgsd and *Pyrus* ×*myloslavensis* Czarna & Antkowiak Grown under Natural Environmental Conditions

Marzanna Hęś [1], Wojciech Antkowiak [2,*], Kinga Stuper-Szablewska [3], Krzysztof Dziedzic [4], Marta Jessa [1] and Paulina Ratajczak [1]

1 Department of Gastronomy Science and Functional Foods, Faculty of Food Science and Nutrition, Poznań University of Life Sciences, Wojska Polskiego 28, 60-637 Poznań, Poland; marzanna.hes@up.poznan.pl (M.H.)

2 Department of Botany, Faculty of Agronomy, Horticulture and Bioengineering, Poznań University of Life Sciences, Wojska Polskiego 28, 60-637 Poznań, Poland

3 Department of Chemistry, Faculty of Forestry and Wood Technology, Poznań University of Life Sciences, Wojska Polskiego 28, 60-637 Poznań, Poland; kinga.stuper@up.poznan.pl

4 Department of Food Technology of Plant Origin, Faculty of Food Science and Nutrition, Poznań University of Life Sciences, Wojska Polskiego 28, 60-637 Poznań, Poland; krzysztof.dziedzic@up.poznan.pl

* Correspondence: wojciech.antkowiak@up.poznan.pl

Abstract: Analyses were conducted on extracts from the fruits of *P. pyraster* and *P.* ×*myloslavensis*. Extraction with 80% methanol was performed at room temperature. The total phenolic content was determined by spectrophotometry using the Folin–Ciocalteu reagent, with gallic acid as the reference standard. Phenolic compounds and organic acids were identified on a liquid chromatograph. The antioxidative activity of the extracts was tested in relation to linoleic acid incubation of the emulsions for 19 h based on the neutralization of the DPPH radical (2,2-diphenyl-1-picrylhydrazyl) and the ABTS cation radical (2,2′-azino-bis[3-ethylbenzothiazoline-6-sulfonic acid]) as well as by the ferric reducing antioxidant power (FRAP) assay. The analyses showed that the extract from *P. pyraster* fruits is characterized by a higher content of phenolic compounds and a higher antioxidative potential compared with that from *P.* ×*myloslavensis*. In extracts of both pear species, seven phenolic compounds and four organic acids were identified. The total fiber content in pears of *P. pyraster* and *P.* ×*myloslavensis* was determined at 36.45 g and 24.74 g/100 g d.m. of the pear fruits, of which most comprised the insoluble fraction (32.49 g and 20.86/100 g, respectively). The results of the conducted research are highly significant, as they confirm that pears contain many valuable nutrients and biologically active compounds, including antioxidants and dietary fiber. Adding pear extracts to food products may offer a way to boost their health benefits while also broadening the variety of items that have appealing sensory characteristics. Moreover, research has shown that fruit extracts can help to prolong the shelf life of food products by safeguarding them against lipid oxidation and the decline in their nutritional value.

Keywords: *Pyrus pyraster*; *Pyrus* ×*myloslavensis*; antioxidative activity; fiber

1. Introduction

As the World Health Organization (WHO) reports, nutrition is the key factor affecting public health. Some evidence confirms that nutrition can have a positive effect on public health worldwide. This suggests that a suitable diet may prevent or even support the treatment of certain diet-dependent diseases [1]. The recommendation is to consume five servings of vegetables and fruit daily. Such an amount is crucial for the maintenance of a healthy diet by enriching it with essential nutrients, e.g., vitamins, minerals, dietary fiber, and, most of all, antioxidants. It has been shown that the consumption of adequate servings of vegetables and fruit may strengthen the immune system, as well as prevent cardiovascular disease, diabetes, certain cancers, and obesity [2]. Evidence has been shown that confirms an inverse proportional relationship between the incidence of obesity and the antioxidative capacity of the diet [3]. Fruit and vegetables are the main sources of antioxidants, such as polyphenolics, vitamin C, and carotenoids [4]. Antioxidants may inhibit the development of various diseases due to their properties. One of them is the ability to suppress the action of free radicals and prevent their oxidation. Neutralization of free radicals protects cells in the living organism against their detrimental action [5].

Materials rich in nutrients and bioactive compounds showing health-promoting properties also include pear fruits. For example, these fruits contain dietary fiber, minerals (magnesium, calcium, copper, iron, manganese, potassium, phosphorus, sodium, zinc, and iodine) and vitamins (C, E, B1, B2, B3, B6, K, and folic acid) [6–8]. Pears constitute a rich source of polyphenols (epicatechins, rutin, and chlorogenic acid) as well as anti-inflammatory triterpenes (ursolic and oleanolic acids). These are contained in both the skin and flesh of fruits; however, higher concentrations of these compounds can be found in the outer layers. The components mentioned above exert a beneficial effect on the human organism. Polyphenols are attributed antioxidative properties, while triterpenes exhibit an anti-inflammatory capacity [7]. Studies have shown that pear fruits can be characterized by an approx. 2-fold greater free radical scavenging capacity compared with peaches and apricots, while this is lower than the capacity of bilberry and cranberry [9]. Fruits of different pear varieties, e.g., Asian (=Japanese, Nashi, Chinese) *P. pyrifolia* (Burm.f.) Nakai, are rich in pectins. Galacturonic acid is their main structural component. Conducted analyses [10] showed that fruits of the Asian pears 'Shinseiki', 'Chojuro' and 'Hosui' contain from 0.61% to 0.78% pectins. These are beneficial for human health, since pectins constitute soluble (fermentable) dietary fiber, which, among other things, prevents constipation, binds harmful metals, and reduces the risk of cardiovascular disease by reducing the levels of LDL cholesterol.

Currently, in many countries, new species of plants with edible fruits rich in biologically active substances are sought after [11,12]. One of the pear species that deserves particular attention is the European wild pear, *P. pyraster* (L.) Burgsd. (=*P. communis* var. *pyraster* L.). It is found in southern, central, and western Europe. *Pyrus pyraster* is a type of tree that, when exposed to water shortage, tolerates dry substrates and high temperatures. In view of the documented global effects of climate change, the role of thermophilic tree species in forests will be increasing thanks to their high biocenotic value and adaptability to changing environmental conditions. Pears could be used on a broader scale both in forestry and in agricultural areas, replacing more sensitive tree species [13].

A new pear species is a spontaneous hybrid of the common pear *P. communis* L. (=*P. communis* subsp. *communis* L., *P. domestica* Medik.) and the willow-leafed pear *P. salicifolia* Pall., which is alien to Polish flora but which has been planted in European parks since 1780. This hybrid taxon, named *Pyrus* ×*myloslavensis*, was recently discovered and described by Antkowiak et al. in 2008 [14], who performed a comparative characterization (with reference to the parent taxa) concerning its flower and fruit morphology, as well as of its leaf morphology and anatomy. The goal of subsequent studies on *P.* ×*myloslavensis* was to determine whether the intermediate nature of this taxon would be evident in the pollen and seed anatomy and morphology [15]. In addition, the authors evaluated self-compatibility and crossability based on the observation of pollen tube growth. The investigations were carried out using light, fluorescence, and electron microscopy techniques and flow cytometry. *P.* ×*myloslavensis* is a highly interesting object of botanical and pomological research, and the data in this article are new and completely original.

Pear fruits, including those of uncultivated species, deserve greater attention as a valuable raw material for health-promoting food. There are few studies describing the antioxidative action of the polyphenols contained in pear, which may be successfully used in the production of preparations used in the development of new food products. Thus, the aim of the conducted investigations was to determine the phenolic content and antioxidative activity of extracts from *P. pyraster* and *P.* ×*myloslavensis* fruits. Lipid-free model systems (examining the scavenging of 2,2-diphenyl-1-picrylhydrazyl (DPPH), 2,2′ azino-bis(3-ethylbenzothiazoline-6-sulfonic acid) (ABTS) radicals, and the ferric reducing antioxidant power (FRAP)) as well as systems containing emulsified linoleic acid were carried out. Furthermore, the total content of dietary fiber and its fractions was also determined in the pear fruits.

2. Materials and Methods

2.1. Chemicals

The following chemicals were used: 2,2-diphenyl-1-picrylhydrazyl (DPPH), 2,2′ azino-bis(3-ethylbenzothiazoline-6-sulfonic acid) (ABTS), Folin–Ciocalteu reagent (FCR), 2,4,6-tris(2-pyridyl)-s-triazine (TPTZ), gallic acid (GAE), (±)-6-hydroxy-2,5,7,8-tetramethylchromane-2-carboxylic acid (Trolox), linoleic acid, 2-(N-Morpholino)ethanesulfonic acid (MES), 2-amino-2-(hydroxymethyl)-1,3-propanediol (TRIS) thermostable (Sigma-Aldrich, Darmstadt, Germany), α-amylase, pepsin, amyloglucosidase (Megazyme, Ireland); acetone, acetic acid, sodium acetate × 3 H_2O, ethanol, hydrochloric acid, iron(III) chloride hexahydrate, sodium carbonate, potassium persulfate, dipotassium hydrogen phosphate and potassium hydroxide were purchased from POCh (Polskie Odczynniki Chemiczne, eng. Polish Chemical Reagents, Gliwice, Poland). All chemicals and solvents used were of analytical grade.

2.2. Preparation of Extracts

Analyses were conducted on pear fruits collected at the beginning of October 2021 from 5 *P.* ×*myloslavensis* trees growing in the garden of the Department of Botany, Poznań University of Life Sciences and 4 *P. pyraster* trees growing in Poznań county. The relevant environmental parameters of the experimental garden of the Department of Botany of the University of Life Sciences in Poznan, in which the *P.* ×*myloslavensis* trees were grown, are shown in Table 1. *P. pyraster* trees grew in the wild in mid-field woodlots, several kilometers away from *P.* ×*myloslavensis*. Pear trees of both species were not fertilized or watered.

Table 1. Air temperature, precipitation, and soil characteristics of *P.* ×*myloslavensis* cultivation in 2021.

Month of Measurement	Air Temperature (°C)	Precipitation (mm)	Soil Characteristics (mg dm^{-3} Soil)	
January	−0.6	80.2	$N\text{-}NH_4 + N\text{-}NO_3$	12
February	−0.7	33.0	P	96
March	3.9	30.0	K	60
April	6.1	65.0	Ca	2598
May	12.0	66.4	Mg	155
June	19.9	22.6	pH_{H2O}	6.85
July	20.7	19.4		
August	17.5	57.2		
September	15.1	32.0	EC (mS cm^{-1})	0.11
October	9.6	34.8		
November	5.5	73.8		
December	−0.8	29.4		
Average temperature	9.0			
Total precipitation		543.8		

In soil, the content of the following minerals was determined in an extract of 0.03 M CH_3COOH by the following methods: $N\text{-}NH_4$—by the distillation method according to Bremmer with Starck modification; $N\text{-}NO_3$—by the potentiometric method using ionoselenium electrodes; P—colorimetrically with ammonium vanadomolybdate;

K, Ca, Mg—by the atomic absorption method; pH in H_2O—potentiometrically; salt concentration—conductometrically.

The harvested fruits were cut into pieces, and the seeds and pedicels were removed, while overripe fruits were discarded. The fruits were frozen at −80 °C, lyophilized, and ground to produce a powder of uniform grain size. The material was then extracted with 80% methanol in a 1:5 ratio (25 g powder: 125 mL of methanol). Samples were shaken in a water bath at room temperature for 24 h, and then the extract was filtered and subsequently centrifuged for 10 min at 3000 RPM. The supernatant was transferred to a round-bottom flask, and methanol was distilled in a Rotavapor R-215 rotary evaporator (Chemland, Stargard, Poland). The resulting extract was frozen at −30 °C for 24 h and then lyophilized.

2.3. *Preparation of the Linoleic Acid Emulsion*

A total of 0.5 mL of Tween 20 was added to 5 mL of 1 M K_2HPO_4 and mixed for 5 min using a magnetic stirrer. Then, after 10 min, 0.5 mg of linoleic acid and 0.4 g of KOH were added and stirred for 2 min. In the next step, 11 mL of 1 M K_2HPO_4 was added and stirred for 10 min. After the addition of 150 mL of distilled water, the emulsion was stirred for 5 min and adjusted to a pH of 7.2 [16].

2.4. *Analytical Methods*

2.4.1. Determination of Total Phenolic Compound Content

The FCR (Folin–Ciocalteu reagent) method [17] was used to determine the total level of phenolic substances in the extract. A sample of the extract (0.2 mL containing 0.2–0.5 mg of extract) was mixed with 8.3 mL of distilled water and 0.5 mL of the FCR. The resulting test material was blended with 1 mL of saturated sodium carbonate solution. Subsequently, the mixture was incubated for 30 min at room temperature, and then a reading of the absorbance was carried out at 750 nm (Specord 40). The results are presented in mg of gallic acid equivalent per gram of dry matter extract (mg of GAE/g of d.m. extract).

2.4.2. Analysis of Phenolic Compound Content

Phenolic compounds and organic acids were identified in a 'Acquity UPLC H-class chromatography system (Waters, Milford, MA, USA). Chromatographic separation was performed on a UPLC®BEH C18 column (100 mm × 2.1 mm; particle size 1.7 µm) at 35 °C. The result was expressed in mg/g d.m. extract [18,19].

2.4.3. Determination of Dietary Fiber Content

The contents of insoluble and soluble dietary fiber were determined using the "AOAC Method 991.43 Total, Soluble, and Insoluble Dietary Fiber in Foods" [20]. The analysis involves the use of enzymes such as thermostable α-amylase (Megazyme, Bray, Ireland), pepsin and amyloglucosidase (Sigma-Aldrich, Darmstadt, Germany), which simulate the environment of the human digestive tract. The total dietary fiber content was calculated as the sum of the soluble and insoluble fractions. The results were expressed in mg/1 g d.m. pear. Analyses were performed using the Fibertec System 1023 apparatus by Tecator.

2.5. *Antioxidative Activity of the Extract*

2.5.1. Scavenging of DPPH Radicals

The capacity of the prepared extract to scavenge the stable free radical 2,2-diphenyl-1-picrylhydrazyl (DPPH) was monitored using the method of Sanchez-Moreno et al. [21]. DPPH (1 mM, 0.25 mL) was dissolved in pure methanol and added to 0.1 mL of pear extracts (0.075–0.15 mg) with 2 mL of methanol. The decrease in absorbance of the resulting solution was determined at 517 nm after 30 min. The results are presented in mg of Trolox/g of d.m. extract.

2.5.2. Scavenging of ABTS Radicals

The radical cation was generated by combining a 7 mM ABTS$^{\bullet+}$ stock solution with 2.45 mM potassium persulfate in a 1:0.5 ratio (v/v) and allowing the mixture to react for 12–16 h until the reaction reached completion and the absorbance stabilized. The ABTS$^{\bullet+}$ solution was then diluted with ethanol to achieve an absorbance of 0.700 ± 0.02 at 734 nm for the measurements. The photometric assay involved adding 3.0 mL of ABTS$^{\bullet+}$ solution to 0.03 mL of the tested samples (concentration range: 0.015–0.15 mg/mL) and mixing for 1 min. Measurements were taken at 734 nm immediately after 6 min of reaction [22]. The results are expressed in mg Trolox per gram of dry matter extract.

2.5.3. Ferric Reducing Antioxidant Power (FRAP)

The FRAP reagent was prepared by combining 300 mM acetate buffer, 10 mL of TPTZ in 40 mM HCl, and 20 mM $FeCl_3 \times 6H_2O$ in a 10:1:1 ratio at 37 °C. Then, 3.0 mL of the FRAP reagent was mixed with 0.1 mL of the extracts at varying concentrations (0.05 mg/mL). The mixture was incubated in the dark at 37 °C for 4 min. The absorbance was recorded at 593 nm [23]. The results are expressed in mM Fe^{2+}/L.

2.5.4. Emulsion System

The ability to inhibit linoleic acid autoxidation was determined using the method developed by Lingnert et al. [16]. This method uses a spectrophotometer to assess the increase in conjugated dienes in 10 mM linoleic acid emulsion at pH 7.2. A dose of 0.2 mL of pear extract was introduced at concentrations of 0.25 and 0.5% into the prepared emulsion. The absorbance of the sample was subsequently measured at wavelength λ = 234 nm immediately after the antioxidants were added. After 19 h of incubation at 37 °C and without access to light another measurement was done. The content of dienes was calculated using the molar absorptivity coefficient according to the Beer–Lambert law: $A = \varepsilon \times c \times l$, where A is the absorbance; ε is the molar absorptivity coefficient ($\varepsilon = 2.56 \times 10^4\ dm^3 \cdot M^{-1} \cdot cm^{-1}$); c is the concentration of peroxides with conjugated dienes (M); and l is the thickness of the absorbing layer (tray thickness = 1 cm).

3. Statistical Analysis

All statistical calculations were performed using Statistica software, ver. 13.3 (StatSoft Inc., Tulsa, OK, USA). The data were expressed as the mean ± standard deviation (SD) of 2 series and 3 independent measurements for each sample ($n = 6$). To determine whether a random sample came from a population with a normal distribution, the Student's t-test was applied. In the case where the hypothesis on the normal distribution was rejected, non-parametric statistics were used in further analyses. Correlation and regression analyses between variables were performed. The dependence between variables was presented using Pearson's linear correlation coefficient.

4. Results and Discussion

4.1. Extraction Efficiency and Total Phenolic Content

The extraction efficiency for fruits of wild pear *P. pyraster* and *P.* ×*myloslavensis* was 3.6% and 5.4%, respectively. Liaudanskas et al. [24] reported that a significant effect on the course of extraction in known European pear cultivars is exerted by time, temperature, and the concentration of the solvent. These authors stated that when ethanol was applied at various concentrations (20–96%), the best extraction capacity was found for the 70% solution. In turn, Patricia and Syaputri [25], when studying contents of bioactive compounds in cultivars of European and Asian pears, used both 80% methanol and ethanol in the extraction process. In both varieties, a better process yield was obtained when ethanol was used instead of methanol. In a study by Guiné et al. [26] on fruits of 'Joaquina' *P. communis*, the mean extraction efficiency before the drying process was 17.23%, while this efficiency increased with a higher temperature (60 °C and 70 °C) and a longer duration of the extraction process.

Another factor affecting the amount of extracted material was the polarity of the solvent. Dai and Mumper [27] showed that the use of methanol as a solvent is more advantageous in the case of phenolic compounds with a lower molecular mass, while it was acetone in the case of polyphenols with a higher molecular mass. Methanol is a polar substance; thus, it effectively extracts the phenolic compounds found in the polar, soluble part. Apart from the polarity and the type of the solvent, the efficiency of extraction is also affected by the type of analyzed material.

Du et al. [28] showed that even when working on one pear cultivar, the results may differ depending on the part of the fruit collected for analysis. The results were compared in terms of the content of phenolic compounds in the endocarp, exocarp, and hypanthium. In all cases, the highest contents were recorded in the exocarp (121.22–153.44 mg GAE/100 g), whereas the lowest were recorded in the hypanthium (56.78–61.22 mg GAE/100 g). Öztürk et al. [29], in their studies, also analyzed different fruit parts of the same pear cultivar. Their investigations showed that a greater proportion of phenolic compounds is found in the skin rather than in the flesh itself.

In this study, the extract from fruits of the wild pear *P. pyraster* was characterized by a 3-fold greater content of total phenolic compounds compared with that of *P.* ×*myloslavensis* fruits (197 mg and 61.7 mg GAE/g d.m. extract, respectively) (Table 2). Kopera and Mitek [30], using 80% methanol extracts from fruits of three cultivars of the Asian pear *P. pyrifolia* ('Shinseiki', 'Hosui', and 'Chojuro'), showed the total content of phenolic compounds to be 102–141 mg GAE/g of d.m. extract, with a process efficiency of 11.9–13.0%. In turn, Wang et al. [31], reported a total content of polyphenols of 1.89–3.14 mg GAE/g d.m. extract in 70% ethanol extracts from the fruits of five *P. communis* cultivars ('Beurre Bosc', 'Josephinede Malines', 'Rico', 'Packham's Triumph', and 'Winter Nelis'). It was stated that the total phenolic content in extracts from fruits of *P. pyraster* and *P.* ×*myloslavensis* also exceeds that of 'Ya Li' *P. bretschneideri* (57.5 mg GAE/g d.m. extract), 'Blanquilla' *P. communis* (13 mg GAE/g d.m. extract), and 'Conference' *P. communis* (2.5 mg GAE/g d.m. extract [32]).

Table 2. Total content of phenolic compounds and the antioxidative activity of extracts from fruits of *Pyrus pyraster* and *Pyrus* ×*myloslavensis*.

Extract from Pear Fruits	TPC * (mg GAE/g d.m.)	DPPH * (mg Trolox/g d.m.)	ABTS * (mg Trolox/g d.m.)	FRAP * (mmol Fe^{2+}/L)
P. pyraster	197.15	289.81	11.33	18.20
P. ×*myloslavensis*	61.71	134.39	0.69	0.60

* Student's *t*-test; normal distribution with dependent estimation of variance (independent *t*-test by groups; normally distributed); statistical differences significant at $p = 0.001$.

4.2. Determination of the Content of Phenolic Compounds and Organic Acids Using HPLC

In extracts from the fruits of *P. pyraster* and *P.* ×*myloslavensis*, a total of seven phenolic compounds (quercetin, catechin, epicatechin, ferrulic, chlorogenic, synapic, and vanillic acids) and four organic acids (citric, maleic, shikimic, and fumaric acids) were detected (Table 3). In the extracts, chlorogenic acid was found in the greatest amounts (0.1225 and 0.0762 mg/g d.m. extract from the fruits of *P. pyraster* and *P.* ×*myloslavensis*, respectively). This acid is a phenolic acid, classified in the group of hydroxycinnamic acids. In turn, among the organic acids, citric acid is found in the greatest amounts (0.0813 and 0.0440 mg/g d.m. extracts, respectively).

Carbonaro et al. [33] showed that among all phenolic compounds, chlorogenic acid is a compound commonly found in pear fruits. Kopera and Mitek [30] provided chemical characteristics of fruits from three of the *P. pirifolia* Asian pear cultivars ('Shinseiki', 'Hosui', and 'Chojuro'). Those authors detected no statistically significant differences between these cultivars in terms of their phenolic acid content. They showed that the fruits of all the cultivars, both in the flesh and the skin, contain p-hydroxycinnamic acid and catechins.

In the cultivars 'Shinseiki' and 'Hosui', the whole fruits also contained cinnamic and pyrocatechuic acid. In addition, var. 'Hosui' also contained *p*-coumaric acid.

Table 3. Contents of phenolic compounds and organic acids in extracts of *Pyrus pyraster* and *Pyrus* ×*myloslavensis*.

	P. pyraster	*P.* ×*myloslavensis*
Phenolic compounds (mg/g d.m.)		
Chlorogenic acid	0.1225 **	0.0762 **
(−) Epicatechin	0.0235 *	0.0133 *
Quercetin	0.0193 ***	0.0086 ***
Synapic acid	0.0182 **	0.0076 **
Vanillic acid	0.0172 ***	0.0016 ***
(+) Catechin	0.0100 *	0.0075 *
Ferrulic acid	0.0090 **	0.0018 **
Organic acids (mg/g d.m.)		
Citric acid	0.0813 ***	0.0440 ***
Maleic acid	0.0127 **	0.0067 **
Fumaric acid	0.0023 ***	0.0013 ***
Shikimic acid	0.0016 *	0.0008 *

* Student's *t*-test; normal distribution with dependent estimation of variance (independent *t*-test by groups; normally distributed); $p = 0.001$; ** Normal distribution; Student's *t*-test with independent estimation of variance (independent *t*-test by groups, with separate variance estimates); $p = 0.001$; *** Non-parametric statistic; Mann–Whitney test; $p = 0.05$.

4.3. DPPH Radical Scavenging Capacity

A greater DPPH radical scavenging capacity was recorded for the extract from fruits of *P. pyraster* (289.8 mg Trolox/g d.m. extract) compared with that of the extract from *P.* ×*myloslavensis* fruits (134.4 mg/g d.m. extract) (Table 2). Statistical analysis showed that the DPPH radical scavenging capacity in extracts from fruits of *P. pyraster* and *P.* ×*myloslavensis* is strongly positively correlated with their total phenolic content ($p < 0.05$; r = 0.99; Figure 1). Wang et al. [31] showed that DPPH radical scavenging capacity of ethanol extracts from fruits of five *P. communis* cultivars ('Beurre Bosc', 'Josephinede Malines', 'Packham's Triumph', 'Winter Nelis', and 'Rico') ranged from 5.72 to 3.75 mg ascorbic acid/g d.m. extract. 'Beurre Bosc' of the *P. communis* cultivar showed the greatest DPPH radical scavenging capacity, while this was lowest in *P. communis* var. 'Winter Nelis'. Ieguchi et al. [34] determined the DPPH radical scavenging capacity in 29 pear cultivars grown in Japan and Europe; for many cultivars, the results were comparable to those reported by Wang et al. [31]. The potential of the antioxidative capacity is influenced not only by the cultivar but also by the growing conditions, the degree of ripeness of the fruit, and even the conditions of its storage [35]. In turn, Manzoor et al. [36] determined the content of phenolic compounds and the DPPH radical scavenging capacity in the pulp and skin of pear fruits. They showed that both the content of polyphenols and the DPPH scavenging capacity are higher in the skin (49.71%) than in the pulp (27.89%). A strong dependence between total content of phenolic compounds and the DPPH radical scavenging capacity was also confirmed in the study by Li et al. [8]. Choi et al. [37] investigated five different ethanol–aqueous extracts of pear fruits and found that their DPPH radical scavenging capacity ranged from 17.4 to 30.3%. These results were lower compared with the activity of *P. pyraster* and *P.* ×*myloslavensis* when methanol was the extractant.

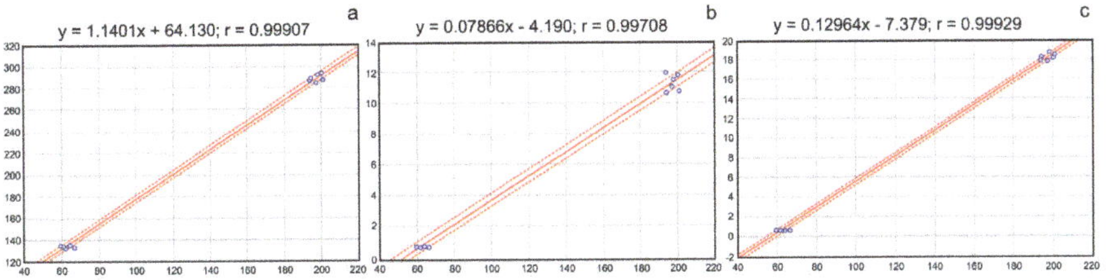

Figure 1. Correlation between the total phenolic content of *P. pyraster* and *P.* ×*myloslavensis* fruit extracts (mg GAE/g d.m.) (X-axis) and (**a**) DPPH, (**b**) the ABTS radical scavenging activity (mg Trolox/g s.m.), and (**c**) the FRAP reducing potential (mM Fe^{2+}) (Y-axis).

4.4. ABTS Cation Radical Scavenging Capacity

The extract from *P. pyraster* fruits showed an ABTS cation radical scavenging capacity of 11.33 mg Trolox/g d.m. of extract, while in the case of *P.* ×*myloslavensis*, it was 0.69 mg Trolox/g d.m. of extract (Table 2). The positive value of Pearson's linear correlation coefficient ($p < 0.05$; r = 0.99; Figure 1) shows that there is a relationship between the ABTS cation radical scavenging capacity and the content of phenolic compounds in extracts from *P.* ×*myloslavensis* and *P. pyraster*. Also, Batista et al. [38] found a strong correlation (r = 0.99) between the antioxidative capacity of dried fruits of *P. communis* var. 'Bartolomeu' and 'Amendoa' and the total content of phenolic compounds. Erbil et al. [39] recorded the ABTS cation radical scavenging capacity in fruits of five Turkish cultivars of *P. communis* ('Egirsah', 'Gugum', 'Deveci', 'Kizil' and 'Banda') and found the highest activity in var. 'Kizil'. In turn, Hu et al. [40] tested fruits of four Australian cultivars of sweet cherry, *Prunus avium* L. ('Bing', 'Ron's', 'Merchant' and 'Lapins'), and found their ABTS radical scavenging capacity to range from 0.51 to 0.37 mg ascorbic acid/g d.m. of extract. The highest ABTS scavenging capacity was recorded in var. 'Merchant', while it was lowest in var. 'Lapins'. Nowak et al. [41] investigated the ABTS cation radical scavenging capacity in the fruit, skin, and leaves of plums extracted using various concentrations of both ethanol and methanol. Their analyses showed that the use of methanol as a solvent was the most effective. They also stated that in the fruit itself, the cation radical scavenging capacity was lower than in the other parts of the plant tested.

4.5. FRAP Reducing Potential

The extract from the fruits of *P. pyraster* showed reducing potential at 18.2 mM Fe(II). The activity of *P.* ×*myloslavensis* fruits was much lower, a value of 0.61 mM Fe(II) (Table 2). Based on the regression and correlation ($p < 0.05$; r = 0.99) it was stated that the capacity to reduce Fe(III) ions to Fe(II) ions by antioxidants contained in extracts from the fruits of *P. pyraster* and *P.* ×*myloslavensis* is very strongly positively correlated with the total phenolic content. Wang et al. [31] determined the reducing potential using the FRAP test for fruits of five *P. communis* cultivars ('Beurre Bosc', 'Josephinede Malines', 'Packham's Triumph', 'Winter Nelis', and 'Rico') and showed a significantly different ($p < 0.05$) capacity for reducing Fe(III) ions to Fe(II) ions in the extracts tested. The reducing potential of the extracts ranged from 2.15 to 4.37 mg ascorbic acid/g d.m. of extract. The highest capacity was found for var. 'Josephinede Maline', while it was lowest for 'Winter Nelis'. Jamuna et al. [42] recorded comparable antioxidative activity using the FRAP test for fruits of *P. communis*, with a reduction potential amounting to 3 mg of ascorbic acid/g d.m. of extract. Bahorun et al. [43], when analyzing lettuce, broccoli, carrot, cauliflower, onion, white cabbage, napa cabbage, chilli peppers, and tomato, showed a strong correlation (r = 0.83) between the investigated factors. Additionally, Dudonne et al. [44] reported an extraordinarily strong correlation between the content of phenols and the antioxidative capacity in the FRAP test (r = 0.906). In the same study conducted on chamomile (*Matricaria recutita* L.; 0.12 mM Fe (II)) and musk mallow (*Abelmoschus moschatus* Medik.; 0.08 mM Fe (II)), the results were lower compared with those obtained for extracts from the fruits of *P. pyraster* and *P.* ×*myloslavensis*. Slightly different data was reported by Sagbas et al. [45], who showed a moderate correlation between total phenolic content and the antioxidative capacity in the FRAP test for the pear *P. elaeagrifolia* Pall. (r = 0.489). In the analysis presented by Gouws et al. [46], a moderate correlation was also obtained between phenolic compounds and the capacity to reduce Fe(III) ions (r = 0.519) in fruits of the Indian fig opuntia.

4.6. Antioxidative Activity in Linoleic Acid Emulsion

The capacity of extracts from fruits of *P. pyraster* and *P.* ×*myloslavensis* to inhibit autoxidation of emulsified linoleic acid was tested at 0.25 and 0.5% addition levels (Table 4). After 19 h of thermal incubation of samples with various levels of pear fruit extracts added, the increase in peroxides with conjugated dienes was by 13.5% up to 80% lower compared with the control. A greater antioxidative activity has been observed for the extract from *P. pyraster* fruits than for of *P.* ×*myloslavensis*. The increased content of peroxides with conjugated dienes in the emulsion system after thermal incubation is very strongly negatively correlated with the concentration of the extracts ($p < 0.05$; r = −0.94 and $p < 0.05$; r = −0.84 for *P. pyraster* and *P.* ×*myloslavensis*, respectively). Along with the increase in concentration, a lower increment was recorded in the content of peroxides with conjugated dienes.

Table 4. The effect of extracts from fruits of *Pyrus pyraster* and *Pyrus* ×*myloslavensis* on changes in the content of conjugated dienes in a linoleic acid emulsion (10^{-5} M).

Extract from Pear Fruits	Content of Conjugated Dienes (10^{-5} M)	
	After Emulsion Preparation	After 19 h Emulsion Incubation
P. pyraster 0.25%	1.11	1.73 *
P. pyraster 0.5%	1.02	1.65 ***
P. ×*myloslavensis* 0.25%	0.85	4.53 *
P. ×*myloslavensis* 0.5%	0.86	2.40 ***
Control	1.00	5.59

* Student's *t*-test; normal distribution with dependent estimation of variance (independent *t*-test by groups, normally distributed); $p = 0.001$; *** Non-parametric statistics; Mann–Whitney test; $p = 0.05$.

Hęś et al. [47] determined antioxidative activity of the aloe vera extract in the linoleic acid emulsion. Their analyses showed that the higher the concentration of Aloe vera extract added to the emulsion system, the lower the content of conjugated dienes after thermal incubation. Thus, an addition of the extract at 0.1–0.5 mg/mL to the emulsion system resulted in a 90% lower increase in conjugated dienes after thermal incubation compared to the control. In turn, addition of the extract at 0.05 mg/mL and 0.06 mg/mL inhibited (by 40% and 57%, respectively) the increase in conjugated dienes in the emulsion system following thermal incubation. Consequently, the results of both the experiment reported by Hęś et al. [47] and this study conclude that the concentration of the extract added to the emulsion system influences changes in the content of conjugated dienes after thermal incubation. The activity of pear extracts can be affected by the methods used for extracting phenolic compounds, the selection of solvents, and the particular part of the fruit from which the sample is taken (e.g., pear skin or pulp). Nevertheless, there is a lack of extensive research in the current literature on the antioxidative potential of pears in model systems, especially in emulsified fat systems, oils, and solid fats. More studies are needed to gain a deeper understanding of the role pear extracts play in these contexts and to investigate their potential for preserving the quality and extending the shelf life of fat-rich food products.

4.7. Dietary Fiber Content

Extracts from the fruits of *P. pyraster* have been characterized by higher contents of total fiber (36.45 g/100 g sample) and its insoluble fraction (32.49 g/100 g d.m. sample) compared with extracts from the fruits of *P.* ×*myloslavensis* (24.74 and 20.86 g/100 g d.m. sample, respectively). On the contrary, pears contain comparable amounts of the soluble fraction (approximately 3.70 g d.m. sample) (Table 5). It is worth noting that the insoluble dietary fiber fraction contains lignin, cellulose, and partially hemicellulose. Research so far has shown that the lignin fraction contains phenolic compounds, which are characterized by high antioxidative activity, as demonstrated through antioxidative activity analyses [48]. In our research, we observed the same trend—the antioxidative activity and the insoluble dietary fiber (IDF) fraction are higher in the *P. pyraster* sample.

Table 5. Dietary fiber content in the fruits of *Pyrus pyraster* and *Pyrus ×mylosloslavensis* (g/100 g lyophilized pear sample).

	P. pyraster	*P. ×myloslavensis*
TDF	36.45 ***	24.74 ***
IDF	32.49 ***	20.86 ***
SDF	3.75 *	3.67 *

TDF—total dietary fiber; IDF—insoluble dietary fiber; SDF—soluble dietary fiber. * Student's *t*-test; normal distribution with dependent estimation of variance (*t*-test independent by groups; normally distributed); $p = 0.05$; *** Non-parametric statistics; Mann–Whitney test; $p > 0.05$.

Witkowska et al. [49] determined the total fiber content in Polish (sweet cherries, pears, apples, plums, strawberries, and black currants) and imported fruits (bananas, peaches, oranges, lemons, kiwi, and tangerines). They stated that in Polish fruits, the total content of dietary fiber in sweet cherries and red currants amounted to 1.14 g/100 g and 4.55 g/100 g. In turn, in imported fruits, the total fiber content in tangerines was 1.36 g/100 g and in lemons 2.89 g/100 g, respectively. Therefore, these authors reported a significantly lower total fiber content in the tested fruits compared with that in the lyophilizate produced from the fruits of *P. pyraster* and *P. ×myloslavensis*. The ratio of soluble fiber to total fiber in tested samples was 0.10:1 and 0.15:1, respectively, which is lower compared with the 0.39:1 ratio obtained from pear pomace [50,51]. Some tropical fruits show higher fiber contents compared with fruits grown in a temperate climate [52]. Moreno-Hernández et al. [53] analyzed fiber fractions in soursop *Annona muricata* L. from Mexico. The SDF level was greater (9.31 g/100 g sample) than in the extracts of the fruits of the investigated pears (Table 5). However, the total fiber content turned out to be greater in the tested pear fruits, as in soursop, it was 19.82 g/100 g of sample.

5. Conclusions

Results of research conducted worldwide confirm the health-promoting effect of consuming products rich in polyphenols and their extensive effect on the sensory attributes of food. It has been shown that materials rich in bioactive compounds exhibiting health-promoting properties also include fruits of *P. pyraster* and *P. ×myloslavensis*. Extracts from fruits of these pear species showed both high antioxidative potential and considerable dietary fiber content; thus, they may find applications in the food, pharmaceutical, and cosmetics industries. In the form of concentrate, gel, juice, or powder, they may be used as additives to numerous foodstuffs, enhancing their nutritive and health-promoting value while also improving their sensory attractiveness. Extracts from pear fruits may be added to jellies, jams, water, tea, juices, sweets, vitamin-enriched fruit smoothies, non-alcoholic beverages, isotonic drinks (with electrolytes), nutritional drinks and yogurts, as well as cream or cottage cheeses and ice-creams. An added aspect confirming the benefits resulting from the use of pear fruits in the food industry relates to their confirmed antioxidative activity in a linoleic acid emulsion. This property may be used to extend the shelf life of foodstuffs containing emulsified fat. The high content of active compounds in pear extracts makes them effective and safe food additives. However, the available literature is still lacking studies concerning the antioxidative potential of pear fruits in model systems, particularly in emulsified fat systems.

For this reason, exploring the most suitable applications and evaluating their effectiveness in food products with varying levels of processing is crucial. This could lead to the productive use of pears rich in bioactive compounds. The results obtained in this document clearly demonstrate the potential of pear extract to inhibit lipid autoxidation in food products, which justifies the need for further research. Now our attention is focused on the choice of antioxidants best suited for that purpose.

Author Contributions: Conceptualization, W.A. and M.H.; methodology, M.H.; validation, K.S.-S. and K.D.; formal analysis, K.D. and K.S.-S.; investigation, M.H., M.J. and P.R.; resources, W.A. and M.H.; data curation, M.J., P.R., K.D. and K.S.-S.; writing—original draft preparation, W.A. and M.H.; writing—review and editing, M.H. and W.A.; visualization, K.D.; supervision, M.H. and W.A.; project administration, M.H.; funding acquisition, M.H. and W.A. All authors have read and agreed to the published version of the manuscript.

Funding: This publication was financed by the Polish Minister of Science and Higher Education as part of the Strategy of the Poznan University of Life Sciences for 2024–2026 within the field of improving scientific research and development work in priority research areas.

Institutional Review Board Statement: Not applicable.

Data Availability Statement: The datasets used and/or analyzed during the current study are available from the first author upon reasonable request.

Conflicts of Interest: The authors declare no conflicts of interest.

References

1. Medina-Vera, I.; Gómez-de-Regil, L.; Gutiérrez-Solis, A.L.; Lugo, R.; Guevara-Cruz, M.; Pedraza-Chaverri, J.; Avila-Nava, A. Dietary strategies by foods with antioxidant effect on nutritional management of dyslipidemias: A systematic review. *Antioxidants* **2021**, *10*, 225. [CrossRef] [PubMed]
2. Arias, A.; Feijoo, G.; Moreira, M.T. Exploring the potential of antioxidants from fruits and vegetables and strategies for their recovery. *Innov. Food Sci. Emerg. Technol.* **2022**, *77*, 102974. [CrossRef]
3. Davinelli, S.; Scapagnini, G. The Pharma-Nutritional Role of Antioxidant Phytochemicals in Health and Disease. *Antioxidants* **2022**, *11*, 1081. [CrossRef]
4. Meitha, K.; Pramesti, Y.; Suhandono, S. Reactive oxygen species and antioxidants in postharvest vegetables and fruits. *Int. J. Food Sci.* **2020**, *2020*, 8817778. [CrossRef]
5. Heydari, A.; Parvini, F.; Fard, N.A. Functional foods and antioxidant effects: Emphasizing the role of probiotics. In *Current Topics in Functional Food*; Shiomi, N., Savitskaya, A., Eds.; IntechOpen: London, UK, 2022; p. 280. [CrossRef]
6. Hong, S.Y.; Lansky, E.; Kang, S.S.; Yang, M. A review of pears (*Pyrus* spp.), ancient functional food for modern times. *BMC Complement. Med. Ther.* **2021**, *21*, 219. [CrossRef]
7. Li, X.; Wang, T.; Zhou, B.; Gao, W.; Cao, J.; Huang, L. Chemical composition and antioxidant and anti-inflammatory potential of peels and flesh from 10 different pear varieties (Pyrus spp.). *Food Chem.* **2014**, *152*, 531–538. [CrossRef]
8. Li, X.; Zhang, J.Y.; Gao, W.Y.; Wang, Y.; Wang, H.Y.; Cao, J.G.; Huang, L.Q. Chemical composition and anti-inflammatory and antioxidant activities of eight pear cultivars. *J. Agric. Food Chem.* **2012**, *60*, 8738–8744. [CrossRef]
9. Beliveau, R.; Gingras, D. Dieta w walce z rakiem. Profilaktyka i wspomaganie terapii przez odżywianie [Diet in the fight against cancer. In *Prevention and Support of Therapy through Nutrition*; Wydawnicza Delta WZ: Warsaw, Poland, 2010; pp. 140–143. (In Polish)
10. Kopera, M.; Mitek, M.; Świeca, E. Zawartość substancji pektynowych w owocach trzech odmian gruszy azjatyckiej (*Pyrus pyrifolia*). [The content of pectin substances in fruit of three asian pear cultivars (*Pyrus pyrifolia*)]. *Żywn. Nauka Technol. Jakość/Food. Sci. Technol. Quality. Suplement* **2004**, *3*, 133. (In Polish)
11. San-Martín-Hernández, C.; Martínez-Téllez, M.A.; Valenzuela-Amavizca, O.N.; Aispuro-Hernández, E.; Sánchez-Sánchez, M.; Hernández-Camarillo, E.; López-Martínez, L.X.; Eber Addí Quintana-Obregón, E.A. *Byrsonima crassifolia* L. Kunth a bio-resource with potential: Overview and opportunities. *Folia Hort.* **2023**, *35*, 61–75. [CrossRef]
12. Cortiella, M.G.; Castro, R.I.; Parra-Palma, C.; Méndez-Yáñez, A.; Ramos, P.; Morales-Quintana, L. Analysis of the contents of *Ugni molinae* Turcz fruits across the ripening stages. *Folia Hort.* **2024**, *36*, 119–134. [CrossRef]
13. Antkowiak, W.; Cedro, A.; Prajs, B.; Wolko, Ł.; Michalak, M. Success of wild pear *Pyrus pyraster* (L.) Burgsd. in colonization of steep sunny slopes: An interdisciplinary study in the Bielinek Reserve (NW Poland). *Pol. J. Ecol.* **2012**, *60*, 57–76.
14. Antkowiak, W.; Czarna, A.; Wawrzyńska, M. *Pyrus* ×*myloslavensis* (*P. communis* L. × *P. salicifolia* Pall.)-A new spontaneons pear hybrid. *Dendrobiology* **2008**, *60*, 45–49.
15. Antkowiak, W.; Maciejewska-Rutkowska, I.; Jędrzejczyk, I.; Wojciechowski, A. Morphological, anatomical and cytological characteristics of spontaneous hybrid *Pyrus* ×*myloslavensis*. *Dendrobiology* **2016**, *75*, 23–30. [CrossRef]
16. Lingnert, H.; Vallentin, K.; Eriksson, C.E. Measurement of antioxidative effect in model system. *J. Food Process. Preserv.* **1979**, *3*, 87–103. [CrossRef]

17. Horwitz, W. *Official Methods of Analysis of the Association of Official Analytical Chemists*; Association of Official Analytical Chemists: Washington, DC, USA, 1970; pp. 15.049–15.055.
18. Telichowska, A.; Kobus-Cisowska, J.; Szulc, P.; Liga, j.M.; Stuper-Szablewska, K.; Szwajgier, D.; Bujak, H. Comparative analysis of infusions with the addtion *P. padus* bark: Assessment of the antioxidant potential and their inhibitory effect on enzymes associated with oxidative stress. *Sustainability* **2021**, *13*, 3913. [CrossRef]
19. Przybylska-Balcerek, A.; Szablewski, T.; Szwajkowska-Michałek, L.; Świerk, D.; Cegielska-Radziejewska, R.; Krejpcio, Z.; Suchowilska, E.; Tomczyk, Ł.; Stuper-Szablewska, K. *Sambucus nigra* extracts-natural antioxidants and antimicrobial compounds. *Molecules* **2021**, *26*, 2910. [CrossRef]
20. Helrich, K. *Official Methods of Analysis of the Association of Official Analytical Chemists*, 15th ed.; Association of Official Analytical Chemists United States: Arlington, VA, USA, 1990; ISBN 0-935584-42-0.
21. Sanchez-Moreno, C.; Larrauri, J.A.; Saura-Calixto, F. A procedure to measure the antiradical efficiency of polyphenols. *J. Sci. Food Agric.* **1998**, *76*, 270–276. [CrossRef]
22. Re, R.; Pellegrini, N.; Proteggente, A.; Pannala, A.; Yang, M.; Rice-Evans, C. Antioxidant activity applying an improved ABTS radical cation decolorization assay. *Free Radic. Biol. Med.* **1999**, *26*, 1231–1237. [CrossRef]
23. Fernandes, R.P.P.; Trindade, M.A.; Tonin, F.G. Evaluation of antioxidant capacity of 13 plant extracts by three different methods: Luster analyses applied for selection of the natural extracts with high er to replace synthetic antioxidant in lambburgers. *J. Food Sci. Technol.* **2016**, *53*, 451–460. [CrossRef]
24. Liaudanskas, M.; Zymonė, K.; Viškelis, J.; Klevinskas, A.; Janulis, V. Determination of the phenolic composition and antioxidant activity of pear extracts. *J. Chem.* **2017**, *2017*, 7856521. [CrossRef]
25. Patricia, V.M.; Syaputri, F.N. Antioxidant activities from two varieties of pear peel extracts using DPPH and CUPRAC methods. *J. Phys. Conf. Ser.* **2021**, *1764*, 012013. [CrossRef]
26. Guiné, R.P.; Barroca, M.J.; Gonçalves, F.J.; Alves, M.; Oliveira, S.; Correia, P.M. Effect of drying on total phenolic compounds, antioxidant activity, and kinetics decay in pears. *Int. J. Fruit Sci.* **2015**, *15*, 173–186. [CrossRef]
27. Dai, J.; Mumper, R.J. Plant phenolics: Extraction, analysis and their antioxidant and anticancer properties. *Molecules* **2010**, *15*, 7313–7352. [CrossRef] [PubMed]
28. Du, B.; Cheng, C.; Chen, Y.; Wu, J.; Zhu, F.; Yang, Y.; Peng, F. Phenolic profiles and antioxidant activities of exocarp, endocarp, and hypanthium of three pear cultivars grown in China. *J. Food Bioact.* **2021**, *14*, 75–80. [CrossRef]
29. Öztürk, A.; Demirsoy, L.; Demirsoy, H.; Asan, A.; Gül, O. Phenolic compounds and chemical characteristics of pears (*Pyrus communis* L.). *Int. J. Food Prop.* **2015**, *18*, 536–546. [CrossRef]
30. Kopera, M.; Mitek, M. Charakterystyka chemiczna owoców wybranych odmian gruszy azjatyckiej (*Pyrus pyrifolia*). [Chemical characteristics of some selected cultivars of asian pear (*Pyrus pyrifolia*)]. *Żywn. Nauka Technol. Jakość/Food. Sci. Technol. Quality. Suplement* **2003**, *10*, 68–75. (In Polish)
31. Wang, Z.; Barrow, C.J.; Dunshea, F.R.; Suleria, H.A. A comparative investigation on phenolic composition, characterization and antioxidant potentials of five different australian grown pear varieties. *Antioxidants* **2021**, *10*, 151. [CrossRef] [PubMed]
32. Djendoubi, N.; Boudhrioua, N.M.; Kechaou, N.; Courtois, F.; Bonazzi, C. Influence of air drying temperature on kinetics, physicochemical properties, total phenolic content and ascorbic acid of pears. *Food Bioprod. Process.* **2012**, *90*, 433–441. [CrossRef]
33. Carbonaro, M.; Mattera, M.; Nicoli, S.; Bergamo, P.; Cappelloni, M. Modulation of antioxidant compounds in organic vs conventional fruit (peach, *Prunus persica* L., and pear, *Pyrus communis* L.). *J. Agric. Food Chem.* **2002**, *50*, 5458–5462. [CrossRef]
34. Ieguchi, T.; Katayama, H.; Nomura, K.; Takaoka, M.; Uematsu, C. Pear (*Pyrus* L.) genetic resources from Northern Japan: Evaluation of antioxidant capacity. *Acta Hortic.* **2015**, *1094*, 539–548. [CrossRef]
35. Łysiak, G.P.; Rutkowski, K.; Walkowiak-Tomczak, D. Effect of storage conditions on storability and antioxidant potential of pears cv. Conference. *Agriculture* **2021**, *11*, 545. [CrossRef]
36. Manzoor, M.; Anwar, F.; Bhatti, I.A.; Jamil, A. Variation of phenolics and antioxidant activity between peel and pulp parts of pear (*Pyrus communis* L.) fruit. *Pak. J. Bot.* **2013**, *45*, 1521–1525.
37. Choi, J.J.; Lim, S.H.; Choi, J.H.; Park, J.H.; Nam, S.H.; Lee, H.C. Antioxidant activity of *Pyrus pyrifolia* fruit in different cultivars and parts. *Korean J. Food Preser.* **2013**, *20*, 222–226. [CrossRef]
38. Batista, S.; Guiné, R.; Barroca, M.J.; Gonçalves, F.; González-SanJosé, M.L.; Rivero-Pérez, M.D. Evaluación de la composición fenólica y de la actividad antioxidante de peras frescas y secadas al sol (*Pyrus communis* L.). In Proceedings of the III Congreso Nacional de Ciência y Tecnología de los Alimentos, Burgos, Spain, 24–26 May 2005; pp. 947–950.
39. Erbil, N.; Murathan, Z.T.; Arslan, M.; Ilcim, A.; Sayin, B. Antimicrobial, antioxidant and antimutagenic activities of five Turkish pear cultivars. *Erwerbs. Obstbau.* **2018**, *60*, 203–209. [CrossRef]
40. Hu, T.; Subbiah, V.; Wu, H.; BK, A.; Rauf, A.; Alhumjdhi, F.A.; Suleria, H. Determination and characterization of phenolic compound from Australia-Grown sweet cherries (*Prunus avium* L.) and their potential antioxidant properties. *Am. Chem. Soc.* **2021**, *6*, 34687–34699. [CrossRef]
41. Nowak, A.; Szatan, D.; Zielonka-Brzezicka, J.; Florkowska, K.; Muzykiewicz, A.; Klimowicz, A. Antioxidant activity of selected parts of *Prunus domestica* L. harvested at two ripening stages. *Pomeranian J. Life Sci.* **2020**, *66*, 65–69. [CrossRef]
42. Jamuna, K.; Ramesh, C.; Srinivasa, T.; Raghu, R. In vitro antioxidant studies in some common fruit. *Int. J. Pharm. Pharm. Sci.* **2011**, *3*, 60–63.

43. Bahorun, T.; Luximon-Ramma, A.; Crozier, A.; Aruoma, O.I. Total phenol, flavonoid, proanthocyanidin and vitamin C levels and antioxidant activities of Mauritian vegetables. *J. Sci. Food Agric.* **2004**, *84*, 1553–1561. [CrossRef]
44. Dudonne, S.; Vitrac, X.; Coutiere, P.; Woillez, M.; Mérillon, J.M. Comparative study of antioxidant properties and total phenolic content of 30 plant extracts of industrial interest using DPPH, ABTS, FRAP, SOD, and ORAC assays. *J. Agric. Food Chem.* **2009**, *57*, 1768–1774. [CrossRef] [PubMed]
45. Sagbas, H.I.; Ilhan, G.; Ercisli, S.; Anjum, M.A.; Holubec, V. Characterization of oleaster-leafed pear (*Pyrus elaeagrifolia* Pall. subsp. *elaeagrifolia*) fruits in Turkey. *Agronomy* **2021**, *11*, 430. [CrossRef]
46. Gouws, C.A.; Georgousopoulou, E.; Mellor, D.D.; Naumovski, N. The effect of juicing methods on the phytochemical and antioxidant characteristics of the purple prickly pear (*Opuntia ficus-indica*)—Preliminary findings on juice and pomace. *Beverages* **2019**, *5*, 28. [CrossRef]
47. Hęś, M.; Dziedzic, K.; Le Thanh-Blicharz, J.; Kmiecik, D.; Górecka, D. Antioxidant activity of true aloe (*Aloe vera*) extract in model systems. *Nauka Przyr. Technol.* **2016**, *10*, 53. [CrossRef]
48. Dziedzic, K.; Górecka, D.; Szwengiel, A.; Sulewska, H.; Kreft, I.; Gujska, E.; Walkowiak, J. The Content of Dietary Fiber and Polyphenols in Morphological Parts of Buckwheat (*Fagopyrum tataricum*). *Plant Foods Hum. Nutr.* **2018**, *73*, 82–88. [CrossRef]
49. Witkowska, A.; Borawska, M.; Lozowska, A.; Siwicka, J. Zawartość błonnika pokarmowego całkowitego w wybranych owocach [Total dietary fiber content in selected fruits]. *Brom. Chem. Toks.* **1998**, *31*, 115–118. (In Polish)
50. Garcia-Amezquita, L.E.; Tejada-Ortigoza, V.; Serna-Saldivar, S.O.; Welti-Chanes, J. Dietary fiber concentrates from fruit and vegetable by-products: Processing, modification, and application as functional ingredients. *Food Bioproc. Tech.* **2018**, *11*, 1439–1463. [CrossRef]
51. Grigelmo-Miguel, N.; Martín-Belloso, O. Comparison of dietary fibre from by-products of processing fruits and greens and from cereals. *LWT—Food Sci. Technol.* **1999**, *32*, 503–508. [CrossRef]
52. Pereira-Netto, A.B. Tropical fruits as natural, exceptionally rich, sources of bioactive compounds. *Int. J. Fruit Sci.* **2018**, *18*, 231–242. [CrossRef]
53. Moreno-Hernández, C.L.; Sáyago-Ayerdi, S.G.; García-Galindo, H.S.; Mata-Montes De Oca, M.; Montalvo-González, E. Effect of the application of 1-methylcyclopropene and wax emulsions on proximate analysis and some antioxidants of soursop (*Annona muricata* L.). *Sci. World J.* **2014**, *2014*, 896853. [CrossRef] [PubMed]

agriculture

MDPI

Article

Effect of Short-Term Storage in Modified Atmosphere Packaging (MAP) and Controlled Atmosphere (CA) on Total Polyphenol Content and Antioxidant Activity in Juices from Haskap Berry (*Lonicera caerulea* L.)

Barbara Anna Kowalczyk [1,*], Monika Bieniasz [2] and Jan Błaszczyk [2]

[1] Department of Ornamental Plants and Garden Art, Faculty of Biotechnology and Horticulture, Agricultural University in Krakow, Al. 29-Listopada 54, 31-425 Cracow, Poland

[2] Department of Horticulture, Faculty of Biotechnology and Horticulture, Agricultural University in Krakow, Al. 29-Listopada 54, 31-425 Krakow, Poland; monika.bieniasz@urk.edu.pl (M.B.); jan.blaszczyk@urk.edu.pl (J.B.)

* Correspondence: barbara.kowalczyk@urk.edu.pl

Abstract: This article focuses on analysing the properties of six varieties of haskap berry (honeyberry) as a valuable raw material for producing health-promoting juices. Significant differences in the content of bioactive compounds were observed between juices derived from fruits of the same species. This study demonstrated that controlled atmosphere (CA) conditions (20% CO_2 and 5% O_2) and modified atmosphere packaging (MAP) in Xtend bags affected juice quality by minimising nutritional losses. The analysis of polyphenol content in the juices revealed significant differences between varieties and years (2021 and 2022), primarily due to varying weather conditions. In 2022, the polyphenol content of the varieties 'Usłada', 'Candy Blue', 'Boreal Beauty', and 'Boreal Beast' was from 69% to twice as high compared to values recorded in 2021. CA and MAP storage conditions were found to be more effective than normal atmosphere (NA) in preserving bioactive components, and thus the antioxidant activity of the fruits, as measured by the DPPH method. The variety 'Sinij Utes' had the highest total polyphenol contents and their lowest loss during storage. Conversely, the variety 'Boreal Beauty' contained the lowest polyphenol levels both after harvest and storage. This study confirmed the importance of proper storage conditions for maintaining the antioxidant properties of haskap berries.

Keywords: polyphenols; DPPH; storage conditions; controlled atmosphere; modified atmosphere; superfruits

1. Introduction

Fruit juices are a popular alternative for consumers seeking fresh fruit substitutes, appreciated for their freshness, taste, and nutritional value. Increased awareness and growing consumer expectations regarding modern fruit and vegetable processing technologies drive the development of the fresh juice industry. The contemporary food market relies heavily on innovative products that not only promote health but also reduce the risk of certain diseases. While consumers continue to value familiar and well-liked ingredients, they are increasingly attentive to the health benefits of products made from novel and less commonly used raw materials [1].

Haskap berry (*Lonicera caerulea* L.) is a species that has recently gained popularity in Poland. This perennial leafy shrub from the family Caprifoliaceae was originally cultivated in Russia, Japan, and China [2–8]. It is now also grown in many other countries, including Poland, the Czech Republic, Slovakia, Lithuania, Belarus, Slovenia, the United States [2,9,10], Switzerland [11], Ukraine [12], Estonia, and Romania [3,6]. Controlled cross-breeding and intensive selection in breeding centres have resulted in the development of new varieties that lack the bitterness originally present in the fruit. Advantages of this species include early fruiting (before strawberries), resistance to diseases and pests, frost tolerance, suitability for mechanical harvesting [13–15], and adaptability to a broader pH range of the substrate [13,16]. In Poland, the harvest period typically begins in June (sometimes May) and lasts, depending on the variety, until mid-August [17].

After harvesting, the berries are mainly used for juice, jam, or wine production [5,6,18–20]. Due to their content of vitamins, phenols, organic acids, and above all, high levels of bioactive compounds, haskap berries are gaining popularity and are considered a superfruit [2,6,13]. Phenolic compounds in haskap berries have been extensively studied, identified, and described, and their high concentration is considered one of the most important characteristics of this species. Berries are rich in anthocyanins, flavonoids, phenolic acids, and tannins, which collectively contribute to their health-promoting properties. Anthocyanins, such as cyanidin, delphinidin and malvidin, are powerful antioxidants that support healthy vision, exert anti-inflammatory effects, and protect the cardiovascular system. Flavonoids, including quercetin and catechins, provide antioxidant and anti-inflammatory benefits, helping to protect cells from oxidative stress. Phenolic acids, such as caffeic and coumaric acids, additionally enhance vascular endothelium function, positively affecting circulatory system condition. Tannins, on the other hand, show antibacterial properties and support the health of the digestive system, alleviating gastrointestinal complaints. The joint action of these phenolic compounds makes berries a valuable component of a health-promoting diet [7,13,15]. The levels of polyphenols primarily depend on factors such as variety, geographic location [21], climatic conditions, cultivation techniques [2,11,22], and harvest period [21]. Foods rich in polyphenols are known to prevent neurodegenerative diseases, age-related disorders, and other civilisational diseases, while also exerting antimutagenic and anticancer properties [5,8,9,11,19,23]. In Japan, on Hokkaido Island, haskap berry juice producers market their product as a 'golden remedy for eternal youth and longevity' [16]. Factors such as fruit origin, variety, climate conditions, maturity level, and post-harvest conditions significantly affect juice quality, including bioactive compound contents [10,24,25].

Haskap berries (*Lonicera caerulea* L.) deteriorate rapidly at room temperature [5,17], with a shelf life of approximately two days for consumption or processing [17]. After this period, damaged berries are often infected with the most common pathogen of soft fruits, grey mould (*Botrytis cinerea*) [26]. Harvesting on plantations can extend over a month, which is particularly challenging for smaller growers who do not use harvesters. Smaller batches are frequently frozen, which increases production costs [27,28]. Freezing causes cell rupture in the berries, triggering enzymatic reactions that may degrade anthocyanins and other phenolic compounds. This degradation is particularly pronounced during thawing due to interactions with oxidising enzymes [28].

Extending the limited storage period to at least 14 days would allow for processing of larger fruit batches. This can be achieved by slowing metabolic processes through more advanced storage methods of fruits and vegetables, such as controlled atmosphere (CA), which involves reducing the temperature to −1 to +2 °C at 95% relative humidity [3,29,30]. Another effective approach is the use of modified atmosphere packaging (MAP) in the form of Xtend bags [3,26,31,32]. The controlled atmosphere (CA) method reduces oxygen levels to approx. 5% and increases carbon dioxide levels to 20%. This adjustment lowers respiration rates in fruits and vegetables, slows ripening and aging processes, and extends their freshness. Similarly, the Xtend bag method, a form of modified atmosphere packaging (MAP) modifies the internal packaging atmosphere to desired gas proportions (e.g., 5–10% O_2, 15–20% CO_2, and 70–80% N_2), lowering metabolic activity, limiting pathogen growth, and preventing spoilage. This technique is commonly applied to seasonal fruits and vegetables such as berries, tomatoes, or lettuce. Both methods have proven effective in extending shelf life and preserving the nutritional quality of agricultural products, contributing to reduced food waste and improved food security [30,33,34].

Considering the growing scale of haskap berry cultivation in Poland and its increased availability for industry, this fruit is emerging as an attractive raw material to produce both single-ingredient and mixed juices. While previous research has primarily examined the effects of frozen storage on the health benefits of juices, the present study aimed to assess the impact of short-term storage under controlled and modified atmosphere conditions on the total polyphenol content and antioxidant activity in one of the final products, i.e., haskap berry juice. Determining health-promoting substances in dry or fresh matter does not always reflect the actual content of bioactive compounds in consumer-ready products.

2. Materials and Methods

2.1. Materials

The research material for the two-year experiment consisted of juices derived from six varieties of haskap berry (*Lonicera caerulea* L.): three Canadian varieties, 'Boreal Beast', 'Boreal Beauty', and 'Boreal Blizzard', and three Russian varieties, 'Candy Blue', 'Sinij Utes', and 'Usłada'. The Boreal series is a new line of haskap varieties developed through hybridisation of germplasm from Russia, Japan, and the Kuril Islands [35].

The berries were sourced from a commercial honeysuckle plantation located in southern Poland, in the Lesser Poland Voivodeship, approx. 20 km from Krakow (coordinates: altitude 360 m above sea level; latitude 50°14′06″ N; longitude 19°53′15″ E). The plants were spaced 1.2 m apart in rows and 4 m apart between rows on a gentle slope. Agricultural fabric was placed beneath the plants, and an irrigation system was installed with watering frequency adjusted to weather conditions. All agronomic practices during the experiment were carried out in accordance with the current guidelines for haskap berry cultivation.

2.1.1. Harvest Timing

The harvest period of haskap berries depended on the variety (Table 1). The Russian varieties, which ripened first, were harvested in the second week of June; the Canadian varieties 'Boreal Beauty' and 'Boreal Beast' were harvested in the third week of July; while the variety 'Boreal Blizzard' was harvested in the first week of August. All fruits were harvested at full ripeness, as they are non-climacteric.

Table 1. Flowering and fruiting dates of six haskap berry varieties.

		June				July				August			
	Week	**1**	**2**	**3**	**4**	**1**	**2**	**3**	**4**	**1**	**2**	**3**	**4**
'Usłada'													
'Sinij Utes'													
'Candy Blue'													
'Boreal Beauty'													
'Boreal Beast'													
'Boreal Blizzard'													

2.1.2. Weather Conditions

Temperatures remained stable during flowering, growth, and ripening. Precipitation varied, with higher rainfall recorded in May, June, and July of 2021—months critical for berry growth and ripening (Figure 1).

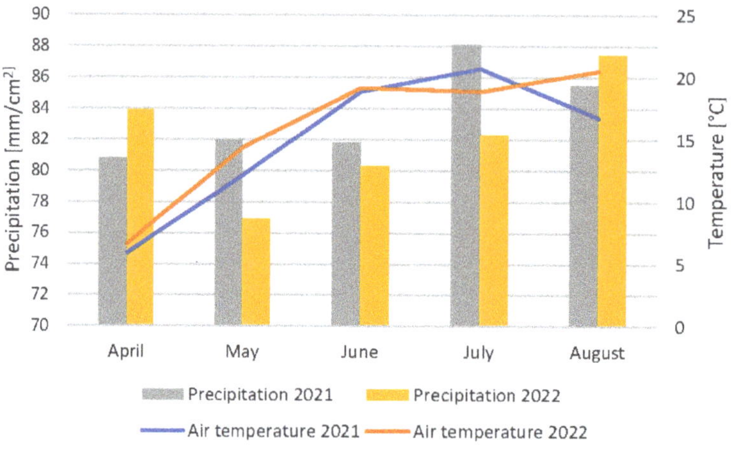

Figure 1. Weather patterns during flowering, growth, and ripening of berries in 2021 and 2022.

2.2. Storage Treatments

Berries of each variety (four replicates, approx. 300 g each) were stored for 7 and 14 days in separate containers under the following conditions:

1. Normal atmosphere (NA): Stored in a cold room (temperature 2 °C, relative humidity 90–92%).
2. Controlled atmosphere (CA): 20% CO_2 and 5% O_2 (temperature 2 °C, relative humidity 90–92%).
3. Modified atmosphere packaging (MAP): Xtend packaging (StePac LA Ltd., Johnson Matthey, Tefen, Israel/USA (Chicago, IL 60181)).

2.3. Juice Preparation

Directly after harvest, and on day 7 and 14 of storage, the bioactive compound contents and antioxidant activity of the berries were determined. For this purpose, a 100 g sample from each of the four replicates was pressed using a laboratory manual basket press. The resulting juice was analysed for total polyphenol contents using the Folin–Ciocalteu method and for their ability to scavenge DPPH radicals.

2.4. The Content of Bioactive Compounds and Antioxidant Activity

2.4.1. Total Polyphenols Content (TPC)

The quantification of phenolic compounds in the extracts was performed under laboratory conditions by reaction with Folin–Ciocalteu reagent. To this end, a juice sample (0.25 mL) was mixed with 0.25 mL of 25% Na_2CO_3; 0.125 mL of Folin–Ciocalteu reagent (Sigma-Aldrich, Darmstadt, Germany), diluted twice with water prior to the analysis; and 2.25 mL of water, which was incubated for 15 min. The absorbance was measured at 760 nm using a JASCO V-530 UV-Vis spectrophotometer (Tokyo, Japan). The final results were expressed as mg of gallic acid (Sigma-Aldrich, Darmstadt, Germany) per litre (gallic acid equivalents, GAE mg L^{-1}) [36,37].

2.4.2. Radical Scavenging Capacity (RSC)—A DPPH Assay

The radical scavenging capacity of the extracts was determined under laboratory conditions by measuring the reduction of the synthetic, stable free radical 2,2-diphenyl-1-picrylhydrazyl (DPPH). The colorimetric method measures the change in absorbance of a DPPH solution at 517 nm, reflecting the antioxidant activity of the sample. Briefly, 2.8 mL of 0.1 mM DPPH (Sigma-Aldrich) solution in 96% ethanol was mixed with 0.2 mL of juice sample. DPPH absorbance was measured after 10 min, with RSC results expressed in micromoles of Trolox (Sigma-Aldrich) equivalents per litre (TE μmol L^{-1}) [38].

2.5. Fruit Firmness

Fruit firmness (N) was measured using a TA 500 Lloyd Texture Analyzer equipped with a 6.35 mm diameter tip (AMETEK Test & Calibration Instruments; Fareham, Hampshire, UK).

2.6. Statistical Analysis

The results were processed using a one-way analysis of variance (ANOVA). Fisher's least significant difference (LSD) test was applied to determine the significance of differences between the means at a significance level of $\alpha = 0.05$. All the statistical calculations were performed using Statistica 13.3 software.

3. Results and Discussion

The use of modern storage methods, such as modified atmosphere packaging (MAP) and controlled atmosphere (CA), as well as the analysis of their effects on juice quality, provides valuable information that can help producers optimise storage processes and extend product freshness without compromising nutritional value. Additionally, such research can support the development of more sustainable and efficient fruit storage and processing methods, addressing global challenges associated with food security and sustainable development.

3.1. Polyphenol Content

The total polyphenol content of the juices was analysed using a spectrophotometer and the Folin–Ciocalteu method. In 2021, significant variations in polyphenol content were observed between the varieties tested. The highest concentrations were recorded for the variety 'Sinij Utes', while the lowest were found in the variety 'Boreal Beauty' (Table 2). Shevchuk et al. [12] conducted studies on the same varieties from the Borealis group, demonstrating that they were among the richest in polyphenols compared to other varieties studied. The latter authors conducted their analyses on raw mass rather than extracted juice, obtaining polyphenol content of 908 mg/L g for 'Boreal Blizzard', 849 mg/L for 'Boreal Beast', and 813 mg/L for 'Boreal Beauty'. Corresponding juice samples from their study showed significantly higher values, reaching 2340, 1463, and 1036 mg/L, respectively.

Wang et al. [39] have argued that the method of juice extraction from haskap berries significantly affects their characteristics. Similarly, Senica et al. [40] have found that the concentration of phenolic compounds and tannins changes based on a complex combination of temperature, light, and other factors affecting berry metabolism. Kalisz et al. [23] highlighted the role of post-harvest conditions and processing methods in shaping these differences. The latter authors found that polyphenol content in haskap berry juice ranged from 200.8 to 416.0 mg/L, likely due to pre-freezing the fruit and subsequent pasteurisation of the juice at 85 °C. Research by Khattab et al. [28] showed a reduction in biological activity after juice preservation by heating [41] and freezing [28], respectively. Comparable polyphenol levels were obtained by Szajdek et al. [42] in juices without maceration from chokeberries, bilberries, and blackcurrants, amounting to 1926, 1237, and 2624 mg/L, respectively.

Table 2. Total polyphenol content (TPC, mg/L GAE) and antioxidant activity (DPPH, % inhibition) of post-harvest juice by variety in 2021.

	Total Polyphenols [mg GAE/L]	Antioxidant Activity [%]
'Usłada'	1708.66 ± 6.47 d	79.93 ± 0.49 b
'Sinij Utes'	2506.57 ± 2.36 a	77.52 ± 0.07 cd
'Candy Blue'	1895.16 ± 5.84 c	78.15 ± 0.04 c
'Boreal Beauty'	1036.20 ± 2.84 f	76.43 ± 0.25 e
'Boreal Beast'	1463.77 ± 7.79 e	81.77 ± 0.54 a
'Boreal Blizzard'	2340.74 ± 31.84 b	76.95 ± 0.13 de

Values are given as means ± standard deviations, followed by letters a–f to indicate statistical significance. The values marked with the same letters in the same column, depending on the number of days of storage, are not statistically different at the $\alpha < 0.05$ level.

In the current study, after seven days of storage, a significant percentage decrease in polyphenol content was observed in the variety 'Boreal Blizzard' (an average of 25.6%) regardless of the method used (Figure 2). The largest reduction occurred during the second week of storage, except for the variety 'Boreal Blizzard', which showed a significant decrease in the first week. After 14 days, the highest percentage losses were recorded for the variety 'Candy Blue', with an average reduction of 39.4%. Previous studies have shown that polyphenol content in fruits tends to decrease during storage. For instance, research on red raspberries demonstrated a gradual decline in total polyphenol content during fruit ripening and storage, mainly due to changes in the activity of biosynthetic enzymes and an increase in the content of other compounds, such as anthocyanins. Similar trends have been reported in apples and various berries, with storage conditions playing a critical role in the stability of polyphenols [43,44].

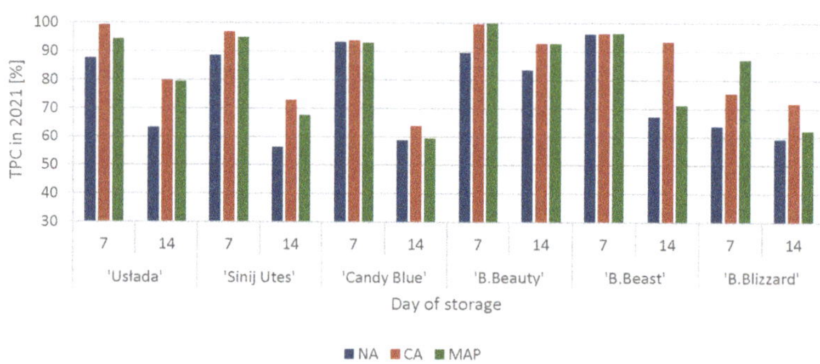

Figure 2. Percent change in polyphenol content (mg/L GAE) after seven days and after fourteen days of berry storage, relative to post-harvest values for individual varieties and storage methods in 2021.

The stability of polyphenols during storage depends on many factors, including temperature, light, and the presence of oxygen. Polyphenols are sensitive to oxidative stress and enzymatic activity, which can lead to their degradation during storage [45].

After the first and second weeks, the highest polyphenol content in honeysuckle juice was recorded for the variety 'Sinij Utes' (2380 mg/L) under controlled atmosphere (CA) conditions, and the lowest in the variety 'Boreal Beauty' in normal atmosphere (NA) storage (928 mg/L) (Table 3). Fruits continue to undergo biochemical changes after harvest, including polyphenol synthesis. The polyphenol content in fruits is also influenced by the composition of the storage atmosphere, particularly CO_2 concentration, which can either inhibit or promote polyphenol synthesis depending on conditions [3].

Table 3. Total polyphenols content (TPC, mg/L GAE) and antioxidant activity (DPPH, % inhibition) in juice after harvest, after seven and fourteen days of berry storage in 2021, depending on variety and storage method.

Cultivars	Storage Conditions	Total Polyphenols [mg GAE/L]	Antioxidant Activity [%]	Total Polyphenols [mg GAE/100 mL]	Antioxidant Activity [%]
		7 days	7 days	14 days	14 days
'Usłada'	Na	1498.13 ± 1.03 i	68.81 ± 0.04 f	1082.61 ± 1.52 h	63.24 ± 0.04 g
	Ca	1693.56 ± 2.19 g	77.21 ± 0.07 b	1364.72 ± 0.82 e	74.58 ± 0.06 a
	MAP	1610.66 ± 1.55 h	75.46 ± 0.14 c	1356.66 ± 6.23 e	69.66 ± 0.05 d
'Sinij Utes'	Na	2219.28 ± 5.31 c	68.78 ± 0.05 f	1411.02 ± 4.25 d	61.68 ± 0.11 h
	Ca	2426.57 ± 2.14 a	75.42 ± 0.10 c	1853.21 ± 3.07 a	69.60 ± 0.09 d
	MAP	2380.02 ± 2.37 b	73.57 ± 0.03 d	1659.06 ± 3.86 b	71.60 ± 0.19 c
'Candy Blue'	Na	1767.37 ± 3.24 f	72.73 ± 0.15 e	999.5 ± 3.65 j	65.64 ± 0.21 e
	Ca	1776.62 ± 5.04 e	73.44 ± 0.23 d	1209.11 ± 3.94 f	69.33 ± 0.08 d
	MAP	1764.09 ± 2.23 f	73.48 ± 0.11 d	1124.78 ± 1.59 g	69.55 ± 0.02 d
'Boreal Beauty'	Na	927.76 ± 2.25 l	68.90 ± 0.15 f	865.58 ± 1.68 l	63.15 ± 0.11 g
	Ca	1031.40 ± 2.94 k	75.95 ± 0.24 c	961.67 ± 5.19 k	71.49 ± 0.03 c
	MAP	1035.28 ± 4.41 k	75.51 ± 0.12 c	962.08 ± 0.64 k	71.73 ± 0.07 c
'Boreal Beast'	Na	1407.35 ± 1.02 j	75.61 ± 0.22 c	1040.87 ± 1.59 i	69.37 ± 0.09 d
	Ca	1409.35 ± 1.61 j	78.78 ± 0.20 a	1367.23 ± 5.26 e	72.54 ± 0.07 b
	MAP	1410.85 ± 1.91 j	78.78 ± 0.14 a	984.49 ± 4.90 jk	74.28 ± 0.07 a
'Boreal Blizzard'	Na	1495.29 ± 2.05 i	64.78 ± 0.09 h	1385.24 ± 1.24 de	61.36 ± 0.05 h
	Ca	1764.92 ± 4.93 f	66.30 ± 0.21 g	1673.07 ± 5.11 b	63.85 ± 0.10 f
	MAP	2039.25 ± 2.95 d	68.87 ± 0.23 f	1455.62 ± 4.73 c	63.24 ± 0.03 g

Values are given as means ± standard deviations, followed by letters a–l to indicate statistical significance. The values marked with the same letters in the same column, depending on the number of days of storage, are not statistically different at the $\alpha < 0.05$ level.

In 2022, the total polyphenol content in juices also showed significant variation across varieties; however, the differences between varieties were significantly smaller than in 2021 (Table 4). The highest polyphenol concentration was recorded in the variety 'Usłada' (2898 mg/L), while the lowest was recorded in 'Boreal Beauty' (2120 mg/L).

Table 4. Total polyphenols content (TPC, mg/L GAE) and antioxidant activity (DPPH, % inhibition) in juice post-harvest by variety in 2022.

	Total Polyphenols [mg GAE/L]	Antioxidant Activity [%]
'Usłada'	2898.76 ± 3.63 a	79.74 ± 0.13 b
'Sinij Utes'	2710.15 ± 2.91 b	77.53 ± 0.13 c
'Candy Blue'	2544.46 ± 2.84 c	78.02 ± 0.32 c
'Boreal Beauty'	2120.75 ± 9.35 f	76.24 ± 0.28 e
'Boreal Beast'	2467.37 ± 8.40 d	81.78 ± 0.14 a
'Boreal Blizzard'	2448.64 ± 1.54 e	76.96 ± 0.05 d

Values are given as means ± standard deviations, followed by letters a–f to indicate statistical significance. The values marked with the same letters in the same column, depending on the number of days of storage, are not statistically different at the $\alpha < 0.05$ level.

The lowest polyphenol concentration after both seven and fourteen days of storage was recorded in the 'Boreal Beauty' cultivar under NA (Table 5). The cultivar with the highest polyphenol content, similar to the previous year, was 'Sinij Utes'.

Table 5. Total polyphenol content (TPC, mg/L GAE) and antioxidant activity (DPPH, % inhibition) in juice after harvest, after seven and fourteen days of berry storage in 2022 depending on variety and storage method.

Cultivars	Storage Conditions	Total Polyphenols [mg GAE/L]	Antioxidant Activity [%]	Total Polyphenols [mg GAE/L]	Antioxidant Activity [%]
		7 days	7 days	14 days	14 days
'Usłada'	Na	1485.86 ± 2.29 e	74.01 ± 0.16 b	1250.60 ± 4.32 e	69.98 ± 0.06 e
	Ca	2230.19 ± 4.01 c	77.46 ± 0.08 a	1592.26 ± 4.95 c	75.55 ± 0.14 a
	MAP	1925.93 ± 2.51 c	77.31 ± 0.17 a	1353.04 ± 3.52 d	75.50 ± 0.34 a
'Sinij Utes'	Na	1924.21 ± 1.74 c	74.09 ± 0.05 b	1591.35 ± 2.38 c	70.04 ± 0.01 e
	Ca	2208.92 ± 3.96 b	76.69 ± 0.16 b	1757.58 ± 1.92 a	75.59 ± 0.28 a
	MAP	1922.67 ± 3.80 c	76.51± 0.12 b	1653.99 ± 4.12 b	73.95 ± 0.16 b
'Candy Blue'	Na	1133.70 ± 2.99 h	72.32 ± 0.23 g	924.64 ± 1.43 h	69.12 ± 0.17 e
	Ca	1889.79 ± 6.04 d	75.56 ± 0.09 c	1351.92 ± 1.53 d	75.47 ± 0.10 a
	MAP	1413.54 ± 4.84 f	74.06 ± 0.01 e	1252.52 ± 2.03 e	73.25 ± 0.11 b
'Boreal Beauty'	Na	1112.99 ± 5.33 i	72.25 ± 0.04 g	916.86 ± 3.25 h	68.95 ± 0.12 e
	Ca	1260.68 ± 5.48 g	73.43 ± 0.07 f	1194.44 ± 3.29 f	72.02 ± 0.01 c
	MAP	1250.73 ± 2.74 g	73.50 ± 0.09 f	1196.23 ± 2.72 f	71.48 ± 0.06 c
'Boreal Beast'	Na	1410.06 ± 0.45 f	72.03 ± 0.02 g	1350.58 ± 1.89 d	69.09 ± 0.14 e
	Ca	1480.46 ± 5.99 e	74.64 ± 0.08 d	1357.90 ± 3.84 d	73.61 ± 0.28 b
	MAP	1415.36 ± 7.01 f	74.04 ± 0.01 e	1253.43 ± 8.66 e	73.32 ± 0.21 b
'Boreal Blizzard'	Na	1252.60 ± 3.68 g	73.36 ± 0.14 f	920.81 ± 6.07 h	70.59 ± 0.08 e
	Ca	1488.64 ± 7.27 d	75.57 ± 0.13 c	1109.42 ± 1.11 g	74.05 ± 0.14 b
	MAP	1894.66 ± 3.68 d	76.41 ± 0.09 b	1255.87 ± 2.33 e	75.59 ± 0.05 a

Values are given as means ± standard deviations, followed by letters a–i to indicate statistical significance. The values marked with the same letters in the same column, depending on the number of days of storage, are not statistically different at the $\alpha < 0.05$ level.

The total polyphenol content in berry juice differed between 2021 and 2022. The varieties 'Usłada', 'Candy Blue', 'Boreal Beauty', and 'Boreal Beast' demonstrated increases ranging from 69% to more than double the polyphenol content in 2022 compared to 2021 (Figure 3). This variation can be attributed to differences in weather conditions between the two growing seasons, particularly due to increased rainfall during growth and ripening phases in 2021 (Table 1). For instance, Wu et al. [46] demonstrated that intense ultraviolet radiation, high growth temperatures, and low rainfall promoted the formation and accumulation of polyphenols and anthocyanins in plants. Similarly, Senica et al. [40] noted that additional stress caused by reduced temperatures led to the accumulation of secondary metabolites, such as polyphenols.

In 2022, the most significant changes in polyphenol content occurred during the first week of storage (Figure 4). The greatest losses were observed in the variety 'Candy Blue' stored in normal atmosphere (55.4%), followed by modified atmosphere packaging (MAP) (44.5%), and the Borealis group under CA conditions (40%). The lowest percentage loss was recorded in the cultivar 'Sinij Utes' (25.5%). After 14 days of storage, the worst results were recorded for the variety 'Candy Blue' in NA (65.7%) and 'Usłada' in CA (45%) and MAP (53.3%). These findings were consistent with those of López et al. [47], who reported that post-harvest handling of fruits and vegetables, including storage under controlled or modified atmospheres and minimal processing, can influence the synthesis or degradation of phenolic compounds.

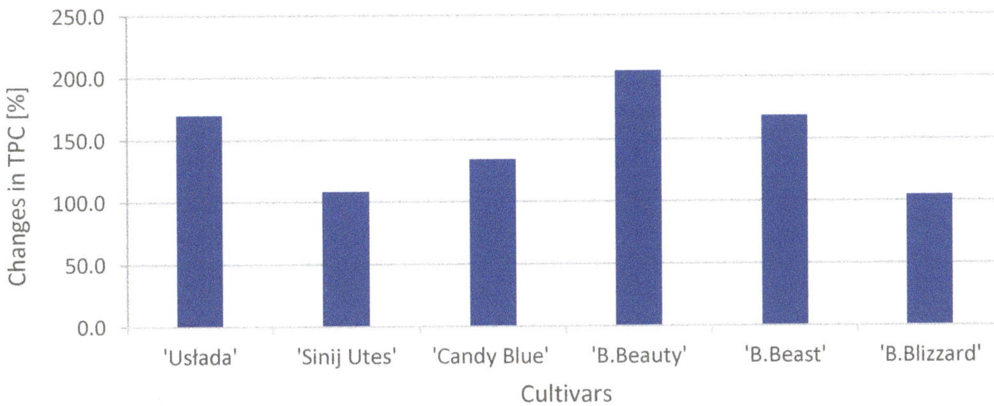

Figure 3. Percentage change in total polyphenol content in 2022 compared to 2021.

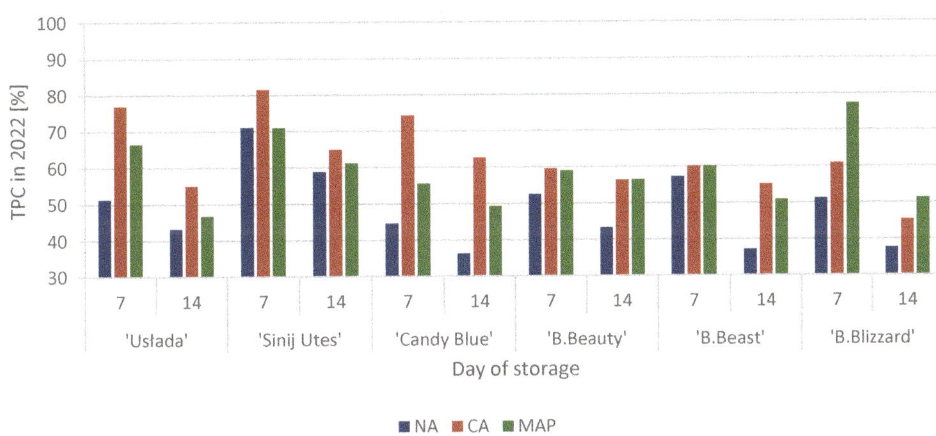

Figure 4. Percentage change in polyphenol content (TPC, mg/L GAE) after seven and fourteen days of berry storage relative to post-harvest values in 2022 based on variety and storage method.

The lowest polyphenol concentration after both storage periods was recorded in the variety 'Boreal Beauty' under NA conditions (Table 5). The variety with the highest polyphenol content, as in the previous year, was 'Sinij Utes'.

3.2. Antioxidant Activity (DPPH)

Antioxidant activity, assessed using DPPH tests, was comparable between individual varieties in 2021 and 2022 (Tables 2 and 4). Interestingly, juice samples with the highest polyphenol content did not always show the highest antioxidant activity. After the first week of storage, the results varied depending on the variety and storage conditions (Tables 2–5), reflecting findings by Dziedzic et al. [3]. Conversely, Kalisz et al. [48] reported a strong correlation between total polyphenol content and antioxidant activity in processed products. Polyphenols are largely responsible for antioxidant properties of fruits, but other factors can also influence their effects. Therefore, maintaining appropriate storage conditions for haskap berries is essential to preserve their nutritional value. Polyphenols are key compounds determining the antioxidant properties of honeysuckle berries, further emphasising the importance of optimised storage conditions.

In 2021, the DPPH test showed the highest inhibition percentage in the variety 'Boreal Beast' (81.77%) and the lowest in 'Boreal Beauty' (76.43%) and 'Boreal Blizzard' (76.95%) (Table 2). Lee et al. [49] demonstrated in vitro antioxidant activity of 86.1% for Korean haskap berries and 92.9% for Chinese berries at a concentration of 300 μg/mL against oxidative stress induced by tert-butyl hydroperoxide in HepG2 human liver cancer cells.

After one week of storage, the most significant declines in the antioxidant activity were observed in the varieties 'Boreal Blizzard' and 'Usłada' under normal atmosphere (NA) (decreases of 15.82% and 13.70%, respectively) and under controlled atmosphere (CA) conditions (13.84% and 13.70%, respectively). Modified atmosphere packaging (MAP) caused a lower decrease in antioxidant activity (10.51%) in 'Boreal Blizzard' (Figure 5), while the smallest losses occurred in the variety 'Candy Blue' under NA (6.77%), and in 'Boreal Beauty' under CA and MAP conditions (0.96% and 0.38%, respectively).

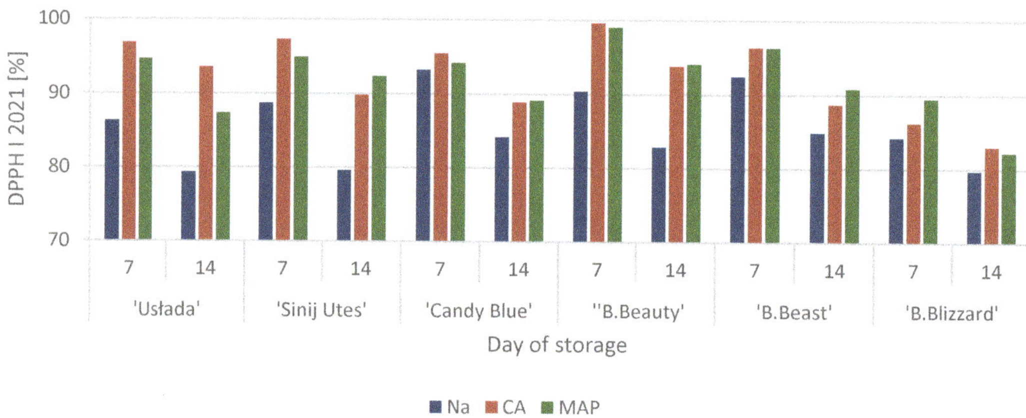

Figure 5. Percentage changes in radical scavenging capacity in juice after seven and fourteen days of storage in 2021 depending on the variety.

After two weeks of storage, the smallest changes in radical scavenging capacity, compared to the first week, were noted in 'Boreal Blizzard' under NA and CA (4.45% and 3.19%, respectively) and in 'Sinij Utes' under MAP conditions (2.54%) (Figure 5). The most significant changes occurred in 'Sinij Utes' and 'Candy Blue' under NA conditions (9.15% and 9.10%, respectively). The lowest percentage changes in antioxidant activity relative to post-harvest values were recorded for varieties 'Boreal Beast' and 'Candy Blue' under NA (15.17% and 15.87%), in 'Boreal Beauty' and 'Usłada' under CA (6.23% and 6.48%), and in 'Boreal Beauty' under MAP conditions (5.92%). In contrast, the most pronounced alteration was observed in the variety 'Usłada' under NA conditions (20.70%).

In 2022, the DPPH test results were consistent with the previous year, with 'Boreal Beast' showing the highest inhibition percentage (81.78%), while 'Boreal Beauty' (76.24%) and 'Boreal Blizzard' (76.96%) showed the lowest (Table 3). Zehfus et al. [50], studying five Canadian honeysuckle varieties, including 'Boreal Blizzard', classified it among varieties with lower antioxidant capacity. After seven days of storage, the highest radical scavenging capacity was observed in 'Usłada' under CA and MAP conditions (77.46% and 77.31%), while the lowest was recorded in 'Boreal Beast' in NA (72.03%) (Table 5). After 14 days, the variety 'Usłada' showed a 5.22% decline in antioxidant capacity during NA storage (Figure 6). The lowest value was recorded in 'Boreal Beauty' (68.95%), and the highest in 'Sinij Utes' under CA and MAP conditions (75.55% and 75.50%).

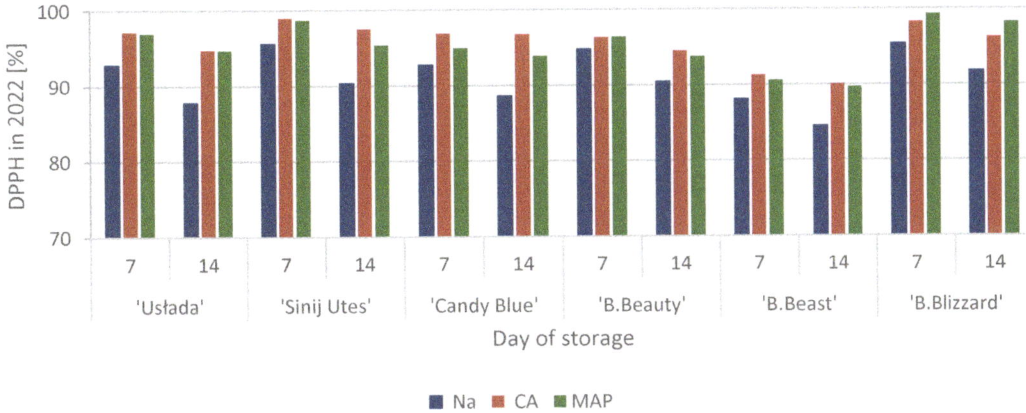

Figure 6. Percentage changes in radical scavenging capacity in juice after seven and fourteen days of storage in 2022 depending on the variety.

Changes in antioxidant potential mirrored the variations in total polyphenol content (TPC) and followed similar trends. According to the existing literature, the most significant changes in TAC and DPPH occur during freezing [51], thermal processing, and oxygen exposure [42,48]. Research by Grobelna et al. [1] also indicated that a four-month storage period of haskap berry juice resulted in a decrease in both polyphenol content and antioxidant activity.

The scientific literature indicates that biochemical processes, such as the synthesis of polyphenolic compounds, continue in fruits after harvest, influencing their radical scavenging capacity. It is also well established that polyphenol content in stored fruits and vegetables depends on atmospheric composition, particularly CO_2 concentration [3]. The current study found that controlled atmosphere (CA) and modified atmosphere (MAP) storage conditions were more favourable than normal atmosphere (NA), especially in the context of haskap berry juice production, as previously demonstrated by Krupa et al. [32]. This is reflected in the amplitude of changes in polyphenol content (Figure 7A–C) and antioxidant activity (Figure 7D–F) under different storage conditions, averaged for 2021 and 2022.

3.3. *Qualitative Measurements*

Depending on the variety, stage of maturity, and growing region, honeysuckle berries are characterised by firmness ranging from 2.9 to 4.9 N at harvest [3,29]. In both years, differences in fruit firmness rates between fruit varieties may explain changes in polyphenol and anthocyanin contents during storage under different conditions, further supporting the advantages of CA and MAP over NA (Figure 8). However, determining the optimal storage method remains complex, as different fruit varieties respond uniquely to varying conditions, significantly affecting the quality of haskap berry juice produced from a single variety. In 2022, the most significant changes in firmness were observed after 14 days in the variety 'Boreal Blizzard', regardless of storage method.

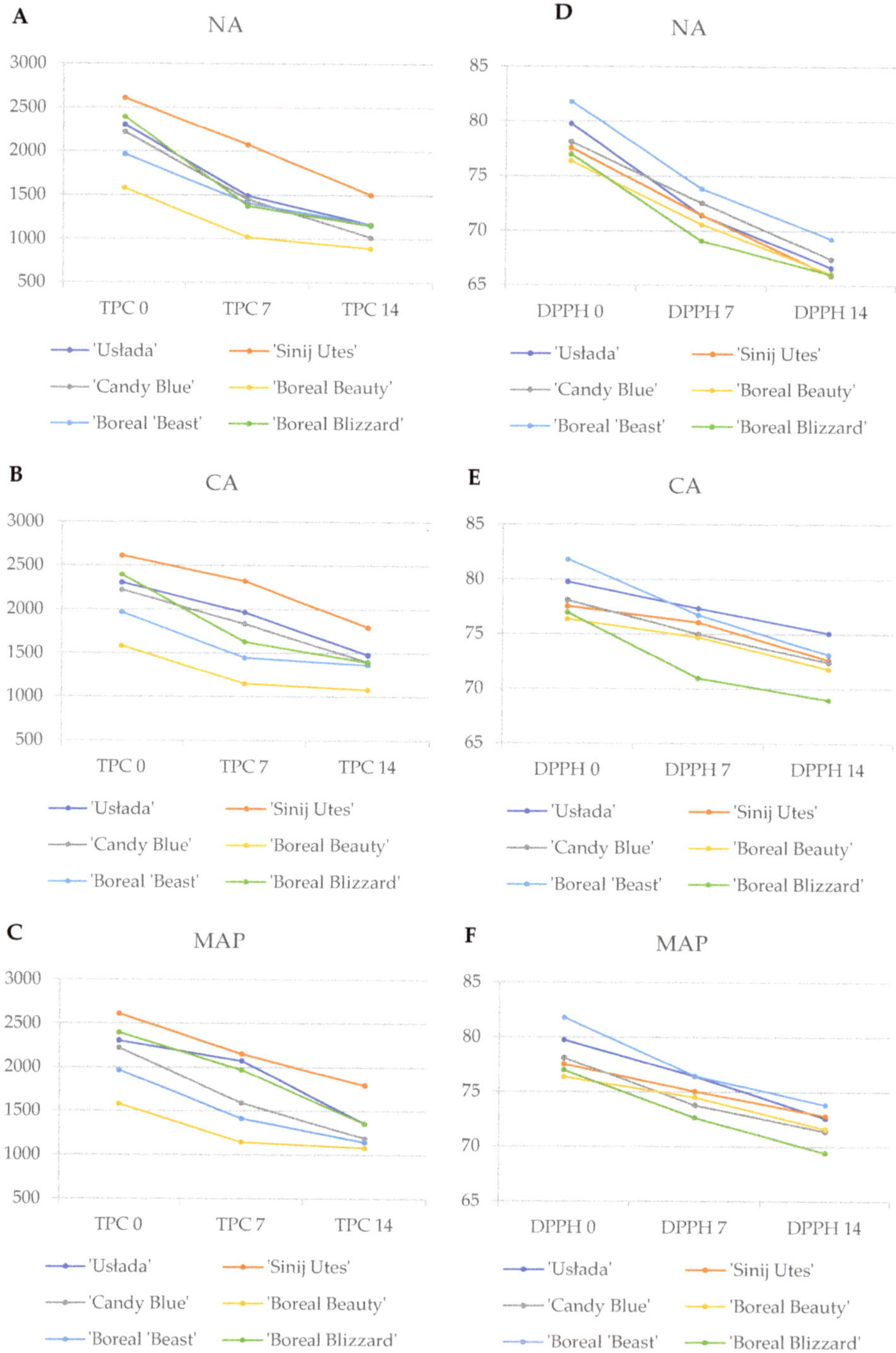

Figure 7. Amplitude of changes in polyphenol content (**A**–**C**) and antioxidant activity (**D**–**F**) under various storage conditions, averaged for 2021 and 2022.

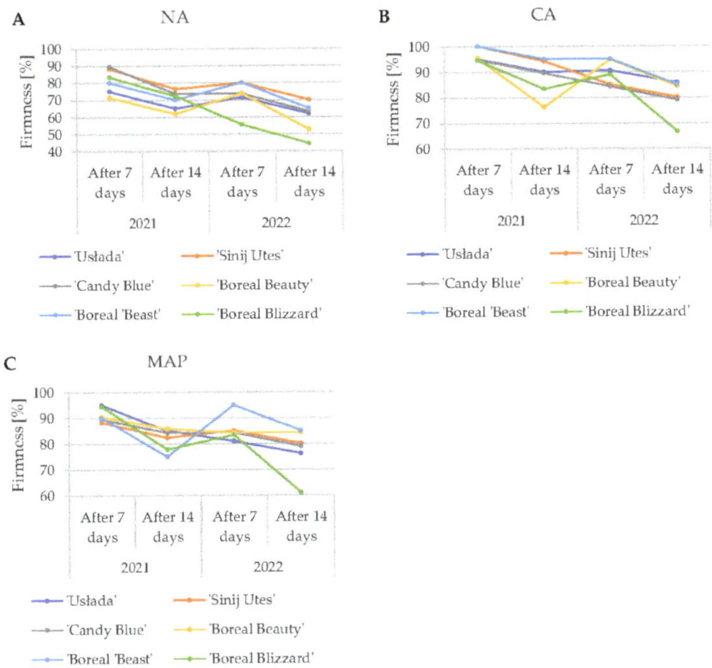

Figure 8. Percentage changes in fruit firmness based on storage method (NA—(**A**), CA—(**B**), or MAP—(**C**)) in 2021 and 2022.

Cluster analysis involving total polyphenol content and antioxidant activity identified two groups in 2021 (Figure 9) and 2022 (Figure 10). The composition of these groups varied between the years, supporting earlier findings that total polyphenol content does not always correlate with high antioxidant activity.

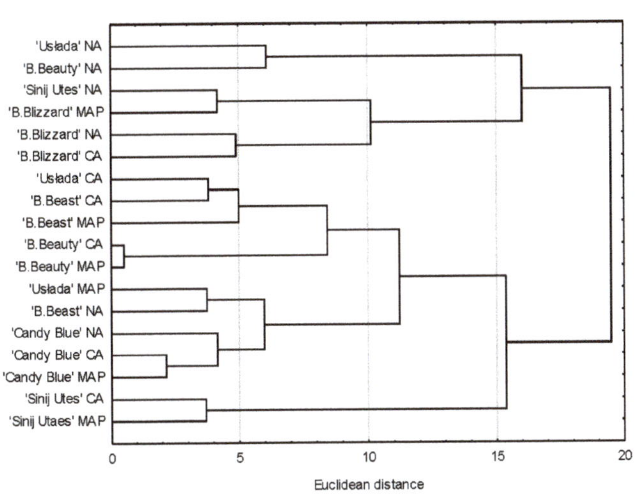

Figure 9. Similarity of storage method combinations and varieties in terms of total polyphenol content and antioxidant activity in 2021.

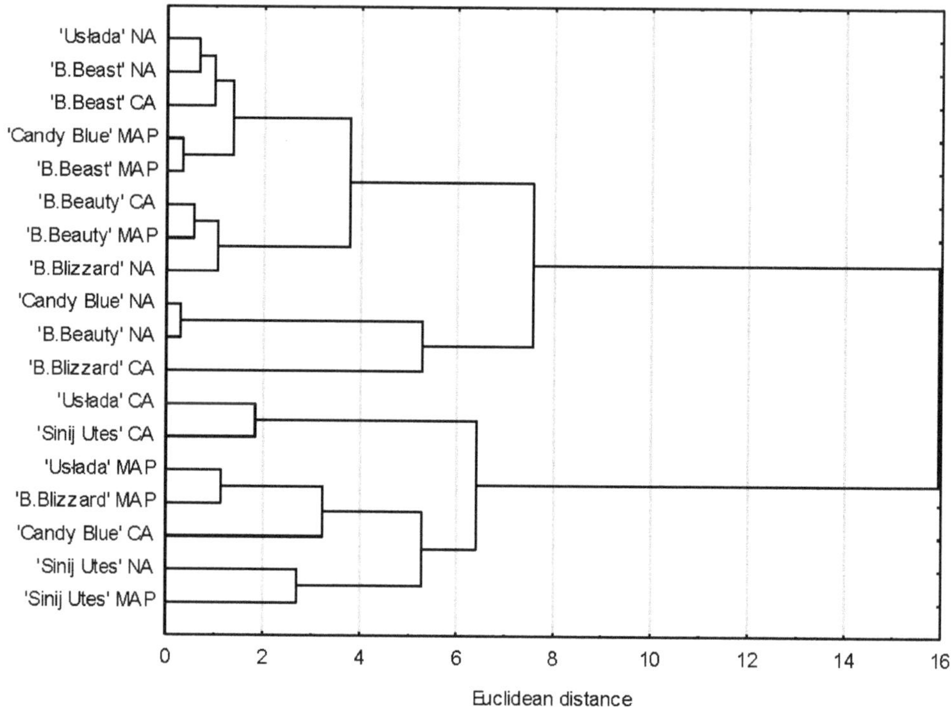

Figure 10. Similarity of storage method combinations and varieties in terms of total polyphenol content and antioxidant activity in 2022.

4. Conclusions

This study evaluated the impact of short-term storage of haskap berry juices under modified atmosphere (MAP) and controlled atmosphere (CA) conditions on their polyphenol content and antioxidant activity. The results confirmed that storage conditions play a significant role in influencing these parameters, which are critical for preserving the quality of food products. The use of advanced storage methods, such as CA and MAP, was shown to effectively stabilise polyphenol content in juices, as evidenced by differences in fruit respiration rates and biochemical changes.

The observed variability in polyphenol content between 2021 and 2022 reflects the impact of weather conditions on the synthesis of phenolic compounds in fruits. Rainfall and temperature play a key role in these processes, which is important for planning harvests and raw material storage.

DPPH test results demonstrated that high polyphenol content does not necessarily translates into the highest antioxidant activity.

The stability of polyphenols in juices depends on storage conditions such as temperature, humidity, and atmospheric composition. Optimising these factors is essential for preserving the nutritional and sensory qualities of the products, which is vital for haskap berry juice producers aiming to offer high-quality, health-promoting products to the food market.

Differences between varieties and storage conditions highlight the need for further research into polyphenol stabilisation mechanisms and their functionality in haskap berry juices, and the relationship between antioxidant activity and the quality and nutritional value of the fruit. Such studies can significantly increase practical value by providing insights on how to improve shelf life and preserve the health-promoting properties of fruit juices.

Author Contributions: B.A.K.: conceptualisation, data care, experimental design, methodology, writing, preparation of original draft, software, project administration; M.B.: reviewing, methodology; J.B.: funding acquisition, methodology. All authors have read and agreed to the published version of the manuscript.

Funding: This research received no external funding.

Institutional Review Board Statement: Not applicable for studies not involving humans or animals.

Data Availability Statement: The original contributions presented in this study are included in the article. Further inquiries can be directed to the corresponding author.

Conflicts of Interest: The authors declare no conflicts of interest. The funders had no role in the design of the study; in the collection, analyses, or interpretation of data; in the writing of the manuscript; or in the decision to publish the results.

References

1. Grobelna, A.; Kalisz, S.; Kieliszek, M. The Effect of the Addition of Blue Honeysuckle Berry Juice to Apple Juice on the Selected Quality Characteristics, Anthocyanin Stability, and Antioxidant Properties. *Biomolecules* **2019**, *9*, 744. [CrossRef] [PubMed]
2. Česonienė, L.; Labokas, J.; Jasutienė, I.; Šarkinas, A.; Kaškonienė, V.; Kaškonas, P.; Kazernavičiūtė, R.; Pažereckaitė, A.; Daubaras, R. Bioactive Compounds, Antioxidant, and Antibacterial Properties of *Lonicera caerulea* Berries: Evaluation of 11 Cultivars. *Plants* **2021**, *10*, 624. [CrossRef]
3. Dziedzic, E.; Błaszczyk, J.; Bieniasz, M.; Dziadek, K.; Kopeć, A. Effect of Modified (MAP) and Controlled Atmosphere (CA) Storage on the Quality and Bioactive Compounds of Blue Honeysuckle Fruits (*Lonicera caerulea* L.). *Sci. Hortic.* **2020**, *265*, 109226. [CrossRef]
4. Li, F.; Zhao, H.; Xu, R.; Zhang, X.; Zhang, W.; Du, M.; Liu, X.; Fan, L. Simultaneous Optimization of the Acidified Water Extraction for Total Anthocyanin Content, Total Phenolic Content, and Antioxidant Activity of Blue Honeysuckle Berries (*Lonicera caerulea* L.) Using Response Surface Methodology. *Food Sci. Nutr.* **2019**, *7*, 2968–2976. [CrossRef] [PubMed]
5. Luo, J.; Fan, Z.; Yang, X.; Bao, Y.H.; Liang, M.; Guo, Y. Anthocyanins and Antioxidant Activity of *Lonicera caerulea* Berry Wine during Different Processes. *Food Sci. Technol.* **2021**, *42*, e25121. [CrossRef]
6. Negreanu-Pirjol, B.S.; Oprea, O.C.; Negreanu-Pirjol, T.; Roncea, F.N.; Prelipcean, A.M.; Craciunescu, O.; Iosageanu, A.; Artem, V.; Ranca, A.; Motelica, L.; et al. Health Benefits of Antioxidant Bioactive Compounds in the Fruits and Leaves of *Lonicera caerulea* L. and *Aronia melanocarpa* (Michx.) Elliot. *Antioxidants* **2023**, *12*, 951. [CrossRef]
7. Piekarska, J.; Szczypka, M.; Gorczykowski, M.; Sokół-łętowska, A.; Kucharska, A.Z. Evaluation of Immunotropic Activity of Iridoid-Anthocyanin Extract of Honeysuckle Berries (*Lonicera caerulea* L.) in the Course of Experimental Trichinellosis in Mice. *Molecules* **2022**, *27*, 1949. [CrossRef]
8. Raudonė, L.; Liaudanskas, M.; Vilkickytė, G.; Kviklys, D.; Žvikas, V.; Viškelis, J.; Viškelis, P. Phenolic Profiles, Antioxidant Activity and Phenotypic Characterization of *Lonicera caerulea* L. Berries, Cultivated in Lithuania. *Antioxidants* **2021**, *10*, 115. [CrossRef]
9. Oszmiański, J.; Kucharska, A.Z. Effect of Pre-Treatment of Blue Honeysuckle Berries on Bioactive Iridoid Content. *Food Chem.* **2018**, *240*, 1087–1091. [CrossRef]
10. Sharma, A.; Lee, H.J. *Lonicera caerulea*: An Updated Account of Its Phytoconstituents and Health-Promoting Activities. *Trends Food Sci. Technol.* **2021**, *107*, 130–149. [CrossRef]
11. Auzanneau, N.; Weber, P.; Kosińska-Cagnazzo, A.; Andlauer, W. Bioactive Compounds and Antioxidant Capacity of *Lonicera caerulea* Berries: Comparison of Seven Cultivars over Three Harvesting Years. *J. Food Compos. Anal.* **2018**, *66*, 81–89. [CrossRef]
12. Shevchuk, L.; Tereshchenko, Y.; Vintskovska, Y.; Levchuk, L.; Babenko, S.; Hrynyk, R. Yield and Content of Biologically Active Substances in Blue Honeysuckle Fruit (*Lonicera caerulea* L.) Grown in the Forest Steppe of Ukraine. *Agron. Res.* **2022**, *20*, 814–826. [CrossRef]
13. Becker, R.; Szakiel, A. Phytochemical Characteristics and Potential Therapeutic Properties of Blue Honeysuckle *Lonicera caerulea* L. (*Caprifoliaceae*). *J. Herb. Med.* **2019**, *16*, 100237. [CrossRef]
14. Bors, B. Breeding the Boreal Series of Haskap (*Lonicera Caerulea*). 2017, p. 2017. Available online: https://research-groups.usask.ca/fruit/documents/haskap/cshs-Poster-2017.pdf (accessed on 4 June 2024).
15. De Silva, A.B.K.H.; Rupasinghe, H.P.V. Polyphenols Composition and Anti-Diabetic Properties in Vitro of Haskap (*Lonicera caerulea* L.) Berries in Relation to Cultivar and Harvesting Date. *J. Food Compos. Anal.* **2020**, *88*, 103402. [CrossRef]

16. Celli, G.B.; Ghanem, A.; Brooks, M.S.L. Haskap Berries (*Lonicera caerulea* L.)—A Critical Review of Antioxidant Capacity and Health-Related Studies for Potential Value-Added Products. *Food Bioprocess Technol.* **2014**, *7*, 1541–1554. [CrossRef]
17. Ochmian, I.; Grajkowski, J.; Skupień, K. Field Performance, fruit chemical composition and firmness under cold storage and simulated "shelf-life" conditions of three blue honey suckle cultigens (*Lonicera caerulea*). *J. Fruit Ornam. Plant Res.* **2008**, *16*, 83–91. Available online: https://www.inhort.pl/files/journal_pdf/journal_2008/full9%202008.pdf (accessed on 18 July 2024).
18. Levon, V.; Grygorieva, O.; Antoniewska-Krzeska, A.; Šramková, K.F.; Zhurba, M. Variation of Fruits Morphometric Parameters and Iridoid Content of *Lonicera caerulea* L. Germplasm Collection. *Agrobiodivers. Improv. Nutr. Health Life Qual.* **2022**, *6*, 262–271. [CrossRef]
19. Khattab, R.; Brooks, M.S.-L.; Ghanem, A. Phenolic Analyses of Haskap Berries (*Lonicera caerulea* L.): Spectrophotometry Versus High Performance Liquid Chromatography. *Int. J. Food Prop.* **2016**, *19*, 1708–1725. [CrossRef]
20. Rupasinghe, H.P.V.; Arumuggam, N.; Amararathna, M.; De Silva, A.B.K.H. The Potential Health Benefits of Haskap (*Lonicera caerulea* L.): Role of Cyanidin-3-O-Glucoside. *J. Funct. Foods.* **2018**, *44*, 24–39. [CrossRef]
21. Gerbrandt, E.M.; Bors, R.H.; Meyer, D.; Wilen, R.; Chibbar, R.N. Fruit Quality of Japanese, Kuril and Russian Blue Honeysuckle (*Lonicera caerulea* L.) Germplasm Compared to Blueberry, Raspberry and Strawberry. *Euphytica* **2020**, *216*, 59. [CrossRef]
22. Gazdik, Z.; Reznicek, V.; Adam, V.; Zitka, O.; Jurikova, T.; Krska, B.; Matuskovic, J.; Plsek, J.; Saloun, J.; Horna, A.; et al. Use of Liquid Chromatography with Electrochemical Detection for the Determination of Antioxidants in Less Common Fruits. *Molecules* **2008**, *13*, 2823–2836. [CrossRef] [PubMed]
23. Kalisz, S.; Polak, N.; Jodłowska, M. Wpływ Odmiany Surowca Na Wybrane Wyróżniki Jakościowe Soków z Jagody Kamczackiej. *Przemysł Spożywczy* **2022**, *76*, 26–32. [CrossRef]
24. Małodobry, M.; Bieniasz, M.; Dziedzic, E. Evaluation of the Yield and Some Components in the Fruit of Blue Honeysuckle (*Lonicera caerulea* Var. Edulis Turcz. Freyn.). *Folia Hortic.* **2010**, *22*, 45–50. [CrossRef]
25. Koort, A.; Moor, U.; Põldma, P.; Kaiser, C.; Starast, M. Comparison of Regular Atmospheric Storage versus Modified Atmospheric Packaging on Postharvest Quality of Organically Grown Lowbush and Half-Highbush Blueberries. *Sustainability* **2018**, *10*, 3916. [CrossRef]
26. Yan, H.; Mi, Y.; Li, Y.; Zang, H.; Guo, L.; Huo, J.; Man, Z.; Chen, Z.; Zhang, B.; Sang, M.; et al. First Report of Postharvest Fruit Rot Caused by Botrytis Cinerea on Blue Honeysuckle (*Lonicera caerulea*) Fruit in China. *Plant Dis.* **2024**, *108*, 527. [CrossRef]
27. Jurikova, T.; Ercişli, S.; Rop, O.; Mlcek, J.; Balla, S.; Zitny, R.; Sochor, J.; Hegedusova, A.; Benedikova, D.; Ďurišová, L. The Evaluation of Anthocyanin Content of Honeyberry (*Lonicera kamtschatica*) Clones during Freezing in Relation to Antioxidant Activity and Parameters of Nutritional Value. *Zemdirb. Agric.* **2014**, *101*, 215–220. [CrossRef]
28. Khattab, R.; Celli, G.B.; Ghanem, A.; Brooks, M.S.L. Effect of Frozen Storage on Polyphenol Content and Antioxidant Activity of Haskap Berries (*Lonicera caerulea* L.). *J. Berry Res.* **2015**, *5*, 231–242. [CrossRef]
29. Leisso, R.; Jarrett, B.; Richter, R.; Miller, Z. Fresh Haskap Berry Postharvest Quality Characteristics and Storage Life. *Can. J. Plant Sci.* **2021**, *101*, 1051–1063. [CrossRef]
30. Ziv, C.; Fallik, E. Postharvest Storage Techniques and Quality Evaluation of Fruits and Vegetables for Reducing Food Loss. *Agronomy* **2021**, *11*, 1133. [CrossRef]
31. Błaszczyk, J.; Bieniasz, M.; Nawrocki, J.; Kopeć, M.; Mierzwa-Hersztek, M.; Gondek, K.; Zaleski, T.; Knaga, J.; Bogdał, S. The Effect of Harvest Date and Storage Conditions on the Quality of Remontant Strawberry Cultivars Grown in a Gutter System under Covers. *Agriculture* **2022**, *12*, 1193. [CrossRef]
32. Krupa, T.; Klimek, K.; Zaraś-Januszkiewicz, E. Nutritional Values of Minikiwi Fruit (*Actinidia arguta*) after Storage: Comparison between DCA New Technology and ULO and CA. *Molecules* **2022**, *27*, 4313. [CrossRef] [PubMed]
33. Bishop, D. Controlled Atmosphere Storage. In *Cold and Chilled Storage Technology*; Springer: Boston, MA, USA, 1997; pp. 53–92. [CrossRef]
34. Fang, Y.; Wakisaka, M. A Review on the Modified Atmosphere Preservation of Fruits and Vegetables with Cutting-Edge Technologies. *Agriculture* **2021**, *11*, 992. [CrossRef]
35. Bors, B.; Sawchuk, E.; Dawson, J.; Sawatzky, R. *Breeding and Selecting Haskap for Nutraceutical and Agronomic Suitability*; Final Report on Saskatchewan Ministry of Agriculture: Saskatoon, SK, Canada, 2015; pp. 1–84. Available online: https://research-groups.usask.ca/fruit/documents/haskap/20080042.pdf (accessed on 26 July 2024).
36. Singleton, V.L.; Orthofer, R.; Lamuela-Raventós, R.M. Analysis of Total Phenols and Other Oxidation Substrates and Antioxidants by Means of Folin-Ciocalteu Reagent. *Methods Enzymol.* **1999**, *299*, 152–178. [CrossRef]
37. Sánchez-Rangel, J.C.; Benavides, J.; Heredia, J.B.; Cisneros-Zevallos, L.; Jacobo-Velázquez, D.A. The Folin–Ciocalteu Assay Revisited: Improvement of Its Specificity for Total Phenolic Content Determination. *Anal. Methods* **2013**, *5*, 5990–5999. [CrossRef]
38. Brand-Williams, W.; Cuvelier, M.E.; Berset, C. Use of a Free Radical Method to Evaluate Antioxidant Activity. *LWT Food Sci. Technol.* **1995**, *28*, 25–30. [CrossRef]
39. Wang, Z.; Svyantek, A.; Miller, Z.; Jarrett, B.; Kapus, A. Haskap Juicing Method Effects on Haskap Juice Quality. *Appl. Sci.* **2023**, *13*, 10784. [CrossRef]

40. Senica, M.; Bavec, M.; Stampar, F.; Mikulic-Petkovsek, M. Blue Honeysuckle (*Lonicera caerulea* Subsp. Edulis (Turcz. Ex Herder) Hultén.) Berries and Changes in Their Ingredients across Different Locations. *J. Sci. Food Agric.* **2018**, *98*, 3333–3342. [CrossRef]
41. Piasek, A.; Kusznierewicz, B.; Grzybowska, I.; Malinowska-Pańczyk, E.; Piekarska, A.; Azqueta, A.; Collins, A.R.; Namieśnik, J.; Bartoszek, A. The Influence of Sterilization with EnbioJet® Microwave Flow Pasteurizer on Composition and Bioactivity of Aronia and Blue-Berried Honeysuckle Juices. *J. Food Compos. Anal.* **2011**, *24*, 880–888. [CrossRef]
42. Szajdek, A.; Dąbkowska, E.; Borowska, E.J. Wpływ obróbki enzymatycznej miazgi owoców jagodowych na zawartość polifenoli i aktywność przeciwutleniającą soku. *Żywność* **2006**, *4*, 59–67. Available online: https://wydawnictwo.pttz.org/magazine-archive/agnieszka-szajdek-ewa-dabkowska-eulalia-j-borowska-wplyw-obrobki-enzymatycznej-miazgi-owocow-jagodowych-na-zawartosc-polifenoli-i-aktywnosc-przeciwutleniajaca-soku/ (accessed on 18 July 2024).
43. Kobori, R.; Yakami, S.; Kawasaki, T.; Saito, A.; Popovi´c, J.; Djordjevi´c, P.-D.; Fernando, L.; De Oliveira, C. Changes in the Polyphenol Content of Red Raspberry Fruits during Ripening. *Horticulturae* **2021**, *7*, 569. [CrossRef]
44. Łysiak, G. The Sum of Active Temperatures as a Method of Determining the Optimum Harvest Date of "Šampion" and "Ligol" Apple Cultivars Impact of Transport Conditions on Storage and Shelf Life of Fruit View Project Impact of Climatic Conditions on the Quality of Frui. *Acta Sci. Pol. Hortorum Cultus.* **2012**, *11*, 3–13. Available online: https://bibliotekanauki.pl/articles/11542008.pdf (accessed on 24 March 2022).
45. Zhang, Y.; Truzzi, F.; D'amen, E.; Dinelli, G. Effect of Storage Conditions and Time on the Polyphenol Content of Wheat Flours. *Processes* **2021**, *9*, 248. [CrossRef]
46. Wu, X.; Zhang, S.; Li, X.; Zhang, F.; Fan, Y.; Liu, Q.; Wan, X.; Lin, T. Postharvest UV-B Radiation Increases Enzyme Activity, Polysaccharide and Secondary Metabolites in Honeysuckle (*Lonicera japonica* Thunb.). *Ind. Crops Prod.* **2021**, *171*, 113907. [CrossRef]
47. López, N.; Puértolas, E.; Condón, S.; Álvarez, I.; Raso, J. Application of Pulsed Electric Fields for Improving the Maceration Process during Vinification of Red Wine: Influence of Grape Variety. *Eur. Food Res. Technol.* **2008**, *227*, 1099–1107. [CrossRef]
48. Kalisz, S.; Polak, N.; Cacak-Pietrzak, G.; Cendrowski, A.; Kruszewski, B. Impact of Production Methods and Storage Time on the Bioactive Compounds and Antioxidant Activity of Confitures Made from Blue Honeysuckle Berry (*Lonicera caerulea* L.). *Appl. Sci.* **2023**, *13*, 12999. [CrossRef]
49. Lee, Y.S.; Cho, I.J.; Kim, J.W.; Lee, S.K.; Ku, S.K.; Lee, H.J. Evaluation of in Vitro Anti-Oxidant and Anti-Inflammatory Activities of Korean and Chinese *Lonicera Caerulea*. *Nutr. Res. Pract.* **2018**, *12*, 486–493. [CrossRef]
50. Zehfus, L.R.; Eskiw, C.; Low, N.H. Phenolic Profiles and Antioxidant Activities of Saskatchewan (Canada) Bred Haskap (*Lonicera caerulea*) Berries. *BioRxiv.* **2021**. [CrossRef]
51. Khattab, R.; Ghanem, A.; Brooks, M.S.-L. Stability of Haskap Berry (*Lonicera caerulea* L.) Anthocyanins at Different Storage and Processing Conditions. *J. Field Robot.* **2016**, *5*, 67. [CrossRef]

agriculture

MDPI

Article

Assessment of Physiological Traits of *Fragaria vesca* Genotypes Under Water Deficit Conditions

Marta Rokosa, Małgorzata Mikiciuk * and Piotr Ptak

Department of Bioengineering, Faculty of Environmental Management and Agriculture, West Pomeranian University of Technology in Szczecin, Słowackiego 17, 71-434 Szczecin, Poland; marta.rokosa@gmail.com (M.R.); piotr.ptak@zut.edu.pl (P.P.)

* Correspondence: malgorzata.mikiciuk@zut.edu.pl

Abstract: Drought is one of the key challenges of climate change. The basic global problem related to the increasing water deficit is that the vast majority of crops are species and varieties that are the result of breeding work that did not anticipate such a rapid decrease in water availability in the soil. The main objective of the conducted research was to compare the physiological and biochemical response to water deficit of plants of the species *Fragaria vesca*—two cultivated varieties, and one collected from the natural environment. A two-year pot experiment was conducted in a polyethylene tunnel. The substrate moisture level was monitored using tensiometer readings. Measurements of gas exchange parameters, chlorophyll "a" fluorescence, content of photosynthetic pigments in leaves, index of relative water content in leaves, total fruit yield, single fruit mass and content of K, Ca, Mg, Na, Cu, Zn, Mn, Mo and the ratio of mono- to divalent cations in leaves, roots and plant crowns were taken three times each year during the experiments. Based on one-way and two-way analysis of variance, statistically significant differences were observed between wild-growing plants and cultivated varieties under control conditions, particularly in terms of chlorophyll fluorescence values and the content of photosynthetic pigments. A significant main effect of the soil moisture level was identified for most measured parameters across the majority of assessment time points. However, a significant interaction effect between soil moisture level and genotype was less frequently observed. Significant changes in response to water deficit varied depending on the parameter and genotype, ranging from 2.5% to 106.1%. For the content of chemical elements, the changes reached up to 157.1%. The results suggest that plants obtained from natural environments exhibit better adaptation to water deficit conditions, making them suitable for use in breeding programs aimed at developing varieties resistant to soil water deficits. However, the study's limitations, particularly the absence of molecular analyses regarding the plants' adaptive mechanisms, should be taken into consideration.

Keywords: wild strawberry; drought; chlorophyll fluorescence; gas exchange; chemical composition

1. Introduction

Phenomena of drought and desertification, exacerbated by climate change, adversely impact global food security and human health and may even pose a threat to global peace [1–3]. Climate models project that drought severity could double across 30% of the global land area under rigorous mitigation scenarios, leading to substantial implications for water supply and demand deficits [4]. Water deficit stress poses a significant challenge to agriculture, making plant resilience to water scarcity of critical economic importance [5]. In

Poland's climate, periodic meteorological droughts [6] disrupt the water balance of affected areas, leading to soil droughts that impair crop productivity [7]. In temperate climates, drought is among the primary factors limiting crop yields [4,8]. A fundamental global challenge associated with the increasing water deficit is that the majority of cultivated crops, including those of the *Fragaria* genus, consist of species and varieties developed through selective breeding, often without adequate consideration for the current significant reductions in soil water availability [9].

The economically most significant species of the *Fragaria* genus is the cultivated strawberry (*Fragaria* × *ananassa* Duch.), whose fruits are among the most widely consumed globally [10,11]. In Poland, plants of the *Fragaria* genus also hold considerable economic importance [8,12]. In 2020, the share of *Fragaria* plants in the total fruit yield from fruit bushes and berry plantations reached 26.0% [13]. The wild strawberry (*Fragaria vesca* L.) is primarily cultivated on an amateur scale. Although *F. vesca* is not commercially grown, its valuable traits as a research model have been extensively highlighted, particularly due to its short generation time, well-defined cycles of seasonal flowering and vegetative growth, and single-gene traits controlling aspects such as flowering and fruit color [14].

Plants of the *Fragaria* genus are highly sensitive to water deficiency due to their large leaf surface area, high water content in fruits, and shallow root systems. This sensitivity has been highlighted in numerous studies, which report a decline in physiological parameters in plants exposed to water deficit conditions [3,15–21]. However, this trait is undesirable in commercial cultivation, as even minor deviations from optimal irrigation levels can result in reduced yields, diminished fruit quality, decreased profitability of crops, and failure to meet consumer and market demands. Consequently, a key objective for breeders is to develop varieties with enhanced tolerance to water deficits in the soil.

An essential preliminary step in breeding programs aimed at enhancing plant resistance to water deficit through improved water-use efficiency is to assess the degree of adaptation of various genotypes (cultivars) to limited water availability. For example, the development of drought-resistant wheat and cotton varieties has shown promise in maintaining agricultural productivity despite environmental stress [22,23]. Differences in physiological, morphological, and morphogenetic traits at the cultivar level have been documented in numerous studies [3,11,24–27]. Wild plant varieties possess genetic resources that enhance their resilience to abiotic stresses, as demonstrated in studies on species such as potato [28] and barley [29]. However, there is a lack of studies evaluating the differences in physiological traits and responses to soil water deficit between cultivated and wild *Fragaria* species.

Therefore, the aim of this study was to compare the physiological responses of *Fragaria vesca* plants—two cultivars ('Rugia' and 'Baron von Solemacher') and a wild strawberry specimen obtained from a natural habitat, to identify the genotypes with greater resistance. This will enable the determination of whether wild-growing plants exhibit lower sensitivity to drought stress, and consequently, whether their use in breeding programs aimed at developing water deficit-resistant varieties is justified. This objective was achieved by testing the following hypotheses: (1) there are physiological differences between cultivated varieties and wild-growing plants, (2) water deficit negatively affects the physiological traits of *Fragaria vesca* plants, and (3) wild-growing plants exhibit lower sensitivity to water deficit in the substrate compared to cultivated varieties.

2. Materials and Methods

2.1. Materials

The plants used in this study included *Fragaria vesca* cultivars 'Rugia' and 'Baron von Solemacher' (seedlings obtained from TOP-PLANT, Szczecin, Poland) as well as wild strawberry seedlings collected from a natural habitat in a forested area near the village of Ciemnik, close to Ińsko, Poland (53°23′ N, 15°34′ E; 105 m a.s.l.). The cultivated varieties used in the study are popular in Poland. Both varieties are high-yielding and produce fruit from June to September. However, 'Baron von Solemacher' exhibits lower frost resistance and greater susceptibility to diseases compared to 'Rugia' [30]. These characteristics made them suitable candidates for experimental analysis.

2.2. Methods

A two-year pot experiment was conducted during the period of 2018–2019, employing a completely randomized design with 15 replications (one plant per replication). The study was carried out in the vegetation hall of the Faculty of Environmental Management and Agriculture at the West Pomeranian University of Technology in Szczecin, Poland (53°25′ N, 14°32′ E; 25 m a.s.l.). Plastic containers with a capacity of 10 dm^3 were used, filled with a substrate consisting of a 1:1 mixture (by volume) of sandy silty clay soil, collected from the arable humus horizon (0–30 cm) at the Agricultural Experimental Station of the West Pomeranian University of Technology in Szczecin, located in Lipnik (Poland), and sand. The physicochemical properties of the soil used in the experiment are presented in Table 1. To allow for acclimatization to the new conditions, the plants were planted one year prior to the start of measurements (in June) of 3 plants per pot. The seedlings had three fully developed leaves.

Table 1. The physicochemical properties of the soil used in the experiment.

pH		Fine Soil Fractions [%]	S_0 [g·cm^{-3}]	Content of P	Content of K	Content of Mg	Salinity [g·NaCl·dm^{-3}]	W_{tv}
H_2O	KCl			[mg · 100 g^{-1}]				
6.70	5.98	21.00	1.25	5.87	12.66	5.98	0.45	37.10

Explanations: S_0—bulk density, W_{tv}—total water capacity volume.

Water potential was maintained at −10 kPa to −15 kPa under control conditions (Ctrl) and −30 kPa to −35 kPa under water deficit conditions (Def) in the substrate. Each pot was watered using individual drippers when the water potential in the substrate increased above −15 kPa under control conditions or above −35 kPa under water deficit conditions. Irrigation was carried out to a value of −10 kPa in the control variant and to −30 kPa under water deficit conditions. The adopted water potential range of −30 to −35 kPa was considered as stress conditions for the studied species, based on the findings of various authors [3]. The need for irrigation was determined based on readings from contact soil tensiometers placed at a depth of 20 cm in the pots of each experimental variant. Tensiometric measurements have been successfully employed in other studies investigating water deficit in plants [31–35].

During the vegetation period (from May to October), the pots were placed in the covered section of the vegetation hall. In the remaining months, they were kept in an unheated greenhouse, with a temperature maintained above 0 °C.

Fertilization was carried out twice during the growing season: prior to planting the plants, during pot filling, and after the plants had finished fruiting. In the second year of the study, fertilization took place in early spring, before the start of vegetation, and after the completion of fruiting. A multi-component fertilizer specifically designed for strawberries was used, which, according to the manufacturer (Agrecol Sp. z o.o., Mesznary 2, Poland), contains the following: 15.0% N (6.5% nitrate N, 8.5% ammonium N), 6.0% P_2O_5 (soluble in neutral ammonium citrate and in water, including 5.2% water-soluble P_2O_5),

11.0% water-soluble K_2O, 0.04% total B, 0.1% total Cu, 0.2% total Fe, 0.17% total Mn, 0.01% total Mo, and 0.035% total Zn. The fertilizer was applied at a rate of 4 g per pot each time.

Morphological observations of the plants were conducted systematically throughout the growing season, and emerging stolons were removed. Additionally, plants were regularly inspected for the presence of pests. When pest populations exceeded the economic threshold, chemical treatments were applied as follows:

- Decis Mega 50 EW, targeting aphids (*Aphidoidea*) at a dosage of 0.25 L/ha.
- Mospilan 20 SP, targeting thrips (*Thysanoptera*, aphids (*Aphidoidea*), and greenhouse whitefly (*Trialeurodes vaporariorum*) at a concentration of 4 g per 10 L of water, applied at 3 L per 100 m^2.

2.3. Measurements

In each year of the study, three times during the plant growing season, in the second ten days of July, August, and September, the following measurements and analyses were conducted:

- the content of photosynthetic pigments was assessed by the method of Arnon et al. [36] modified by Lichtenthaler and Wellburn [37] for chlorophyll "a", "b" and total chlorophyll, and the method of Hager and Mayer-Berthenrath [38] for carotenoids;
- the relative leaf water content index (RWC) by the method of Yamasaki and Dillenburg [39];
- the gas exchange parameters, i.e., assimilation intensity CO_2 net (P_n), transpiration intensity (E), stomatal conductivity H_2O (g_s) and substomatal CO_2 concentration (c_i), were measured using a TPS-2 portable gas analyzer with a PLC-4 (PP Systems, Amesbury, MA, USA); the photosynthetic water utilization factor (WUE) was determined on the basis of the P_n/E quotient;
- the fluorescence parameters of chlorophyll "a" were determined using a Handy PEA spectrofluorometer (Hansatech Ltd., Kings Lynn, UK), based on the standard procedure of the apparatus (3 × 650 nm LED, maximum actinic light intensity 3000 μmol m^{-2} s^{-1}). The measurement was carried out in each variant on 15 randomly selected, fully grown bean leaves (the repetition was the measurement on one leaf), in a place previously darkened for 20 min, using factory-made clips (the irradiation area was 4 mm). The following parameters were measured: initial (zero) fluorescence, index of excitation energy loss in power antennas (F_0), maximum fluorescence, after reduction of acceptors in PS II and after darkroom adaptation (F_M), variable fluorescence, determined after dark adaptation (the parameter depends on maximum quantum efficiency of PS II) ($F_V = F_M - F_0$), maximum potential efficiency of the photochemical reaction in PS II, determined after dark adaptation, after reduction of acceptors in PS II (F_V/F_M), chlorophyll fluorescence growth time from the beginning of the measurement to reaching maximum (T_{FM}), and the area above the fluorescence induction curve of chlorophyll "a" between points F_0 and F_M proportional to the size of the pool of reduced plastoquinone electron acceptors in PSII (A_M (Area)).

In each growing season, the total yield of fresh strawberry fruit was determined by weight, expressed in grams per pot, as well as the weight of a single fruit. Due to the fact that wild strawberries produce fruit throughout almost the entire growing season, harvesting was carried out successively as the fruit reached collective maturity, spanning the period from June to August each year of experiment.

After each experiment, the fresh and dry mass of the root system was determined. The content of K, Ca, Mg, Na, Cu, Zn, Mn, and Mo, as well as the ratio of monovalent to divalent elements in the leaves, roots, and plant crowns, was also measured. The content of calcium, magnesium, sodium, and microelements in plant tissues was determined by atomic absorption spectrometry (AAS) using an iCE 3000 series spectrophotometer (ThermoScientific, Waltham, MA, USA) in wet mineralized material, prepared with a mixture of nitric acid (HNO_3) and hydrochloric acid (HCl) in a 1:1 ratio. Phosphorus content was determined colorimetrically using a Marcel Mini spectrophotometer. Based on the obtained results, the equivalent ratio of monovalent to divalent elements, i.e., K: (Ca + Mg), was additionally calculated.

2.4. Meteorological Data

Szczecin is located in a sub-Atlantic climate zone, where the primary influencing factors include latitude, topography, distance from the sea, and the presence of large industrial and production facilities. These factors result in relatively high annual average air temperatures and a low annual temperature amplitude. Additionally, Szczecin typically experiences mild winters and cool summers.

The key meteorological elements relevant to the conducted research, along with comparisons to multi-year averages, are presented in Table 2. The data used in this study were sourced from the nearest meteorological station, IMGW Szczecin-Dąbie (Przestrzenna 10 Street).

Table 2. Meteorological conditions during the growing seasons of the studied plants from 2017 to 2019, compared to multi-year averages for the Szczecin-Dąbie Meteorological Station.

Years	**Months**							**Average/Sum**
	IV	**V**	**VI**	**VII**	**VIII**	**IX**	**X**	
	Mean air temperature (°C)							
2017	7.4	14.1	17.2	17.7	18.2	13.6	11.2	14.2
Deviation	−1.8	0.5	0.4	−1.2	−0.3	−0.7	1.7	−0.2
2018	12.3	16.6	18.5	20.0	20.0	15.4	10.3	16.2
Deviation	3.1	3.0	1.7	1.1	1.5	1.1	0.8	1.8
2019	10.1	12.1	21.5	18.8	20.1	14.5	10.7	15.4
Deviation	0.9	−1.5	4.7	−0.1	1.6	0.2	1.2	1.0
1991–2020	9.2	13.6	16.8	18.9	18.5	14.3	9.5	14.4
	Sunshine duration (h)							
2017	143.2	240.3	215.5	197.1	229.1	130.6	96.4	11,252.2
% of the norm	71.9	98.3	89.0	80.1	99.5	81.6	96.8	88.1
2018	257.6	349.8	253.6	301.0	269.2	204.0	170.36	1805.8
% of the norm	129.3	143.1	104.7	122.3	116.9	127.5	171.3	127.0
2019	285.7	198.2	345.5	222.3	239.1	168.5	-	1459.3
% of the norm	143.4	81.1	142.7	90.3	103.8	105.3	x	102.6
1991–2020	199.3	244.5	242.2	246.2	230.3	160.0	99.6	1422.1
	Total precipitation (mm)							
2017	42.2	90.8	132.7	192.5	43.8	30.4	93.9	626.3
% of the norm	135.3	162.7	224.5	252.6	72.6	63.7	215.9	167.5
2018	26.8	22.5	15.0	92.8	21.4	16.3	20.2	215.0
% of the norm	85.9	40.3	25.4	121.8	35.5	34.2	46.4	57.5
2019	10.7	68.7	70.8	23.5	41.8	63.2	46.1	324.8
% of the norm	34.3	123.1	119.8	30.8	69.3	132.5	106.0	86.9
1991–2020	31.2	55.8	59.1	76.2	60.3	47.1	43.5	373.8

Table 2. *Cont.*

Years	Months							Average/Sum
	IV	V	VI	VII	VIII	IX	X	
				Relative air humidity (%)				
2017	73.8	70.9	74.5	79.4	76.5	82.1	85.6	77.5
Deviation	3.0	−0.6	1.6	5.0	0.6	0.9	0.1	1.5
2018	69.3	63.7	66.4	71.5	67.8	74.2	79.1	70.3
Deviation	−1.5	−7.8	−6.5	−2.9	−8.1	−7.0	−6.4	−5.7
2019	60.0	71.2	65.5	70.0	69.7	78.2	84.3	71.3
Deviation	−10.8	−0.3	−7.4	−4.4	−6.2	−3.0	−1.2	−4.8
1991–2020	70.8	71.5	72.9	74.4	75.9	81.2	85.5	76.0

—no data available; x—irrelevant data. Source: IMGW Station in Szczecin-Dąbie.

In the first year of the study (2017), the average air temperature during the growing season was 0.2 °C lower than the multi-year average. The coldest month in the analyzed period was April (7.4 °C), while the warmest was August (18.2 °C). Sunshine duration during the growing season in 2017 was 21.9% below the multi-year average. No month recorded sunshine levels above the average values.

In the second year of the study (2018), the average air temperature during the growing season was 1.8 °C higher than the multi-year average. The lowest average temperature was observed in October (10.3 °C), while the highest was recorded in July and August (20.0 °C). Sunshine duration during the growing season was 27.0% above the multi-year average, with all months exceeding average values.

The third year of the study (2019) was characterized by an average air temperature during the growing season that was 1.0 °C higher than the multi-year average. The coldest month was April (10.1 °C), and the warmest was June (21.5 °C). Sunshine duration during the analyzed period was 2.6% above the multi-year average. April, June, August, and September recorded higher sunshine values compared to the same months in the multi-year period, while the remaining months had sunshine levels below the long-term average.

2.5. Statistical Analysis

The results of the study were statistically analyzed in STATISTICA version 13.3 (StatSoft, Poland). A one-way analysis of variance (ANOVA) was employed to compare genotypes grown under control conditions, while a two-way ANOVA was used to evaluate genotype × substrate moisture interactions in a completely randomized design. The significance of differences between means was determined using Duncan's multiple range test at a significance level of $\alpha = 0.05$. Identical single-letter labels were assigned to means that did not differ statistically. The analyses were conducted separately for each measurement period.

In the case of two-factor analysis of variance, means denoted by the same lowercase letter did not differ significantly within the context of a significant interaction effect. Similarly, means denoted by the same uppercase letter showed no significant differences with respect to the significant main effect (substrate moisture level; SML). The main effect of genotype (G) was not analyzed in the two-factor analysis due to the assumption that responses to water stress vary among genotypes and may follow different directions of change. Comparisons of genotype values were performed using one-way analysis of variance exclusively under control conditions.

2.6. Radar Charts

The radar charts present the values of selected analyzed parameters. The differences between the values for plants grown under control and water-deficient soil moisture conditions were used to calculate the parameter values shown in the charts.

For the S parameters—i.e., those for which a smaller decrease under water deficit conditions, compared to controlmoisture conditions, indicated better adaptation to stress (E, g_s, P_n, c_i, WUE, F_0, F_M, F_V, F_V/F_M, chla, chlb, chl total, carot, RWC, fruit fresh mass yield, single fruit mass)—the following formula was used:

$$S = \frac{S_{Def} \times 100}{S_{Ctrl}} \quad (1)$$

where:

S_{Def}—parameter value in conditions of deficit soil moisture;
S_{Ctrl}—parameter value in conditions of control soil moisture.

For the D parameters—i.e., those for which a smaller increase under water deficit conditions, compared to control moisture conditions, indicated better adaptation to stress (T_{FM}, A_M)—the following formula was used:

$$D = 100 - \left(\frac{D_{Def} \times 100}{D_{Ctrl}} - 100 \right) \quad (2)$$

where:

D_{Def}—parameter value in conditions of deficit soil moisture;
D_{Ctrl}—parameter value in conditions of control soil moisture.

The charts were constructed based on a series of analyses testing the significance of differences between means for individual object types, using one-way analysis of variance (ANOVA). For means that did not differ significantly from each other, as well as for values showing trends divergent from the general patterns—i.e., for S parameters, where values under deficit conditions were significantly higher than those under control conditions, and for D parameters, where values under deficit conditions were significantly lower than under control conditions—a value of 100 was assigned on the radar graphs.

Comparisons of the studied objects were made by calculating the area of the polygon formed by the chart line.

3. Results

3.1. Evaluation of Differences in Physiological Traits, Chemical Composition, and Yield Among Genotypes

3.1.1. Gas Exchange Parameters

The results showed that, for each measurement term, the highest values of transpiration intensity (E) under control substrate moisture conditions were characteristic of different varieties (Table 3). On the first measurement date of each study year, the 'Rugia' variety exhibited a significantly lower E value compared to the 'Baron von Solemacher' variety (by an average of 13.6%) and plants obtained from the natural environment (by an average of 13.0%). For the remaining dates, except for the third measurement date of the second study year, the 'Rugia' variety most frequently exhibited the highest E value (by an average of 35.3%). On the final measurement date of the experiment, this variety again had the lowest E value among the tested genotypes (by 8.4% lower than 'Baron von Solemacher' and by 42.1% lower than wild plants).

Table 3. Gas exchange parameters in leaves of *F. vesca* genotypes.

Parameters	Genotypes	Measurement Terms					
		1st Year			2nd Year		
		July	August	September	July	August	September
E [mmol $H_2O \cdot m^{-2} \cdot s^{-1}$]	B	1.105 ^{a*}	0.502 ^{b}	0.967 ^{a}	0.707 ^{a}	0.841 ^{ab}	0.573 ^{b}
	R	1.017 ^{b}	0.645 ^{a}	1.036 ^{a}	0.571 ^{b}	0.992 ^{a}	0.525 ^{c}
	N	1.120 ^{a}	0.653 ^{a}	0.726 ^{b}	0.687 ^{a}	0.737 ^{b}	0.907 ^{a}
g_s [mol $H_2O \cdot m^{-2} \cdot s^{-1}$]	B	0.094 ^{a}	0.137 ^{a}	0.061 ^{ab}	0.064 ^{a}	0.043 ^{b}	0.063 ^{b}
	R	0.087 ^{a}	0.107 ^{b}	0.064 ^{a}	0.048 ^{b}	0.053 ^{a}	0.049 ^{b}
	N	0.086 ^{a}	0.150 ^{a}	0.050 ^{b}	0.060 ^{a}	0.038 ^{b}	0.120 ^{a}
P_n [mmol $H_2O \cdot m^{-2} \cdot s^{-1}$]	B	6.747 ^{a}	2.200 ^{b}	4.727 ^{a}	3.760 ^{a}	1.520 ^{b}	2.513 ^{b}
	R	6.847 ^{a}	1.780 ^{b}	3.740 ^{ab}	3.487 ^{a}	3.873 ^{a}	4.107 ^{a}
	N	7.307 ^{a}	3.253 ^{a}	3.427 ^{b}	3.760 ^{a}	4.260 ^{a}	4.893 ^{a}
c_i [µmol $CO_2 \cdot mol^{-1}$]	B	311.00 ^{a}	421.33 ^{ab}	315.00 ^{a}	298.20 ^{a}	355.73 ^{a}	498.47 ^{a}
	R	299.93 ^{a}	426.20 ^{a}	302.53 ^{a}	280.87 ^{a}	289.40 ^{b}	371.47 ^{b}
	N	279.80 ^{b}	411.73 ^{b}	304.53 ^{a}	298.87 ^{a}	244.33 ^{c}	449.80 ^{a}
WUE	B	6.201 ^{a}	3.484 ^{b}	5.167 ^{a}	5.409 ^{a}	1.922 ^{c}	4.656 ^{b}
	R	6.747 ^{a}	3.812 ^{b}	3.799 ^{a}	6.253 ^{a}	4.028 ^{b}	5.432 ^{b}
	N	6.519 ^{a}	5.168 ^{a}	5.173 ^{a}	5.328 ^{a}	6.533 ^{a}	20.713 ^{a}

Explanations: B—'Baron von Solemacher' cultivar; R—'Rugia' cultivar; N—plants obtained from a natural habitat. *—means marked with the same letters do not differ significantly at the significance level of $\alpha = 0.05$.

Similar differences were observed for the stomatal conductance of H_2O (g_s). In the first term of the first year of the study, no significant differences were found between the genotypes. The 'Rugia' cultivar exhibited the lowest g_s value in the second term of the first year (by an average of 25.3%), the first term of the second year (by an average of 22.5%), and, together with the 'Baron von Solemacher' cultivar, in the third term of the second year (by an average of 53.3%). However, in the second term of the second year, the highest g_s value was observed for the 'Rugia' cultivar among the analyzed genotypes (by an average of 23.6%).

A significant differentiation between the analyzed genotypes, depending on the measurement date, was also observed for net CO_2 assimilation intensity (P_n). On the first measurement date of each year, no differences in P_n values were found among the genotypes. Plants collected from the natural environment exhibited the highest P_n value on the second measurement date of the first year (by an average of 38.8%), and, along with the 'Rugia' cultivar, on the second and third measurement dates of the second year (by an average of 53.1%).

Similar to other gas exchange parameters, the substomatal CO_2 concentration (c_i) also varied across the individual measurement dates. Across all dates, the 'Baron von Solemacher' cultivar most frequently exhibited the highest c_i value among the tested genotypes, particularly in the second term of the second year of the study, where the value was 18.6% higher than that of 'Rugia' and 31.3% higher than that of wild plants.

No differences were observed between the studied genotypes in terms of the photosynthetic water use efficiency (WUE) in the first and third terms of the first year, and in the first term of the second year of the study. In the remaining terms, the highest WUE values were observed in plants collected from the natural habitat (by an average of 53.2%).

3.1.2. Chlorophyll "a" Fluorescence Parameters

In most of the measurement terms, no differences were observed in the initial (zero) fluorescence value (F_0) between the tested genotypes (Table 4). The exception occurred on the second measurement term of the second year, when the lowest F_0 value was observed in plants collected from the natural environment (by an average of 14.0%).

Table 4. Chlorophyll "a" fluorescence parameters in leaves of *F. vesca* genotypes.

Parameters	Genotypes	Measurement Terms					
		1st Year			2nd Year		
		July	August	September	July	August	September
F_0	B	600.786 [a*]	616.857 [a]	587.429 [a]	259.143 [a]	261.857 [a]	266.857 [a]
	R	594.375 [a]	612.857 [a]	598.250 [a]	268.625 [a]	273.125 [a]	278.188 [a]
	N	600.429 [a]	594.000 [a]	582.929 [a]	258.571 [a]	229.857 [b]	250.643 [a]
F_M	B	2804.571 [a]	2523.143 [b]	2917.929 [ab]	1246.571 [b]	1222.714 [b]	1270.500 [a]
	R	2795.000 [a]	2680.688 [ab]	2821.563 [b]	1346.688 [a]	1398.500 [a]	1323.500 [a]
	N	2998.429 [a]	2782.643 [a]	3067.071 [a]	1333.143 [a]	992.571 [c]	1258.643 [a]
F_V	B	2203.786 [a]	1906.286 [b]	2330.500 [ab]	987.429 [b]	960.857 [b]	1003.643 [a]
	R	2200.625 [a]	2068.250 [ab]	2223.313 [b]	1078.063 [a]	1125.375 [a]	954.313 [a]
	N	2398.000 [a]	2188.643 [a]	2484.143 [a]	1074.571 [a]	762.714 [c]	1008.000 [a]
F_V/F_M	B	0.782 [a]	0.754 [b]	0.796 [ab]	0.792 [a]	0.785 [b]	0.780 [ab]
	R	0.784 [a]	0.771 [ab]	0.785 [b]	0.799 [a]	0.804 [a]	0.772 [b]
	N	0.793 [a]	0.779 [a]	0.809 [a]	0.805 [a]	0.743 [ab]	0.801 [a]
T_{FM}	B	373.571 [b]	240.714 [b]	305.000 [ab]	578.571 [b]	277.857 [a]	535.000 [a]
	R	404.375 [b]	236.875 [b]	279.375 [b]	493.751 [c]	324.375 [a]	518.750 [a]
	N	507.143 [a]	273.571 [a]	339.286 [a]	671.429 [a]	317.144 [a]	560.000 [a]
A_M	B	52,530.786 [b]	31,023.786 [b]	45,809.500 [b]	28,356.286 [b]	16,458.857 [b]	22,092.714 [b]
	R	54,069.125 [b]	35,609.250 [ab]	42,670.813 [b]	30,417.688 [b]	23,104.000 [a]	23,176.813 [b]
	N	68,608.929 [a]	39,444.500 [a]	57,824.214 [a]	40,694.786 [a]	17,188.714 [b]	32,145.643 [a]

Explanations: B—'Baron von Solemacher' cultivar; R—'Rugia' cultivar; N—plants obtained from a natural habitat. *—means marked with the same letters do not differ significantly at the significance level of $\alpha = 0.05$.

For maximum fluorescence (F_M), no differences were observed between the tested genotypes during the first term of the first year and the third term of the second year of the study. For the remaining measurements, the highest F_M values were most frequently observed in plants collected from the natural environment (e.g., in the second term of the first year, where the value was 9.3% higher than that of 'Baron von Solemacher' and the first term of the second year—6.5%); however, these results were not consistent.

Similar to F_M, no differences in variable fluorescence (F_V) were observed between the tested genotypes in the first term of the first year and the third term of the second year of the study. In the remaining terms, the 'Baron von Solemacher' cultivar was most frequently characterized by the lowest F_V value, in comparison with the 'Rugia' cultivar and plants collected from the natural environment, particularly in the first term of the second year of the study, where the value was 8.4% lower than that of 'Rugia' and 8.1% lower than that of wild plants.

No differences were found between the compared genotypes in terms of the ratio of variable to maximum fluorescence (F_V/F_M) during the first term of both the first and the second years of the study. Although the results in the other terms were not consistent, it can be concluded that the 'Baron von Solemacher' cultivar most frequently exhibited the lowest F_V/F_M values (e.g., in the second term of the first year the value was 3.2% lower than that of wild plants, and in the second term of the second year the value was 2.4% lower than that of 'Rugia'), while plants collected from the natural environment were characterized by the highest values (e.g., in the third term of the second year the value was 3.6% higher than that of 'Rugia').

In all measurement terms, except for the second and third terms of the second year, where no significant differences were observed, the highest fluorescence growth time (T_{FM}) was recorded in plants collected from the natural environment (by an average of 17.5%). However, it should be noted that during the third term of the first year, the results did not indicate this unambiguously.

For the area above the fluorescence induction curve (A_M), the highest values were most frequently observed in plants collected from the natural habitat (by an average of 25.3%). The exceptions were the second term of the first year, where the results were ambiguous, and the second term of the second year, when the highest A_M value was recorded for the 'Rugia' cultivar (by an average of 27.2%).

3.1.3. Content of Photosynthetic Pigments

In most measurement terms during both years of the study, the highest chlorophyll "a" content in the leaves was observed in strawberries collected from the natural environment—by an average of 29.8% (Table 5). In contrast, the lowest values were typically recorded for the 'Rugia' cultivar (e.g., in the third term of the first year the value was 27.0% lower than that of 'Baron von Solemacher' and 41.4% lower than that of wild plants).

Table 5. Content of photosynthetic pigments in leaves of *F. vesca* genotypes [mg · g^{-1} FM].

Parameters	Genotypes	Measurement Terms					
		1st Year			2nd Year		
		July	August	September	July	August	September
Chlorophyll "a"	B	1.711 ^{ab*}	0.851 ^{b}	1.427 ^{b}	1.327 ^{b}	1.952 ^{a}	1.857 ^{b}
	R	1.360 ^{b}	1.090 ^{b}	1.042 ^{c}	1.355 ^{b}	1.528 ^{b}	1.641 ^{b}
	N	1.995 ^{a}	1.420 ^{a}	1.777 ^{a}	2.116 ^{a}	1.780 ^{ab}	2.201 ^{a}
Chlorophyll "b"	B	0.622 ^{a}	0.565 ^{a}	0.531 ^{b}	0.564 ^{a}	0.627 ^{a}	0.627 ^{ab}
	R	0.523 ^{a}	0.562 ^{a}	0.523 ^{b}	0.654 ^{a}	0.486 ^{b}	0.591 ^{b}
	N	0.702 ^{a}	0.700 ^{a}	0.855 ^{a}	0.846 ^{a}	0.508 ^{ab}	0.721 ^{a}
Chlorophyll "a" to chlorophyll "b" ratio	B	2.747 ^{a}	1.528 ^{b}	2.684 ^{a}	2.385 ^{a}	3.113 ^{b}	2.963 ^{a}
	R	2.632 ^{a}	1.932 ^{a}	2.074 ^{a}	2.208 ^{a}	3.154 ^{b}	2.780 ^{b}
	N	2.839 ^{a}	2.028 ^{a}	2.107 ^{a}	2.523 ^{a}	3.504 ^{a}	3.056 ^{a}
Total chlorophyll	B	2.333 ^{ab}	1.415 ^{b}	1.958 ^{b}	2.333 ^{ab}	1.415 ^{b}	1.958 ^{b}
	R	1.882 ^{b}	1.652 ^{b}	1.565 ^{c}	1.882 ^{b}	1.652 ^{b}	1.565 ^{c}
	N	2.697 ^{a}	2.119 ^{a}	2.631 ^{a}	2.697 ^{a}	2.119 ^{a}	2.631 ^{a}
Carotenoids	B	0.959 ^{ab}	0.672 ^{b}	0.866 ^{b}	0.683 ^{b}	0.966 ^{a}	0.933 ^{b}
	R	0.728 ^{b}	0.748 ^{b}	0.639 ^{c}	0.696 ^{b}	0.902 ^{ab}	0.825 ^{b}
	N	1.179 ^{a}	1.009 ^{a}	1.170 ^{a}	1.050 ^{a}	0.772 ^{b}	1.120 ^{a}

Explanations: B—'Baron von Solemacher' cultivar; R—'Rugia' cultivar; N—plants obtained from a natural habitat. *—means marked with the same letters do not differ significantly at the significance level of $\alpha = 0.05$.

In all three measurement periods, no statistically significant differences were found between the studied genotypes regarding chlorophyll "b" content in the leaves. In the third term of the first year, the highest chlorophyll "b" content was observed in the leaves of wild strawberry plants collected from the natural environment (by an average of 31.5%). In the second term of the second year, the 'Baron von Solemacher' cultivar exhibited a higher chlorophyll "b" content compared to the 'Rugia' cultivar (by 22.5%). In the third term of the same year, the leaves of wild strawberry plants from the natural environment had a higher chlorophyll "b" content, by 18.0%, than those of the 'Rugia' cultivar.

In the first and third terms of the first year, as well as in the first term of the second year of the study, no differences were observed between the compared genotypes regarding the ratio of chlorophyll "a" to chlorophyll "b". For the remaining measurement terms, the highest values of this parameter were found in plants collected from the natural environment (e.g., in the second term of the second year the value was 11.2% higher than that of 'Baron von Solemacher', and 10.0% higher than that of 'Rugia').

The content of total chlorophyll in the leaves of the tested *F. vesca* genotypes exhibited similar trends to those observed for the previously discussed assimilation pigments. In all measurement terms, except for the first term of the first year and second term of the second year, the highest total chlorophyll content was found in the leaves of plants collected from a natural environment (by an average of 30.3%). In the first term of the first year, the 'Baron von Solemacher' cultivar also showed a similarly high value of this parameter.

Additionally, the content of carotenoids in the leaves of the studied genotypes was highest in plants collected from the natural environment in most measurement terms (by an average of 30.3%). The exception was the first measurement term of the first year, where the 'Baron von Solemacher' cultivar exhibited an equally high carotenoid content, and the second measurement term of the second year, when this cultivar surpassed by 20.1% the wild strawberry collected from a natural environment in carotenoid content.

3.1.4. Relative Water Content in Leaves (RWC)

In the first term of the second year of the study, no significant differences in relative water content (RWC) were observed among the genotypes studied. Additionally, the results obtained in the other terms did not allow for a clear identification of plants with the highest or lowest RWC values (Table 6). Only in the second term of the first year was the highest RWC observed in the leaves of wild strawberry plants collected from the natural environment (15.8% greater than 'Baron von Solemacher', and 15.2% greater than 'Rugia').

Table 6. Relative water content (RWC) in leaves of *F. vesca* genotypes [%].

Genotypes	Measurement Terms					
	1st Year			2nd Year		
	July	August	September	July	August	September
B	54.445 ^{b*}	54.765 ^{b}	56.308 ^{ab}	45.247 ^{a}	42.064 ^{a}	46.258 ^{b}
R	62.644 ^{a}	55.159 ^{b}	61.177 ^{a}	39.875 ^{a}	40.545 ^{ab}	58.035 ^{a}
N	58.285 ^{ab}	65.073 ^{a}	54.813 ^{b}	37.150 ^{a}	30.509 ^{b}	52.958 ^{ab}

Explanations: B—'Baron von Solemacher' cultivar; R—'Rugia' cultivar; N—plants obtained from a natural habitat. *—means marked with the same letters do not differ significantly at the significance level of $\alpha = 0.05$.

3.1.5. Fruit Yield and Weight of a Single Wild Strawberry Fruit

Throughout the two-year experiment, flowering, which was absent in plants collected from their natural habitats, led to the production of fruit yield. In the second year, the 'Baron von Solemacher' cultivar produced 29.7% higher yield of fresh fruit mass than the 'Rugia' cultivar. However, in both years of the study, no significant differences were observed between the 'Rugia' and 'Baron von Solemacher' cultivars in terms of single fruit mass (Table 7).

Table 7. Total fruit yield from one plant and weight of a single fruit during the vegetation period (June–September) in *F. vesca* genotypes [g].

Parameters	Genotypes	1st Year	2nd Year
Fresh Fruit Mass Yield	B	23.79 ^{a*}	33.11 ^{a}
	R	22.20 ^{a}	23.27 ^{b}
	N	-	-
Single Fruit Mass	B	0.65 ^{a}	0.55 ^{a}
	R	0.59 ^{a}	0.54 ^{a}
	N	-	-

Explanations: B—'Baron von Solemacher' cultivar; R—'Rugia' cultivar; N—plants obtained from a natural habitat. *—means marked with the same letters do not differ significantly at the significance level of $\alpha = 0.05$.

3.1.6. Fresh and Dry Mass of the Root System

Upon completion of the two-year experiment, the fresh and dry mass of the root system was measured. No significant differences were observed among the tested *F. vesca* genotypes regarding both fresh and dry root mass (Table 8).

Table 8. Fresh and dry mass of the root system of *F. vesca* genotypes [g].

Genotypes	Fresh Mass of the Root System	Dry Mass of the Root System
B	11.81 ^{a*}	6.83 ^{a}
R	13.30 ^{a}	6.83 ^{a}
N	17.80 ^{a}	10.47 ^{a}

Explanations: B—'Baron von Solemacher' cultivar; R—'Rugia' cultivar; N—plants obtained from a natural habitat. *—means marked with the same letters do not differ significantly at the significance level of $\alpha = 0.05$.

3.1.7. Chemical Composition

Analysis of the average elemental content in the tested *F. vesca* genotypes revealed no significant differences in the concentrations of K, Ca, Mg, Na, Zn, Mn, and Fe (Table 9). However, the 'Rugia' cultivar exhibited 25.7% lower Mo content compared to the other genotypes and 31.7% higher Cu content than plants collected from the natural environment.

Table 9. Chemical composition of leaves, crowns and roots of *F. vesca* genotypes.

Plant Organ	Genotypes	K [g·kg^{-1}]	Ca [g·kg^{-1}]	Mg [g·kg^{-1}]	Na [g·kg^{-1}]	Cu [mg·kg^{-1}]	Zn [mg·kg^{-1}]	Mn [mg·kg^{-1}]	Mo [mg·kg^{-1}]	Fe [mg·kg^{-1}]
Leaves	B	97.6 ^{a*}	66.9 ^{a}	13.5 ^{b}	4.8 ^{a}	37.3 ^{a}	172.5 ^{ab}	201.8 ^{a}	11.7 ^{a}	1.9 ^{a}
	R	106.9 ^{a}	69.7 ^{a}	15.7 ^{a}	2.8 ^{a}	52.2 ^{a}	236.4 ^{a}	205.4 ^{a}	6.6 ^{b}	1.8 ^{a}
	N	96.5 ^{a}	51.9 ^{b}	17.2 ^{a}	2.7 ^{a}	45.6 ^{a}	146.8 ^{b}	139.2 ^{b}	12.8 ^{a}	0.7 ^{b}
Crowns	B	82.9 ^{a}	115.6 ^{b}	3.9 ^{b}	28.0 ^{ab}	87.9 ^{ab}	155.2 ^{b}	31.4 ^{b}	17.6 ^{a}	0.2 ^{a}
	R	92.2 ^{a}	91.4 ^{b}	3.4 ^{b}	37.4 ^{a}	117.1 ^{a}	141.3 ^{b}	30.8 ^{b}	16.4 ^{a}	0.2 ^{a}
	N	50.4 ^{b}	159.7 ^{a}	5.4 ^{a}	10.0 ^{b}	74.4 ^{b}	212.9 ^{a}	61.8 ^{a}	17.3 ^{a}	0.2 ^{a}
Roots	B	25.9 ^{a}	80.7 ^{a}	13.4 ^{a}	6.1 ^{a}	92.5 ^{b}	853.5 ^{ab}	373.3 ^{b}	12.4 ^{a}	6.3 ^{b}
	R	24.5 ^{a}	63.4 ^{a}	13.4 ^{a}	5.9 ^{a}	111.4 ^{a}	677.8 ^{b}	388.0 ^{b}	9.4 ^{a}	6.4 ^{b}
	N	12.9 ^{b}	60.2 ^{a}	9.8 ^{b}	5.9 ^{a}	71.8 ^{c}	883.8 ^{a}	940.8 ^{a}	14.6 ^{a}	13.7 ^{a}
Average	B	69.9 ^{a}	89.9 ^{a}	9.8 ^{a}	32.6 ^{a}	73.7 ^{ab}	375.4 ^{a}	189.0 ^{a}	14.2 ^{a}	2.6 ^{a}
	R	74.5 ^{a}	74.8 ^{a}	10.8 ^{a}	15.4 ^{a}	93.6 ^{a}	351.8 ^{a}	208.1 ^{a}	10.8 ^{b}	2.8 ^{a}
	N	53.3 ^{a}	90.6 ^{a}	10.8 ^{a}	6.2 ^{a}	63.9 ^{b}	414.5 ^{a}	380.6 ^{a}	14.9 ^{a}	4.9 ^{a}

Explanations: B—'Baron von Solemacher' cultivar; R—'Rugia' cultivar; N—plants obtained from a natural habitat. *—means marked with the same letters do not differ significantly at the significance level of $\alpha = 0.05$.

An analysis of element content within specific plant organs revealed that wild strawberries collected from the natural environment exhibited lower concentrations of Ca (by an average of 24.0%), Mn (by 31.6%), and Fe (by 62.1%) in the leaves compared to the 'Rugia' and 'Baron von Solemacher' cultivars, as well as 37.9% lower Zn concentration than the

'Rugia' cultivar. The Baron von Solemacher variety showed the lowest Mg content in the leaves (by an average of 17.8%), while 'Rugia' had the lowest Mo content (by an average of 46.0%). No significant differences were observed among the genotypes in terms of K, Na, and Cu content in the leaves.

The genotypes studied showed no significant differences in Mo and Fe content within the crowns. However, plants collected from the natural environment exhibited higher concentrations of Ca (by an average of 35.2%), Mg (by 32.4%), Zn (by 30.4%), and Mn (by 49.7%) in the crowns compared to both cultivars, as well as lower levels of Na (by 274%) and Cu (by 57.4%) relative to the 'Rugia' cultivar. Additionally, these wild-sourced plants contained lower levels of K in the crowns than the other genotypes (by an average of 48.8%).

In the roots, the genotypes did not differ in Ca, Na, and Mo content. Nonetheless, plants from the natural environment displayed lower levels of K (by an average of 48.8%) and Mg (by 26.9%) but higher levels of Mn (by 147.2%) and Fe (by 115.8%) in the roots compared to the cultivars. These wild plants also had 30.4% higher Zn content in the roots than the 'Rugia' cultivar, while exhibiting the lowest Cu content (by an average of 29.0%), in contrast to the 'Rugia' cultivar, which showed the highest Cu concentration in the roots.

3.2. Evaluation of Differences in Physiological Traits, Chemical Composition, and Yield Among Genotypes in Various Soil Moisture Conditions

3.2.1. Gas Exchange Parameters

The two-factor analysis of variance (ANOVA) revealed a significant main effect of the substrate moisture level (SML) factor on various physiological parameters. Specifically, transpiration intensity (E) showed significant effects during the first four measurement terms, stomatal conductivity to H_2O (g_s) during the first five measurement terms, net CO_2 assimilation intensity (P_n) across all measurement terms, substomatal CO_2 concentration (c_i) during the first four and the final measurement dates, and the photosynthetic water use efficiency (WUE) during the first three and the fifth measurement dates (Table S1). In general, these parameters exhibited a decline under water deficit conditions, with the exception of g_s and c_i on the second measurement date, as well as WUE on the first measurement date, where an increase in values was observed (respectively, by 88.5%, 4.1% and 66.4%). The average decrease was 36.7% for E, 30.0% for g_s, 23.7% for P_n, 18.2% for c_i, and 57.6% for WUE (Table 10).

A significant effect of the Genotype × SML interaction on transpiration intensity (E) was observed in all measurement terms except the second. For stomatal conductivity to H_2O (gs) and substomatal CO_2 concentration (c_i), significant effects were detected in the first, fourth, and fifth terms; for net CO_2 assimilation intensity (P_n), in the last three terms; and for photosynthetic water use efficiency (WUE), in the first and last three measurement terms (Table S1). Post hoc analysis revealed a more pronounced reduction in E under water deficit conditions in the cultivars 'Rugia' and 'Baron von Solemacher' (respectively, by an average of 48.7% and 40.4%, while in wild plants, it was 26.2%). For g_s, the cultivars exhibited a greater reduction in the first measurement term (58.5% in 'Baron von Solemacher', 50.6%—'Rugia', and 23.3%—wild plants). However, in the fourth term, 'Rugia' demonstrated a smaller change compared to plants from the natural environment (no statistically significant change compared to a 25.0% reduction), while in the fifth term, 'Baron von Solemacher' exhibited a smaller reduction compared to other plants (50.9% and no statistically significant change). In the case of P_n, the results were relatively inconsistent. For example, in 'Baron von Solemacher', an increase in P_n was observed in the last two measurement terms (by 106.1% and 79.6%). However, in the remaining terms, P_n values generally decreased, with the greatest reductions occurring in plants from the natural environment (by an average of 47.6%). Regarding c_i, plants from the natural environment exhibited the smallest decrease in value in the first measurement term (10.7% compared to an average of 26.9%). In the fourth term, c_i values increased in 'Rugia' by 14.0% similarly to the fifth term in plants from the natural environment (by 30.6%). In contrast, 'Baron von Solemacher' showed a 2.5% decrease in c_i values in the fourth term. Summarizing the changes in WUE values across the measurement terms, the most unfavorable changes were observed in 'Rugia', which predominantly responded to water deficit by decreasing this parameter (by an average of 44.5%) (Table 10).

Table 10. Gas exchange parameters in leaves of *F. vesca* genotypes in various soil moisture conditions.

Parameters	Genotypes	SML	Measurement Terms					
			1st Year			2nd year		
			July	August	September	July	August	September
E [mmol $H_2O{\cdot}m^{-2}{\cdot}s^{-1}$]	B	Ctrl	1.105 ^{ab}	1.041 ^{a}	0.967 ^{a}	0.707 ^{a}	0.841 ^{bc}	0.573 ^{c}
		Def	0.417 ^{d}	0.645 ^{a}	0.727 ^{b}	0.465 ^{c}	1.097 ^{a}	0.499 ^{cd}
	R	Ctrl	1.017 ^{b}	1.021 ^{a}	1.036 ^{a}	0.571 ^{b}	0.992 ^{ab}	0.252 ^{e}
		Def	0.405 ^{d}	0.502 ^{a}	0.579 ^{b}	0.565 ^{b}	0.578 ^{d}	0.431 ^{d}
	N	Ctrl	1.120 ^{a}	1.059 ^{a}	0.726 ^{b}	0.687 ^{a}	0.735 ^{cd}	0.907 ^{a}
		Def	0.789 ^{c}	0.653 ^{a}	0.619 ^{b}	0.530 ^{bc}	0.733 ^{cd}	0.801 ^{b}
	Average	Ctrl	1.081 ^{A}	1.040 ^{A}	0.910 ^{A}	0.655 ^{A}	0.856 ^{A}	0.577 ^{A}
		Def	0.537 ^{B}	0.600 ^{B}	0.642 ^{B}	0.520 ^{B}	0.803 ^{A}	0.574 ^{A}
g_s [mol $H_2O{\cdot}m^{-2}{\cdot}s^{-1}$]	B	Ctrl	0.094 ^{a}	0.137 ^{a}	0.061 ^{a}	0.064 ^{a}	0.043 ^{bc}	0.063 ^{a}
		Def	0.039 ^{c}	0.254 ^{a}	0.045 ^{a}	0.036 ^{c}	0.050 ^{ab}	0.082 ^{a}
	R	Ctrl	0.087 ^{a}	0.107 ^{a}	0.064 ^{a}	0.048 ^{b}	0.053 ^{a}	0.055 ^{a}
		Def	0.043 ^{c}	0.182 ^{a}	0.038 ^{a}	0.046 ^{b}	0.026 ^{e}	0.049 ^{a}
	N	Ctrl	0.086 ^{a}	0.150 ^{a}	0.050 ^{a}	0.060 ^{a}	0.038 ^{cd}	0.113 ^{a}
		Def	0.066 ^{b}	0.305 ^{a}	0.042 ^{a}	0.045 ^{b}	0.032 ^{de}	0.120 ^{a}
	Average	Ctrl	0.089 ^{A}	0.131 ^{B}	0.058 ^{A}	0.058 ^{A}	0.045 ^{A}	0.083 ^{A}
		Def	0.049 ^{B}	0.247 ^{A}	0.042 ^{B}	0.042 ^{B}	0.036 ^{B}	0.077 ^{A}
P_n [mmol $H_2O{\cdot}m^{-2}{\cdot}s^{-1}$]	B	Ctrl	6.747 ^{a}	2.200 ^{a}	4.727 ^{a}	3.760 ^{a}	1.520 ^{d}	2.513 ^{c}
		Def	4.313 ^{a}	1.667 ^{a}	3.873 ^{a}	2.787 ^{b}	3.133 ^{bc}	4.513 ^{b}
	R	Ctrl	6.847 ^{a}	1.780 ^{a}	3.740 ^{a}	3.487 ^{a}	3.873 ^{ab}	4.107 ^{bc}
		Def	5.380 ^{a}	1.773 ^{a}	2.607 ^{a}	2.060 ^{c}	2.300 ^{cd}	3.667 ^{bc}
	N	Ctrl	7.307 ^{a}	3.253 ^{a}	3.427 ^{a}	3.533 ^{a}	4.260 ^{a}	7.780 ^{a}
		Def	5.800 ^{a}	2.553 ^{a}	2.673 ^{a}	3.840 ^{a}	1.787 ^{d}	4.893 ^{b}
	Average	Ctrl	6.697 ^{A}	2.411 ^{A}	3.964 ^{A}	3.593 ^{A}	3.218 ^{A}	5.320 ^{A}
		Def	5.164 ^{B}	1.998 ^{A}	3.051 ^{B}	2.896 ^{B}	2.407 ^{B}	3.838 ^{B}
c_i [μmol $CO_2{\cdot}mol^{-1}$]	B	Ctrl	311.000 ^{a}	421.333 ^{a}	315.000 ^{a}	298.200 ^{ab}	355.733 ^{a}	498.467 ^{a}
		Def	220.667 ^{c}	441.600 ^{a}	257.133 ^{a}	290.800 ^{c}	316.133 ^{ab}	361.333 ^{a}
	R	Ctrl	299.933 ^{ab}	426.200 ^{a}	302.533 ^{a}	280.867 ^{bc}	289.400 ^{bc}	371.467 ^{a}
		Def	225.933 ^{c}	434.200 ^{a}	282.533 ^{a}	320.133 ^{a}	300.533 ^{b}	323.800 ^{a}
	N	Ctrl	279.800 ^{b}	411.733 ^{a}	304.533 ^{a}	298.867 ^{ab}	244.333 ^{c}	449.800 ^{a}
		Def	250.000 ^{c}	434.733 ^{a}	286.467 ^{a}	255.400 ^{c}	319.000 ^{ab}	339.800 ^{a}
	Average	Ctrl	296.911 ^{A}	419.756 ^{B}	307.356 ^{A}	292.644 ^{A}	296.489 ^{A}	439.911 ^{A}
		Def	232.200 ^{B}	436.844 ^{A}	275.378 ^{B}	278.778 ^{A}	311.889 ^{A}	341.644 ^{B}
WUE	B	Ctrl	6.201 ^{c}	3.484 ^{a}	5.167 ^{a}	5.409 ^{bc}	1.922 ^{c}	4.656 ^{c}
		Def	11.071 ^{b}	1.668 ^{a}	0.196 ^{a}	6.593 ^{b}	2.938 ^{bc}	9.285 ^{bc}
	R	Ctrl	6.747 ^{c}	3.812 ^{a}	3.799 ^{a}	6.253 ^{ab}	4.028 ^{b}	20.713 ^{a}
		Def	13.763 ^{a}	1.806 ^{a}	0.231 ^{a}	3.979 ^{c}	3.760 ^{b}	9.805 ^{b}
	N	Ctrl	6.519 ^{c}	5.168 ^{a}	5.173 ^{a}	5.328 ^{bc}	6.533 ^{a}	5.432 ^{bc}
		Def	7.561 ^{c}	2.569 ^{a}	0.243 ^{a}	7.685 ^{a}	2.554 ^{bc}	9.967 ^{b}
	Average	Ctrl	6.489 ^{B}	4.155 ^{A}	4.713 ^{A}	5.663 ^{A}	4.161 ^{A}	10.267 ^{A}
		Def	10.798 ^{A}	2.014 ^{B}	0.223 ^{B}	6.089 ^{A}	3.084 ^{B}	9.686 ^{A}

Explanations: SML—substrate moisture level; Ctrl—control, Def—deficit; B—'Baron von Solemacher' cultivar; R—'Rugia' cultivar; N—plants obtained from a natural habitat. Uppercase letters indicate means that significantly differ in the presence of a significant main effect of the SML factor, while lowercase letters denote means that significantly differ in the presence of a significant interaction effect (SML × genotype).

3.2.2. Chlorophyll "a" Fluorescence Parameters

A significant main effect of the SML factor on F_0 was observed exclusively on the fourth measurement date. For F_M and F_V, significant effects were noted in the second and last three measurement terms, while for T_{FM}, they were observed during the first two and last three terms. Similarly, A_M demonstrated significant effects on the last three dates. However, no significant differences were found for F_V/F_M in relation to the SML factor on any measurement date (Table S2). A general decrease in the values of F_0 (by 3.8%), F_V (by an average of 7.5%), and F_M (by an average of 12.1%) was observed across the analyzed dates, with the exception of the second measurement date, where F_V and F_M values were higher under water deficit conditions (respectively, by 6.6% and 5.1%). In contrast, T_{FM} and A_M values increased under water deficit conditions (respectively, by an average of 29.7% and 27.8%) (Table 11).

Table 11. Chlorophyll "a" fluorescence parameters in leaves of *F. vesca* genotypes in various soil moisture conditions.

Parameters	Genotypes	SML	Measurement Terms					
			1st Year			2nd Year		
			July	August	September	July	August	September
F_0	B	Ctrl	600.786 ^{a}	616.857 ^{a}	587.429 ^{a}	259.143 ^{a}	261.857 ^{a}	266.857 ^{ab}
		Def	598.333 ^{a}	620.000 ^{a}	585.933 ^{a}	252.200 ^{a}	266.733 ^{a}	274.867 ^{ab}
	R	Ctrl	594.375 ^{a}	610.143 ^{a}	598.250 ^{a}	268.625 ^{a}	273.125 ^{a}	278.188 ^{a}
		Def	608.929 ^{a}	612.438 ^{a}	594.000 ^{a}	258.214 ^{a}	267.500 ^{a}	271.857 ^{ab}
	N	Ctrl	600.429 ^{a}	594.000 ^{a}	582.929 ^{a}	258.571 ^{a}	229.857 ^{a}	250.643 ^{b}
		Def	603.867 ^{a}	593.200 ^{a}	593.600 ^{a}	247.200 ^{a}	223.067 ^{a}	211.933 ^{c}
	Average	Ctrl	598.341 ^{A}	607.977 ^{A}	589.932 ^{A}	262.409 ^{A}	255.773 ^{A}	265.818 ^{A}
		Def	603.591 ^{A}	907.727 ^{A}	591.114 ^{A}	252.409 ^{B}	252.091 ^{A}	252.455 ^{A}
F_M	B	Ctrl	2804.571 ^{a}	2523.143 ^{a}	2917.929 ^{a}	1246.571 ^{c}	1222.714 ^{b}	1270.500 ^{a}
		Def	2936.867 ^{a}	2729.667 ^{a}	2959.667 ^{a}	1272.400 ^{bc}	1305.333 ^{ab}	1179.800 ^{a}
	R	Ctrl	2701.214 ^{a}	2680.688 ^{a}	2821.563 ^{a}	1346.688 ^{a}	1398.500 ^{a}	1232.500 ^{a}
		Def	2795.000 ^{a}	2733.000 ^{a}	2998.643 ^{a}	1201.357 ^{c}	1237.857 ^{b}	1162.857 ^{a}
	N	Ctrl	2998.429 ^{a}	2782.643 ^{a}	3067.071 ^{a}	1333.143 ^{ab}	992.571 ^{c}	1258.643 ^{a}
		Def	3005.067 ^{a}	2928.533 ^{a}	2922.600 ^{a}	1237.867 ^{c}	887.333 ^{d}	1097.000 ^{a}
	Average	Ctrl	2862.773 ^{A}	2663.000 ^{B}	2930.341 ^{A}	1310.523 ^{A}	1213.409 ^{A}	1525.909 ^{A}
		Def	2885.136 ^{A}	2798.523 ^{A}	2959.432 ^{A}	1238.023 ^{B}	1141.364 ^{B}	1146.182 ^{B}
F_V	B	Ctrl	2203.786 ^{a}	1906.286 ^{a}	2330.500 ^{a}	987.429 ^{cb}	960.857 ^{b}	1003.643 ^{a}
		Def	2338.533 ^{a}	2109.667 ^{a}	2373.733 ^{a}	1020.200 ^{ab}	1038.600 ^{ab}	904.933 ^{a}
	R	Ctrl	2200.625 ^{a}	2068.250 ^{a}	2223.313 ^{a}	1078.063 ^{a}	1125.375 ^{a}	954.313 ^{a}
		Def	2092.286 ^{a}	2122.857 ^{a}	2404.643 ^{a}	943.143 ^{c}	970.357 ^{b}	891.000 ^{a}
	N	Ctrl	2398.000 ^{a}	2188.643 ^{a}	2484.143 ^{a}	1074.571 ^{a}	762.714 ^{c}	1008.000 ^{a}
		Def	2401.200 ^{a}	2335.333 ^{a}	2329.000 ^{a}	990.667 ^{bc}	664.267 ^{d}	885.067 ^{a}
	Average	Ctrl	2264.432 ^{A}	2055.023 ^{B}	2340.409 ^{A}	1048.114 ^{A}	957.636 ^{A}	987.091 ^{A}
		Def	2281.545 ^{A}	2190.795 ^{A}	2368.318 ^{A}	985.614 ^{B}	889.273 ^{B}	893.727 ^{B}
F_V/F_M	B	Ctrl	0.782 ^{a}	0.754 ^{a}	0.796 ^{a}	0.792 ^{ab}	0.785 ^{a}	0.780 ^{a}
		Def	0.796 ^{a}	0.769 ^{a}	0.800 ^{a}	0.801 ^{a}	0.795 ^{a}	0.765 ^{a}
	R	Ctrl	0.784 ^{a}	0.771 ^{a}	0.785 ^{a}	0.799 ^{a}	0.804 ^{a}	0.772 ^{a}
		Def	0.765 ^{a}	0.775 ^{a}	0.800 ^{a}	0.784 ^{b}	0.783 ^{a}	0.764 ^{a}
	N	Ctrl	0.793 ^{a}	0.779 ^{a}	0.809 ^{a}	0.805 ^{a}	0.743 ^{a}	0.801 ^{a}
		Def	0.797 ^{a}	0.796 ^{a}	0.796 ^{a}	0.800 ^{a}	0.737 ^{a}	0.805 ^{a}
	Average	Ctrl	0.786 ^{A}	0.768 ^{A}	0.796 ^{A}	0.799 ^{A}	0.779 ^{A}	0.783 ^{A}
		Def	0.787 ^{A}	0.780 ^{A}	0.798 ^{A}	0.795 ^{A}	0.772 ^{A}	0.778 ^{A}

Table 11. *Cont.*

Parameters	Genotypes	SML	Measurement Terms					
			1st Year			2nd Year		
			July	August	September	July	August	September
T_{FM}	B	Ctrl	373.571 [a]	240.714 [a]	305.000 [a]	578.571 [d]	277.857 [a]	535.000 [a]
		Def	506.667 [a]	266.667 [a]	341.333 [a]	726.667 [b]	500.000 [a]	611.333 [a]
	R	Ctrl	404.375 [a]	236.875 [a]	279.375 [a]	493.751 [e]	324.375 [a]	518.750 [a]
		Def	507.143 [a]	251.429 [a]	309.286 [a]	821.429 [a]	491.429 [a]	628.571 [a]
	N	Ctrl	507.143 [a]	273.571 [a]	339.286 [a]	671.429 [c]	317.144 [a]	560.000 [a]
		Def	620.000 [a]	276.000 [a]	356.667 [a]	806.667 [a]	462.000 [a]	713.333 [a]
	Average	Ctrl	427.273 [B]	249.773 [B]	306.591 [A]	577.273 [B]	307.273 [B]	537.045 [B]
		Def	545.455 [A]	265.000 [A]	336.364 [A]	784.091 [A]	484.318 [A]	651.591 [A]
A_M	B	Ctrl	52,530.79 [a]	31,023.79 [a]	45,809.50 [bc]	28,356.29 [a]	16,458.86 [c]	22,092.71 [a]
		Def	62,576.73 [a]	36,214.73 [a]	52,129.73 [ab]	38,018.53 [a]	29,509.93 [a]	31,889.93 [a]
	R	Ctrl	54,069.13 [a]	35,609.25 [a]	42,670.81 [c]	30,417.69 [a]	23,104.00 [b]	23,176.81 [a]
		Def	52,151.14 [a]	34,992.36 [a]	50,287.57 [abc]	38,595.50 [a]	27,784.21 [a]	29,535.43 [a]
	N	Ctrl	68,608.93 [a]	39,444.50 [a]	57,824.21 [a]	40,694.79 [a]	17,188.72 [c]	32,145.64 [a]
		Def	73,210.20 [a]	43,184.27 [a]	50,782.27 [abc]	46,048.60 [a]	19,009.80 [c]	36,060.13 [a]
	Average	Ctrl	58,205.96 [A]	35,370.55 [A]	48,491.02 [A]	33,031.77 [B]	19,107.50 [B]	25,685.59 [B]
		Def	62,884.55 [A]	38,201.77 [A]	51,084.02 [A]	40,939.64 [A]	25,381.25 [A]	32,562.43 [A]

Explanations: SML—substrate moisture level; Ctrl—control, Def—deficit; B—'Baron von Solemacher' cultivar; R—'Rugia' cultivar; N—plants obtained from a natural habitat. Uppercase letters indicate means that significantly differ in the presence of a significant main effect of the SML factor, while lowercase letters denote means that significantly differ in the presence of a significant interaction effect (SML × genotype).

A significant effect of the genotype × SML interaction on F_0 was observed exclusively in the last measurement term. For F_M and F_V, significant effects were noted in the fourth and fifth measurement term, while for F_V/F_M and T_{FM}, they were observed in the fourth term. Significant effects for A_M were found in the third, fifth, and sixth measurement terms (Table S2). Regarding F_0, the most pronounced changes under water deficit conditions occurred in plants obtained from the natural environment (reduction of 15.4%). For F_M and F_V, the greatest decreases on the fourth and fifth measurement dates were observed in the 'Rugia' cultivar (respectively, by an average of 11.1% and 13.1%). A similar trend was evident for F_V/F_M, where a more substantial decrease was noted in 'Rugia' (by 1.9%). In the case of T_{FM}, the most adverse changes under water deficit conditions were recorded in 'Rugia' (an increase of 66.4%), but in the case of A_M it was 'Baron von Solemacher' (an increase of 79.3%) (Table 11).

3.2.3. Content of Photosynthetic Pigments

A significant main effect of the SML factor on chlorophyll "a" content was observed in the second, third, fourth, and fifth measurement terms. For chlorophyll "b," significant effects were found in the first three and fifth terms. The chlorophyll "a" to "b" ratio showed significant effects in the second and fifth terms, while total chlorophyll content was significantly affected in the second, third, and fifth terms. Carotenoid content exhibited significant effects in the first five measurement terms (Table S3). The results, however, demonstrated notable variability. In the first, second, fourth, and fifth measurement terms, under water deficit conditions, an increase was observed in the contents of chlorophyll "a" (by an average of 23.5%), chlorophyll "b" (by an average of 28.2%), total chlorophyll (by an average of 27.5%), and carotenoids (by an average of 21.3%), accompanied by a decrease in the chlorophyll "a" to "b" ratio (by an average of 7.8%). Conversely, on the third measurement date, decreases were recorded for chlorophyll "a" (by 11.2%), chlorophyll "b" (by 19.3%), total chlorophyll (by 13.7%), and carotenoid content (by 9.0%) (Table 12).

Table 12. Content of photosynthetic pigments in leaves of *F. vesca* genotypes in various soil moisture conditions [mg · g^{-1} FM].

Parameters	Genotypes	SML	Measurement Terms					
			1st Year			2nd Year		
			July	August	September	July	August	September
Chlorophyll "a"	B	Ctrl	1.711 a	0.851 d	1.427 b	1.327 a	1.952 a	1.857 bc
		Def	1.727 a	1.409 b	1.386 b	1.852 a	2.372 a	1.717 bc
	R	Ctrl	1.360 a	1.090 c	1.042 c	1.355 a	1.528 a	1.641 bc
		Def	1.629 a	1.274 bc	1.460 b	1.553 a	2.143 a	2.009 ab
	N	Ctrl	1.995 a	1.420 b	1.777 a	2.116 a	1.780 a	2.201 a
		Def	2.194 a	1.629 a	0.923 d	2.250 a	2.034 a	1.685 bc
	Average	Ctrl	1.689 A	1.120 B	1.415 A	1.599 B	1.754 B	1.900 A
		Def	1.850 A	1.437 A	1.256 B	1.885 A	2.183 A	1.804 A
Chlorophyll "b"	B	Ctrl	0.622 a	0.565 c	0.531 b	0.564 a	0.627 a	0.627 abc
		Def	0.685 a	0.709 b	0.553 b	0.705 a	0.768 a	0.627 abc
	R	Ctrl	0.523 a	0.562 c	0.523 b	0.606 a	0.486 a	0.591 bc
		Def	0.741 a	0.652 bc	0.601 b	0.654 a	0.695 a	0.693 ab
	N	Ctrl	0.702 a	0.700 b	0.387 c	0.846 a	0.508 a	0.721 a
		Def	0.860 a	1.037 a	0.855 a	0.824 a	0.637 a	0.537 c
	Average	Ctrl	0.616 B	0.609 B	0.637 A	0.688 A	0.541 B	0.646 A
		Def	0.762 A	0.800 A	0.514 B	0.712 A	0.700 A	0.619 A
Chlorophyll "a" to chlorophyll "b" ratio	B	Ctrl	2.747 a	1.528 b	2.684 a	2.385 a	3.113 bc	2.963 ab
		Def	2.525 a	1.986 a	2.506 a	2.607 a	3.091 bc	2.730 c
	R	Ctrl	2.632 a	1.932 a	2.074 a	2.208 a	3.154 bc	2.780 c
		Def	2.190 a	1.950 a	2.429 a	2.565 a	3.083 c	2.898 bc
	N	Ctrl	2.839 a	2.028 a	2.107 a	2.523 a	3.504 a	3.056 ab
		Def	2.557 a	1.573 b	2.386 a	2.732 a	3.196 b	3.136 a
	Average	Ctrl	2.739 A	1.829 A	2.288 A	2.372 A	3.257 A	2.933 A
		Def	2.424 B	1.836 A	2.440 A	2.635 A	3.123 B	2.921 A
Total chlorophyll	B	Ctrl	2.333 a	1.415 a	1.958 b	1.890 a	2.579 a	2.483 abc
		Def	2.411 a	2.117 a	1.938 b	2.556 a	3.139 a	2.343 abc
	R	Ctrl	1.882 a	1.652 a	1.565 c	2.008 a	2.014 a	2.231 c
		Def	2.369 a	1.926 a	2.061 b	2.159 a	2.837 a	2.701 ab
	N	Ctrl	2.697 a	2.119 a	2.631 a	2.962 a	2.288 a	2.921 a
		Def	3.053 a	2.666 a	1.309 d	3.073 a	2.670 a	2.222 c
	Average	Ctrl	2.304 A	1.729 B	2.051 A	2.286 A	2.294 B	2.545 A
		Def	2.611 A	2.236 A	1.769 B	2.596 A	2.882 A	2.422 A
Carotenoids	B	Ctrl	0.959 a	0.672 a	0.866 b	0.683 a	0.966 a	0.933 bc
		Def	1.023 a	0.983 a	0.920 b	0.917 a	1.155 a	0.912 bc
	R	Ctrl	0.728 a	0.748 a	0.639 c	0.696 a	0.772 a	0.825 c
		Def	0.992 a	0.866 a	0.846 b	0.808 a	1.058 a	1.047 ab
	N	Ctrl	1.179 a	1.009 a	1.170 a	1.050 a	0.902 a	1.120 a
		Def	1.403 a	1.204 a	0.667 c	1.108 a	1.047 a	0.902 bc
	Average	Ctrl	0.955 B	0.810 B	0.891 A	0.810 B	0.880 B	0.959 A
		Def	1.139 A	1.018 A	0.811 B	0.944 A	1.087 A	0.954 A

Explanations: SML—substrate moisture level; Ctrl—control, Def—deficit; B—'Baron von Solemacher' cultivar; R—'Rugia' cultivar; N—plants obtained from a natural habitat. Uppercase letters indicate means that significantly differ in the presence of a significant main effect of the SML factor, while lowercase letters denote means that significantly differ in the presence of a significant interaction effect (SML × genotype).

A significant effect of the genotype × SML interaction on the content of chlorophyll "a" and chlorophyll "b" was observed in the second, third, and final measurement terms. For the chlorophyll "a" to "b" ratio, significant effects were noted in the second, fifth, and sixth terms, while for total chlorophyll and carotenoid content, significant effects were recorded in the third and sixth terms (Table S3). For chlorophyll "a," all genotypes responded to water deficit on the second measurement date with an increase in its content, with the most pronounced change observed in the 'Baron von Solemacher' cultivar (by 65.6%). However, in the third and sixth terms, plants from the natural environment exhibited a decrease in chlorophyll "a" content (by an average of 35.8%). A similar pattern was observed for chlorophyll "b," where an increase in content was noted in the earlier terms (by an average of 64.9%), but a 25.5% decrease occurred in wild plants in the final term. Consequently, the largest reductions in total chlorophyll content were recorded in wild plants (by an average of 37.1%). For the chlorophyll "a" to "b" ratio, a 30.0% increase was observed in cultivated plants in the second measurement term, while a 22.4% decrease occurred in wild plants. In the fifth term, the most substantial reduction in this ratio was recorded in wild plants (by 8.8%), whereas in the sixth term, the greatest decrease was observed in the 'Baron von Solemacher' cultivar (by 7.9%). Regarding carotenoids, a decrease in content was observed exclusively in wild plants during the third and sixth terms (by an average of 31.2%), whereas the 'Rugia' cultivar demonstrated the most significant increase in carotenoid content (by an average of 29.7%) (Table 12).

3.2.4. Relative Water Content in Leaves (RWC)

A significant main effect of the SML factor on relative water content (RWC) was observed in the first, second, and fifth measurement terms (Table S4). In all instances, RWC values decreased under water deficit conditions by an average of 14.4% (Table 13).

Table 13. Relative water content (RWC) in leaves of *F. vesca* genotypes in various soil moisture conditions [%].

Genotypes	SML	Measurement Terms					
		1st Year			2nd Year		
		July	August	September	July	August	September
B	Ctrl	54.445 ^{bc}	54.765 ^{a}	56.308 ^{a}	45.245 ^{a}	42.064 ^{a}	46.258 ^{a}
	Def	57.062 ^{bc}	55.160 ^{a}	63.152 ^{a}	45.247 ^{a}	29.790 ^{a}	42.738 ^{a}
R	Ctrl	62.644 ^{a}	55.159 ^{a}	61.177 ^{a}	39.875 ^{a}	40.545 ^{a}	58.035 ^{a}
	Def	53.366 ^{c}	43.781 ^{a}	52.717 ^{a}	38.022 ^{a}	31.226 ^{a}	50.866 ^{a}
N	Ctrl	58.285 ^{b}	65.073 ^{a}	54.813 ^{a}	37.150 ^{a}	30.509 ^{a}	52.958 ^{a}
	Def	56.126 ^{bc}	60.660 ^{a}	56.865 ^{a}	35.471 ^{a}	18.812 ^{a}	47.108 ^{a}
Average	Ctrl	58.458 ^{A}	58.332 ^{A}	57.432 ^{A}	40.757 ^{A}	37.706 ^{A}	52.417 ^{A}
	Def	55.518 ^{B}	53.200 ^{B}	57.578 ^{A}	37.285 ^{A}	26.609 ^{B}	46.904 ^{A}

Explanations: SML—substrate moisture level; ctrl—Control, Def—deficit; B—'Baron von Solemacher' cultivar; R—'Rugia' cultivar; N—plants obtained from a natural habitat. Uppercase letters indicate means that significantly differ in the presence of a significant main effect of the SML factor, while lowercase letters denote means that significantly differ in the presence of a significant interaction effect (SML × genotype).

A significant effect of the genotype × SML interaction on RWC was observed only in the first measurement term (Table S4). During this time, the most substantial decrease in RWC was recorded in the 'Rugia' plants (by 14.8%) (Table 13).

3.2.5. Fruit Yield and Weight of a Single Wild Strawberry Fruit

In plants obtained from the natural environment, flowering was not observed, and consequently, they were excluded from the analysis of fruit mass as they did not produce any fruits. A significant main effect of the SML factor was identified across all study years for both Fresh Fruit Mass Yield and Single Fruit Mass (Table S5). In each case, plants subjected to water deficit conditions produced fruits with lower mass. This represented an average decrease of 62.9% in Fresh Fruit Mass Yield and an average decrease of 35.6% in Single Fruit Mass (Table 14).

However, no significant effect of the SML × genotype interaction was observed in any year of the study for either Fresh Fruit Mass Yield or Single Fruit Mass (Table S5).

Table 14. Total fruit yield from one plant and weight of a single fruit during the vegetation period (June–September) in *F. vesca* genotypes in various soil moisture conditions [g].

Parameters	Genotypes	SML	1st Year	2nd Year
Fresh Fruit Mass Yield	B	Ctrl	23.79 a	33.11 a
		Def	5.29 a	14.83 a
	R	Ctrl	22.20 a	23.27 a
		Def	5.64 a	14.62 a
	N	Ctrl	-	-
		Def	-	-
	Average	Ctrl	23.00 A	29.19 A
		Def	5.46 B	14.72 B
Single Fruit Mass	B	Ctrl	0.65 a	0.55 a
		Def	0.36 a	0.38 a
	R	Ctrl	0.59 a	0.54 a
		Def	0.39 a	0.38 a
	N	Ctrl	-	-
		Def	-	-
	Average	Ctrl	0.62 A	0.55 A
		Def	0.37 B	0.38 B

Explanations: SML—substrate moisture level; Ctrl—control, Def—deficit; B—'Baron von Solemacher' cultivar; R—'Rugia' cultivar; N—plants obtained from a natural habitat. Uppercase letters indicate means that significantly differ in the presence of a significant main effect of the SML factor, while lowercase letters denote means that significantly differ in the presence of a significant interaction effect (SML × genotype).

3.2.6. Fresh and Dry Mass of the Root System

Neither the main effect of SML nor the SML × genotype interaction was significant for the fresh and dry mass of the root system of *F. vesca* (Table S6 and Table 15).

Table 15. Fresh and dry mass of the root system of *F. vesca* genotypes in various soil moisture conditions [g].

Genotypes	SML	Fresh Mass of the Root System	Dry Mass of the Root System
B	Ctrl	11.81 a	6.83 a
	Def	6.74 a	4.40 a
R	Ctrl	12.30 a	6.53 a
	Def	8.26 a	4.40 a
N	Ctrl	17.80 a	10.47 a
	Def	23.80 a	15.31 a
Average	Ctrl	13.64 a	7.66 a
	Def	16.61 a	10.46 a

Explanations: SML—substrate moisture level; Ctrl—control, Def—deficit; B—'Baron von Solemacher' cultivar; R—'Rugia' cultivar; N—plants obtained from a natural habitat. Uppercase letters indicate means that significantly differ in the presence of a significant main effect of the SML factor, while lowercase letters denote means that significantly differ in the presence of a significant interaction effect (SML × genotype).

3.2.7. Chemical Composition

For the content of individual elements in leaves, a significant main effect of the SML factor was observed for potassium (K), magnesium (Mg), zinc (Zn), and manganese (Mn) (Table S7). In each case, the content of these elements increased under drought conditions (respectively, by 8.5%, 10.9%, 32.8% and 20.6%) (Table 15). A significant effect of the SML × genotype interaction was detected for calcium (Ca), manganese (Mn), and iron (Fe) (Table S7). The most pronounced changes were observed in plants obtained from the natural environment, which exhibited higher contents of Zn (by 96.3%), Mn (by 62.2%), and Fe (by 157.1%) (Table 16).

For the element content in the crowns, a significant main effect of the SML factor was observed for sodium (Na) and iron (Fe) (Table S7), with the content of both elements increasing under drought conditions (respectively, by 68.9% and 50.0%) (Table 16). A significant effect of the SML × genotype interaction was identified only for calcium (Ca) (Table S7). A significant decrease in Ca content was observed exclusively in plants collected from the natural environment (by 18.5%). In contrast, cultivated varieties showed an increase in Ca content (by an average of 21.9%); however, this increase was not statistically significant.

Table 16. Chemical composition of leaves, crowns and roots of *F. vesca* genotypes in various soil moisture conditions.

Plant Organ	Genotypes	SML	K [g·kg^{-1}]	Ca [g·kg^{-1}]	Mg [g·kg^{-1}]	Na [g·kg^{-1}]	Cu [mg·kg^{-1}]	Zn [mg·kg^{-1}]	Mn [mg·kg^{-1}]	Mo [mg·kg^{-1}]	Fe [mg·kg^{-1}]
Leaves	B	Ctrl	97.6 ^{a}	66.9 ^{ab}	13.5 ^{a}	64.8 ^{a}	37.3 ^{a}	172.5 ^{bc}	201.8 ^{ab}	11.7 ^{a}	1.9 ^{b}
		Def	100.4 ^{a}	70.2 ^{a}	16.6 ^{a}	4.9 ^{a}	53.4 ^{a}	225.0 ^{ab}	241.8 ^{a}	12.8 ^{a}	2.6 ^{a}
	R	Ctrl	106.9 ^{a}	69.7 ^{a}	15.7 ^{a}	2.8 ^{a}	52.2 ^{a}	236.4 ^{ab}	205.4 ^{ab}	6.6 ^{a}	1.8 ^{b}
		Def	118.0 ^{a}	68.4 ^{a}	17.0 ^{a}	5.5 ^{a}	56.0 ^{a}	228.0 ^{ab}	186.4 ^{b}	6.7 ^{a}	1.2 ^{bc}
	N	Ctrl	96.5 ^{a}	51.9 ^{bc}	17.2 ^{a}	2.7 ^{a}	45.6 ^{a}	146.8 ^{c}	139.2 ^{c}	12.8 ^{a}	0.7 ^{c}
		Def	108.7 ^{a}	42.2 ^{c}	18.2 ^{a}	4.6 ^{a}	46.0 ^{a}	288.2 ^{a}	225.8 ^{ab}	12.4 ^{a}	1.8 ^{b}
	Average	Ctrl	100.5 ^{B}	62.5 ^{A}	15.6 ^{B}	20.5 ^{A}	45.6 ^{A}	186.1 ^{B}	180.7 ^{B}	10.3 ^{A}	1.4 ^{A}
		Def	109.0 ^{A}	60.6 ^{A}	17.3 ^{A}	5.0 ^{A}	51.8 ^{A}	247.1 ^{A}	218.0 ^{A}	10.6 ^{A}	1.8 ^{A}
Crowns	B	Ctrl	82.9 ^{a}	115.6 ^{bc}	3.9 ^{a}	28.0 ^{a}	87.9 ^{a}	155.2 ^{a}	31.4 ^{a}	17.6 ^{a}	0.2 ^{a}
		Def	92.2 ^{a}	129.7 ^{b}	4.2 ^{a}	49.8 ^{a}	84.1 ^{a}	160.8 ^{a}	49.8 ^{a}	14.6 ^{a}	0.5 ^{a}
	R	Ctrl	92.2 ^{a}	91.4 ^{c}	3.4 ^{a}	37.4 ^{a}	117.1 ^{a}	141.3 ^{a}	30.8 ^{a}	16.4 ^{a}	0.2 ^{a}
		Def	104.3 ^{a}	120.3 ^{bc}	3.9 ^{a}	56.6 ^{a}	85.6 ^{a}	164.1 ^{a}	34.5 ^{a}	13.9 ^{a}	0.3 ^{a}
	N	Ctrl	50.4 ^{a}	159.7 ^{a}	5.4 ^{a}	10.0 ^{a}	74.4 ^{a}	212.9 ^{a}	61.8 ^{a}	17.3 ^{a}	0.2 ^{a}
		Def	54.9 ^{a}	130.2 ^{b}	5.1 ^{a}	20.8 ^{a}	64.0 ^{a}	245.6 ^{a}	77.2 ^{a}	18.6 ^{a}	0.3 ^{a}
	Average	Ctrl	75.2 ^{A}	122.2 ^{A}	4.3 ^{A}	25.1 ^{B}	93.1 ^{A}	169.8 ^{A}	41.3 ^{A}	17.1 ^{A}	0.2 ^{B}
		Def	83.8 ^{A}	126.7 ^{A}	4.4 ^{A}	42.4 ^{A}	77.9 ^{A}	190.2 ^{A}	53.8 ^{A}	15.7 ^{A}	0.3 ^{A}
Roots	B	Ctrl	25.9 ^{a}	80.7 ^{a}	13.4 ^{a}	6.1 ^{a}	92.5 ^{bc}	853.5 ^{bc}	373.3 ^{a}	12.4 ^{a}	6.3 ^{a}
		Def	26.9 ^{a}	88.3 ^{a}	12.1 ^{a}	8.6 ^{a}	121.2 ^{a}	844.4 ^{b}	422.2 ^{a}	19.0 ^{a}	7.5 ^{a}
	R	Ctrl	24.5 ^{a}	63.4 ^{a}	13.4 ^{a}	5.9 ^{a}	111.4 ^{a}	677.8 ^{c}	388.0 ^{a}	9.4 ^{a}	6.4 ^{a}
		Def	29.0 ^{a}	103.3 ^{a}	13.1 ^{a}	8.9 ^{a}	107.6 ^{ab}	913.2 ^{b}	309.0 ^{a}	14.0 ^{a}	7.3 ^{a}
	N	Ctrl	12.9 ^{a}	60.2 ^{a}	9.8 ^{b}	5.9 ^{a}	71.8 ^{d}	883.8 ^{b}	940.8 ^{a}	14.6 ^{a}	13.7 ^{a}
		Def	20.9 ^{a}	77.6 ^{a}	12.9 ^{a}	9.6 ^{a}	93.0 ^{c}	1232.3 ^{a}	926.8 ^{a}	12.4 ^{a}	12.6 ^{a}
	Average	Ctrl	20.7 ^{A}	67.2 ^{B}	12.1 ^{A}	6.0 ^{B}	91.9 ^{B}	801.6 ^{B}	581.2 ^{A}	12.1 ^{A}	9.0 ^{A}
		Def	25.9 ^{A}	90.6 ^{A}	12.7 ^{A}	9.0 ^{A}	108.3 ^{A}	979.8 ^{A}	525.9 ^{A}	15.3 ^{A}	8.9 ^{A}

Explanations: SML—substrate moisture level; Ctrl—control, Def—deficit; B—'Baron von Solemacher' cultivar; R—'Rugia' cultivar; N—plants obtained from a natural habitat. Uppercase letters indicate means that significantly differ in the presence of a significant main effect of the SML factor, while lowercase letters denote means that significantly differ in the presence of a significant interaction effect (SML × genotype).

For the element content in roots, a significant main effect of the SML factor was observed for Ca, Na, Zn, and copper (Cu) (Table S7), with the content of each element increasing under drought conditions (respectively, by 34.8%, 50.0%, 22.2% and 17.8%) (Table 15). A significant effect of the SML × genotype interaction was identified for Mg, Cu, and Zn (Table S7). A significant increase in Mg content in roots under water deficit conditions was observed exclusively in plants obtained from the natural environment (by 31.6%). These plants, along with the 'Rugia' cultivar, also showed the most pronounced increase in Zn content (by 34.7% in 'Rugia' and by 39.4% in wild plants). In contrast, an increase in Cu content was observed in the 'Baron von Solemacher' cultivar (by 31.0%) and in plants obtained from the natural environment (by 29.5%) (Table 16).

3.2.8. Radar Charts

Figure 1 illustrates an analysis of changes in the mean values of parameters used to assess the physiological characteristics of *F. vesca* plants, grouped by specific measurement dates and genotypes. Based on the graph's surface area, the 'Baron von Solemacher' cultivar showed the lowest adaptation to water deficit conditions in the substrate and the strongest adverse reaction to drought stress (as indicated by the smallest area under the curve) during the first and second terms of the first study year. The 'Rugia' cultivar exhibited the smallest adaptation in the third measurement terms of both study years, while plants collected from the natural environment showed the least adaptation in the first and second terms of the second year.

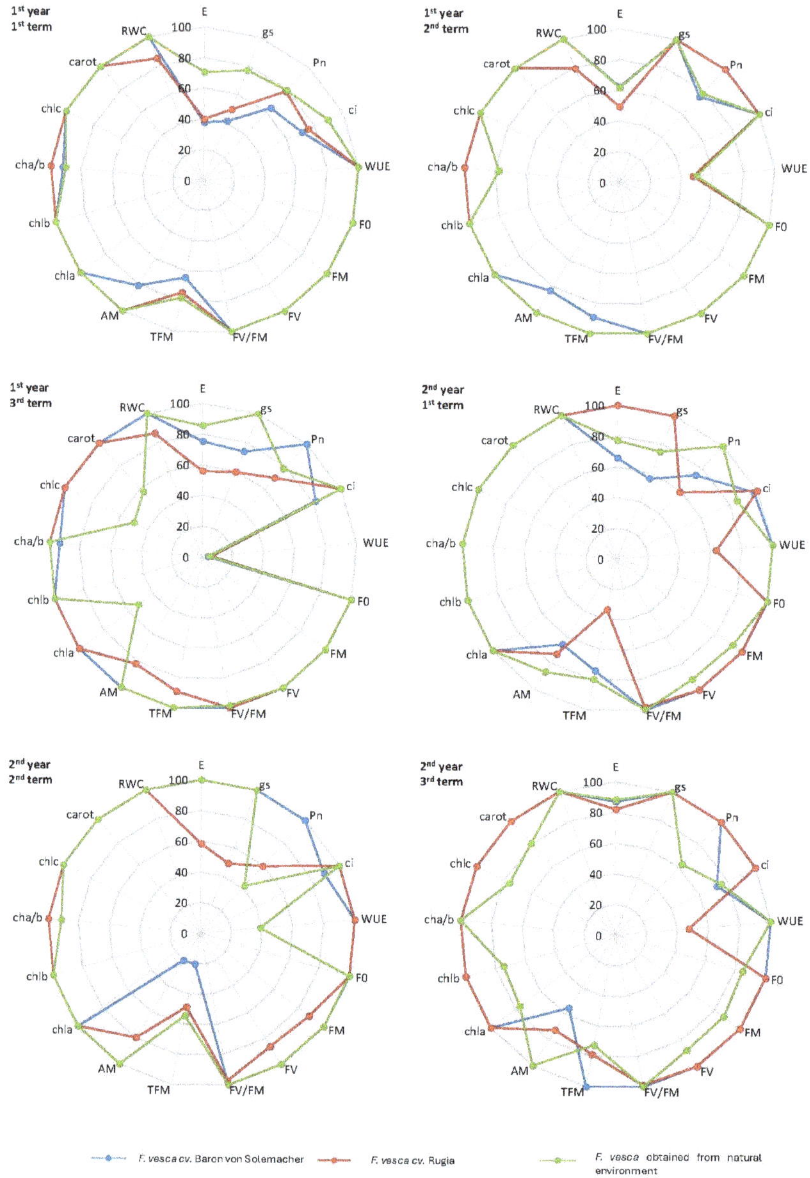

Figure 1. Changes in the mean values of selected physiological parameters of *F. vesca* genotypes under water deficit conditions in the substrate compared to mean values under controlmoisture conditions. Explanations: E—H_2O transpiration rate, g_s—stomatal H_2O conductance, P_n—net CO_2 assimilation rate, c_i—substomatal CO_2 concentration, WUE—photosynthetic water use coefficient, F_0—initial fluorescence, F_M—maximum fluorescence, F_V—variable fluorescence, F_V/F_M—maximum, potential efficiency of the photochemical reaction in PS II, T_{FM}—chlorophyll fluorescence growth time, A_M—area above the chlorophyll fluorescence induction curve, chla—chlorophyll "a" content, chlb—chlorophyll "b" content, chla/b—chlorophyll "a" to "b" content ratio, chlc—total chlorophyll content, carot—carotenoid content, RWC—relative leaf water content index.

Conversely, the Baron von Solemacher variety demonstrated the highest level of adaptation to drought conditions (reflected by the largest area under the curve) on the third date of the first study year and on the second and third dates of the second study year. Plants obtained from the natural environment exhibited the greatest adaptation on the first measurement date of both years, while the Rugia variety showed the greatest adaptation on the second date of the first study year.

4. Discussion

Physiological traits of cultivated plant varieties, including those within the *Fragaria* genus, exhibit substantial variability due to both genetic and environmental factors. Research indicates that differences in gas exchange parameters, chlorophyll fluorescence, and yield are pronounced across *Fragaria* species and varieties, highlighting the critical role of genetic diversity in these traits [40–42]. These findings align with the results of our study. While differences in physiological traits among *Fragaria* varieties are well documented, it is essential to consider that environmental factors can also profoundly influence these characteristics, sometimes overshadowing genetic effects. This complexity underscores the need for continued research into genotype–environment interactions that shape plant physiological responses.

Physiological characteristics of *F. vesca* plants growing in natural environments, including chlorophyll fluorescence and assimilation pigment content, surpassed those observed in cultivated varieties in our study, likely reflecting their adaptation to natural conditions. These plants thrive in diverse and often challenging habitats, which enhances their capacity for physiological adaptation [43,44]. The mechanisms of drought stress adaptation in plants, both cultivated and wild-growing, share fundamental similarities. However, the scope and effectiveness of these mechanisms can vary significantly due to the distinct environmental pressures experienced by wild plants [45]. Wild plants exhibited a less pronounced reduction in transpiration in response to the same drought stress compared to cultivated plant varieties, suggesting a lower sensitivity to this type of stress.

The observed differences in gas exchange parameters between wild and cultivated *F. vesca* plants under water stress conditions can be attributed to the inherent genetic diversity and adaptive traits present in wild varieties. Wild *F. vesca* evolved in natural environments where water stress is a common challenge, leading to the development of physiological and biochemical mechanisms that enhance drought tolerance. In contrast, cultivated varieties have been selectively bred for traits that may not prioritize drought resistance, resulting in a more pronounced reduction in gas exchange under similar stress conditions [46,47]. The greater genetic variability present in natural populations, due to uncontrolled interbreeding and mutations, compared to cultivated varieties, results in a broader range of adaptive responses to environmental stressors [48]. This genetic variability is crucial for the expression of drought-responsive genes, which facilitate improved gas exchange under water-limited conditions. For example, studies on wild relatives of other crops, such as peanut (*Arachis hypogaea* L.), have shown that they possess alleles that confer resistance to abiotic stresses, including drought, which can be leveraged to improve cultivated varieties [49].

Cultivated varieties of *F. vesca* are often subjected to agricultural practices that prioritize yield over stress tolerance. Such selective breeding may reduce the expression of genes associated with drought tolerance, rendering these plants more susceptible to water stress. Consequently, cultivated varieties may show more pronounced declines in gas exchange parameters under water stress, as they lack the robust adaptive traits present in their wild counterparts [50].

Similar treatments may trigger opposite changes in other adaptive mechanisms. The observed greater reduction in leaf photosynthetic pigment content in *F. vesca* plants from natural environments under soil moisture deficit, as compared to cultivated varieties, may result from selective breeding practices favoring higher yields. Cultivated varieties may possess enhanced mechanisms for maintaining pigment levels under stress, such as improved water use efficiency and osmotic regulation [51–53]. Although plants may initially increase photosynthetic pigment production under mild stress, severe water deficits can inhibit this response, ultimately leading to a net reduction in pigment content [51,52]. Wild *F. vesca* plants, frequently exposed to fluctuating moisture levels and other abiotic stresses, may exhibit a more pronounced physiological response to additional stressors, such as drought [54]. In contrast, cultivated varieties, typically grown in more controlled environments, can better withstand water stress without a significant reduction in pigment content [54].

The analysis of mineral composition indicated that wild plants exhibited more substantial variation in the content of specific elements depending on soil moisture conditions, suggesting greater adaptive flexibility. Under control conditions, these plants had a higher calcium content, while under moisture deficit, levels of elements such as potassium and manganese increased. The increase in potassium content is associated with plant survival mechanisms during drought stress [55]. Elevated potassium levels may also function as a tolerance mechanism, limiting the absorption of toxic ions by the roots [56,57]. Manganese, a critical component of photosynthetic proteins and enzymes, plays a central role in the water-splitting reaction that supplies electrons for subsequent stages of photosynthesis. Its deficiency can significantly impair the activity of photosystem II (PSII) [58]. Therefore, the observed increase in manganese content in tissues of plants exposed to drought conditions likely reflects the activation of physiological processes aimed at mitigating the adverse effects of water deficit on plant metabolism.

In contrast, cultivated varieties exhibited fewer or less pronounced differences in mineral content under varying moisture conditions, affecting only select elements. For instance, the 'Baron von Solemacher' cultivar showed increased sodium and copper content under drought stress, whereas the 'Rugia' displayed greater stability in its mineral composition regardless of soil moisture levels. Sodium contributes to osmotic adjustment in plant tissues under conditions of low external water potential. In C4 species, sodium also plays a role in both the light and dark phases of photosynthesis; for example, it aids in the transport of pyruvate to the chloroplast mesophyll and supports the structural integrity of chloroplast granules [59]. Copper, meanwhile, is crucial at the cellular level, influencing transcriptional and protein transport mechanisms, oxidative phosphorylation, and iron accumulation, all of which significantly shape the plant's response to drought stress [60].

Other research indicates that wild plants develop more versatile adaptation mechanisms to changing and challenging environmental conditions, a trait supported by their greater genetic diversity. Studies on various plant genotypes, for instance, reveal that wild plants produce higher levels of antioxidant enzymes and osmoprotectants, which enhance their survival under stress [61]. Wild plants also demonstrated more marked differences in elemental content depending on substrate moisture levels. Their leaves, grown under control conditions, contained more calcium (Ca), whereas under drought conditions, they showed elevated concentrations of potassium (K), sodium, manganese, and iron. This suggests an ability to dynamically adjust the internal transport and allocation of minerals in response to drought [62], a response characteristic of wild plants that must adapt to naturally fluctuating environmental conditions where water availability varies. In comparison, cultivated varieties, which have been selectively bred primarily for high yield efficiency and uniformity, may not exhibit such a broad range of adaptive responses under stress [63].

Elemental content differences between the roots of wild plants and cultivated varieties were even more pronounced. The roots of wild plants displayed higher concentrations of potassium, magnesium, sodium, copper, and zinc, indicating an enhanced ability to effectively absorb and store minerals under deficit conditions. While increased sodium and copper accumulation was observed in the roots of the 'Baron von Solemacher' cultivar and increased sodium and zinc in the roots of the 'Rugia', this response to water deficiency was more extensive in wild plants, with a broader range of mineral accumulation and redistribution [64]. Natural populations exhibit greater genetic diversity, enabling a wider spectrum of responses to environmental stressors, such as low soil moisture. This genetic variability is critical for the efficient uptake and translocation of essential minerals, as demonstrated in studies emphasizing the role of genetic biofortification in wild crop relatives [65]. Furthermore, wild plants frequently display considerable phenotypic plasticity, allowing them to modify root and leaf characteristics in response to fluctuating moisture levels, as shown in research on *Plantago lanceolata* [66]. Species adapted to low-nutrient soils also develop specialized root morphologies and nutrient acquisition strategies that enhance mineral uptake under stress conditions [67].

Cultivars have likely been selectively bred with an emphasis on yield rather than resilience to environmental variability, which may limit their capacity to adapt to low soil moisture conditions. This focus on yield can lead to reduced genetic diversity and a narrower range of functional traits, potentially impairing their ability to accumulate essential minerals in challenging environments.

Under conditions of water deficiency (drought stress), numerous physiological and molecular processes are either activated or inhibited to adjust plant growth and development, thereby enhancing stress tolerance. The terms "water deficiency" or "water stress" refer to a state in which a plant's water content is reduced to levels that impair its physiological functions, such as growth, stomatal conductance, and photosynthesis rate. Determining water deficit in absolute terms is challenging due to the complex interplay among atmospheric, plant, and soil factors. Furthermore, a plant's response to stress can be influenced by conditions preceding the stress event itself. Jacques et al. [68] highlight the concept of "stress memory," suggesting that plants possess mechanisms to retain information from drought-related stress, potentially leading to enhanced tolerance in subsequent growing seasons. This phenomenon was also observed in our study, where the plant response to drought stress was less pronounced in the second year of the experiment.

The physiological advantages of wild *Fragaria vesca* result from evolutionary adaptations, extensive genetic diversity, and complex metabolic profiles, which are often diminished in cultivated forms. While cultivated varieties may display characteristics advantageous for commercial production, such as uniformity and increased fruit yield, they frequently lack the resilience and metabolic diversity inherent to wild populations. This trade-off underscores the significance of wild species in breeding programs, as they offer genetic material that can enhance stress tolerance and nutritional quality in cultivated varieties.

Although wild *F. vesca* demonstrates superior drought tolerance, it is important to recognize that cultivated varieties may possess other valuable traits, such as higher yield and improved fruit quality. This highlights the value of integrating wild genetic resources into breeding efforts to increase the resilience of cultivated strawberries while preserving desirable agronomic characteristics. Evidence suggests that naturally occurring plants exhibit considerable heritable genetic variability for traits associated with stress resistance, implying that wild *F. vesca* may contain untapped genetic resources valuable for improving cultivated varieties [69]. Balancing drought tolerance with other agronomic traits remains a crucial focus for future research and development in *Fragaria* breeding.

Our study, however, has several notable limitations. First, as the experiments were conducted under controlled pot conditions, the results may not fully translate to field conditions, where numerous additional factors influence physiological responses. Despite this, our aim was to isolate and examine a single factor, minimizing the influence of others as much as possible. Second, the absence of molecular and proteomic analyses prevented us from definitively identifying the mechanisms underlying plant adaptation to stress. Such analyses would have provided more detailed insights and likely corroborated our findings. Third, the lack of evapotranspiration data means we cannot unequivocally quantify total water loss by the plants and their environment, which includes both transpiration (water loss through stomata) and soil surface evaporation. Nonetheless, tensiometric measurements are widely regarded as a reliable method with high accuracy, offering an efficient alternative to the more labor-intensive gravimetric method. Finally, some measurements may appear slightly inconsistent due to the extended duration of the study and the influence of uncontrolled factors, such as wind, pests, and humidity, which were not accounted for in our analyses.

5. Conclusions

- This study enabled a comparative analysis of the physiological responses of *Fragaria vesca* plants, including two cultivated varieties ('Rugia' and 'Baron von Solemacher') and a wild strawberry population obtained from a natural habitat. The key findings are as follows:
- *F. vesca* genotypes exhibit differences in physiological traits. Wild plants display distinct physiological patterns compared to cultivated varieties, including higher chlorophyll fluorescence values and greater assimilation pigment content.
- Water deficiency in the substrate significantly impacts the physiological characteristics of *F. vesca* plants. Drought conditions lead to reductions in gas exchange parameters, water use efficiency, chlorophyll "a" fluorescence parameters, relative water content in leaves, and the mass of fruits. Conversely, drought increases the content of elements such as K, Ca, Mg, Na, Cu, Zn, and Mn in various plant organs.
- *F. vesca* genotypes differ in their physiological responses to water deficit in the substrate. Plants from the natural environment demonstrate lower sensitivity, in terms of physiological responses, to water deficit compared to cultivated varieties.

Supplementary Materials: The following supporting information can be downloaded at: https://www.mdpi.com/article/10.3390/agriculture15010070/s1, Table S1: Results of two-factor analysis of variance for gas exchange parameters, Table S2: Results of two-factor analysis of variance for chlorophyll "a" fluorescence parameters, Table S3: Results of two-factor analysis of variance for content of photosynthetic pigments, Table S4: Results of two-factor analysis of variance for relative water content (RWC), Table S5: Results of two-factor analysis of variance for total fruit yield from one plant and weight of a single fruit, Table S6: Results of two-factor analysis of variance for fresh and dry mass of the root system, Table S7: Results of two-factor analysis of variance for chemical composition of leaves, crowns and roots,

Author Contributions: Conceptualization, M.R. and M.M.; methodology, M.R. and M.M.; software, M.R. and P.P.; validation, M.R. and M.M.; formal analysis, M.R.; investigation, M.R. and P.P.; resources, M.R., M.M. and P.P.; data curation, M.R.; writing—original draft preparation, M.R. and M.M.; writing—review and editing, M.R. and M.M.; visualization, M.R.; supervision, M.M.; project administration, M.R. and M.M.; funding acquisition, M.R. and M.M. All authors have read and agreed to the published version of the manuscript.

Funding: This research received no external funding.

Institutional Review Board Statement: Not applicable.

Informed Consent Statement: Not applicable.

Data Availability Statement: The datasets used and analyzed during the current study are available from the corresponding author on reasonable request.

Conflicts of Interest: The authors declare no conflicts of interest.

References

1. Hamdan, M.; Shafar, J. Effect of water irrigation techniques on growth and yield of rock melon (*Cucumis melo Linn* cv. Glamour). *Adv. Plant Sci. Technol.* **2017**, *24*, 80–84.
2. Ghaderi, N.; Normohammadi, S.; Javadi, T. Morpho-physiological responses of strawberry (*Fragaria × ananassa*) to exogenous salicylic acid application under drought stress. *J. Agric. Sci. Technol.* **2015**, *17*, 167–178.
3. Klamkowski, K.; Treder, W.; Wójcik, K. Effects of long-term water stress on leaf gas exchange, growth and yield of three strawberry cultivars. *J. Agric. Sci. Technol.* **2015**, *17*, 167–178.
4. Naumann, G.; Alfieri, L.; Wyser, K.; Mentaschi, L.; Betts, R.A.; Carrao, H.; Spinoni, J.; Vogt, J.; Feyen, L. Global Changes in Drought Conditions Under Different Levels of Warming. *Geophys. Res. Lett.* **2018**, *45*, 3285–3296. [CrossRef]
5. Shao, H.-B.; Bu, L.-Y.; Jaleel, C.; Zhao, C.-X. Water-deficit stress-induced anatomical changes in higher plants. *Comptes Rendus Biol.* **2008**, *331*, 215–225. [CrossRef] [PubMed]
6. Doroszewski, A.; Jadczyszyn, J.; Kozyra, J.; Pudełko, R.; Stuczyński, T.; Mizak, K.; Łopatka, A.; Koza, P.; Górski, T.; Wróblewska, E. Podstawy systemu monitoringu suszy rolniczej (Fundamentals of the agricultural drought monitoring system). *Water Environ. Rural Areas* **2015**, *2*, 77–91.
7. Boczoń, A.; Wróbel, M. Wpływ suszy na pobór wody przez sosnę zwyczajną (*Pinus sylvestris* L.) o różnej pozycji w drzewostanie (The influence of drought on the water uptake by Scots pines (*Pinus sylvestris* L.) at different positions in the tree stand). *For. Res. Work* **2015**, *76*, 370–376.
8. Treder, W.; Wójcik, K.; Tryngiel-Grać, A.; Krzewińska, D.; Klamkowski, K. Rozwój nawodnień roślin sadowniczych w świetle badań ankietowych (Development of irrigation of orchard plants reflected by survey investigations). *Infrastruct. Ecol. Rural Areas* **2011**, *5*, 61–69.
9. Dietz, K.; Zörb, C.; Geilfus, C. Drought and crop yield. *Plan Biol.* **2021**, *23*, 881–893. [CrossRef]
10. Liu, F.; Savic, S.; Jensen, C.; Shahnazari, A.; Jacobsen, S.; Stikic, R.; Andersen, M. Water relations and yield of lysimeter-grown strawberries under limited irrigation. *Sci. Hortic.* **2007**, *111*, 128–132. [CrossRef]
11. Żebrowska, J.; Dyduch, M. Quantitative assessment of the morphogenetic efficiency in strawberry *Fragaria × ananassa* Duch. *Folia Hortic.* **2009**, *21*, 53–64. [CrossRef]
12. Statistics Poland, Agriculture Department. *Crop Production in 2016*; Statistical Publishing Establishment: Warsaw, Poland, 2017.
13. Statistics Poland, Agriculture Department. *Production of Agricultural and Horticultural Crops in 2020*; Statistical Publishing Establishment: Warsaw, Poland, 2021.
14. Alsheikh, M.; Suso, H.; Robson, M.; Battey, N.; Wetten, A. Appropriate choice of antibiotic and *Agrobacterium* strain improves transformation of antibiotic-sensitive *Fragaria vesca* and *F. v. semperflorens*. *Plant Cell Rep.* **2002**, *20*, 1173–1180. [CrossRef]
15. Caulet, R.-P.; Gradinariu, G.; Iurea, D.; Morariu, A. Influence of furostanol glycosides treatments on strawberry (*Fragaria × ananassa* Duch.) growth and photosynthetic characteristics under drought conditions. *Sci. Hort.* **2014**, *169*, 179–188. [CrossRef]
16. Klamkowski, K.; Treder, K. wpływ deficytu wody na wymianę gazową liści, wzrost i plonowanie dwóch odmian truskawki uprawianych pod osłonami (Effect of water deficit on gas exchange parameters, growth and yield of two strawberry cultivars grown under greenhouse conditions). *Infrastruct. Ecol. Rural Areas* **2011**, *5*, 105–113.
17. Sun, C.; Li, X.; Hu, Y.; Zhao, P.; Xu, T.; Sun, J.; Gao, X. Proline, sugars, and antioxidant enzymes respond to drought stress in the leaves of strawberry plants. *Hortic. Sci. Technol.* **2015**, *33*, 625–632. [CrossRef]
18. Yin, B.; Wang, Y.; Liu, P.; Hu, J.; Zhen, W. Effects of vesicular-arbuscular mycorrhiza on the protective system in strawberry leaves under drought stress. *Front. Agric. China* **2010**, *4*, 165–169. [CrossRef]
19. Ibrahim, M.H.; Nulit, R.; Sakimin, S.Z. Influence of drought stress on growth, biochemical changes and leaf gas exchange of strawberry (*Fragaria × ananassa* Duch.) in Indonesia. *AIMS Agric. Food.* **2022**, *7*, 37–60.
20. Jiang, L.; Song, R.; Wang, X.; Wang, J.; Wu, C. Transcriptomic and Metabolomic Analyses Provide New Insights into the Response of Strawberry (*Fragaria × ananassa* Duch.) to Drought Stress. *Horticulturae* **2024**, *10*, 734. [CrossRef]
21. Mikiciuk, G.; Mikiciuk, M.; Hawrot-Paw, M. Influence of superabsorbent polymers on the chemical composition of strawberry (*Fragaria × ananassa* Duch.) and biological activity in the soil. *Folia Hortic.* **2015**, *27*, 63–69. [CrossRef]
22. Krishna, K.V.; Manasa, N.; Krishna, N.V. Drought Resistance in Wheat (*Triticum aestivum* L.). *J. Adv. Biol. Biotechnol.* **2024**, *27*, 409–415. [CrossRef]
23. Sun, F.; Chen, Q.; Chen, Q.; Jiang, M.; Qu, Y. Yield-based drought tolerance index evaluates the drought tolerance of cotton germplasm lines in the interaction of genotype-by-environment. *PeerJ* **2023**, *11*, e14367. [CrossRef] [PubMed]

24. Bartczak, M.; Lisiecka, J.; Knaflewski, M. Correlation between selected parameters of planting material and strawberry yield. *Folia Hortic.* **2010**, *22*, 9–12. [CrossRef]
25. Błaszczyk, J. Influence of harvest date and storage conditions on the content of chlorophyll pigments in pear peels. *Folia Hortic.* **2012**, *24*, 91–95. [CrossRef]
26. Ghaderi, N.; Siosemardeh, A. Response to drought stress of two strawberry cultivars (cv. Kurdistan and Selva). *Hortic. Environ. Biotechnol.* **2011**, *52*, 6–12. [CrossRef]
27. Zahedi, S.M.; Hosseini, M.S.; Fahadi Hoveizeh, N.; Kadkhodaei, S.; Vaculík, M. Physiological and biochemical responses of commercial strawberry cultivars under optimal and drought stress conditions. *Plants* **2023**, *12*, 496. [CrossRef]
28. Docimo, T.; Paesano, A.; D'Agostino, N.; D'Amelia, V.; Garramone, R.; Carputo, D.; Aversano, R. Exploring CDF gene family in wild potato under salinity stress unveils promising candidates for developing climate-resilient crops. *Sci. Rep.* **2024**, *14*, 24619. [CrossRef] [PubMed]
29. Najafipour, B.; Mirlohi, A.; Majidi, M.M.; Saeidi, G.; Abtahi, M. Wild barley genomic resources for drought adaptability and quality improvement. *J. Cereal Sci.* **2023**, *114*, 103802. [CrossRef]
30. Żurawicz, E. *Truskawka i Poziomka (Strawberry and Wild Strawberry)*; Państwowe Wydawnictwo Rolnicze i Leśne: Warsaw, Poland, 2005.
31. Kaya, C.; Şenbayram, M.; Akram, N.A.; Ashraf, M.; Alyemeni, M.N.; Ahmad, P. Sulfur-enriched leonardite and humic acid soil amendments enhance tolerance to drought and phosphorus deficiency stress in maize (*Zea mays* L.). *Sci. Rep.* **2020**, *10*, 6432. [CrossRef]
32. Wang, X.; Fu, J.; Min, Z.; Zou, D.; Liu, H.; Wang, J.; Zheng, H.; Jia, L.; Yang, L.; Xin, W.; et al. Response of rice with overlapping growth stages to water stress by assimilates accumulation and transport and starch synthesis of superior and inferior grains. *Int. J. Mol. Sci.* **2022**, *23*, 11157. [CrossRef] [PubMed]
33. Frost, C.J.; Nyamdari, B.; Tsai, C.J.; Harding, S.A. The tonoplast-localized sucrose transporter in Populus (PtaSUT4) regulates whole-plant water relations, responses to water stress, and photosynthesis. *PLoS ONE* **2012**, *7*, e44467. [CrossRef]
34. Küçükyumuk, C.; Yıldız, H.; Meriç, M.K. The response of Braeburn apple to regulated deficit irrigation. *J. Agric. Sci.* **2020**, *26*, 154–163. [CrossRef]
35. Jones, H.G. Monitoring plant and soil water status: Established and novel methods revisited and their relevance to studies of drought tolerance. *J. Exp. Bot.* **2007**, *58*, 119–130. [CrossRef] [PubMed]
36. Arnon, D.I.; Allen, M.B.; Whatley, F.R. Photosynthesis by isolated chloroplasts. IV. General concept and comparison of three photochemical reactions. *Biochim. Biophys. Acta* **1956**, *20*, 449–461. [CrossRef]
37. Lichtenthaler, H.; Wellburn, A. Determinations of total carotenoids and chlorophyll a and b of leaf extracts in different solvents. *Biochem. Soc. Trans.* **1966**, *11*, 591–592. [CrossRef]
38. Hager, A.; Mayer-Berthenrath, T. Die Isolierung und quantative Bestimung der Carotenoide und Chlorophyll von Blatern, Algen und isolierten Chloroplasten mit Hilfe Dunnschichtchromatographischer Methoden. *Planta* **1966**, *69*, 198–217. [CrossRef] [PubMed]
39. Yamasaki, S.; Rebello Dillenburg, I. Measurements of leaf relative water content in Araucaria angustifolia. *Rev. Bras. Fisiol. Veg.* **1999**, *11*, 69–75.
40. Cameron, J.S.; Hartley, C.A. Gas exchange characteristics of *Fragaria chiloensis* genotypes. *Hort. Sci.* **1990**, *25*, 327–329. [CrossRef]
41. Gao, S.; Yan, Q.; Chen, L.; Song, Y.; Li, J.; Fu, C.; Dong, M. Effects of ploidy level and haplotype on variation of photosynthetic traits: Novel evidence from two *Fragaria* species. *PLoS ONE* **2017**, *12*, e0179899. [CrossRef]
42. Lapshin, V.I.; Yakovenko, V.V.; Shcheglov, S.N.; Podorojny, V.N. A methodical approach for evaluating the variability of productivity and fruit quality in the genetic collections of strawberry (*Fragaria* × *ananassa* Duch.). *Vavilovskij Žurnal Genet. I Sel.* **2019**, *23*, 675–682. [CrossRef]
43. Kiprono, D.; Thagun, C.; Kodama, Y. Light-dependent chloroplast relocation in wild strawberry (*Fragaria vesca*). *Plant Signal. Behav.* **2024**, *19*, 2342744. [CrossRef]
44. Na, Y.W.; Jeong, H.J.; Lee, S.Y.; Choi, H.G.; Kim, S.H.; Rho, I.R. Chlorophyll fluorescence as a diagnostic tool for abiotic stress tolerance in wild and cultivated strawberry species. *Hortic. Environ. Biotechnol.* **2014**, *55*, 280–286. [CrossRef]
45. Malambane, G.; Madumane, K.; Sewelo, L.T.; Batlang, U. Drought stress tolerance mechanisms and their potential common indicators to salinity, insights from the wild watermelon (*Citrullus lanatus*): A review. *Front. Plant Sci.* **2023**, *13*, 1074395. [CrossRef]
46. Skender, A.; Ajdinović, T.; Bećirspahić, D. The comparison of phenotypic characteristics of current varieties and wild species of *Fragaria*. *Genetika* **2015**, *47*, 45–52. [CrossRef]
47. Ulrich, D.; Olbricht, K. Diversity of metabolite patterns and sensory characters in wild and cultivated strawberries 1. *J. Berry Res.* **2014**, *4*, 11–17. [CrossRef]
48. Wei, Y.; Kong, Y.; Li, H.; Yao, A.; Han, J.; Zhang, W.; Li, X.; Han, D. Genome-Wide Characterization and Expression Profiling of the AP2/ERF Gene Family in *Fragaria vesca* L. *Int. J. Mol. Sci.* **2024**, *25*, 7614. [CrossRef] [PubMed]

49. Brasileiro, A.; Morgante, C.; Araujo, A.; Leal-Bertioli, S.; Silva, A.; Martins, A.; Vinson, C.; Santos, C.; Bonfim, O.; Togawa, R.; et al. Transcriptome profiling of wild Arachis from water-limited environments uncovers drought tolerance candidate genes. *Plant Mol. Biol. Rep.* **2015**, *33*, 1876–1892. [CrossRef]
50. Schulze, J.; Erhardt, A.; Stoll, P. Reduced clonal reproduction indicates low potential for establishment of hybrids between wild and cultivated strawberries (*Fragaria vesca* × *F.* × *ananassa*). *Ecol. Res.* **2013**, *28*, 43–52. [CrossRef]
51. Rugienius, R.; Bendokas, V.; Siksnianas, T.; Stanys, V.; Sasnauskas, A.; Kazanaviciute, V. Characteristics of *Fragaria vesca* yield parameters and anthocyanin accumulation under water deficit stress. *Plants* **2021**, *10*, 557. [CrossRef]
52. Rokosa, M.; Mikiciuk, M. Assessment of physiological and morphological traits of plants of the genus *Fragaria* under conditions of water deficit-a study review. *Acta Sci. Pol. Hortorum Cultus* **2020**, *19*, 21–39. [CrossRef]
53. Vinskienė, J.; Bendokas, V.; Stanys, V.; Sasnauskas, A.; Rugienius, R. The effect of osmotic stress, lighting spectrum and temperature on growth and gene expression related to anthocyanin biosynthetic pathway in wild strawberry (L.). *Folia Hortic.* **2023**, *35*, 419–431. [CrossRef]
54. Nezhadahmadi, A.; Faruq, G.; Rashid, K. The impact of drought stress on morphological and physiological parameters of three strawberry varieties in different growing conditions. *Pak. J. Agric. Sci.* **2015**, *52*, 79–92.
55. Al-Yasi, H.; Attia, H.; Alamer, K.; Hassan, F.; Ali, E.; Elshazly, S.; Siddique, K.; Hessini, K. Impact of drought on growth, photosynthesis, osmotic adjustment, and cell wall elasticity in Damask rose. *Plant Physiol. Biochem.* **2020**, *150*, 133–139. [CrossRef] [PubMed]
56. Perin, E.; Messias, R.; Galli, V.; Borowski, J.; Souza, E.; Avila, L.; Bamberg, A.; Rombaldi, C. Mineral content and antioxidant compounds in strawberry fruit submitted to drough stress. *Food Sci. Technol. Int.* **2019**, *39*, 245–254. [CrossRef]
57. Keutgen, A.J.; Pawelzik, E. Quality and nutritional value of strawberry fruit under long term salt stress. *Food Chem.* **2008**, *107*, 1413–1420. [CrossRef]
58. Buchanan, B.; Gruissem, W.; Jones, R. Photosynthesis. In *Biochemistry and Molecular Biology of Plants*; Wiley Blackwell: Hoboken, NJ, USA; American Society of Plants Biologists: Rockville, MD, USA, 2000; pp. 568–629. ISBN 978-0-470-71421-8.
59. Martinez, J.P.; Ledent, J.F.; Bajji, M.; Kinet, J.M.; Lutts, S. Effect of water stress on growth, Na^+ and K^+ accumulation and water use efficiency in relation to osmotic adjustment in two populations of *Atriplex halimus* L. *Plant Growth Regul.* **2003**, *41*, 63–73. [CrossRef]
60. Yurela, I. Copper in plants. *Braz. J. Plant Physiol.* **2005**, *17*, 145–156. [CrossRef]
61. Gupta, B.; Shrestha, J. Abiotic stress adaptation and tolerance mechanisms in crop plants. *Front. Plant Sci.* **2023**, *14*, 1278895. [CrossRef]
62. Munné-Bosch, S. Vitamin E function in stress sensing and signaling in plants. *Dev. Cell* **2019**, *48*, 290–292. [CrossRef] [PubMed]
63. Lata, C.; Prasad, M. Role of DREBs in regulation of abiotic stress responses in plants. *J. Exp. Bot.* **2011**, *62*, 4731–4748. [CrossRef]
64. Craine, J.M.; Nippert, J.B.; Elmore, A.J.; Skibbe, A.M.; Hutchinson, S.L.; Brunsell, N.A. Timing of climate variability and grassland productivity. *Proc. Natl. Acad. Sci. USA* **2012**, *09*, 3401–3405. [CrossRef]
65. Maqbool, A.; Bakhsh, A.; Aksoy, E. Effects of natural variations on biofortification. In *Wild Germplasm for Genetic Improvement in Crop Plants*; Academic Press: Cambridge, MA, USA, 2021; pp. 115–138.
66. Dawson, W.; Bòdis, J.; Bucharova, A.; Catford, J.; Duncan, R.; Fraser, L.; Groenteman, R.; Kelly, R.; Moore, J.; Pärtel, M.; et al. Root traits vary as much as leaf traits and have consistent phenotypic plasticity among 14 populations of a globally widespread herb. *Funct. Ecol.* **2024**, *4*, 926–941. [CrossRef]
67. Bowsher, A.W.; Miller, B.J.; Donovan, L.A. Evolutionary divergences in root system morphology, allocation, and nitrogen uptake in species from high-versus low-fertility soils. *Funct. Plant Biol.* **2015**, *43*, 129–140. [CrossRef] [PubMed]
68. Jacques, C.; Salon, C.; Barnard, R.L.; Vernoud, V.; Prudent, M. Drought stress memory at the plant cycle level: A review. *Plants* **2021**, *10*, 1873. [CrossRef] [PubMed]
69. Omondi, E.O.; Lin, C.Y.; Huang, S.M.; Liao, C.A.; Lin, Y.P.; Oliva, R.; van Zonneveld, M. Landscape genomics reveals genetic signals of environmental adaptation of African wild eggplants. *Ecol. Evol.* **2024**, *14*, e11662. [CrossRef]

agriculture

MDPI

Article

Increasing Productivity and Fruit Quality of 'Mutsu' Apple Orchard by Dwarfing Treatments

Maria Małachowska *, Tomasz Majak, Tomasz Krupa and Kazimierz Tomala

Department of Pomology and Horticulture Economics, Institute of Horticultural Sciences, Warsaw University of Life Sciences (SGGW-WULS), ul. Nowoursynowska 159 C, 02-787 Warsaw, Poland
* Correspondence: maria_malachowska@sggw.edu.pl; Tel.: +48-22-593-2080

Abstract: The aim of this 2022 study was to evaluate the effect of tree growth-limiting treatments on the tree yield and quality of 'Mutsu' apples. The experiment was established on 7-year-old trees on M.9 rootstock in a commercial orchard in Wilga near Warsaw. Growth-limiting treatments included unilateral root pruning, spraying the trees with Regalis Plus 10 WG at various times, and spraying with Flordimex 480 SL. Eight combinations were used, with four replicates of 20 trees per repetition. The measurements included fruit set, length of this year's shoots, yield per tree, average fruit weight, and the size structure of the yield. The distinctive physiological status of the apples was assessed directly after harvest, directly after 8 months of storage under CA conditions (1.5% CO_2, 1.5% O_2, 1 °C, >92% RH) and after an additional 7 days of shelf-life. Spraying trees with Regalis Plus 10 WG from the balloon stage onwards, irrespective of the treatment with root pruning, was most effective in both inhibiting long-stem growth and increasing tree yield (by almost two times) by increasing the number of apples per tree. The growth response of long-stemmed apple trees to both unilateral root pruning and Ethephon spray was significantly lower than it was to Regalis Plus 10 WG spray and had relatively little effect on their yield. Regalis Plus 10 WG resulted in a clear reduction in average fruit weight (by about 100 g), which, in the case of the large-fruited cultivar 'Mutsu', should be seen as an advantage. Its application from the balloon stage onwards promoted higher apple firmness at harvest and after simulated handling preceded by long-term storage.

Keywords: apple tree; Regalis Plus 10 WG; Ethephon; root pruning; growth; yield; apple quality

1. Introduction

Apple production in Poland is developing dynamically and is the most important sector of the fruit-growing industry. Thanks to modern storage technologies, apples are available on the market year-round. As Europe's leading apple producer, Poland faces many challenges. Despite their global success, some apple cultivars are not cultivated [1,2] due to their specific climatic requirements. The yield of cultivated apple cultivars, as well as other fruit species grown in temperate climates, is determined by the weather conditions during the peri-bloom period and depends on the occurrence of spring frosts [3–5]. According to the Central Statistical Office, the apple harvest in Poland in 2022 is 4.2 million tonnes, while the World Apple and Pear Association (WAPA) has estimated the harvest at 4.495 million tonnes [6]. Despite the increase in apple production, export volumes are not increasing accordingly. The challenges of the Russian embargo continue to affect Polish producers, forcing them to look for new markets and implement advanced agrotechnical practices to ensure high-quality apples that meet global standards [7].

Until recently, Polish producers focused mainly on the 'Idared' cultivar for the eastern market. Currently, 'Gala', 'Golden Delicious', and 'Red Delicious' are the leading cultivars. Depending on the genetic characteristics of the cultivar, the problem of adequate growth and the yield of the trees occurs to varying degrees. Bilateral root pruning is performed to maintain the balance between growth and fruiting [8]. However, drought occurring after such treatment can result in both russeting and excessive fruit drop [9–11]. Therefore, combining unilateral root pruning with another method of regulating tree growth can alleviate the stress caused by bilateral root pruning [12,13].

Moderate shoot growth and good insolation of the tree crown are critical factors in ensuring high yields, as are proper apple size and coloration. The production of "Extra" Class fruit is facilitated by bioregulators for growth [14–17]. For this purpose, calcium prohexadione (contained in Regalis Plus 10 WG) is used to prevent the formation of active gibberellin GA_1 from inactive gibberellin GA_{20}. In addition, there is increased production of GA_4 gibberellin, which reduces fruit russeting [18–20].

Intense shoot growth adversely affects flower bud formation and fruit setting after flowering. This is because the strongly growing shoots compete with the fruit for nutrients and contribute to fruit drop [21–24]. In contrast, calcium prohexadione causes a decrease in auxins and an increase in cytokinins [25,26]. Cytokinins are considered promoters of flower bud set initiation [26–28]. An inhibitory effect of calcium prohexadione on ACC (1-aminocyclopropane-1-carboxylic acid) oxidase activity was also observed. The reaction requires 2-oxyglutaric acid and ascorbic acid [29,30]. Perhaps because of the similar structure, calcium prohexadione is incorporated in place of the compounds mentioned above, with the resulting disruption of the production of ethylene—a phytohormone produced in the plant in significant quantities, especially during periods of high stress. Stress can be caused by various factors, including but not limited to adverse weather conditions, such as lack of or excess water in the soil. Intense ethylene production contributes to faster ageing and the formation of a cut-off layer and, consequently, to the dropping of buds and leaves [30–32]. To counteract the negative effects of stress and to achieve high-quality yields, the growth of this year's shoots must be skillfully controlled.

In a high-apple-production environment, obtaining the highest possible percentage of premium quality fruit in the crop is becoming a priority. Such apples are accepted in the world's most demanding markets. Among the green-skinned apple cultivars, 'Mutsu' is the most noteworthy [33]. However, trees of this cultivar require the skillful management of shoot growth, which is significant for the size and regularity of yield. The strong tree growth of this cultivar has a negative effect on flower bud formation and fruit quality [34,35]. Therefore, crown contour pruning and root pruning are recommended, alongside growth regulators, including calcium prohexadione [14,36,37]. This compound was shown to effectively inhibit the growth of this year's shoots, with no negative effect on either fruit yield and quality or the initiation of tree flowering [38]. For apples' high quality and storability, their physiological stage at harvest is also important [39,40]. Adequate action in these areas makes it possible to obtain fruit that is well-developed and meets high quality standards, even after long-term storage [41–43].

The present study aimed to evaluate the effectiveness of unilateral root pruning, the application of Regalis Plus 10 WG, and the spraying of Ethephon on apple tree growth reduction and to find out the effects of these treatments on tree yield and quality of 'Mutsu' apples at harvest, directly after storage, and after simulated marketing.

2. Materials and Methods

The study material consisted of trees and fruit of the 'Mutsu' cultivar on a private orchard farm in the municipality of Wilga. The orchard was established in the spring of 2015 with annual trees on M.9 rootstock at a spacing of 3.5 × 1 m on sandy loam soil with a slight acid reaction. Herbicide fallow was maintained in the tree rows and turf in the inter-rows.

Tree protection treatments against disease and pests were carried out based on pheromone trap monitoring and control according to the current requirements of Integrated Plant Protection, using registered plant protection products authorized for use in 2022. A soil analysis for mineral content and pH was also carried out.

Over-crown sprinkling was used to protect the trees from frost. In the spring of 2022, temperature drops below 0 °C at 1 m were recorded several times. Thanks to the use of over-crown sprinkling, no damage to flower buds or flowers was recorded.

The study was carried out in eight treatments of the experiment. Four repetitions were distinguished in each treatment, with 20 trees per repetition. The experiment involved two rows of trees, well apart from each other, to avoid the risk of spray drift on trees in other treatments. In one of these rows, unilateral root pruning was performed using an oblique cut. Regalis Plus 10 WG was applied in two separate treatments, differing in the timing of spraying the trees.

The following treatments were used in the experiment:

1. Control (no growth-limiting treatments)—C;
2. Unilateral root pruning with an oblique knife—RP;
3. Regalis Plus 10 WG applied from the balloon stage (BBCH 60–69)—RB;
4. Regalis Plus 10 WG applied at standard times, according to the label (BBCH 71–73)—RS;
5. Flordimex 480 SL applied 4 times, from the end of flowering, at 2-week intervals, at doses of 250, 200, 150, and 100 mL·ha^{-1}—Ethephon;
6. Unilateral root pruning with an angled knife + Regalis Plus 10 WG applied from the balloon stage (BBCH 60–69)—RP + RB;
7. One-sided root pruning with an angled knife + Regalis Plus 10 WG applied at standard dates, according to the label (BBCH 71–73)—RP + RS;
8. Unilateral root pruning with an angled knife + spraying the trees 4 times with Flordimex 480 SL from the end of flowering, at 2-week intervals, at doses of 250, 200, 150, and 100 mL·ha^{-1}—RP + Ethephon.

2.1. Growth-Restricting Treatments

Root pruning: Unilateral root pruning with an angled knife was carried out on 1 April 2022 at a distance of 50 cm from the trunk and to a depth of 40 cm.

Regalis Plus 10 WG: The first spraying of the trees with Regalis Plus 10 WG was carried out at two different times, depending on the treatment of the experiment. In one treatment, the first application was carried out at the 'balloon' stage (BBCH 60–69)—3 May 2022, while in the other treatment, the first spraying of the trees was carried out 2 weeks later, i.e., at the standard date according to the label (BBCH 71–73)—17 May 2022, when this year's growth reached 4–7 cm in length and had fully developed 4–5 leaves. Another tree spraying in both treatments was performed after 2 weeks.

Flordimex 480 SL: Spraying of trees with Flordimex 480 SL was carried out four times. The first treatment at 250 mL·ha^{-1} was made when 80% of the flower petals had fallen. Subsequent spraying was carried out at 2-week intervals, using decreasing product doses, on successive dates, of 200, 150, and 100 mL·ha^{-1}. The trees were sprayed under high sun operation in the morning at a temperature of at least 16 °C, using 750 L of water per ha for the treatment.

2.2. Fruit Storage Conditions

To assess the quality of the fruit after storage, one pallet box of fruit (300 kg) was taken each within a given combination of experiments. Directly after harvesting, they were placed in a cold chamber for rapid cooling, and after 7 days, the target atmospheric composition (1.5% CO_2 and 1.5% O_2) was established. The apples were stored for 8 months at a temperature of 1 °C and a relative humidity above 92%. To reduce the development of apple rot agents (grey mold, wet rot), the fruit-filled chamber was fogged with a biological plant protection product (Polyversum®, at a dose of 150 g·500 m^{-3}; Biopreparaty Ltd., Unetice, Czech Republic). The chamber was fogged for 1 h using a K-22 BIO fogger (Pulsfog®, Überlingen, Germany).

2.3. Indicators Determined in the Orchard

Number of fruitlets per 100 inflorescences: In eight treatments, on 20 trees in each treatment (four repetitions of five trees each), one branch was marked, and inflorescences were counted before full flowering. The selected branches were at a similar height. After the June drop, the number of lingering fruitlets was recorded. These data were used to calculate the number of fruitlets per 100 inflorescences. The shoots used to calculate the number of fruitlets were always selected along the line of the rows, alternating between the north and south sides of the trees.

The average length of this year's increments: Five representative trees per treatment were each selected in September 2022 after shoot growth had finished. Within each tree, the lengths of 20 randomly taken shoots were measured, with 10 from the lower part of the crown and a further 10 shoots from the upper part of the crown. The shoots used to measure growth were taken along the row lines, with samples collected alternately from the northern and southern sides of the trees. On each side, five shoots were measured from the lower part of the crown and another five from the upper part, ensuring balanced sampling across both sides of the tree. Results are given in centimeters [cm].

Yield per tree: To determine the value of this indicator, fruit was harvested and weighed separately from each of the five trees in the treatment. The results are given in $kg \cdot tree^{-1}$.

Average fruit weight: Results were obtained by dividing the yield per tree (kg) by the number of apples picked. The results are given in grams [g].

Size structure of apple yield: The percentage of apples in the five size classes in a given treatment was determined after sorting the fruit in a pallet box (300 kg) within a given experimental treatment. A total of eight pallet crates of fruit were calibrated. Calibration was carried out using a Dutch Greefa MSE 2000 (Greefa Machinebouw B.V, Tricht, The Netherlands) sorter equipped with a water-based fruit unloader to prevent bruising and damage to the apples when unloading from the crates.

2.4. Physiological Status of Fruit at Harvest Directly After Storage and After 7 Days of Shelf-Life

The apples were harvested on 29 September 2022. On the same day, their physiological status was assessed in the Department of Horticulture and Horticultural Economics laboratory of the Warsaw University of Life Sciences. All determinations were performed on 40 fruits representing a given treatment (four repetitions of 10 apples per repetition). The evaluation of apple quality after 8 months of storage and after an additional 7 days of storage at room temperature (20 °C). These evaluations were carried out on 40 apples per treatment, again in four repetitions of 10 fruits, with samples taken from the trees in all eight treatments.

Ethylene concentration in the seed chambers directly after fruit harvest (IEC): Measurements were taken using a specialized gas chromatograph (Clarus 690, Perkin Elmerm, Inc., Waltham, MA, USA) after 1 cm^3 of air was taken from the apple seed chambers using a syringe. Results are given in $\mu L \cdot L^{-1}$ [44].

Starch test directly after fruit picking (SI): After cutting 10 apples in replication, the halves were sprayed with Lugol's liquid. An applied solution of iodine in potassium iodide (I_2 in KI) stained the starch-containing flesh dark blue. The resulting color of the flesh of the assessed fruits was compared with a 10-stage reference table [45].

Background skin color: Measurements of the background color of the apple skin were taken with a Minolta CR-400 colorimeter. This device sends a beam of light and compares it with the light reflected from the surface under examination. The results of the determinations are given in the CIE L*a*b* system, where L* denotes the brightness from 100 to 0, the a* value is for the transition from red to green, and b* is for the transition from yellow to blue.

Flesh firmness (FF): This parameter was determined with an Instron firmness meter (Instron 5542, Instron, Norwood, MA, USA), using an 11 mm diameter stylus for measurements. The skin was removed from the apples on two opposite sides before measurements were taken. During firmness determinations, a pin was driven into the apple flesh to a depth of 8 mm [46]. The results are given in N.

Titratable acidity (TA): This parameter was determined as described by Malachowska and Tomala [44]. For this purpose, a piece of flesh with skin was cut from each of the 10 fruits in the repetition. The juice was squeezed from the samples thus collected, then 10 mL of juice was taken and placed in a volumetric flask and diluted with 100 mL of distilled water. The resulting solution was titrated with 0.1 mol sodium hydroxide (NaOH) until pH = 8.1 using an automated titrator (TitroLine 5000, Xylem Analytics Germany GmbH, Weilheim, Germany). After conversion, the results are given as a percentage of malic acid [% malic acid].

Soluble solids content (SSC): The apple juice prepared for the titratable acidity test was also used to determine SSC content. For this purpose, a few drops of its juice were applied to a digital refractometer (Atago, Palette PR-32, Atago, Co., Ltd., Tokyo, Japan). The results are given in degrees Brix [°Bx].

2.5. Statistical Analysis

The Shapiro–Wilk test was carried out to check whether the results showed a normal distribution. A one-way ANOVA analyzed the results using Tukey's HSD post hoc test for normal distribution. Differences between average values were reported as significant at $p = 0.05$ and $p = 0.01$. Statistical analyses were performed using Statistica 13.3 (Statsoft Inc., Tulsa, OK, USA).

The strength of the relationship between the variables was assessed using Pearson's linear correlation coefficients.

3. Results

3.1. Indicators Determined in the Orchard

Spraying the trees with Regalis Plus 10 WG, irrespective of both the date of the first treatment and the treatment with prior root pruning, increased the number of fruitlets per 100 inflorescences by more than twofold compared to treatments without Regalis Plus 10 WG (Table 1).

Table 1. Effect of tree growth retardation method on the number of fruitlets per 100 inflorescences: shoot length, yield, and fruit weight of 'Mutsu' cultivar.

Treatment	Number of Fruitlets per 100 Inflorescences	Shoot Length [cm]		Yield [kg·tree^{-1}]	Fruit Weight [g]
		Top of the Tree	Bottom of the Tree		
C	47.8 b [1]	69.9 h	56.3 g	18.3 a	393 b
RP	44.0 ab	59.7 g	52.6 f	21.5 b	367 b
RB	126.3 e	18.0 b	15.0 a	37.9 h	296 a
RS	118.1 cd	26.1 d	23.9 c	32.1 f	286 a
Ethephon	44.8 ab	58.1 f	49.0 e	25.5 d	378 b
RP + RB	120.5 d	15.6 a	14.6 a	34.0 g	284 a
RP + RS	115.0 c	22.7 c	19.9 b	27.9 e	284 a
RP + Ethephon	42.4 a	53.2 e	42.7 d	23.3 c	365 b

[1] Tukey's HSD test; normal distribution (normality was checked using the Shapiro–Wilk test-$p \leq 0.05$); means followed by the same letter within a column are not significantly different. Explanation of abbreviations: C—control; RP—one-sided root pruning (with an oblique knife); RB—Regalis Plus 10 WG (applied from the balloon stage—BBCH 60–69); RS—Regalis Plus 10 WG (applied at standard terms—BBCH 71–73, according to the label); Ethephon—spraying trees with Flordimex 480 SL; RP + RB—one-sided root pruning (with an oblique knife) + Regalis Plus 10 WG (applied from the balloon stage—BBCH 60–69); RP + RS—one-sided root pruning (with an oblique knife) + Regalis Plus 10 WG (applied at standard terms—BBCH 71–73, according to the label); RP + Ethephon—one-sided root pruning (with an oblique knife) + Spraying trees with Flordimex 480 SL; n = 4.

Trees sprayed with Regalis Plus 10 WG from the 'balloon' stage (BBCH 60–69) were characterized by approximately four times shorter growth than control trees. This relationship was recorded in both the upper and lower parts of the tree crowns, regardless of the treatment with prior root pruning. In contrast, root pruning 'solo' and spraying the trees with Ethephon did little to limit long shoot growth.

The varying number of fruitlets per 100 inflorescences was reflected in the yield of the trees. The highest yield of 37.9 kg per tree was recorded after applying Regalis Plus 10 WG from the 'balloon' stage (BBCH 60–69). The average yield per tree in this treatment was two times higher than in the control treatment (Figure 1). Interestingly, Regalis Plus 10 WG applied 'solo' increased yield more compared to spraying the trees with this preparation after root pruning. In contrast, root pruning alone or spraying the trees with Ethephon alone resulted in a slight increase in yield relative to the control treatment.

(a) (b)

Figure 1. Fruit load in 'Mutsu' cultivar trees: (**a**) control (no growth-limiting treatments)—C; (**b**) Regalis Plus 10 WG applied from the balloon stage (BBCH 60–69)—RB.

Limiting tree growth with Regalis Plus 10 WG consistently reduced the average weight of the apples. In contrast, in the Ethephon program and the case of unilateral root, pruning both 'solo' and in treatment with Ethephon, the apples did not differ from the control fruit regarding the value of the trait in question. In these treatments, the average fruit weight was more than 360 g, while it fluctuated around 300 g in the other four.

3.2. Physiological State of Apples Directly After Harvest

Table 2 contains the results describing the physiological state of the apples directly after harvest. In the experiment, tree growth-limiting treatments had no significant effect on the ethylene concentration in the seed chambers (IEC) or the background color of the fruit skin as expressed by the a* and b* values of the trichromatic color stimulus. In contrast, it was found that apples from control trees had the lowest starch test (SI) value, while those from trees sprayed with the Ethephon preparation had the highest.

Table 2. The effect of tree growth retardation method on the fruit quality of 'Mutsu' apple directly after harvest.

Treatment	IEC [μL·L^{-1}]	SI [-]	Background Color			FF [N]	SSC [°Bx]	TA [%]
			L*	a*	b*			
C	0.081 a [1]	6.6 a	66.3 ab	−15.1 a	35.1 a	76.9 a	11.5 a	0.544 b
RP	0.041 a	8.5 c	68.6 b	−15.5 a	36.2 a	77.9 ab	12.3 b	0.504 ab
RB	0.067 a	7.6 b	65.4 ab	−15.1 a	35.2 a	87.4 c	13.0 c	0.479 a
RS	0.054 a	7.4 b	67.4 b	−15.2 a	35.3 a	81.8 b	12.9 c	0.494 a
Ethephon	0.049 a	9.4 d	66.9 ab	−15.3 a	34.7 a	82.3 b	12.8 bc	0.475 a
RP + RB	0.036 a	7.4 b	65.2 a	−15.0 a	34.8 a	88.7 c	12.9 c	0.487 a
RP + RS	0.064 a	7.3 b	68.3 b	−15.2 a	35.4 a	81.7 b	13.2 c	0.491 a
RP + Ethephon	0.058 a	7.4 b	68.2 b	−15.5 a	35.4 a	88.7 c	13.1 c	0.488 a

[1] Tukey's HSD test; normal distribution (normality was checked using the Shapiro–Wilk test-$p \leq 0.05$); means followed by the same letter within a column are not significantly different. Explanation of abbreviations: C—control; RP—one-sided root pruning (with an oblique knife); RB—Regalis Plus 10 WG (applied from the balloon stage—BBCH 60–69); RS—Regalis Plus 10 WG (applied at standard terms—BBCH 71–73, according to the label); Ethephon—spraying trees with Flordimex 480 SL; RP + RB—one-sided root pruning (with an oblique knife) + Regalis Plus 10 WG (applied from the balloon stage—BBCH 60–69); RP + RS—one-sided root pruning (with an oblique knife) + Regalis Plus 10 WG (applied at standard terms—BBCH 71–73, according to the label); RP + Ethephon—one-sided root pruning (with an oblique knife) + spraying trees with Flordimex 480 SL; IEC—internal ethylene content; SI—starch test; L*, a*, and b*—the color of apples in CIE L*a*b* coordinates; FF—flesh firmness; SSC—soluble solids content; TA—titratable acidity; n = 4.

There were also significant differences in tree scarification treatment for flesh firmness (FF). Statistically, apples harvested from trees sprayed with Regalis Plus 10 WG from the 'balloon' stage onwards (in both root pruning treatments), as well as fruit from trees where root pruning was combined with an Ethephon program, had the highest firmness. The apples harvested from the control trees had the least-firm flesh. The fruit from this treatment also had the lowest soluble solids content. A significantly higher value for this trait was recorded in fruit from trees after root pruning. At the same time, in the other treatments of the experiment, this indicator generally took on the highest values.

As with the previously discussed indicators, tree growth scarification treatments significantly affected apple acidity; in this case, almost all tree growth-limiting treatments significantly reduced the malic acid content of the apples compared to the control sample.

3.3. Quality of Apples Directly After Storage and After Shelf-Life

The data in Table 3 showed that the apple skin's background color directly after storage did not differ in terms of the L*, a*, and b* values in any tree growth scarification treatments. In contrast, fruit from trees with pruned roots had a higher L* parameter value after shelf-life than trees sprayed with Ethephon. Apples after the Ethephon application were generally characterized by a lower a* value, describing the background color of the skin than apples from the other treatments of the experiment.

In the experiment, there were also statistically proven differences in apple firmness under the influence of tree growth retardation treatments. Directly after storage, it was found that root pruning, spraying the trees with Regalis Plus 10 WG at standard times, as well as the application of the Ethephon program, irrespective of the treatment with root pruning, decreased apple firmness compared to both the control treatment and root pruning in combination with spraying the trees with Regalis Plus 10 WG from the 'balloon' stage. On the other hand, after simulated handling, the highest firmness was ensured by applying Regalis Plus 10 WG from the 'balloon' stage onwards, regardless of the tree root pruning treatment. In contrast, root pruning alone, as well as the use of the Ethephon program, including treatment with root pruning, resulted in the lowest apple firmness after simulated handling.

Table 3. The effect of tree growth retardation method on fruit quality of 'Mutsu' apples directly after storage and after 7 days of shelf life.

Treatments	Background Color			FF [N]	SSC [°Bx]	TA [%]
	L*	a*	b*			
Directly after storage						
C	66.6 a [1]	−13.3 a	35.7 a	72.8 b	14.8 ab	0.403 a
RP	69.3 a	−13.6 a	36.4 a	66.4 a	14.9 ab	0.417 ab
RB	64.4 a	−13.8 a	35.6 a	69.1 ab	14.3 a	0.445 b
RS	66.1 a	−13.1 a	34.6 a	66.1 a	14.6 ab	0.406 a
Ethephon	65.5 a	−13.0 a	34.4 a	67.4 a	14.5 ab	0.415 ab
RP + RB	66.2 a	−14.3 a	36.8 a	75.1 b	14.4 a	0.443 b
RP + RS	67.0 a	−13.3 a	35.3 a	68.6 ab	14.7 ab	0.425 ab
RP + Ethephon	66.1 a	−13.4 a	37.4 a	66.8 a	15.4 b	0.398 a
After 7 days of shelf life						
C	71.9 ab	−10.6 a	41.3 b	56.3 b	15.0 a	0.375 a
RP	72.9 b	−10.7 a	41.3 b	47.8 a	14.5 a	0.366 a
RB	71.5 ab	−11.0 a	41.2 b	61.7 c	14.8 a	0.357 a
RS	72.2 ab	−10.6 a	41.3 b	53.4 b	15.0 a	0.375 a
Ethephon	69.3 a	−10.1 a	39.1 a	49.5 a	14.7 a	0.379 a
RP + RB	70.4 ab	−10.8 a	40.7 ab	62.0 c	15.0 a	0.373 a
RP + RS	72.4 ab	−10.6 a	41.4 b	54.8 b	15.1 a	0.376 a
RP + Ethephon	71.4 ab	−10.4 a	40.2 ab	48.4 a	14.9 a	0.374 a

[1] Tukey's HSD test; normal distribution (normality was checked using the Shapiro–Wilk test-$p \leq 0.05$); means followed by the same letter within a column are not significantly different. Explanation of abbreviations: C—control; RP—one-sided root pruning (with an oblique knife); RB—Regalis Plus 10 WG (applied from the balloon stage—BBCH 60–69); RS—Regalis Plus 10 WG (applied at standard terms—BBCH 71–73, according to the label); Ethephon—spraying trees with Flordimex 480 SL; RP + RB—one-sided root pruning (with an oblique knife) + Regalis Plus 10 WG (applied from the balloon stage—BBCH 60–69); RP + RS—one-sided root pruning (with an oblique knife) + Regalis Plus 10 WG (applied at standard terms—BBCH 71–73, according to the label); RP + Ethephon—one-sided root pruning (with an oblique knife) + spraying trees with Flordimex 480 SL; L*, a*, and b*—the color of apples in CIE L*a*b* coordinates; FF—flesh firmness; SSC—soluble solids content; TA—titratable acidity; n = 4.

Treatments that restricted tree growth also significantly affected SSC and TA, but only directly after apple storage. It was then found that root pruning in treatment with the Ethephon program provided higher soluble solids content in apples than spraying the trees with Regalis Plus 10 WG from the 'balloon' stage, regardless of the root pruning treatment. In contrast, the opposite relationship occurred for acidity.

3.4. Values of Correlation Coefficients Between Selected Parameters

Table 4 shown the values of the linear correlation coefficients r between the different parameters determined in the harvest period. Statistical analysis showed a highly significant positive correlation between fruit firmness, tree yield, and SSC but a negative correlation with TA. This means that higher tree yields promoted higher firmness in apples with a higher SSC/TA ratio. In addition, a very strong negative correlation was found between the number of fruitlets per 100 inflorescences and the length of growth at both the top and bottom of the tree crown ($r = -0.96$ and -0.99, $p < 0.01$), demonstrating that fruit is a fundamental factor in limiting the growth of tree shoots. These relationships were reflected in the high yield of the trees ($r = 0.88$, $p < 0.01$) while yielding fruit of lower weight ($r = -0.96$, $p < 0.01$), and such fruit is particularly desirable for the large-fruited cultivar 'Mutsu'.

Table 4. Correlation coefficients among various analyzed parameters in the period around the harvest.

	A	B	C	D	E	F	G	H	I	J	K	L	M
A	1.00												
B	−0.96 **	1.00											
C	−0.99 **	0.98 **	1.00										
D	0.88 **	−0.92 **	−0.89 **	1.00									
E	−0.96 **	0.98 **	0.95 **	−0.87 **	1.00								
F	−0.05	0.13	0.07	−0.11	0.14	1.00							
G	−0.35 *	0.24	0.35 *	−0.04	0.17	−0.37 *	1.00						
H	−0.36 *	0.31	0.37 *	−0.46 *	0.22	−0.01	0.18	1.00					
I	0.28	−0.23	−0.26	0.23	−0.20	0.07	−0.17	−0.57 **	1.00				
J	−0.15	0.12	0.11	−0.18	0.12	−0.06	0.03	0.60 **	−0.62 **	1.00			
K	0.38 *	−0.55 **	−0.46 *	0.61 **	−0.43 *	−0.09	−0.08	−0.40 *	0.25	−0.36 *	1.00		
L	0.50 **	−0.67 **	−0.53 **	0.63 **	−0.68 **	−0.11	0.18	0.13	0.05	−0.01	0.60 **	1.00	
M	−0.29	0.44 *	0.31	−0.50 **	0.43 *	0.34 *	−0.39 *	0.02	0.13	0.05	−0.49 **	−0.58 **	1.00

Explanation of the designations used: A—number of fruitlets/100 inflorescences; B—shoot length in top tree crown; C—shoot length in bottom tree crown; D—yield per tree; E—fruit weight; F—internal ethylene content; G—starch test; H—L* value; I—a* value; J—b* value; K—flesh firmness; L—soluble solids content; M—titratable acidity; L* value, a* value, and b* value—the color of apples in CIE L*a*b* coordinates. *—correlation significant at $\alpha = 0.05$, **—correlation significant at $\alpha = 0.01$; n = 30.

Table 5 presents the values of the linear correlation coefficients r between the individual parameters that describe the growth and yield of the trees and the quality of the fruit after storage, as well as between the particular characteristics of the quality determined directly after storage. The number of fruitlets per 100 inflorescences and the yield per tree was positively correlated with TA ($r = 0.53$, $r = 0.58$; $p < 0.01$). This means apples from trees with more abundant yields contained more organic acids. On the other hand, TA was negatively correlated with shoot length at the top and bottom of the crown ($r = -0.56$ and -0.58, $p < 0.01$), as well as mean fruit weight ($r = -0.50$, $p < 0.01$). This means that larger fruits and those from trees with more intense shoot growth had a lower TA. On the other hand, the relationships between the individual characteristics of apple quality were unclear. In the statistical analysis, a negative correlation was found between SSC and TA ($r = -0.59$, $p < 0.01$), which indicates that fruits containing more sugars were characterized by lower acidity. At the same time, a significant positive correlation was found between SSC and the value of b* ($r = 0.61$, $p < 0.01$).

Table 5. Correlation coefficients between various analyzed parameters directly after storage.

	A	B	C	D	E	F	G	H	I	J	K
F	−0.25	0.24	0.21	−0.35 *	0.20	1.00					
G	−0.24	0.27	0.29	−0.25	0.20	−0.15	1.00				
H	−0.11	0.03	0.02	−0.07	0.11	0.46 *	−0.67 **	1.00			
I	−0.08	0.09	0.05	0.03	0.15	−0.01	−0.09	0.14	1.00		
J	−0.44 *	0.36 *	0.40 *	−0.46 *	0.38 *	0.36 *	−0.12	0.61 **	−0.02	1.00	
K	0.53 **	−0.56 **	−0.58 **	0.58 **	−0.50 **	−0.13	−0.31	−0.07	0.10	−0.59 **	1.00

Explanation of the designations used: A—number of fruitlets/100 inflorescences; B—shoot length in top tree crown; C—shoot length in bottom tree crown; D—yield per tree; E—fruit weight; F—L* value; G—a* value; H—b* value; I—flesh firmness; J—soluble solids content; K—titratable acidity. F–K (consider parameters directly after storage); L* value, a* value, and b* value—the color of apples in CIE L*a*b* coordinates. *—correlation significant at $\alpha = 0.05$, **—correlation significant at $\alpha = 0.01$; n = 30.

Table 6 presents the values of the linear correlation coefficients r between the individual parameters that describe the growth and yield of the trees and the quality of the fruit after simulated handling, as well as between the individual distinguishing characteristics of the quality determined after simulated handling. Directly after harvest, a positive correlation was found with both the number of fruitlets per 100 inflorescences ($r = 0.61$, $p < 0.01$) and with the yield per tree ($r = 0.63$, $p < 0.01$). However, regarding the average fruit weight and the length of long shoots, a negative relationship was noted with the firmness of apples after simulated handling. These data indicate that moderate growth and high yielding of trees ensure higher firmness of apples. However, with an increase in the weight of apples and the length of this year's growth, the firmness of the fruit after simulated handling decreases.

Table 6. Correlation coefficients between various analyzed parameters as well as after shelf life.

	A	B	C	D	E	F	G	H	I	J	K
F	0.05	0.01	−0.06	−0.14	−0.01	1.00					
G	−0.26	0.26	0.31	−0.23	0.21	−0.09	1.00				
H	0.33 *	−0.21	−0.31	0.09	−0.21	0.56 **	−0.22	1.00			
I	0.61 **	−0.54 **	−0.63 **	0.63 **	−0.42 *	−0.17	−0.32	0.11	1.00		
J	0.22	−0.19	−0.21	0.05	−0.18	−0.01	−0.06	−0.13	0.19	1.00	
K	−0.09	0.08	0.09	−0.10	0.05	0.06	−0.08	−0.10	−0.16	0.07	1.00

Explanation of the designations used: A—number of fruitlets/100 inflorescences; B—shoot length in top tree crown; C—shoot length in bottom tree crown; D—yield per tree; E—fruit weight; F—L* value; G—a* value; H—b* value; I—flesh firmness; J—soluble solids content; K—titratable acidity. F–K (consider parameters after shelf life); L* value, a* value, and b* value—the color of apples in CIE L*a*b* coordinates. *—correlation significant at $\alpha = 0.05$, **—correlation significant at $\alpha = 0.01$; n = 30.

3.5. Size Structure of Apple Yield

Treatments limiting tree growth modified the apple size structure of the yield (Figure 2). With the 'Mutsu' cultivar, fruits with a diameter of up to 90 mm are the most desirable on the market. Single-sided root pruning in combination with spraying of trees with Regalis Plus 10 WG from the 'balloon' phase (BBCH 60–69) provided the highest percentage share in the yield of this expected fruit category. Similar results were obtained by combining tree root pruning with Regalis Plus 10 WG on dates consistent with the label. In both of these combination experiments, the percentage of fruits with a diameter of more than 90 mm was small and amounted to 5 and 8%, respectively. However, after using Regalis Plus 10 WG without root cutting, the percentage of such fruits was twice as high. The structure of apple yield of the trees after root pruning and after application of the Ethephon program was similar to that of the control combination, but the percentage of apples with a diameter greater than 90 mm was alarmingly high and oscillated around 40%.

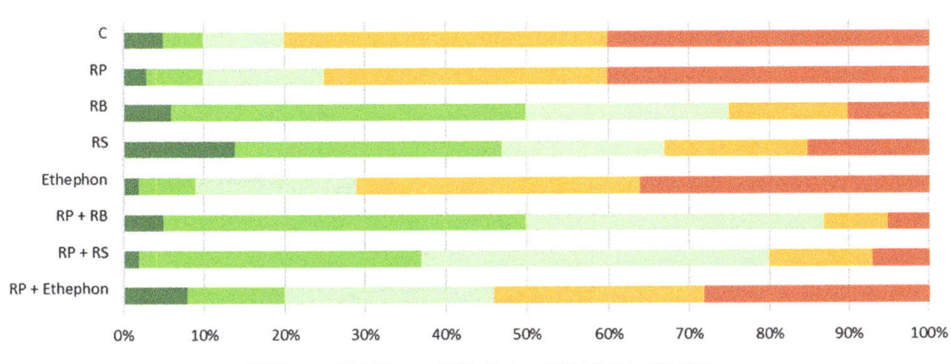

Figure 2. Size structure of 'Mutsu' apples yield in relation to tree growth retardation method (fruits divided into 5 apple size classes). Explanation of abbreviations: C—control; RP—one-sided root pruning (with an oblique knife); RB—Regalis Plus 10 WG (applied from the balloon stage—BBCH 60–69); RS—Regalis Plus 10 WG (applied at standard terms—BBCH 71–73, according to the label); Ethephon—spraying trees with Flordimex 480 SL; RP + RB—one-sided root pruning (with an oblique knife) + Regalis Plus 10 WG (applied from the balloon stage—BBCH 60–69); RP + RS—one-sided root pruning (with an oblique knife) + Regalis Plus 10 WG (applied at standard terms—BBCH 71–73, according to the label); RP + Ethephon—one-sided root pruning (with an oblique knife) + spraying trees with Flordimex 480 SL.

The dendrogram in Figure 3 showed the degree of similarity between the treatment of tree growth retarding based on the measurements taken in the orchard and considering the results of the evaluation of the physiological state of the fruit directly after harvest. Based on these data, a hierarchical aggregation of eight tree dwarf combinations evaluated in the present experiment was shown. There was a large difference between the cluster formed by the control tree, RP, Ethephon, and RP + Ethephon, and the cluster comprised of the four remaining experiment treatments (RB, RP + RB, RS, and RP + RS). Interestingly, the most similar objects within the first cluster were RP and Ethephon. Therefore, it can be assumed that these treatments can be used interchangeably, depending on the agrotechnical situation in the orchard. In the second cluster formed by all Regalis Plus 10 WG combinations, the most similar objects were RB and RP + RB and RS and RP + RS. It can be assumed that trees and fruits react similarly to both root pruning in combination with Regalis Plus 10 WG and to using Regalis Plus 10 WG alone. The height of the dendrogram, which indicates the order in which the clusters were combined, can be used to select the best treatment to retard tree growth and facilitate the management of their growth for an optimal effect.

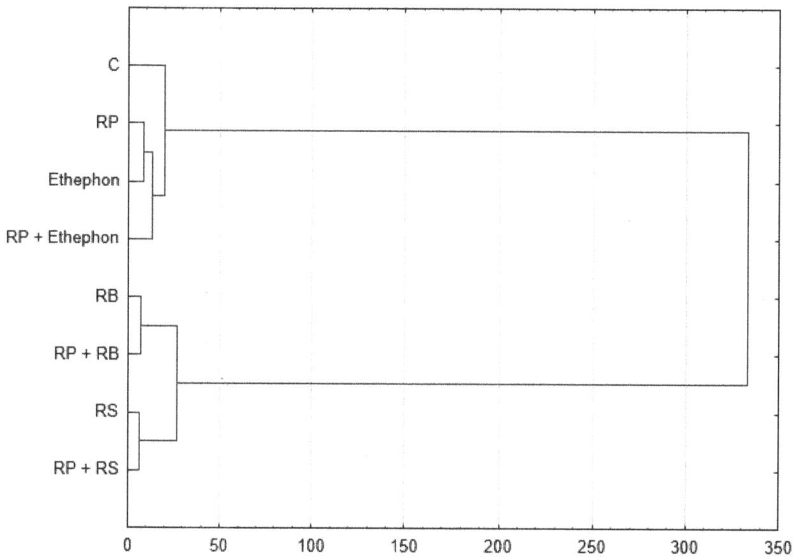

Figure 3. Dendrogram for different tree growth-limiting treatments based on all the results available directly after harvest. Explanation of abbreviations: C—control; RP—one-sided root pruning (with an oblique knife); RB—Regalis Plus 10 WG (applied from the balloon stage—BBCH 60–69); RS—Regalis Plus 10 WG (applied at standard terms—BBCH 71–73, according to the label); Ethephon—spraying trees with Flordimex 480 SL; RP + RB—one-sided root pruning (with an oblique knife) + Regalis Plus 10 WG (applied from the balloon stage—BBCH 60–69); RP + RS—one-sided root pruning (with an oblique knife) + Regalis Plus 10 WG (applied at standard terms—BBCH 71–73, according to the label); RP + Ethephon—one-sided root pruning (with an oblique knife) + spraying trees with Flordimex 480 SL.

An analogous dendrogram considering only the quality characteristics of the shelf-life apples is shown in Figure 4. In this case, there was a big difference between the cluster combining RB and RP + RB and the other cluster combining the six remaining experience combinations. Apples from the RB and RP + RB combinations characterized the most similar quality. This means that in the planned spraying of trees with Regalis Plus 10 WG from the balloon phase, it is unnecessary to prune the roots beforehand. The following two most similar combinations were RP and RP + Ethephon.

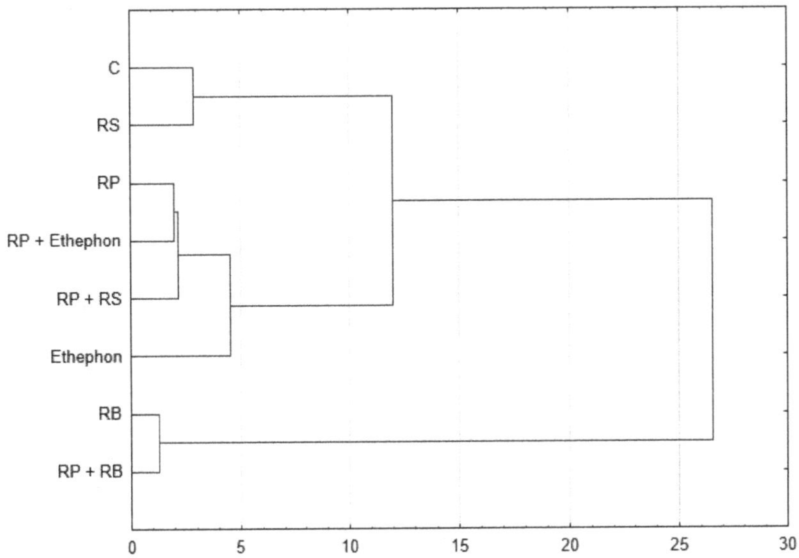

Figure 4. Dendrogram for different tree growth-limiting treatments based on apple quality after shelf-life. Explanation of abbreviations: C—control; RP—one-sided root pruning (with an oblique knife); RB—Regalis Plus 10 WG (applied from the balloon stage—BBCH 60–69); RS—Regalis Plus 10 WG (applied at standard terms—BBCH 71–73, according to the label); Ethephon—spraying trees with Flordimex 480 SL; RP + RB—one-sided root pruning (with an oblique knife) + Regalis Plus 10 WG (applied from the balloon stage—BBCH 60–69); RP + RS—one-sided root pruning (with an oblique knife) + Regalis Plus 10 WG (applied at standard terms—BBCH 71–73, according to the label); RP + Ethephon—one-sided root pruning (with an oblique knife) + spraying trees with Flordimex 480 SL.

4. Discussion

In the coming years, the demand for apples will depend even more on their quality than it does today. The condition to obtain high-quality apples is the proper selection and timely performance of agrotechnical treatments in the orchard. Regarding such treatments, it is important to control the intensity of tree growth. Trees' moderate growth is conducive to the formation of flower buds and regular fruiting. However, growth that is too strong or weak creates problems with fruit size and yield quality. Undoubtedly, the best method of regulating tree growth is annual/regular fruiting. In the experiment described in this study, a spectacular reduction in shoot growth was observed with a simultaneous increase in the number of apples on trees sprayed with Regalis Plus 10 WG. Such an impact is by all means desirable, especially in the case of the 'Mutsu' cultivar, which tends to shed its buds excessively.

It is known that limiting the strength of tree growth modifies the direction of the flow of various components to shoots and fruits. As a result, the fruit is better nourished and has better quality. By reducing the length of this year's growth, tree growth retardant treatments also increased the penetration of sunlight into the crown [47]. Therefore, they will contribute to better flower buds for the next year and higher tree yields, as well as promote a more economical and rational use of pesticides to protect trees against diseases and pests [48]. In addition, they contribute to improving the color of apples [49–52].

Ferree and Knee [53] noted the high effectiveness of root pruning in reducing the growth of tree shoots. On the other hand, Wajja-Musukwe et al. [54] believe that root pruning has only a minor effect on tree growth. The results of the experiment described in this study confirm a slight growth reaction of long shoots of apple trees of the 'Mutsu' cultivar to unilateral root pruning. It should be noted that the effect of limiting tree growth due to root pruning can be modified under the influence of the timing of this treatment and the weather conditions during the growing season [54].

Other experiments have shown that root pruning, in addition to regulating the growth and abundance of tree flowering, reduced leaf and fruit size while increasing yield [55]. The fineness of apples in the case of the large-fruited cultivar 'Mutsu' is a positive effect, also noted in the experiment described in this study. Schupp [56], on the other hand, believes that the yield decreases due to pruning the roots, and the number of flowers in the following year decreases. In our experiment, slightly fewer fruitlets per 100 inflorescences after unilateral root pruning were observed on 'Mutsu' trees than in the control combination. However, tree yield after unilateral root pruning was significantly higher than control trees.

Some researchers say limiting tree growth with Regalis Plus 10 WG is more spectacular than it is with cutting the roots. Using the maximum dose of this preparation (2.5 $kg \cdot ha^{-1}$) per season can reduce shoot growth by 50%, resulting from limited gibberellin synthesis after using calcium prohexadione [57,58]. Greene [59] also observed such a relationship on trees of the 'Mutsu' and 'McIntosh' cultivars. In this study, Regalis Plus 10 WG increased the number of fruitlets per 100 inflorescences by approximately 2.5 times compared to control trees. This was reflected in the yield of trees, which was about two times higher. Thus, previous observations by Zadravec et al. [60], who reported that Regalis had no significant effect on the tree yield and fruit quality of the 'Gala' cultivar, were not confirmed. Therefore, it can be inferred that the application of Pro-Ca at a later stage may be less effective, as vegetative growth is more advanced [61].

All methods of limiting the growth of trees used in the experiment resulted in a significant shortening of long shoots, but only slightly affected the IEC and the background color of apples directly after harvest. On the other hand, retarding the tree growth resulted in a decrease in apple acidity and an increase in SSC. Similar relationships were previously found in the pears of cv. 'Shinseiki' [62] and tomatoes [63], while no relationships were found in pears of cv. 'Pass Crassane' [64].

In the presented study, using Ethephon caused the most significant decrease in the firmness of apples after the simulated handling period, which repeats the previous observations of Smith et al. [65]. In other studies, the results were inconclusive, which suggests that the efficacy of Ethephon depends on various factors, including the timing of treatment, the course of weather conditions, the degree of fruit ripeness at harvest, and the conditions of their storage [66–70]. For example, the Ethephon used for the late thinning of primordia effectively increased the average weight of the fruit [71], which is particularly important in the cultivation of small-fruit cultivars. According to Koen et al. [72], using Ethephon was the safest and most predictable alternative to labor-intensive manual thinning.

To sum up, all the methods to regulate the growth and fruiting of trees presented in this study influenced the production of high-quality fruit to a greater or lesser extent. When choosing the proper dwarfing method for trees, the fruit grower should rely on his own experience and knowledge about the growth and yield of individual cultivars in the orchard.

5. Conclusions

One-sided spring pruning of the roots with an oblique knife, together with the application of Regalis Plus 10 WG from the balloon phase, is the most effective method of inhibiting the growth of annual shoots and reducing the share of overgrown fruit in the yield. Apples from such trees show a slow rate of firmness decline during storage and simulated handling. Limiting shoot growth and excessive apple growth by pruning the roots or spraying trees with Ethephon gives much worse results compared to using Regalis Plus 10 WG. In addition, trees sprayed with Regalis Plus 10 WG are suppressed for naturally occurring stronger growth of the upper shoots compared to the bottom part of the tree crown.

Author Contributions: Conceptualization, M.M. and K.T.; methodology, M.M., T.M. and K.T.; formal analysis, M.M. and K.T.; investigation, T.M. and K.T.; writing—original draft preparation, M.M., T.M., T.K. and K.T.; writing—review and editing, M.M., T.M., T.K. and K.T.; visualization, M.M. All authors have read and agreed to the published version of the manuscript.

Funding: This research received no external funding.

Institutional Review Board Statement: Not applicable.

Data Availability Statement: All data are present in the manuscript.

Conflicts of Interest: The authors declare no conflicts of interest.

References

1. Daugaard, H.; Callesen, O. The effect of rootstock on yield and quality of 'Mutsu'apples. *J. Hortic. Sci. Biotechnol.* **2002**, *77*, 248–251. [CrossRef]
2. Peşteanu, A. Influence of Regalis Plus product on the growth and fruiting of apple trees. *Ann. Univ. Craiova* **2022**, *XXVII*, 291–296. [CrossRef]
3. Łysiak, G.P.; Kurlus, R.; Michalska, A. Increasing the frost resistance of 'Golden Delicious', 'Gala' and 'Šampion' apple cultivars. *Folia Hortic.* **2016**, *28*, 25–135. [CrossRef]
4. Bakos, J.L.; Ladányi, M.; Szalay, L. Frost hardiness of flower buds of 16 apricot cultivars during dormancy. *Folia Hortic.* **2024**, *36*, 81–93. [CrossRef]
5. Kurlus, R.; Świerczyński, S.; Rutkowski, K.; Ratajkiewicz, H.; Malinowska, A.; Wyrwał, A. Exogenus 'Ga_3' and 'Ga_{4+7}' effects on phenological indices, frost hardiness and quality properties of 'English morello' sour cherry (*Prunus cerasus* L.). *Acta Sci. Polonorum. Hortorum Cultus* **2017**, *16*, 99–109. [CrossRef]
6. The Apple Market in the EU: Vol. 1: Production, Areas and Yields. 2022. Available online: https://agriculture.ec.europa.eu/system/files/2022-10/apples-production_en.pdf (accessed on 27 July 2024).
7. Kacprzak, T.; Himstedt, K. The Influence of the Russian Embargo on the Economic Situation of Apple Producers in the Eastern Part of the Masovia Province. *Zesz. Nauk. SGGW Warsz. Probl. Roln. Świat.* **2019**, *19*, 54–64. (In Polish) [CrossRef]
8. Elias, T.S. *The Complete Trees of North America. Field Guide and Natural History*; Van Nostrand Reinhold Company & Times Mirror Magazines. Inc.: New York, NY, USA, 1980.
9. Schupp, J.; Ferree, D. Influence of time of root pruning on growth, mineral nutrition, net photosynthesis and transpiration of young apple trees. *Sci. Hortic.* **1990**, *42*, 299–306. [CrossRef]
10. Maggs, D. Growth-rates in relation to assimilate supply and demand: I. Leaves and roots as limiting regions. *J. Exp. Bot.* **1964**, *15*, 574–583. [CrossRef]
11. Geisler, D.; Ferree, D. The influence of root pruning on water relations, net photosynthesis, and growth of young 'Golden Delicious' apple trees. *J. Am. Soc. Hortic. Sci.* **1984**, *109*, 827–831. [CrossRef]
12. Fang, Y.; Xu, B.; Turner, N.; Li, F. Grain yield, dry matter accumulation and remobilization, and root respiration in winter wheat as affected by seeding rate and root pruning. *Eur. J. Agron.* **2010**, *33*, 257–266. [CrossRef]
13. Ma, S.; Xu, B.; Li, F.; Liu, W.; Huang, Z. Effects of root pruning on competitive ability and water use efficiency in winter wheat. *Field Crops Res.* **2010**, *105*, 56–63. [CrossRef]
14. Cline, J.A.; Embree, C.G.; Hebb, J.; Nichols, D.S. Performance of prohexadione-calcium on shoot growth and fruit quality of apple—Effect of spray surfactants. *Can. J. Plant Sci.* **2008**, *88*, 165–174. [CrossRef]
15. Miller, S.S. Prohexadione-calcium controls vegetative shoot growth in apple. *J. Tree Fruit Prod.* **2002**, *3*, 11–28. [CrossRef]
16. Milyaev, A.; Kofler, J.; Moya, Y.; Lempe, J.; Stefanelli, D.; Hanke, M.; Flachowsky, H.; Wirén, N.; Wünsche, J. Profiling of phytohormones in apple fruit and buds regarding their role as potential regulators of flower bud formation. *Tree Physiol.* **2022**, *42*, 2319–2335. [CrossRef] [PubMed]
17. Beauvieux, R.; Wenden, B.; Dirlewanger, E. Bud Dormancy in Perennial Fruit Tree Species: A Pivotal Role for Oxidative Cues. *Front. Plant Sci.* **2018**, *9*, 657. [CrossRef] [PubMed]
18. Kang, S.; Kim, J.; Hamayun, M.; Hwang, I.; Khan, A.; Kim, Y.; Lee, J.; Lee, I. Influence of prohexadione-calcium on growth and gibberellins content of Chinese cabbage grown in Alpine region of South Korea. *Sci. Hortic.* **2010**, *125*, 88–92. [CrossRef]
19. Kim, Y.; Khan, A.; Hamayun, M.; Kim, J.; Lee, J.; Hwang, I.; Yoon, C.; Lee, I. Effects of Prohexadione Calcium on growth and gibberellins contents of Chrysanthemum morifolium R. cv Monalisa White. *Sci. Hortic.* **2010**, *123*, 423–427. [CrossRef]
20. Amarante, C.; Silveira, J.; Steffens, C.; Freitas, S.; Mitcham, E.; Miqueloto, A. Post-bloom and preharvest treatment of 'Braeburn' apple trees with prohexadione-calcium and GA_{4+7} affects vegetative growth and postharvest incidence of calcium-related physiological disorders and decay in the fruit. *Sci. Hortic.* **2020**, *261*, 108919. [CrossRef]
21. Thorp, T.; Aspinall, D.; Sedgley, M. Influence of shoot age on floral development and early fruit set in avocado (*Persea americana* Mill.) cv. Hass. *J. Hortic. Sci.* **1993**, *68*, 645–651. [CrossRef]
22. Samach, A.; Smith, H. Constraints to obtaining consistent annual yields in perennials. II: Environment and fruit load affect induction of flowering. *Plant Sci.* **2013**, *207*, 168–176. [CrossRef]

23. Martín-Fontecha, E.; Tarancón, C.; Cubas, P. To grow or not to grow, a power-saving program induced in dormant buds. *Curr. Opin. Plant Biol.* **2018**, *41*, 102–109. [CrossRef]
24. Jain, R.; Rathod, S.; Banjare, M.; Nidhi, R.; Sood, N.; Sharma, A. Physiological Aspects of Flowering, Fruit Setting, Fruit Development and Fruit Drop, Regulation and their Manipulation: A Review. *Int. J. Environ. Clim. Change* **2023**, *13*, 205–224. [CrossRef]
25. Ramírez, H.; Alonso, S.; Benavides, A.; Antonio, A. Prohexadione-Ca Modifies Growth and Endogenous Hormones in the Shoot Apex in Apple Trees. *Acta Hortic.* **2006**, *727*, 117–124. [CrossRef]
26. Ramírez, H.; Sanchez-Canseco, J.; Zamora-Villa, V.; Rancaño-Arrioja, J. Effect of Prohexadione calcium, 6-benzyl amino purine and 6-furfuryladenine on vegetative growth and fruit quality in apple. *Phyton* **2017**, *86*, 282–289. [CrossRef]
27. Li, Y.; Zhang, D.; An, N.; Fan, S.; Zuo, X.; Zhang, X.; Zhang, L.; Gao, C.; Han, M.; Xing, L. Transcriptomic analysis reveals the regulatory module of apple (*Malus* × *domestica*) floral transition in response to 6-BA. *BMC Plant Biol.* **2019**, *19*, 93. [CrossRef]
28. Schwarz, I.; Scheirlinck, M.; Otto, E.; Bartrina, I.; Schmidt, R.; Schmülling, T. Cytokinin regulates the activity of the inflorescence meristem and components of seed yield in oilseed rape. *J. Exp. Bot.* **2020**, *71*, 7146–7159. [CrossRef]
29. Vaughan-Hirsch, J.; Li, D.; Martinez, A.; Roden, S.; Pattyn, J.; Taira, S.; Shikano, H.; Miyama, Y.; Okano, Y.; Voet, A.; et al. A 1-aminocyclopropane-1-carboxylic-acid (ACC) dipeptide elicits ethylene responses through ACC-oxidase mediated substrate promiscuity. *Front. Plant Sci.* **2022**, *13*, 995073. [CrossRef]
30. Houben, M.; Poel, B. 1-Aminocyclopropane-1-Carboxylic Acid Oxidase (ACO): The Enzyme That Makes the Plant Hormone Ethylene. *Front. Plant Sci.* **2019**, *10*, 695. [CrossRef]
31. Gamalero, E.; Glick, B. Bacterial Modulation of Plant Ethylene Levels. *Plant Physiol.* **2015**, *169*, 13–22. [CrossRef]
32. Rademacher, W. Plant growth regulators: Backgrounds and uses in plant production. *J. Plant Growth Regul.* **2015**, *34*, 845–872. [CrossRef]
33. Sosna, I.; Spiak, Z.; Sosna, I.; Gudarowska, E. Effect of rootstock on leaf nutrient concentration and productive value of 'Mutsu' apple trees. *J. Elem.* **2020**, *25*, 1581–1593. [CrossRef]
34. Jackson, J.E.; Palmer, J.W. Light distribution within hedgerow orchards. *J. Appl. Ecol.* **1972**, *9*, 341–357. [CrossRef]
35. Swarbrick, T. The Seasonal elongation growth of apple varieties on some vegetative rootstocks and its possible relation to fruit bud formation. *J. Pomol. Hortic. Sci.* **1928**, *7*, 100–129. [CrossRef]
36. Hoying, S.A.; Robinson, T.L. Effects of chain saw girdling and root pruning of apple trees. *Acta Hortic.* **1992**, *322*, 167–172. [CrossRef]
37. Marandici, S. Increasing productivity of apple orchard by using the growth regulator based on prohexadion-calcium. *An. Univ. Craiova Ser. Biol. Hortic. Tehnol. Preluc. Prod. Agric. Ing. Med.* **2016**, *21*, 141–144.
38. Medjdoub, R.; Val, J.; Blanco, A. Prohexadione–Ca inhibits vegetative growth of 'Smoothee Golden Delicious' apple trees. *Sci. Hortic.* **2004**, *101*, 243–253. [CrossRef]
39. Skic, A.; Szymańska-Chargot, M.; Kruk, B.; Chylińska, M.; Pieczywek, P.M.; Kurenda, A.; Zdunek, A.; Rutkowski, K.P. Determination of the Optimum Harvest Window for Apples Using the Non-Destructive Biospeckle Method. *Sensors* **2016**, *16*, 661. [CrossRef]
40. Tomala, K. Orchard Factors Affecting Fruit Storage Quality And Prediction Of Harvest Date Of Apples. *Acta Hortic.* **1999**, *485*, 373–382. [CrossRef]
41. Łysiak, G. The determination of harvest index of 'Sampion' apples intended for long storage. *Acta Sci. Pol. Hortorum Cultus* **2011**, *10*, 273–282.
42. Bulens, I.; Van de Poel, B.; Hertog, M.P.; De Proft, M.P.; Geeraerd, A.H.; Nicolai, B.M. Influence of harvest time and 1-MCP application on postharvest ripening and ethylene biosynthesis of 'Jonagold' apple. *Postharvest Biol. Technol.* **2012**, *72*, 11–19. [CrossRef]
43. Lafer, G. Optimaler Erntetermin—Optimale Ergebnisse. *Besseres Obst* **1998**, *43*, 15–20.
44. Małachowska, M.; Tomala, K. Apple Quality during Shelf-Life after Long-Term Storage and Simulated Transport. *Agriculture* **2023**, *13*, 2045. [CrossRef]
45. Tomala, K.; Grzęda, M.; Guzek, D.; Głąbska, D.; Gutkowska, K. Analysis of Possibility to Apply Preharvest 1-Methylcyclopropene (1-MCP) Treatment to Delay Harvesting of Red Jonaprince Apples. *Sustainability* **2020**, *12*, 4575. [CrossRef]
46. Tomala, K.; Guzek, D.; Głąbska, D.; Małachowska, M.; Widłak, Ł.; Krupa, T.; Gutkowska, K. Assessment of the quality of 'Red Jonaprince' apples during storage after delayed harvesting and 1-methylcyclopropene (1-MCP) preharvest and postharvest treatment. *Agronomy* **2023**, *13*, 1730. [CrossRef]
47. Vermeulen, P. Crown depth as a result of evolutionary games: Decreasing solar angle should lead to shallower, not deeper crowns. *New Phytol.* **2014**, *202*, 1249–1256. [CrossRef]
48. Dayatilake, G.A.; Wunschke, J.N.; Wood, P.; McArtney, S.; Manktelow, D.; Gurnsey, S.; Tustin, D.S. The use of prohexadione-ca for improved crop management. *Acta Hortic.* **2005**, *694*, 315–319. [CrossRef]
49. Jackson, J.; Palmer, J. Effects of Shade on the Growth and Cropping of Apple Trees. II. Effects on Components of Yield. *J. Hortic. Sci.* **1977**, *52*, 253–266. [CrossRef]
50. Pitchers, B.; Do, F.; Pradal, C.; Dufour, L.; Lauri, P. Apple tree adaptation to shade in agroforestry: An architectural approach. *Am. J. Bot.* **2021**, *108*, 732–743. [CrossRef]

51. Solomakhin, A.; Blanke, M. Coloured hailnets alter light transmission, spectra and phytochrome, as well as vegetative growth, leaf chlorophyll and photosynthesis and reduce flower induction of apple. *Plant Growth Regul.* **2008**, *56*, 211–218. [CrossRef]
52. Łysiak, G. The base colour of fruit as an indicator of optimum harvest date for two apple cultivars (*Malus domestica* Borkh.). *Folia Hort.* **2012**, *24*, 81–89. [CrossRef]
53. Ferree, D.C.; Knee, M. Influence of root pruning and rootstocks on growth and performance of 'Golden Delicious' apple. *HortScience* **1997**, *32*, 645–648. [CrossRef]
54. Wajja-Musukwe, T.; Wilson, J.; Sprent, J.; Ong, C.; Deans, J.; Okorio, J. Tree growth and management in Ugandan agroforestry systems: Effects of root pruning on tree growth and crop yield. *Tree Physiol.* **2008**, *28*, 233–242. [CrossRef] [PubMed]
55. Schupp, J.; Ferree, D. Effect of Root Pruning at Different Growth Stages on Growth and Fruiting of Apple Trees. *HortScience* **1987**, *22*, 387–390. [CrossRef]
56. Schupp, J.R. Root pruning of apple trees. *Fruit Notes* **1990**, *55*, 2–3.
57. Kim, H.; Lee, I.; Hamayun, M.; Kim, J.; Won, J.; Hwang, I.; Kim, K. Effect of Prohexadione Calcium on Growth Components and Endogenous Gibberellins Contents of Rice (*Oryza sativa* L.). *J. Agron. Crop Sci.* **2007**, *193*, 445–451. [CrossRef]
58. Grossmann, K.; Konig-Kranz, S.; Kwiatkowski, J. Phytohormonal changes in intact shoots of wheat and oilseed rape treated with the acylcyclohexanedione growth retardant prohexadione calcium. *Physiol. Plant.* **1994**, *90*, 139–143. [CrossRef]
59. Greene, D.W. The effect of prohexadione-calcium on fruit set and chemical thinning of apple trees. *HortScience* **2007**, *42*, 1361–1365. [CrossRef]
60. Zadravec, P.; Čmelik, Z.; Tojnko, S.; Unuk, T.; Schlauer, B. Vegetative Growth, Yield and Fruit Quality of 'Gala' Apple Treated with Regalis (Prohexadione-Ca). *Acta Hortic.* **2008**, *774*, 287–290. [CrossRef]
61. Musacchi, S.; Sheick, R.; Mia, M.J.; Serra, S. Studies on physiological and productive effects of multi-leader training systems and Prohexadione-Ca applications on apple cultivar 'WA 38'. *Sci. Hortic.* **2023**, *312*, 111850. [CrossRef]
62. Carrà, B.; Fachinello, J.; Abreu, E.; Pasa, M.; Spagnol, D.; Giovanaz, M.; Silva, C. Control of the vegetative growth of 'Shinseiki' pear trees by prohexadione calcium and root pruning. *Pesq. Agropec. Bras.* **2017**, *52*, 177–185. [CrossRef]
63. Gao, Y.; Zhou, X.; Liang, H.; Ji, Y.; Liu, M. Effects of Comparative Metabolism on Tomato Fruit Quality under Different Levels of Root Restriction. *HortScience* **2023**, *58*, 885–892. [CrossRef]
64. Shehaj, M.; Hodaj, B.; Rama, P.; Lepaja, K. Effect of prohexadione-calcium (Regalis) on shoot growth and fruit set in pear var. 'Passe Crassane'. *J. Int. Sci. Publ. Agric. Food* **2014**, *2*, 298–304.
65. Smith, S.; Johnson, D.; Mori, J. The effect of the growth regulator Ethephon on the quality of discovery apples during simulated marketing. *J. Hortic. Sci.* **1985**, *60*, 305–310. [CrossRef]
66. McArtney, S.; Unrath, D.; Obermiller, J.; Green, A. Naphthaleneacetic Acid, Ethephon, and Gibberellin A_4 + A_7 Have Variable Effects on Flesh Firmness and Return Bloom of Apple. *Horttechnology* **2007**, *17*, 32–38. [CrossRef]
67. Whale, S.; Singh, Z.; Behboudian, M.; Janes, J.; Dhaliwal, S. Fruit quality in 'Cripp's Pink' apple, especially colour, as affected by preharvest sprays of aminoethoxyvinylglycine and Ethephon. *Sci. Hortic.* **2008**, *115*, 342–351. [CrossRef]
68. Sun, Y.; Shi, Z.; Jiang, Y.; Zhang, X.; Li, X.; Li, F. Effects of preharvest regulation of ethylene on carbohydrate metabolism of apple (Malus domestica Borkh cv. Starkrimson) fruit at harvest and during storage. *Sci. Hortic.* **2021**, *276*, 109748. [CrossRef]
69. Jędrszczyk, E.; Skowera, B.; Kędzior, R.; Gawęda, M. The influence of Ethephon application to processing tomato plants on yield structure in relation to weather conditions during the growing period. *Folia Hortic.* **2017**, *29*, 75–81. [CrossRef]
70. Zhang, Z. Prohexadione calcium regulates wheat tolerance to drought stress by maintaining water balance and promoting antioxidant metabolism and photosynthesis. *Plant Soil Environ.* **2024**, *70*, 673–681. [CrossRef]
71. Jones, K.M.; Koen, T.B.; Oakford, M.J. Thinning 'Red Fuji' apples using Ethephon at two timings. *J. Hortic. Sci.* **2015**, *65*, 381–384. [CrossRef]
72. Koen, T.B.; Jones, K.M.; Longley, S.B. Spray thinning strategies for 'Red Delicious' apple using naphthalene acetic acid and Ethephon. *J. Hortic. Sci.* **1988**, *63*, 31–35. [CrossRef]

agriculture

MDPI

Article

Analysis of the Impact of Treatments Stimulating Branching on the Quality of Maiden Apple Trees

Magdalena Kapłan [1], Kamila E. Klimek [2,*] and Kamil Buczyński [1]

[1] Institute of Horticulture Production, University of Life Science, 28 Głęboka Street, 20-612 Lublin, Poland; magdalena.kaplan@up.lublin.pl (M.K.); kamil.buczynski@up.lublin.pl (K.B.)
[2] Department of Applied Mathematics and Computer Sciences, University of Life Science, 28 Głęboka Street, 20-612 Lublin, Poland
* Correspondence: kamila.klimek@up.lublin.pl

Abstract: Nursery material intended for establishing intensive apple orchards should be characterised by a dominant and straight leader with an appropriate number of shoots that develop at the right height and are regularly spaced along the leader. The use of well-branched trees can lead to fruiting in the first year after planting. However, many apple varieties have difficulty forming lateral shoots due to strong apical dominance. The aim of the study was to assess the effectiveness of treatments stimulating the branching of maiden apple trees of the 'Gloster' variety. The research was carried out in 2017–2019 at a private nursery farm located in eastern Poland. The studied trees were subjected to a mechanical branching stimulation treatment, which consisted of pinching off 4–5 of the youngest leaves located below the growth cone, and chemical branching stimulation treatments, which consisted of applying growth regulator mixtures in the form of an aqueous solution, i.e., BA+GA_3 and BA+GA_{4+7}. The conducted studies showed that the type of branching-stimulating treatment had a significant effect on the height and trunk diameter of the maiden trees, the number of lateral shoots, the average length of one shoot and the sum of the lengths of all sylleptic shoots. The maiden trees treated with BA+GA_3 were characterised by the best quality among the analysed combinations. Maiden apple trees treated with BA+GA_3 were the tallest (2017—167.7 cm; 2018—175.3 cm; 2019—164.4 cm), produced the largest number of shoots (2017—6.5 pcs; 2018—6.8 pcs; 2019—6.3 pcs) and had the largest sum of lateral shoot lengths (2017—148.0 cm; 2018—155.4 cm; 2019—140.6 cm) among the evaluated combinations. The number of treatments and the concentration of applied growth regulators had a significant effect on the structure of the crown of the maiden apple trees of the 'Gloster' cultivar.

Keywords: sylleptic shoots; apical dominance; growth regulators; leaf pinching

1. Introduction

Apple trees are one of the main horticultural crops in the world [1]. It is estimated that the global planted area of this crop in 2021 was 4,825,629.0 ha, and the yield reached approximately 93.1 million tonnes [2]. At the same time, intensive orchard production characterised by high yields due to high tree planting densities requires high-quality planting material with well-developed crowns [3]. The presence of apple tree shoots is crucial for their fruiting [4–7]. High-quality apple tree planting material should be characterised by a dominant, straight central canopy with an adequate number of shoots that develop to the right height and are regularly spaced along the canopy. These shoots should reach the right length and bifurcation angle. Trees with these characteristics root quickly after planting, effectively fill the space in the rows and efficiently capture light in the early stages of growth [8]. Growers typically choose dwarf and semi-dwarf trees for their quick production of apple fruit and quick return on investment, making these plants more desirable. The use of well-branched trees can lead to fruiting as early as the first year after planting. Branching treatments, including mechanical as well as chemical methods, are used to produce well-branched trees [9,10].

Strongly branched apple trees promote faster crown formation, which facilitates pruning and reduces maintenance costs [11]. Furthermore, well-branched apple trees are crucial for commercial orchards to achieve maximum yields. However, many varieties find it difficult to form these lateral branches under commercial nursery conditions due to strong apical dominance. There are several methods to aid the formation of these branches in nurseries [12]. Shoot branching, which is formed by the growth of axillary buds, is a key element of the aboveground architecture of plants. An important quality parameter in apple seedling production is the number of shoots formed. Unfortunately, at present, the molecular mechanism regulating bud activation and cell cycle processes in apple trees is still poorly understood [13]. The axillary buds of apple trees, located in the leaf angles, usually remain dormant due to the correlative inhibition exerted by the apical buds [14]. The number of branches is largely determined by the formation, release and growth of axillary buds. Humans have attempted to regulate shoot branching according to their needs to improve agricultural production [15,16].

Diversity in plant structure, the result of evolution and artificial selection, is controlled by complex, integrated networks of genetic and environmental signals [17]. Various plant hormones that interact with environmental signals play a major role in regulating branching [18]. These factors can act alone or together to influence the formation of plant architecture. However, a full understanding of the mechanisms controlling or regulating lateral branch development is still a long way off. Although many genes involved in the regulation of these branches have recently been identified, their overarching regulators and target effects are largely unknown. The consideration of factors influencing lateral branch formation is the foundation of rational plant breeding and the cultivation of optimal plant types [16].

Currently, the most common methods used to branch apple trees in nurseries are chemical methods based primarily on growth regulators as well as mechanical methods, such as apical leaf pinching, which is aimed at breaking apical dominance and stimulating lateral shoot development.

Plant growth regulators are naturally biosynthesised chemical compounds that interact with plant physiological processes. Their synthetic counterparts initiate a variety of biochemical and physiological processes related to plant growth and development [19]. Exogenous growth regulators based on benzyladenine and gibberellins show strong potential for use in apple branching [20]. Both benzyladenine and BA+GA_{4+7} and BA+GA_3 are effective in temporarily interrupting apical dominance [21]. Growth regulators based on 6-benzyladenine (6-BA), used alone or in combination with gibberellin GA_{4+7} (6-BA+GA_{4+7}) or GA_3 (6-BA+GA_3), are commonly used to increase the intensity of lateral shoot development in juvenile trees [22].

The aim of this study was to evaluate the effectiveness of treatments to stimulate the branching of 'Gloster' apple trees. The effects of different modes, the number of branching treatments and the concentrations of applied growth regulators on the growth and quality of nursery stock were evaluated.

2. Materials and Methods

The study was carried out in 2017–2019 at a private nursery farm located in the village of Karczmiska in the municipality of Karczmiska, Opolski County, Lubelskie Province (GPS: 51.247144, 22.033154). The experiment evaluated the growth and quality of maiden apple trees of the 'Gloster' cv. grafted on M.9 RN29 rootstock after the application of branching stimulation treatments. The experiment was set up in a randomised block design and included 11 combinations with 5 replications (one replication was a plot with 15 plants).

2.1. Treatments

The following combinations were used in the experiment:

1. Control—trees not subjected to branching stimulation treatments;
2. Pinching out of 4–5 youngest leaves—1 x season;
3. Pinching out of 4–5 youngest leaves—2 x season;
4. 4.5 mL of Globaryll 100 SL + 0.45 g of Florgib/1 L of water—1 x season;
5. 4.5 mL of Globaryll 100 SL + 0.45 g of Florgib/1 L of water—2 x season;
6. 6.8 mL of Globaryll 100 SL + 0.63 g of Florgib/1 L of water—1 x season;
7. 6.8 mL of Globaryll 100 SL + 0.63 g of Florgib/1 L of water—2 x season;
8. 4.5 mL of Globaryll 100 SL + 45 mL of Gibb Plus 11 SL/1 L of water—1 x season;
9. 4.5 mL of Globaryll 100 SL + 45 mL of Gibb Plus 11 SL/1 L of water—2 x season;
10. 6.8 mL of Globaryll 100 SL + 67.5 mL of Gibb Plus 11 SL/1 L of water—1 x season;
11. 6.8 mL of Globaryll 100 SL + 67.5 mL of Gibb Plus 11 SL/1 L of water—2 x season.

The experiment was carried out every summer when the apple trees were in their second year and last year in the nursery and had reached a height of about 75 cm. Depending on the combination, treatments were carried out once or twice per season; the first treatment occurred when the trees reached a height of about 75 cm and the second was 7 days later.

The chemical branching stimulation treatments consisted of an aqueous solution application of Globaryll 100 SL (BA—6-benzyladenine—100 g/L (9.50%), Florgib (GA_3—gibberellic acid—204 g/kg (20.4%)) and Gibb Plus 11 SL (GA_{4+7}—gibberellin GA_{4+7}—10 g/L (0.97%)). BA—6-benzyladenine—is a substance belonging to the cytokinin group, which stimulates the growth and metabolism of plants. This hormone regulates many biochemical processes at the cellular level of plants, but it has a particularly strong influence on cell division. Florgib is a growth regulator based on gibberellic acid (GA_3; stimulates elongation of cells and shoot elongation), which has a unique formula of rapidly dissolving tablets. The gibberellin GA_{4+7} causes excessive cell elongation. The aqueous solution was sprayed on the six youngest well-developed lateral buds together with the leaves just below the growth cone. The adjuvant Superam 10 AL was added to the preparation of the solution at a dose of 0.5 mL per 1 L of water each time in order to prolong the durability of the effect of the mixtures. This formulation was only used in combinations where growth regulators were applied. During the experiment, regular protection of the nursery against diseases, pests and weeds was carried out according to the recommendations of the orchard protection programme.

2.2. Measurements

In autumn, after the end of growth, the following measurements and observations were made: diameter of the rootstock stems at 10 cm above the soil surface, diameter of the whorl stems at 30 cm from the ground, height of the trees, number of lateral shoots, length of sylleptic shoots and the sum of the lengths of the sylleptic shoots. The diameters of rootstocks and whorls were measured using callipers with an accuracy of 0.1 mm. The height of the trees from the point of budding to the apical bud of the main shoot was measured using a scaler to the nearest 1 cm. On each oculant, lateral shoots were counted and their length measured.

The results obtained in the experiment were statistically analysed using the one-factor analysis of variance method. Inference was based on significance $p < 0.05$. All statistical analyses were performed using Statistica 13.

The average air temperature in the months from April to October in the consecutive years of the study was higher than the multi-year average (Table 1). Analysis of the above weather parameter showed that the temperature in all months in 2017–2019 was higher than for the multi-year average. It was shown that, in the summers of 2018 and 2019, the average air temperature reached the same level, i.e., 15.7 °C, and was higher than the multi-year average by 2.5 °C. Optimal environmental conditions, such as light, temperature and precipitation, are the most important factors influencing plant growth [22]. Excessive deviations in at least one of the above-mentioned environmental parameters may have a negative effect on plant growth and development and may disturb aromatic vegetative development. Koyuncu and Ersoy [23], studying different apple varieties in nurseries grown under controlled greenhouse and field conditions, observed that trees grown under protected conditions had higher vegetative growth and smaller trunk cross-sectional areas than trees grown under orchard conditions.

Table 1. Average monthly air temperatures and total precipitation according to the agrometeorological station in Lublin during the months of April to October in 2017–2019.

	Air Temperature, °C							
	IV	V	VI	VII	VIII	IX	X	Mean from IV to X °C
2017	8.0	14.1	19.6	18.9	20.2	13.5	8.5	14.7
2018	7.5	16.7	18.8	20.6	20.8	15.5	10.0	15.7
2019	9.5	13.4	21.5	19.4	20.3	14.5	11.0	15.7
Multi-year Average	7.4	13.0	16.3	18.0	17.2	12.6	7.6	13.2
	Total Precipitation, mm							
	IV	V	VI	VII	VIII	IX	X	∑ Precipitation
2017	55.0	29.1	282.0	107.9	48.0	80.0	90.0	692.0
2018	40.0	56.0	65.0	124.0	72.0	68.0	36.0	461.0
2019	49.0	93.0	37.0	38.0	102.0	52.0	29.0	400.0
Multi-year Average	39.0	60.7	65.9	82.0	70.7	53.7	40.1	400.0

IV—April; V—May; VI—June; VII—July; VIII—August; IX—September; X—October.

Total rainfall in the months of April to October differed significantly between study years. It was found that the driest year was 2019, with a total rainfall of 400 mm. In 2018, rainfall reached 461 mm. In 2017, the rainfall level was the highest and differed significantly from the multi-year average, reaching 692 mm.

Analysis of weather conditions would indicate that the effectiveness of the treatments used in 2019, when the total rainfall was lower than the multi-year average, could have been lower than in other years of the study. Water deficit during the period of intensive tree growth may have a negative impact on growth and quality parameters.

3. Results and Discussion

Tables of Total Rainfall

The height of 'Gloster' apple trees in autumn after growth was significantly affected by the type of branching stimulation treatment and the year of the study (Table 2). It was found that the trees treated with BA+GA_3 each year were significantly taller than the others. A significant effect of the type of branching treatment on the height of apple trees of the 'Rubinola' and 'Topaz' cultivars was shown by Laňar et al. in their study [24]. Different results were obtained by Gąstoł et al. [20] when evaluating apple trees of the 'Boskoop' cultivar, where preparations based on growth regulators did not affect tree growth. In the present study, it was shown that in 2018, the perianths of the studied apple cultivar were significantly higher than in the other years of the study; this relationship was not shown in the case of trees treated with BA+GA_{4+7}. Similar relationships were shown in the work by Gąstoł et al. [20] and Kaplan et al. [25].

It was shown that the applied branching stimulation treatments and the year of the study had no significant effect on the diameter of the rootstock stems. The above study confirmed the previous results of Kaplan et al. [26], whose aim was to evaluate the effect of the branching treatment on the growth and quality of 'Gloster' and 'Jonagold' apple trees' perianths. Research conducted in 2017–2019 by Kaplan et al. [25] on the apple cultivars 'Alwa' and 'Najdared' showed that the diameter of rootstock trunks was significantly affected by branching treatments.

The diameter of the trunks of the whorled trees depended significantly on the type of treatment given. Control trees and trees pinched annually were significantly thicker than those treated with BA+GA_{4+7}. There was no significant effect of test year on the trait studied. Carra et al. [22], evaluating the effect of 6-benzyladenine (6-BA) and 6-benzyladenine + gibberellin 4 and 7 (6-BA+GA_{4+7}) on the growth and quality of apple trees of the cultivars 'Early Red One' and 'Fuji', showed no significant effect of branching treatments on the stem diameter of trees. Doric et al. [27] observed differences in the trunk diameter of grafted young apple trees of the 'Gala' cultivar after three sprayed applications of 6-BA at concentrations of 250 and 350 mg L^{-1}. The application of this treatment resulted in a decrease in the trunk diameter compared to untreated trees, while, in the case of the 'Jonagold' cultivar, the trend was the opposite. Steiner et al. [28] showed a reduction in stem diameter in 'Golden Delicious' apple trees treated with branching treatments. Rossi et al. [29] showed that a single spray of 6-BA+GA_{4+7} at different concentrations (0, 500, 1000 and 1500 mg L^{-1}) reduced the trunk diameter of young grafted 'Catarina' apple plants, but the difference was not statistically significant. Literature data indicate an ambiguous effect of growth regulators on the stem thickness of trees subjected to branching treatments, so the mechanism of the effect of this type of preparation on tree diameter has not been precisely determined.

Table 2. Effect of branching stimulation treatments on growth and quality of 'Gloster' cv. apple trees in 2017–2019.

	2017	2018	2019	*p*-value
	Height of maiden apple trees in autumn, cm			
Control	153.8 ± 9.7 Bb *	163.4 ± 10.1 Ba	153.2 ± 9.5 Bb	0.1064
Pinching leaves	154.3 ± 8.6 Bb	161.2 ± 9.0 Ba	151.2 ± 8.4 Bb	0.0001
BA+GA_3 **	167.7 ± 5.2 Ab	175.3 ± 5.4 Aa	164.4 ± 5.1 Ac	0.0001
BA+GA_{4+7} ***	155.4 ± 24.0 Ba	162.4 ± 25.1 Ba	152.3 ± 23.6 Ba	0.1064
p-value	0.0001	0.0001	0.0001	
	Diameter of rootstocks, mm			
Control	24.0 ± 2.9 Aa	25.1 ± 3.0 Aa	24.3 ± 3.0 Aa	0.0711
Pinching leaves	22.6 ± 2.2 Aa	23.7 ± 2.3 Aa	23.0 ± 2.2 Aa	0.1841
BA+GA_3	23.8 ± 3.0 Aa	24.9 ± 3.2 Aa	24.2 ± 3.1 Aa	0.2166
BA+GA_{4+7}	22.9 ± 2.2 Aa	23.9 ± 2.3 Aa	23.2 ± 2.2 Aa	0.0641
p-value	0.0810	0.0810	0.0810	
	Diameter of maiden trees, mm			
Control	17.2 ± 2.2 Aa	17.8 ± 2.3 Aa	17.6 ± 2.3 Aa	0.9377
Pinching leaves	16.7 ± 7.7 Aa	17.2 ± 7.9 Aa	17.1 ± 7.8 Aa	0.4447
BA+GA_3	15.8 ± 1.4 ABa	16.4 ± 1.5 ABa	16.2 ± 1.4 ABa	0.1852
BA+GA_{4+7}	14.5 ± 1.4 Ba	15.0 ± 1.5 Ba	14.8 ± 1.5 Ba	0.2494
p-value	0.0078	0.0078	0.0078	
	Number of lateral shoots, pcs			
Control	1.4 ± 1.1 Cb	1.3 ± 1.1 Cab	2.7 ± 4.8 BCa	0.8974
Pinching leaves	1.8 ± 1.4 Ca	1.9 ± 1.5 Ca	1.7 ± 1.4 Ca	0.0228
BA+GA_3	6.5 ± 1.4 Aa	6.8 ± 1.5 Aa	6.3 ± 1.4 Aa	0.2439
BA+GA_{4+7}	3.9 ± 1.5 Ba	4.0 ± 1.5 Ba	3.7 ± 1.4 Ba	0.6263
p-value	0.0001	0.0001	0.0001	
	Total length of lateral shoots, cm			
Control	3.3 ± 2.9 Da	3.4 ± 3.1 Da	3.1 ± 2.8 Da	0.8738
Pinching leaves	24.9 ± 23.9 Ca	26.1 ± 25.1 Ca	23.6 ± 22.7 Ca	0.8549
BA+GA_3	148.0 ± 50.3 Aa	155.4 ± 52.9 Aa	140.6 ± 47.8 Aa	0.3428
BA+GA_{4+7}	106.7 ± 54.0 Ba	112.1 ± 56.7 Ba	101.4 ± 51.3 Ba	0.6151
p-value	0.0001	0.0001	0.0001	
	Average length of one shoot, cm			
Control	1.7 ± 1.3 Da	1.8 ± 1.3 Da	1.7 ± 1.3 Da	0.9898
Pinching leaves	10.3 ± 9.5 Ca	10.4 ± 9.6 Ca	10.1 ± 9.3 Ca	0.9851
BA+GA_3	22.4 ± 4.8 Ba	22.7 ± 4.8 Ba	22.1 ± 4.7 Ba	0.8242
BA+GA_{4+7}	26.6 ± 8.8 Aa	26.9 ± 8.9 Aa	26.2 ± 8.7 Aa	0.9236
p-value	0.0001	0.0001	0.0001	

* Significant difference: A, B, C and D indicate significant differences in the column using capital letters, while a, b and c indicate significant differences in the row using small letters, for significant differences at α = 0.05. ** BA+GA_3—regardless of the dose of applied growth regulators and the number of treatments. *** BA+GA_{4+7}—regardless of the dose of applied growth regulators and the number of treatments

The number of lateral shoots ranged from 1.3 to 6.8 and was significantly dependent on the branching treatment and the type of growth regulators used. In the first two years of the study, it was shown that control and pinch-treated trees produced significantly fewer lateral shoots than chemically treated branching trees. In 2019, this relationship was shown in the combination of leaf pinching and a chemical treatment. It was shown, not depending on the year of the study, that the whorls treated with BA+GA_3 produced significantly more shoots than those treated with BA+GA_{4+7}. Laňar et al. [24] also found that the number of lateral shoots was significantly influenced by the method of branching treatment and the type of growth regulators used. No significant effect of the year of study on the trait studied was shown in the present study. A similar trend was shown by Gąstoł et al. [20]. The sum of lateral shoot lengths ranged from 3.1 to 155.4 cm and varied significantly. Significant differences were found between all the combinations used. Significantly, the smallest sum of lateral shoot lengths was obtained in control trees and the largest in BA+GA_3 treated trees. Trees subjected to mechanical branching treatments formed a significantly lower sum of lateral shoot lengths than those treated with chemical treatments. It was shown that the type of mixture used, in the case of chemical branching, had a significant effect on the trait under study. Trees treated with the BA+GA_3 solution produced a significantly higher sum of lateral shoot lengths than those treated with BA+GA_{4+7}. There was no significant effect of the year of testing on the trait studied. Kumawat et al. [8], evaluating the effect of branching stimulation treatments on the quality of 'Oregon Spur' apple trees, showed that trees with a higher number of lateral shoots produced a higher total sum of lateral shoot lengths, a relationship that was confirmed in the present study.

The length of lateral shoots is directly related to the productivity of trees in the orchard and is crucial in determining the quality of nursery stock [8]. Statistical analysis for the parameter determining the length of one shoot showed identical relationships to those for the sum of lateral shoot lengths. Although the control trees formed significantly fewer lateral shoots than those treated with growth regulators, their length was also, significantly, the smallest among the evaluated trees. This does not confirm the observations of other authors [8,27,30,31].

On trees treated with BA+GA_3, the mean shoot length was significantly lower than after the application of BA+GA_{4+7}. As confirmed in other studies [8,27,31], the production of a greater number of lateral shoots by a tree can have an effect on reducing the length of these shoots; this was fully confirmed in the present study.

The height of the whorls was significantly modified by the type of branching-stimulating treatment (Table 3). It was shown that trees treated with BA+GA_3 were significantly taller than the others. A significant effect of the number of applications on the studied trait was shown only for trees treated with pinching. Laňar et al. [24], evaluating apple trees of the 'Rubinola' and 'Topaz' cultivars, showed a significant effect of the number of benzyladein (BA) treatments on tree height.

The diameter of rootstock trunks significantly depended on the type of branching treatment and the number of applications. The highest values of this parameter were obtained in maiden trees treated once and twice with BA+GA_3 and in those treated twice with BA+GA_{4+7}. The rootstock trunks of maiden trees treated once were significantly thicker than those treated twice. This trend was not confirmed in trees treated with BA+GA_{4+7}.

The trunk diameter of perianths treated once with branching treatments were significantly larger in the pinching combination than in those treated with BA+GA_{4+7}. With two treatments, it was shown that the trunks of perianths sprayed with BA+GA_3 were significantly thicker than the others. There was no clear effect of the number of treatments on the traits under study.

The number of lateral shoots and the sum of lateral shoot lengths in the BA+GA_3-treated trees were significantly higher than in the other combinations, regardless of the number of treatments. The number of treatments did not show any significant effect on the number of lateral shoots in maiden trees treated with BA+GA_3. In the case of trees treated with BA+GA_3 and BA+GA_{4+7}, it was found that trees sprayed twice formed a significantly higher number of lateral shoots and total lateral shoot lengths than trees treated once.

This trend was confirmed in trees treated with pinching by analysing the sum of lateral shoot lengths. The effect of the number of treatments on the number of lateral shoots was confirmed by Laňar et al. in their study [24].

Table 3. Influence of the number of branching stimulation treatments on the growth and quality of 'Gloster' apple trees, irrespective of the year of the study.

	Number of Treatments	Pinching Leaves	$BA+GA_3$	$BA+GA_{4+7}$	*p*-Value
Height of maiden apple trees in autumn, cm	1 **	161.9 ± 9.3 Ab *	168.3 ± 6.3 Aa	154.5 ± 7.4 Ac	0.0001
	2 ***	149.2 ± 4.1 Bc	169.9 ± 7.5 Aa	158.9 ± 33.8 Ab	0.0001
	p-value	0.0001	0.1549	0.2799	
Diameter of rootstocks, mm	1	23.7 ± 2.6 Ab	25.1 ± 3.2 Aa	23.3 ± 2.6 Ab	0.0002
	2	22.5 ± 1.7 Bb	23.5 ± 2.9 Ba	23.4 ± 1.8 Aa	0.0112
	p-value	0.0018	0.0021	0.7077	
Diameter of maiden trees, mm	1	18.5 ± 10.7 Aa	16.2 ± 1.4 Aab	14.5 ± 1.7 Bb	0.0001
	2	15.5 ± 1.3 Bb	16.0 ± 1.5 Aa	15.0 ± 1.1 Ab	0.0001
	p-value	0.0148	0.4698	0.0302	
Number of lateral shoots, pcs	1	2.0 ± 1.4 Ac	6.0 ± 1.2 Ba	2.8 ± 1.0 Bb	0.0001
	2	1.6 ± 1.4 Ac	7.1 ± 1.4 Aa	5.0 ± 1.0 Ab	0.0001
	p-value	0.1668	0.0001	0.0001	
Total length of lateral shoots, cm	1	21.0 ± 20.7 Bc	122.5 ± 34.8 BAb	63.8 ± 35.4 Bb	0.0001
	2	28.8 ± 26.1 Ac	173.5 ± 50.8 Aa	149.7 ± 29.1 Ab	0.0001
	p-value	0.0432	0.0001	0.0001	
Average length of one shoot, cm	1	7.2 ± 6.2 Bb	20.5 ± 4.0 Ba	22.5 ± 9.8 Ba	0.0001
	2	13.3 ± 11.0 Ac	24.3 ± 4.6 Ab	30.6 ± 5.1 Aa	0.0001
	p-value	0.0001	0.0001	0.0001	

* Significant difference: A, B and C indicate significant differences in the column using capital letters, while a, b and c indicate significant differences in the row using small letters for significant differences at $\alpha = 0.05$. ** Single number of branching procedures, regardless of dose and year of study. *** Double number of branching procedures, regardless of dose and year of study.

The average shoot length in trees treated with pinching was significantly lower than in the other combinations, regardless of the number of applications. Trees treated twice with branching-stimulating treatments formed significantly longer shoots than those treated once.

The size of the dose had no significant effect on the height of apple trees (Table 4). Trees treated with $BA+GA_3$ were significantly taller than those treated with $BA+GA_{4+7}$. Laňar et al. [24] showed that doubling the dose of Progerbalin ®LG had a significantly negative effect on the height of apple trees.

The stem diameter of rootstocks of trees treated with the lower dose of $BA+GA_3$ was significantly smaller than those treated with the higher dose. At the higher concentration, trees treated with $BA+GA_3$ had significantly thicker trunks than those that received applications of $BA+GA_{4+7}$.

The diameter of the trunks of trees treated with the lower dose of $BA+GA_3$ was significantly smaller than those treated with the higher dose, while the opposite trend was shown for the application of $BA+GA_{4+7}$. Trees treated with $BA+GA_3$ were significantly thicker than those that received applications of $BA+GA_{4+7}$, and this was independent of the application rate.

The number of lateral shoots on trees treated with a higher dose of the $BA+GA_3$ mixture was significantly higher than those that received applications of the higher dose of $BA+GA_{4+7}$. The effect of the concentration of growth regulators applied on the number of lateral shoots was confirmed in numerous studies [8,20,22,24].

The sum of lateral shoot lengths in trees treated with the $BA+GA_3$ solution at a higher concentration was significantly higher than those that received applications of the lower dose. The sum of lateral shoot lengths on maiden trees treated with $BA+GA_3$ was significantly greater than those that received applications of the lower dose. Trees treated with the BA+GA_3 solution produced a significantly higher sum of lateral shoot lengths than those treated with BA+GA_{4+7}.

Table 4. Effect of the branching stimulation solution dose on the growth and quality of 'Gloster' apple trees, irrespective of study year.

	Concentration Value	BA+GA_3	BA+GA_{4+7}	*p*-Value
Height of maiden apple trees in autumn, cm	small **	168.9 ± 9.2 Aa *	154.6 ± 8.3 Ab	0.0001
	large ***	169.3 ± 7.0 Aa	156.7 ± 24.5 Ab	0.0001
	p-value	0.6368	0.3252	
Diameter of rootstocks, mm	small	23.4 ± 2.2 Ba	23.8 ± 2.0 Aa	0.1244
	large	24.3 ± 3.1 Aa	23.3 ± 2.2 Ab	0.0017
	p-value	0.0033	0.0725	
Diameter of maiden trees, mm	small	15.6 ± 1.5 Ba	15.2 ± 1.4 Ab	0.0172
	large	16.1 ± 1.4 Aa	14.8 ± 1.5 Bb	0.0001
	p-value	0.0025	0.0104	
Number of lateral shoots, pcs	small	5.0 ± 1.8 Ba	3.5 ± 1.4 Bb	0.0001
	large	6.6 ± 1.4 Aa	3.9 ± 1.5 Ab	0.0001
	p-value	0.0001	0.0127	
Total length of lateral shoots, cm	small	115.6 ± 48.4 Ba	97.9 ± 42.1 Ab	0.0001
	large	148.6 ± 50.2 Aa	106.7 ± 53.8 Ab	0.0001
	p-value	0.0001	0.1128	
Average length of one shoot, cm	small	22.5 ± 6.5 Ab	27.3 ± 8.8 Aa	0.0001
	large	22.2 ± 4.6 Ab	26.5 ± 8.8 Aa	0.0001
	p-value	0.7094	0.4569	

* Significant difference: A and B indicate significant differences in the column using capital letters, while a and b indicate significant differences in the row using small letters for significant differences at $\alpha = 0.05$. ** small: lower dose of applied growth regulators, i.e., BA (Globaryll 100 SL 4.5 mL) + GA_3 (0.45 g Florgib) and BA (Globaryll 100 SL 4.5 mL) + GA_{4+7} (Gibb Plus 11 SL 45 mL/1 L of water), regardless of the number of applications and the year of testing. *** large: higher dose of applied growth regulators, i.e., BA (Globaryll 100 SL 6.8 mL) + GA_3 (0.63 g Florgib) and BA (Globaryll 100 SL 6.8 mL) + GA_{4+7} (Gibb Plus 11 SL 67.5 mL/1 L of water), regardless of the number of applications and the year of testing.

Mean shoot length was significantly dependent on the type of solution applied. The whorls treated with BA+GA_{4+7} formed significantly longer shoots than those that received applications of BA+GA_3.

The PCA showed the influence of the concentration of the applied preparations on the number, the average length and the sum of the lengths of the lateral shoots. The above-mentioned parameters defining the quality of the crown structure were analysed jointly for the types of growth regulators used and their concentrations. The sum of the total variable PCs for the analysed parameters defining the structure of the crown in maiden apple trees of the 'Gloster' cultivar was 100% (for PC1 98.14% and for PC2 2.86%, respectively) (Figure 1). Analysing the quality of the trees expressed by the number, average length and the sum of the lengths of lateral shoots, a very high similarity and dependence was observed between BA+GA_3 and BA+GA_{4+7} in low and high doses, regardless of the number of applications. The presented analysis clearly indicates that the factor that most strongly determines the effectiveness of using growth regulators is the number of applications, not the type of preparation used.

Figure 2 shows a comprehensive analysis of the growth and quality of the maiden 'Gloster' trees. The above analysis was developed based on the following parameters: growth of the maiden trees; diameter of the rootstock and maiden trunks; and number, average length and sum of the lengths of side shoots. As can be seen from the presented dendrogram (Figure 2a), for the parameters determining the growth and quality of maiden apple trees of the 'Gloster' variety after a single application of treatments to stimulate branching, there is one cluster focusing on the pinching procedure and another focusing on the mixture of growth regulators: BA+GA_{4+7} and an outlier in the form of the mixture BA+GA_3. In the case of the dendrogram (Figure 2b), determining the growth and quality of maiden apple trees of the 'Gloster' variety after two applications of branching-stimulating treatments, slightly different relationships were observed, maiden trees of the 'Gloster' variety after two applications of branching-stimulating treatments created one cluster consisting of the mixture of the growth regulators BA+GA_{4+7} and BA+GA_3 and an outlier in the form of the pinching procedure.

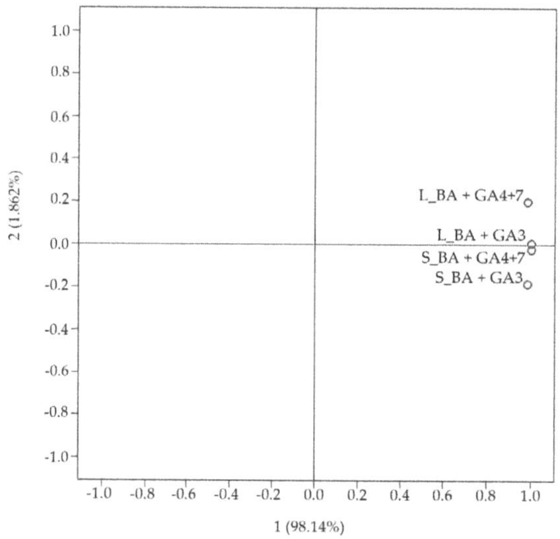

Figure 1. PCA analysis for the effect of the applied concentrations of growth regulators, regardless of their type, on the quality of the crown structure of maiden apple trees of the 'Gloster' variety.

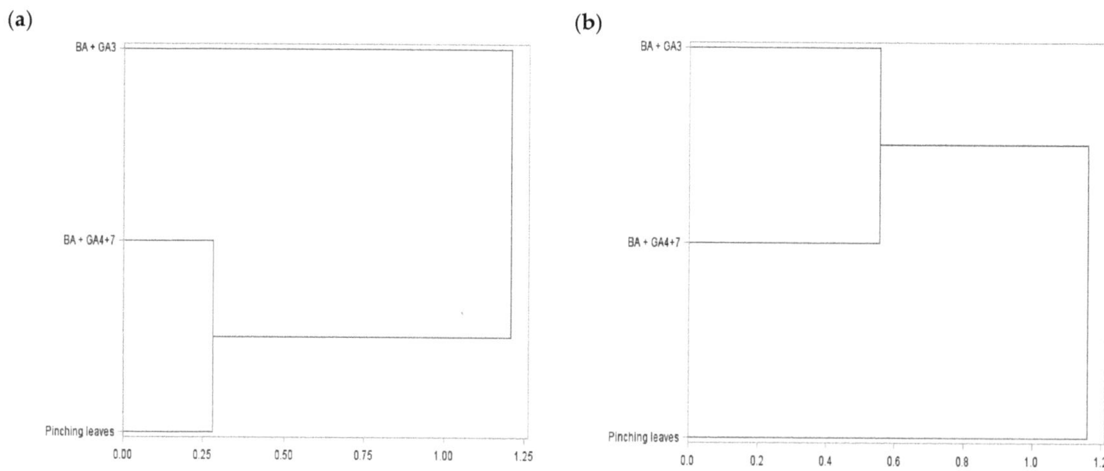

Figure 2. Branching tree diagram for the growth and quality of maiden apple trees of the 'Gloster' variety (**a**) for a single number of treatments, regardless of the concentration of applied growth regulator and the year of research; (**b**) for twice the number of treatments, regardless of the concentration of applied growth regulator and the year of research.

4. Conclusions

1. The type of branching-stimulating treatment of the apple cv. 'Gloster' had a significant effect on the height, stem diameter, number of lateral shoots, average length of one shoot and the sum of the lengths of all sylleptic shoots.

2. The year of the study had a significant effect on the height and number of lateral shoots of the studied apple cultivar.

3. Maiden apple trees treated with BA+ GA_3 were, significantly, the tallest (2017—167.7 cm; 2018—175.3 cm; 2019—164.4 cm), produced the largest number of shoots (2017—6.5 pcs; 2018—6.8 pcs; 2019—6.3 pcs) and had the largest sum of lateral shoot lengths (2017—148.0 cm; 2018—155.4 cm; 2019—140.6 cm) among the evaluated combinations.

4. The number of treatments and the concentration of applied growth regulators had a significant effect on the crown structure of 'Gloster' apple trees. Trees chemically treated for branching formed a significantly higher number and total length of lateral shoots after two applications and a higher concentration of a growth regulator than after a single application and a lower dose.

Author Contributions: Conceptualization, M.K. and K.E.K.; methodology, M.K.; software, K.E.K.; validation, M.K., K.E.K. and K.B.; formal analysis, K.B.; investigation, M.K.; resources, K.E.K.; data curation, K.B.; writing—original draft preparation, M.K., K.E.K. and K.B.; writing—review and editing, M.K. and K.E.K.; visualization, K.E.K.; supervision, M.K. and K.E.K.; project administration, K.E.K.; funding acquisition, M.K. All authors have read and agreed to the published version of the manuscript.

Funding: The cost was incurred from financial resources from the IDUB University Development Strategy for 2024–2026 in the discipline of Agriculture and Horticulture as part of the task "Stage: 1, payment from funds: SUBB.RNN.24.019".

Data Availability Statement: The data presented in this study are available on request from the corresponding author.

Conflicts of Interest: The authors declare no conflict of interest. The funders had no role in the design of the study; in the collection, analyses or interpretation of data; in the writing of the manuscript; or in the decision to publish the results.

References

1. Li, Q.; Gao, Y.; Wang, K.; Feng, J.; Sun, S.; Lu, X.; Liu, Z.; Zhao, D.; Li, L.; Wang, D. Transcriptome Analysis of the Effects of Grafting Interstocks on Apple Rootstocks and Scions. *Int. J. Mol. Sci.* **2023**, *24*, 807. [CrossRef] [PubMed]
2. Food and Agriculture Organization of the United Nations (FAO). Available online: https://www.fao.org/Faostat/En/#data/QCL (accessed on 7 July 2024).
3. Volk, G.M.; Chao, C.T.; Norelli, J.; Brown, S.K.; Fazio, G.; Peace, C.; McFerson, J.; Zhong, G.-Y.; Bretting, P. The Vulnerability of US Apple (Malus) Genetic Resources. *Genet. Resour. Crop. Evol.* **2015**, *62*, 765–794. [CrossRef]
4. Yang, Y.-Y.; Ren, Y.-R.; Zheng, P.-F.; Zhao, L.-L.; You, C.-X.; Wang, X.-F.; Hao, Y.-J. Cloning and Functional Identification of a Strigolactone Receptor Gene MdD14 in Apple. *Plant Cell Tiss. Organ. Cult.* **2020**, *140*, 197–208. [CrossRef]
5. Kapłan, M.; Baryła, P. The Effect of Growth Regulators on the Quality of Two-Year-Old Apple Trees of 'Šampion' and 'Jonica' Cultivars. *Acta Sci. Pol. Hortorum Cultus* **2006**, *5*, 79–89.
6. Sosna, I.; Licznar-Małańczuk, M. Growth, Yielding and Tree Survivability of Several Apricot Cultivars on Myrobalan and "Wangenheim Prune" Seedlings. *Acta Sci. Pol. Hortorum Cultus* **2012**, *11*, 27–37.
7. Świerczyński, S.; Borowiak, K.; Bosiacki, M.; Urbaniak, M.; Malinowska, A. Estimation of the growth of 'Vanda' maiden sweet cherry trees on three rootstocks and after aplication of foliar fertilization in a nursery. *Acta Sci. Pol. Hortorum Cultus* **2019**, *18*, 109–118. [CrossRef]
8. Kumawat, K.L.; Raja, W.H.; Singh, D.B.; Chand, L.; Mir, J.I.; Rai, K.M.; Kirmani, S.N. Effects of Plant Growth Regulators Applications on Induction of Lateral Branching in Oregon Spur Apple Nursery Trees. *Ind. J. Hort.* **2020**, *77*, 72. [CrossRef]
9. Coşkun, N.; Pırlak, L. Effect of Plant Growth Promoting Rhizobacteria (PGPR) on Branching and Growing of Apple Sapling. *Erwerbs-Obstbau* **2017**, *59*, 309–313. [CrossRef]
10. Klimek, K.; Kapłan, M.; Najda, A. Effect of Growth Regulators on Quality of Apple Tree Maidens. *Acta Agroph.* **2018**, *25*, 277–288. [CrossRef]
11. Verma, P.; Sharma, N.C.; Sharma, D.P.; Thakur, H. Induction of Lateral Branches in Apple (*Malus* × *domestica* Borkh.) by Notching. *Erwerbs-Obstbau* **2023**, *65*, 2219–2228. [CrossRef]

12. Vanderzande, S.; Hias, N.; Edge-Garza, D.; Costes, E.; Davey, M.W.; Keulemans, J. Sylleptic Branching in Winter-Headed Apple (Malus × Domestica) Trees: Accession-Dependent Responses and Their Relationships with Other Tree Architectural Characteristics. *Tree Genet. Genomes* **2016**, *12*, 87. [CrossRef]
13. Li, G.; Tan, M.; Cheng, F.; Liu, X.; Qi, S.; Chen, H.; Zhang, D.; Zhao, C.; Han, M.; Ma, J. Molecular Role of Cytokinin in Bud Activation and Outgrowth in Apple Branching Based on Transcriptomic Analysis. *Plant Mol. Biol.* **2018**, *98*, 261–274. [CrossRef] [PubMed]
14. Olesen, T.; Muldoon, S.J. Branch Development in Custard Apple (Cherimoya Annona Cherimola Miller × Sugar Apple *A. Squamosa* L.) in Relation to Tip-Pruning and Flowering, Including Effects on Production. *Trees* **2009**, *23*, 855–862. [CrossRef]
15. Matysiak, K.; Adamczewski, K. Regulatory Wzrostu i Rozwoju Roślin-Kierunki Badań w Polsce i Na Świecie. *Prog. Plant Prot.* **2009**, *49*, 1810–1816.
16. Yan, Y.; Zhao, N.; Tang, H.; Gong, B.; Shi, Q. Shoot Branching Regulation and Signaling. *Plant Growth Regul.* **2020**, *92*, 131–140. [CrossRef]
17. Hill, J.L.; Hollender, C.A. Branching out: New Insights into the Genetic Regulation of Shoot Architecture in Trees. *Curr. Opin. Plant Biol.* **2019**, *47*, 73–80. [CrossRef]
18. Li, J.; Zhao, Y.; Zhang, Y.; Ye, F.; Hou, Z.; Zhang, Y.; Hao, L.; Li, G.; Shao, J.; Tan, M. Genome-Wide Analysis of MdPLATZ Genes and Their Expression during Axillary Bud Outgrowth in Apple (*Malus domestica* Borkh.). *BMC Genom.* **2023**, *24*, 329. [CrossRef]
19. Sabagh, A.E.; Mbarki, S.; Hossain, A.; Iqbal, M.A.; Islam, M.S.; Raza, A.; Llanes, A.; Reginato, M.; Rahman, M.A.; Mahboob, W.; et al. Potential Role of Plant Growth Regulators in Administering Crucial Processes Against Abiotic Stresses. *Front. Agron.* **2021**, *3*, 648694. [CrossRef]
20. Gąstoł, M.; Domagała-Świątkiewicz, I.; Bijak, M. The Effect of Different Bioregulators on Lateral Shoot Formation in Maiden Apple Trees. *Folia Hortic.* **2012**, *24*, 147–152. [CrossRef]
21. Rufato, L.; Marchioretto, L.D.R.; Orlandi, J.C.; Michelon, M.F.; De Rossi, A.; Sander, G.F.; Macedo, T.A.D. Lateral branch induction at nursery with growth regulators in 'Maxi Gala' apple trees grafted on four rootstocks. *Sci. Hortic.* **2019**, *253*, 349–357. [CrossRef]
22. Carra, B.; Cabrera, D.; Rodriguez, P.; Dini, M. Improving Feathering in Different Nursery Apple Trees by Plant Growth Regulators. *Rev. Bras. Frutic.* **2023**, *45*, e965. [CrossRef]
23. Karamürsel, Ö.F. Nursery Growing of Some Apple Varieties Using Different Grafting Methods in Greenhouse and Orchard. *Afr. J. Biotechnol.* **2011**, *10*, 2518. [CrossRef]
24. Laňar, L.; Mészáros, M.; Kyselová, K.; Náměstek, J.; Sus, J.; Bělíková, H.; Čonka, P. Branching of Nursery Apples and Plums Using Various Branching Inducing Methods. *J. Cent. Eur. Agric.* **2020**, *21*, 113–123. [CrossRef]
25. Kapłan, M.; Klimek, K.; Borkowska, A.; Buczyński, K. Effect of Growth Regulators on the Quality of Apple Tree Whorls. *Appl. Sci.* **2023**, *13*, 11472. [CrossRef]
26. Kapłan, M.; Jurkowski, G.; Krawiec, M.; Borowy, A.; Wójcik, I.; Palonka, S. Wpływ Zabiegów Stymulujących Rozgałęzianie Na Jakość Okulantów Jabłoni. *Ann. Hortic.* **2018**, *27*, 5–20. [CrossRef]
27. Doric, M.; Magazin, N.; Milić, B.; Keserovic, Z. Enhancing Feathering of One-Year-Old Gala and Jonagold Apple Trees through Application of 6-Benzylaminopurine and Gibberellins. *Bulg. J. Agric. Sci.* **2015**, *21*, 631–637.
28. Steiner, M.; Hrotkó, K.; Vegvari, G. Performance of Hormonal Content and Branching of Apple Nursery Trees after BA (6-Benzyladenine) Application. *Acta Hortic.* **2013**, 419–423. [CrossRef]
29. Rossi, A.D.; Rufato, L.; Giacobbo, C.L.; Gomez, F.R.C.; Fachinello, J.C. Use of Promalin® on One-Year Old Trees of the Apple CV. "Catarina". *Acta Hortic.* **2004**, *636*, 145–149. [CrossRef]
30. Cowgill, W.; Beese, M.; Magron, R.; Autio, W.R.; Clements, J.M.; Robinson, T. Studies and Recommendations for Branching Young Apple Trees. *Fruit Notes* **2014**, *79*, 1–8.
31. Radivojevic, D.; Momirovic, I.; Milivojevic, J.; Velickovic, M.; Oparnica, C.; Lukic, M. The Influence of BA and BA+GA4+7 on Formation of Sylleptic Shoots on One-Year-Old Apple Nursery Trees. *J. Agric. Sci. BGD* **2015**, *60*, 89–95. [CrossRef]

agriculture

MDPI

Article

Influence of Biostimulants and Microbiological Preparations on the Yield and the Occurrence of Diseases and the European Corn Borer (*Ostrinia nubilalis* Hbn, Lepidoptera, Crambidae) on Sweet Corn (*Zea mays* L. Var. *saccharata*)

Elżbieta Wojciechowicz-Żytko [1,*], Edward Kunicki [2] and Jacek Nawrocki [1]

1 Department of Botany, Physiology and Plant Protection, University of Agriculture in Krakow, Al. 29-Listopada 54, 31-425 Cracow, Poland; jacek.nawrocki@urk.edu.pl

2 Department of Horticulture, University of Agriculture in Krakow, Al. 29-Listopada 54, 31-425 Cracow, Poland; edward.kunicki@urk.edu.pl

* Correspondence: elzbieta.wojciechowicz-zytko@urk.edu.pl

Abstract: The aim of this work was to determine the influence of chosen biostimulants and microbiological preparations on the yield of sweet corn and the occurrence of *Ostrinia nubilalis* Hbn, and diseases. In both years of the study, the preparations used in this experiment did not have a statistically significant effect on marketable yield; however, in 2017, the highest weight was observed in the cobs of plants treated with Rizocore and Polyversum WP while the lowest in the cobs treated with RhizoVital 42. The biostimulant Asahi SL and the biological fungicide Serenade ASO proved to be the most effective in protecting sweet corn against cob and shoot infections by fungi of the genus *Fusarium*. All the preparations reduced the development of the common smut in corn, especially on the cobs. There were no statistically significant differences in cob infection by the *O. nubilalis* in the combinations treated with different preparations, although the lowest number of cobs damaged by pest in both years were observed on plots treated with Serenade ASO and RhizoVital 42, while the highest on plots treated with Goëmar BM.

Keywords: corn; yield; biostimulants; pests; diseases

1. Introduction

Sweet corn is one of the eight most consumed vegetables in the world. The world leader in corn production is the United States, where this crop covers an area of about 120,000 hectares and its consumption is estimated at 10 kg per capita. In Poland, sweet corn is grown both for the fresh market and for the processing industry, with an annual area of about 10,000 ha and an annual consumption of 0.5 kg per capita [1].

The gradually increasing consumption of this vegetable is due not only to its high nutritional value, but also to its health-promoting properties. Sweet corn is a rich source of vitamin E, selenium, antioxidants (lutein and zeaxanthin), and dietary fibre, among others. Moreover, its grains do not contain gluten, so sweet corn is a valuable component of gluten-free diets [1].

The dynamically increasing size of this vegetable crop has prompted the search for methods to improve its productivity per unit area. One of these is the use of biostimulants, which, unlike standard chemicals, have no negative impact on the environment [2]. Biostimulants belong to a group of compounds containing active substances that, when applied in small doses, improve plant metabolism by stimulating the synthesis and activity of natural hormones, increasing nutrient uptake, stimulating root system development and increasing resistance to unfavourable environmental factors. Biostimulants may include hormones, enzymes, proteins, ammo acids, vitamins, trace elements and other biologically active compounds [3–5]. Recently, many articles have been published on the use of biostimulants, which indicate their positive or negative effects on plant growth, chemical composition, and the yield of different vegetables [6–10], but the number of articles on their usefulness in sweet corn cultivation remains quite limited. Among others, Das et al. [11] conducted a field experiment to investigate the effect of organic and standard sources of fertilizer on the yield and quality of sweet corn in West Bengal, India; Al-Temimi and Al-Hilfy [12] investigated the application of a combination of certain mineral fertilizers and seaweed extract on the number of cobs per plant, number of rows per cob, number of grains per row, and the grain field; Tadros et al. [13] used a foliar application of amino acid biostimulants at a chosen growth stage of the sweet corn plants.

Sweet corn can be infected by a number of pathogens during the growing season, the most damaging of which are fungi of the genus *Fusarium* Link. These pathogens can cause two forms of disease: corn cob fusariosis, as well as root rot and stem base rot. *Fusarium* spp. fungi can infect various parts of plants, starting from the seedling stage, but the most harmful infections affect corn cobs. Infected cobs become deformed and the seeds may get damaged. A coating of white, yellowish, or reddish (salmon-coloured) mycelium, the colour of which depends on the predominant fungus species, develops on the infected tissues. Fungi of the genus *Fusarium* adversely affect human health, as some species produce harmful mycotoxins [14,15]. *Fusarium* spp. are polyphagous fungi and can also occur on other plants such as cereals and many vegetables. They are easily transmitted by conidial spores produced in large numbers and survive in the soil for several years in the form of spores—chlamydospores [16–18].

Fusarium root rot and stem base rot, often abbreviated as 'stem fusariosis', is a serious disease capable of causing the death of entire plants. The first symptoms of the disease are visible on the ground side of the stems, on which rot spots can be seen, and sometimes the wilting and drying of the leaves from the bottom upwards can be observed. In the final stage of the disease, further localized drying and dying of entire plants may occur [16–18].

Common smut in corn, caused by *Ustilago maydis* (DC.) Corda, can be quite often observed mainly on the cobs in sweet corn plantations. This pathogen does not produce mycotoxins and therefore does not pose a threat to human health in the same way as the fungi *Fusarium* spp. The main sources of infestation can be infected seed and corn residues in the soil. Significant damage, due to the high level of infection, is associated with damage caused by pests such as thrips, aphids, the fruit fly, and the European corn borer. The most important symptom of the disease is growths in the form of nodules with a mass of compact, black-grey spores inside. Earlier, the growths are bright and hard, but as the spores mature inside, the inside of the nodules darkens, and their surface wrinkles and may crack. Early infected plants may not produce cobs at all. Infection by the pathogen at a later stage of cob development, reduces their yield and quality [16,17,19].

A disease of lesser importance is the small leaf spot of corn caused by the fungus *Kabatiella zeae* Narita and Y. Hirats (syn. *Aureobasidium zeae*). The main source of the pathogen is post-harvest corn, and infection is promoted by wet leaves during rainy weather, or long-lasting dew. Optimal conditions for the causative agent are high humidity above 75% and temperatures of 18–25 °C. Symptoms on the leaves are small, initially chlorotic spots, which enlarge over time and become necrotic inside. At high severity, the spots can cover the entire leaves and cause a decrease in yield quantity and quality [14].

The most effective preparations for the protection of sweet corn against disease are strobilurin and triazole fungicides, which are used either preventively or curatively. Unfortunately, the number of synthetic plant protection products currently available for vegetable protection is decreasing annually, due to the non-renewal of the registration of certain chemical preparations in the European Union member states.

This is in line with one of the strategies of the European Green Deal, which promotes the use of biological plant protection methods and aims to reduce the number of registered pesticides [14]. Consequently, in line with the principles of integrated plant protection, research should focus on the evaluation of the effectiveness of biological and biotechnical preparations, as well as supportive measures to enhance plant health, including new fertilizer formulations.

Corn is also attacked by a number of insects, one of the most harmful being the European corn borer (*Ostrinia nubilalis* Hubner). This is a widespread pest known to feed on 250 different kinds of plants, including sweet corn and grain corn. The larvae eat leaves, mine all parts of the stalks and ears, and feed on cobs, causing serious yield loss [20,21].

Insecticides are commonly applied to protect corn plants against damage [22,23]. The research for innovative and alternative methods is an increasingly growing reality [24]. The use of cultural practices, host plant resistance, biological control and biostimulants has been shown to improve many agronomic characteristics of plants and protect them against pests and diseases [25–29].

Seaweed (marine algae) extracts are the new type of products currently used in plant cultivations. Microalgae are a rich source of bioactive substances that can find many applications in agriculture [30,31]. Primary and secondary metabolites are of great importance in medicine and agriculture [32,33]. Seaweeds are classified as the most important organism that can be used in plant production as biostimulants of plant growth [26,29,34–39] and as biostimulants or microbiological preparations for the control of pests and diseases [26,27,32,33,40,41].

Another substance that can be used in alternative pest control is silicon. This element acts as a mechanical barrier limiting the damage caused by pests. It also strengthens the induced chemical defence of plants after insect feeding [42–44]. In general, silicon is involved in plant resistance against insect-caused damage via two major defence mechanisms: physical defence and induced biochemical (chemical) defence [45–47].

The aim of this work was to determine the influence of chosen biostimulants and microbiological preparations on the yield of sweet corn and the occurrence of pests and diseases.

2. Material and Methods

2.1. Research Site

The experiment was carried out in 2016 and 2017 at the experimental station of the Agricultural University in Mydlniki near Krakow, located in southern Poland (50°04′ N, 19°51′ E). According to Koppen's classification, the climate is humid continental (Dfb). According to the classification of the Food and Agriculture Organization of the United Nations [48], the soil type is a Fluvic Cambisol (Humic).

2.2. Methods

Seeds of the sweet corn (*Zea mays* L. ssp. *saccharata* Kcke) cultivar Sweet Wonder F_1 (Agri Frimeko) were sown in the middle of May, at 75 × 20 cm spacing and covered for three weeks with non-woven fabric (Agryl PP, 19 g m^{-2}). The plants were treated four times (foliar spray) at weekly intervals with biostimulants and microbiological preparations (Table 1). The control were unsprayed plants.

Table 1. Doses and dates of application in the experiment in years 2016–2017.

Treatment	Dose	Date of Application	
		2016	2017
Tytanit (TY)	0.04%		
Optisil (OSI)	0.1%		
Asahi SL (AS)	0.5 dm^3 ha^{-1}	18.07	17.07
Goëmar BM 86 (BM)	1.5 dm^3 ha^{-1}	25.07	24.07
Polyversum WP (POL)	0.05%	1.08	31.07
Rhizo Vital 42 (RHV)	0.5 dm^3 ha^{-1}	8.08	7.08
Rizocore (RI)	60 g ha^{-1}		
Serenade ASO (SE)	8 dm^3 ha^{-1}		

Each treatment combination was performed in 4 replications. Each plot consisted of 40 plants sown in 4 rows where 2 of them served as protective rows. Standard cultivation practices were performed, including weed control, fertilization and irrigation. The doses of fertilizers were calculated on the basis of soil analyses using the universal method according to Nowosielski [49] to achieve nutrient contents of 120 mg N (NH_4 + NO_3), 60 mg P, 200 mg K, 70 mg Mg, and 1500 mg Ca in 1 dm^3 of soil.

The biostimulants and microbiological preparations used in this experiment are as follows:

AS-Asahi SL (UPL, formerly Arysta LifeScience, Warszawa, Poland) is a biostimulator based on three active substances from the group of nitrophenols naturally occurring in plants. It also improves the tolerance of crops to stress factors unfavourable to plant growth and development. The content of active substances is as follows: sodium para-nitrophenolate—3 g/dm^3 (0.3%), sodium ortho-nitrophenolate—2 g/dm^3 (0.2%), 5-sodium nitroguaiacolate—1 g/dm^3 (0.1%);

TY—Tytanit® (Intermag, Okusz, Poland) is a liquid mineral growth stimulator. It increases the content of chlorophyll in the leaves, which intensifies photosynthesis. It significantly improves the efficiency of the pollination process and increases the uptake of nutrients from the soil, especially potassium (K), nitrogen (N), magnesium (Mg), calcium (Ca), and iron (Fe). It strengthens the natural resistance of plants to stress factors. Furthermore, it contains titanium (Ti) calculated as TiO_2 (8.5 g/dm^3; 0.8% m/m);

OSI—Optysil® (Intermag, Olkusz, Poland) is a liquid silicon anti-stress agent that activates the natural immune systems of plants and stimulates their growth and development. It increases plants' tolerance to unfavourable growing conditions, e.g., drought, and reduces the impact of biotic stress caused by pathogens and/or pest attack. It contains silicon (Si) calculated as SiO_2 (200 g/dm^3; 16.5% m/m);

BM—Goëmar BM 86 (UPL, formerly Arysta LifeScience, Warszawa, Poland) is a liquid preparation recommended for the cultivation of vegetables whose edible part is fruit. It activates mineral nutrition and ensures optimal flowering. It contains biologically active filtrate GA 142 obtained from *Ascophyllum nodosum* (sea alga), with the following composition: boron—2.03% (m/m), molybdenum—0.02% (m/m), magnesium—4.8% (m/m);

POL—Polyversum WP (Biopreparaty Praque, Czech Republic) is a biological agent based on the fungus-like organism *Pythium oligandrum* (concentration 10^6 oospores/g). It is a parasite of some species of oomycetes and pathogenic fungi. *Pythium oligandrum* decomposes pathogen hyphae through enzymatic decomposition, while stimulating the immune mechanisms of the protected plant by introducing phytohormones, phosphorus and sugars into them. The dose used is 0.05%;

RhV—RhizoVital 42 (ABiTEP, Berlin, Germany) is a biostimulator that contains the beneficial bacterium *Bacillus amyloliquefaciens* FZB42, naturally occurring in the soil (concentration above 2.5×10^{10} spores per mL). It supports the plant in building a strong and efficient root system, improving its tolerance towards stress (e.g., water deficiency, salinity) caused by unfavourable climatic conditions and field management. The bacteria germinate and colonize the young growing roots and release hormones and enzymes, which stimulate plant growth and mobilize nutrients, which increases plant growth and promotes a higher yield. The preparation is used at a dose of 0.5 L/ha.

RI—Rizocore (Biogard, Grassobbio, Italy) is a microbiological soil preparation with a unique composition of three ingredients: *Bacillus megaterium* (concentration 10^4 CFU/g), *Trichoderma harzianum* NAT11 (concentration 10^{10} CFU/g), and mycorrhizal fungi *Glomus* spp. (5%). The preparation improves the microbiological composition of the soil, increases the vigor and health of seedlings and young plants, causes better development of the roots in young plants and the growth and health of plants, and increases the resistance of mature plants in defence processes in the presence of pathogens; mycorrhiza also increases the absorbent surface of roots. The preparation is used at a dose of 60 g/ha.

SE—Serenade ASO (Bayer AG, Leverkusen, Germany) is biological preparation based on the beneficial soil bacterium *Bacillus subtilis* QST 713 (concentration 13.96 g/L, 1.34%,

1.042×10^{12} CFU/L). It is intended for use against diseases caused by fungi and bacteria. The bacterium occurs naturally and has fungicidal and fungistatic properties. It works by disrupting the development of mycelium as a result of contact with the pathogen on the plant surface and producing substances that disrupt the functioning of fungal cell membranes. *Bacillus subtilis* QST 713 also competes with pathogens for living space and nutrients as well as induces the systemic immunity of the plant. Its effect on bacterial diseases is to block the synthesis of the bacterial cell wall. The preparation was used at a dose of 8 L/ha.

A study on the effect of applied preparations on the health of the shoot base, leaves, and cobs of sweet corn was performed just before harvest. The severity of the disease symptoms was evaluated on 30 randomly selected leaves, shoots, and cobs of maize from each plot, respectively. During direct observations of the plants, the severity of the disease symptoms was visually assessed according to the following scale: 0—no symptoms, 1—light infestation (spot area of up to 5% of the leaf blade), 2—medium infestation (spot area of 6–25%), 3—heavy infestation (spot area of 26–50%), 4—very heavy infestation (more than 51% of the area). The infestation scale was also applied to the area of rot spots on a 30 cm section, measured from the base, of shoots infected by *Fusarium*-induced stem base rot and the area of cobs attacked by the nodular head [17]. Based on the obtained data, an infestation index was calculated for each combination replicate. The infestation index expressed as a percentage was calculated according to the formula:

$$Ip = \sum (a \times b) \times 100/N \times n$$

where Ip—infestation index; a—number of plants infested at a given scale level; b—scale level; N—total number of plants analysed; n—the highest scale level.

The effects of the preparations on the plant crop and the occurrence of *O. nubilalis* were estimated after harvesting the cobs at the end of August. Fifteen cobs from each combination were weighed and measured, and the number of cobs damaged by *O. nubilalis* Hbn was noted.

2.3. Meteorological Data

The data source is a meteorological station located on the research site.

There was less rainfall in each month of the growing season (May–August) of 2016 than in the corresponding period of 2017 (Figure 1). However, the distribution of rainfall on individual days was different—in 2017 rain events were less frequent, but more intense, while 2016 had more regular rainfall. This means that in the second year of research there were many periods of drought during the growing season. Such a pattern was unfavourable for the development of pathogens, including fungal ones. The highest air temperatures in 2016 were noted in June and July, whereas in 2017 they were noted in May and July.

2.4. Statistical Analysis

Data were analysed using a one-way analysis of variance ANOVA with Statistica software v. 13.3 (Tibco Software Inc., Palo Alto, CA, USA); calculations were conducted for each season separately. The values expressed as a percentage were transformed according to the Bliss function ($y = \arcsin \sqrt{x}$). Tukey's HSD test for the results of the analyses concerning yield and cob parameters, Duncan's MRT test for the results concerning the occurrence of disease symptoms, and Fisher's LSD test for cob infestation by *O. nubilalis* were used to determine the significance of differences between mean values at the $p = 0.05$ significance level.

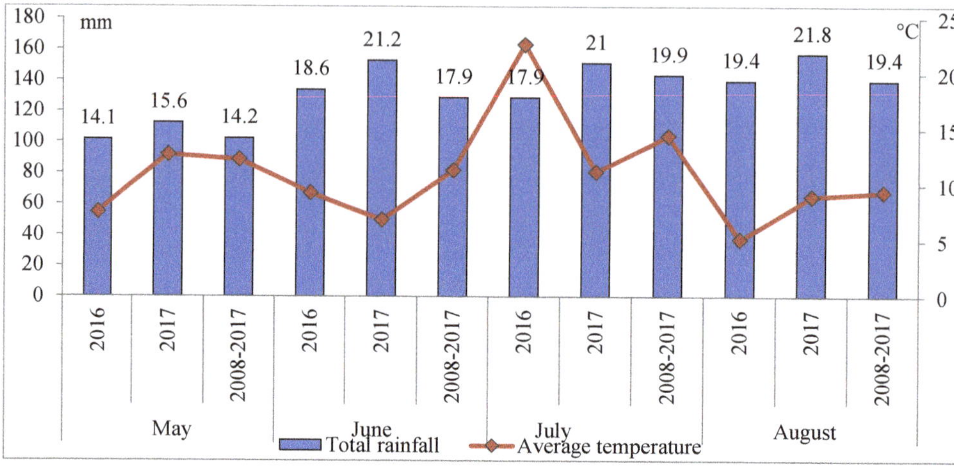

Figure 1. Rainfall and air temperature in 2016–2017—Mydlniki.

3. Results

Considering the cob parameters studied, it should be noted that the results are inconclusive (Table 2). On average, the highest corn cob weight was obtained and recorded in the first year of the study, while the cob diameter and length were slightly bigger in 2017. In 2017, the cobs of plants treated with Rizocore and Polyversum WP had the heaviest weight (but, similar to the control and plants treated with Asahi SL), while the cobs treated with RhizoVital 42 were the lightest, but their weight was similar to those treated with Tytanit and Optysil. No statistically significant differences were found for the cob weight in 2017, the cob diameter in both years, and the cob length in 2016. The control and RhizoVital 42-treated plants produced the shortest cobs in 2017, but their length was comparable to that of the cobs from the other treatments.

Table 2. Effect of the treatment on corn cob parameters in years 2016–2017.

Treatment	Weight (g)		Diameter (cm)		Length (cm)	
	2016	**2017**	**2016**	**2017**	**2016**	**2017**
Control	249 bc *	228 a	4.6 a	4.8 a	21.3 a	21.8 a
Asahi SL	253 bc	229 a	4.6 a	4.8 a	21.4 a	22.0 ab
Tytanit	246 ab	207 a	4.6 a	4.8 a	20.9 a	21.8 ab
Optysil	244 ab	207 a	4.6 a	4.7 a	21.2 a	21.9 ab
Serenade ASO	250 bc	211 a	4.5 a	4.9 a	21.2 a	21.9 ab
RhizoVital 42	241 a	207 a	4.6 a	4.9 a	21.2 a	21.8 a
Rizocore	260 c	217 a	4.7 a	4.8 a	20.8 a	22.8 b
Polyversum WP	262 c	230 a	4.6 a	4.8 a	21.1 a	22.2 ab

* means that for each year marked with the same letter they do not differ significantly at $p = 0.05$; Tukey's HSD test.

Figure 2 demonstrates that in both years of the study, the preparations used in this experiment did not have a statistically significant effect on the marketable yield. In 2016, it ranged from 14.97 (RhizoVital 42) to 17.00 t ha^{-1} (Rizocore), while in the following year, from 13.30 (Goëmar BM) to 15.33 t ha^{-1} (Polyversum WP).

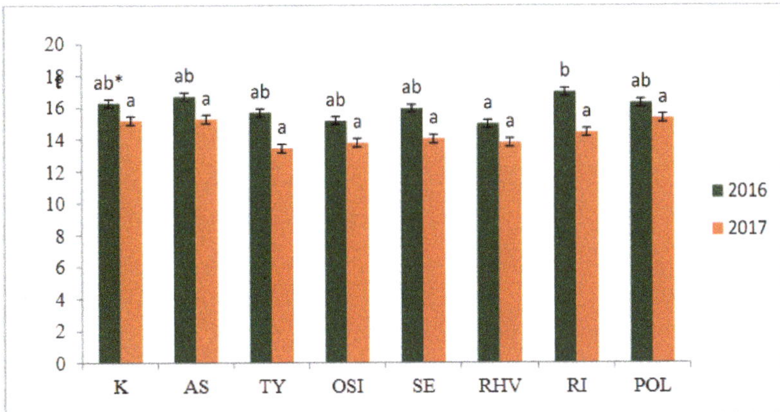

Figure 2. Marketable yield (t ha^{-1}) of sweet corn depending on preparation in years 2016–2017. * means that for each year marked with the same letter they do not differ significantly at $p = 0.05$, Tukey's HSD test. K-Control, AS—Asahi SL, TY—Tytanit, OSI—Optysil, SE—Serenade ASO, RHV—RhizoVital 42, RI—Rizocore, POL—Polyversum.

In the first year of the study, all the preparations used effectively limited the development of fungi of the *Fusarium* genus on sweet corn cobs compared to the control. The most effective were Goëmar BM 86, Asahi SL and Serenade ASO, and the weakest protective effect was demonstrated by RhizoVital 42, Tytanit and Optysil (Table 3). In the next year, the following products most effectively protected corn against *Fusarium* spp. infection: Asahi SL, Serenade ASO, Goëmar BM 86, and Polyversum WP. The remaining preparations were ineffective and limited the development of *Fusarium* rot on corn cobs to a small extent only.

Table 3. Indices of infection of sweet corn cobs with *Fusarium* fungi [%].

Treatment	2016	2017
Tytanit	7.84 d *	5.23 bcd
Optysil	6.47 cd	6.76 cd
Asahi SL	4.31 abc	0.63 a
Goëmar BM 86	2.70 a	1.83 ab
Polyversum WP	5.62 bcd	2.84 abc
RhizoVital 42	8.45 d	7.36 cd
Rizocore	6.32 cd	4.84 bcd
Serenade ASO	3.40 ab	0.70 a
Control	12.03 e	10.41 d

* means that for each year followed by the same letter, they do not differ significantly at $p = 0.05$; Duncan's MRT test.

During the 2016 growing season, the infection of the base of corn shoots in all the combinations was greater than in the next year of research (Table 4). In 2016, the most effective protective effect on shoots was observed on plants sprayed with Asahi SL and Serenade ASO, and the weakest in combinations with Tytanit, RhizoVital 42, and Polyversum WP. In 2017, similar results were obtained, although Polyversum WP turned out to be one of the most effective preparations.

In the first year of the study, all the preparations used effectively reduced infections with the causative agent of common smut in corn (Table 5). The most effective were Goëmar BM 86; RhizoVital 42; and Asahi SL. In 2017, corn infection by *Ustilago maydis* was at a very low level; in some combinations practically no disease symptoms were observed. Tytanit did not significantly improve the health of corn cobs; the number of infected cobs was only slightly lower than in the control.

Table 4. Indices of infection of the base of sweet corn shoots with *Fusarium* fungi [%].

Treatment	2016	2017
Tytanit	4.85 d *	3.70 cd
Optysil	2.96 c	2.59 bc
Asahi SL	1.18 a	0.08 a
Goëmar BM 86	2.83 bc	1.60 bc
Polyversum WP	3.59 cd	0.83 ab
RhizoVital 42	3.69 cd	2.61 bc
Rizocore	2.35 abc	1.63 bc
Serenade ASO	1.50 ab	0.46 ab
Control	8.86 e	6.68 d

* means that for each year followed by the same letter, they do not differ significantly at $p = 0.05$; Duncan's MRT test.

Table 5. Intensity of infection of sweet corn cobs with the fungus *Ustilago maydis*—the cause of common smut in corn—infection indices (%).

Treatment	2016	2017
Tytanit	6.61 c *	3.37 bc
Optysil	5.14 bc	4.10 c
Asahi SL	3.75 ab	0.08 a
Goëmar BM 86	3.05 a	0.08 a
Polyversum WP	4.76 abc	0.92 ab
RhizoVital 42	3.37 ab	0.00 a
Rizocore	4.55 abc	0.63 a
Serenade ASO	4.41 abc	0.16 a
Control	9.67 d	6.10 c

* means that for each year followed by the same letter, they do not differ significantly at $p = 0.05$; Duncan's MRT test.

In both years of the experiments, among the applied agents, only Tytanit and Optysil fertilizers did not significantly reduce leaf infection by *Kabatiella zeae* compared to the control (Table 6). The most effective preparation was Goëmar BM 86; Asahi SL, RhizoVital 42 and Serenade ASO quite effectively protected corn leaves against the pathogen causing small leaf spot in both growing seasons.

The results indicate that the level of infestation by the European corn borer differed between the study years. In 2016, a greater number of damaged cobs was recorded compared to 2017 (Table 7).

There were no statistically significant differences in cob infection by the *O. nubilalis* in the combinations treated with different biostimulants and microbiological preparations, although the least number of cobs damaged by pests in both years was observed in plots treated with Serenade (2.0 in 2016 and 2.5 in 2017) and RhizoVital (2.75 in both years) while the highest (5.0 in 2016 and 4.0 in 2017) in plots treated with Goëmar BM. On control plots, an average of 4.75 in 2016 and 3.0 in 2017 damaged cobs were recorded (Table 7, Figure 3).

Table 6. Infection of sweet corn leaves with the fungus *Kabatiella zeae* (syn. *Aureobasidium zeae*)—the cause of small corn leaf spot—infection indices [%].

Treatment	2016	2017
Tytanit	18.05 d *	14.03 cd
Optysil	14.78 cd	11.74 bcd
Asahi SL	8.06 ab	5.59 a
Goëmar BM 86	7.14 a	4.79 a
Polyversum WP	9.39 ab	7.60 ab
RhizoVital 42	8.66 ab	5.63 a
Rizocore	11.79 bc	8.58 abc
Serenade ASO	9.18 ab	5.61 a
Control	17.07 d	18.00 d

* means that for each year followed by the same letter, they do not differ significantly at $p = 0.05$, Duncan's MRT test.

Table 7. Mean number of cobs infested with *O. nubilalis* depending on the treatment in years 2016–2017.

Treatment	2016	2017
Control	4.75 a *	3.00 a
Asahi SL	3.75 a	3.00 a
Tytanit	3.75 a	3.25 a
Optysil	3.75 a	3.25 a
Goëmar BM 86	5.00 a	4.00 a
Serenade ASO	2.00 a	2.50 a
RhizoVital 42	2.75 a	2.75 a
Rizocore	2.75 a	3.50 a
Polyversum WP	4.25 a	2.50 a

* means that the same letters within a column indicate no significant differences at $p = 0.05$; Fisher's LSD test.

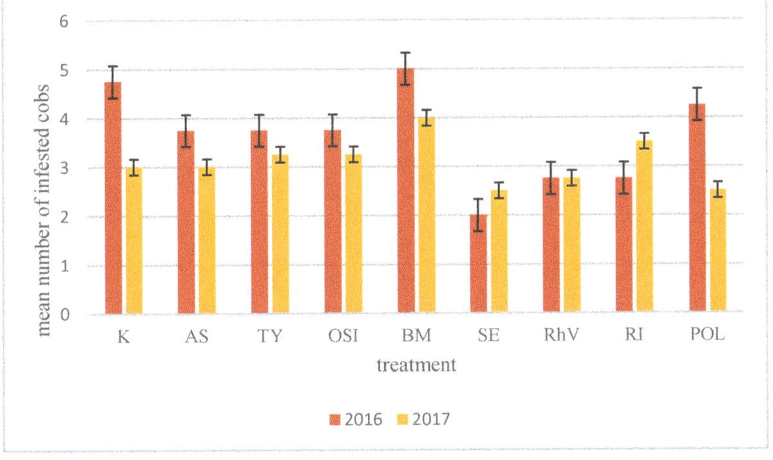

Figure 3. Mean number of cobs damaged by *O. nubilalis*. K—Control, AS—Asahi SL, TY—Tytanit, OSI—Optysil, BM—Goëmar BM 86, SE—Serenade ASO, RHV—RhizoVital 42, RI—Rizocore, POL—Polyversum.

4. Discussion

Seaweeds are a popular choice among farmers for encouraging plant growth and improving crop yields in terms of both quality and quantity. In our experiment in both years of the study, no statistically significant differences were found in the marketable yield of corn.

Waligóra and Majchrzak [50] reported that in 2016–2017, the yield of the tested cultivars in Poland ranged from 7.8 to 21.4 t ha^{-1} depending on the cultivar and the year of the study, and a yield of 15 t ha^{-1} is considered profitable, which corresponds with the results achieved in this experiment. Das et al. [10] used vermicompost, an extract from the seaweed *Ascophyllum nodosum*, and a mineral fertilizer, and observed that the application of a full dose of the mineral fertilizer alone or in combination with vermicompost resulted in a higher yield of sweet corn as compared to treatments solely with the seaweed, which confirms the results obtained in this experiment, i.e., the lack of a significant effect of Goëmar BM 86 on sweet corn yield. Similarly, Al-Temimi and Al-Hilfy [12] found out that the application of a seaweed extract resulted in the highest number of cobs per plant, rows per cob, and grains per row, and a higher grain yield only when combined with mineral fertilizers. In the research conducted by Tadros et al. [13], the goal was to determine the optimal concentration and timing of foliar applications of amino acid biostimulants to achieve more efficient use in sweet corn cultivation. The plants were sprayed with four different concentrations: 1, 2, 3, and 4 mL/L of amino acids biostimulant liquid (commercial product PerfectoseTM), at three different growth stages: the seventh leaf, the tasselling and milk stage, and while the control plants were sprayed with distilled water only. The results indicated that the corn plant height; leaf water potential; chlorophyll content; and nitrogen, phosphorus, potassium, and protein content were significantly affected by the foliar application treatments, while the cob length was not significant. Similar effects of biostimulants on yield have been observed in other vegetables. Kunicki et al. [7] noted that spraying with Aminoplant had no effect on the yield of two cultivars of spinach. In an experiment conducted by Grabowska and Kunicki [6], there was no significant effect of Aminoplant and Goëmar Goteo preparations on the yield of broccoli, despite their positive effect on increasing the aboveground weight of seedlings. Similarly, the biostimulants Asahi SL, Biozyme, and Goëmar BM 86 did not contribute to an increase in the yield of the tomato grown in the open field [10], whereas an experiment with carrots showed an increase in the yield after the application of Aminoplant at a dose of 1.5 dm^3 ha^{-1}, but only in the first year of the study [9]. The study by Xu et al. [51] on the effectiveness of different foliar sprays, two forms of silicon and two kinds of organic nitrogenous compounds, showed that the treatments improved the marketable yield of tomatoes.

The use of preparations based on seaweed extracts, mainly containing laminarin, can significantly improve the physiological parameters and condition of sweet corn. Seaweed extract based preparations will keep the plants in good condition and less susceptible to pathogen infections [52,53]. In our experiment, Goëmar BM 86 was quite effective in protecting corn against *Fusarium* spp. and at the same time was most effective in reducing the development of the pathogens responsible for small leaf spot and nodular head.

The formulation Rizocore, which contains both beneficial microorganisms and arbuscular mycorrhizal fungi, did not show during the trials any significant efficacy in reducing dangerous pathogens, especially soil-borne pathogens such as *Fusarium* spp. The lack of effectiveness could be due to a short growing season, which might be insufficient for the mycorrhizal process to develop, or because of unfavourable soil conditions, especially soil drought, during the summer period. Subaedah et al. [54] showed that mycorrhiza applied on sweet corn had a positive effect on improving the yield and quality of sweet corn. The lower effectiveness of Rizocore may have been duem not only to a too-short growing season for sweet corn, but also due to soil moisture that was insufficient in promoting the development of beneficial soil microorganisms. The development of arbuscular mycorrhizal fungi requires a longer period than a few months, especially under varying weather conditions. The development of bacteria of the genus *Bacillus* contained in the fungicides

Serenade ASO and RhizoVital 42 is faster than that of fungi under optimal conditions, and they can develop, not only in the rhizosphere, but also in the phyllosphere of the protected plants. This may be one of the reasons why these beneficial bacteria are more effective in protecting sweet corn from pathogens, compared to the microbial complex contained in Rizocore. This points to the need for continued research in future years, under different weather conditions, to gain more information on this subject.

Antagonistic bacteria, actinomycetes, oomycetes as well as fungi can effectively reduce the occurrence of some pathogens, including *Fusarium* fungi. Ridout et al. [15] showed that the presence of some beneficial bacteria as well as fungi on corn grains can reduce the number of pathogens of the genus *Fusarium* as well as reduce the content of some mycotoxins produced by these fungi. In a study, the *Pythium oligandrum* fungus-like organism contained in Polyversum WP effectively inhibited, especially in the second year, the occurrence of disease symptoms of *Fusarium* head rot on cobs, as well as on sweet corn stems.

The tested fertilizer Tytanit proved ineffective in reducing sweet corn pathogens in both years of the experiments. Pangaribuan et al. [55] and Mohammed et al. [56] showed that the quality and yield of sweet corn cobs were positively affected by urea as well as organic fertilizers. In addition, the applied fertilizers had a beneficial effect on stimulating beneficial microbial communities in the soil. Similar results were reported by Fahrurrozi et al. [57] on several sweet corn cultivars. Organic fertilization can affect soil structure, improving conditions for sweet corn growth [58]. By contrast, Canatoy [59] showed that the organic fertilization of sweet corn was less effective than mineral fertilization in improving the quality and quantity of cob yield.

The variation in the occurrence of the symptoms of common smut in corn between the years of the experiments was due, among other things, to different weather conditions during the growing season. In 2017, the symptoms were not severe, which was perhaps also related to the better health of the seeds used in the experiment. The high relevance of sweet corn cultivar and meteorological conditions prevailing during the growing season in the degree of plant infection by *Ustilago maydis* was shown by Waligóra et al. [19] and Frommer et al. [60] Szőke et al. [61] found in their research, that the use of auxin, cytokinin, and gibberellin reduced the negative effects of *Ustilago maydis* infection on sweet corn and that these phytohormones may be a new method of protecting plants against common smut in corn.

In 2016, there were longer periods of higher air and soil moisture during the growing season compared to 2017, which resulted from lower but more frequent rainfalls. Air humidity is one of the most important factors influencing the growth of bacteria and fungi, both those that are beneficial and those that are pathogenic to plants. Lower air humidity during the summer of 2017 may have had a limiting effect on the development of fungi that is pathogenic to sweet corn. The disease symptoms were less severe in 2017, with no common smut symptoms observed on corn in some plots. In addition to weather conditions, the presence of sources of infection with a particular pathogen is essential for pathogen development. Fungi of the genus *Fusarium*, as polyphages, can develop from many sources, not only from crop residues or infected seeds, but also from other, neighbouring crops. Therefore, they also pose a very high threat to corn every year. In the case of *Ustilago maydis*, there are fewer inoculum sources [16]. Dry, hot days and scarcity of water in the soil adversely affect the application of biopreparations, especially those containing living organisms. The development of beneficial microorganisms is inhibited under such conditions, leading to a significant reduction in their effectiveness.

The high efficacy of Serenade ASO in the protection of sweet corn against pathogens during our tests confirms the information contained, among other things, on the label of the product. The *Bacillus subtilis* bacterium, which is the main component of this biofungicide, can directly affect certain plant pathogens by producing important antibiotic-like substances and thus inhibiting the growth of other microorganisms. Additionally, this beneficial bacterium plays an important role in stimulating induced systemic resistance to pathogenic microorganisms. The biosynthesis of phenylpropanoids, flavonoids, and flavonols is important in this process [62]. Currently, the use of plant growth-promoting bacteria [PGPB], which include *B. subtilis* contained in the fungicide Serenade ASO and *B. megaterium* in Rizocore, is gaining popularity in agricultural production. Different strains of beneficial bacterial species have been found to have a great diversity of action mechanisms affecting crop plants in manifold ways. Katsenios et al. [63] showed the significant effect of some tested PGPB strains, including *B. subtilis*, in sweet corn crops on yield increase as well as on grain quality by increasing protein and fibre content in the grains.

When using biological preparations, it is important to bear in mind several factors that determine their effectiveness. The timing of their application is very important; biopreparations should be applied as early as possible so that the beneficial microorganisms have time to develop on the surface of plant organs (phyllosphere and rhizosphere) and activate their multidirectional action on pathogens. Protective treatments should be carried out under conditions ensuring the proper development and survival of beneficial microorganisms in the environment; crops should be sprayed during cloudy weather with moderate temperatures and high air and soil humidity. Biopreparations should not be applied in bright sunshine and periods of drought with high air temperatures because such conditions are unfavourable for the development of beneficial microorganisms and, at the same time, inhibit or even block their effect on pathogens [64]. The low effectiveness of biopreparations is most often the result of their application late in the growing season, sometimes already on infected plants, or under unfavourable climatic conditions. The fungi most dangerous to sweet corn are those of the genus *Fusarium*, which can infect plants already in the early stages of plant development, even in the seedling stage [16]. The use of biopreparations on infected plants is pointless in such a case. Although seaweeds are registered as plant growth stimulants, little is known about their effect against plant pests. In our experiment, no statistically significant differences were found in cob infection by the *O. nubilalis* in the combinations treated with different preparations; however, less pest-damaged cobs were noted in plots treated with biostimulants and microbiological preparations compared to the control, except Goëmar BM (seaweeds), where the highest infestation was observed. Similar information was provided by Warner [65], who studied the influence of seaweeds on the occurrence of orchards' pests. The study revealed that seaweed treatment had no effect on fruit damage caused by *Hoplocampa testudinea, Conotrachelus nenuphar, Cydia pomonella, Grapholita molesta, Grapholita prunivora*, or *tortricids* and also noted that those extracts could slightly limit the occurrence of spider mites compared to the control. Yu et al. [32] observed the potential of seaweeds and their isolated components, such as polysaccharides, phenolics, proteins, terpenes, lipids, and halogenated compounds against eggs, larvae, pupae, and adult mosquitoes. They also found out that seaweed compounds affected the growth, development, and reproduction of adults and larvae. Also, Bianco et al. [66] and Cetin et al. [67] concluded that seaweed contained substances with larvicidal properties against mosquitoes. According to Ishii et al. [33], compounds from red alga (seaweeds) showed strong toxicity against *Sithophilus zeamais*, and termites. Asharaja and Sahayaray [68], while working on the biocidal activity of the water extracts (100, 200, 400, 600 and 800 ppm) of brown algae *Sargassum wightii* and *Padina pavonica* against *Dysdercus cingulatus*, noted that extracts from seaweeds controlled the nymphs and adults of this pest on cotton, negatively influencing their biological parameters by, e.g., increasing nymphal mortality, significantly prolonging the mating period, and highly reducing fecundity and hatchability.

Thawfeeq et al. [69] and Thawfeeq et al. [70] studied the insecticidal activity of brown seaweed (*Sargassum tenerrimum*) and red seaweed (*Gracilaria corticata*) against *Aphis craccivora* adults on cowpea. Methanolic extracts of both seaweeds showed insecticidal activity against aphids (at 0.1, 0.2, 0.4, 0.8 and 1.6%) and caused a mortality ratio of over 60% within 72 h after the treatment against *A. craccivora*. The authors concluded that seaweeds could be an alternative ecological pesticide in the control of aphids. Similarly, *Osmundae pinnatifida* showed insecticidal activity against five different pests *Callosobruchus analis, Sitophilus oryzae, Tribolium castaneum, Rhyzopertha dominica*, and *Trogoderma granarium* [71].

In addition, seaweed extracts from *Plocamium cartilagineum* could be used to manage several agricultural pests, such as aphids, bugs, weevils, or moths [72]. Gonzalez-Castro et al. [41] found out that seaweed extract caused a 100% mortality rate in *Diaphorina citri* (Asian citrus psyllid) after 24 h of exposure and suggested that seaweed extracts offered an applicable alternative method for pest control. However, Machado et al. [27] stated that the promising results of the application of seaweed extracts to pest control obtained under laboratory or greenhouse conditions do not always reflect the actual weather conditions affecting plantations.

In our studies no effect of silicon on the plant infestation by *O. nubilalis* was found, whereas Alhousari and Greger [44], who tested two types of mechanisms (physical or mechanical barriers and biochemical mechanisms) limiting pest feeding, found out that Si played an important role in strengthening direct and indirect protection against many insects. They noted that Si acted in plants both below and above ground, inducing lignin accumulation in the roots, increasing toughness and, eventually, resistance to pest attack. Han et al. [45] showed that silicon application at 0.16 and 0.32 g Si/kg soil made a susceptible rice variety resistant to the *Cnaphalocrocis medinalis*. The authors noted that silicon significantly prolonged larval development and reduced the larval survival and pupation rate. High doses of silicon negatively influenced the development of third instar larvae and the weight of pupae. In addition, silicon reduced the intrinsic rate of increase, the finite rate of increase, and the net reproduction rate of the *C. medinalis* population. According to Lang et al. [46], the modification of plants by adding silicone increased the plants' physical barrier and changed the insect behaviours, causing enhanced resistance in Si-amended plants. Si application to rice had an effect on plant defence responses induced by the feeding of *Nilaparvata lugens*. Si supplement improved the silicification of leaf sheaths. It was noted that silicon affected direct and indirect plant defence against chewing and sucking insects by influencing the development time, larvae survival, and population increase rate. The authors suggested that, due to its protective plant defence effects, silicon could be an eco-friendly alternative to chemical pesticides in plant protection.

5. Conclusions

The presented study on the effects of biostimulants and microbiological preparations on the yield and occurrence of selected diseases and pests on sweet corn has shown inconclusive results. Thus, further investigations are necessary in the future in order to clarify the impact of such substances on the quantity and quality of the yield of sweet corn and the occurrence of *O. nubilalis*.

The biostimulant Asahi SL and the biological fungicide Serenade ASO proved to be the most effective in protecting sweet corn against cob and shoot infections by fungi of the genus *Fusarium* in both years of the study. The biostimulant Goëmar BM 86 and the biofungicide Polyversum WP were effective in 2017. All the preparations reduced the development of the common smut in corn, especially on the cobs. Tytanit was the least effective fertilizer, and it did not significantly affect the health of sweet corn compared to the control in the second year of the study. The fertilizer Titanite and the biostimulant Optysil were completely ineffective in protecting sweet corn against small leaf spots in both years of the experiment. The biostimulants Goëmar BM 86 and Asahi SL were the most effective, and the biopreparations RhizoVital 42 and Serenade ASO were moderately effective in this regard.

No statistically significant differences in cob infection by the *O. nubilalis* were noted between the respective plots, although the lowest number of cobs damaged by pest were observed in both years on plots treated with Serenade and RhizoVital while the highest on plots treated with Goëmar BM.

Further research is needed on the effects of different application techniques, doses or mixtures of the biostimulants and microbiological preparations under alternative environmental conditions.

Author Contributions: All the authors contributed to this work. E.W.-Ż. designed and performed the experiments, analysed the data, and wrote the paper; J.N. designed and performed the experiments, analysed the data, and wrote the paper; and E.K. designed and performed the experiments, analysed the data, and wrote the paper. All authors have read and agreed to the published version of the manuscript.

Funding: This research was supported by the Polish Ministry of Science and Higher Education and under the research subsidy granted to the University of Agriculture in Krakow.

Institutional Review Board Statement: Not applicable.

Data Availability Statement: The data presented in this study are available on request from the corresponding author.

Conflicts of Interest: The authors declare no conflicts of interest.

References

1. Warzecha, R. Sweet corn—A valuable crop for vegetable processing and the fresh market. *Agro Serwis* **2019**, *10*, 52–54.
2. Paradikovic, N.; Vinkovic, T.; Vrcek, I.; Zuntar, I.; Bojic, M.; Medic, M. Effect of natural biostimulants on yield and nutritional quality: An example of sweet yellow pepper (*Capsicum annuum* L.) plants. *Sci. Food Agric.* **2011**, *91*, 2146–2152. [CrossRef] [PubMed]
3. Khan, W.; Rayirath, U.P.; Subramanian, S.; Jithesh, M.N.; Rayorath, P.; Hodges, D.M.; Critchley, A.; Craigie, J.; Norrie, J.; Prithiviraj, B. Seaweed extracts as biostimulants of plant growth and development. *J. Plant Growth Reg.* **2009**, *28*, 386–399. [CrossRef]
4. Calvo, P.; Nelson, L.; Kloepper, J.W. Agricultural uses of plant biostimulants. *Plant Soil.* **2014**, *383*, 3–41. [CrossRef]
5. Ojeda-Barrios, D.L.; Cruz-Alvarez, O.; Sánchez-Chavez, E.; Ciscomani-Larios, J.P. Effect of foliar application of zinc on annual productivity, foliar nutrients, bioactive compounds and oxidative metabolism in pecan. *Folia Hort.* **2023**, *35*, 179–192. [CrossRef]
6. Grabowska, A.; Kunicki, E. The influence of selected biopreparations on the yield of broccoli in spring cultivation. *Zesz. Probl. Postep. Nauk. Roln* **2009**, *539*, 193–197.
7. Kunicki, E.; Grabowska, A.; Sękara, A.; Wojciechowska, R. The effect of cultivar type, time of cultivation, and biostimulant treatment on the field of spinach (*Spinacia oleracea* L.). *Folia Hort.* **2010**, *22/2*, 9–13. [CrossRef]
8. Guinan, K.J.; Sujeeth, N.; Copeland, R.B.; Jones, P.W.; O'Brien, N.M.; Sharma, H.S.S.; Prouteau, P.F.J.; O'Sullivan, J.T. Discrete roles for extracts of *Ascophyllum nodosum* in enhancing plant growth and tolerance to abiotic and biotic stresses. *Acta Hort.* **2013**, *1009*, 127–135. [CrossRef]
9. Grabowska, A.; Kunicki, E.; Sękara, A.; Kalisz, A.; Wojciechowska, R. The effect of cultivar and biostimulant treatment on the carrot yield and its quality. *Veget. Crop. Res. Bull.* **2012**, *77*, 37–48. [CrossRef]
10. Grabowska, A.; Kunicki, E.; Sękara, A.; Kalisz, A.; Jezdinský, A.; Gintrowicz, K. The effect of biostimulants on the quality parameters of tomato grown for the processing industry. *Agrochimica* **2015**, *59*, 203–217. [CrossRef]
11. Das, A.; Murmu, K.; Bandopadhyay, P.; Roy, M. Seaweed and its role in enhancing yield and antioxidant properties in sweet corn. *Res. Sq.* **2022**, 1–17. Available online: https://www.researchgate.net/publication/360059422_Seaweed_and_its_role_in_enhancing_yield_and_antioxidant_properties_in_sweet_corn/fulltext/636c6041431b1f5300863bd7/Seaweed-and-its-role-in-enhancing-yield-and-antioxidant-properties-in-sweet-corn.pdf (accessed on 15 June 2024).
12. Al-Temimi, A.H.M.; Al-Hilfy, I.H.H. Effect of the application of biostimulants on the agronomic parameters of corn varieties. *Rev. Bras. Cienc. Agrar. Recife* **2021**, *16*, e9257. [CrossRef]
13. Tadros, M.J.; Omari, H.J.; Turk, M.A. The morphological, physiological and biochemical responses of sweet corn to foliar application of amino acids biostimulants sprayed at three growth stages. *AJCS* **2019**, *13*, 412–417.
14. Korbas, M. Common smuts of maize and other diseases—Harmfulness and possibilities of their control. *Prog. Plant Prot.* **2006**, *46*, 354–357.
15. Ridout, M.E.; Godfrey, B.; Newcombe, G. Effects of antagonists on mycotoxins of seedborne *Fusarium* spp. in sweet corn. *Toxins* **2019**, *1*, 438. [CrossRef]
16. Koike, T.S.; Gladders, P.; Paulus, O.A. *Vegetable Diseases: A Colour Handbook*; Manson Publishing: London, UK, 2007.
17. Bereś, P.K.; Siekaniec, Ł.; Kontowski, Ł. Occurrence of maize (*Zea mays* L.) diseases caused by fungi on long-term monoculture fields in south-eastern Poland in 2010–2021. *Prog. Plant Prot.* **2022**, *62*, 117–127. [CrossRef]
18. Filho, R.C.; Guimarães, M.; de Abreu, V.P.; Rocha, G.A.; de Carvalho Menezes, R.; Dias, V.D.; da Cunha, M.G. *Fusarium sacchari* associated with stem rot in sweet corn in Brazil. *Pesq. Agropec. Trop. Goiânia* **2023**, *53*, e74263. [CrossRef]
19. Waligóra, H.; Sawinska, Z.; Skrzypczak, W.; Idziak, R.; Weber, A.; Głowicka-Wołoszyn, R. Susceptibility of sweet corn varieties to *Ustilago maydis* (DC). *Fragm. Agron.* **2016**, *33*, 76–86.
20. Hudon, M.; LeRoux, E.J.; Harcourt, D.G. Seventy years of European corn borer (*Ostrinia nubilalis*) research in North America. *Agric Zool Rev.* **1989**, *3*, 53–96.
21. Capinera, J.L. European corn borer Ostrinia nubilalis (Hübner) (Insecta: Lepidoptera: Crambidae) Distribution—Life Cycle and Description—Host Plants—Natural Enemies—Weather—Damage—Management; Publication Number EENY-156. 2000. Available online: https://edis.ifas.ufl.edu/publication/IN313 (accessed on 15 June 2024).

22. Nault, B.A.; Kennedy, G.G. Timing insecticide applications for managing European corn borer (Lepidoptera: Pyralidae) infestations in potato. *Crop Prot.* **1996**, *15*, 465–471. [CrossRef]
23. Toninato, A.; Burkness, E.C.; Hutchison, W.D. Insecticidal Control of Corn Earworm in Minnesota Sweet Corn. *Arthropod Manag. Tests* **2021**, *46*, 1–2. [CrossRef]
24. Liu, J.; Chen, J.; Ye, S.; Ding, Y.; Guo, S.; Ningning, F. Inhibitory action of ginkgolic acid against pathogenic fungi and characterisation of its inhibitory activities on Nigrospora oryzae. *Folia Hortic.* **2023**, *35*, 49–59. [CrossRef]
25. Rice, M.E.; Pilcher, C.D. Potential benefits and limitations of transgenic Bt corn for management of the European corn borer. *Am. Entomol.* **1998**, *44*, 75–78. [CrossRef]
26. Tuhy, Ł.; Chowańska, J.; Chojnacka, K. Seaweed extracts as biostimulants of plant growth: Review. *Chemik* **2013**, *67*, 636–641.
27. Machado, L.P.; de Godoy, M.C.; Gasparoto; Alves, N.; Filho, S.; Pavarini, R. Seaweeds in the Control of Plant Diseases and Insects. In *Seaweeds as Plant Fertilizer, Agricultural Biostimulants and Animal Fodder*; CRC Press: Boca Raton, FL, USA, 2019; pp. 98–124. ISBN 9780429487156. [CrossRef]
28. Pereira, R.V.; Filgueiras, C.C.; Dória, J.; Peñaflor, M.F.G.V.; Willett, D.S. The Effects of Biostimulants on Induced Plant Defense. *Front. Agron.* **2021**, *3*, 630596. [CrossRef]
29. Martínez-Gutiérrez, A.; Zamudio-González, B.; Tadeo-Robledo, M.; Espinosa-Calderón, A.; Cardoso-Galvão, J.C.; Vázquez-Carrillo, M.G. Yield of corn hybrids in response to foliar fertilization with biostimulants. *Rev. Mex. Cienc. Agríc* **2022**, *13*, 2. [CrossRef]
30. Aziz, A.; Poinssot, B.; Daire, X.; Adrian, M.; Bézier, A.; Lambert, B.; Joubert, J.-M.; Pugin, A. Laminarin elicits defense responses in grapevine and induces protection against *Botrytis cinerea* and *Plasmopara viticola*. *Mol. Plant Microbe Interact.* **2003**, *16*, 1118–1128. [CrossRef]
31. Chandía, N.P.; Matsuhiro, B. Characterization of a fucoidan from *Lessonia vadosa* (Phaeophyta) and its anticoagulant and elicitor properties. *Int. J. Biol. Macromol.* **2008**, *1*, 235–240. [CrossRef]
32. Yu, K.-X.; Jantan, I.; Ahmad, R.; Wong, C.-L. The major bioactive components of seaweeds and their mosquitocidal potential. *Parasitol Res* **2014**, *113*, 3121–3141. [CrossRef]
33. Ishii, T.; Nagamine, T.; Nguyen, B.C.Q.; Tawata, S. Insecticidal and Repellent Activities of Laurinterol from the Okinawan Red Alga *Laurencia nidifica*. *Rec. Nat. Prod.* **2017**, *11*, 63–68.
34. El-Naggar, A.H.; Osman, M.E.H.; El-Sheekh, M.; Gheda, S.F. Influence of the aqueous extracts of *Ulva lactuca* and *Chlorella kessleri* on growth and yield of *Vicia faba*. *Algol. Stud.* **2005**, *116*, 213–229. [CrossRef]
35. Salah El Din, R.; Elbakry, A.; Ghazi, S.; Abdel Hamid, O. Effect of seaweed extract on the growth and yield of faba bean (*Vicia faba* L.). *Egypt. J. Phycol.* **2008**, *9*, 25–38. [CrossRef]
36. Zodape, S.; Kawarkhe, V.; Patolia, J.; Warade, A. Effect of liquid seaweed fertilizer on yield and quality of okra (*Abelmoschus esculentus* L.). *J. Sci. Ind. Res* **2009**, *67*, 1115–1117.
37. Zodape, S.T.; Mukhopadhyay, S.; Eswaran, K.; Reddy, M.P.; Chikara, J. Enhanced yield and nutritional quality in green gram (*Phaseolus radiata* L) treated with seaweed (*Kappaphycus alvarezii*) extract. *J. Sci. Ind. Res.* **2010**, *69*, 468–471.
38. Kavipriya, R.; Dhanalakshmi, P.K.; Jayashree, S.; Thangaraju, N. Seaweed extract as a biostimulant for legume crop, green gram. *J. Ecobiotech.* **2011**, *3*, 16–19.
39. Zodape, S.T.; Gupta, A.; Bhandari, S.C.; Rawat, U.S.; Chaudhary, D.R.; Eswaran, K.; Chikara, J. Foliar application of seaweed sap as biostimulant for enhancement of yield and quality of tomato (*Lycopersicon esculentum* Mill). *J. Sci. Ind. Res* **2011**, *70*, 215–219.
40. Baloch, G.N.; Tariq, S.; Ehteshamul-Haque, S.; Athar, M.; Sultana, V.; Ara, J. Management of root diseases of eggplant and watermelon with the application of asafoetida and seaweeds. *J Appl Bot Food* **2013**, *86*, 138–142. [CrossRef]
41. González-Castro, A.; Muñoz-Ochoa, M.; Hernandez-Carmona, G.; López-Vivas, J. Evaluation of seaweed extracts for the control of the Asian citrus psyllid *Diaphorina citri*. *J. Appl. Phycol.* **2019**, *31*, 3815–3821. [CrossRef]
42. Fauteux, F.; Rémus-Borel, W.; Menzies, J.G.; Bélanger, R.R. Silicon and plant disease resistance against pathogenic fungi. *FEMS Microbiol. Lett.* **2005**, *249*, 1–6. [CrossRef]
43. Ali, J.G.; Agrawal, A.A. Specialist versus generalist insect herbivores and plant defense. *Trends Plant Sci.* **2012**, *17*, 293–302. [CrossRef]
44. Alhousari, F.; Greger, M. Silicon and Mechanisms of Plant Resistance to Insect Pests. *Plants* **2018**, *7*, 33. [CrossRef] [PubMed]
45. Han, Y.; Lei, W.; Wen, L.; Hou, M. Silicon-mediated resistance in a susceptible rice variety to the rice leaf folder, *Cnaphalocrocis medinalis* Guenée (Lepidoptera: Pyralidae). *PLoS ONE* **2015**, *10*, e0120557. [CrossRef] [PubMed]
46. Lang, Y.; Han, Y.Q.; Li, P.; Li, F.; Ali, S.; Hou, M.L. Silicon amendment is involved in the induction of plant defense responses to a phloem feeder. *Sci. Rep.* **2017**, *7*, 4232.
47. Liang, Y.; Nikolic, M.; Belanger, R.; Gong, H.; Song, A. Silicon and Insect Pest Resistance. In *Silicon in Agriculture*; Springer: Dordrecht, The Netherlands, 2015; pp. 197–207. [CrossRef]
48. FAO. *World Reference Base for Soil Resources*; FAO: Rome, Italy, 2014.
49. Nowosielski, O. *Principles for Developing Fertilizer Recommendations in Horticulture*; PWRiL: Warszawa, Poland, 1978; pp. 250–280. (In Polish)
50. Waligóra, H.; Majchrzak, L. Yield of sweet corn cultivars under conditions in Wielkopolska, Poland. *Fragm. Agron.* **2019**, *36*, 8–14.
51. Xu, X.; Lei, X.; Liao, S.; Li, Y.; Sun, Y. Foliar application of potassium silicate, potassium fulvate and betaine improve summer-time tomato yield by promoting plant nitrogen and potassium uptake. *Folia Hort.* **2022**, *34*, 125–138. [CrossRef]

52. Skwarek, M.; Pipiak, P. Maize pathogenic diseases and the role of biostimulators in their control. *Technol. I Jakość Wyr.* **2020**, *65*, 129–143.
53. Villaver, J.P. Physiological and yield parameters of sweet corn (*Zea mays* L. var. *rogusa*) in response to biofertilizer application under low elevation conditio. *Int. J. Agric. Techno* **2022**, *18*, 2307–2314.
54. Subaedah, S.; Edy, E.; Numba, S.; Sabahannur, S.; Fausiah, R. Effectiveness of arbuscular mycorrhizal fungi and NPK fertilizer in increasing the production of sweet corn. *Plant Asian J. Plant Sci.* **2023**, *21*, 685–692. [CrossRef]
55. Pangaribuan, D.H.; Hendarto, K.; Elzhivago, S.R.; Yulistiani, A. The effect of organic fertilizer and urea fertilizer on growth, yield and quality of sweet corn and soil health. *Asian J. Agri. Biol.* **2018**, *6*, 335–344.
56. Mohammed, A.A.; Majid, Z.M.; Kasnazany, S.A.S.; Rashis, Z.O.; Hama, B.M. Evaluation of growth, yield and some quality characteristics of sweet corn under effect of different nutrient sources. *J. Zankoy Sulaimani* **2018**, 561–568. [CrossRef]
57. Fahrurrozi, F.; Muktamar, Z.; Dwatmadji, D.; Setyowati, N.; Sudjatmiko, S.; Chozin, M. Growth and yield responses of three sweet corn (*Zea mays* L. var. *saccharata*) varieties to local-based liquid organic fertilizer. *Int. J. Adv. Sci. Eng. Inf. Techno* **2016**, *6*, 318–323. [CrossRef]
58. Intansari, R.S.I. The effectiveness of organic fertilizer granules for increasing sweet corn production on acid dryland in Bogor district. *J. Soilsc. Agric* **2022**, *1*, 40–52. [CrossRef]
59. Canatoy, R.C. Growth and yield response of sweet corn (*Zea mays* L. var. *saccharata*) as affected by tillage operations and fertilizer applications. *Int. J. Educ.* **2018**, *6*, 265–276.
60. Frommer, D.; Veres, S.; Radócz, L. Susceptibility of stem infected sweet corn hybrids to common smut disease. *Acta Agrar. Debrec.* **2018**, *7*, 55–57. [CrossRef]
61. Szőke, L.; Moloi, M.J.; Kovács, G.E.; Biró, G.; Radócz, L.; Hájos, M.T.; Kovács, B.; Rácz, D.; Danter, M.; Toth, B. The Application of Phytohormones as Biostimulants in Corn Smut Infected Hungarian Sweet and Fodder Corn Hybrids. *Plants* **2021**, *10*, 1822. [CrossRef]
62. Shao, M.; Chen, Y.; Gong, Q.; Miao, S.; Li, C.; Sun, Y.; Qin, D.; Guo, X.; Yan, X.; Cheng, P.; et al. Biocontrol endophytes *Bacillus subtilis* R31 influence the quality, transcriptome and matabolome of sweet corn. *PeerJ* **2023**, *11*, e14967. [CrossRef]
63. Katsenios, N.; Andreou, V.; Sparangis, P.; Djordjevic, N.; Giannoglou, M.; Chanioti, S.; Kasimatis, C.N.; Kakabouki, I.; Leonidakis, D.; Danalatos, N.; et al. Assessment of plant growth promoting bacteria strains on growth, yield and quality of sweet corn. *Sci. Rep.* **2022**, *12*, 11598. [CrossRef] [PubMed]
64. Pięta, D.; Patkowska, E.; Pastucha, A. The protective effect of biopreparations applied as the dressing for common bean (*Phaseolus vulgaris* L.) and pea (*Pisum sativum* L.). *Acta Sci. Pol. Hort. Cult.* **2005**, *4*, 59–67.
65. Warner, G. Seaweeds tested for pest control. Insects and mites. *Good Fruit Grow* **2012**, *1*, 1–5.
66. Bianco, E.; Pires, L.; Santos, G.; Dutra, K.; Reis, T.; Vasconcelos, E.; Cocentino, A.; Navarro, D. Larvicidal activity of seaweeds from northeastern Brazil and of a halogenated sesquiterpene against the dengue mosquito (*Aedes aegypti*). *Ind. Crop. Prod.* **2013**, *43*, 270–275. [CrossRef]
67. Cetin, H.; Gokoglu, M.; Oz, E. Larvicidal Activity of the Extract of Seaweed, *Caulerpa scalpelliformis*, against *Culex pipiens*. *J. Am. Mosq. Control Assoc.* **2010**, *26*, 433–435. [CrossRef] [PubMed]
68. Asharaja, A.; Sahayaraj, K. Screening of insecticidal activity of brown macroalgal extracts against *Dysdercus cingulatus* (Fab.) (Hemiptera: Pyrrhocoridae). *J. Biopest.* **2013**, *6*, 193–203. [CrossRef]
69. Thawfeeq, A.; Srinivasan, G.; Shanthi, M.; Mini, M.L. Insecticidal activity of brown and red seaweed extracts against cowpea bean aphid, *Aphis craccivora* Koch. *J. Pharm Innov.* **2022**, *11*, 4223–4225.
70. Thawfeeq, A.J.; Shanthi, M.; Mini, M.L.; Srinivasan, G. Multifaceted effects of seaweed extracts against cowpea aphid, *Aphis craccivora* Koch, by evaluating four macroalgae. *J. Appl Phycol.* **2023**, *35*, 1–10.
71. Rizvi, M.A.; Shameela, M. Biological activity and Elementology of Benthic Algae from Karachi coast. *Pak. J. Bot.* **2003**, *35*, 717–729.
72. Argadoña, V.; Del Pozo, T.; San-Martín, A.; Rovirosa, J. Insecticidal activity of *Plocamium cartilagineum* monoterpenes. *J. Chil. Chem. Soc.* **2000**, *45*, 371–376.

agriculture

MDPI

Article

The Effect of Chosen Biostimulants on the Yield of White Cabbage

Edward Kunicki [1], Ewa Capecka [1] and Elżbieta Wojciechowicz-Żytko [2,*]

[1] Department of Horticulture, University of Agriculture in Krakow, al. 29-Listopada 54, 31-425 Cracow, Poland; edward.kunicki@urk.edu.pl (E.K.); ewa.capecka@urk.edu.pl (E.C.)
[2] Department of Botany, Physiology and Plant Protection, University of Agriculture in Krakow, al. 29-Listopada 54, 31-425 Cracow, Poland
* Correspondence: elzbieta.wojciechowicz-zytko@urk.edu.pl

Abstract: An open-field study evaluated the effects of four biostimulants (Asahi, Optysil, Optycal, and Tytanit) on the yield and chemical composition of two white cabbage cultivars ('Caraflex' and 'Alfredo'). Although the biostimulants did not significantly impact the marketable yield, all treatments led to a significant increase in ascorbic acid content. The influence on dry matter, sugars, phenols, and antioxidant activity varied by cultivar but generally did not result in inferior outcomes compared to the control. The effect on nitrate levels also varied, with Optysil and Asahi showing some adverse effects depending on the cultivar.

Keywords: *Brassica oleracea* var. *capitata* f. *alba*; dry matter; ascorbic acid; soluble sugars; phenolic compounds; nitrates

1. Introduction

White cabbage is one of the six main vegetables cultivated in Poland, with an 18% share in the production of vegetables grown in the field, which makes it a crop of high economic value in that country [1]. The yield and its quality depend largely on environmental factors and agricultural technology, but currently relying only on the latter does not allow for achieving the best results. To keep cabbage production economically profitable, there is an urgent need for other solutions which could improve yields. One of them is using preparations, especially biostimulants, enhancing the resistance of plants to unfavourable environmental factors. Being natural and environmental friendly substances, such preparations currently play an important role in horticultural practise [2,3]. According to the European Biostimulants Industry Council (EBIC) [4], the largest market for biostimulants in 2012 was Europe, where they were applied on over 6.2 million hectares. The definition and concept of plant biostimulants are constantly being refined, but generally they are considered to be substances containing organic and mineral compounds or their mixtures whose function, when they are applied to plants or the rhizosphere, is to stimulate natural processes to enhance nutrient uptake and efficiency, tolerance to biotic and abiotic stresses, and crop quality [5–8]. Biostimulants may include hormones, enzymes, proteins, amino acids, vitamins, trace elements, and other biologically active compounds [5,9,10]. Although biostimulants contain nutrients, they work differently from fertilizers [11–14]. Their influence on plants does not result from their ability to directly regulate metabolism and their action can be multidirectional. The crucial feature distinguishing biostimulants from bioregulators and hormones is that biostimulants improve plant metabolic processes without changing their natural pathway [15].

Biostimulating formulations available commercially make use of a wide range of diverse active compounds. Preparations such as those containing Ti, Si, Ca, or nitrophenols are known to promote good effects in plant production.

Titanium activates metabolic processes during the growth and development of plants and boosts the photosynthesis and absorption of nutrients [16]. Many studies have demonstrated a positive effect of titanium application in the form of, e.g., sprays, on plant crops. Titanium was found to enhance pollen grain vigour, increase the rate of nutrient uptake, and to improve the health condition of plants [17]. Dobromilska [18] showed that the application of Tytanit spray improved the yield of tomato plants.

Silicon is an element which can enhance plant tolerance to biotic and abiotic stresses by making plants rigid and strong. It can be introduced to plants in root or foliar application using different sources, i.e., sodium silicate, calcium silicate, calcium silicate. The fortification of plants with this element has improved their resistance to diseases, insect attacks, and unfavourable climatic conditions [19–25].

Cabbage is a crop with very high requirements for calcium. The sufficient supply of this element is a necessary precondition for increasing cabbage yield [26]. Calcium also plays a crucial role in preventing tipburn, a frequent disorder of cabbage heads caused by Ca deficiency in the youngest parts of plants [27,28]. Thus, the use of preparations containing calcium activators is an option allowing for the obtaining of a high quality yield of cabbage crops. Calcium can be applied to plants in different forms, with prohexadione-calcium being of particular relevance [29].

Phenolic compounds, as effective antioxidants at low concentrations, influence various biochemical and physiological processes in plants. Components of Asahi SL, a commercially available mixture of sodium *ortho*- and *para*-nitrophenolates with sodium 5-nitroguaiacolate, are easily metabolized by plants into other phenolics which are involved in mitochondrial energy-generating processes [15]. Applied exogenously, they strengthen the cell walls of plants, and increase the resistance of plants to temperature and water stress, mechanical damage, and disease infections. They have been shown to have a positive effect on limiting the severity of diseases caused by *Alternaria* [30]. In addition, they play an important role in the metabolism of nitrogen in plants and in cytoplasmatic transport, contributing to increased plant productivity [31].

The aim of this study was to determine the effect of selected biostimulants applied in the open-field cultivation of two early cultivars of white-headed cabbage ('Caraflex' and 'Alfredo') on the yield and chemical composition of the crop.

2. Material and Methods

2.1. Research Site

The open-field experiment on the cultivation of white cabbage was carried out in 2016–2017, with harvest planned at the end of summer. The experimental plots were located in the Vegetable Experimental Station in Kraków-Mydlniki (50°08′ N, 19°85′ E). The soil type is Fluvic Cambisoil (Humic), according to the classification of the Food and Agriculture Organization of the United Nations [32], containing 3.4% organic matter, with pH = 6.9.

2.2. Material

Two white cabbage cultivars (hybrids F_1) from Bejo Zaden B.V. were tested: 'Alfredo', which forms spherical heads weighing 2–4 k and whose vegetation period is about 90 days, and 'Caraflex', with spherical-conical heads weighing 1–2 kg and a vegetation period of about 80 days.

Four biostimulants were applied: Asahi, Optysil, Optycal, and Tytanit.

Asahi SL (ARYSTA LifeScience, Warsaw, Poland) is composed of three phenolic compounds: sodium *para*-nitrophenolate PNP (0.3%), sodium *ortho*-nitrophenolate ONP (0.2%), sodium 5-nitroguaiacolate 5NG (0.1%), and water.

Optycal (INTERMAG Company, Oklusz, Poland) contains 350 g CaO per 1 kg of preparation.

Optysil (INTERMAG Company, Olkusz, Poland) contains 200 g SiO_2 in 1 dm^3 of solution.

Tytanit (INTERMAG Company, Olkusz, Poland) contains 8.5 g of Ti in 1 dm^3 of solution.

The control plants were plants not treated with biostimulants.

2.3. Methods

Transplants were produced in multipots containing 96 cells, each with a volume of 0.053 dm^3. They were planted out at a spacing of 75 × 40 cm on plots 6 m^2 containing 20 plants. The experiment was designed as a randomized block with four replications. The agronomic treatments were applied in accordance with the requirements of the species [33]. The biostimulants were applied four times in a foliar spraying with the addition of Faster adjuvant (containing sodium salts of alkyl derivatives of C10–C13 benzenesulphonic acids). The doses of biostimulants used were as follows: Asahi and Optysil 0.5 dm^3 ha^{-1}, Optycal 1.5 kg ha^{-1}, Tytanit 0.3 dm^3 ha^{-1}. The trial schedule is presented in Table 1.

Table 1. The course of experiments in 2016 and 2017.

Operation	2016	2017
Seed sowing	27 May	22 May
Transplant planting-out	26 June	27 June
Application of biostimulants	8 and 25 July; 1 and 8 August	13, 20, and 27 July; 4 August
Harvest		
'Caraflex'	8 September	4 September
'Alfredo'	19 September	20 September

The results of the treatments were determined on the basis of obtained yields of marketable heads of the two cabbage cultivars, their dry matter and the content of sugars, vitamin C, total phenolics, nitrates, and DPPH scavenging ability.

2.4. Chemical Analyses

The analyses were carried out according to standard methods.

Dry matter was estimated after drying fresh plant material at 105 °C to constant weight.

Soluble sugar content was evaluated using the anthrone colorimetric method. The optical density of ethanol extracts after reaction with anthrone reagent was measured at 625 nm on a Helios Beta spectrophotometer (Thermo Fisher Scientific Inc., Carlsbad, CA, USA). The content of soluble sugars was calculated using the calibration curve of glucose as standard.

The content of L-ascorbic acid was assessed using the Tillman's method. Plant material was mixed with CH_3COOH, extracted for 30 min, and then titrated with 2.6-dichlorophenolindophenol (Tillman's reagent, Merck Group, Wien, Austria).

Total phenolic content was determined in 80% methanol extracts of plant material by the photometric method with the Folin–Ciocalteu reagent, using gallic acid as standard.

The radical scavenging activity (RSA) of 80% methanol extracts of plant material was tested with DPPH as a stable free radical. The decrease in absorbance was monitored at 517 nm using a UV-VIS spectrophotometer (Thermo Electron Corporation, Cambridge, UK). The RSA of plant extracts was expressed as the percentage of stable free radical DPPH (2,2-diphenyl-1-picrylhydrazyl) neutralization after 15 min from the start of the reaction.

Nitrate content was determined calorimetrically by the method of Cataldo et al. [34].

2.5. Meteorological Data

There was less rainfall in each month of the growing season (May–August) of 2016 than in the corresponding period of 2017 (Figure 1). In 2017, rain events were less frequent but more intense, while 2016 had more regular rainfall. The highest air temperatures in 2016 were noted in June and July, as opposed to in 2017—in May and July.

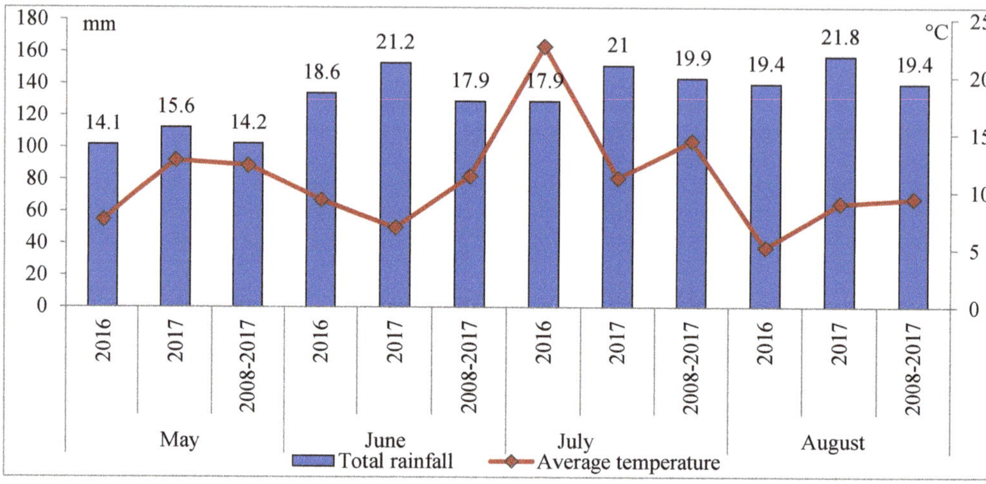

Figure 1. Rainfall and air temperature in 2016–2017 in Mydlniki. The data source is a meteorological station located on the research site.

2.6. Statistical Analysis

Statistical analysis was performed by a two-way ANOVA and Tukey *t*-test at $p \geq 0.05$ using Statistica 13.3 (StatSoft). The first factor was the type of biostimulant and the second was the cabbage cultivar. All results were expressed as the means from two observation years. Homogenous groups were marked by the same letters.

3. Results

The obtained yield ranged from 40.98 to 56.05 t ha^{-1} (Table 2). In both years of the study, only the cultivar had significant influence on the marketable yield of cabbage—the yield of 'Alfredo' was 20% higher than that of 'Caraflex'. The effect of biostimulants was statistically insignificant; however, the marketable yield of 'Caraflex' was observed to be higher after the application of Asahi and Optysil compared to the control plants.

Table 2. The effect of cultivar and biostimulant on the marketable yield of cabbage (t ha^{-1}).

Cultivar	Biostimulant					Mean
	Control	Asahi	Optycal	Optysil	Tytanit	
'Alfredo'	56.05 b	53.95 ab	53.11 ab	53.35 ab	48.76 ab	53.04 B
'Caraflex'	43.21 ab	47.17 ab	42.67 ab	47.01 ab	40.98 a	44.21 A
Mean	49.63 A	50.56 A	47.89 A	50.18 A	44.87 A	-

Values marked with the same letters are not significantly different at $p < 0.05$. Lowercase letters indicate the significance of the interaction between the factors (biostimulant and cultivar), while uppercase letters indicate the significance of the biostimulant (in the last row) or cultivar (in the last column).

Independently of the cultivar, less dry matter was observed in plants sprayed with Tytanit than in those treated with Asahi (Table 3). The differences in dry matter between the control and treated plants were not always statistically significant. The 'Alfredo' plants, which generally produce less dry matter than 'Caraflex', had the lowest share of dry matter (6.84%) when treated with Tytanit, while the highest share was observed for the 'Caraflex' plants treated with Asahi (7.62%).

Table 3. The effect of cultivar and biostimulant on the dry matter in cabbage heads (% f.m.).

Cultivar	Biostimulant					Mean
	Control	Asahi	Optycal	Optysil	Tytanit	
'Alfredo'	7.29 ab	7.52 ab	7.16 ab	7.16 ab	6.84 a	7.19 A
'Caraflex'	7.52 ab	7.62 b	7.35 ab	7.56 b	7.20 ab	7.45 B
Mean	7.40 AB	7.57 B	7.25 AB	7.36 AB	7.02 A	-

Values marked with the same letters are not significantly different at $p < 0.05$. Lowercase letters indicate the significance of the interaction between the factors (biostimulant and cultivar), while uppercase letters indicate the significance of the biostimulant (in the last row) or cultivar (in the last column).

Similarly to the case of dry matter, the content of soluble sugars (Table 4) in the heads of 'Alfredo' was lower (by 12%) compared to 'Caraflex'. A positive impact of Asahi and Optysil application (an increase in sugar content as compared to the control) was noted only for 'Alfredo'. The use of Tytanit and Optycal had no effect on sugar content in "Alfredo', while for 'Caraflex' this was the case for every biostimulant tested.

Table 4. The effect of cultivar and biostimulant on the content of soluble sugars in cabbage heads (% f.m.).

Cultivar	Biostimulant					Mean
	Control	Asahi	Optycal	Optysil	Tytanit	
'Alfredo'	2.90 abc	3.02 d	2.84 ab	2.99 de	2.74 a	2.90 A
'Caraflex'	3.34 e	3.29 e	3.20 de	3.13 de	3.24 de	3.24 B
Mean	3.12 AB	3.15 B	3.02 AB	3.06 AB	2.99 A	-

Values marked with the same letters are not significantly different at $p < 0.05$. Lowercase letters indicate the significance of the interaction between the factors (biostimulant and cultivar), while uppercase letters indicate the significance of the biostimulant (in the last row) or cultivar (in the last column).

Heads of 'Caraflex' contained 8% more ascorbic acid than heads of 'Alfredo' (Table 5). Regardless of the cultivar, all biostimulants caused an increase in ascorbic acid content in comparison with the control. However, there were differences between the cultivars when it comes to the response to the biostimulants, as reflected in the level of ascorbic acid. For 'Caraflex', the increase in the content of this compound was noted after every treatment, especially with Optysil. By contrast, the response of 'Alfredo' to this biostimulant was the weakest, and a significant increase as compared to the control was brought about only by Asahi and Optycal.

Table 5. The effect of cultivar and biostimulant on the content of ascorbic acid in cabbage heads (mg 100 g^{-1} f.m.).

Cultivar	Biostimulant					Mean
	Control	Asahi	Optycal	Optysil	Tytanit	
'Alfredo'	40.39 a	42.73 bc	43.73 cd	40.85 ab	41.53 ab	41.85 A
'Caraflex'	45.86 d	49.61 e	49.51 e	52.83 f	48.58 e	49.28 B
Mean	43.13 A	46.17 BC	46.62 C	46.84 C	45.06 B	-

Values marked with the same letters are not significantly different at $p < 0.05$. Lowercase letters indicate the significance of the interaction between the factors (biostimulant and cultivar), while uppercase letters indicate the significance of the biostimulant (in the last row) or cultivar (in the last column).

'Caraflex' was also found to have significantly (49%) higher total phenolic compounds than 'Alfredo' (Table 6). With respect to the effect of biostimulants on the total phenolic content, clearly the best results were achieved by Optycal, Optysil, and Tytanit, regardless of the cultivar. The response of each cultivar to the treatments was nevertheless different. For 'Alfredo', none of the biostimulants contributed to a significant change in the phenolic content, whereas for 'Caraflex', Asahi caused the greatest increase.

Table 6. The effect of cultivars and biostimulant on total phenolic compounds in cabbage heads (mg g^{-1} f.m.).

Cultivar	Biostimulant					Mean
	Control	Asahi	Optycal	Optysil	Tytanit	
'Alfredo'	0.381 ab	0.344 a	0.437 b	0.371 ab	0.409 ab	0.388 A
'Caraflex'	0.529 c	0.605 de	0.595 cde	0.516 c	0.544 cd	0.578 B
Mean	0.455 A	0.474 A	0.516 B	0.493 AB	0.477 AB	-

Values marked with the same letters are not significantly different at $p < 0.05$. Lowercase letters indicate the significance of the interaction between the factors (biostimulant and cultivar), while uppercase letters indicate the significance of the biostimulant (in the last row) or cultivar (in the last column).

As shown in Table 7, 'Caraflex' exhibited 2.4-fold higher antioxidant activity than 'Alfredo'. Neither of the tested preparations affected this trait.

Table 7. The effect of biostimulant and cultivar on DPPH radical scavenging activity (%).

Cultivar	Biostimulant					Mean
	Control	Asahi	Optycal	Optysil	Tytanit	
'Alfredo'	2.86 a	2.03 a	3.05 a	2.09 a	2.24 a	2.45 A
'Caraflex'	6.18 b	5.55 b	6.26 b	6.26 b	5.28 b	5.91 B
Mean	4.52 AB	3.79 A	4.65 B	4.18 A	3.76 A	-

Values marked with the same letters are not significantly different at $p < 0.05$. Lowercase letters indicate the significance of the interaction between the factors (biostimulant and cultivar), while uppercase letters indicate the significance of the biostimulant (in the last row) or cultivar (in the last column).

The nitrate content of 'Caraflex' exceeded that of 'Alfredo' by 110.79 mg per kg f.m. (Table 8). The highest level of nitrates was noted after using Optysil (729.83 mg)—it was significantly higher than the level found in other combinations, where it ranged from 595.55 mg (Tytanit) to 641.40 mg (Optycal). Optysil most strongly affected the nitrate content in 'Alfredo'—it exceeded significantly that of the control, while the other biostimulants were observed to have either no effect or even to reduce the nitrate content in this cultivar. 'Caraflex' had the highest nitrate content after Asahi, well above the control, while the other biostimulants did not affect this trait.

Table 8. The effect of cultivar and biostimulant on the content of nitrates in cabbage heads (mg NO_3 kg^{-1} f. m.).

Cultivar	Biostimulant					Mean
	Control	Asahi	Optycal	Optysil	Tytanit	
'Alfredo'	610.65 bc	482.43 a	561.20 ab	733.70 de	561.20 ab	589.84 A
'Caraflex'	667.01 cd	758.71 e	721.59 de	725.96 de	629.89 bc	700.63 B
Mean	638.83 A	620.57 A	641.40 A	729.83 B	595.55 A	-

Values marked with the same letters are not significantly different at $p < 0.05$. Lowercase letters indicate the significance of the interaction between the factors (biostimulant and cultivar), while uppercase letters indicate the significance of the biostimulant (in the last row) or cultivar (in the last column).

4. Discussion

The present investigation on the application of four plant biostimulants containing nitrophenols, Ca, Si, and Ti, in the open-field cultivation of white cabbage aimed to determine whether such treatment contributed to yield improvement. The studies on the effect of biostimulants on crop production have so far produced inconsistent results. In the experiment described in this paper, the marketable yield of cabbage was not affected by the examined preparations, which agreed with the results obtained by Heckman [11], who tested two biostimulants (mixtures of humic acids and plant extracts) and found no positive effect on the yield of spring cabbage. On the other hand, Hassan and Obaid [35] obtained a higher yield of 'Galaxy' head cabbage sprayed with a silicon solution at a concentration of 3 mL dm^{-3}, and Kováčik et al. [36] observed that the use of Mg-Tytanit resulted in an increased yield of oilseed rape seeds and that crops treated with biostimulant three times gave higher yields than those treated twice. According to Hara and Sonoda [26], cabbage head yield increased with the increase in Ca, Mg, or S concentration in the culture solution from 0 to 100 ppm Ca. In turn, sprays of $CaCl_2$ or calcium chelate had no effect on cauliflower productivity [27,28], but in the case of Chinese cabbage the largest head weight was obtained when the plants were sprayed at first with Tytanit and after 4 days with Wapnovit, a preparation containing calcium [37].

The Tytanit preparation did not have any positive effect on the average potato tuber weight [38]. Neither was Tytanit found to significantly influence the yield of tested strawberry cultivars [17]. At the same time, however, there were numerous reports of the beneficial effects of biostimulants on plant growth. Czeczko and Mikos-Bielak [31] observed that spraying plants with Asahi at a concentration of 0.1% significantly increased the yield of leek, celeriac, and potato. Similarly, Dobrzański et al. [39] reported a positive impact of this preparation on the yield of onion. Asahi (0.1%) and Tytanit (0.05%) preparations were found to enhance the marketable yield of carrot roots by reducing the share of small and unmarketable roots in the total yield [40]. A study carried out on celery demonstrated a significant effect of foliar fertilization with various Tytanit concentrations. The highest yield was observed after a single spray with a 3.6% solution while double spraying resulted in insignificant changes in the yield [41]. The application of titanium at an annual dose of 960 g Ti·ha^{-1} caused a significant increase in the marketable yield of greenhouse tomato culture on rockwool [42]. Three- and fourfold spraying (at 10 day intervals) of the cherry tomato cultivar 'Picolino' with 0.03% Tytanit significantly increased the yield and limited the formation of non-commercial fruit [18]. Borkowski et al. [43] reported that tomatoes 'Admiro' treated with silicon (twice with 1% of liquid potassium silicate Silvit using 0.5 dm^3 solution per plant) gave a higher yield than the control, but more fruit with blossom-end rot was observed in the treated sample. According to Xu et al. [25], foliar spraying with biostimulants improved the tomato yield by increasing the uptake of nitrogen and potassium. In case of herbs, foliar application with 0.05% Tytanit increased the yield of thyme (*Thymus vulgaris*) by 16.0% and contributed to a reduction in the share of stalks in herb [44]. The growth stimulators (Asahi SL 0.1% and Tytanit 0.05%) applied in foliar spraying twice in the vegetation period contributed to a better growth of sweet basil plants and a significant (by 28–31% in comparison with the control) increase in herb yield [45].

The literature's reports also vary as to the chemical composition of crops obtained with the use of various substances supporting the growth and development of plants, and this applies to both nutritional and non-nutritional ingredients that together determine the quality of vegetables. In this experiment, all used preparations, especially Optysil, had a clearly positive influence only on the content of ascorbic acid in cabbage heads, which was confirmed by Hassan and Obaid [35], Silva et al. [46], and Aleyashe and Obaid [47], for white and Chinese cabbage. As for dry matter, sugars, phenolic content, and antioxidant activity, their impact was uncertain, frequently different depending on the cultivar, but in the majority of cases comparable to the control. Similarly, divergent results were obtained by other authors investigating the effect of biostimulants on the yield quality of vegetables or fodder plants. Truba and Sosnowski [48] found out that Tytanit changed the content of fibrous fractions in plant cell walls of *Medicago* and *Trifolium*. Wadas and Kalinowski [49] noted that Tytanit did not affect the content of dry matter and nitrogen, phosphorus, potassium, calcium, magnesium, or sulphur in potato tubers. Kleiber and Markiewicz [42] showed that tomato fruit treated with Tytanit in doses of 80 and 480 g Ti·ha^{-1} had the highest vitamin C content compared to the control and other Ti concentrations applied, but no significant effect of this preparation was identified for the content of dry matter and sugars in fruit, or active acidity. Tomatoes treated with the Asahi SL biostimulator were characterized by the highest content of dry matter, P, K, Mg, Na, and Zn. In addition, they were distinguished by the highest amount of bioactive components (vitamin C and polyphenols) and the highest antioxidant potential [50]. Czeczko and Mikos-Bielak [31] reported that Asahi increased the content of reducing sugars and phenolic compounds in the tested vegetables (potato, tomato, celery, leek), thus improving their antioxidant properties, but in most cases lowered the content of protein and vitamin C. In another study, foliar application of calcium in the form of Pro-Ca preparation improved total soluble sugar content in the heads of Chinese cabbage [29].

The biostimulants investigated in the present work affected the content of nitrates in the cabbage heads; for Optysil, the effect was unfavourable since its application led to their significant increase, especially in 'Alfredo'. On the other hand, according to D'Imperio et al. [51], in other brassica crops, namely tatsoi (*Brassica rapa* L., Tatsoi group) and mizuna (*Brassica rapa* L., Mizuna group), the nitrate content was reduced by Si treatments, which stands in contrast to the results obtained in the present experiment. Other studies on nitrogen content in plant crops also reported various effects of stimulant treatments. The application of Asahi SL at a concentration of 0.1% significantly reduced the total nitrogen levels in eggplant fruit (cv. 'Black Beauty') [52]. In the present investigation, the content of nitrates was higher in 'Caraflex' but lower in 'Alfredo' compared to the control after the application of Asahi. This preparation was also found to significantly reduce N (as % of air-dry weight) in sweet basil herb, while Tytanit did not cause the change [45].

5. Conclusions

- The constant need to improve the yield quantity and quality of cabbage, which is one of the most important vegetable crops, forces the search for new agrotechnics. The use of biostimulants can be helpful in this regard. These types of substances improve the condition of plants, are environmentally friendly, and therefore play an important role in agricultural practise. Many studies have been carried out, but the reaction of plants to these substances is not always the same. The use of biostimulants in plant production is expected to positively impact the yields and quality of crops. However, biostimulants may bring about no yield improvement or their effect may be both favourable and unfavourable. As a result of our research, the treatment of two cabbage cultivars with the investigated biostimulants (Asahi, Optycal, Optysil, and Tytanit) did not affect the yield in comparison with the control. In general, the yield of cabbage and the content of selected compounds in the cabbage heads were found to more strongly depend on the cultivar than on the use of biostimulants. A higher marketable yield was achieved from the cultivar 'Alfredo' than from 'Caraflex'. However, 'Caraflex' had a higher content of dry matter, soluble sugars, ascorbic acid, total phenolic compounds, and nitrates as well as showing a higher antioxidant activity. The biostimulants, in turn, were found to have a clearly positive effect only on the content of ascorbic acid. Their influence on the content of dry matter, soluble sugars, phenolic compounds, and on the antioxidant activity of cabbage heads was not clear. Optysil significantly increased the level of nitrates in the plants of both tested cultivars, while the other biostimulants either had no effect on this feature or every preparation affected it differently in each cultivar.
- Current research has confirmed the belief that the effect of biostimulants on cabbage yield exists, but that this is not positive in every case; therefore, further studies on other ways of biostimulant use, e.g., different doses, times of application, together with a wider range of tested cultivars, are necessary.

Author Contributions: All authors contributed to this work. E.K. designed and performed the experiments, analyzed the data, and wrote the paper. E.C. designed and performed the experiment, analyzed the data, and wrote the paper. E.W.-Ż. designed the experiment, analyzed the data, and wrote the paper. All authors have read and agreed to the published version of the manuscript.

Funding: This research was supported by the Ministry of Science and Higher Education of Poland as a part of a research subsidy to the University of Agriculture in Krakow.

Institutional Review Board Statement: Not applicable.

Data Availability Statement: The data presented in this study are available on request from the corresponding author.

Conflicts of Interest: The authors declare no conflicts of interest.

References

1. Statistical Yearbook of Agriculture. 2017. Available online: https://stat.gov.pl/obszary-tematyczne/roczniki-statystyczne/roczniki-statystyczne/rocznik-statystyczny-rolnictwa-2017,6,11.html (accessed on 15 June 2022).
2. Abd Alla, M.A.; El-Greadly, N.H.M.; Emam, Y.T. Effect of foliar spray with biostimulants on growth, head yield, phytohormones and nutrients of cabbage (*Brassica oleracea* var. capitata). *Egypt. J. Hort.* **2015**, *42*, 707–719.
3. Brown, P.; Saa, S. Biostimulants in agriculture. *Front. Plant Sci.* **2015**, *6*, 671. [CrossRef] [PubMed]
4. European Biostimulants Industry Council 2013. Economic Overview of the Biostimulants Sector in Europe, 17 April 2013. Available online: http://www.biostimulants.eu/wp-content/uploads/2013/04/Biostimulant_economics_17April2013.pdf (accessed on 15 June 2022).
5. Calvo, P.; Nelson, L.; Kloepper, J.W. Agricultural uses of plant biostimulants. *Plant Soil* **2014**, *383*, 3–41. [CrossRef]
6. Bulgari, R.; Cocetta, G.; Trivellini, A.; Vernieri, P.; Ferrante, A. Biostimulants and crop response: A review. *Biol. Agric. Hortic.* **2015**, *31*, 117. [CrossRef]
7. Pereira, R.V.; Filgueiras, C.C.; Dória, J.; Peñaflor, M.F.G.V.; Willett, D.S. The Effects of Biostimulants on Induced Plant Defense. *Front. Agron.* **2021**, *3*, 630596. [CrossRef]
8. Buga, N.; Petek, M. Use of Biostimulants to Alleviate Anoxic Stress in Waterlogged Cabbage (*Brassica oleracea* var. capitata)—A Review. *Agriculture* **2023**, *13*, 2223. [CrossRef]
9. Ojeda-Barrios, D.L.; Cruz-Alvarez, O.; Sánchez-Chavez, E.; Ciscomani-Larios, J.P. Effect of foliar application of zinc on annual productivity, foliar nutrients, bioactive compounds and oxidative metabolism in pecan. *Folia Hort.* **2023**, *35*, 179–192. [CrossRef]
10. EL-Bauome, H.; Doklega, S.; Saleh, S.; Mohamed, A.; Suliman, A.; Abd El-Hady, M. Effects of melatonin on lettuce plant growth, antioxidant enzymes and photosynthetic pigments under salinity stress conditions. *Folia Hort.* **2024**, *36*, 1–17. [CrossRef]
11. Heckman, J.R. Evaluating phosphorous fertilization and commercial biostimulants for producing cabbage. *HortTechnology* **1995**, *5*, 298–300. [CrossRef]
12. European Biostimulants Industry Council. 2012a; EBIC and Biostimulants in Brief. Available online: https://biostimulants.eu/ (accessed on 15 June 2022).
13. European Biostimulants Industry Council. 2012b; What Are Biostimulants? Available online: https://www.biostimulants.eu/about/what-arebiostimulants/ (accessed on 15 June 2022).
14. European Biostimulants Industry Council 2024. Plant Biostimulants Contribute to Europe's Agricultural Policy Objectives. Available online: https://biostimulants.eu/ (accessed on 22 July 2024).
15. Posmyk, M.M.; Szafrańska, K. Biostimulators: A new trend towards solving an old problem. *Front. Plant Sci.* **2016**, *7*, 1–6. [CrossRef]
16. Lyu, S.; Wei, X.; Chen, J.; Wang, C.; Wang, X.; Pan, D. Titanium as a beneficial element for crop production. *Front Plant Sci.* **2017**, *8*, 597. [CrossRef] [PubMed]
17. Michalski, P. The effect of Tytanit on the yield structure and the fruit size of strawberry 'Senga Sengana' and 'Elsanta'. *Annales UMCS Agricult.* **2008**, *63*, 109–118. [CrossRef]
18. Dobromilska, R. The influence of Tytanit application on the growth and yield of small-fruited tomato. *Rocz. AR Pozn.* **2007**, *41*, 451–454.
19. Fauteux, F.; Rémus-Borel, W.; Menzies, J.G.; Bélanger, R.R. Silicon and plant disease resistance against pathogenic fungi. *FEMS Microbiol. Lett.* **2005**, *249*, 1–6. [CrossRef] [PubMed]
20. Ali, J.G.; Agrawal, A.A. Specialist versus generalist insect herbivores and plant defense. *Trends Plant Sci.* **2012**, *17*, 293–302. [CrossRef]
21. Vasanthi, N.; Saleena, L.M.; Raj, S.A. Silicon in Day Today Life. *World Appl. Sci. J.* **2012**, *17*, 1425–1440.
22. Meena, V.D.; Dotaniya, M.L.; Coumar, V.; Rajendiran, A.S.; Kundu, S.; Subba Rao, A. A case for silicon fertilization to improve crop yields in tropical soils. *Proc. Natl. Acad. Sci. India* **2014**, *84*, 505–518. [CrossRef]
23. Weerahewa, D.; Somapala, K. Role of silicon on enhancing disease resistance in tropical fruits and vegetables: A review. *OUSL J.* **2016**, *11*, 135–162. [CrossRef]
24. Alhousari, F.; Greger, M. Silicon and Mechanisms of Plant Resistance to Insect Pests. *Plants* **2018**, *7*, 33. [CrossRef]
25. Xu, X.; Lei, X.; Liao, S.; Li, Y.; Sun, Y. Foliar application of potassium silicate, potassium fulvate and betaine improve summer-time tomato yield by promoting plant nitrogen and potassium uptake. *Folia Hort.* **2022**, *34*, 125–138. [CrossRef]
26. Hara, T.; Sonoda, Y. The role of macronutrients in cabbage head formation. II. Contribution to cabbage-head formation of calcium, magnesium or sulfur supplied at different growth stages. *Soil Sci. Plant Nutr.* **1981**, *27*, 45–54. [CrossRef]
27. Palzkill, D.A.; Tibbitts, T.W.; Williams, P.H. Enhancement of calcium transport to inner leaves of cabbage for prevention of tipburn. *H. Amer. Soc. Hort. Sci.* **1976**, *101*, 645–648. [CrossRef]
28. Rosen, C.J. Leaf tipburn in cauliflower as affected by cultivar, calcium sprays, and nitrogen nutrition. *HortScience* **1990**, *25*, 660–663. [CrossRef]

29. Kang, S.M.; Kimb, J.T.; Hamayun, M.; Hwang, I.C.; Khan, A.L.; Kim, Y.H.; Lee, J.H.; Lee, I.J. Influence of prohexadione-calcium on growth and gibberellins content of Chinese cabbage grown in alpine region of South Korea. *Sci. Hortic.* **2010**, *125*, 88–92. [CrossRef]
30. Sawicka, B.; Skiba, D. Influence of foliar nutrition on plant sanitary conditions in vegetation period of potato. *Annales UMCS Sec. E.* **2009**, *64*, 39–51. [CrossRef]
31. Czeczko, R.; Mikos-Bielak, M. Effects of Asahi bio-stimulator application in the cultivation of different vegetable species. *Annales UMCS Sec. E.* **2004**, *59*, 1073–1079.
32. FAO. *World Reference Base for Soil Resources*; FAO: Rome, Italy, 2014.
33. Wize, A. *Uprawa Kapusty Białej, Czerwonej i Włoskiej (Growing White, Red and Savoy Cabbages)*; Plantpress: Kraków, Poland, 2004.
34. Cataldo, D.A.; Haroon, M.; Schrader, L.E.; Youngs, V.L. Rapid colorimetric determination of nitrate in plant tissue by nitration of salicylic acid. *Commun. Soil Sci. Plant Anal.* **1975**, *6*, 71–80. [CrossRef]
35. Hassan, I.F.; Obaid, A. The effect of spraying with silicon and moringa leaf extract on some growth and yield indicators of the cabbage plant (*Brassica oleracea* var. capitata). *Euphrates J. Agric. Sci.* **2020**, *12*, 356–364.
36. Kováčik, P.; Šimanský, V.; Wierzbowska, J.; Renčo, M. Impact of foliar application of the biostimulator Mg-Tytanit on the formation of winter oilseed rape phytomass and titanium kontent. *J. Elem.* **2016**, *21*, 1235–1251.
37. Borkowski, J.; Kowalczyk, W.; Felczyńska, A.; Szczechura, W. Effect of spraying with Tytanit and Wapnowit on the yield and healthiness of Chinese cabbage. *Zeszyt. Nauk. Inst. Ogrod.* **2017**, *25*, 187–195.
38. Kalinowski, K.; Wadas, W. Effect of Tytanit® on the yield and yield components of very early-maturing potato cultivars. *J. Cent. Eur. Agric.* **2017**, *18*, 441–459. [CrossRef]
39. Dobrzański, A.; Anyszka, Z.; Pałczyński, J. Response of onion and carrot to Asahi SL biostimulator used with herbicides. In *Biostimulators in Modern Agriculture*; Dąbrowski, Z.T., Ed.; Editorial House Wieś Jutra, Limited Warsaw, 7–20; Vegetable Crops: Skierniewice, Poland, 2008.
40. Kwiatkowski, A.C.; Kołodziej, B.; Woźniak, A. Yield and quality parameters of carrot (*Daucus carota* L.) roots depending on growth stimulators and stubble crops. *Acta Sci. Pol. Hortorum Cultus* **2013**, *12*, 55–68.
41. Malinowska, E.; Kalembasa, S. The yield and content of Ti, Fe, Mn, Cu in celery leaves (*Apium graveolens* L. var. *dulce* Mill. Pers.) as a result of Tytanit application. *Acta Sci. Pol. Hortorum Cultus* **2012**, *11*, 69–80.
42. Kleiber, T.; Markiewicz, B. Application of 'Tytanit' in greenhouse tomato growing. *Acta Sci. Pol. Hortorum Cultus* **2013**, *12*, 117–126.
43. Borkowski, J.; Felczyńska, A.; Górecki, R. Effect of silicon fertilization on the growth, yield and healthiness of tomato. *Zesz. Nauk. Inst. Ogrod.* **2014**, *22*, 195–202.
44. Król, B. The effect of foliar fertilization with Tytanit and Ekolist in thyme culture. *Annales UMCS Sec. E* **2009**, *64*, 1–6. [CrossRef]
45. Kwiatkowski, A.C.; Juszczak, J. The response of sweet basil (*Ocimum basilicum* L.) to the application of growth stimulators and forecrops. *Acta Agrobot.* **2011**, *64*, 69–76. [CrossRef]
46. Silva, D.L.; Prado, R.M.; Tenesaca, L.F.T.; Silva, J.L.F.; Mattiuz, B.H. Silicon attenuates calcium deficiency by increasing ascorbic acid content, growth and quality of cabbage leaves. *Sci. Rep.* **2021**, *11*, 1770. [CrossRef]
47. Aleyashe, H.H.; Obaid, M.H. The effect of adding whey and spraying with silicone on the chemical of chinese cabbage plant *Brassica campestris* var. *pekinensis*. *IOP Conf. Ser. Earth Environ. Sci.* **2023**, *1259*, 012051. [CrossRef]
48. Truba, M.; Sosnowski, J. The Effect of Tytanit on Fibre Fraction Content in *Medicago x varia* T. Martyn and *Trifolium pratense* L. Cell Walls. *Agriculture* **2022**, *12*, 191. [CrossRef]
49. Wadas, W.; Kalinowski, K. Effect of Tytanit® on the dry matter and macroelement contents in potato tuber. *J. Cent. Eur. Agric.* **2018**, *19*, 557–570. [CrossRef]
50. Ambroszczyk, A.M.; Liwińska, E.; Bieżanowska-Kopeć, R. Variation of the nutritional and health-promoting value of tomato fruit depending on the growth stimulants used. In *The Role of Technological Processes in Shaping Food Quality*; Polskiego Towarzystwa Technologów Żywności Oddział Małopolski: Kraków, Poland, 2016; pp. 173–182.
51. D'Imperio, M.; Renna, M.; Cardinali, A.; Buttaro, D.; Santamaria, P.; Serio, F. Silicon biofortification of leafy vegetables and its bioaccessibility in the edible parts. *J. Sci. Food Agric.* **2016**, *96*, 751–756. [CrossRef] [PubMed]
52. Majkowska-Gadomska, J.; Wierzbicka, B. Effect of the biostimulator Asahi on the mineral content of eggplants (*Solanum melongena* L.) grown in an unheated plastic tunnel. *J. Elem.* **2013**, *18*, 269–276.

agriculture

MDPI

Review

Harnessing Beneficial Microbes for Drought Tolerance: A Review of Ecological and Agricultural Innovations

Grzegorz Mikiciuk [1], Tymoteusz Miller [2,3,*], Anna Kisiel [2], Danuta Cembrowska-Lech [4], Małgorzata Mikiciuk [5], Adrianna Łobodzińska [4,6] and Kamila Bokszczanin [7]

1 Department of Horticulture, Faculty of Environmental Management and Agriculture, West Pomeranian University of Technology in Szczecin, 71-434 Szczecin, Poland; grzegorz.mikiciuk@zut.edu.pl
2 Institute of Marine and Environmental Sciences, University of Szczecin, 71-415 Szczecin, Poland; anna.kisiel@usz.edu.pl
3 Faculty of Data Science and Information, INTI International University, Nilai 71800, Negeri Sembilan, Malaysia
4 Institute of Biology, University of Szczecin, 71-415 Szczecin, Poland; danuta.cembrowska-lech@usz.edu.pl (D.C.-L.); 2532@stud.usz.edu.pl (A.Ł.)
5 Department of Bioengineering, Faculty of Environmental Management and Agriculture, West Pomeranian University of Technology in Szczecin, 71-434 Szczecin, Poland; malgorzata.mikiciuk@zut.edu.pl
6 Doctoral School of the University of Szczecin, 71-412 Szczecin, Poland
7 Department of Pomology and Horticulture Economics, Institute of Horticultural Sciences SGGW, Nowoursynowska 159 Str., 02-787 Warsaw, Poland; kamila_bokszczanin@sggw.edu.pl
* Correspondence: tymoteusz.miller@usz.edu.pl

Abstract: Drought is an increasingly critical global challenge, significantly impacting agricultural productivity, food security, and ecosystem stability. As climate change intensifies the frequency and severity of drought events, innovative strategies are essential to enhance plant resilience and sustain agricultural systems. This review explores the vital role of beneficial microbes in conferring drought tolerance, focusing on Plant Growth-Promoting Rhizobacteria (PGPR), mycorrhizal fungi, endophytes, actinomycetes, and cyanobacteria. These microorganisms mitigate drought stress through diverse mechanisms, including osmotic adjustment, enhancement of root architecture, modulation of phytohormones, induction of antioxidant defenses, and regulation of stress-responsive gene expression. Ecological and agricultural innovations leveraging these beneficial microbes have demonstrated significant potential in bolstering drought resilience. Strategies such as soil microbiome engineering, bioaugmentation, and the integration of microbial synergies within pest management frameworks enhance ecosystem resilience and agricultural sustainability. Additionally, advancements in agricultural practices, including seed coating, soil amendments, the development of microbial consortia, and precision agriculture technologies, have validated the effectiveness and scalability of microbial interventions in diverse farming systems. Despite promising advancements, several challenges hinder the widespread adoption of microbial solutions. Environmental variability can affect microbial performance, necessitating the development of robust and adaptable strains. Scale-up and commercialization hurdles, economic constraints, and regulatory and safety considerations also pose significant barriers. Furthermore, the complex interactions between microbes, plants, and their environments require a deeper understanding to optimize microbial benefits consistently. Future research should focus on integrating cutting-edge technologies such as genomics, synthetic biology, and precision agriculture to refine and enhance microbial interventions. Collaborative efforts among academia, industry, and government are essential to bridge the gap between research and practical implementation. By addressing these challenges and harnessing microbial innovations, it is possible to develop resilient and sustainable agricultural systems capable of thriving in an increasingly water-scarce world.

Keywords: drought tolerance; beneficial microbes; plant growth-promoting rhizobacteria; mycorrhizal fungi; soil microbiome engineering; sustainable agriculture; climate change

1. Introduction

1.1. Background on Drought and Its Global Impact

Drought, a prolonged period of deficient precipitation relative to the statistical multi-year average for a region, represents one of the most pervasive and challenging environmental stressors affecting both natural ecosystems and human societies. Drought can be classified into various types based on its causative factors and manifestations, including meteorological drought, characterized by a significant decrease in precipitation; agricultural drought, which refers to the lack of moisture affecting crop production; hydrological drought, indicating reduced water levels in rivers; and socioeconomic drought, reflecting adverse impacts on communities and economies due to insufficient water resources [1–3].

The economic and ecological consequences of drought are profound and multifaceted. Economically, droughts can lead to substantial losses in agricultural productivity, increased food prices, and heightened costs associated with water scarcity management. Water-dependent industries such as agriculture, energy, and manufacturing often experience disrupted operations and financial strain during prolonged drought periods. Ecologically, droughts contribute to the degradation of habitats, loss of biodiversity, and diminished ecosystem services. Prolonged water shortages can lead to the decline of plant and animal populations, altering species composition and ecosystem dynamics [4–6].

In recent decades, the frequency and severity of drought events have escalated, a trend closely linked to anthropogenic climate change. Rising global temperatures exacerbate evaporation rates and alter precipitation patterns, increasing the likelihood of prolonged and intense droughts in many regions. Climate models project that drought-prone areas will continue to expand, with significant implications for water resource management, agricultural sustainability, and ecosystem resilience. Understanding these trends is crucial for developing effective strategies to mitigate the adverse effects of drought and enhance the resilience of both human and natural systems [7–10].

1.2. Importance of Drought Tolerance in Agriculture and Ecology

Drought tolerance is a critical attribute for both agricultural productivity and the sustainability of natural ecosystems. In the agricultural sector, the ability of crops to withstand periods of water scarcity directly influences food security. As global populations grow and climate variability increases, the demand for resilient agricultural systems becomes imperative to ensure consistent food supply and mitigate the risks of crop failure. Enhancing drought tolerance in crops can lead to more stable yields, reduced dependency on irrigation, and improved economic outcomes for farmers, particularly in regions susceptible to recurrent droughts [11–15].

Beyond agriculture, drought tolerance plays a vital role in maintaining the health and functionality of natural ecosystems. Natural ecosystems, including forests, grasslands, and wetlands, rely on adequate water availability to support diverse biological communities and ecosystem processes. Drought-induced stress can lead to reduced plant growth, increased susceptibility to pests and diseases, and even widespread mortality of key species. These changes not only diminish biodiversity but also disrupt ecosystem services such as carbon sequestration, water purification, and soil stabilization. Preserving the resilience of natural ecosystems against drought is essential for maintaining ecological balance and ensuring the provision of vital services that support both human well-being and environmental health [16–21].

1.3. Role of Microbes in Plant Health and Stress Tolerance

Microbes, encompassing a vast array of bacteria, fungi, and other microorganisms, play an integral role in plant health and stress tolerance. Plant–microbe interactions are fundamental to various physiological processes, including nutrient acquisition, growth regulation, and defense against pathogens. Beneficial microbes, often referred to as plant growth-promoting microorganisms (PGPMs), can enhance plant resilience to abiotic stresses such as drought by facilitating water uptake, modulating hormone levels, and improving soil structure. These microorganisms establish symbiotic relationships with plants, contributing to the host's ability to adapt and thrive under adverse environmental conditions [22–26].

Historically, the application of microbes in agriculture dates back to practices such as the use of nitrogen-fixing bacteria in legume cultivation and the incorporation of mycorrhizal fungi to enhance nutrient uptake. These early successes laid the groundwork for modern biotechnological approaches aimed at harnessing microbial potential to address contemporary agricultural challenges. Advances in microbiology and molecular biology have expanded our understanding of the complex interactions between plants and their associated microbial communities, revealing the intricate mechanisms through which microbes confer stress tolerance. This historical perspective underscores the enduring significance of microbial applications in agriculture and highlights the potential for innovative microbial-based strategies to mitigate the impacts of drought [27–31].

1.4. Objectives and Scope of Review

The primary objective of this review is to provide a comprehensive synthesis of current knowledge and advancements in the utilization of beneficial microbes to enhance drought tolerance in agricultural and ecological contexts. By examining both ecological and agricultural innovations, this article aims to elucidate the multifaceted roles that microorganisms play in mitigating the effects of drought and promoting sustainable water management practices. The review will explore the diverse types of beneficial microbes, the underlying mechanisms by which they confer drought resilience, and the practical applications and innovations that leverage microbial interactions to improve plant and ecosystem performance under water-limited conditions.

Key themes covered in this review include the classification and functional roles of various beneficial microbes, the biological and biochemical pathways involved in microbe-mediated drought tolerance, and the integration of microbial strategies into ecological and agricultural practices. Additionally, the review will address the challenges and limitations associated with deploying microbial solutions at scale, as well as future research directions and technological advancements that hold promise for enhancing the efficacy and reliability of microbial interventions. By consolidating insights from ecological and agricultural research, this review seeks to provide a holistic understanding of how beneficial microbes can be harnessed to achieve greater resilience against drought, thereby contributing to the broader goals of sustainable agriculture and environmental stewardship.

2. Literature Review Methodology

2.1. Overview of the Methodology

The literature review in this chapter provides a comprehensive examination of current research surrounding the use of beneficial microbes in promoting drought tolerance in agricultural and ecological contexts. The methodology employed in this review involves a structured and systematic approach to identify, select, and analyze relevant peer-reviewed articles, academic publications, and authoritative sources from various disciplines, including microbiology, agriculture, and environmental sciences.

2.2. Search Strategy

A systematic search was conducted across the following academic databases to gather relevant portions of the literature:

- PubMed: For articles focusing on the biological and ecological aspects of beneficial microbes and their role in drought tolerance.
- Web of Science: To access a broad range of multidisciplinary studies on agriculture, environmental sciences, and microbial ecology.
- Scopus: For comprehensive coverage of peer-reviewed literature in the fields of microbiology, agriculture, and biotechnology.
- Google Scholar: To supplement the search with the gray literature and access a wider array of scientific publications, including conference proceedings and dissertations.

The search terms used included combinations of key phrases such as "drought tolerance", "beneficial microbes", "plant growth-promoting rhizobacteria (PGPR)", "mycorrhizal fungi", "endophytes", "soil microbiome", and "sustainable agriculture". Boolean operators (AND, OR) were applied to refine the search. The search was limited to articles published between 2000 and 2024.

To ensure comprehensiveness, backward and forward citation tracking was employed on selected key studies to uncover additional relevant portions of the literature.

2.3. Inclusion and Exclusion Criteria

The following inclusion criteria were applied:

- Studies published in peer-reviewed journals from 2000 onwards.
- Research that focused on the use of beneficial microbes in agricultural and environmental drought tolerance.
- Both experimental studies and review articles were included.
- Only studies available in English were considered.

Exclusion criteria included the following:

- Articles that focused on non-agricultural or non-ecological contexts without a direct link to drought resilience.
- Studies that did not include original research data or lacked scientific rigor.

2.4. Data Extraction and Analysis

Key information was extracted from selected articles, including study objectives, methodology, microbial species studied, outcomes related to drought tolerance, and limitations. These data were then organized thematically to identify recurring trends, gaps in the literature, and areas of innovation in the field of microbial solutions to drought stress.

2.5. Quality Assessment

Each study was critically assessed for its methodological quality, including sample size, experimental design, statistical analysis, and the validity of conclusions drawn. Studies were ranked based on their contribution to advancing knowledge in microbial-assisted drought tolerance and were categorized according to their relevance to the review's objectives.

2.6. Synthesis of Findings

The literature was synthesized into several core themes:

- Mechanisms of microbe-mediated drought tolerance.
- Ecological innovations utilizing beneficial microbes.
- Agricultural applications and case studies demonstrating practical impacts.
- Challenges and future research directions.

The synthesis highlighted both the current advancements in microbial-based drought resilience strategies and the limitations that still need to be addressed in future research.

3. Types of Beneficial Microbes for Drought Tolerance

3.1. Plant Growth-Promoting Rhizobacteria (PGPR)

Plant Growth-Promoting Rhizobacteria (PGPR) constitute a diverse group of bacteria that colonize plant roots and confer numerous benefits to their host plants, particularly under stress conditions such as drought. Prominent examples of PGPR include genera such as *Pseudomonas, Bacillus, Rhizobium,* and *Azospirillum.* These bacteria are characterized by their ability to thrive in the rhizosphere—the narrow region of soil directly influenced by root secretions—and establish beneficial interactions with plants. PGPR enhance plant growth through various mechanisms, which are both direct and indirect. Direct mechanisms involve the synthesis of phytohormones like indole-3-acetic acid (IAA). Additionally, PGPR can facilitate nutrient acquisition by fixing atmospheric nitrogen, solubilizing phosphates, and mobilizing other essential minerals, thereby improving soil fertility and plant health [32–37] (Figure 1).

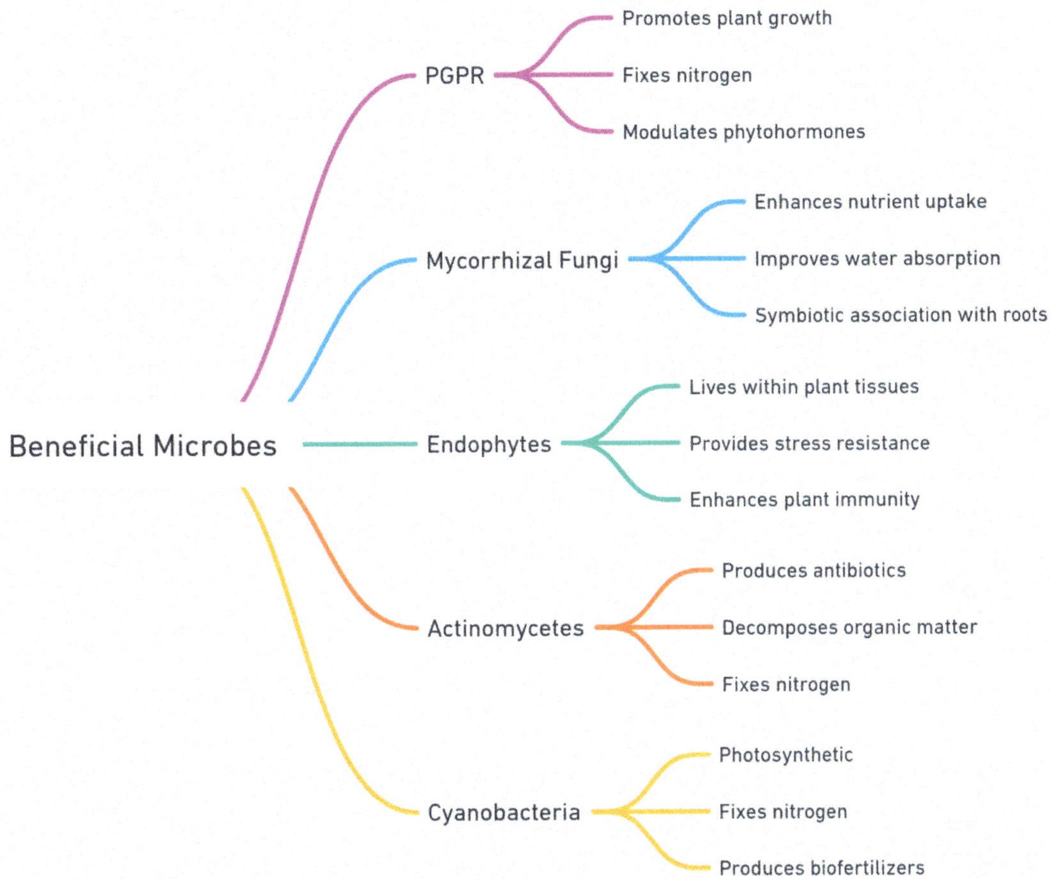

Figure 1. Mechanisms of microbe-mediated drought tolerance.

Indirect mechanisms by which PGPR confer drought tolerance include the production of exopolysaccharides, which improve soil structure and water retention, thereby reducing the impact of water stress. In addition, PGPR can induce systemic resistance in plants by triggering the expression of stress-responsive genes, enhancing the plant's innate ability to cope with drought-induced oxidative stress. PGPR also protect the plant from stress by reducing ethylene levels through the production of the enzyme ACC deaminase. These multifaceted interactions make PGPR a cornerstone in developing sustainable agricultural practices aimed at increasing crop drought resistance [33–36] (Table 1).

Table 1. Beneficial microbes and their characteristics.

Microbe Type	Examples	Key Characteristics	Role in Drought Tolerance
Plant Growth-Promoting Rhizobacteria (PGPR)	*Pseudomonas* spp., *Bacillus subtilis*, *Azospirillum brasilense* [38]	Nitrogen fixation, phosphate solubilization, root colonization, production of phytohormones [39]	Enhances root growth, improves nutrient uptake, induces osmotic adjustment, modulates hormone levels [40]
Mycorrhizal Fungi	*Glomus intraradices* (arbuscular mycorrhizae), *Rhizopogon* spp. (ectomycorrhizae) [41]	Symbiotic association with plant roots, extensive hyphal networks, nutrient exchange [42]	Increases water uptake capacity, improves soil structure, enhances nutrient availability, reduces plant water stress [43]
Endophytes	*Bacillus amyloliquefaciens*, *Fusarium* spp., *Trichoderma harzianum* [44]	Internal colonization of plant tissues, production of bioactive compounds, resistance to pathogens [45]	Produces osmolytes and antioxidants, enhances root architecture, stimulates plant defense mechanisms [46]
Actinomycetes	*Streptomyces* spp., *Micromonospora* spp. [47]	Filamentous growth, production of antibiotics and enzymes, decomposition of organic matter [48]	Improves soil fertility, enhances nutrient cycling, promotes plant growth under drought conditions [49]
Cyanobacteria	*Anabaena* spp., *Nostoc* spp., *Spirulina* spp. [50]	Photosynthetic, nitrogen-fixing, formation of biofilms, secretion of extracellular polysaccharides [51]	Enhances soil moisture retention, fixes atmospheric nitrogen, improves soil structure, provides organic matter [52]
Other Beneficial Microorganisms	*Trichoderma* spp., *Penicillium* spp., *Rhizobium* spp. [53]	Diverse metabolic capabilities, symbiotic relationships, production of growth-promoting substances [54]	Suppresses soil pathogens, enhances nutrient uptake, stimulates plant growth, improves stress resilience [55]

3.2. *Mycorrhizal Fungi*

Mycorrhizal fungi form symbiotic associations with the roots of most terrestrial plants, playing a pivotal role in enhancing plant water uptake and overall drought tolerance. The two primary types of mycorrhizae are arbuscular mycorrhizae (AM) and ectomycorrhizae (EM), each differing in their structural associations with plant roots and their functional contributions. Arbuscular mycorrhizal fungi, belonging to the phylum Glomeromycota, penetrate the cortical cells of plant roots, forming highly branched structures known as arbuscules. These structures facilitate the exchange of nutrients, where the fungi assist in the uptake of water and minerals, particularly phosphorus, in exchange for carbohydrates produced by the plant through photosynthesis. Ectomycorrhizal fungi, primarily associated with woody plants and trees, form a sheath around the roots and extend their hyphae into the surrounding soil, thereby increasing the surface area for water and nutrient absorption [56–60].

The benefits of mycorrhizal associations in drought conditions are manifold. Mycorrhizal fungi enhance the plant's ability to access water from a larger soil volume, effectively increasing the plant's water uptake capacity during periods of scarcity. They also improve soil structure by binding soil particles into aggregates, which enhances soil porosity and reduces water runoff and evaporation. Additionally, mycorrhizal fungi contribute to the plant's stress tolerance by modulating hormonal pathways and enhancing the plant's antioxidant defenses, thereby mitigating the adverse effects of drought-induced oxidative stress. The presence of mycorrhizal fungi has been shown to improve plant survival rates, biomass accumulation, and overall growth performance under drought conditions, underscoring their critical role in sustainable agriculture and ecosystem resilience [61–64].

3.3. Endophytes

Endophytic microorganisms, commonly referred to as endophytes, reside within the tissues of plants without causing any apparent harm. These microbes, which include bacteria and fungi, establish intimate associations with their host plants, conferring various benefits that enhance plant growth and stress resilience. Endophytes can colonize various plant parts, including roots, stems, leaves, and even seeds, forming a protective and symbiotic relationship that is particularly advantageous under drought conditions. By inhabiting the internal tissues of plants, endophytes can directly influence the plant's physiological processes, thereby enhancing its ability to withstand water deficit stress [45,65,66].

The role of endophytes in enhancing drought resilience is multifaceted. They can produce phytohormones such as gibberellins and cytokinins, which regulate plant growth and development, facilitating better root architecture and increased water uptake efficiency. Endophytes also contribute to osmotic adjustment by synthesizing osmoprotectants like proline and trehalose, which help maintain cellular water balance and protect cellular structures from dehydration. Furthermore, endophytic microbes can bolster the plant's antioxidant defense systems, reducing the accumulation of reactive oxygen species (ROS) that are typically elevated during drought stress. This antioxidative protection mitigates cellular damage and maintains cellular integrity, thereby enhancing the plant's overall drought tolerance. The ability of endophytes to modulate gene expression related to stress responses further underscores their potential as natural agents for improving plant resilience to drought [66–69].

3.4. Other Beneficial Microorganisms

Beyond PGPR, mycorrhizal fungi, and endophytes, several other groups of microorganisms contribute uniquely to drought tolerance in plants. Actinomycetes, a group of Gram-positive bacteria known for their filamentous growth, are renowned for their ability to produce a wide array of bioactive compounds, including antibiotics and enzymes that can enhance plant stress resilience. These microorganisms improve soil structure and fertility through the decomposition of organic matter, thereby enhancing water retention and availability to plants. Additionally, actinomycetes can form symbiotic relationships with plants, similar to PGPR, promoting root growth and nutrient uptake [47–49,70].

Cyanobacteria, also known as blue-green algae, are another group of beneficial microorganisms that play a significant role in drought tolerance. These photosynthetic bacteria can fix atmospheric nitrogen, thereby enriching soil fertility and reducing the need for chemical fertilizers. Cyanobacteria contribute to soil stability and water retention through the secretion of extracellular polysaccharides, which help in the formation of soil aggregates. Their ability to enhance soil moisture content makes them valuable allies in arid and semi-arid regions prone to drought [51,52,71,72].

Furthermore, certain species of fungi and bacteria exhibit unique capabilities such as the production of volatile organic compounds (VOCs) that can modulate plant growth and stress responses. These VOCs can induce systemic tolerance in plants, preparing them to better withstand drought conditions. Additionally, some microorganisms possess the ability to degrade soil pollutants, thereby improving soil health and creating a more conducive environment for plant growth under stress [73,74].

In summary, the diversity of beneficial microorganisms extends beyond the well-studied PGPR, mycorrhizal fungi, and endophytes. Actinomycetes, cyanobacteria, and other specialized microbes offer unique contributions to drought tolerance, each enhancing plant resilience through distinct mechanisms. The integration of these diverse microbial groups into agricultural and ecological practices holds immense potential for developing robust strategies to combat drought stress and ensure sustainable plant productivity in the face of increasing water scarcity.

4. Mechanisms of Microbe-Mediated Drought Tolerance

The ability of plants to withstand drought conditions is significantly enhanced by their interactions with beneficial microbes. These microorganisms employ a variety of sophisticated mechanisms to bolster plant resilience, ensuring sustained growth and productivity even under water-limited scenarios. The following sections delve into the primary mechanisms through which microbes mediate drought tolerance, encompassing osmotic adjustment, root system enhancement, phytohormone production, exopolysaccharides (EPSs), antioxidant defense induction, and the regulation of gene expression associated with stress responses [75–77] (Table 2) (Figure 2).

Table 2. Mechanisms of microbe-mediated drought tolerance.

Mechanism	Microbial Actions	Examples
Osmotic Adjustment	Production of osmolytes (proline, trehalose), secretion of extracellular polymeric substances (EPS) [78]	*Bacillus subtilis* producing trehalose and EPS, *Pseudomonas putida* producing proline [79]
Root Architecture Enhancement	Synthesis of auxins (IAA), promotion of root elongation and branching, stimulation of root hair development [80]	*Bacillus amyloliquefaciens* producing IAA, mycorrhizal fungi enhancing root surface area [81]
Phytohormone Production	Modulation of phytohormones (auxins, gibberellins, cytokinins, ethylene), production of ACC deaminase to reduce ethylene levels [82]	*Azospirillum brasilense* producing auxins, *Bacillus subtilis* producing ACC deaminase [83]
Antioxidant Defense Induction	Induction of antioxidant enzymes (SOD, CAT, POD), production of antioxidants (ascorbate, glutathione), secretion of antioxidant metabolites [84]	*Trichoderma harzianum* inducing SOD and CAT in plants, *Pseudomonas fluorescens* producing antioxidants [85]
Gene Expression Regulation	Activation of drought-responsive genes, epigenetic modifications, regulation of transcription factors [86]	*Bacillus subtilis* activating drought-responsive genes, mycorrhizal fungi inducing expression of stress-related genes [87]

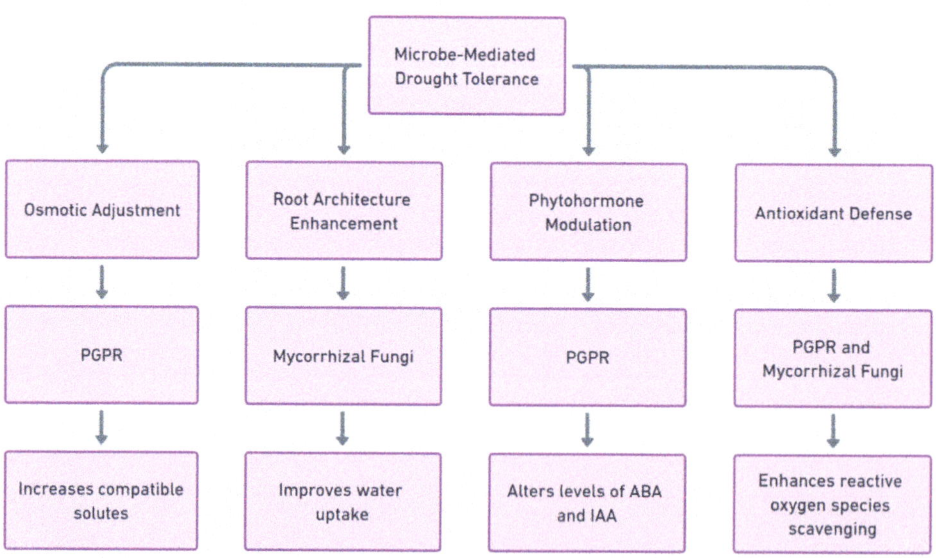

Figure 2. Ecological innovations in sustainable agriculture.

4.1. Osmotic Adjustment and Water Retention

One of the fundamental challenges plants face during drought is maintaining cellular hydration and osmotic balance. Beneficial microbes contribute to osmotic adjustment by facilitating the production of osmolytes—small organic molecules such as proline, glycine betaine, and trehalose. These osmolytes accumulate within plant cells, lowering the osmotic potential and enabling the retention of water despite external water scarcity. By synthesizing and regulating these compounds, microbes help plants maintain turgor pressure, which is essential for cellular functions and overall plant rigidity [75,79,88]. High proline content is involved in the protection of the cell membrane and the maintenance of the water status in cells during limited water supply (Ortiz et al., 2015). The use of *Pseudomonas putida* GAP-P45 strain improved proline accumulation in maize plants subjected to drought stress [89,90].

In addition to osmolyte production, microbes play a crucial role in improving soil structure, which directly influences water retention. Microbial activities, particularly those of PGPR and mycorrhizal fungi, lead to the formation of soil aggregates through the secretion of extracellular polymeric substances (EPSs) and the binding of soil particles. These aggregates enhance soil porosity and water-holding capacity, reducing water runoff and evaporation. Improved soil structure not only facilitates better water infiltration but also creates microenvironments that retain moisture around plant roots, ensuring a more consistent water supply during drought periods. Consequently, the synergistic effects of osmotic adjustment and enhanced soil structure collectively contribute to the plant's ability to endure prolonged drought stress [91–93]. Bacterial strains *Proteus penneri* Pp1, *P. aeruginosa* Pa2 and *Alcaligenes faecalis* AF3 can produce EPSs and maintain soil moisture, composition, biomass, root and shoot length, and leaf area in plants [77].

4.2. Enhancement of Root System Architecture

A robust and extensive root system is paramount for effective water absorption, particularly in drought-prone environments. Beneficial microbes significantly influence root system architecture by promoting root elongation and branching. PGPR, such as species of *Pseudomonas* and *Bacillus*, produce phytohormones like auxins, which stimulate root growth and encourage the formation of lateral roots. This increased root branching not only expands the root surface area but also enhances the plant's ability to explore a larger soil volume in search of water and nutrients [17,80,94,95].

Moreover, microbes contribute to the development of root hairs, which further amplify the root surface area and improve water uptake efficiency. Enhanced root surface area allows for more effective absorption of water from the soil, particularly from deeper layers that may retain moisture longer during drought conditions. Additionally, the improved root architecture facilitates better anchorage and stability of the plant, reducing the risk of lodging and promoting overall plant health. Through these modifications, beneficial microbes enable plants to optimize their root systems for maximum water acquisition, thereby enhancing their drought tolerance and ensuring sustained growth and productivity [91,96–98]. Bacterial inoculation of wheat plants improved lateral root formation and enhanced root growth, thereby enhancing water uptake under drought conditions [99]. Inoculation of maize plants with *Pseudomonas putida* improved leaf water potential, RWC, and plant biomass after exposure to drought stress [89].

4.3. Production of Phytohormones

Phytohormones are critical regulators of plant growth and stress responses, and beneficial microbes are adept at modulating their levels within plant tissues. Among the key phytohormones involved in drought tolerance are auxins, gibberellins, cytokinins, and ethylene. PGPR synthesize auxins, such as indole-3-acetic acid (IAA), which promote root growth and development, enhancing the plant's capacity to absorb water. Gibberellins produced by microbes stimulate stem elongation and leaf expansion, contributing to overall plant vigor and resilience [100–102].

Cytokinins, another class of phytohormones, are involved in cell division and differentiation, and their regulation by microbes can influence shoot growth and leaf senescence, thereby optimizing resource allocation during drought stress. Ethylene, a hormone associated with stress responses, is modulated by microbial activity through the production of the enzyme ACC deaminase. This enzyme breaks down 1-aminocyclopropane-1-carboxylate (ACC), the precursor to ethylene, thereby reducing ethylene levels in plants. Lower ethylene concentrations mitigate stress-induced growth inhibition and senescence, allowing plants to maintain growth and physiological functions under drought conditions [95,103–105]. The effect of ACC deaminase-producing rhizobacteria under drought stress was reported in wheat, maize, millet, rice, and tomato [106–108].

The intricate modulation of phytohormones by beneficial microbes ensures that plants can finely tune their growth and stress responses, enhancing their ability to cope with water deficits. By influencing hormone signaling pathways, microbes enable plants to maintain growth, optimize resource use, and activate stress mitigation mechanisms, thereby significantly contributing to drought tolerance [81,82,109–111].

4.4. Induction of Antioxidant Defense Systems

Drought stress often leads to the overproduction of reactive oxygen species (ROS) in plants, resulting in oxidative stress that can damage cellular components and impair plant function. Beneficial microbes play a pivotal role in inducing and enhancing a plant's antioxidant defense systems, thereby mitigating the detrimental effects of oxidative stress. Microbial interactions stimulate the production of antioxidant enzymes such as superoxide dismutase (SOD), catalase (CAT), and peroxidases (POD), which scavenge excess ROS and protect cellular structures from oxidative damage [84,98,112].

In addition to enzyme induction, microbes can also enhance the synthesis of non-enzymatic antioxidants, including ascorbate, glutathione, and tocopherols. These molecules work synergistically with antioxidant enzymes to neutralize ROS and maintain cellular redox balance. By bolstering the antioxidant defense mechanisms, beneficial microbes help plants maintain cellular integrity and function under drought conditions, reducing the impact of oxidative stress and promoting overall plant health and resilience [92,113–115].

Furthermore, certain microbes produce secondary metabolites with antioxidant properties, which can directly scavenge ROS or act as signaling molecules to trigger antioxidant responses in plants. This dual role of microbes in both enzyme induction and metabolite production ensures a comprehensive defense against oxidative stress, thereby enhancing the plant's capacity to survive and thrive during periods of water scarcity [75,93,116,117]. *Streptomyces* strains increased MDA, H_2O_2, and total sugar content along with APX activity, while decreasing CAT and GPX activities under stress conditions in tomato [117]. Tomato plants grafted with *B. subtilis* Rhizo SF 48 increased the antioxidant activity of SOD and APX enzymes [106].

4.5. Regulation of Gene Expression Related to Stress Response

The ability of plants to respond to drought stress at the molecular level is significantly influenced by beneficial microbes, which can modulate the expression of stress-responsive genes. Microbial interactions can activate or upregulate the expression of genes involved in various protective mechanisms, thereby enhancing a plant's intrinsic ability to cope with drought conditions. For instance, microbes can induce the expression of genes responsible for osmolyte biosynthesis, antioxidant enzyme production, and heat shock proteins, all of which play crucial roles in mitigating drought-induced damage [118–120].

Moreover, microbes can influence epigenetic modifications, such as DNA methylation and histone acetylation, which alter gene expression patterns without changing the underlying DNA sequence. These epigenetic changes can lead to the activation of drought-responsive pathways, enabling plants to better manage water stress. By regulating gene expression, beneficial microbes help plants fine-tune their physiological and biochemical responses to drought, ensuring a more effective and coordinated stress response [81,83,121–123].

Advances in genomics and transcriptomics have revealed the intricate networks through which microbes influence plant gene expression. Studies have shown that microbial inoculation can lead to differential expression of hundreds of genes related to stress tolerance, growth regulation, and metabolic processes. This comprehensive regulation at the genetic level underscores the profound impact that beneficial microbes have on enhancing plant resilience to drought. By orchestrating complex gene expression programs, microbes enable plants to adapt dynamically to fluctuating environmental conditions, thereby ensuring sustained growth and productivity even under adverse water availability [101,124–126].

In summary, the mechanisms through which beneficial microbes mediate drought tolerance are multifaceted and interdependent. From facilitating osmotic adjustment and improving soil structure to enhancing root architecture, modulating phytohormones, inducing antioxidant defenses, and regulating gene expression, these microorganisms provide a comprehensive suite of tools that empower plants to withstand and thrive in drought conditions. Understanding these mechanisms not only highlights the critical role of microbes in plant resilience but also paves the way for the development of innovative strategies to harness microbial potential for sustainable agriculture and ecosystem management in the face of increasing water scarcity.

5. Ecological Innovations Utilizing Beneficial Microbes

The integration of beneficial microbes into ecological management practices represents a frontier in enhancing drought resilience and sustaining ecosystem health. Ecological innovations that harness microbial communities not only mitigate the adverse effects of drought but also promote overall ecosystem stability and productivity. This section explores key ecological strategies, including soil microbiome engineering, bioaugmentation and inoculant applications, integrated pest management synergies, and sustainable agricultural practices that incorporate microbial benefits. Each subsection delves into the methodologies employed, the resulting impacts on ecosystem resilience, and illustrative case studies that highlight the effectiveness of these approaches [27,127–129] (Table 3).

Table 3. Ecological Innovations Leveraging Beneficial Microbes.

Innovation Type	Techniques/Methods	Impact on Ecosystem Resilience
Soil Microbiome Engineering	Metagenomic analysis, tailored microbial consortia introduction, CRISPR–Cas mediated microbial modifications [130]	Improved soil structure, enhanced nutrient cycling, increased microbial diversity, enhanced plant resilience [131].
Bioaugmentation	Introduction of specific beneficial microbial strains, application of liquid or granular inoculants, slow-release formulations [132]	Enhanced soil fertility, improved water retention, increased plant growth and drought tolerance, restoration of degraded soils [133]
Integrated Pest Management Synergies	Combining beneficial microbes with biological control agents, simultaneous application of PGPR and entomopathogenic fungi, integrated microbial and cultural practices [134]	Reduced pest outbreaks, enhanced plant defense mechanisms, improved crop health and yield, minimized reliance on chemical pesticides [135]
Sustainable Agricultural Practices	Conservation tillage with microbial amendments, crop rotation incorporating microbial inoculants, use of organic amendments (compost, biochar) to support microbial communities [136]	Enhanced soil moisture retention, improved soil health and structure, increased nutrient availability, sustained plant productivity under drought conditions [137]

5.1. Soil Microbiome Engineering

Soil microbiome engineering involves the deliberate manipulation of soil microbial communities to enhance their functional capabilities and promote ecosystem resilience under drought conditions. This innovative approach leverages advanced techniques in microbiology, molecular biology, and bioinformatics to alter the composition, diversity, and activity of soil microorganisms. Key techniques include the use of metagenomic analysis to identify beneficial microbial taxa, the application of microbial consortia tailored to specific environmental conditions, and the implementation of biotechnological tools such as CRISPR–Cas systems for precise genetic modifications of microbial strains [112,130,131,138].

One prominent method in soil microbiome engineering is the introduction of tailored microbial consortia that are designed to perform specific functions, such as nitrogen fixation, phosphorus solubilization, and the production of stress-alleviating compounds. By selecting and combining microbes with complementary traits, researchers can create synergistic communities that enhance soil fertility, improve water retention, and increase plant resilience to drought. Additionally, the use of bioinformatics and machine learning algorithms facilitates the prediction and optimization of microbial interactions, enabling the development of robust and adaptable soil microbiomes [56,138–140].

The impact of soil microbiome engineering on ecosystem resilience is profound. Enhanced microbial communities contribute to improved soil structure and nutrient cycling, which are critical for maintaining plant health and productivity during drought periods. Moreover, engineered microbiomes can increase the stability of soil ecosystems by promoting biodiversity and preventing the dominance of pathogenic or opportunistic species. This increased stability ensures that ecosystems are better equipped to withstand and recover from environmental stressors, thereby sustaining ecosystem services such as carbon sequestration, water filtration, and habitat provision [141–144].

Case studies illustrate the successful application of soil microbiome engineering in various agricultural and natural ecosystems. For instance, in arid regions of the Mediterranean, the introduction of drought-tolerant microbial consortia has led to significant improvements in soil moisture retention and crop yields. Similarly, in degraded lands subjected to prolonged drought, engineered microbiomes have facilitated the restoration of soil health and the re-establishment of native plant communities. These examples underscore the potential of soil microbiome engineering as a transformative tool for enhancing ecological resilience and ensuring sustainable land management in the face of increasing drought frequency and severity [130,131,144,145].

5.2. Bioaugmentation and Inoculant Applications

Bioaugmentation, the process of introducing beneficial microorganisms into an environment to enhance its functional capabilities, and inoculant applications, the strategic deployment of microbial agents to specific sites, are pivotal strategies in leveraging microbes for drought tolerance. These approaches involve the selection, cultivation, and application of microbial strains that possess traits conducive to improving plant resilience and soil health under water-limited conditions [146–148].

Methods of bioaugmentation typically begin with the isolation and characterization of native or exotic microbial strains that exhibit drought-alleviating properties. These strains are then cultured under controlled conditions to produce high-density inoculants, which can be applied to soil or directly to plant roots. Inoculant applications may take various forms, including liquid suspensions, powder formulations, or encapsulated agents, each tailored to ensure effective delivery and sustained microbial activity in the target environment. Advanced delivery systems, such as slow-release formulations and biofilm carriers, are often employed to enhance the persistence and efficacy of introduced microbes [149–152].

The success of bioaugmentation and inoculant applications is evidenced by numerous case studies across different climatic and agricultural settings. In the semi-arid regions of sub-Saharan Africa, the application of PGPR inoculants has resulted in increased maize and sorghum yields by enhancing root growth and improving water uptake efficiency. Similarly, in the drought-prone vineyards of California, the introduction of mycorrhizal fungi has not only improved grapevine health and productivity but also contributed to the sustainability of viticultural practices by reducing the reliance on irrigation. These instances demonstrate the tangible benefits of bioaugmentation and inoculant applications in fostering drought resilience and supporting agricultural sustainability [149,153,154].

Furthermore, bioaugmentation has been successfully employed in the restoration of degraded ecosystems subjected to severe drought stress. In the Australian outback, the application of drought-tolerant cyanobacteria has facilitated the establishment of vegetation in arid soils, promoting soil stabilization and preventing erosion. In another example, the introduction of actinomycetes into degraded grasslands has enhanced soil fertility and moisture retention, leading to the recovery of native plant species and the restoration of ecosystem functions. These success stories highlight the versatility and effectiveness of bioaugmentation and inoculant applications as ecological innovations for combating drought [155–157].

5.3. Integrated Pest Management and Microbe Synergies

Integrated Pest Management (IPM) is a holistic approach to controlling pests that combines biological, chemical, cultural, and mechanical strategies to achieve sustainable pest control with minimal environmental impact. The incorporation of beneficial microbes into IPM frameworks represents a synergistic strategy that enhances both drought tolerance and pest resistance in crops. By leveraging microbial interactions, IPM can address multiple stressors concurrently, promoting plant health and reducing the reliance on synthetic pesticides [158–160].

Beneficial microbes, such as PGPR, mycorrhizal fungi, and entomopathogenic fungi, play critical roles in IPM by enhancing plant defenses against pests and pathogens while simultaneously improving drought resilience. For example, PGPR can induce systemic resistance in plants, making them less susceptible to insect infestations and microbial infections. Mycorrhizal fungi enhance nutrient uptake and stress tolerance, which indirectly strengthens plant defenses against pests. Additionally, certain entomopathogenic fungi can directly target and suppress pest populations, reducing the need for chemical interventions [161–163].

Combining drought tolerance with pest resistance through microbial synergies offers several advantages. Plants that are more resilient to water stress are better equipped to withstand pest attacks, as healthy, well-nourished plants can allocate more resources to defense mechanisms. Moreover, the use of beneficial microbes can lead to the suppression of pest populations through competitive exclusion, antibiosis, and the induction of plant-mediated defenses. This multifaceted approach not only enhances plant health and productivity but also contributes to the sustainability of agricultural ecosystems by reducing chemical inputs and minimizing environmental pollution [140,164,165].

Case studies demonstrate the effectiveness of integrating beneficial microbes into IPM strategies. In rice cultivation, the application of PGPR has been shown to enhance both drought tolerance and resistance to the rice stem borer, a major pest species. Similarly, in tomato production, the use of mycorrhizal fungi in conjunction with biological control agents has resulted in improved plant vigor, reduced pest damage, and increased yields under water-limited conditions. These examples illustrate the potential of microbe-mediated synergies in IPM to provide comprehensive solutions for managing multiple agricultural challenges simultaneously [166–169].

5.4. Sustainable Agricultural Practices Incorporating Microbes

Sustainable agricultural practices that incorporate beneficial microbes are essential for promoting long-term soil health, enhancing crop resilience, and ensuring the sustainability of food production systems in the face of increasing drought stress. Practices such as conservation tillage, crop rotation, and the use of organic amendments provide conducive environments for beneficial microbial communities to thrive, thereby amplifying their positive effects on plant growth and drought tolerance [27,86,170].

Conservation tillage, which minimizes soil disturbance, plays a significant role in preserving soil structure and microbial habitats. By reducing plowing and other disruptive practices, conservation tillage maintains soil aggregates and promotes the establishment of stable microbial communities. This leads to enhanced water retention, improved nutrient cycling, and increased microbial activity, all of which contribute to greater plant resilience during drought periods. Additionally, conservation tillage practices reduce soil erosion and surface runoff, further conserving soil moisture and maintaining ecosystem integrity [170–172].

Crop rotation, the practice of alternating different crops in the same field across seasons, is another sustainable practice that benefits microbial communities and enhances drought tolerance. Diverse crop rotations introduce a variety of root exudates and organic residues into the soil, fostering a rich and dynamic microbial ecosystem. This diversity supports beneficial interactions, such as nitrogen fixation, phosphorus solubilization, and the suppression of soil-borne pathogens, which collectively enhance plant health and resilience to drought. Moreover, crop rotation breaks pest and disease cycles, reducing the need for chemical interventions and promoting a balanced and sustainable agricultural system [28,149,150,173].

The application of organic amendments, such as compost, manure, and biochar, further supports beneficial microbial populations by providing essential nutrients and improving soil physical properties. Organic amendments increase soil organic matter content, which enhances soil moisture retention and provides a steady supply of nutrients for both plants and microbes. The addition of biochar, a stable form of carbon, not only improves soil structure and water-holding capacity but also serves as a habitat for beneficial microbes, promoting their persistence and activity in the soil. These amendments create an environment that is conducive to microbial growth and function, thereby amplifying their beneficial effects on plant drought tolerance and overall soil health [152,154,170].

Integrating these sustainable agricultural practices with the application of beneficial microbes creates synergistic effects that enhance the resilience and productivity of agricultural systems. For example, in Mediterranean climates prone to extended dry periods, the combination of conservation tillage, crop rotation, and the application of PGPR has resulted in improved soil moisture retention, increased crop yields, and enhanced plant health. Similarly, in the semi-arid regions of India, the use of organic amendments in conjunction with mycorrhizal fungi has led to significant improvements in soil fertility, water use efficiency, and crop resilience to drought stress. These integrated approaches demonstrate the potential of combining sustainable practices with microbial innovations to achieve resilient and sustainable agriculture in the face of climate-induced water scarcity [28,167,174,175].

In conclusion, ecological innovations that utilize beneficial microbes offer comprehensive strategies for enhancing drought tolerance and sustaining ecosystem health. Through soil microbiome engineering, bioaugmentation and inoculant applications, integrated pest management synergies, and the incorporation of microbes into sustainable agricultural practices, these innovations provide multifaceted solutions to the challenges posed by drought. The successful implementation of these strategies not only improves plant resilience and agricultural productivity but also contributes to the broader goals of environmental sustainability and ecosystem resilience.

6. Agricultural Innovations and Applications

The application of beneficial microbes in agriculture has spurred a range of innovative practices aimed at enhancing crop resilience and productivity under drought conditions. These agricultural innovations encompass advanced techniques for microbial delivery, the development of synergistic microbial communities, the integration of precision technologies for monitoring and optimization, and the implementation of successful case studies that demonstrate the practical benefits of microbe-based drought mitigation. This section explores these innovations in detail, highlighting the technologies and methodologies employed, their efficacy, adoption rates, and the tangible outcomes achieved across various agricultural settings [112,176–178] (Table 4).

Table 4. Case studies of successful microbe-based drought mitigation.

Location	Crop Type	Microbial Strains Used	Interventions Applied	Outcomes Achieved
Sub-Saharan Africa [179]	Maize, Sorghum	*Azospirillum brasilense*, *Bacillus subtilis*	PGPR inoculation, microbial consortia application	Increased crop yields by 20%, improved water use efficiency by 15%, enhanced root development
Mediterranean Europe [174]	Grapevines	*Glomus intraradices*, *Trichoderma harzianum*	Mycorrhizal fungi introduction, bioaugmentation	Improved soil moisture retention, increased grape quality and yield, reduced soil erosion
Southeast Asia [180]	Rice	*Azospirillum brasilense*, *Rhizobium* spp.	PGPR inoculation, microbial consortia application	Enhanced water use efficiency, increased rice yields by 25%, improved nutrient uptake
California (USA) [92]	Grapevines	*Glomus intraradices*, *Bacillus subtilis*	Mycorrhizal fungi introduction, PGPR inoculation	Increased grapevine health and productivity, improved soil structure, reduced irrigation needs
Australia [181]	Native Vegetation	*Cyanobacteria* spp., *Actinomycetes* spp.	Cyanobacteria restoration, actinomycetes inoculation	Restoration of native plant communities, enhanced soil moisture retention, improved soil fertility
Southeast Asia [182]	Rice	*Trichoderma harzianum*, *Pseudomonas fluorescens*	Microbial consortia application, bioaugmentation	Increased grain yield by 30%, enhanced plant resilience to drought stress, improved soil health

6.1. Seed Coating and Soil Amendments with Beneficial Microbes

Seed coating and soil amendments represent two pivotal methods for introducing beneficial microbes into agricultural systems, facilitating enhanced plant growth and drought resilience from the earliest stages of plant development. Seed coating involves the application of microbial inoculants directly onto seeds before planting, ensuring that beneficial microorganisms are present from the outset of germination. This method utilizes a variety of technologies and materials to create effective coatings that protect the microbes and ensure their viability and efficacy. Common materials used for seed coatings include biopolymers, such as alginate and chitosan, which form protective matrices that encapsulate the microbes, providing a stable environment during seed storage and handling. Additionally, carriers like clay minerals, humic substances, and inert fillers are employed to enhance the adhesion and distribution of microbes on seed surfaces [183–185].

Soil amendments, on the other hand, involve the incorporation of microbial inoculants into the soil through various formulations, including liquid suspensions, granular formulations, and biochar-based carriers. These amendments are designed to improve soil microbial diversity and activity, thereby enhancing soil fertility, structure, and water retention capacity. Technologies such as slow-release formulations and encapsulation techniques are utilized to prolong the persistence and effectiveness of introduced microbes in the soil environment [97,180,186,187].

The efficacy of seed coating and soil amendment techniques has been demonstrated in numerous studies, showcasing significant improvements in plant growth parameters, yield stability, and drought tolerance. For instance, the application of PGPR-coated seeds has been shown to enhance root development, increase biomass accumulation, and improve water uptake efficiency, thereby enabling plants to better withstand periods of water scarcity.

Similarly, soil amendments with mycorrhizal fungi have led to enhanced soil moisture retention, reduced soil erosion, and increased nutrient availability, all of which contribute to improved plant resilience under drought conditions [183,188–190].

Adoption rates of these technologies vary across different agricultural sectors and regions, influenced by factors such as cost, accessibility, and farmer awareness. However, the growing recognition of the benefits of microbial applications in sustainable agriculture has led to increased uptake, particularly in regions prone to recurrent droughts and water scarcity. The development of user-friendly formulations and the establishment of support systems for farmers, including extension services and training programs, have further facilitated the widespread adoption of seed coating and soil amendment practices [189,191,192].

Overall, seed coating and soil amendments with beneficial microbes represent effective agricultural innovations that provide a foundation for enhancing crop resilience to drought. By ensuring the presence and activity of beneficial microorganisms from the earliest stages of plant growth, these techniques contribute to the establishment of robust and sustainable agricultural systems capable of withstanding the challenges posed by water-limited environments.

6.2. Microbial Consortia for Crop Improvement

The use of microbial consortia, comprising multiple beneficial microorganisms, represents a sophisticated approach to crop improvement that leverages the synergistic interactions among different microbial species to enhance plant growth and drought tolerance. Unlike single-strain inoculants, microbial consortia are designed to perform a range of complementary functions, thereby providing a more comprehensive and resilient solution to drought stress [191,193,194].

The synergistic effects of multiple microbes within a consortium are achieved through the combination of various functional traits, such as nutrient acquisition, hormone production, pathogen suppression, and stress alleviation. For example, a consortium might include PGPR that fix atmospheric nitrogen, mycorrhizal fungi that enhance phosphorus uptake, and endophytic bacteria that produce osmolytes and antioxidants. This multifaceted approach ensures that plants receive a broad spectrum of benefits, enhancing their overall resilience to drought conditions [94,195–197].

Formulating effective microbial consortia poses several challenges, including ensuring the compatibility and stability of different microbial species, maintaining their viability during storage and application, and optimizing their interactions within the soil–plant environment. Overcoming these challenges requires a thorough understanding of microbial ecology and the specific needs of target crops. Advanced formulation techniques, such as co-encapsulation and biofilm-based carriers, are employed to protect and support the diverse microbial populations within consortia. Additionally, the use of genetic and metabolic profiling tools allows for the selection of microbial strains that are not only beneficial individually but also complementary when combined, thereby enhancing the overall efficacy of the consortium [183,184,198,199].

Solutions to formulation challenges also involve the development of tailored delivery systems that ensure the targeted release and sustained activity of microbial consortia in soil. For instance, encapsulation in biodegradable polymers can protect microbes from environmental stresses and facilitate their gradual release, ensuring prolonged activity and interaction with plant roots. Furthermore, the incorporation of stabilizing agents, such as trehalose and other osmoprotectants, can enhance microbial survival during storage and application, thereby maintaining the integrity and functionality of the consortia [88,97,200,201].

The benefits of microbial consortia for crop improvement are evidenced by numerous studies that demonstrate enhanced plant growth, increased yield, and improved drought tolerance compared to single-strain inoculants. For example, in maize cultivation, the application of a microbial consortium comprising nitrogen-fixing bacteria, mycorrhizal fungi, and drought-tolerant PGPR has resulted in significant increases in root biomass, grain yield, and water use efficiency under drought conditions. Similarly, in wheat production, consortia that include endophytic bacteria and mycorrhizal fungi have been shown to enhance plant vigor, reduce water stress indicators, and improve overall crop performance during periods of limited water availability [179,185,192,202].

In conclusion, microbial consortia offer a powerful tool for crop improvement by harnessing the collective benefits of multiple beneficial microorganisms. Despite the challenges associated with their formulation and application, ongoing advancements in microbial ecology, biotechnology, and formulation technologies are paving the way for the widespread adoption of consortia-based approaches in agriculture. By providing a holistic and resilient solution to drought stress, microbial consortia hold significant promise for enhancing the sustainability and productivity of agricultural systems in the face of increasing water scarcity.

6.3. Precision Agriculture and Microbe Monitoring

The advent of precision agriculture has revolutionized the way farmers manage their crops, enabling the integration of advanced technologies to optimize agricultural practices and improve crop outcomes. In the context of microbial applications for drought tolerance, precision agriculture facilitates the precise monitoring and management of microbial activity, ensuring that beneficial microbes are effectively utilized to enhance plant resilience and productivity. The use of sensors, Internet of Things (IoT) devices, and data-driven approaches plays a crucial role in tracking microbial dynamics and optimizing their application in real-time [171,172,203,204].

Sensors and IoT technologies are employed to monitor various environmental and soil parameters that influence microbial activity and plant health. Soil moisture sensors, temperature probes, and pH sensors provide continuous data on soil conditions, enabling farmers to make informed decisions about irrigation, fertilization, and microbial inoculant applications. Additionally, microbial sensors, such as biosensors that detect specific microbial metabolites or genetic markers, can provide insights into the presence and activity of beneficial microbes in the soil–plant system. This real-time monitoring allows for the timely and targeted application of microbial inoculants, ensuring that they are deployed when and where they are most needed to mitigate drought stress [160,205–207].

Data-driven approaches, including machine learning algorithms and predictive analytics, are integral to optimizing microbial applications in precision agriculture. By analyzing large datasets generated by sensors and IoT devices, these approaches can identify patterns and correlations between environmental conditions, microbial activity, and plant performance. This information can be used to develop predictive models that forecast drought events, microbial responses, and crop outcomes, enabling proactive and adaptive management strategies. For example, machine learning models can predict the optimal timing and dosage of microbial inoculants based on soil moisture levels, temperature trends, and crop growth stages, thereby maximizing the efficacy of microbial applications and minimizing resource wastage [156,157,169,208].

Precision agriculture also facilitates the customization of microbial applications to specific field conditions and crop requirements. Spatial variability in soil properties, moisture levels, and microbial populations can be addressed through site-specific management practices that tailor microbial inoculants to the unique needs of different areas within a field. This targeted approach not only enhances the effectiveness of microbial interventions but also reduces the environmental impact by avoiding over-application and ensuring that microbes are utilized efficiently [153,155,208,209].

The integration of precision agriculture with microbial applications has demonstrated significant benefits in various agricultural settings. In drought-prone regions of Australia, precision irrigation systems combined with real-time microbial monitoring have led to improved water use efficiency, increased crop yields, and enhanced plant health. Similarly, in the rice paddies of Southeast Asia, the use of IoT-enabled sensors to track soil moisture and microbial activity has facilitated the precise application of mycorrhizal fungi, resulting in improved water retention, reduced irrigation needs, and increased rice productivity under water-limited conditions [149,150,153,210,211].

In summary, precision agriculture and microbe monitoring represent cutting-edge innovations that enhance the effectiveness and sustainability of microbial applications in agriculture. By leveraging advanced technologies and data-driven approaches, these practices enable the precise management of beneficial microbes, optimizing their contributions to drought tolerance and crop performance. The continued advancement and integration of precision agriculture technologies hold immense potential for transforming agricultural practices, ensuring food security, and promoting sustainable farming in an era of increasing water scarcity and climate variability.

6.4. Case Studies of Successful Microbe-Based Drought Mitigation

The practical implementation of microbe-based drought mitigation strategies has yielded numerous success stories across diverse agricultural regions and crop systems. These case studies provide tangible evidence of the benefits of leveraging beneficial microbes to enhance crop resilience and productivity under drought conditions. By examining detailed examples from various contexts, this subsection highlights the versatility and effectiveness of microbial interventions in real-world agricultural settings [152–154,212].

One notable example comes from the semi-arid regions of sub-Saharan Africa, where smallholder farmers have adopted the use of PGPR inoculants to improve maize and sorghum yields during drought periods. In these regions, the application of PGPR strains such as *Azospirillum brasilense* and *Bacillus subtilis* has been shown to enhance root growth, increase nutrient uptake, and improve water use efficiency. As a result, farmers have reported significant increases in crop yields, reduced reliance on irrigation, and enhanced soil fertility, contributing to greater food security and economic stability in drought-prone communities [147,213,214].

In the Mediterranean climate zones, vineyards have successfully implemented mycorrhizal inoculation to mitigate the impacts of prolonged dry spells. The introduction of arbuscular mycorrhizal fungi (*Glomus intraradices*) has led to improved grapevine health, increased water uptake, and enhanced resistance to soil erosion. This microbial intervention has not only improved grape yields and quality but also contributed to the sustainability of viticultural practices by reducing the need for chemical fertilizers and irrigation. The success of mycorrhizal inoculation in Mediterranean vineyards underscores the potential of microbial solutions to enhance the resilience and productivity of perennial crops in water-limited environments [134,138,139,215,216].

Another compelling case study is found in the rice paddies of India, where the application of a microbial consortium comprising nitrogen-fixing bacteria, mycorrhizal fungi, and drought-tolerant PGPR has significantly improved rice yields under water-stressed conditions. The consortium has facilitated enhanced nutrient uptake, improved soil moisture retention, and increased plant resilience to drought, leading to higher grain yields and greater stability in rice production. This integrated microbial approach has been particularly beneficial in regions where traditional irrigation practices are constrained by water scarcity, providing a sustainable solution to maintain rice productivity [129,130,217,218].

In the United States, research conducted in drought-prone vineyards of California has demonstrated the effectiveness of microbial inoculants in improving grapevine resilience and productivity. The application of mycorrhizal fungi in conjunction with PGPR resulted in increased grape yields, improved fruit quality, and enhanced plant health under water-limited conditions [219,220]. These findings have been instrumental in promoting the adoption of microbial-based practices among Californian viticulturists, contributing to the sustainability and economic viability of the wine industry in the face of increasing drought stress [127,128,221,222].

In the agricultural landscapes of Australia, the restoration of degraded lands subjected to severe drought has been successfully achieved through the application of drought-tolerant cyanobacteria and actinomycetes. These microorganisms have facilitated the establishment of vegetation in arid soils, improved soil structure and moisture retention, and promoted the recovery of native plant species. The restoration efforts have not only enhanced the ecological resilience of the regions but also provided valuable insights into the potential of microbial interventions for rehabilitating degraded ecosystems and promoting sustainable land management practices [86,103,223,224].

Evidence from Case Studies and Meta-Analyses

To strengthen the claims regarding the efficacy of microbial interventions in enhancing drought resilience, it is essential to present robust statistical data from empirical studies. The following case studies and meta-analyses provide quantitative evidence of the positive impact of beneficial microbes on plant drought tolerance.

Pinus edulis and Ectomycorrhizal Fungi

Gehring et al. (2017) [225] conducted a long-term study on *Pinus edulis* (pinyon pine) to examine the role of ectomycorrhizal fungi in drought tolerance. The study found that specific combinations of tree genotypes and fungal communities significantly improved drought resilience. Over a decade of drought conditions, drought-tolerant trees associated with particular ectomycorrhizal fungi exhibited 25% higher growth rates and had one-third the mortality rate compared to drought-intolerant trees. This indicates that selecting compatible plant–fungal partnerships can substantially enhance tree survival and growth under water-limited conditions.

Pseudomonas argentinensis in Alfalfa

Alwutayd et al. (2023) [226] investigated the effects of the root endophytic bacterium *Pseudomonas argentinensis* strain SA190 on alfalfa plants under drought stress. The inoculated alfalfa showed enhanced performance, including increased biomass and improved root architecture. Specifically, the treated plants exhibited a significant increase in root length and dry weight compared to non-inoculated controls. The bacterium modulated root development and gene expression through the abscisic acid (ABA) pathway, which is crucial for plant responses to drought. This study demonstrates the potential of root endophytes to improve drought tolerance in forage crops.

Populus Trees and Root Microbiome

Research by Xie et al. (2021) [227] on *Populus* trees revealed that drought stress alters the root microbial community composition. The study identified specific microbes, such as *Streptomyces rochei* and *Bacillus* species, that were enriched under drought conditions and associated with enhanced plant growth and drought tolerance. In controlled experiments, inoculation with these beneficial microbes led to a 20% increase in biomass and improved physiological parameters, including higher chlorophyll content and water use efficiency, compared to non-inoculated plants under drought stress.

Meta-Analyses

Microbial Mitigation of Drought Stress

Shaffique et al. (2022) [223] conducted a comprehensive review highlighting the role of various microbes in mitigating drought stress in plants. The meta-analysis demonstrated that microbial inoculation could lead to significant improvements in plant growth parameters under drought conditions. On average, inoculated plants showed a 30% increase in shoot biomass, a 25% increase in root biomass, and a 20% improvement in water use efficiency compared to non-inoculated controls. The mechanisms involved include osmotic adjustment, phytohormonal modulation, and enhanced antioxidant enzyme activities.

Rhizosphere Microbiota Management

Vidal et al. (2022) [175] emphasized the importance of managing rhizosphere microbiota to increase plant drought tolerance. Their review of multiple studies found that plants associated with beneficial rhizosphere microbes experienced an average yield increase of 15–40% under drought conditions. These microbes enhanced drought tolerance through plant growth-promoting characteristics such as nitrogen fixation, phosphate solubilization, and the production of phytohormones like auxins and cytokinins. The authors suggested that employing multi-omics approaches could further elucidate these interactions and improve the development of microbial interventions.

Integration of Quantitative Data

The quantitative data from these studies underscore the significant impact that beneficial microbes can have on plant drought tolerance. For instance, the 25% higher growth rates and reduced mortality in *Pinus edulis* demonstrate the long-term benefits of specific plant–microbe combinations [225]. Similarly, the increase in biomass and improved physiological parameters in *Populus* trees and alfalfa plants highlight the potential for microbial applications to enhance crop performance under drought stress [226,227].

These findings are further supported by meta-analyses showing substantial average increases in growth and yield parameters across various plant species and microbial treatments [175,223]. The statistical significance of these results strengthens the argument for integrating beneficial microbes into agricultural practices to combat drought-related challenges.

Incorporating robust statistical data from case studies and meta-analyses provides compelling evidence of the efficacy of microbial interventions in enhancing drought resilience. These studies offer quantifiable improvements in plant growth, yield, and physiological responses, reinforcing the potential of beneficial microbes as a viable strategy for sustainable agriculture under changing climatic conditions.

These case studies collectively illustrate the diverse applications and significant benefits of microbe-based drought mitigation strategies across different agricultural systems and climatic regions. They highlight the adaptability and effectiveness of beneficial microbes in enhancing crop resilience, improving soil health, and ensuring sustainable agricultural productivity under drought conditions. The successful implementation of these strategies underscores the critical role of microbial innovations in addressing the challenges posed by water scarcity and climate variability, paving the way for more resilient and sustainable agricultural practices globally.

7. Challenges and Limitations

While the utilization of beneficial microbes for drought tolerance presents significant promise, several challenges and limitations must be addressed to fully realize their potential in agricultural and ecological applications. These challenges span environmental variability, scale-up and commercialization hurdles, regulatory and safety considerations, and the intricate understanding of microbe–plant–environment interactions. Addressing these issues is crucial for the effective implementation and widespread adoption of microbial solutions aimed at enhancing drought resilience [98,109,228,229] (Table 5).

Table 5. Challenges and strategies for enhancing microbial efficacy.

Challenge	Description	Strategies to Overcome
Environmental Variability	The performance and efficacy of beneficial microbes are highly influenced by environmental factors such as soil type, temperature, moisture levels, pH, and the presence of native microbial communities. These variables can lead to inconsistent outcomes in microbial activity and plant drought tolerance across different regions and seasons [230].	- Selection and Engineering of Resilient Strains: Identify and develop microbial strains with inherent tolerance to a wide range of environmental conditions [231]. - Protective Formulations: Utilize encapsulation techniques and protective carriers to shield microbes from extreme conditions [232]. - Site-Specific Application: Tailor microbial inoculants to specific environmental contexts through localized testing and customization [233]. - Integration with Agronomic Practices: Combine microbial applications with practices like conservation tillage and organic amendments to create favorable soil environments [113].

Table 5. *Cont.*

Challenge	Description	Strategies to Overcome
Scale-Up and Commercialization	Transitioning microbial technologies from controlled laboratory settings to large-scale field applications involves challenges related to maintaining microbial viability, ensuring consistent performance, and managing production costs. Additionally, scaling up requires robust formulation and delivery systems that can withstand storage and transportation conditions [234].	- Cost-Effective Production Methods: Develop scalable and economically viable microbial cultivation and formulation processes [235]. - Robust Formulation Techniques: Implement advanced encapsulation and stabilization methods to preserve microbial viability during storage and transport [236]. - Quality Control and Standardization: Establish stringent quality control protocols to ensure consistency and reliability of microbial products [237] - Economic Incentives and Support: Provide financial incentives, subsidies, and support programs to encourage investment and adoption by farmers and agricultural businesses [204].
Regulatory and Safety Concerns	The deployment of beneficial microbes must comply with stringent regulatory standards to ensure safety for humans, non-target organisms, and the environment. Regulatory frameworks vary across regions, and navigating these can be complex and time-consuming. Additionally, there is a need to address potential non-target effects and ensure that microbial applications do not disrupt existing ecosystems [203].	- Comprehensive Risk Assessments: Conduct thorough evaluations of microbial strains to assess potential risks and non-target effects (Imathiu I in. 2020). - Adherence to Regulatory Standards: Ensure all microbial products meet the regulatory requirements of target markets through rigorous testing and documentation (Rossi I in. 2021). - Transparent Communication: Maintain transparency with stakeholders about the safety and benefits of microbial interventions to build trust and acceptance [195]. - Development of Best Practices: Establish standardized guidelines and protocols for the safe and effective use of beneficial microbes in agriculture [238].
Complex Microbe-Plant-Environment Interactions	The interactions between microbes, plants, and their environment are intricate and dynamic, making it challenging to predict and optimize microbial behavior and plant responses consistently. Factors such as plant genotype, microbial community composition, and fluctuating environmental conditions add layers of complexity that are not fully understood [83].	- Interdisciplinary Research Initiatives: Foster collaboration among microbiologists, plant scientists, ecologists, and data scientists to gain a holistic understanding of microbe–plant–environment interactions [194]. - Advanced Omics Technologies: Utilize genomics, transcriptomics, proteomics, and metabolomics to dissect the molecular mechanisms underlying microbe-mediated drought tolerance [239]. - Longitudinal Field Studies: Implement long-term studies to monitor microbial and plant dynamics across multiple growing seasons and environmental conditions [193]. - Modeling and Simulation:Develop predictive models using machine learning and bioinformatics to anticipate microbial behavior and optimize interventions based on environmental variables [191].

7.1. Environmental Variability and Microbe Efficacy

The efficacy of beneficial microbes in conferring drought tolerance to plants is highly dependent on the prevailing environmental conditions, which can vary significantly across different regions and seasons. Factors such as soil type, temperature, moisture levels, pH, and the presence of native microbial communities influence the performance and survival of introduced beneficial microbes. For instance, high soil temperatures or extreme pH levels can inhibit microbial activity and reduce their ability to colonize plant roots effectively. Additionally, fluctuations in soil moisture can impact the stability and persistence of microbial inoculants, leading to inconsistent outcomes in plant drought tolerance [180,181,240,241].

To enhance the reliability of microbial interventions under varying environmental conditions, several strategies can be employed. One approach is the selection and engineering of microbial strains that are inherently more resilient to environmental stresses. By identifying and utilizing microbes with traits such as heat tolerance, acid resistance, or desiccation resilience, researchers can develop inoculants that maintain their functionality under adverse conditions. Another strategy involves the use of protective formulations, such as encapsulation in biopolymers or the incorporation of osmoprotectants, which shield microbes from environmental extremes and enhance their survival rates. Additionally, site-specific application techniques that consider local soil and climate conditions can optimize the performance of microbial inoculants, ensuring that they are deployed in environments where they are most likely to succeed [88,242–244].

Furthermore, integrating microbial applications with other agronomic practices, such as conservation tillage and organic amendments, can create more favorable conditions for microbial activity and persistence. By fostering a conducive soil environment, these practices can support the establishment and maintenance of beneficial microbial communities, thereby enhancing the overall efficacy of microbial interventions. Continuous monitoring and adaptive management, facilitated by precision agriculture technologies, also play a vital role in addressing environmental variability by enabling real-time adjustments to microbial applications based on changing conditions [204,235,245,246].

7.2. Scale-Up and Commercialization Issues

Transitioning beneficial microbial technologies from laboratory settings to large-scale field applications presents a range of challenges. One of the primary hurdles is ensuring that microbial formulations developed under controlled conditions retain their efficacy and viability when applied in diverse and dynamic agricultural environments. Laboratory experiments often involve optimized conditions that may not accurately reflect the complexities of real-world fields, leading to discrepancies in microbial performance and plant responses [162,206,247].

Scaling up microbial applications requires the development of robust production and formulation processes that can produce high-quality inoculants at commercial volumes. This involves optimizing fermentation conditions, ensuring consistent microbial viability, and developing scalable encapsulation and delivery systems that protect microbes during storage and transportation. Additionally, maintaining the genetic and functional stability of microbial strains during large-scale production is essential to preserve their beneficial traits and ensure reliable performance in the field [156,158,161,248,249].

Economic considerations also play a critical role in the commercialization of microbial solutions. The costs associated with microbial production, formulation, and application can be significant, particularly for smallholder farmers and regions with limited financial resources. Developing cost-effective production methods, such as utilizing low-cost substrates and implementing energy-efficient processes, can help reduce the overall expenses and make microbial technologies more accessible. Furthermore, demonstrating the return on investment through comprehensive field trials and economic analyses can encourage adoption by showcasing the tangible benefits and cost savings associated with enhanced drought tolerance and increased crop yields [94,202,236,245].

Another challenge in commercialization is the variability in regulatory frameworks across different regions and countries. Navigating the regulatory landscape requires a thorough understanding of local agricultural regulations, certification processes, and safety standards. Collaborations between researchers, industry stakeholders, and regulatory bodies are essential to streamline the approval and registration processes, ensuring that beneficial microbial products meet the necessary standards for safety and efficacy [196,234,250,251].

7.3. Regulatory and Safety Considerations

The deployment of beneficial microbes in agriculture must adhere to stringent regulatory and safety standards to ensure that they do not pose risks to human health, non-target organisms, or the environment. Compliance with agricultural regulations involves obtaining necessary approvals and certifications from relevant authorities, which typically require comprehensive data on the safety, efficacy, and environmental impact of microbial products. This regulatory scrutiny is essential to prevent the introduction of harmful pathogens, invasive species, or unintended ecological disruptions that could arise from the widespread use of microbial inoculants [161,184,236,252].

Ensuring that non-target effects are minimized is a critical aspect of regulatory and safety considerations. Beneficial microbes must be carefully evaluated to assess their interactions with native microbial communities, soil fauna, and plant species. Potential risks, such as the horizontal gene transfer of antibiotic resistance genes or the disruption of existing microbial symbioses, must be thoroughly investigated through rigorous testing and risk assessment protocols. Developing standardized guidelines and best practices for microbial application can help mitigate these risks and promote the safe use of beneficial microbes in agriculture [157,162,166,173].

Moreover, public perception and acceptance of microbial technologies play a significant role in their regulatory and commercial success. Transparent communication about the benefits and safety of microbial interventions, coupled with education and outreach efforts, can help build trust and alleviate concerns among farmers, consumers, and other stakeholders. Addressing misconceptions and providing evidence-based information on the environmental and economic advantages of beneficial microbes can facilitate broader acceptance and support for their integration into sustainable agricultural practices [143,148,154,178].

To streamline regulatory processes and enhance safety, fostering international collaborations and harmonizing regulatory standards can be beneficial. By establishing consistent frameworks and sharing best practices across countries, the agricultural sector can accelerate the development and deployment of safe and effective microbial solutions, thereby promoting global agricultural sustainability and resilience to drought.

7.4. Understanding Microbe–Plant–Environment Interactions

The interactions between microbes, plants, and the surrounding environment are inherently complex and multifaceted, posing significant challenges to fully understanding and optimizing the benefits of microbial interventions for drought tolerance. Microbial communities are influenced by a myriad of factors, including soil chemistry, plant genotype, climate conditions, and agricultural practices, making it difficult to predict and manipulate their behavior consistently. The dynamic nature of these interactions necessitates a comprehensive and integrative approach to study and harness microbial functions in diverse agricultural settings [131,136,218,253].

One of the primary complexities lies in deciphering the specific mechanisms through which beneficial microbes confer drought tolerance to plants. While numerous studies have identified key microbial traits and pathways involved in stress alleviation, the intricate networks of gene expression, metabolic interactions, and signaling pathways that underpin these benefits remain only partially understood. Advanced molecular and omics-based techniques, such as genomics, transcriptomics, proteomics, and metabolomics, are essential for unraveling the detailed interactions and identifying the critical factors that drive microbe-mediated drought resilience [104,115,223,254,255].

Another significant gap in current knowledge is the long-term sustainability and stability of introduced microbial communities in the soil–plant ecosystem. Understanding how beneficial microbes interact with native microbial populations, compete for resources, and adapt to changing environmental conditions is crucial for ensuring the sustained efficacy of microbial interventions. Longitudinal studies that monitor microbial dynamics over multiple growing seasons and under varying climatic scenarios are necessary to assess the persistence and functional stability of beneficial microbes in the field [96,126,256,257].

Additionally, the influence of plant genotype on microbe–plant interactions represents an important area of research. Different plant varieties may exhibit varying degrees of responsiveness to microbial inoculants, influenced by their root architecture, exudate profiles, and inherent stress tolerance mechanisms. Identifying plant traits that synergize with microbial benefits can inform the development of tailored microbial solutions that are compatible with specific crop varieties, thereby enhancing their effectiveness and reliability [81,83,117,258].

Addressing these complexities requires interdisciplinary collaboration among microbiologists, plant scientists, agronomists, ecologists, and data scientists. Integrating diverse perspectives and expertise can facilitate a more holistic understanding of microbe–plant–environment interactions and drive the development of innovative strategies to optimize microbial benefits for drought tolerance. Furthermore, bridging the gaps in current knowledge through targeted research initiatives and collaborative projects will be instrumental in advancing the field and overcoming the challenges associated with harnessing beneficial microbes for sustainable agriculture and ecosystem management [73,75,259,260].

In summary, while the application of beneficial microbes for drought tolerance holds immense potential, addressing the associated challenges and limitations is essential for achieving consistent and scalable success. Environmental variability, scale-up and commercialization hurdles, regulatory and safety considerations, and the complexities of microbe–plant–environment interactions represent significant barriers that must be navigated through strategic research, technological innovation, and collaborative efforts. By overcoming these challenges, the agricultural sector can fully leverage the advantages of microbial interventions, fostering resilient and sustainable systems capable of withstanding the increasing pressures of water scarcity and climate change.

7.5. Economic Impact and Socio-Economic Barriers

The application of beneficial microbes to enhance drought tolerance holds significant economic implications for the agricultural sector. On the positive side, microbial interventions can lead to cost savings and increased profitability for farmers by reducing the reliance on chemical fertilizers and pesticides, enhancing crop yields, and improving product quality [261,262]. Beneficial microbes promote nutrient uptake and enhance stress tolerance, which can decrease the need for expensive agrochemicals and increase crop resilience under drought conditions [263,264].

Improved drought tolerance can lead to more stable and increased crop yields, enhancing food security and providing greater economic returns for farmers [265]. Long-term benefits also include improved soil health and sustainability, which are critical for the economic viability of agricultural systems facing climate change and resource scarcity [266].

However, several socio-economic barriers may hinder the widespread adoption of these technologies, especially among smallholder farmers and in developing regions. The initial costs associated with purchasing microbial inoculants and implementing new practices can be prohibitive for farmers with limited financial resources [267]. Lack of awareness and understanding of the benefits and application methods of beneficial microbes contributes to hesitancy among farmers [268]. Access to these technologies is often limited in rural and marginalized areas due to inadequate distribution networks and infrastructural challenges [269].

Regulatory hurdles also play a significant role in influencing the economic impact and adoption of microbial solutions. Inconsistent or underdeveloped regulatory frameworks can create uncertainty for both producers and users of microbial products, potentially discouraging investment and innovation in this sector [270]. Moreover, the absence of standardized guidelines and certifications can lead to variability in product quality and efficacy, undermining farmer confidence and the perceived economic benefits [271,272].

Addressing these socio-economic barriers requires a multifaceted approach. Enhancing farmer education and training through extension services and demonstration projects can increase awareness and understanding of microbial technologies [273]. Developing cost-effective production methods and delivery systems, such as microbial seed coatings or encapsulation techniques, can reduce costs and improve accessibility [263]. Policy interventions, including subsidies, incentives, and supportive regulatory frameworks, can stimulate market growth and encourage adoption by lowering financial risks and ensuring product quality [274–276].

Collaborative efforts among researchers, industry stakeholders, policymakers, and farming communities are essential to maximize the economic impact of beneficial microbes. By aligning research objectives with farmer needs and market demands, the development of microbial solutions can be tailored to address specific economic challenges [277]. Furthermore, fostering public–private partnerships can leverage resources and expertise to overcome infrastructural and distribution barriers, making microbial technologies more accessible to farmers across diverse socio-economic contexts [278,279].

In summary, while the economic impact of utilizing beneficial microbes for drought tolerance is potentially substantial, realizing these benefits requires overcoming significant socio-economic barriers. Through strategic interventions and collaborative efforts, the agricultural sector can enhance the adoption of microbial technologies, leading to improved economic outcomes, sustainable practices, and greater resilience against drought and other environmental stresses.

8. Future Perspectives and Research Directions

As the global climate continues to evolve, the urgency to develop sustainable and resilient agricultural systems becomes increasingly paramount. Harnessing beneficial microbes for drought tolerance represents a promising avenue, yet several avenues warrant further exploration to fully realize their potential. Future research must focus on leveraging advancements in genomics and microbiome analysis, employing synthetic biology for microbe engineering, enhancing microbial resilience and functionality, and fostering robust policy frameworks and collaborative efforts. These directions will not only deepen our understanding of microbe–plant interactions but also facilitate the practical implementation of microbial solutions in diverse agricultural contexts [62,280–282].

8.1. Advances in Genomics and Microbiome Analysis

The rapid advancements in genomics and microbiome analysis have revolutionized our ability to study and manipulate microbial communities. High-throughput sequencing technologies, including next-generation sequencing (NGS) and third-generation sequencing platforms, have enabled comprehensive profiling of soil- and plant-associated microbiomes. Metagenomics, in particular, allows for the exploration of the genetic diversity within microbial communities without the need for culturing individual species. This approach provides invaluable insights into the functional potential and ecological roles of microbes involved in drought tolerance [50,52,116,283,284].

Through metagenomic analyses, researchers can identify key microbial taxa and functional genes that contribute to drought resilience. These studies have revealed intricate networks of microbial interactions and metabolic pathways that underpin plant–microbe symbioses. Additionally, transcriptomic and proteomic approaches complement metagenomics by elucidating gene expression patterns and protein functions under drought stress conditions. Such integrative omics strategies facilitate a deeper understanding of how microbial communities respond to environmental stresses and interact with host plants to confer drought tolerance [41,65,285,286].

Furthermore, bioinformatics and machine learning algorithms play a critical role in analyzing the vast datasets generated by high-throughput sequencing. These computational tools enable the identification of biomarkers for beneficial microbes, the prediction of microbial functions, and the modeling of microbe–plant–environment interactions. Insights gained from these analyses can inform the development of targeted microbial inoculants and the optimization of microbial consortia for enhanced drought resilience [33,34,287,288].

Overall, the continued advancement of genomics and microbiome analysis technologies holds significant promise for uncovering the complex mechanisms by which beneficial microbes enhance drought tolerance. By leveraging these tools, researchers can accelerate the discovery of novel microbial traits and interactions, paving the way for innovative applications in sustainable agriculture.

8.2. Synthetic Biology and Microbe Engineering

Synthetic biology offers transformative opportunities for designing and engineering microbes with enhanced capabilities to support plant drought tolerance. By manipulating the genetic makeup of beneficial microorganisms, researchers can create custom strains tailored to specific agricultural needs and environmental conditions. This precision engineering approach allows for the introduction of novel traits, such as the production of stress-alleviating compounds, improved nutrient acquisition, and enhanced root colonization abilities [29,30,57,289].

One of the key applications of synthetic biology in this context is the engineering of microbes to overproduce phytohormones like indole-3-acetic acid (IAA) and gibberellins, which promote root growth and water uptake. Additionally, microbes can be engineered to produce higher levels of osmolytes, such as proline and trehalose, which aid in osmotic adjustment and cellular protection under drought stress. The incorporation of synthetic gene circuits can also enable microbes to sense and respond to environmental cues, ensuring that beneficial traits are expressed precisely when needed [8,9,16,22,290].

Moreover, synthetic biology facilitates the construction of synthetic microbial communities with defined interactions and functional redundancies. By designing consortia with complementary traits, researchers can enhance the overall stability and resilience of microbial interventions. For instance, a synthetic consortium might include nitrogen-fixing bacteria, phosphate-solubilizing fungi, and drought-tolerant PGPR, each contributing distinct functions that collectively enhance plant resilience [30,35,291].

However, the application of synthetic biology in agriculture also presents challenges, including regulatory hurdles, public acceptance, and the potential for unintended ecological impacts. Ethical considerations and stringent safety assessments are essential to ensure that engineered microbes do not disrupt native ecosystems or pose risks to human health. Despite these challenges, the potential benefits of synthetic biology in developing customized microbial solutions for drought tolerance are immense, offering a pathway to more resilient and sustainable agricultural systems.

8.3. Enhancing Microbe Resilience and Functionality

For beneficial microbes to be effective in conferring drought tolerance, their resilience and functionality under stress conditions must be robust. Enhancing microbial survival and activity in harsh environments is a critical area of research that encompasses both natural and engineered strategies. Improving microbial resilience involves optimizing their physiological and metabolic capabilities to withstand extreme temperatures, desiccation, and nutrient limitations commonly associated with drought conditions [19,67,289,292].

One approach to enhancing microbial resilience is the selection and breeding of naturally robust strains through adaptive laboratory evolution (ALE). By subjecting microbes to prolonged stress conditions in controlled environments, researchers can select for strains that exhibit superior stress tolerance traits. These evolved strains often demonstrate enhanced survival rates, metabolic efficiency, and functional stability under drought stress, making them more effective as bioinoculants [13,293,294].

Another strategy involves the application of stress protectants, such as trehalose, proline, and other compatible solutes, which can be incorporated into microbial formulations to improve their desiccation tolerance and overall viability. Encapsulation techniques, including microencapsulation and biofilm formation, provide physical protection to microbes, shielding them from environmental extremes and facilitating sustained activity in the soil [9,10,60,295].

Additionally, the engineering of stress-responsive regulatory networks within microbes can enhance their ability to dynamically respond to drought stress. By incorporating synthetic regulatory elements that activate stress response pathways, researchers can create microbes that upregulate protective mechanisms in real time, thereby maintaining their functionality and beneficial interactions with plants during periods of water scarcity [16,19,67].

Enhancing microbial functionality also involves ensuring efficient colonization and persistence within the plant rhizosphere. This can be achieved by optimizing microbial adhesion properties, root colonization strategies, and mutualistic interactions with host plants. Understanding the factors that influence microbe–plant compatibility and leveraging this knowledge to improve colonization efficacy are essential for maximizing the benefits of microbial interventions [88,200,243,296].

Overall, enhancing microbe resilience and functionality is a multifaceted endeavor that requires a combination of natural selection, genetic engineering, and formulation optimization. By developing more robust and active microbial strains, researchers can ensure the consistent performance of beneficial microbes in promoting drought tolerance and supporting sustainable agricultural practices.

8.4. Policy and Collaborative Efforts for Implementation

The successful implementation of beneficial microbial technologies for drought tolerance extends beyond scientific advancements; it requires robust policy frameworks and collaborative efforts among various stakeholders. Encouraging interdisciplinary research and facilitating partnerships between academia, industry, and government are essential for translating scientific discoveries into practical agricultural solutions [180,184,186].

Policy initiatives play a crucial role in fostering innovation and ensuring the safe and effective deployment of microbial interventions. Governments can support research and development through funding programs, grants, and incentives aimed at advancing microbial technologies. Additionally, the establishment of clear regulatory guidelines and streamlined approval processes can facilitate the commercialization of beneficial microbial products, reducing barriers to market entry and encouraging investment from the private sector [190,193,199,250].

Interdisciplinary research is another key component in addressing the complex challenges associated with harnessing beneficial microbes for drought tolerance. Collaboration among microbiologists, plant scientists, agronomists, ecologists, and data scientists can lead to a more comprehensive understanding of microbe–plant–environment interactions and the development of integrated solutions. Joint research initiatives and cross-disciplinary projects can accelerate the discovery of novel microbial traits, optimize formulation and delivery methods, and enhance the scalability of microbial interventions [102,161,236,245].

Partnerships between academia, industry, and government are also vital for bridging the gap between research and practical application. Industry stakeholders, including agricultural biotechnology companies and biofertilizer manufacturers, can collaborate with academic institutions to translate laboratory findings into commercial products. Government agencies can facilitate these partnerships by providing platforms for knowledge exchange, supporting pilot projects, and promoting the adoption of sustainable agricultural practices through extension services and educational programs [94,191,194,297].

Moreover, international collaborations are essential for addressing the global nature of drought and water scarcity challenges. Sharing knowledge, resources, and best practices across countries and regions can enhance the collective capacity to develop and implement effective microbial solutions [298]. Participating in global networks and consortia can also promote the standardization of methodologies, facilitate the exchange of microbial strains, and harmonize regulatory frameworks, thereby fostering a more cohesive and coordinated approach to combating drought through microbial innovations [102,161,171].

Public engagement and awareness are also critical for the widespread acceptance and adoption of beneficial microbial technologies. Educating farmers, consumers, and policymakers about the benefits and safety of microbial interventions can build trust and support for their integration into agricultural systems. Transparent communication, demonstration projects, and success stories can highlight the tangible advantages of using beneficial microbes, encouraging broader acceptance and utilization [158,208,299].

Despite advancements in understanding stress mitigation through biological treatments, several challenges persist. The use of exogenous compounds like melatonin has shown promise in enhancing plant stress tolerance under salinity conditions, and exogenous glutamine has been effective in improving stress-related gene expression in onions [300,301]. However, the application of nanomaterials, such as zinc oxide and silver, for improving plant stability and growth in controlled conditions, and magnetic field treatments for boosting plant resilience, still face limitations when scaling to field applications [269,302]. These technologies are highly dependent on environmental factors and often require precise control to be effective, limiting their broader application in diverse climates [303]. While biological coatings like chitosan–putrescine nanoparticles show potential for reducing postharvest losses, their commercial viability in large-scale agriculture is still a hurdle due to costs and varied results across different crops [304,305]. Moreover, although living mulches can enhance soil health and microbial communities, their long-term impact on weed suppression and crop yield in different orchard systems remains under-researched [306,307]. Additionally, while advances in microbial biocontrol and diagnostics offer new solutions, more studies are needed to improve the consistency of these applications under varying environmental conditions, especially in drought scenarios [308–310].

In conclusion, policy support and collaborative efforts are indispensable for the successful implementation of beneficial microbial technologies in agriculture. By fostering interdisciplinary research, facilitating partnerships, and promoting supportive policies, stakeholders can collectively advance the development and adoption of microbial solutions that enhance drought tolerance and contribute to sustainable and resilient agricultural systems.

9. Conclusions

Drought poses an escalating threat to global agricultural productivity, food security, and ecosystem stability, exacerbated by the ongoing impacts of climate change. In response to these challenges, the utilization of beneficial microbes has emerged as a promising strategy to enhance plant resilience and sustain agricultural systems under water-limited conditions. This review has elucidated the multifaceted roles that various beneficial microorganisms—such as Plant Growth-Promoting Rhizobacteria (PGPR), mycorrhizal fungi, endophytes, actinomycetes, and cyanobacteria—play in mitigating drought stress. These microbes employ a diverse array of mechanisms, including osmotic adjustment, improvement of root architecture, modulation of phytohormones, induction of antioxidant defenses, and regulation of stress-responsive gene expression, all of which collectively contribute to enhanced plant drought tolerance.

Ecological and agricultural innovations leveraging these beneficial microbes have demonstrated significant potential in bolstering drought resilience. Strategies such as soil microbiome engineering, bioaugmentation, and the integration of microbial synergies within pest management frameworks have shown promising results in enhancing ecosystem resilience and agricultural sustainability. Additionally, advancements in agricultural practices, including seed coating, soil amendments, the development of microbial consortia, and the adoption of precision agriculture technologies, have further validated the effectiveness and scalability of microbial interventions in diverse farming systems.

However, the path to widespread adoption of microbial solutions is not without challenges. Environmental variability can significantly influence microbial performance, necessitating the development of robust and adaptable microbial strains capable of thriving under diverse and fluctuating conditions. The transition from laboratory successes to field applications involves overcoming scale-up and commercialization hurdles, including economic constraints and the need for efficient production and delivery systems. Regulatory and safety considerations also play a critical role, requiring stringent assessments to ensure that microbial applications do not pose risks to human health or non-target organisms. Moreover, the intricate interactions between microbes, plants, and their environments demand a deeper understanding to optimize microbial benefits and ensure consistent outcomes.

Future research must prioritize the integration of cutting-edge technologies such as genomics, synthetic biology, and precision agriculture to refine and enhance microbial interventions. Advances in high-throughput sequencing and metagenomics will continue to unravel the complex dynamics of microbial communities, providing insights into their functional roles and interactions with host plants. Synthetic biology offers the potential to engineer custom microbial strains with tailored traits for improved drought tolerance, while precision agriculture technologies enable the precise monitoring and management of microbial activity in real time, ensuring optimal application and efficacy.

Collaborative efforts among academia, industry, and government are essential to bridge the gap between research and practical implementation. Policy support, interdisciplinary research, and robust partnerships will facilitate the development, regulation, and adoption of microbial technologies, ensuring that their benefits are realized on a global scale. Additionally, fostering public awareness and acceptance through education and transparent communication will be crucial in promoting the safe and effective use of beneficial microbes in agriculture.

In conclusion, the strategic harnessing of beneficial microbes presents a transformative approach to addressing the pressing challenges of drought and water scarcity. By leveraging microbial innovations and overcoming existing barriers, it is possible to develop resilient and sustainable agricultural systems capable of thriving in an increasingly unpredictable climate. The continued advancement and integration of microbial solutions will be instrumental in ensuring food security, maintaining ecosystem health, and promoting environmental sustainability in a world facing the dual pressures of population growth and climate change.

Author Contributions: Conceptualization, G.M., T.M., AK., D.C.-L., M.M. and A.Ł.; investigation, G.M., T.M., A.K., D.C.-L., M.M., K.B. and A.Ł.; resources, G.M., T.M., A.K., D.C.-L., M.M., K.B. and A.Ł.; writing—original draft preparation, G.M., T.M., A.K., D.C.-L., M.M., K.B. and A.Ł.; writing—review and editing, G.M., T.M., A.K., D.C.-L., M.M. and A.Ł. All authors have read and agreed to the published version of the manuscript.

Funding: This research was co-financed by the Minister of Science (Poland) under the "Regional Excellence Initiative" Program for 2024–2027 (RID/SP/0045/2024/01).

Institutional Review Board Statement: Not applicable.

Data Availability Statement: Not applicable.

Conflicts of Interest: The authors declare no conflicts of interest.

References

1. Haile, G.G.; Tang, Q.; Li, W.; Liu, X.; Zhang, X. Drought: Progress in Broadening Its Understanding. *WIREs Water* **2020**, *7*, e1407. [CrossRef]
2. Orimoloye, I.R.; Belle, J.A.; Orimoloye, Y.M.; Olusola, A.O.; Ololade, O.O. Drought: A Common Environmental Disaster. *Atmosphere* **2022**, *13*, 111. [CrossRef]
3. Naumann, G.; Cammalleri, C.; Mentaschi, L.; Feyen, L. Increased Economic Drought Impacts in Europe with Anthropogenic Warming. *Nat. Clim. Change* **2021**, *11*, 485–491. [CrossRef]
4. Savelli, E.; Rusca, M.; Cloke, H.; Di Baldassarre, G. Drought and Society: Scientific Progress, Blind Spots, and Future Prospects. *WIREs Clim. Change* **2022**, *13*, e761. [CrossRef]
5. Lottering, S.; Mafongoya, P.; Lottering, R. Drought and Its Impacts on Small-Scale Farmers in Sub-Saharan Africa: A Review. *S. Afr. Geogr. J.* **2021**, *103*, 319–341. [CrossRef]
6. Salvador, C.; Nieto, R.; Linares, C.; Díaz, J.; Gimeno, L. Effects of Droughts on Health: Diagnosis, Repercussion, and Adaptation in Vulnerable Regions under Climate Change. Challenges for Future Research. *Sci. Total Environ.* **2020**, *703*, 134912. [CrossRef]
7. Patil, R.; Polisgowdar, B.S.; Rathod, S.; Bandumula, N.; Mustac, I.; Srinivasa Reddy, G.V.; Wali, V.; Satishkumar, U.; Rao, S.; Kumar, A.; et al. Spatiotemporal Characterization of Drought Magnitude, Severity, and Return Period at Various Time Scales in the Hyderabad Karnataka Region of India. *Water* **2023**, *15*, 2483. [CrossRef]
8. Palazzolo, N.; Peres, D.J.; Bonaccorso, B.; Cancelliere, A. A Probabilistic Analysis of Drought Areal Extent Using SPEI-Based Severity-Area-Frequency Curves and Reanalysis Data. *Water* **2023**, *15*, 3141. [CrossRef]
9. Nasir, J.; Assefa, E.; Zeleke, T.; Gidey, E. Meteorological Drought in Northwestern Escarpment of Ethiopian Rift Valley: Detection Seasonal and Spatial Trends. *Environ. Syst. Res.* **2021**, *10*, 16. [CrossRef]

10. Ullah, I.; Ma, X.; Ren, G.; Yin, J.; Iyakaremye, V.; Syed, S.; Lu, K.; Xing, Y.; Singh, V.P. Recent Changes in Drought Events over South Asia and Their Possible Linkages with Climatic and Dynamic Factors. *Remote Sens.* **2022**, *14*, 3219. [CrossRef]
11. Ilyas, M.; Nisar, M.; Khan, N.; Hazrat, A.; Khan, A.H.; Hayat, K.; Fahad, S.; Khan, A.; Ullah, A. Drought Tolerance Strategies in Plants: A Mechanistic Approach. *J. Plant Growth Regul.* **2021**, *40*, 926–944. [CrossRef]
12. Ali, S.; Tyagi, A.; Park, S.; Mir, R.A.; Mushtaq, M.; Bhat, B.; Mahmoudi, H.; Bae, H. Deciphering the Plant Microbiome to Improve Drought Tolerance: Mechanisms and Perspectives. *Environ. Exp. Bot.* **2022**, *201*, 104933. [CrossRef]
13. Zia, R.; Nawaz, M.S.; Siddique, M.J.; Hakim, S.; Imran, A. Plant Survival under Drought Stress: Implications, Adaptive Responses, and Integrated Rhizosphere Management Strategy for Stress Mitigation. *Microbiol. Res.* **2021**, *242*, 126626. [CrossRef] [PubMed]
14. Kapoor, D.; Bhardwaj, S.; Landi, M.; Sharma, A.; Ramakrishnan, M.; Sharma, A. The Impact of Drought in Plant Metabolism: How to Exploit Tolerance Mechanisms to Increase Crop Production. *Appl. Sci.* **2020**, *10*, 5692. [CrossRef]
15. Więcław, H. Within-Species Variation among Populations of the *Carex flava* Complex as a Function of Habitat Conditions. *Plant Ecol. Divers.* **2017**, *10*, 443–451. [CrossRef]
16. Decunta, F.A.; Pérez, L.I.; Malinowski, D.P.; Molina-Montenegro, M.A.; Gundel, P.E. A Systematic Review on the Effects of Epichloë Fungal Endophytes on Drought Tolerance in Cool-Season Grasses. *Front. Plant Sci.* **2021**, *12*, 644731. [CrossRef]
17. Aslam, M.M.; Okal, E.J.; Idris, A.L.; Qian, Z.; Xu, W.; Karanja, J.K.; Wani, S.H.; Yuan, W. Rhizosphere Microbiomes Can Regulate Plant Drought Tolerance. *Pedosphere* **2022**, *32*, 61–74. [CrossRef]
18. Leontidou, K.; Genitsaris, S.; Papadopoulou, A.; Kamou, N.; Bosmali, I.; Matsi, T.; Madesis, P.; Vokou, D.; Karamanoli, K.; Mellidou, I. Plant Growth Promoting Rhizobacteria Isolated from Halophytes and Drought-Tolerant Plants: Genomic Characterisation and Exploration of Phyto-Beneficial Traits. *Sci. Rep.* **2020**, *10*, 14857. [CrossRef]
19. Schnabel, F.; Liu, X.; Kunz, M.; Barry, K.E.; Bongers, F.J.; Bruelheide, H.; Fichtner, A.; Härdtle, W.; Li, S.; Pfaff, C.-T.; et al. Species Richness Stabilizes Productivity via Asynchrony and Drought-Tolerance Diversity in a Large-Scale Tree Biodiversity Experiment. *Sci. Adv.* **2021**, *7*, eabk1643. [CrossRef]
20. Rutkowski, K.; Łysiak, G.P. Effect of Nitrogen Fertilization on Tree Growth and Nutrient Content in Soil and Cherry Leaves (*Prunus cerasus* L.). *Agriculture* **2023**, *13*, 578. [CrossRef]
21. Oszust, K.; Szpilska, K.; Gryta, A.; Panek, J.; Pylak, M.; Lipa, T.; Frąc, M. New Biotechnological Solutions in Biocontrol and Molecular Diagnostics of *Neofabraea* spp. in Apples—A Review. *Postharvest Biol. Technol.* **2023**, *204*, 112442. [CrossRef]
22. Koza, N.; Adedayo, A.; Babalola, O.; Kappo, A. Microorganisms in Plant Growth and Development: Roles in Abiotic Stress Tolerance and Secondary Metabolites Secretion. *Microorganisms* **2022**, *10*, 1528. [CrossRef] [PubMed]
23. Harman, G.; Khadka, R.; Doni, F.; Uphoff, N. Benefits to Plant Health and Productivity From Enhancing Plant Microbial Symbionts. *Front. Plant Sci.* **2021**, *11*, 610065. [CrossRef] [PubMed]
24. Więcław, H.; Bosiacka, B.; Hrivnák, R.; Dajdok, Z.; Mesterházy, A.; Koopman, J. Morphological Variability of *Carex buekii* (Cyperaceae) as a Function of Soil Conditions: A Case Study of the Central European Populations. *Sci. Rep.* **2022**, *12*, 11761. [CrossRef] [PubMed]
25. Silva, A.R.L.; Cavalcante, Í.H.L.; Silva, M.A.; Paiva Neto, V.B.; Amariz, R.A.; Amorim, L.Y.A. Does the Sunblock Alleviate Abiotic Stress in Mango Trees Grown in the Tropical Semiarid? *Folia Hortic.* **2022**, *34*, 211–221. [CrossRef]
26. Seymour, Z.J.; Mercedes, J.F.; Fang, J.-Y. Effect of Heat Acclimation on Thermotolerance of In Vitro Strawberry Plantlets. *Folia Hortic.* **2024**, *36*, 135–147. [CrossRef]
27. Ray, P.; Lakshmanan, V.; Labbé, J.L.; Craven, K.D. Microbe to Microbiome: A Paradigm Shift in the Application of Microorganisms for Sustainable Agriculture. *Front. Microbiol.* **2020**, *11*, 622926. [CrossRef]
28. O'Callaghan, M.; Ballard, R.A.; Wright, D. Soil Microbial Inoculants for Sustainable Agriculture: Limitations and Opportunities. *Soil Use Manag.* **2022**, *38*, 1340–1369. [CrossRef]
29. Sanjuán, J.; Nápoles, M.C.; Pérez-Mendoza, D.; Lorite, M.J.; Rodríguez-Navarro, D.N. Microbials for Agriculture: Why Do They Call Them Biostimulants When They Mean Probiotics? *Microorganisms* **2023**, *11*, 153. [CrossRef]
30. Tripathi, A.N.; Meena, B.R.; Pandey, K.K.; Singh, J. Microbial Bioagents in Agriculture: Current Status and Prospects. In *New Frontiers in Stress Management for Durable Agriculture*; Springer Singapore: Singapore, 2020; pp. 331–368.
31. Paliwoda, D.; Mikiciuk, G. Use of Rhizosphere Microorganisms in Plant Production—A Review Study. *J. Ecol. Eng.* **2020**, *21*, 292–310. [CrossRef]
32. Santoyo, G.; Urtis-Flores, C.A.; Loeza-Lara, P.D.; Orozco-Mosqueda, M.d.C.; Glick, B.R. Rhizosphere Colonization Determinants by Plant Growth-Promoting Rhizobacteria (PGPR). *Biology* **2021**, *10*, 475. [CrossRef] [PubMed]
33. Basu, A.; Prasad, P.; Das, S.N.; Kalam, S.; Sayyed, R.Z.; Reddy, M.S.; El Enshasy, H. Plant Growth Promoting Rhizobacteria (PGPR) as Green Bioinoculants: Recent Developments, Constraints, and Prospects. *Sustainability* **2021**, *13*, 1140. [CrossRef]
34. Riaz, U.; Murtaza, G.; Anum, W.; Samreen, T.; Sarfraz, M.; Nazir, M.Z. Plant Growth-Promoting Rhizobacteria (PGPR) as Biofertilizers and Biopesticides. In *Microbiota and Biofertilizers*; Springer International Publishing: Cham, Switzerland, 2021; pp. 181–196.
35. Vocciante, M.; Grifoni, M.; Fusini, D.; Petruzzelli, G.; Franchi, E. The Role of Plant Growth-Promoting Rhizobacteria (PGPR) in Mitigating Plant's Environmental Stresses. *Appl. Sci.* **2022**, *12*, 1231. [CrossRef]
36. Mikiciuk, G.; Sas-Paszt, L.; Mikiciuk, M.; Derkowska, E.; Trzciński, P.; Ptak, P.; Chylewska, U.; Statkiewicz, M.; Lisek, A. Physiological Response of Three Grapevine Cultivars Grown in North-Western Poland to Mycorrhizal Fungi. *S. Afr. J. Enol. Vitic.* **2018**, *40*, 1–14. [CrossRef]

37. Dakora, F.D.; Phillips, D.A. Root Exudates as Mediators of Mineral Acquisition in Low-Nutrient Environments. In *Food Security in Nutrient-Stressed Environments: Exploiting Plants' Genetic Capabilities*; Springer: Dordrecht, The Netherlands, 2002; pp. 201–213.
38. Fourneau, E.; Pannier, M.; Riah, W.; Personeni, E.; Morvan-Bertrand, A.; Bodilis, J.; Pawlak, B. A "Love Match" Score to Compare Root Exudate Attraction and Feeding of the Plant Growth-Promoting Rhizobacteria Bacillus Subtilis, Pseudomonas Fluorescens, and Azospirillum Brasilense. *Front. Microbiol.* **2024**, *15*, 1473099. [CrossRef]
39. Wang, Y.; Pei, Y.; Wang, X.; Dai, X.; Zhu, M. Antimicrobial Metabolites Produced by the Plant Growth-Promoting Rhizobacteria (PGPR): Bacillus and Pseudomonas. *Adv. Agrochem* **2024**, *3*, 206–221. [CrossRef]
40. Essalimi, B.; Esserti, S.; Rifai, L.A.; Koussa, T.; Makroum, K.; Belfaiza, M.; Rifai, S.; Venisse, J.S.; Faize, L.; Alburquerque, N.; et al. Enhancement of Plant Growth, Acclimatization, Salt Stress Tolerance and Verticillium Wilt Disease Resistance Using Plant Growth-Promoting Rhizobacteria (PGPR) Associated with Plum Trees (*Prunus domestica*). *Sci. Hortic.* **2022**, *291*, 110621. [CrossRef]
41. Mota, I.; Sánchez-Sánchez, J.; Pedro, L.G.; Sousa, M.J. Composition Variation of the Essential Oil from Ocimum Basilicum L. Cv. Genovese Gigante in Response to Glomus Intraradices and Mild Water Stress at Different Stages of Growth. *Biochem. Syst. Ecol.* **2020**, *90*, 104021. [CrossRef]
42. Zou, Y.-N.; Wu, Q.-S.; Kuča, K. Unravelling the Role of Arbuscular Mycorrhizal Fungi in Mitigating the Oxidative Burst of Plants under Drought Stress. *Plant Biol.* **2021**, *23*, 50–57. [CrossRef]
43. Tang, H.; Hassan, M.U.; Feng, L.; Nawaz, M.; Shah, A.N.; Qari, S.H.; Liu, Y.; Miao, J. The Critical Role of Arbuscular Mycorrhizal Fungi to Improve Drought Tolerance and Nitrogen Use Efficiency in Crops. *Front. Plant Sci.* **2022**, *13*, 919166. [CrossRef]
44. Sheteiwy, M.S.; Abd Elgawad, H.; Xiong, Y.; Macovei, A.; Brestic, M.; Skalicky, M.; Shaghaleh, H.; Alhaj Hamoud, Y.; El-Sawah, A.M. Inoculation with *Bacillus amyloliquefaciens* and Mycorrhiza Confers Tolerance to Drought Stress and Improve Seed Yield and Quality of Soybean Plant. *Physiol. Plant.* **2021**, *172*, 2153–2169. [CrossRef] [PubMed]
45. Byregowda, R.; Prasad, S.R.; Oelmüller, R.; Nataraja, K.N.; Prasanna Kumar, M.K. Is Endophytic Colonization of Host Plants a Method of Alleviating Drought Stress? Conceptualizing the Hidden World of Endophytes. *Int. J. Mol. Sci.* **2022**, *23*, 9194. [CrossRef] [PubMed]
46. Dubey, A.; Saiyam, D.; Kumar, A.; Hashem, A.; Abd_Allah, E.F.; Khan, M.L. Bacterial Root Endophytes: Characterization of Their Competence and Plant Growth Promotion in Soybean (*Glycine max* (L.) Merr.) under Drought Stress. *Int. J. Environ. Res. Public Health* **2021**, *18*, 931. [CrossRef] [PubMed]
47. Li, H.; Guo, Q.; Jing, Y.; Liu, Z.; Zheng, Z.; Sun, Y.; Xue, Q.; Lai, H. Application of Streptomyces Pactum Act12 Enhances Drought Resistance in Wheat. *J. Plant Growth Regul.* **2020**, *39*, 122–132. [CrossRef]
48. Chukwuneme, C.F.; Babalola, O.O.; Kutu, F.R.; Ojuederie, O.B. Characterization of Actinomycetes Isolates for Plant Growth Promoting Traits and Their Effects on Drought Tolerance in Maize. *J. Plant Interact.* **2020**, *15*, 93–105. [CrossRef]
49. Xie, F.; Andrews, B.; Asenjo, J.A.; Goodfellow, M.; Pathom-aree, W. Atacama Desert Actinomycetes: Taxonomic Analysis, Drought Tolerance and Plant Growth Promoting Potential. *World J. Microbiol. Biotechnol.* **2024**, *40*, 283. [CrossRef]
50. El-Shazoly, R.M.; Aloufi, A.S.; Fawzy, M.A. The Potential Use of Arthrospira (*Spirulina platensis*) as a Biostimulant for Drought Tolerance in Wheat (*Triticum aestivum* L.) for Sustainable Agriculture. *J. Plant Growth Regul.* **2024**, *in press*. [CrossRef]
51. GR, S.; Yadav, R.K.; Chatrath, A.; Gerard, M.; Tripathi, K.; Govindsamy, V.; Abraham, G. Perspectives on the Potential Application of Cyanobacteria in the Alleviation of Drought and Salinity Stress in Crop Plants. *J. Appl. Phycol.* **2021**, *33*, 3761–3778. [CrossRef]
52. Abbasi, S.; Mohammadi, P. Improving Drought Tolerance and Glycyrrhizin Content of Licorice Plant by EPS-Producing Cyanobacteria. *Ind. Crops Prod.* **2023**, *197*, 116578. [CrossRef]
53. Cornejo-Ríos, K.; Osorno-Suárez, M.d.P.; Hernández-León, S.; Reyes-Santamaría, M.I.; Juárez-Díaz, J.A.; Pérez-España, V.H.; Peláez-Acero, A.; Madariaga-Navarrete, A.; Saucedo-García, M. Impact of Trichoderma Asperellum on Chilling and Drought Stress in Tomato (*Solanum lycopersicum*). *Horticulturae* **2021**, *7*, 385. [CrossRef]
54. Scudeletti, D.; Crusciol, C.A.C.; Bossolani, J.W.; Moretti, L.G.; Momesso, L.; Servaz Tubaña, B.; de Castro, S.G.Q.; De Oliveira, E.F.; Hungria, M. Trichoderma Asperellum Inoculation as a Tool for Attenuating Drought Stress in Sugarcane. *Front. Plant Sci.* **2021**, *12*, 645542. [CrossRef]
55. Battaglia, M.E.; Martinez, S.I.; Covacevich, F.; Consolo, V.F. *Trichoderma harzianum* Enhances Root Biomass Production and Promotes Lateral Root Growth of Soybean and Common Bean under Drought Stress. *Ann. Appl. Biol.* **2024**, *185*, 36–48. [CrossRef]
56. Hussain, M.; Shakoor, N.; Adeel, M.; Ahmad, M.A.; Zhou, H.; Zhang, Z.; Xu, M.; Rui, Y.; White, J.C. Nano-Enabled Plant Microbiome Engineering for Disease Resistance. *Nano Today* **2023**, *48*, 101752. [CrossRef]
57. Kavadia, A.; Omirou, M.; Fasoula, D.; Ioannides, I.M. The Importance of Microbial Inoculants in a Climate-Changing Agriculture in Eastern Mediterranean Region. *Atmosphere* **2020**, *11*, 1136. [CrossRef]
58. Branco, S.; Schauster, A.; Liao, H.; Ruytinx, J. Mechanisms of Stress Tolerance and Their Effects on the Ecology and Evolution of Mycorrhizal Fungi. *New Phytol.* **2022**, *235*, 2158–2175. [CrossRef]
59. Cheng, S.; Zou, Y.-N.; Kuča, K.; Hashem, A.; Abd_Allah, E.F.; Wu, Q.-S. Elucidating the Mechanisms Underlying Enhanced Drought Tolerance in Plants Mediated by Arbuscular Mycorrhizal Fungi. *Front. Microbiol.* **2021**, *12*, 809473. [CrossRef]
60. Koopman, J.; Więcław, H.; Cembrowska-Lech, D. Intersectional Hybrids of *Carex remota* with *C. otrubae* and *C. vulpina* (Cyperaceae) in Europe. *Bot. J. Linn. Soc.* **2023**, *203*, 350–369. [CrossRef]
61. Jajoo, A.; Mathur, S. Role of Arbuscular Mycorrhizal Fungi as an Underground Saviuor for Protecting Plants from Abiotic Stresses. *Physiol. Mol. Biol. Plants* **2021**, *27*, 2589–2603. [CrossRef]

62. Wahab, A.; Muhammad, M.; Munir, A.; Abdi, G.; Zaman, W.; Ayaz, A.; Khizar, C.; Reddy, S.P.P. Role of Arbuscular Mycorrhizal Fungi in Regulating Growth, Enhancing Productivity, and Potentially Influencing Ecosystems under Abiotic and Biotic Stresses. *Plants* **2023**, *12*, 3102. [CrossRef]
63. Marro, N.; Grilli, G.; Soteras, F.; Caccia, M.; Longo, S.; Cofré, N.; Borda, V.; Burni, M.; Janoušková, M.; Urcelay, C. The Effects of Arbuscular Mycorrhizal Fungal Species and Taxonomic Groups on Stressed and Unstressed Plants: A Global Meta-analysis. *New Phytol.* **2022**, *235*, 320–332. [CrossRef]
64. Mikiciuk, G.; Sas-Paszt, L.; Mikiciuk, M.; Derkowska, E.; Trzciński, P.; Głuszek, S.; Lisek, A.; Wera-Bryl, S.; Rudnicka, J. Mycorrhizal Frequency, Physiological Parameters, and Yield of Strawberry Plants Inoculated with Endomycorrhizal Fungi and Rhizosphere Bacteria. *Mycorrhiza* **2019**, *29*, 489–501. [CrossRef]
65. Salvi, P.; Mahawar, H.; Agarrwal, R.; Kajal; Gautam, V.; Deshmukh, R. Advancement in the Molecular Perspective of Plant-Endophytic Interaction to Mitigate Drought Stress in Plants. *Front. Microbiol.* **2022**, *13*, 981355. [CrossRef] [PubMed]
66. Hosseyni Moghaddam, M.S.; Safaie, N.; Soltani, J.; Hagh-Doust, N. Desert-Adapted Fungal Endophytes Induce Salinity and Drought Stress Resistance in Model Crops. *Plant Physiol. Biochem.* **2021**, *160*, 225–238. [CrossRef] [PubMed]
67. Verma, S.K.; Sahu, P.K.; Kumar, K.; Pal, G.; Gond, S.K.; Kharwar, R.N.; White, J.F. Endophyte Roles in Nutrient Acquisition, Root System Architecture Development and Oxidative Stress Tolerance. *J. Appl. Microbiol.* **2021**, *131*, 2161–2177. [CrossRef] [PubMed]
68. Abideen, Z.; Cardinale, M.; Zulfiqar, F.; Koyro, H.-W.; Rasool, S.G.; Hessini, K.; Darbali, W.; Zhao, F.; Siddique, K.H.M. Seed Endophyte Bacteria Enhance Drought Stress Tolerance in Hordeum Vulgare by Regulating, Physiological Characteristics, Antioxidants and Minerals Uptake. *Front. Plant Sci.* **2022**, *13*, 980046. [CrossRef]
69. Koopman, J.; Dajdok, Z.; Więcław, H.; Marinetto, E.; Grulich, V.; Řepka, R.; Jimenez-Mejas, P. Global Distribution of *Carex buekii* (Cyperaceae) Reappraised. *Phytotaxa* **2018**, *358*, 139–161. [CrossRef]
70. Oyedoh, O.P.; Yang, W.; Dhanasekaran, D.; Santoyo, G.; Glick, B.R.; Babalola, O.O. Rare Rhizo-Actinomycetes: A New Source of Agroactive Metabolites. *Biotechnol. Adv.* **2023**, *67*, 108205. [CrossRef]
71. Sneha, G.R.; Govindasamy, V.; Singh, P.K.; Kumar, S.; Abraham, G. Priming of Seeds with Cyanobacteria Improved Tolerance in Wheat (*Triticum aestivum* L.) during Post-Germinative Drought Stress. *J. Appl. Phycol.* **2024**, *36*, 1233–1246. [CrossRef]
72. Więcław, H.; Šumberová, K.; Bosiacka, B.; Hrivnák, R.; Dajdok, Z.; Mesterházy, A.; Minuzzo, C.; Martinetto, E.; Koopman, J. Ecology, Threats and Conservation Status of *Carex buekii* (Cyperaceae) in Central Europe. *Sci. Rep.* **2019**, *9*, 11162. [CrossRef]
73. Yasmin, H.; Rashid, U.; Hassan, M.N.; Nosheen, A.; Naz, R.; Ilyas, N.; Sajjad, M.; Azmat, A.; Alyemeni, M.N. Volatile Organic Compounds Produced by *Pseudomonas pseudoalcaligenes* Alleviated Drought Stress by Modulating Defense System in Maize (*Zea mays* L.). *Physiol. Plant.* **2021**, *172*, 896–911. [CrossRef]
74. Chieb, M.; Gachomo, E.W. The Role of Plant Growth Promoting Rhizobacteria in Plant Drought Stress Responses. *BMC Plant Biol.* **2023**, *23*, 407. [CrossRef]
75. Gontia-Mishra, I.; Sapre, S.; Deshmukh, R.; Sikdar, S.; Tiwari, S. Microbe-Mediated Drought Tolerance in Plants: Current Developments and Future Challenges. In *Plant Microbiomes for Sustainable Agriculture*; Springer: Berlin/Heidelberg, Germany, 2020; pp. 351–379.
76. Koopman, J.; Kalinowski, P.; Stech, M.; Więcław, H. *Carex* × *kneuckeri*, a Hybrid New for Central Europe and Neotypification of This Name. *Preslia* **2019**, *91*, 161–177. [CrossRef]
77. Naseem, H.; Ahsan, M.; Shahid, M.A.; Khan, N. Exopolysaccharides Producing Rhizobacteria and Their Role in Plant Growth and Drought Tolerance. *J. Basic Microbiol.* **2018**, *58*, 1009–1022. [CrossRef]
78. Yadav, K.; Yadav, P.; Singh, K. Microbial-Mediated Molecular Mechanisms to Cope Up Salinity and Drought Stress in Plants. In *Microbial Based Land Restoration Handbook, Volume 2*; CRC Press: Boca Raton, FL, USA, 2022; pp. 31–64. ISBN 9781003147077.
79. Jain, D.; Phurailatpam, L.; Mishra, S. Microbes-Mediated Mitigation of Drought Stress in Plants: Recent Trends and Future Challenges. In *Advances in Plant Microbiome and Sustainable Agriculture*; Springer: Berlin/Heidelberg, Germany, 2020; pp. 199–218.
80. Narayanasamy, S.; Uthandi, S. Microbial Mitigation of Drought Stress: Potential Mechanisms and Challenges. In *Microbial Management of Plant Stresses*; Elsevier: Amsterdam, The Netherlands, 2021; pp. 185–201.
81. Jatav, S.S.; Parihar, M.; Patra, A.; Singh, S.K.; Chitara, M.K.; Mohapatra, K.K.; Rana, K. Soil Microbes in Plant Growth Promotion and for Mitigation of Abiotic Stress of Drought. In *Soil Microbiomes for Sustainable Agriculture*; Springer: Berlin/Heidelberg, Germany, 2021; pp. 175–201.
82. Asghar, M.A.; Ahmad, B.; Raza, A.; Adil, B.; Hassan Javed, H.; Farooq, M.U.; Ghafoor, A.; Iftikhar Hussain, M.; Shafiq, I.; Karim, H.; et al. Shade and Microbes Enhance Drought Stress Tolerance in Plants by Inducing Phytohormones at Molecular Levels: A Review. *J. Plant Ecol.* **2022**, *15*, 1107–1117. [CrossRef]
83. Singh, D.; Thapa, S.; Yadav, J.; Singh, D.; Chakdar, H.; Kumar, M.; Saxena, A.K. Deciphering the Mechanisms of Microbe Mediated Drought Stress Alleviation in Wheat. *Acta Physiol. Plant.* **2023**, *45*, 81. [CrossRef]
84. Kour, D.; Yadav, A.N. Bacterial Mitigation of Drought Stress in Plants: Current Perspectives and Future Challenges. *Curr. Microbiol.* **2022**, *79*, 248. [CrossRef]
85. Kour, D.; Khan, S.S.; Kaur, T.; Kour, H.; Singh, G.; Yadav, A.; Yadav, A.N. Drought Adaptive Microbes as Bioinoculants for the Horticultural Crops. *Heliyon* **2022**, *8*, e09493. [CrossRef]
86. Gupta, A.; Bano, A.; Rai, S.; Mishra, R.; Singh, M.; Sharma, S.; Pathak, N. Mechanistic Insights of Plant-Microbe Interaction towards Drought and Salinity Stress in Plants for Enhancing the Agriculture Productivity. *Plant Stress* **2022**, *4*, 100073. [CrossRef]

87. Yadav, B.; Jogawat, A.; Rahman, M.S.; Narayan, O.P. Secondary Metabolites in the Drought Stress Tolerance of Crop Plants: A Review. *Gene Rep.* **2021**, *23*, 101040. [CrossRef]
88. Andreo-Jimenez, B.; Schilder, M.T.; Nijhuis, E.H.; te Beest, D.E.; Bloem, J.; Visser, J.H.M.; van Os, G.; Brolsma, K.; de Boer, W.; Postma, J. Chitin- and Keratin-Rich Soil Amendments Suppress Rhizoctonia Solani Disease via Changes to the Soil Microbial Community. *Appl. Environ. Microbiol.* **2021**, *87*, e00318-21. [CrossRef]
89. Sandhya, V.; Ali, S.Z.; Grover, M.; Reddy, G.; Venkateswarlu, B. Effect of Plant Growth Promoting *Pseudomonas* spp. on Compatible Solutes, Antioxidant Status and Plant Growth of Maize under Drought Stress. *Plant Growth Regul.* **2010**, *62*, 21–30. [CrossRef]
90. Ortiz, N.; Armada, E.; Duque, E.; Roldán, A.; Azcón, R. Contribution of Arbuscular Mycorrhizal Fungi and/or Bacteria to Enhancing Plant Drought Tolerance under Natural Soil Conditions: Effectiveness of Autochthonous or Allochthonous Strains. *J. Plant Physiol.* **2015**, *174*, 87–96. [CrossRef] [PubMed]
91. Gamalero, E.; Glick, B.R. Recent Advances in Bacterial Amelioration of Plant Drought and Salt Stress. *Biology* **2022**, *11*, 437. [CrossRef] [PubMed]
92. Hanaka, A.; Ozimek, E.; Reszczyńska, E.; Jaroszuk-Ściseł, J.; Stolarz, M. Plant Tolerance to Drought Stress in the Presence of Supporting Bacteria and Fungi: An Efficient Strategy in Horticulture. *Horticulturae* **2021**, *7*, 390. [CrossRef]
93. Camaille, M.; Fabre, N.; Clément, C.; Ait Barka, E. Advances in Wheat Physiology in Response to Drought and the Role of Plant Growth Promoting Rhizobacteria to Trigger Drought Tolerance. *Microorganisms* **2021**, *9*, 687. [CrossRef]
94. Singh, S.K.; Srikanth, G.S.; Puranik, S.; Shukla, L. Chemical Talk within Plant Holobiont: A Fascinating Conversation. In *Plant-Microbe Interaction-Recent Advances in Molecular and Biochemical Approaches*; Elsevier: Amsterdam, The Netherlands, 2023; pp. 165–203.
95. Rajput, S.; Sengupta, P.; Kohli, I.; Varma, A.; Singh, P.K.; Joshi, N.C. Role of Piriformospora Indica in Inducing Soil Microbial Communities and Drought Stress Tolerance in Plants. In *New and Future Developments in Microbial Biotechnology and Bioengineering*; Elsevier: Amsterdam, The Netherlands, 2022; pp. 93–110.
96. Kour, D.; Rana, K.L.; Kaur, T.; Sheikh, I.; Yadav, A.N.; Kumar, V.; Dhaliwal, H.S.; Saxena, A.K. Microbe-Mediated Alleviation of Drought Stress and Acquisition of Phosphorus in Great Millet (*Sorghum bicolour* L.) by Drought-Adaptive and Phosphorus-Solubilizing Microbes. *Biocatal. Agric. Biotechnol.* **2020**, *23*, 101501. [CrossRef]
97. Zhang, K.; Khan, Z.; Yu, Q.; Qu, Z.; Liu, J.; Luo, T.; Zhu, K.; Bi, J.; Hu, L.; Luo, L. Biochar Coating Is a Sustainable and Economical Approach to Promote Seed Coating Technology, Seed Germination, Plant Performance, and Soil Health. *Plants* **2022**, *11*, 2864. [CrossRef]
98. Singh, D.P.; Singh, V.; Gupta, V.K.; Shukla, R.; Prabha, R.; Sarma, B.K.; Patel, J.S. Microbial Inoculation in Rice Regulates Antioxidative Reactions and Defense Related Genes to Mitigate Drought Stress. *Sci. Rep.* **2020**, *10*, 4818. [CrossRef]
99. Mahdi Dar, Z.; Masood, A.; Hussain Mughal, A.; Asif, M.; Ahamd Malik, M. Review on Drought Tolerance in Plants Induced by Plant Growth Promoting Rhizobacteria. *Int. J. Curr. Microbiol. Appl. Sci.* **2018**, *7*, 412–422. [CrossRef]
100. Singh, S.; Kumar, V.; Datta, S.; Ramamurthy, P.C.; Singh, J. Molecular Mechanism and Signaling Pathways Interplay between Plant Hormones during Plant-Microbe Crosstalk. In *Microbial Management of Plant Stresses*; Elsevier: Amsterdam, The Netherlands, 2021; pp. 93–105.
101. Kaya, C. Microbial Modulation of Hormone Signaling, Proteomic Dynamics, and Metabolomics in Plant Drought Adaptation. *Food Energy Secur.* **2024**, *13*, e513. [CrossRef]
102. Gupta, A.; Singh, U.B.; Sahu, P.K.; Paul, S.; Kumar, A.; Malviya, D.; Singh, S.; Kuppusamy, P.; Singh, P.; Paul, D.; et al. Linking Soil Microbial Diversity to Modern Agriculture Practices: A Review. *Int. J. Environ. Res. Public Health* **2022**, *19*, 3141. [CrossRef]
103. Pandey, P.; Tripathi, A.; Dwivedi, S.; Lal, K.; Jhang, T. Deciphering the Mechanisms, Hormonal Signaling, and Potential Applications of Endophytic Microbes to Mediate Stress Tolerance in Medicinal Plants. *Front. Plant Sci.* **2023**, *14*, 1250020. [CrossRef] [PubMed]
104. Dhar, N.; Gopalan, N.S.R.; Nikhil, P.T.; Mohapatra, S. Role of Phytohormones in Plant-Microbial Interaction. In *Auxins, Cytokinins and Gibberellins Signaling in Plants*; Springer: Berlin/Heidelberg, Germany, 2022; pp. 313–336.
105. Bora, S.S.; Anshu; Deka, P.; Barooah, M. Drought-Tolerant Plant Growth-Promoting Rhizobacteria Associated with Millets. In *Millet Rhizosphere*; Springer Nature Singapore: Singapore, 2023; pp. 83–96.
106. Gowtham, H.G.; Singh, B.; Murali, M.; Shilpa, N.; Prasad, M.; Aiyaz, M.; Amruthesh, K.N.; Niranjana, S.R. Induction of Drought Tolerance in Tomato upon the Application of ACC Deaminase Producing Plant Growth Promoting Rhizobacterium Bacillus Subtilis Rhizo SF 48. *Microbiol. Res.* **2020**, *234*, 126422. [CrossRef] [PubMed]
107. Chandra, D.; Srivastava, R.; Gupta, V.V.S.R.; Franco, C.M.M.; Sharma, A.K. Evaluation of ACC-Deaminase-Producing Rhizobacteria to Alleviate Water-Stress Impacts in Wheat (*Triticum aestivum* L.) Plants. *Can. J. Microbiol.* **2019**, *65*, 387–403. [CrossRef] [PubMed]
108. Zhang, Y.; Du, H.; Xu, F.; Ding, Y.; Gui, Y.; Zhang, J.; Xu, W. Root-Bacteria Associations Boost Rhizosheath Formation in Moderately Dry Soil through Ethylene Responses. *Plant Physiol.* **2020**, *183*, 780–792. [CrossRef] [PubMed]
109. Ma, Y. Abiotic Stress Responses and Microbe-Mediated Mitigation in Plants. *Agronomy* **2023**, *13*, 1844. [CrossRef]
110. Bhagat, N.; Raghav, M.; Dubey, S.; Bedi, N. Bacterial Exopolysaccharides: Insight into Their Role in Plant Abiotic Stress Tolerance. *J. Microbiol. Biotechnol.* **2021**, *31*, 1045–1059. [CrossRef]
111. Ault, T.R. On the Essentials of Drought in a Changing Climate. *Science* **2020**, *368*, 256–260. [CrossRef]

112. Thakur, N.; Nigam, M.; Mann, N.A.; Gupta, S.; Hussain, C.M.; Shukla, S.K.; Shah, A.A.; Casini, R.; Elansary, H.O.; Khan, S.A. Host-Mediated Gene Engineering and Microbiome-Based Technology Optimization for Sustainable Agriculture and Environment. *Funct. Integr. Genom.* **2023**, *23*, 57. [CrossRef]
113. Sinha, T.; Nandi, K.; Das, R.; Prasad, S.N.; Pradhan, M.; Maurya, S.; Nandi, A. Microbe-Mediated Biotic and Abiotic Stress Tolerance in Crop Plants. In *Microbes and Microbial Biotechnology for Green Remediation*; Elsevier: Amsterdam, The Netherlands, 2022; pp. 93–116.
114. Dastogeer, K.M.G.; Zahan, M.I.; Rhaman, M.S.; Sarker, M.S.A.; Chakraborty, A. Microbe-Mediated Thermotolerance in Plants and Pertinent Mechanisms—A Meta-Analysis and Review. *Front. Microbiol.* **2022**, *13*, 833566. [CrossRef]
115. Ummara, U.; Noreen, S.; Afzal, M.; Zafar, Z.U.; Akhter, M.S.; Iqbal, S.; Hefft, D.I.; Kazi, M.; Ahmad, P. Induced Systemic Tolerance Mediated by Plant-Microbe Interaction in Maize (*Zea mays* L.) Plants under Hydrocarbon Contamination. *Chemosphere* **2022**, *290*, 133327. [CrossRef]
116. Elnajar, M.; Aldesuquy, H.; Abdelmoteleb, M.; Eltanahy, E. Mitigating Drought Stress in Wheat Plants (*Triticum aestivum* L.) through Grain Priming in Aqueous Extract of *Spirulina platensis*. *BMC Plant Biol.* **2024**, *24*, 233. [CrossRef] [PubMed]
117. Fadiji, A.E.; Orozco-Mosqueda, M.d.C.; Santos-Villalobos, S.d.l.; Santoyo, G.; Babalola, O.O. Recent Developments in the Application of Plant Growth-Promoting Drought Adaptive Rhizobacteria for Drought Mitigation. *Plants* **2022**, *11*, 3090. [CrossRef]
118. Joshi, H.; Bisht, N.; Mishra, S.K.; Prasad, V.; Chauhan, P.S. *Bacillus amyloliquefaciens* Modulate Carbohydrate Metabolism in Rice-PGPR Cross-Talk Under Abiotic Stress and Phytohormone Treatments. *J. Plant Growth Regul.* **2023**, *42*, 4466–4483. [CrossRef]
119. Galicia-Campos, E.; García-Villaraco, A.; Montero-Palmero, M.B.; Gutiérrez-Mañero, F.J.; Ramos-Solano, B. Bacillus G7 Improves Adaptation to Salt Stress in *Olea europaea* L. Plantlets, Enhancing Water Use Efficiency and Preventing Oxidative Stress. *Sci. Rep.* **2023**, *13*, 22507. [CrossRef] [PubMed]
120. Hasanuzzaman, M.; Bhuyan, M.H.M.; Zulfiqar, F.; Raza, A.; Mohsin, S.; Mahmud, J.; Fujita, M.; Fotopoulos, V. Reactive Oxygen Species and Antioxidant Defense in Plants under Abiotic Stress: Revisiting the Crucial Role of a Universal Defense Regulator. *Antioxidants* **2020**, *9*, 681. [CrossRef]
121. Lee, S.; Kim, J.-A.; Song, J.; Choe, S.; Jang, G.; Kim, Y. Plant Growth-Promoting Rhizobacterium Bacillus Megaterium Modulates the Expression of Antioxidant-Related and Drought-Responsive Genes to Protect Rice (*Oryza sativa* L.) from Drought. *Front. Microbiol.* **2024**, *15*, 1430546. [CrossRef]
122. Li, W.; Wang, H.; Lv, G.; Wang, J.; Li, J. Regulation of Drought Stress on Nutrient Cycle and Metabolism of Rhizosphere Microorganisms in Desert Riparian Forest. *Sci. Total Environ.* **2024**, *954*, 176148. [CrossRef]
123. Islam, M.R.; Islam, R.; Dutta, T.; Jhelom, F.R. Genetic and Microbial Insights into Drought Stress Alleviation in Tomato (*Solanum lycopersicum* L.). *Biol. Life Sci. Forum* **2023**, *27*, 22. [CrossRef]
124. Krishnamoorthy, R.; Anandham, R.; Senthilkumar, M.; Venkatramanan, V. Adaptation Mechanism of Methylotrophic Bacteria to Drought Condition and Its Strategies in Mitigating Plant Stress Caused by Climate Change. In *Exploring Synergies and Trade-offs between Climate Change and the Sustainable Development Goals*; Springer Singapore: Singapore, 2021; pp. 145–158.
125. Kaya, C.; Uğurlar, F.; Adamakis, I.-D.S. Epigenetic and Hormonal Modulation in Plant–Plant Growth-Promoting Microorganism Symbiosis for Drought-Resilient Agriculture. *Int. J. Mol. Sci.* **2023**, *24*, 16064. [CrossRef]
126. Nisha; Batra, N.; Sharma, V.; Sharma, R.; Sharma, A. Abiotic Stress Responses and Strategies of Microbes Mediated Mitigation for Sustainable Agriculture. *Vegetos* **2022**, *36*, 20–27. [CrossRef]
127. Lewis, W.H.; Tahon, G.; Geesink, P.; Sousa, D.Z.; Ettema, T.J.G. Innovations to Culturing the Uncultured Microbial Majority. *Nat. Rev. Microbiol.* **2021**, *19*, 225–240. [CrossRef] [PubMed]
128. Vishwakarma, K.; Kumar, N.; Shandilya, C.; Mohapatra, S.; Bhayana, S.; Varma, A. Revisiting Plant–Microbe Interactions and Microbial Consortia Application for Enhancing Sustainable Agriculture: A Review. *Front. Microbiol.* **2020**, *11*, 560406. [CrossRef] [PubMed]
129. Poppeliers, S.W.; Sánchez-Gil, J.J.; de Jonge, R. Microbes to Support Plant Health: Understanding Bioinoculant Success in Complex Conditions. *Curr. Opin. Microbiol.* **2023**, *73*, 102286. [CrossRef] [PubMed]
130. Jansson, J.K.; McClure, R.; Egbert, R.G. Soil Microbiome Engineering for Sustainability in a Changing Environment. *Nat. Biotechnol.* **2023**, *41*, 1716–1728. [CrossRef]
131. Bano, S.; WU, X.; Zhang, X. Towards Sustainable Agriculture: Rhizosphere Microbiome Engineering. *Appl. Microbiol. Biotechnol.* **2021**, *105*, 7141–7160. [CrossRef]
132. Nwankwegu, A.S.; Zhang, L.; Xie, D.; Onwosi, C.O.; Muhammad, W.I.; Odoh, C.K.; Sam, K.; Idenyi, J.N. Bioaugmentation as a Green Technology for Hydrocarbon Pollution Remediation. Problems and Prospects. *J. Environ. Manag.* **2022**, *304*, 114313. [CrossRef]
133. Gao, D.; Zhao, H.; Wang, L.; Li, Y.; Tang, T.; Bai, Y.; Liang, H. Current and Emerging Trends in Bioaugmentation of Organic Contaminated Soils: A Review. *J. Environ. Manag.* **2022**, *320*, 115799. [CrossRef]
134. Wangithi, C.; Muriithi, B.W.; Diiro, G.; Dubois, T.; Mohamed, S.; Lattorff, M.G.; Ngowi, B.V.; Abdel-Rahman, E.M.; Adan, M.; Kassie, M. Synergies of Integrated Pest and Pollinator Management in Avocado Farming in East Africa: An Ex-Ante Economic Analysis. *PLoS ONE* **2022**, *17*, e0271241. [CrossRef]
135. Ratnadass, A.; Avelino, J.; Fernandes, P.; Letourmy, P.; Babin, R.; Deberdt, P.; Deguine, J.-P.; Grechi, I.; Naudin, K.; Rhino, B.; et al. Synergies and Tradeoffs in Natural Regulation of Crop Pests and Diseases under Plant Species Diversification. *Crop Prot.* **2021**, *146*, 105658. [CrossRef]

136. Piñeiro, V.; Arias, J.; Dürr, J.; Elverdin, P.; Ibáñez, A.M.; Kinengyere, A.; Opazo, C.M.; Owoo, N.; Page, J.R.; Prager, S.D.; et al. A Scoping Review on Incentives for Adoption of Sustainable Agricultural Practices and Their Outcomes. *Nat. Sustain.* **2020**, *3*, 809–820. [CrossRef]
137. Kapoore, R.V.; Wood, E.E.; Llewellyn, C.A. Algae Biostimulants: A Critical Look at Microalgal Biostimulants for Sustainable Agricultural Practices. *Biotechnol. Adv.* **2021**, *49*, 107754. [CrossRef] [PubMed]
138. Arif, I.; Batool, M.; Schenk, P.M. Plant Microbiome Engineering: Expected Benefits for Improved Crop Growth and Resilience. *Trends Biotechnol.* **2020**, *38*, 1385–1396. [CrossRef] [PubMed]
139. Kaul, S.; Choudhary, M.; Gupta, S.; Dhar, M.K. Engineering Host Microbiome for Crop Improvement and Sustainable Agriculture. *Front. Microbiol.* **2021**, *12*, 635917. [CrossRef] [PubMed]
140. Kumar, A.; Dubey, A. Rhizosphere Microbiome: Engineering Bacterial Competitiveness for Enhancing Crop Production. *J. Adv. Res.* **2020**, *24*, 337–352. [CrossRef]
141. Lau, S.-E.; Teo, W.F.A.; Teoh, E.Y.; Tan, B.C. Microbiome Engineering and Plant Biostimulants for Sustainable Crop Improvement and Mitigation of Biotic and Abiotic Stresses. *Discov. Food* **2022**, *2*, 9. [CrossRef]
142. Phour, M.; Sindhu, S.S. Mitigating Abiotic Stress: Microbiome Engineering for Improving Agricultural Production and Environmental Sustainability. *Planta* **2022**, *256*, 85. [CrossRef]
143. Yang, S.; Liu, H.; Xie, P.; Wen, T.; Shen, Q.; Yuan, J. Emerging Pathways for Engineering the Rhizosphere Microbiome for Optimal Plant Health. *J. Agric. Food Chem.* **2023**, *71*, 4441–4449. [CrossRef]
144. Rodríguez, R.; Barra, P.J.; Larama, G.; Carrion, V.J.; de la Luz Mora, M.; Hale, L.; Durán, P. Microbiome Engineering Optimized by Antarctic Microbiota to Support a Plant Host under Water Deficit. *Front. Plant Sci.* **2023**, *14*, 1241612. [CrossRef]
145. Zhang, Y.; Trivedi, P.; Xu, J.; Roper, M.C.; Wang, N. The Citrus Microbiome: From Structure and Function to Microbiome Engineering and Beyond. *Phytobiomes J.* **2021**, *5*, 249–262. [CrossRef]
146. Li, H.; Li, Y.; Bao, M.; Li, S. Solid Inoculants as a Practice for Bioaugmentation to Enhance Bioremediation of Hydrocarbon Contaminated Areas. *Chemosphere* **2021**, *263*, 128175. [CrossRef]
147. Lu, X.-L.; Ding, K.; Dong, X.-X.; Li, G.; Ma, J. A Bioaugmentation Strategy for Promoting the Humification Process during Composting by Microbial Inoculants: A Review. In *Organic Fertilizers-New Advances and Applications [Working Title]*; IntechOpen: Rijeka, Croatia, 2023.
148. Adetunji, C.O.; Anani, O.A. Bioaugmentation: A Powerful Biotechnological Techniques for Sustainable Ecorestoration of Soil and Groundwater Contaminants. In *Microbial Rejuvenation of Polluted Environment*; Springer: Berlin/Heidelberg, Germany, 2021; pp. 373–398.
149. Brzeszcz, J.; Kapusta, P.; Steliga, T.; Turkiewicz, A. Hydrocarbon Removal by Two Differently Developed Microbial Inoculants and Comparing Their Actions with Biostimulation Treatment. *Molecules* **2020**, *25*, 661. [CrossRef] [PubMed]
150. Tondera, K.; Chazarenc, F.; Chagnon, P.-L.; Brisson, J. Bioaugmentation of Treatment Wetlands—A Review. *Sci. Total Environ.* **2021**, *775*, 145820. [CrossRef] [PubMed]
151. Gorodylova, N.; Seron, A.; Michel, K.; Joulian, C.; Delorme, F.; Soulier, C.; Bresch, S.; Garreau, C.; Giovannelli, F.; Michel, C. Zeolite-Supported Biofilms as Inoculants for the Treatment of MCPA-Polluted Soil and Sand by Bioaugmentation: A Microcosm Study. *Appl. Soil Ecol.* **2022**, *180*, 104614. [CrossRef]
152. Balseiro-Romero, M.; Wick, L.Y.; Vila, J.; Grifoll, M.; Ortega-Calvo, J.J. Drivers for Efficient Bioaugmentation and Clean-Up of Contaminated Soil. In *Soil Remediation Science and Technology*; Springer: Berlin/Heidelberg, Germany, 2024; pp. 261–291.
153. Dehnavi, S.M.; Ebrahimipour, G. Comparative Remediation Rate of Biostimulation, Bioaugmentation, and Phytoremediation in Hydrocarbon Contaminants. *Int. J. Environ. Sci. Technol.* **2022**, *19*, 11561–11586. [CrossRef]
154. Song, Y.Q.; Shahir, S.; Abd Manan, F. Bacterial Inoculant-Assisted Phytoremediation of Heavy Metal-Contaminated Soil: Inoculant Development and the Inoculation Effects. *Biologia* **2021**, *76*, 2675–2685. [CrossRef]
155. López, A.M.Q.; dos Santos Silva, A.L. Biostimulation and Bioaugmentation. In *Genomics Approach to Bioremediation*; Wiley: Hoboken, NJ, USA, 2023; pp. 53–68.
156. Agrawal, A.; Maharana, J.K.; Patel, A.K. Microbial Inoculants and Their Potential Application in Bioremediation. In *Microbes Based Approaches for the Management of Hazardous Contaminants*; Wiley: Hoboken, NJ, USA, 2024; pp. 321–344.
157. Mawang, C.-I.; Azman, A.-S.; Fuad, A.-S.M.; Ahamad, M. Actinobacteria: An Eco-Friendly and Promising Technology for the Bioaugmentation of Contaminants. *Biotechnol. Rep.* **2021**, *32*, e00679. [CrossRef]
158. Baker, B.P.; Green, T.A.; Loker, A.J. Biological Control and Integrated Pest Management in Organic and Conventional Systems. *Biol. Control* **2020**, *140*, 104095. [CrossRef]
159. Deguine, J.-P.; Aubertot, J.-N.; Flor, R.J.; Lescourret, F.; Wyckhuys, K.A.G.; Ratnadass, A. Integrated Pest Management: Good Intentions, Hard Realities. A Review. *Agron. Sustain. Dev.* **2021**, *41*, 38. [CrossRef]
160. Akanmu, A.O.; Babalola, O.O.; Venturi, V.; Ayilara, M.S.; Adeleke, B.S.; Amoo, A.E.; Sobowale, A.A.; Fadiji, A.E.; Glick, B.R. Plant Disease Management: Leveraging on the Plant-Microbe-Soil Interface in the Biorational Use of Organic Amendments. *Front. Plant Sci.* **2021**, *12*, 700507. [CrossRef]
161. Fahad, S.; Saud, S.; Akhter, A.; Bajwa, A.A.; Hassan, S.; Battaglia, M.; Adnan, M.; Wahid, F.; Datta, R.; Babur, E.; et al. Bio-Based Integrated Pest Management in Rice: An Agro-Ecosystems Friendly Approach for Agricultural Sustainability. *J. Saudi Soc. Agric. Sci.* **2021**, *20*, 94–102. [CrossRef]

162. Dannon, H.F.; Dannon, A.E.; Douro-Kpindou, O.K.; Zinsou, A.V.; Houndete, A.T.; Toffa-Mehinto, J.; Elegbede, I.A.T.M.; Olou, B.D.; Tamò, M. Toward the Efficient Use of Beauveria Bassiana in Integrated Cotton Insect Pest Management. *J. Cotton Res.* **2020**, *3*, 24. [CrossRef]
163. Niu, B.; Wang, W.; Yuan, Z.; Sederoff, R.R.; Sederoff, H.; Chiang, V.L.; Borriss, R. Microbial Interactions Within Multiple-Strain Biological Control Agents Impact Soil-Borne Plant Disease. *Front. Microbiol.* **2020**, *11*, 585404. [CrossRef] [PubMed]
164. Benmrid, B.; Ghoulam, C.; Zeroual, Y.; Kouisni, L.; Bargaz, A. Bioinoculants as a Means of Increasing Crop Tolerance to Drought and Phosphorus Deficiency in Legume-Cereal Intercropping Systems. *Commun. Biol.* **2023**, *6*, 1016. [CrossRef] [PubMed]
165. Poveda, J. Cyanobacteria in Plant Health: Biological Strategy against Abiotic and Biotic Stresses. *Crop Prot.* **2021**, *141*, 105450. [CrossRef]
166. Bhattacharjee, P.; Nath, S.; Bhattacharjee, S.; Yadav, M.R. Exploring the Synergy of Microbes and Plants: A Promising Strategy for Managing Moisture Stress in Medicinal and Aromatic Crops. In *Microbial Biostimulants for Plant Growth and Abiotic Stress Amelioration*; Elsevier: Amsterdam, The Netherlands, 2024; pp. 367–395.
167. Sani, M.N.H.; Yong, J.W.H. Harnessing Synergistic Biostimulatory Processes: A Plausible Approach for Enhanced Crop Growth and Resilience in Organic Farming. *Biology* **2021**, *11*, 41. [CrossRef]
168. Lin, P.-A.; Paudel, S.; Lai, P.-C.; GC, R.K.; Yang, D.-H.; Felton, G.W. Integrating Water and Insect Pest Management in Agriculture. *J. Pest Sci.* **2024**, *97*, 521–538. [CrossRef]
169. Gvozdenac, S.; Dedić, B.; Mikić, S.; Ovuka, J.; Miladinović, D. Impact of Climate Change on Integrated Pest Management Strategies. In *Climate Change and Agriculture*; Wiley: Hoboken, NJ, USA, 2022; pp. 311–372.
170. Sammauria, R.; Kumawat, S.; Kumawat, P.; Singh, J.; Jatwa, T.K. Microbial Inoculants: Potential Tool for Sustainability of Agricultural Production Systems. *Arch. Microbiol.* **2020**, *202*, 677–693. [CrossRef]
171. Bamdad, H.; Papari, S.; Lazarovits, G.; Berruti, F. Soil Amendments for Sustainable Agriculture: Microbial Organic Fertilizers. *Soil Use Manag.* **2022**, *38*, 94–120. [CrossRef]
172. Bertola, M.; Ferrarini, A.; Visioli, G. Improvement of Soil Microbial Diversity through Sustainable Agricultural Practices and Its Evaluation by -Omics Approaches: A Perspective for the Environment, Food Quality and Human Safety. *Microorganisms* **2021**, *9*, 1400. [CrossRef]
173. Elnahal, A.S.M.; El-Saadony, M.T.; Saad, A.M.; Desoky, E.-S.M.; El-Tahan, A.M.; Rady, M.M.; AbuQamar, S.F.; El-Tarabily, K.A. The Use of Microbial Inoculants for Biological Control, Plant Growth Promotion, and Sustainable Agriculture: A Review. *Eur. J. Plant Pathol.* **2022**, *162*, 759–792. [CrossRef]
174. Poveda, J.; Baptista, P. Filamentous Fungi as Biocontrol Agents in Olive (*Olea europaea* L.) Diseases: Mycorrhizal and Endophytic Fungi. *Crop Prot.* **2021**, *146*, 105672. [CrossRef]
175. Vidal, C.; González, F.; Santander, C.; Pérez, R.; Gallardo, V.; Santos, C.; Aponte, H.; Ruiz, A.; Cornejo, P. Management of Rhizosphere Microbiota and Plant Production under Drought Stress: A Comprehensive Review. *Plants* **2022**, *11*, 2437. [CrossRef] [PubMed]
176. Silverstein, M.R.; Segrè, D.; Bhatnagar, J.M. Environmental Microbiome Engineering for the Mitigation of Climate Change. *Glob. Change Biol.* **2023**, *29*, 2050–2066. [CrossRef] [PubMed]
177. Curiel-Alegre, S.; Velasco-Arroyo, B.; Rumbo, C.; Khan, A.H.A.; Tamayo-Ramos, J.A.; Rad, C.; Gallego, J.L.R.; Barros, R. Evaluation of Biostimulation, Bioaugmentation, and Organic Amendments Application on the Bioremediation of Recalcitrant Hydrocarbons of Soil. *Chemosphere* **2022**, *307*, 135638. [CrossRef]
178. Bakina, L.G.; Chugunova, M.V.; Polyak, Y.M.; Mayachkina, N.V.; Gerasimov, A.O. Bioaugmentation: Possible Scenarios Due to Application of Bacterial Preparations for Remediation of Oil-Contaminated Soil. *Environ. Geochem. Health* **2021**, *43*, 2347–2356. [CrossRef]
179. Galindo, F.S.; Rodrigues, W.L.; Fernandes, G.C.; Boleta, E.H.M.; Jalal, A.; Rosa, P.A.L.; Buzetti, S.; Lavres, J.; Teixeira Filho, M.C.M. Enhancing Agronomic Efficiency and Maize Grain Yield with Azospirillum Brasilense Inoculation under Brazilian Savannah Conditions. *Eur. J. Agron.* **2022**, *134*, 126471. [CrossRef]
180. Yasuda, M.; Dastogeer, K.M.G.; Sarkodee-Addo, E.; Tokiwa, C.; Isawa, T.; Shinozaki, S.; Okazaki, S. Impact of Azospirillum Sp. B510 on the Rhizosphere Microbiome of Rice under Field Conditions. *Agronomy* **2022**, *12*, 1367. [CrossRef]
181. Ingrey, S.D.; Pearson, L.A.; Kalaitzis, J.A.; Neilan, B.A. Australian Bush Medicines Harbour Diverse Microbial Endophytes with Broad-spectrum Antibacterial Activity. *J. Appl. Microbiol.* **2021**, *131*, 2244–2256. [CrossRef]
182. Mathur, P.; Roy, S. Insights into the Plant Responses to Drought and Decoding the Potential of Root Associated Microbiome for Inducing Drought Tolerance. *Physiol. Plant.* **2021**, *172*, 1016–1029. [CrossRef]
183. Etesami, H. Potential Advantage of Rhizosheath Microbiome, in Contrast to Rhizosphere Microbiome, to Improve Drought Tolerance in Crops. *Rhizosphere* **2021**, *20*, 100439. [CrossRef]
184. Loiko, N.; Islam, M.N. Plant–Soil Microbial Interaction: Differential Adaptations of Beneficial vs. Pathogenic Bacterial and Fungal Communities to Climate-Induced Drought. *Agronomy* **2024**, *14*, 1949. [CrossRef]
185. Kang, J.; Peng, Y.; Xu, W. Crop Root Responses to Drought Stress: Molecular Mechanisms, Nutrient Regulations, and Interactions with Microorganisms in the Rhizosphere. *Int. J. Mol. Sci.* **2022**, *23*, 9310. [CrossRef] [PubMed]
186. Muhammad, M.; Waheed, A.; Wahab, A.; Majeed, M.; Nazim, M.; Liu, Y.-H.; Li, L.; Li, W.-J. Soil Salinity and Drought Tolerance: An Evaluation of Plant Growth, Productivity, Microbial Diversity, and Amelioration Strategies. *Plant Stress* **2024**, *11*, 100319. [CrossRef]

187. Liu, J.; Chen, J.; Ye, S.; Ding, Y.; Guo, S.; Fan, N. Inhibitory Action of Ginkgolic Acid against Pathogenic Fungi and Characterisation of Its Inhibitory Activities on *Nigrospora oryzae*. *Folia Hortic.* **2023**, *35*, 49–59. [CrossRef]
188. Zhang, H.; Sun, X.; Dai, M. Improving Crop Drought Resistance with Plant Growth Regulators and Rhizobacteria: Mechanisms, Applications, and Perspectives. *Plant Commun.* **2022**, *3*, 100228. [CrossRef]
189. Sudheer, S.; Bai, R.G.; Usmani, Z.; Sharma, M. Insights on Engineered Microbes in Sustainable Agriculture: Biotechnological Developments and Future Prospects. *Curr. Genom.* **2020**, *21*, 321–333. [CrossRef]
190. Allison, S.D. Microbial Drought Resistance May Destabilize Soil Carbon. *Trends Microbiol.* **2023**, *31*, 780–787. [CrossRef]
191. Azeem, M.; Sun, T.-R.; Jeyasundar, P.G.S.A.; Han, R.-X.; Li, H.; Abdelrahman, H.; Shaheen, S.M.; Zhu, Y.-G.; Li, G. Biochar-Derived Dissolved Organic Matter (BDOM) and Its Influence on Soil Microbial Community Composition, Function, and Activity: A Review. *Crit. Rev. Environ. Sci. Technol.* **2023**, *53*, 1912–1934. [CrossRef]
192. Yadav, V.K.; Raghav, M.; Sharma, S.K.; Bhagat, N. Rhizobacteriome: Promising Candidate for Conferring Drought Tolerance in Crops. *J. Pure Appl. Microbiol.* **2020**, *14*, 73–92. [CrossRef]
193. Fulcher, M.R.; Tancos, M.A.; Mueller, R.C.; Tannières, M. Importance of Pathobiomes to the Success of Microbial Weed Biocontrol Agents. *Biol. Control* **2024**, *192*, 105498. [CrossRef]
194. Ahmed, T.; Noman, M.; Gardea-Torresdey, J.L.; White, J.C.; Li, B. Dynamic Interplay between Nano-Enabled Agrochemicals and the Plant-Associated Microbiome. *Trends Plant Sci.* **2023**, *28*, 1310–1325. [CrossRef] [PubMed]
195. Allan, J.; Belz, S.; Hoeveler, A.; Hugas, M.; Okuda, H.; Patri, A.; Rauscher, H.; Silva, P.; Slikker, W.; Sokull-Kluettgen, B.; et al. Regulatory Landscape of Nanotechnology and Nanoplastics from a Global Perspective. *Regul. Toxicol. Pharmacol.* **2021**, *122*, 104885. [CrossRef] [PubMed]
196. Rossi, S.; Antal, A.; Bestmann, S.; Bikson, M.; Brewer, C.; Brockmöller, J.; Carpenter, L.L.; Cincotta, M.; Chen, R.; Daskalakis, J.D.; et al. Safety and Recommendations for TMS Use in Healthy Subjects and Patient Populations, with Updates on Training, Ethical and Regulatory Issues: Expert Guidelines. *Clin. Neurophysiol.* **2021**, *132*, 269–306. [CrossRef]
197. Sofo, A.; Zanella, A.; Ponge, J. Soil Quality and Fertility in Sustainable Agriculture, with a Contribution to the Biological Classification of Agricultural Soils. *Soil Use Manag.* **2022**, *38*, 1085–1112. [CrossRef]
198. Kumawat, K.C.; Razdan, N.; Saharan, K. Rhizospheric Microbiome: Bio-Based Emerging Strategies for Sustainable Agriculture Development and Future Perspectives. *Microbiol. Res.* **2022**, *254*, 126901. [CrossRef]
199. Diatta, A.A.; Min, D.; Jagadish, S.V.K. Drought Stress Responses in Non-Transgenic and Transgenic Alfalfa—Current Status and Future Research Directions. *Adv. Agron.* **2021**, *170*, 35–100.
200. Afzal, I.; Javed, T.; Amirkhani, M.; Taylor, A.G. Modern Seed Technology: Seed Coating Delivery Systems for Enhancing Seed and Crop Performance. *Agriculture* **2020**, *10*, 526. [CrossRef]
201. Qiu, Y.; Amirkhani, M.; Mayton, H.; Chen, Z.; Taylor, A.G. Biostimulant Seed Coating Treatments to Improve Cover Crop Germination and Seedling Growth. *Agronomy* **2020**, *10*, 154. [CrossRef]
202. Sheoran, S.; Kaur, Y.; Kumar, S.; Shukla, S.; Rakshit, S.; Kumar, R. Recent Advances for Drought Stress Tolerance in Maize (*Zea Mays* L.): Present Status and Future Prospects. *Front. Plant Sci.* **2022**, *13*, 872566. [CrossRef]
203. Akbar, S.; Coiera, E.; Magrabi, F. Safety Concerns with Consumer-Facing Mobile Health Applications and Their Consequences: A Scoping Review. *J. Am. Med. Inform. Assoc.* **2020**, *27*, 330–340. [CrossRef]
204. Testa, C.J.; Hu, C.; Shvedova, K.; Wu, W.; Sayin, R.; Casati, F.; Halkude, B.S.; Hermant, P.; Shen, D.E.; Ramnath, A.; et al. Design and Commercialization of an End-to-End Continuous Pharmaceutical Production Process: A Pilot Plant Case Study. *Org. Process Res. Dev.* **2020**, *24*, 2874–2889. [CrossRef]
205. Haddi, K.; Turchen, L.M.; Viteri Jumbo, L.O.; Guedes, R.N.; Pereira, E.J.; Aguiar, R.W.; Oliveira, E.E. Rethinking Biorational Insecticides for Pest Management: Unintended Effects and Consequences. *Pest. Manag. Sci.* **2020**, *76*, 2286–2293. [CrossRef] [PubMed]
206. Karlsson Green, K.; Stenberg, J.A.; Lankinen, Å. Making Sense of Integrated Pest Management (IPM) in the Light of Evolution. *Evol. Appl.* **2020**, *13*, 1791–1805. [CrossRef] [PubMed]
207. Lundin, O.; Rundlöf, M.; Jonsson, M.; Bommarco, R.; Williams, N.M. Integrated Pest and Pollinator Management—Expanding the Concept. *Front. Ecol. Environ.* **2021**, *19*, 283–291. [CrossRef]
208. Snyder, L.D.; Gómez, M.I.; Power, A.G. Crop Varietal Mixtures as a Strategy to Support Insect Pest Control, Yield, Economic, and Nutritional Services. *Front. Sustain. Food Syst.* **2020**, *4*, 60. [CrossRef]
209. Kumar, V.; Nautiyal, C.S. Plant Abiotic and Biotic Stress Alleviation: From an Endophytic Microbial Perspective. *Curr. Microbiol.* **2022**, *79*, 311. [CrossRef]
210. Pereira, S.I.A.; Abreu, D.; Moreira, H.; Vega, A.; Castro, P.M.L. Plant Growth-Promoting Rhizobacteria (PGPR) Improve the Growth and Nutrient Use Efficiency in Maize (*Zea Mays* L.) under Water Deficit Conditions. *Heliyon* **2020**, *6*, e05106. [CrossRef]
211. Wang, H.; Liu, R.; You, M.P.; Barbetti, M.J.; Chen, Y. Pathogen Biocontrol Using Plant Growth-Promoting Bacteria (PGPR): Role of Bacterial Diversity. *Microorganisms* **2021**, *9*, 1988. [CrossRef]
212. Adedeji, A.A.; Häggblom, M.M.; Babalola, O.O. Sustainable Agriculture in Africa: Plant Growth-Promoting Rhizobacteria (PGPR) to the Rescue. *Sci. Afr.* **2020**, *9*, e00492. [CrossRef]
213. Muter, O. Current Trends in Bioaugmentation Tools for Bioremediation: A Critical Review of Advances and Knowledge Gaps. *Microorganisms* **2023**, *11*, 710. [CrossRef]

214. Afridi, M.S.; Ali, S.; Salam, A.; César Terra, W.; Hafeez, A.; Sumaira; Ali, B.; Al Tami, M.S.; Ameen, F.; Ercisli, S.; et al. Plant Microbiome Engineering: Hopes or Hypes. *Biology* **2022**, *11*, 1782. [CrossRef] [PubMed]
215. Neshat, M.; Abbasi, A.; Hosseinzadeh, A.; Sarikhani, M.R.; Dadashi Chavan, D.; Rasoulnia, A. Plant Growth Promoting Bacteria (PGPR) Induce Antioxidant Tolerance against Salinity Stress through Biochemical and Physiological Mechanisms. *Physiol. Mol. Biol. Plants* **2022**, *28*, 347–361. [CrossRef] [PubMed]
216. Łotocka, B.; Wysokińska, E.; Pitera, E.; Szpadzik, E. Ultrastructure of Receptive Stigma and Transmitting Tissue at Anthesis in Two Pear Species. *Acta Agrobot.* **2023**, *76*, 169344. [CrossRef]
217. Das, P.P.; Singh, K.R.; Nagpure, G.; Mansoori, A.; Singh, R.P.; Ghazi, I.A.; Kumar, A.; Singh, J. Plant-Soil-Microbes: A Tripartite Interaction for Nutrient Acquisition and Better Plant Growth for Sustainable Agricultural Practices. *Environ. Res.* **2022**, *214*, 113821. [CrossRef]
218. Alsharif, W.; Saad, M.M.; Hirt, H. Desert Microbes for Boosting Sustainable Agriculture in Extreme Environments. *Front. Microbiol.* **2020**, *11*, 1666. [CrossRef]
219. Wiesnerová, L.; Hřebečková, T.; Jablonský, I.; Koudela, M. Effect of Different Water Contents in the Substrate on Cultivation of *Pleurotus ostreatus* Jacq. P. Kumm. *Folia Hortic.* **2023**, *35*, 25–31. [CrossRef]
220. Łysiak, G.P.; Szot, I. The Possibility of Using Fruit-Bearing Plants of Temperate Climate in the Treatment and Prevention of Diabetes. *Life* **2023**, *13*, 1795. [CrossRef]
221. Trivedi, P.; Mattupalli, C.; Eversole, K.; Leach, J.E. Enabling Sustainable Agriculture through Understanding and Enhancement of Microbiomes. *New Phytol.* **2021**, *230*, 2129–2147. [CrossRef]
222. Ullah, S.; Bano, A.; Ullah, A.; Shahid, M.A.; Khan, N. A Comparative Study of Plant Growth Promoting Rhizobacteria (PGPR) and Sowing Methods on Nutrient Availability in Wheat and Rhizosphere Soil under Salinity Stress. *Rhizosphere* **2022**, *23*, 100571. [CrossRef]
223. Shaffique, S.; Khan, M.A.; Imran, M.; Kang, S.-M.; Park, Y.-S.; Wani, S.H.; Lee, I.-J. Research Progress in the Field of Microbial Mitigation of Drought Stress in Plants. *Front. Plant Sci.* **2022**, *13*, 870626. [CrossRef]
224. Sheteiwy, M.S.; Ali, D.F.I.; Xiong, Y.-C.; Brestic, M.; Skalicky, M.; Hamoud, Y.A.; Ulhassan, Z.; Shaghaleh, H.; AbdElgawad, H.; Farooq, M.; et al. Physiological and Biochemical Responses of Soybean Plants Inoculated with Arbuscular Mycorrhizal Fungi and Bradyrhizobium under Drought Stress. *BMC Plant Biol.* **2021**, *21*, 195. [CrossRef]
225. Gehring, C.A.; Sthultz, C.M.; Flores-Rentería, L.; Whipple, A.V.; Whitham, T.G. Tree Genetics Defines Fungal Partner Communities That May Confer Drought Tolerance. *Proc. Natl. Acad. Sci. USA* **2017**, *114*, 11169–11174. [CrossRef]
226. Alwutayd, K.M.; Rawat, A.A.; Sheikh, A.H.; Almeida-Trapp, M.; Veluchamy, A.; Jalal, R.; Karampelias, M.; Froehlich, K.; Alzaed, W.; Tabassum, N.; et al. Microbe-induced Drought Tolerance by ABA-mediated Root Architecture and Epigenetic Reprogramming. *EMBO Rep.* **2023**, *24*, e56754. [CrossRef]
227. Xie, J.; Dawwam, G.E.; Sehim, A.E.; Li, X.; Wu, J.; Chen, S.; Zhang, D. Drought Stress Triggers Shifts in the Root Microbial Community and Alters Functional Categories in the Microbial Gene Pool. *Front. Microbiol.* **2021**, *12*, 744897. [CrossRef]
228. Akhtar, M.N.; Balodi, R.; Ghatak, A. Microbe-Mediated Mitigation of Plant Stress. In *Microbial Services in Restoration Ecology*; Elsevier: Amsterdam, The Netherlands, 2020; pp. 271–282.
229. Sood, G.; Kaushal, R.; Sharma, M. Alleviation of Drought Stress in Maize (*Zea mays* L.) by Using Endogenous Endophyte *Bacillus subtilis* in North West Himalayas. *Acta Agric. Scand. B Soil Plant Sci.* **2020**, *70*, 361–370. [CrossRef]
230. Bauer, S.; McNamara, J.M.; Barta, Z. Environmental Variability, Reliability of Information and the Timing of Migration. *Proc. R. Soc. B Biol. Sci.* **2020**, *287*, 20200622. [CrossRef]
231. Sangare, N.; Lo-Yat, A.; Le Moullac, G.; Pecquerie, L.; Thomas, Y.; Lefebvre, S.; Gendre, R.L.; Beliaeff, B.; Andréfouët, S. Impact of Environmental Variability on Pinctada Margaritifera Life-History Traits: A Full Life Cycle Deb Modeling Approach. *Ecol. Modell.* **2020**, *423*, 109006. [CrossRef]
232. Watson, J.W.; Boyd, R.; Dutta, R.; Vasdekis, G.; Walker, N.D.; Roy, S.; Everitt, R.; Hyder, K.; Sibly, R.M. Incorporating Environmental Variability in a Spatially-Explicit Individual-Based Model of European Sea Bass. *Ecol. Modell.* **2022**, *466*, 109878. [CrossRef]
233. Castelo, W.F.L.; Martins, M.V.A.; Martínez-Colón, M.; Guerra, J.V.; Dadalto, T.P.; Terroso, D.; Soares, M.F.; Frontalini, F.; Duleba, W.; Socorro, O.A.A.; et al. Disentangling Natural vs. Anthropogenic Induced Environmental Variability during the Holocene: Marambaia Cove, SW Sector of the Sepetiba Bay (SE Brazil). *Environ. Sci. Pollut. Res.* **2021**, *28*, 22612–22640. [CrossRef]
234. Kumar, R.E.; Tiihonen, A.; Sun, S.; Fenning, D.P.; Liu, Z.; Buonassisi, T. Opportunities for Machine Learning to Accelerate Halide-Perovskite Commercialization and Scale-Up. *Matter* **2022**, *5*, 1353–1366. [CrossRef]
235. Usmani, Z.; Sharma, M.; Awasthi, A.K.; Lukk, T.; Tuohy, M.G.; Gong, L.; Nguyen-Tri, P.; Goddard, A.D.; Bill, R.M.; Nayak, S.C.; et al. Lignocellulosic Biorefineries: The Current State of Challenges and Strategies for Efficient Commercialization. *Renew. Sustain. Energy Rev.* **2021**, *148*, 111258. [CrossRef]
236. Saini, R.; Osorio-Gonzalez, C.S.; Hegde, K.; Brar, S.K.; Magdouli, S.; Vezina, P.; Avalos-Ramirez, A. Lignocellulosic Biomass-Based Biorefinery: An Insight into Commercialization and Economic Standout. *Curr. Sustain./Renew. Energy Rep.* **2020**, *7*, 122–136. [CrossRef]
237. Mula, C.; Zybura, N.; Hipp, T. From Digitalized Start-up to Scale-up: Opening the Black Box of Scaling in Digitalized Firms towards a Scaling Process Framework. *Technol. Forecast. Soc. Change* **2024**, *202*, 123275. [CrossRef]

238. Chen, Y.; Ma, X.; Ma, W.; Xu, L.; Lai, C.; Zhang, X.; Guan, Q.; Zhang, Z.; Chen, Y.; Lai, Z.; et al. AGL61/80 Regulates BGAL9, Which Participates in Early Somatic Embryogenesis and Increases Longan Resistance to Heat Stress. *Fruit Res.* **2024**, *4*, e013. [CrossRef]
239. Poveda, J.; Eugui, D. Combined Use of Trichoderma and Beneficial Bacteria (*Mainly bacillus* and *Pseudomonas*): Development of Microbial Synergistic Bio-Inoculants in Sustainable Agriculture. *Biol. Control* **2022**, *176*, 105100. [CrossRef]
240. Paravar, A.; Piri, R.; Balouchi, H.; Ma, Y. Microbial Seed Coating: An Attractive Tool for Sustainable Agriculture. *Biotechnol. Rep.* **2023**, *37*, e00781. [CrossRef]
241. Ebrahimi-Zarandi, M.; Etesami, H.; Glick, B.R. Fostering Plant Resilience to Drought with Actinobacteria: Unveiling Perennial Allies in Drought Stress Tolerance. *Plant Stress* **2023**, *10*, 100242. [CrossRef]
242. Bonanomi, G.; De Filippis, F.; Zotti, M.; Idbella, M.; Cesarano, G.; Al-Rowaily, S.; Abd-ElGawad, A. Repeated Applications of Organic Amendments Promote Beneficial Microbiota, Improve Soil Fertility and Increase Crop Yield. *Appl. Soil Ecol.* **2020**, *156*, 103714. [CrossRef]
243. Imran; Amanullah; Arif, M.; Shah, Z.; Bari, A. Integration of Peach (*Prunus persica* L.) Residues, Beneficial Microbes and Phosphorous Enhance Phenology, Growth and Yield of Soybean. *Russ. Agric. Sci.* **2020**, *46*, 223–230. [CrossRef]
244. Warrad, M.; Hassan, Y.M.; Mohamed, M.S.M.; Hagagy, N.; Al-Maghrabi, O.A.; Selim, S.; Saleh, A.M.; AbdElgawad, H. A Bioactive Fraction from *Streptomyces* sp. Enhances Maize Tolerance against Drought Stress. *J. Microbiol. Biotechnol.* **2020**, *30*, 1156–1168. [CrossRef]
245. Imathiu, S. Benefits and Food Safety Concerns Associated with Consumption of Edible Insects. *NFS J.* **2020**, *18*, 1–11. [CrossRef]
246. Khaliq, A.; Perveen, S.; Alamer, K.H.; Zia Ul Haq, M.; Rafique, Z.; Alsudays, I.M.; Althobaiti, A.T.; Saleh, M.A.; Hussain, S.; Attia, H. Arbuscular Mycorrhizal Fungi Symbiosis to Enhance Plant–Soil Interaction. *Sustainability* **2022**, *14*, 7840. [CrossRef]
247. Ayilara, M.S.; Adeleke, B.S.; Akinola, S.A.; Fayose, C.A.; Adeyemi, U.T.; Gbadegesin, L.A.; Omole, R.K.; Johnson, R.M.; Uthman, Q.O.; Babalola, O.O. Biopesticides as a Promising Alternative to Synthetic Pesticides: A Case for Microbial Pesticides, Phytopesticides, and Nanobiopesticides. *Front. Microbiol.* **2023**, *14*, 1040901. [CrossRef]
248. Morales-Quintana, L.; Moya, M.; Santelices-Moya, R.; Cabrera-Ariza, A.; Rabert, C.; Pollmann, S.; Ramos, P. Improvement in the Physiological and Biochemical Performance of Strawberries under Drought Stress through Symbiosis with Antarctic Fungal Endophytes. *Front. Microbiol.* **2022**, *13*, 939955. [CrossRef]
249. Li, M.; Ren, Y.; He, C.; Yao, J.; Wei, M.; He, X. Complementary Effects of Dark Septate Endophytes and Trichoderma Strains on Growth and Active Ingredient Accumulation of Astragalus Mongholicus under Drought Stress. *J. Fungi* **2022**, *8*, 920. [CrossRef]
250. Chen, Y.; Kang, Y.; Zhao, Y.; Wang, L.; Liu, J.; Li, Y.; Liang, Z.; He, X.; Li, X.; Tavajohi, N.; et al. A Review of Lithium-Ion Battery Safety Concerns: The Issues, Strategies, and Testing Standards. *J. Energy Chem.* **2021**, *59*, 83–99. [CrossRef]
251. Djebaili, R.; Pellegrini, M.; Rossi, M.; Forni, C.; Smati, M.; Del Gallo, M.; Kitouni, M. Characterization of Plant Growth-Promoting Traits and Inoculation Effects on Triticum Durum of Actinomycetes Isolates under Salt Stress Conditions. *Soil Syst.* **2021**, *5*, 26. [CrossRef]
252. Ahmad, H.M.; Fiaz, S.; Hafeez, S.; Zahra, S.; Shah, A.N.; Gul, B.; Aziz, O.; Mahmood-Ur-Rahman; Fakhar, A.; Rafique, M.; et al. Plant Growth-Promoting Rhizobacteria Eliminate the Effect of Drought Stress in Plants: A Review. *Front. Plant Sci.* **2022**, *13*, 875774. [CrossRef]
253. Afridi, M.S.; Fakhar, A.; Kumar, A.; Ali, S.; Medeiros, F.H.V.; Muneer, M.A.; Ali, H.; Saleem, M. Harnessing Microbial Multitrophic Interactions for Rhizosphere Microbiome Engineering. *Microbiol. Res.* **2022**, *265*, 127199. [CrossRef]
254. Yan, S.; Li, J.; Zhang, Q.; Jia, S.; Zhang, Q.; Wang, R.; Ju, M.; Gu, P. Transcriptional Response of Wolfberry to Infestation with the Endophytic *Fusarium nematophilum* Strain NQ8GII4. *Plant Dis.* **2024**, *108*, 1514–1525. [CrossRef]
255. Iqbal, S.; Iqbal, M.A.; Li, C.; Iqbal, A.; Abbas, R.N. Overviewing Drought and Heat Stress Amelioration—From Plant Responses to Microbe-Mediated Mitigation. *Sustainability* **2023**, *15*, 1671. [CrossRef]
256. Singh, D.; Thapa, S.; Singh, J.P.; Mahawar, H.; Saxena, A.K.; Singh, S.K.; Mahla, H.R.; Choudhary, M.; Parihar, M.; Choudhary, K.B.; et al. Prospecting the Potential of Plant Growth-Promoting Microorganisms for Mitigating Drought Stress in Crop Plants. *Curr. Microbiol.* **2024**, *81*, 84. [CrossRef]
257. Gu, Z.; Hu, C.; Gan, Y.; Zhou, J.; Tian, G.; Gao, L. Role of Microbes in Alleviating Crop Drought Stress: A Review. *Plants* **2024**, *13*, 384. [CrossRef]
258. Khan, N.; Ali, S.; Tariq, H.; Latif, S.; Yasmin, H.; Mehmood, A.; Shahid, M.A. Water Conservation and Plant Survival Strategies of Rhizobacteria under Drought Stress. *Agronomy* **2020**, *10*, 1683. [CrossRef]
259. Fincheira, P.; Quiroz, A.; Tortella, G.; Diez, M.C.; Rubilar, O. Current Advances in Plant-Microbe Communication via Volatile Organic Compounds as an Innovative Strategy to Improve Plant Growth. *Microbiol. Res.* **2021**, *247*, 126726. [CrossRef]
260. Inbaraj, M.P. Plant-Microbe Interactions in Alleviating Abiotic Stress—A Mini Review. *Front. Agron.* **2021**, *3*, 667903. [CrossRef]
261. Vessey, J.K. Plant Growth Promoting Rhizobacteria as Biofertilizers. *Plant Soil* **2003**, *255*, 571–586. [CrossRef]
262. Bhattacharyya, P.N.; Jha, D.K. Plant Growth-Promoting Rhizobacteria (PGPR): Emergence in Agriculture. *World J. Microbiol. Biotechnol.* **2012**, *28*, 1327–1350. [CrossRef]
263. Bashan, Y.; de-Bashan, L.E. How the Plant Growth-Promoting Bacterium Azospirillum Promotes Plant Growth—A Critical Assessment. *Adv. Agron.* **2010**, *108*, 77–136.
264. Bashan, Y.; de-Bashan, L.E.; Prabhu, S.R.; Hernandez, J.-P. Advances in Plant Growth-Promoting Bacterial Inoculant Technology: Formulations and Practical Perspectives (1998–2013). *Plant Soil* **2014**, *378*, 1–33. [CrossRef]

265. Glick, B.R. Plant Growth-Promoting Bacteria: Mechanisms and Applications. *Scientifica* **2012**, *2012*, 963401. [CrossRef]
266. Pérez-Montaño, F.; Alías-Villegas, C.; Bellogín, R.A.; del Cerro, P.; Espuny, M.R.; Jiménez-Guerrero, I.; López-Baena, F.J.; Ollero, F.J.; Cubo, T. Plant Growth Promotion in Cereal and Leguminous Agricultural Important Plants: From Microorganism Capacities to Crop Production. *Microbiol. Res.* **2014**, *169*, 325–336. [CrossRef]
267. Van Der Heijden, M.G.A.; Bardgett, R.D.; Van Straalen, N.M. The Unseen Majority: Soil Microbes as Drivers of Plant Diversity and Productivity in Terrestrial Ecosystems. *Ecol. Lett.* **2008**, *11*, 296–310. [CrossRef]
268. Ryan, M.H.; Graham, J.H. Is There a Role for Arbuscular Mycorrhizal Fungi in Production Agriculture? *Plant Soil* **2002**, *244*, 263–271. [CrossRef]
269. Herrmann, L.; Lesueur, D. Challenges of Formulation and Quality of Biofertilizers for Successful Inoculation. *Appl. Microbiol. Biotechnol.* **2013**, *97*, 8859–8873. [CrossRef]
270. Singh, J.S.; Pandey, V.C.; Singh, D.P. Efficient Soil Microorganisms: A New Dimension for Sustainable Agriculture and Environmental Development. *Agric. Ecosyst. Environ.* **2011**, *140*, 339–353. [CrossRef]
271. Malusá, E.; Sas-Paszt, L.; Ciesielska, J. Technologies for Beneficial Microorganisms Inocula Used as Biofertilizers. *Sci. World J.* **2012**, *2012*, 491206. [CrossRef]
272. Bakos, J.L.; Ladányi, M.; Szalay, L. Frost Hardiness of Flower Buds of 16 Apricot Cultivars during Dormancy. *Folia Hortic.* **2024**, *36*, 81–93. [CrossRef]
273. Barrios, E. Soil Biota, Ecosystem Services and Land Productivity. *Ecol. Econ.* **2007**, *64*, 269–285. [CrossRef]
274. Smith, P.; Bustamante, M.; Ahammad, H.; Clark, H.; Dong, H.; Elsiddig, E.A.; Tubiello, F. Agriculture, Forestry and Other Land Use (AFOLU). In *Climate Change 2014: Mitigation of Climate Change. Contribution of Working Group III to the Fifth Assessment Report of the Intergovernmental Panel on Climate Change*; Cambridge University Press: Cambridge, UK, 2014.
275. EL-Bauome, H.A.; Doklega, S.M.; Saleh, S.A.; Mohamed, A.S.; Suliman, A.A.; Abd El-Hady, M.A.M. Effects of Melatonin on Lettuce Plant Growth, Antioxidant Enzymes and Photosynthetic Pigments under Salinity Stress Conditions. *Folia Hortic.* **2024**, *36*, 1–17. [CrossRef]
276. San-Martín-Hernández, C.; Martínez-Téllez, M.Á.; Valenzuela-Amavizca, O.N.; Aispuro-Hernández, E.; Sánchez-Sánchez, M.; Hernández-Camarillo, E.; López-Martínez, L.X.; Quintana-Obregón, E.A. *Byrsonima crassifolia* L. Kunth a Bio-Resource with Potential: Overview and Opportunities. *Folia Hortic.* **2023**, *35*, 61–75. [CrossRef]
277. Compant, S.; Clément, C.; Sessitsch, A. Plant Growth-Promoting Bacteria in the Rhizo- and Endosphere of Plants: Their Role, Colonization, Mechanisms Involved and Prospects for Utilization. *Soil Biol. Biochem.* **2010**, *42*, 669–678. [CrossRef]
278. O'Callaghan, M. Microbial Inoculation of Seed for Improved Crop Performance: Issues and Opportunities. *Appl. Microbiol. Biotechnol.* **2016**, *100*, 5729–5746. [CrossRef]
279. Rutkowski, K.; Łysiak, G.P. Weather Conditions, Orchard Age and Nitrogen Fertilization Influences Yield and Quality of 'Łutówka' Sour Cherry Fruit. *Agriculture* **2022**, *12*, 2008. [CrossRef]
280. Siddique, S.; Naveed, M.; Yaseen, M.; Shahbaz, M. Exploring Potential of Seed Endophytic Bacteria for Enhancing Drought Stress Resilience in Maize (*Zea mays* L.). *Sustainability* **2022**, *14*, 673. [CrossRef]
281. Hwang, H.-H.; Chien, P.-R.; Huang, F.-C.; Yeh, P.-H.; Hung, S.-H.W.; Deng, W.-L.; Huang, C.-C. A Plant Endophytic Bacterium Priestia Megaterium StrainBP-R2 Isolated from the Halophyte Bolboschoenus Planiculmis Enhances Plant Growth under Salt and Drought Stresses. *Microorganisms* **2022**, *10*, 2047. [CrossRef]
282. Diagne, N.; Ngom, M.; Djighaly, P.I.; Fall, D.; Hocher, V.; Svistoonoff, S. Roles of Arbuscular Mycorrhizal Fungi on Plant Growth and Performance: Importance in Biotic and Abiotic Stressed Regulation. *Diversity* **2020**, *12*, 370. [CrossRef]
283. Nawaz, T.; Saud, S.; Gu, L.; Khan, I.; Fahad, S.; Zhou, R. Cyanobacteria: Harnessing the Power of Microorganisms for Plant Growth Promotion, Stress Alleviation, and Phytoremediation in the Era of Sustainable Agriculture. *Plant Stress* **2024**, *11*, 100399. [CrossRef]
284. Sharma, A.; Usha; Gupta, S.; Chaubey, K.K.; Singh, S.V. An Overview of Microbial-Mediated Alleviation of Abiotic Stress Response in Plant. In *Plant Stress Mitigators*; Springer Nature Singapore: Singapore, 2022; pp. 581–596.
285. Verma, H.; Kumar, D.; Kumar, V.; Kumari, M.; Singh, S.K.; Sharma, V.K.; Droby, S.; Santoyo, G.; White, J.F.; Kumar, A. The Potential Application of Endophytes in Management of Stress from Drought and Salinity in Crop Plants. *Microorganisms* **2021**, *9*, 1729. [CrossRef]
286. Castaño, C.; Suarez-Vidal, E.; Zas, R.; Bonet, J.A.; Oliva, J.; Sampedro, L. Ectomycorrhizal Fungi with Hydrophobic Mycelia and Rhizomorphs Dominate in Young Pine Trees Surviving Experimental Drought Stress. *Soil Biol. Biochem.* **2023**, *178*, 108932. [CrossRef]
287. Ullah, S.; Ikram, M.; Sarfaraz, S.; ul Haq, I.; Khan, A.; Murad, Z.; Munsif, F. Influence of Plant Growth Promoting Rhizobacteria (PGPR) on the Growth and Yield of Sunflower (*Helianthus annus* L.) Under Salt Stress. *J. Crop Health* **2024**, *76*, 1221–1234. [CrossRef]
288. Ha-Tran, D.M.; Nguyen, T.T.M.; Hung, S.-H.; Huang, E.; Huang, C.-C. Roles of Plant Growth-Promoting Rhizobacteria (PGPR) in Stimulating Salinity Stress Defense in Plants: A Review. *Int. J. Mol. Sci.* **2021**, *22*, 3154. [CrossRef] [PubMed]
289. Meier, M.A.; Lopez-Guerrero, M.G.; Guo, M.; Schmer, M.R.; Herr, J.R.; Schnable, J.C.; Alfano, J.R.; Yang, J. Rhizosphere Microbiomes in a Historical Maize-Soybean Rotation System Respond to Host Species and Nitrogen Fertilization at the Genus and Subgenus Levels. *Appl. Environ. Microbiol.* **2021**, *87*, e03132-20. [CrossRef]
290. Li, H.-P.; Han, Q.-Q.; Liu, Q.-M.; Gan, Y.-N.; Rensing, C.; Rivera, W.L.; Zhao, Q.; Zhang, J.-L. Roles of Phosphate-Solubilizing Bacteria in Mediating Soil Legacy Phosphorus Availability. *Microbiol. Res.* **2023**, *272*, 127375. [CrossRef] [PubMed]

291. Poudel, M.; Mendes, R.; Costa, L.A.S.; Bueno, C.G.; Meng, Y.; Folimonova, S.Y.; Garrett, K.A.; Martins, S.J. The Role of Plant-Associated Bacteria, Fungi, and Viruses in Drought Stress Mitigation. *Front. Microbiol.* **2021**, *12*, 743512. [CrossRef]
292. Mushtaq, A.; Rawat, S. Plant–Microbe Interactions: An Insight into the Underlying Mechanisms to Mitigate Diverse Environmental Stresses. In *Microbes and Signaling Biomolecules Against Plant Stress*; Springer: Berlin/Heidelberg, Germany, 2021; pp. 127–150.
293. Panda, D.; Mishra, S.S.; Behera, P.K. Drought Tolerance in Rice: Focus on Recent Mechanisms and Approaches. *Rice Sci.* **2021**, *28*, 119–132. [CrossRef]
294. Bashir, S.S.; Hussain, A.; Hussain, S.J.; Wani, O.A.; Zahid Nabi, S.; Dar, N.A.; Baloch, F.S.; Mansoor, S. Plant Drought Stress Tolerance: Understanding Its Physiological, Biochemical and Molecular Mechanisms. *Biotechnol. Biotechnol. Equip.* **2021**, *35*, 1912–1925. [CrossRef]
295. Wiclaw, H.; Podlasiński, M. Morphological Differences between Natural Populations of *Carex viridula* (Cyperaceae): Effects of Soil Conditions. *Ann. Bot. Fenn.* **2013**, *50*, 13–22. [CrossRef]
296. Murali, M.; Singh, S.B.; Gowtham, H.G.; Shilpa, N.; Prasad, M.; Aiyaz, M.; Amruthesh, K.N. Induction of Drought Tolerance in *Pennisetum glaucum* by ACC Deaminase Producing PGPR- *Bacillus amyloliquefaciens* through Antioxidant Defense System. *Microbiol. Res.* **2021**, *253*, 126891. [CrossRef]
297. French, E.; Kaplan, I.; Iyer-Pascuzzi, A.; Nakatsu, C.H.; Enders, L. Emerging Strategies for Precision Microbiome Management in Diverse Agroecosystems. *Nat. Plants* **2021**, *7*, 256–267. [CrossRef]
298. Markuszewski, B.; Bieniek, A.A.; Wachowska, U.; Bieniek, A.; Krzymińska, I. Effect of Biological Treatment Used before Harvesting and Storage Methods on the Quality, Health and Microbial Characteristics of Unripe Hazelnut in the Husk (*Corylus avellana* L.). *PeerJ* **2022**, *10*, e12760. [CrossRef] [PubMed]
299. Dellagnezze, B.M.; Sierra-Garcia, I.N. Immobilization of Microbial Inoculants for Improving Soil Nutrient Bioavailability. In *Microbial Inoculants*; Elsevier: Amsterdam, The Netherlands, 2023; pp. 161–181.
300. Ulukapi, K.; Nasircilar, A.G. The Role of Exogenous Glutamine on Germination, Plant Development and Transcriptional Expression of Some Stress-Related Genes in Onion under Salt Stress. *Folia Hortic.* **2024**, *36*, 19–34. [CrossRef]
301. Abbasi, S.; Sadeghi, A.; Safaie, N. Streptomyces Alleviate Drought Stress in Tomato Plants and Modulate the Expression of Transcription Factors ERF1 and WRKY70 Genes. *Sci. Hortic.* **2020**, *265*, 109206. [CrossRef]
302. Tymoszuk, A.; Szałaj, U.; Wojnarowicz, J.; Kowalska, J.; Antkowiak, M.; Kulus, D. Zinc Oxide and Silver Effects on the Growth, Pigment Content and Genetic Stability of Chrysanthemums Propagated by the Node Culture Method. *Folia Hortic.* **2024**, *36*, 35–66. [CrossRef]
303. Demir, H.; Yalçi, H.K.; Katgici, A. Ameliorative Effects of Microbial Fertiliser on Yield and Quality Parameters of Curly Lettuce and Cucumber with Fertiliser Saving. *Folia Hortic.* **2023**, *35*, 91–106. [CrossRef]
304. Ahmad, R.; Anjum, M.A.; Ercisli, S.; Malik, W.; Sakar, E.; Marc, R.A.; Karunakaran, R. Physico-Chemical Properties Revealed Huge Diversity in 50 Date Palm (*Phoenix dactylifera* L.) Genotypes. *Folia Hortic.* **2023**, *35*, 107–122. [CrossRef]
305. Bahmani, R.; Razavi, F.; Mortazavi, S.N.; Juárez-Maldonado, A.; Gohari, G. Chitosan–Putrescine Nanoparticle Coating Attenuates Postharvest Decay and Maintains ROS Scavenging System Activity of Strawberry Cv. 'Camarosa' during Cold Storage. *Folia Hortic.* **2024**, *36*, 149–160. [CrossRef]
306. Kunc, N.; Hudina, M.; Osterc, G.; Mikulic-Petkovsek, M. Changes in Various Secondary Metabolites by Crossing Modern Rose Cultivars. *Folia Hortic.* **2024**, *36*, 161–185. [CrossRef]
307. Żelazny, W.R.; Licznar-Małańczuk, M. Living Mulch Persistence in an Apple Orchard and Its Effect on the Weed Flora in Temperate Climatic Conditions. *Weed Res.* **2022**, *62*, 85–99. [CrossRef]
308. Bieniek, A.; Lachowicz-Wiśniewska, S.; Bojarska, J. The Bioactive Profile, Nutritional Value, Health Benefits and Agronomic Requirements of Cherry Silverberry (*Elaeagnus multiflora* Thunb.): A Review. *Molecules* **2022**, *27*, 2719. [CrossRef]
309. Bokszczanin, K.Ł.; Przybyłko, S.; Molska-Kawulok, K.; Wrona, D. Tree Root-Associated Microbial Communities Depend on Various Floor Management Systems in an Intensive Apple (*Malus* × *domestica* Borkh.) Orchard. *Int. J. Mol. Sci.* **2023**, *24*, 9898. [CrossRef]
310. Sosna, I.; Fudali, E. Living Mulches In-Rows as an Alternative for Herbicide Fallow in a Pear *Pyrus communis* L. Orchard. *Hortic. Sci.* **2024**, *51*, 160–167. [CrossRef]

agriculture

MDPI

Review

Biochar as a Multi-Action Substance Used to Improve Soil Properties in Horticultural and Agricultural Crops—A Review

Agnieszka Kozioł [1], Dominika Paliwoda [2], Grzegorz Mikiciuk [2,*] and Nadhira Benhadji [1]

1 Institute of Technology and Life Sciences—National Research Institute Falenty, Hrabska Avenue 3, 05-090 Raszyn, Poland; a.koziol@itp.edu.pl (A.K.); nbenhadji@itp.edu.pl (N.B.)

2 Department of Horticulture, Faculty of Environmental Management and Agriculture, West Pomeranian University of Technology in Szczecin, Słowackiego 17, 71-434 Szczecin, Poland; dominika.paliwoda@zut.edu.pl

* Correspondence: grzegorz.mikiciuk@zut.edu.pl

Abstract: With climate change escalating to global proportions over the past decade, along with a growing population, methods are being sought to help the natural and cultivated environment function in an ecologically balanced manner. Over the past few years, there has been a significantly increased interest in research on the use of natural substances for sustainable agriculture and horticulture. One of the most effective solutions to the emerging need is biochar, which has been the subject of environmental research for years due to its potential to increase soil carbon sequestration, reduce greenhouse gas emissions, remediate contaminated soil, and alleviate anthropogenic pressures. There is evidence of improved soil fertility and increased crop yields in agricultural production after biochar application. Our work comprehensively describes the effects of biochar on soil properties, crop productivity, and mitigating environmental stresses, and its remediation potential in heavy metal-contaminated soils. We analyzed a wide range of the literature on the most important properties of biochar for various potential uses. We summarized the results of research work over the past two decades to analyze soil and plant responses to biochar application.

Keywords: biochar; biomass; soil conditioner; plant growth; environment

1. Introduction

The transformation of the natural environment around us has caused progressive pollution of the environment and anthropogenic emission of greenhouse gases, which in turn has resulted in progressive global warming. In addition to the negative impact on the condition of nature, ecosystems, and the Earth's climate, environmental pollution also directly threatens human health [1,2].

Agricultural and horticultural crops are also severely affected by this problem due to abiotic stressors such as too high or low temperature, drought, mineral deficiency, salinity, and intense radiation [3,4]. Environmental degradation affecting agriculture and horticulture also manifests in soil degradation caused by erosion, deterioration of physical properties, or contamination with plant protection products and heavy metals [5,6].

The use of biochar in agriculture is believed to be one of the most promising technologies for climate change mitigation [7–9]. Biochar research has attracted the attention of dozens of teams worldwide [10], and the pool of literature on its versatile applications is extensive [11,12].

Biochar may positively affect soil health, its nutritional status, and the biological activities of the rhizosphere [13]. It can also be used as an additive to the soil to improve its physical, chemical, and biological properties [14,15]. Many authors have shown that the application of biochar in the soil can (1) directly increase the uptake of nutrients by plants, and thus indirectly enhance yields [16]; (2) improve soil properties [17]; (3) reduce greenhouse gas emission [18]; (4) and immobilize heavy metals and organic pollutants [19] (Figure 1).

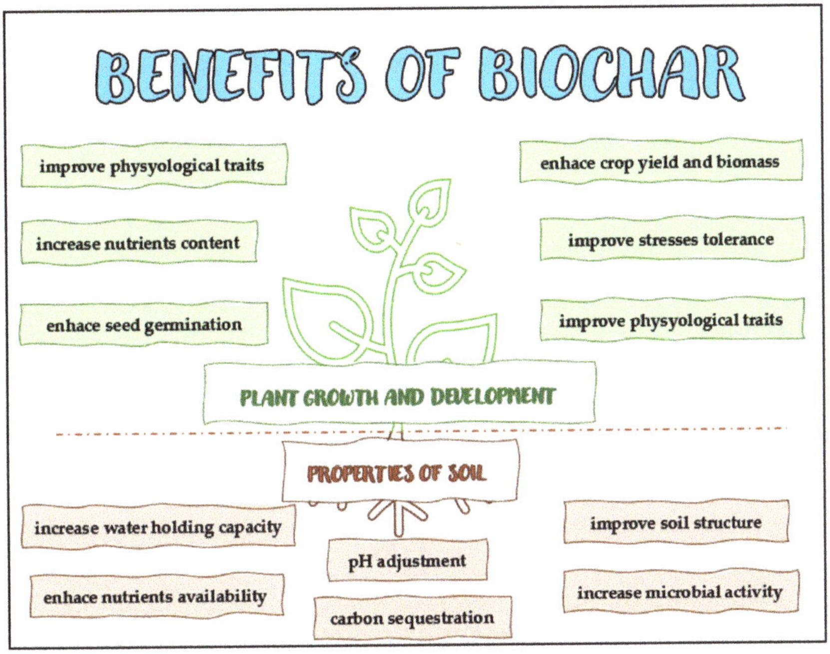

Figure 1. Effect of biochar application on soil properties, plant growth, and development.

The positive effect of biochar on the physiological parameters of plants has also been demonstrated. Indeed, the addition of biochar can significantly increase the rate of photosynthesis, leaf chlorophyll content, and transpiration rate [20,21]. The various advantages of biochar make it a potentially attractive research material for current science and technology [22], especially amid numerous changes in the soil environment.

Among the benefits of using biochar, scientists cite positive agronomic aspects (e.g., maintaining high soil culture and increasing nutrient availability for plants), as well as the ability to mitigate climate change by reducing the negative impact of environmental stresses on cultivated plant species [23–25]. The quality of biochar is mainly influenced by the origin and properties of the substrates as well as the parameters of the pyrolysis process, namely, the temperature, pressure, time, or period of use [26].

This article provides an up-to-date review of current knowledge on the potential uses of biochar as a substance with multifaceted effects both in crop production and its positive impacts on the environment.

2. Literature Review Methodology

2.1. Overview of the Methodology

The literature review in this chapter provides a comprehensive examination of current research surrounding the use of the properties and effects of biochar on soil and plants in horticultural and agricultural crops.

In this article, the literature refers only to biochar obtained in the pyrolysis process.

The methodology employed in this review involves a structured and systematic approach to identify, select, and analyze relevant peer-reviewed articles and academic publications written in English from various disciplines, including agriculture, horticulture, and environmental sciences.

2.2. Search Strategy

A systematic search was conducted across the following academic databases to gather the relevant literature: Google Scholar, Scopus, SpringerLink, PubMed, and Web of Science. The search terms used included combinations of key phrases such as biomass, pyrolysis, gasification, carbonization, soil amendment, soil health, yielding, environmental stresses, remediation, microbial activity, and greenhouse gas emissions. The use of these databases allowed for a thorough study of the issue based not only on the effects of biochar on soil, but also on the environment and crop production. The search was limited to articles published between 2000 and 2024.

2.3. Data Extraction and Analysis of Quality Assessment

Key information about research objectives, methodology, and results related to the impact of biochar on soil, environment, and plant production was extracted from the selected articles. These data were then organized thematically to identify recurring trends in the literature and areas of innovation in the field of biochar use. Each study was critically assessed for its methodological quality, experimental design, statistical analysis, and conclusions drawn.

2.4. Synthesis of Findings

The literature was synthesized into several core themes:

- Effect of biochar on the physical and chemical properties of soils;
- Effect of biochar on soil microbial composition and activity;
- The use of biochar in agricultural and horticultural production;
- The benefits of using biochar for the environment;
- Threats related to the application of biochar to the soil.

The synthesis presents current knowledge about both the positive impact of biochar on soil and plants, as well as the risks resulting from its use.

3. Short Historical Outline of the Use of Biochar

The first references to the use of biochar date back 7000 years, and charcoal, as a precursor to biochar, was known as early as Paleolithic times. Natives living in the area of today's Amazonia transformed the soil through settlement activities. For many generations, it was enriched with charcoal from campfires, burned refuse, animal bones, broken pottery, compost, and manure. As a result, a soil type known as dark Amazonian soil—*Terra preta*—was formed. It is characterized by its dark color, high nutrient content, and ability to retain water [27].

As far back as when the ancient Native American ethnic groups lived in the area, charcoal was used to enrich the soil with nutrients, i.e., potassium, phosphorus, and nitrogen, by combining charcoal with biomass (compost, manure, and agricultural biomass) [28]. Despite the passage of time, these soils are still distinguished by their high productivity and have aroused great interest in the use of biochar in plant cultivation [29]. Throughout Asia, biochar has been used for centuries not only in agriculture, but also in the production of ceramics, ink, and dyes [30].

In the 19th century, its use in agriculture began in Europe and South America. In the 1980s, its use was revisited, and now the properties of biochar are considered more broadly, not only for agricultural purposes but also in waste management and bioenergy [31]. Naturally, biochar finds its way into soils around the world as a result of grassland and forest fires, and lands rich in high levels of biochar are among the most fertile, such as the North American prairies [29].

Nowadays, not only has biochar found applications not only in crop production and environmental studies (to immobilize heavy metals and organic pollutants), but it is widely used in the energy industry and metallurgy, among others [32,33]. Currently, interest in the use of biochar is growing, as evidenced by numerous scientific papers and conferences [34].

4. Biochar—Production and Feedstock Sources

Biochar is a product formed from organic or plant waste subjected to various processes, resulting in a fine-grained porous substance [35], which is known by a uniform chemical composition compared to the starting material [36]. In its composition, it not only contains carbon, but is also rich in ash, hydrogen, oxygen, nitrogen, and sulfur. It can also contain phenols, alcohols, and several aromatic compounds, which ultimately affect its properties [37].

Biochar is mainly produced by the biomass combustion process known as pyrolysis (fast and slow), but there are several methods of obtaining biochar that are constantly being improved to improve the quality of the output product. Based on the literature review that has been carried out, methods such as gasification, torrefaction, hydrothermal carbonization, and electro-modification are used to produce biochar [38], while biocarbon is a solid carbon material that is derived from renewable and sustainable raw materials subjected to thermochemical conversion at high temperatures (>350 °C) in the absence or limited availability of oxygen. It can be used as a component of composite materials and energy storage and conversion [39].

The most popular and cheapest way to obtain biochar is production from the thermal decomposition of biomass, the so-called pyrolysis (slow, intermediate, or fast pyrolysis), carried out under anaerobic conditions at 250–1000 °C (Figure 2). The result is a fine-grained and porous substance with homogeneous chemical composition compared to the starting material [40,41]. Therefore, due to the relatively simple production method and the popularity of its use, this review focuses on biochar obtained as a result of the pyrolysis process.

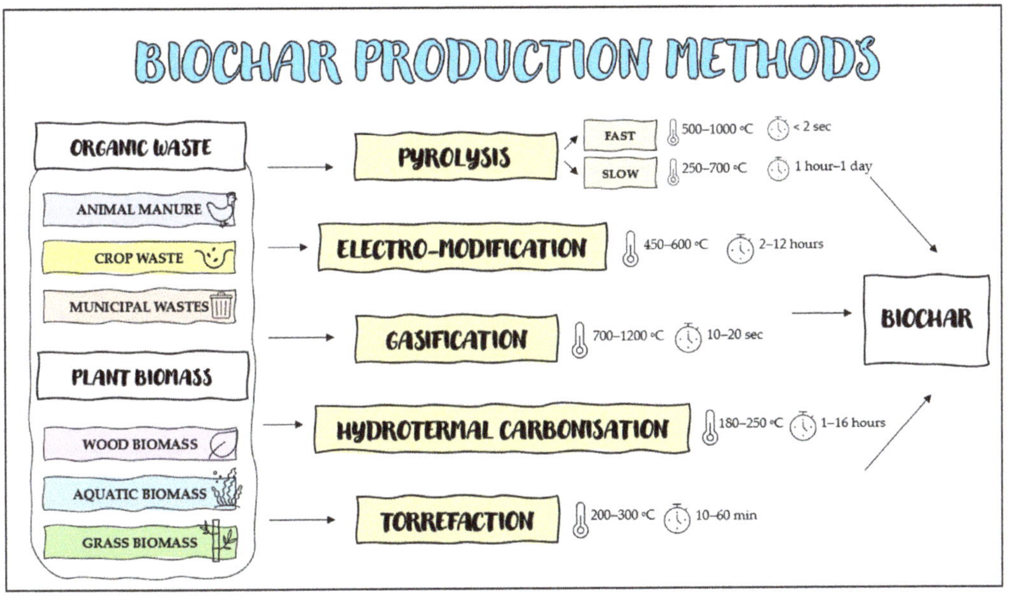

Figure 2. Different methods used for biochar production.

The physicochemical properties and performance of biochar largely depend on the pyrolysis process and the raw material used, which can be agricultural biomass, energy crops (e.g., willow, miscanthus), wood waste (e.g., sawdust, bark), agro-food industry waste, animal manure, sewage sludge, municipal waste, industrial by-products, and aquaculture waste (Table 1) [42–44].

As already mentioned, the physical and chemical properties of biochar (e.g., surface area, pH, or concentration of various elements: carbon, oxygen, nitrogen, phosphorus, potassium, and calcium) depend on the raw materials used and the production temperature [45]. Higher temperatures of the pyrolysis process produce biochar with higher porosity and surface area, a higher C:N ratio, a more alkaline pH, and lower dissolved organic carbon concentrations [46,47].

Temperatures of 350–650 °C rearrange chemical bonds and create new functional groups in biochar (e.g., carboxyl, lactone, lactol, quinine, chromene, anhydride, phenol, ether, pyrone, pyridine, pyridone, and pyrrole). Biochar produced at lower temperatures (300–400 °C) has a more diverse organic character due to the presence of aliphatic compounds and cellulose-type structures. When the pyrolysis temperature increases to 300 °C, biochar contains furans and phenols, but above 300 °C condensed aromatic compounds, aromatic benzenoids and alkylfurans appear. Some authors indicate that biochar produced at temperatures up to 480 °C retains functional groups such as phenolic and carboxylic acid. Biochar produced at high temperatures from 600 to 700 °C is highly hydrophobic in nature, with well-organized carbon layers, but has a lower content of functional groups as a result of dehydration and deoxidation biomass [48–50]. These differences in the properties of biochar are highlighted when it is used as a soil additive [46,47].

Antonangelo et al. [51] report that, as a result of increasing the pyrolysis temperature from 350 to 700 °C, the Cation Exchange Capacity, Specific Surface Area, and microporosity of biochar increased. According to the cited authors, the pyrolysis temperature did not affect the surface functional groups responsible for the retention of heavy metals, which is why biochar pyrolyzed at 700 °C has a much greater buffering capacity and is therefore more promising for improving soil health and reducing pollution bioavailability.

Table 1. Types of biochar according to the raw material used.

Type of Biochar	Pyrolysis Temperature [°C]	Residence Time (Hours)	pH	EC [μS·cm^{-1}]	Product Yield [%]	Ash Content [%]	Elemental Compositions [%] C	N	P	K	Cation Exchange Capacity [cmol$_+$·kg^{-1}]	Bulk Density [g·cm^{-3}]	Specific Surface Area [m^2·g^{-1}]	Diameter of Pore [nm]	Total Pore Volume [cm^3·g^{-1}]	Reference
ORGANIC WASTE																
ANIMAL MANURE																
poultry manure	300	0.3	7.3	8960.0	71.0	n.m.	42.4	5.3	1.26	4.13	69.0	n.m.	n.m.	n.m.	n.m.	[52]
chicken manure	300	0.5	8.1	519.0	n.m.	34.8	n.m.	4.7	0.22	0.51	n.m.	n.m.	n.m.	n.m.	n.m.	[53]
	500	1.0	10.6	662.0	n.m.	38.0	n.m.	1.5	0.14	0.57	n.m.	n.m.	n.m.	n.m.	n.m.	
cow manure	300	n.m.	8.5	n.m.	n.m.	45.0	34.3	2.6	n.m.	n.m.	n.m.	n.m.	3.5	n.m.	n.m.	[54]
	600	n.m.	9.6	n.m.	n.m.	54.5	38.5	2.1	n.m.	n.m.	n.m.	n.m.	13.9	n.m.	n.m.	
swine manure	700	2.0	7.0	n.m.	n.m.	44.0	48.4	2.2	n.m.	n.m.	n.m.	n.m.	319.0	n.m.	0.250	[55]
	400	0.3	10.5	1655.0	70.7	57.6	38.3	1.9	n.m.	n.m.	n.m.	n.m.	7.39	9.5	0.017	[56]
	700	0.5	11.4	1835.0	61.6	71.8	25.0	0.9	n.m.	n.m.	n.m.	n.m.	73.54	3.04	0.056	
CROP WASTE																
husks	400	2.0	8.5	n.m.	43.3	27.0	47.9	1.3	n.m.	n.m.	n.m.	n.m.	39.6	n.m.	n.m.	[57]
	750	n.m.	9.7	900.0	n.m.	48.2	41.0	1.72	n.m.	n.m.	n.m.	0.18	n.m.	n.m.	n.m.	[58]
shells	400–500	2.0	6.4	880.0	n.m.	n.m.	25.3	0.4	0.09	1.39	21.84	0.58	n.m.	n.m.	n.m.	[59]
	400	1.3	10.2	n.m.	39.0	12.2	67.2	3.1	0.41	1.07	n.m.	0.29	2.8	n.m.	n.m.	[60]
	500	1.7	9.8	n.m.	33.0	14.2	69.3	2.8	0.56	1.71	n.m.	0.32	2.9	n.m.	n.m.	
seeds	450	0.7	9.1	n.m.	n.m.	8.2	70.4	3.6	n.m.	n.m.	n.m.	n.m.	3.3	n.m.	0.006	[61]
	550	0.9	9.6	n.m.	n.m.	8.9	73.0	3.7	n.m.	n.m.	n.m.	n.m.	3.8	n.m.	0.007	
	700	1.0	11.3	n.m.	30.0	17.2	82.0	1.3	n.m.	n.m.	n.m.	n.m.	420.3	1.09	0.190	[62]
straws	350	0.3	7.5	1270.0	57.3	11.5	57.5	1.3	0.19	0.10	56.3	n.m.	n.m.	n.m.	n.m.	[63]
	700	2.0	9.5	12,210.0	33.0	40.6	41.0	3.0	n.m.	n.m.	64.8	0.24	n.m.	n.m.	n.m.	[64]
cobs	445	0.9	10.3	2390.0	38.0	10.4	n.m.	n.m.	n.m.	15.40	n.m.	0.15	71.7	n.m.	n.m.	[65]
leaves	300	1.0	5.7	122.0	52.1	5.8	62.5	2.6	n.m	n.m	223.6	n.m.	5.5	n.m.	0.002	[66]
	500	2.0	10.0	n.m.	n.m.	5.2	63.7	4.5	n.m.	n.m.	n.m.	n.m.	17.2	1.42	0.700	[67]
MUNCIPAL WASTE																
sewage sludge	300	11.0	7.2	90.0	43.8	38.2	47.5	n.m.	n.m.	n.m.	44.15	n.m.	n.m.	n.m.	n.m.	[68]
	900	0.5	12.1	n.m.	53.3	88.1	15.9	0.5	2.02	0.87	247.5	n.m.	67.6	3.84	0.099	[69]
solid waste	400	0.5	8.0	n.m.	n.m.	6.1	48.6	1.3	n.m.	n.m.	n.m.	n.m.	20.7	n.m.	0.027	[70]

Table 1. *Cont.*

Type of Biochar	Pyrolysis Temperature [°C]	Residence Time (Hours)	pH	EC [μS·cm^{-1}]	Product Yield [%]	Ash Content [%]	Elemental Compositions [%]				Cation Exchange Capacity [cmol$_+$·kg^{-1}]	Bulk Density [g·cm^{-3}]	Specific Surface Area [m^2·g^{-1}]	Diameter of Pore [nm]	Total Pore Volume [cm^3·g^{-1}]	Reference
							C	N	P	K						
PLANT BIOMASS																
WOOD BIOMASS																
sawdust	500	1.0	7.1	n.m.	19.0	2.4	71.0	0.3	n.m.	n.m.	n.m.	0.43	n.m.	n.m.	n.m.	[71]
chips	500	2.0	6.5	200.0	n.m.	2.6	88.8	0.5	0.03	0.61	n.m.	1.52	6.2	n.m.	n.m.	[72]
stems	450–500	n.m.	9.6	1350.0	n.m.	n.m.	58.5	1.3	n.m.	n.m.	11.0	n.m.	200.5	2.43	0.122	[73]
branches	600	2.2	10.6	n.m.	28.5	9.4	80.0	1.3	0.34	1.14	19.0	n.m.	108.6	1.25	0.058	[74]
GRASS BIOMASS																
alfalfa	650	2.0	8.1	n.m.	27.5	13.6	72.2	4.6	n.m.	n.m.	n.m.	n.m.	405.0	n.m.	n.m.	[75]
cane	400 500	0.7 0.8	5.2 6.2	1460.0 1630.0	21.3 19.5	n.m. n.m.	46.6 48.8	1.0 0.9	n.m. n.m.	0.55 0.61	86.2 44.8	n.m. n.m.	n.m. n.m.	n.m. n.m.	n.m. n.m.	[76]
	250 600	0.4 1.0	5.0 7.7	n.m. n.m.	77.1 22.9	n.m. n.m.	n.m. n.m.	n.m. n.m.	n.m. n.m.	0.39 3.54	n.m. n.m.	0.08 0.07	0.8 14.1	n.m. n.m.	0.056 0.141	[77]
miscanthus	450	2.0	9.9	n.m.	30.6	8.2	73.1	0.4	n.m.	n.m.	n.m.	0.28	90.5	n.m.	0.054	[78]
bamboo	500	1.0	n.m.	n.m.	38.0	3.9	82.1	0.5	n.m.	n.m.	n.m.	0.11	n.m.	n.m.	n.m.	[79]
	300 450 600	n.m. n.m. n.m.	6.7 5.2 7.9	n.m. n.m. n.m.	73.2 26.3 24.0	n.m. n.m. n.m.	66.2 76.9 80.9	0.4 0.2 0.2	0.24 0.36 0.50	0.30 0.35 0.52	n.m. n.m. n.m.	n.m. n.m. n.m.	1.3 18.2 470.4	n.m. n.m. n.m.	n.m. n.m. n.m.	[80]
AQUATIC BIOMASS																
algae	350 550 750	0.9 1.3 2.0	9.2 11.0 11.2	28.0 32.9 44.7	51.0 37.9 33.3	33.3 42.1 54.2	33.9 33.0 27.8	2.1 1.9 1.4	0.54 0.69 0.67	n.m. n.m. n.m.	44.4 34.6 2.3	n.m. n.m. n.m.	1.1 14.6 29.8	15.4 5.01 4.37	0.004 0.018 0.033	[81]

n.m.—not mentioned.

5. Effect of Biochar on the Physical and Chemical Properties of Soils

Studies have shown that biochar can improve the physicochemical and biological properties of soil, creating a suitable environment for plant roots, enabling nutrient uptake and improving plant growth [82,83].

Among other things, its use affects soil water infiltration, Water Holding Capacity, soil aeration and porosity, bulk density, pH, Cation Exchange Capacity, and nutrient cycling. Biochar has also been shown to remain in agricultural fields for up to several years and, therefore, has the advantage of durability in the environment and soil [84–86]. Table 2 presents examples of the impact of biochar on the physical and chemical properties of soils.

The deep-soil application of biochar contributes to an increase in the organic carbon content and influences its circulation. Thanks to its considerable sorption capacity, it stimulates an increase in the soil's content of micro- and macronutrients [87–89]. Biochar also contains aromatic and aliphatic compounds as well as oxidized carbon compounds, which are extremely valuable for soil microorganisms due to the high content of calcium, magnesium, and carbonates, among others [90–92].

This substance can be used as an additive to compost consisting of organic waste with a high nitrogen content, e. g. manure, chicken manure, or sewage sludge [93]. According to Clough et al. [94], biochar changes the dynamics of nitrogen in the soil. Laird et al. [95] obtained a 7% increase in this element after applying biochar. Many authors also report that adding biochar to the soil has a positive effect on the phosphorus content in the soil. It usually occurs in easily digestible forms, but the ability to change the pH of the soil and the strong dependence of phosphorus on pH is of great importance [96,97].

Adding biochar to soils may change reactions and thus improve plant yield by increasing the availability of nutrients. The pH of biochar is generally alkaline (from 7.1 to 10.5) [48,98] and depends on the type of biomass. It was found that biochar produced from wood has a lower pH compared to other biomass formed under similar pyrolysis conditions (by two units on average) [99].

Specific Surface Area is an important physical parameter that, like many characteristics of biochar, varies based on the raw material from which it was produced (Table 1) [100]. These latter cited authors reported that biochar produced from coconut shell (25.8 $m^2 \cdot g^{-1}$) had a lower Specific Surface Area than biochar produced from sugar cane leaves (253.2 $m^2 \cdot g^{-1}$). Weber and Quicker [101] reported that biochar made from sewage sludge had a surface area of 100 $m^2 \cdot g^{-1}$, while the majority of biomasses possess surface areas ranging from 100 to 800 $m^2 \cdot g^{-1}$. In another study, it was found that biochar from corn straw (*Zea mays* L.) had a larger surface area than biochar from poplar or aspen leaves (*Populus* species) [102].

In soils where biochar has been applied, the agronomic performance has improved. This is due precisely to its Specific Surface Area, and thus its effect on the basic functions of soil fertility, its ability to store water and nutrients, aeration, and microbial activity [103,104].

This substance, due to high Specific Surface Area, also promotes the formation of complex bonds with cations and anions (Cation Exchange Capacity), as well as metals and elements on the soil surface, and consequently improves the soil's ability to retain nutrients [105] (Table 2).

Cation Exchange Capacity is an important element characterizing the quality of soil and its ability to retain positively charged ions, such as calcium, magnesium, and potassium, using electrostatic forces. Cations retained by electrostatic forces are exchanged with cations in the soil solution, thanks to which soil with higher Cation Exchange Capacity has better properties for retaining Ca, Mg, and K, which, however, does not indicate increased fertility, especially in the case where the soil has acid cations in the form of hydrogen and aluminum. The physical, chemical, and biological properties of soil have a significant effect on Cation Exchange Capacity [106].

Water Holding Capacity is the amount of water held by the soil solution against the force of gravity. Water Holding Capacity depends on the soil texture and organic matter content and affects the regulation, drainage, and functioning of the soil. The higher the organic matter content, the higher the Water Holding Capacity. It determines the moisture content required in the soil for plant growth. A suitable soil should be rich in air and water. The physicochemical properties of the soil are of great importance, especially in terms of nutrient and water retention, which affect land productivity [107].

Water Holding Capacity increases water availability for plants, and thus land productivity, contributing to reduced water stress and drought. The physicochemical changes in the soil that occur under its influence contribute to the improvement of its aeration, water infiltration rate, and soil drainage [108].

High porosity promotes the high water absorption, sorption capacity, and nutrient retention of biochar. Hernandez-Mena et al. [77] showed that biochar is characterized by high porosity, where micropores and macropores are distinguished.

Biochar, due to high porosity, also affects the soil's management of air and water [109], significantly improving both the water retention capacity and evaporation resistance [110,111]. Using biochar reduces the bulk density and increases the soil's ability to retain moisture [112]. Since the pore volume of biochar is suitable for effective water retention, it proves to be a promising alternative to peat and vermiculite in horticultural crops [113].

Using biochar as a soil additive may affect the bulk density of soils. Adding about 2% (by weight) may already change the bulk density of soils [114]. Mankasingh et al. [115], after biochar application, showed a decrease in this parameter from 1.66 to 1.53 $g \cdot cm^{-3}$. Also, other studies have obtained changes in this index compared to the control without biochar [116].

Therefore, the reduction in the bulk density of biochar-enriched soil may be one of the indicators of soil structure improvement or aggregation and aeration [117].

It is estimated that the decomposition time of biochar is from 100 to 1000 years. Low susceptibility to degradation and microbiological decomposition contribute to the high stability of biochar in soils [118]. However, some components of biochar may be degraded, and, due to the activity of microorganisms, oxidation, and physical factors, there may be mobility deeper into the soil profile [119,120]. In the studies by Kuzyakov et al. [121], rapid decomposition of the used biochar was obtained in the first three months after its introduction into the soil, followed by a slow and partial decomposition throughout the subsequent years from the start of the experiment.

In the soil profiles of the Amazon River, trace amounts of biochar have been found for hundreds of years. The biochar susceptibility to microbiological processes also depends on soil aggregates [122,123]. Mineralization of biochar occurs to a lesser extent under the influence of abiotic and biotic factors, i.e., climate, type of soil, topography, and the activity of microorganisms [124–126].

Table 2. Effects of biochar on the physicochemical properties of soils.

Type of Biochar	Pyrolysis Temperature [°C]	Dose of Biochar	Type of Soil	Effect on Properties of Soil	Reference
corn residue	350–650	1.0, 2.0, and 4.0 g per 5.0 kg of soil	calcareous sandy loam	increasing soil organic matter content by 1.8 $g \cdot kg^{-1}$, lowering soil bulk density by 0.09 $Mg \cdot m^{-3}$, increasing soil water availability to 0.156 cm^3	[127]
willow wood	350	10.9 $Mg \cdot ha^{-1}$ soil	brown soil	increasing soil pH by 1.2 units, decreasing soil bulk density by 7%, increasing soil total nitrogen content from 4.0 to 6.5 $Mg \cdot ha^{-1}$, increasing phosphorus content in soil from 2.0 to 3.0 $Mg \cdot ha^{-1}$	[128]
oil palm empty fruit bunch	450–500	10.0 $Mg \cdot ha^{-1}$ soil, 20.0 $Mg \cdot ha^{-1}$ soil	sandy soil	increasing soil pH by 2.32 units	[129]
coconut shell	600	10.0 $Mg \cdot ha^{-1}$ soil	acidic soil	increasing soil pH by 1.09 units, increasing soil organic matter by 0.06%, lowering soil bulk density by 0.12 $Mg \cdot m^{-3}$, increasing plant available water by 7.32%	[130]
rice husk	500	380.0 $Mg \cdot ha^{-1}$ soil	clay–loamy soil	increasing soil magnesium content by 64.77% and soil iron content by 25.45%, increasing plant available water from 30.53 to 50.13%	[131]
paddy straw, silver grass residue	500	50.0 $Mg \cdot ha^{-1}$ soil	sandy soil	increasing pH by 0.22 units, increasing soil organic matter content from 37 to 76%, increasing soil calcium content by 3.19%, soil magnesium content by 0.33%, and soil potassium content by 0.87%	[132]
red gram, cotton, maize stalk	350–400	2.5 and 5.0 $Mg \cdot ha^{-1}$ soil	sandy loam	increasing soil pH by 0.6 units, lowering bulk density by 0.05 $g \cdot cm^{-3}$, increasing soil Cation Exchange Capacity from 0.40 to 0.80 $cmol_+ \cdot kg^{-1}$	[133]
palm kernel shell	400	20.0 $Mg \cdot ha^{-1}$ soil	loamy soil	increasing soil pH by 1.57 units, increasing soil organic matter content by 332.4 $g \cdot kg^{-1}$, increasing soil Cation Exchange Capacity from 0.78 to 0.83 $cmol_+ \cdot kg^{-1}$	[134]
rice husk	350–400	3.0, 6.0, and 12.0 $Mg \cdot ha^{-1}$ soil	Alfisol	increasing soil pH by 1.21 units, increasing soil organic matter content by 9.41 $g \cdot kg^{-1}$	[135]

6. Effect of Biochar on Soil Microbial Composition and Activity

Biochar is an extremely valuable soil conditioner, as it not only changes numerous soil's physicochemical properties (increases moisture retention and air permeability), but also affects microbial activity [136]. This is confirmed by He et al. [137], who reported that the addition of biochar changes the soil nutrient cycle and supply, which in turn can affect the microbial community (Table 3).

In its structure, biochar has pore spaces, an ideal habitat for microbial and fungal communities [138,139]. Its porous structure promotes the productivity and increased survival of microorganisms in the soil. For example, by comparing biochar from pine bark and biochar from sewage sludge with poultry litter and perlite as a carrier for *Bradyrhizobium* strains, Araujo et al. [140] found that biochar from pine bark was the optimal carrier extending the shelf life of the bacteria to one year. In addition, Hale et al. [141] found that the addition of a mixture of biochar and pine wood bacteria (*Enterobacter cloacae* UW5 strains) to sandy-clay soils increased the bacterial population density by 16%.

The addition of this substance changes the structure of bacterial population in the soil [142]. Bacteria involved in nitrogen fixation (*Rhizobium* sp., *Azospirillum* sp.), nitrification (*Nitrospora* sp., *Nitrobacter* sp.), and methanotrophic bacteria (*Methylobacterium* sp.) showed growth in soil modified with biochar [143].

Soil bacteria are one of the basic factors regulating soil productivity [144]. The presence of soil bacteria promotes nutrient uptake and improves soil aggregation [145]. The addition of biochar can help inhibit or reverse soil degradation processes. Biochar can also promote the growth of mycorrhizal fungi, which contributes to nutrient circulation, uptake, and aggregation in the soil [146,147].

Both biochar and mycorrhiza have the potential to promote soil quality, likely creating synergy between the two elements [148]. Both elements are of increasing interest due to their positive impact on soil fertility, but in the case of biochar, the processes occurring in the soil under its influence are not fully understood.

The interactions of the combination of both components are unspecified and depend on many factors, such as soil structure, organic matter content, and the presence of soil minerals and microorganisms; hence, it is important to continue research to understand the changes taking place [149]. Research on the interaction of biochar with soil microorganisms has been conducted for many years by many researchers, and the resulting reactions to the use of biochar depend largely on the substrate from which it was produced.

Table 3. Effect of biochar on soil microbial composition and activity.

Type of Biochar	Pyrolysis Temperature [°C]	Dose of Biochar	Type of Soil	Effect on Soil Microbial Activity	Reference
eucalyptus wood	not mentioned	1.5, 3.0, and 6.0 $Mg \cdot ha^{-1}$ soil	sandy clay loam	increasing 20–40% of AMF colonization	[150]
Californian pine	not mentioned	15.0 and 30.0 $Mg \cdot ha^{-1}$ soil	silt loam soil	positive influence on promoting an increase effect on bacterial abundance (*Bradyrhizobiaceae*, *Hyphomicrobiaceae*)	[151]
mineral-enriched	450	6.5 $Mg \cdot ha^{-1}$ soil	Red Chromosol soil	promoting the growth of soil microorganism populations by creating suitable habitats for their development	[138]
maize straw	450	2.5, 7.5, and 22.5 $Mg \cdot ha^{-1}$ soil	alluvial soil	biochar application significantly increased the phospholipid fatty acids of AMF	[152]
rice husk	350	3.0, 6.0, and 12.0 $Mg \cdot ha^{-1}$ soil	gray forest soil	alkaline conditions resulting from biochar amendment favor fungal growth over bacterial growth	[153]
amur silvergrass rice straw	500–600	22.2 g per 100 g soil	sandy loam soil	biochar reduced the Pb bioavailability, while rice straw biochar increased the As; biochar addition enhances the dissolved organic matter and AMF growth inside sandy soil	[154]
agricultural wastes	600	0.0025 $Mg \cdot ha^{-1}$ soil	garden soil	increasing the shelf life of *Burkholderia* sp. and *Bacillus* sp.	[155]

7. The Use of Biochar in Agricultural and Horticultural Production

Biochar is considered one of the alternative soil enrichment practices for sustainable agriculture and the environment [7]. It supports sustainable farming practices, which are essential for the future of agriculture. It also improves soil quality and reduces the need for chemical fertilizers and water, promoting more environmentally friendly agriculture [156,157].

Moreover, biochar has been demonstrated to improve plant growth and development, and thus yield (Tables 4–6), under both optimal growing conditions and stressful conditions, such as water deficit and excessive soil salinity (Tables 7 and 8) [158].

In laboratory and field studies, biochar has consistently shown significant beneficial effects on plant growth and development, resulting in increased crop yields and productivity (Table 4) [159,160]. For example, the use of biochar from grass has been shown to increase the yield of onion (*Allium cepa*) [161]. A positive effect of biochar has also been observed by Carpenter and Nair [162], who showed a higher weight of carrots after using this substance.

This substance can provide oxygen for seed germination and increase seedling growth and development by improving soil aeration and reducing soil bulk density (Table 4) [163]. Latini et al. [164] showed an increase in fresh and dry mass in five varieties of durum wheat after the use of biochar resulting from wood chips and wheat straw. Biochar from wheat straw clearly increased the aboveground maize biomass and its root system (including total volume, surface area, and number of maize root tips) [165].

In their research on the impact of biochar on soil properties, Petrova et al. [166] found that biochar can increase soil fertility and retention capacity, which in turn leads to improved plant growth conditions and better yields.

Table 4. Impact of the application of different types of biochar on plant growth, development, and yield.

Type of Biochar	Pyrolysis Temperature [°C]	Dose of Biochar	Type of Soil	Plant Species	Effect on Plant Growth and Development	Reference
rainbow eucalyptus	350	25.5 and 45.5 g per 1.0 kg of soil	inceptisol soil	rice (*Oryza sativa*)	increasing grain yield from 232.0 to 250.0%	[167]
wheat straw	507	15.0, 30.0, 75.0, 150.0, and 225.0 $Mg \cdot ha^{-1}$ soil	loamy sand soil	sunflower (*Helianthus annuus*)	increasing seed germination from 8.0 to 31.0%	[168]
bamboo	500	2.0 and 5.0% per 1.0 kg soil	clay loam soil	cherry tomato (*Solanum lycopersicum*)	improving growth, number of leaves, fresh biomass yield of aboveground and underground parts from 21.0 to 114.0%	[169]
apple branch	450	1.0, 2.0, 4.0, and 6.0% of soil volume	silt–clay soil	durum wheat (*Triticum durum*)	increasing the crop yield by 14.77 to 43.02%	[170]
rice husk	450	50.0 $Mg \cdot ha^{-1}$ soil	fluvo-aquic soil	peanut (*Arachis hypogaea*)	increasing the crop yield by 42.40 and 48.40%	[171]
corn cob	450	20 $t \cdot ha^{-1}$ soil	sandy loam	soy (*Glycine max*. L. Merr)	increasing the rate of seed germination from 24.0 to 27.0%, increasing shoot growth from 22.0 to 27.0%	[172]
walnut shell	550	10.0, 20.0, 40.0, 80.0, and 120.0 $Mg \cdot ha^{-1}$ soil	Petri dish bioassay method	triticale (*Triticosecale* Wittmack), green pea (*Pisum sativum* sp. arvense)	higher germination rate and growth from 69.0 to 97.0%	[173]

Table 4. *Cont.*

Type of Biochar	Pyrolysis Temperature [°C]	Dose of Biochar	Type of Soil	Plant Species	Effect on Plant Growth and Development	Reference
cotton straw	450	1.0 and 3.0% per 15.0 kg of soil	gray desert soil	cotton (*Gossypium hirsutum*)	increasing root dry weight from 7.52 to 13.06 g per plant and shoot dry weight from 95.93 to 136.81 g per plant in harvesting stage	[174]
pinewood	400	5.0, 10.0, 15.0, and 20.0% per 5.7 L pot	Ultisols soil	muscadine grape (*Vitis rotundifolia*)	strengthening the root architecture, increasing the length of roots from 2000.0 to 7000.0 cm and the number of root forks from 3000.0 to 6000.0	[175]

Photosynthesis is a basic and very complex physiological process of plants, determined by many factors, both internal and external. Any reduction in its intensity causes a decrease in the quantity and quality of the crop [176].

Many authors have shown that the use of biochar not only increases the rate of photosynthesis, but also increases the chlorophyll content of leaves, the rate of transpiration, the total content of flavonoids, sugars and glucose, and the Relative Water Content of plants (Figure 3) [177–179]. The beneficial effects of biochar on plants physiological and biochemical traits have been reviewed and are summarized in Table 5.

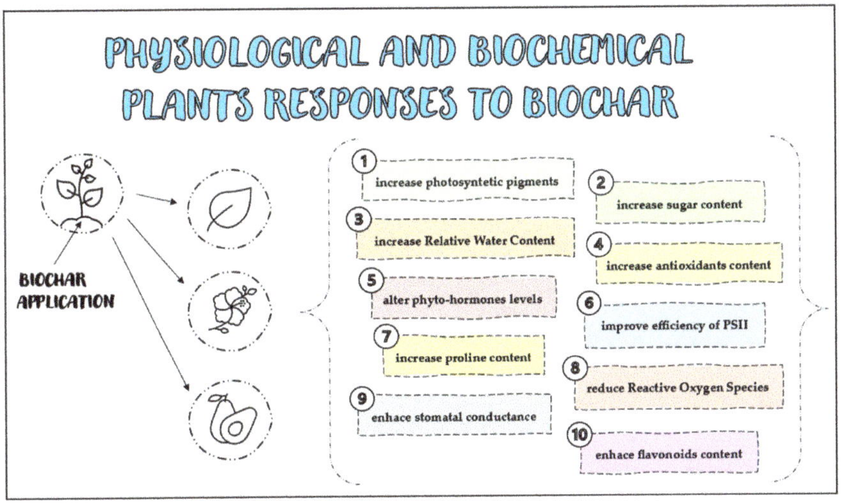

Figure 3. Physiological and biochemical plants responses to biochar.

In studies conducted by Abid et al. [180] and Jabborova et al. [181], the authors noted an increase the total content of chlorophylls and carotenoids in tomato leaves and common basil. Similar results were obtained by Hafez et al. [182], where the addition of biochar stimulated the growth of assimilation dyes in barley. Lévesque et al. [183] demonstrated a 32% increase in the value of the Water Use Efficiency index in photosynthesis due to the use of biochar in tomatoes.

Table 5. Impact of the application of different types of biochar on physiological and biochemical traits of plants.

Type of Biochar	Pyrolysis Temperature [°C]	Dose of Biochar	Type of Soil	Plant Species	Effect on Physiological and Biochemical Traits	Reference
Lantana camara stems	450	10.0 and 30.0 g per 1.0 kg of soil	sandy clay	okra (*Abelmoschus esculentus* Moenc)	increasing net photosynthetic rate and Water Use Efficiency	[184]
cotton stalk	not mentioned	5.0 $Mg \cdot ha^{-1}$ soil	sandy clay loam	cotton (*Gossypium* sp.)	increasing Relative Water Content and chlorophyll stability index, decreasing proline accumulated in leaves	[185]
rice husk	450	1.0, 5.0, and 10.0% per soil	not mentioned	soybean (*Glycine max*)	increasing leaf net photosynthetic rate, chlorophyll index, leaf soluble sugar, and starch contents	[186]
black cherry wood	450	2.0 and 3.0% per 400.0 g of soil	loam soil	basil (*Ocimum basilicum*)	enhancing chlorophyll content, total sugar, flavonoids, and soil enzymes activities	[181]
black cherry wood	450	2.0 and 3.0% per 1.5 kg of soil	not mentioned	ginger (*Zingiber officinale* Rosc.)	increasing Relative Water Content, chlorophyll, and carotenoid content	[187]
olive pomace waste	500	40.0 $Mg \cdot ha^{-1}$ soil	sandy loam	olive (*Olea europaea*)	increasing water potential in leaves, net photosynthetic rate, Water Use Efficiency, and electron transport rate	[188]
rice husk	450	20.0, 30.0, and 50.0 g per 1.0 kg of soil	sandy loam soil	tomato (*Lycopersicon esculentum*)	enhancing stomatal conductance, net photosyntetic rate, transpiration rate, and Water Use Efficiency	[189]

Biochar helps improve crop yields and Nutrient Use Efficiency, which leads to improved soil quality, i.e., improved physicochemical and biological properties [190].

It can increase not only the total content of N, K, and P in the soil, but also all the necessary nutrients for plant growth (Table 6) [139,191–193], which are related to its properties—a large number of oxidized functional groups and porous structure [194].

Table 6. Impact of the application of different types of biochar on nutrient availability.

Type of Biochar	Pyrolysis Temperature [°C]	Dose of Biochar	Type of Soil	Plant Species	Effect on Nutrient Availability	Reference
peanut hull, pine chip	400	11.2 and 22.4 $Mg \cdot ha^{-1}$ soil	loamy sand soil	maize (*Zea mays*)	increasing the availability of N, P, K, Ca, and Mg	[195]
wastewater sludge	550	10.0 $Mg \cdot ha^{-1}$ soil	Chromosol	cherry tomato (*Lycopersicon esculentum*)	increasing nutrient availability (P and N)	[196]
poultry manure	550	10.0 and 20.0 $Mg \cdot ha^{-1}$ soil	tropical acid soil	maize (*Zea mays*)	increasing soil availability of N, P, K, Ca, and Mg	[197]
poultry litter, pinewood	350	40.0 $Mg \cdot ha^{-1}$ soil	Ultisols	winter wheat (*Triticum aestivum*)	increasing uptake of Ca, Mg, Na, P, and K	[198]
peach and vine pruning waste	550	4.0, 10.0, 20.0, 30.0, and 40.0 g per 1 L	heavy alkaline-calcareous soil	kiwifruit (*Actinidia* sp.)	increasing exchangeable Fe in the soil, greater absorption of nutrients by plants	[199]
fruit trees wood waste	450	50.0 $Mg \cdot ha^{-1}$ soil	silty clay loam	tomato (*Lycopersicon esculentum*)	increasing N and P uptake	[200]
bamboo, rice	750–800 (bamboo), 500 (rice)	2.5 and 5% per pot	loamy clay	tea (*Camellia sinensis*)	increasing P, K, and Mg concentrations in plants	[201]
wheat straw	400	12.5 g per 5.0 kg of soil	clay–loam soil	rice (*Oryza sativa*)	increasing N, P, K, and Fe uptake	[202]

Many authors have demonstrated positive effects of biochar on plants growing under conditions of biotic and abiotic stress (Tables 7 and 8), particularly its capacity to increase resistance to adverse environmental stressors and to maintain plant growth [203–206].

Because biochar affects the soil, rhizosphere, pathogens, and plant microbiome, it can assist plants in the rapid regulation of defense-related processes, such as oxidative burst, when encountering abiotic and biotic stress [207].

Biochar is effectively used against a wide range of air- and soil-borne plant pathogens, and the mechanisms by which this substance can protect plants from disease are varied [208,209].

A large number of studies indicate that the addition of biochar can confer induced resistance to pathogenic organisms on plants (Table 7). In a study on the control of Fusarium ear rot in maize, it was found that soil amendment of poultry manure biochar and sawdust biochar through effective management of the resident soil pathogens proved beneficial in inhibiting *Fusarium verticillioides* infection [210]. Also, in studies conducted on tomato (*Solanum lycopersicum*), it was found that after enriching the soil with biochar from greenhouse waste, plant infections by *Botrytis cinerea* pathosystem were reduced by up to 50% [211].

Although various soil pathogens can adversely affect crop growth in the form of diseases, benevolent soil microorganisms can displace the negative effects of harmful bacteria and fungi [212]. The action of biochar results in disease inhibition through the aforementioned induction of systemic resistance in plants, i.e., by increasing the density and activity of beneficial microorganisms such as N_2-fixing bacteria [213], plant growth-promoting rhizobacteria [214], and *Trichoderma* spp. [215].

Table 7. Impact of the application of different types of biochar on plants under biotic stresses.

Type of Biochar	Pyrolysis Temperature [°C]	Dose of Biochar	Type of Soil	Plant Species	Effect on Plants Under Biotic Stresses	Reference
citrus wood	450	1.0 and 3.0 g per 100.0 g potting mixture	coconut fiber (peat)	strawberry (*Fragaria ananassa*)	induction of systemic resistance against *Botrytis cinerea*, *Colletotrichum acutatum*, and *Podosphaera apahanis*	[216]
pinewood	550–600	5.0% per 2.54 L pot	substrate (sphagnum peatmoss, horticultural grade perlite, dolomitic limestone)	red oak (*Quercus rubra*), red maple (*Acer rubrum*)	induction of systemic resistance against *Phytophthora cinnamon* and *Phytophthora cactorum*	[217]
eucalyptus wood, greenhouse waste	350–600	0.5, 1.0, and 3.0% per 100.0 g potting mixture	potting mixture	cucumber (*Cucumis sativus*)	suppressing soil-borne pathogen *Rhizoctonia solani* (belly rot)	[218]
poultry fecal waste, sawdust	485	1.0, 2.0, and 3.0 $kg \cdot m^2$ soil	not mentioned	maize (*Zea mays*)	suppressing the infection caused by *Fusarium verticillioides* (maize ear rot)	[210]
greenhouse waste	350	1.0 and 3.0% per 0.5 L pot	substrate (peat tuff)	tomato (*Solanum lycopersicum*)	mediating systemic resistance against Fusarium crown rot (*Fusarium oxysporum* f. sp. *radicis lycopersici*)	[219]

Abiotic stresses pose a serious threat to crops and global food security. The intensity of abiotic stresses (drought, heat stress, salinity, and soil acidification) continues to increase, negatively affecting crop productivity [220]. Biochar is a carbon-rich product used as an important soil conditioner to improve soil quality and plant performance [221]. Its effect on plant physiological traits and soil properties contributes to the development of a resistance mechanism in plants to environmental stresses [222]. The beneficial effects of biochar abiotic stresses on plants have been reviewed and are summarized in Table 8.

It has been shown repeatedly that salinity stress adversely affects plant growth and development and yield [223]. It causes weaker seed germination, hinders plant nutrient uptake, and contributes to a decrease in the activity of antioxidant enzymes [224–226].

Biochar application brought favorable changes in soil and plant functioning to improve salinity stress tolerance [221,227]. As reported by Parkash and Singh [228] and Kerbab et al. [229], this substance reduces the toxic effects of salinity by increasing the efficiency of the photosynthetic process, regulating water relations in plant cells, accumulating osmolytes, secondary metabolites, and hormones, increasing antioxidant activity, and thereby reducing the production of reactive oxygen species in plants (Table 8). Abo-Elyousr et al. [230] indicated that the application of biochar improves the anatomical and physiological properties of mung bean plants under salt stress conditions. A range of research authors have revealed that in plants exposed to salt stress, stomatal density and stomatal conductance improved after soil application of biochar [158,231,232].

Drought stress is one of the most serious abiotic stresses, affecting crop yield reduction to a great extent [233,234].

In fact, under water deficit conditions, the stomata close, which contributes to a decrease in CO_2 content and thus photosynthetic efficiency [235]. There is a loss of plant cell turgor as a result of reduced leaf water content [236]. Drought is a factor limiting the growth and development of plants, which affects the course of many physiological and biochemical processes in plants [237]. As a result of the inhibition of gas exchange and thus the assimilation of CO_2, there is a restriction of the synthesis of sugars. The change in the distribution of nutrients causes the need to restrict the energy-intensive processes of growth, and in extreme cases, even generative development. Inhibition of growth, weaker fruit binding and reduction in yield size, and sometimes also deterioration of its quality have been observed during drought in various crop species [238,239].

Biochar, due to its porous structure and thus soil Water Holding Capacity, affects water availability to plants under drought stress (Table 8) [240]. For instance, Langeroodi et al. [241] showed an increase in water content in pumpkin leaves growing under water-scarce conditions in response to biochar application. The positive effect of biochar application was also observed by Trupiano et al. [222], who reported an increase in photosynthetic rate, chlorophyll content, and transpiration rate in plants. Biochar also improves the gas exchange capacity of crop leaves and reduces oxidative stress [177].

Moreover, Gharred et al. [242] and Keabetswe et al. [243] reported that the application of biochar under water-deficient conditions improves the nutritional status of plants, affecting biomass growth and production, antioxidant activity, and osmolyte accumulation. It has also been proven that biochar increases nutrient uptake under drought stress conditions (Table 8). This is confirmed by Poormansour et al. [244], who detected an increase in Ca and Mg uptake by faba bean (*Vicia faba*) at different irrigation levels and different doses of biochar.

Under heat stress conditions, the addition of biochar can improve soil bulk density, organic matter content, as well as soil microbial structure and microbial abundance, thereby affecting the size of the plant root system [245].

Table 8. Impact of the application of different types of biochar on plants under abiotic stresses.

Type of Biochar	Pyrolysis Temperature [°C]	Dose of Biochar	Type of Soil	Plant Species	Effect on Plants Under Abiotic Stresses	Reference
			SALINITY STRESS IN CROPS			
dried leaves, sawdust	350	1.0 and 2.0% per 10.0 kg soil in pot	not mentioned	wheat (*Triticum aestivum*)	improving germination and growth, and root and shoot length, increasing leaf water potential and osmotic potential under salinity conditions	[246]
Conocarpus wood	300	12.0 g per 4.0 kg of soil	not mentioned	tomato (*Lycopersicon esculentum*)	improving plant growth and total yield under salinity stress	[247]
softwood, hardwood	not mentioned	0.625 g per 12.0 kg of soil	sandy–clay–loam soil	eggplant (*Solanum melongena*)	minimizing impacts of salinity stress, enhancing stomatal conductance and photosynthesis rate, decreasing leaf temperature and electrolyte leakage in leaf tissues, better root growth, shoot growth, and fruit yield	[228]
wood	250–300	30.0 and 45.0 g per pot	silty clay loam	maize (*Zea mays*)	improving plant height, number of leaves, leaf area, fresh and dry shoot biomass, and fresh and dry root biomass under salinity conditions	[248]
			DROUGHT STRESS IN CROPS			
wheat straw	450	3.0 and 5.0 g per 100.0 g soil	clay loam with sand, silt, and clay	rice (*Oryza sativa*)	increasing total chlorophyll content in leaves under drought conditions	[249]
woody branches of button mangrove	450	3.0% per 100.0 g soil	seed cradle	chickpea (*Cicer arietinum*)	increasing root and shoot length, Relative Water Content, chlorophyll content, and membrane stability index	[250]
hornbeam wood chips	450–500	10.0, 20.0, and 30.0 $g \cdot kg^{-1}$ soil	clay loam	chestnut-leaved oak (*Quercus castaneifolia* C.A.M.)	alleviating effects of water deficit, improving photosynthesis, stomatal conductance, and leaf nutrient concentration (P, K)	[251]
rosehip seeds	500	0.5, 1.0, and 2.0% per pot	silty clayey loam	sugar beet (*Beta vulgaris*)	improving K, Mg, and Mn content in plants under drought stress	[252]
wheat straw	500	27.88 and 37.18 $g \cdot kg^{-1}$ soil	fine loam soil	wheat (*Triticum aestivum*)	mitigating effects of drought stress, improving the number of fertile tillers, spike length, the number of grains per spike, thousand grain weight, economic yield, and water use efficiency	[253]

Table 8. *Cont.*

Type of Biochar	Pyrolysis Temperature [°C]	Dose of Biochar	Type of Soil	Plant Species	Effect on Plants Under Abiotic Stresses	Reference
rice straw	600	30.0 $Mg \cdot ha^{-1}$ soil	clayey sand	rapeseed (*Brassica napus*)	mitigating the hostile effects of drought stress to reduce malondialdehyde, leaf hydrogen peroxide, and electrolyte leakage by increasing superoxide dismutase, peroxidase, and catalase activities	[254]
			TEMPERATURE STRESS IN CROPS			
rapeseed straw	not mentioned	40.0 $g \cdot kg^{-1}$ soil	fluvisols	rice (*Oryza sativa*)	improving protein transport and N assimilation by shoots under heat stress conditions	[245]

8. The Benefits of Using Biochar for the Environment

Biochar has recently received a lot of attention for its potential to slow global warming and improve soil health. It is widely used in horticultural and agricultural production as a soil additive to prevent soil leaching, moisture loss, toxic element accumulation, and fertilizer runoff, among other benefits [255,256].

The substance also finds applications in environmental management, such as in offsetting excess greenhouse gas emissions, carbon sequestration, soil and groundwater remediation, and wastewater treatment (WWT) [257,258].

Contamination of soils with various substances is growing global concern, posing a significant threat to the environment and human health [259]. The increase in the content of heavy metals, polycyclic aromatic hydrocarbons, dioxins, and furans covers an increasing number of affected areas [260]. All these pollutants have a toxic effect on ecosystems by permeating from the soil to plants and soil microorganisms and then to groundwater [261,262]. In response to the existing problems, new, efficient, and at the same time, cheap remediation options are sought, along with the simultaneous revitalization of land and improvement of plant growth [263,264].

The use of biochar to remove heavy metals from soils is a positive and promising strategy, due to its low production costs and exceptional efficiency [265], compared to other heavy metal removal methods, such as chemical precipitation, ion exchange, electrodialysis, or membrane filtration [266]. The presence of functional groups on the surface of biochar gives the potential to adsorb toxic substances such as manganese (Mn) and aluminum (Al) in acidic soils as well as arsenic (As), nickel (Ni), copper (Cu), cadmium (Cd), and lead (Pb) in soils contaminated with heavy metals [267,268]. Table 9 presents reports on the impact of biochar on the remediation of contaminated sites.

Many authors report that adding biochar has a positive effect on cleaning the environment of substances commonly considered toxic, thus limiting their availability to living organisms [269,270]. Modified biochar is considered as a soil treatment agent that can rehabilitate soil contaminated with heavy metals by improving soil properties [271].

Fahmi et al. [272] showed that biochar from empty fruit bunches can help remove lead and cadmium from soils. Also, Gong et al. [273] demonstrated the positive effect of pyrolysis on the stabilization of heavy metals (cadmium, chromium, zinc, lead, and copper) in biomass derived from *Boehmeria nivea* used for phytoremediation. In addition, Koh et al. [274] observed a decrease in the concentration of arsenic and heavy metals in the soil with the addition of biochar, compared to the control variant.

These effects are mainly due to the biochar structure. Indeed, biochar has a large Specific Surface Area, porous structure, and the ability to exchange cations, which makes it an efficient sorbent for various pollutants, both organic and inorganic [275]. This substance has great adsorption capacity due to its permeable structure and asymmetric plates [276,277].

There are also reports indicating the phytostabilizing abilities of plants for the reclamation of heavily contaminated (As and Pb) piles [278]. The latter authors showed that fine particles of biochar allowed *Salix viminalis* to grow on contaminated soil, allowing the species to be used for phytostabilization.

Dai et al. [279] indicated that organic pollutants under the influence of biochar are sorbed twice as fast as in natural organic matter. It has also been reported that during the production of biochar from biomass obtained from areas contaminated with heavy metals, toxic elements volatilized during the pyrolysis process at a higher temperature (700 °C) [280]. Additionally, in soils where biochar formed from contaminated biomass was used, it did not show any adverse effects on microorganisms [281,282]. This reflects a promising effect on the economy, especially in the case of land that requires phytostabilization or is heavily industrialized and exposed to high levels of heavy metals.

Table 9. Effect of biochar in the remediation of contaminated land.

Type of Biochar	Pyrolysis Temperature [°C]	Dose of Biochar	Type of Soil	Effect in the Remediation of Contaminated Land	Reference
oak wood-derived	400	4.0% per 3.5 kg soil	settling pond soil	reducing the phytoavailable contents of Cu, Pb, Ni, and Zn in soil	[283]
rice straw	500	33.0 and 65.0 Mt·ha^{-1} soil	red soil	increasing soil pH and nutrient content, immobilizing Cd	[284]
vegetable wastes	500	5.0 g per 100.0 g soil	sandy loam	achieving Pb immobilization of 87%	[285]
willow	500–700	1.0 g per 50.0 mL soil	soilless	reducing the concentration of Pb, Ni, and Cd in the soil	[286]
rice husk	400–600	0.5, 1.0, and 2.0% per 280.0 g soil	clay loam soil	reducing the bioavailability of Pb from contaminated land	[287]
paulownia	700–800	1.0 g into a 100.0 mL polycarbonate tube	polycarbonate tube	reducing the concentration and minimizing the availability of Cu, Pb, and Cd in the soil	[288]

Heavy metal stress is a serious threat to plants, causing morphological, physiological, and biochemical changes in plants, which most often leads to a reduction in their biomass and yield [289].

As a result of this type of stress, there is an accumulation of reactive oxygen species in plants and cell damage occurs due to the excessive production of free radicals [290]. Heavy metal toxicity interferes with plant metabolism, disrupting the primary process for plant growth and development—photosynthesis [291], as well as indirect processes such as respiration, transpiration, and acquisition of micro- and macronutrients [292,293]. In addition, metal toxicity stunts root growth and reduces root hair formation [294]. Consequently, due to phytotoxicity caused by metal ions, yields and crop quality are also reduced.

The answer to this type of stress may be biochar, which is being studied for environmental remediation, reduction in mobility of contaminants in contaminated soils, and reduction in adverse changes caused by elements such as arsenic, cadmium, chromium, copper, lead, mercury, and nickel in agricultural and horticultural crops [8]. Biochar can capture toxic compounds such as pesticides and herbicides or organic impurities (i.e., agrochemicals, antibiotics, and other hydrocarbons) [295,296], reducing their bioavailability and thus their accumulation in crops such as wheat (*Triticum aestivum*) [297], lettuce (*Lactuca sativa*) [298], and alfalfa (*Medicago sativa*) [299].

Many studies, represented in Table 10, reported that adding biochar to the soil reduced the uptake and toxicity of metals to the plant.

Table 10. Effects of biochar on alleviating heavy metal stress (HMS) in crop production.

Type of Biochar	Pyrolysis Temperature [°C]	Dose of Biochar	Type of Soil	Plant Species	Effect on Alleviating Heavy Metal Stress	Reference
rice straw	500	5.0 and 10.0 g per 100.0 g soil	not mentioned	sedum (*Sedum plumbizincicola*)	reducing the concentration of Cu, Cd, and Pb in soil	[300]
woody biomass	500	1.0, 2.5, 5.0, and 10.0% per 1.0 kg soil	loamy sand	maize (*Zea mays*)	decreasing Pb and Cd toxicity, immobilizing them into more stable forms	[301]

Table 10. *Cont.*

Type of Biochar	Pyrolysis Temperature [°C]	Dose of Biochar	Type of Soil	Plant Species	Effect on Alleviating Heavy Metal Stress	Reference
eucalyptus wood, bamboo, rice husk	550	5.0, 10.0, and 20.0 $Mg \cdot ha^{-1}$ soil	not mentioned	rice (*Oryza sativa*)	alleviating soil Al toxicity	[302]
cotton straw	350–550	46.8 $Mg \cdot ha^{-1}$ soil	not mentioned	cotton (*Gossypium hirsutum*)	reduction in Cd accumulation in plant organs	[303]

Soil acidification is a natural process that occurs globally [304], but intensive agricultural practices such as fertilization and pesticide use can also lower the pH of soils [305]. When the pH value is lowered below 4.5, aluminum, manganese, and iron ions are activated, which, when present in excess in the soil solution, exhibit toxicity to plants [306].

Liming is the standard treatment to raise soil pH and minimize the negative effects of soil acidity on crops grown [307]. Recently, an alternative method in the form of biochar has also emerged as a liming material for controlling soil acidity [308,309].

Biochar has repeatedly been shown to have a positive effect on reducing soil acidity (Table 11), improving the soil pH, increasing exchangeable cations, and decreasing exchangeable aluminum [310,311]. The application of biochar on acidified soils increases the availability of nutrients in the soil and the absorption of nutrients (N, P, K, and Mg) by plants [312].

Evidently, biochar has positive effects on lowering soil acidity, thereby improving plant growth and development (Table 11). Das et al. [313] showed that this substance also accelerates seed germination, increases shoot and root length, weight and shoot biomass in maize, and black gram in acidic soil. Similar results were obtained by Nguyen et al. [314], who demonstrated that biochar added to acidic soil increased elephant grass biomass yield.

Table 11. Effects of biochar on mitigating soil acidification stress in crop production.

Type of Biochar	Pyrolysis Temperature [°C]	Dose of Biochar	Type of Soil	Plant Species	Effect on Mitigating Soil Acidification Stress	Reference
peanut shells	400–800	not mentioned	soilless	cabbage (*Brassica oleracea*)	increasing plant height, biomass production, and root growth, higher soil pH, and reduced Al concentration in the soil	[315]
poultry litter	600–700	15.0 $g \cdot kg^{-1}$ soil	acid soils	maize (*Zea mays*)	higher soil pH, decreasing in exchangeable soil acidity, increasing in the content of bioavailable N, P, K, Ca, and Mg in the soil	[316]
bamboo, rice straw	750–800	2.5 and 5.0% per pot	loamy clay	tea (*Camellia sinensis*)	higher soil pH, increasing concentration of P, K, and Mg in plants, reduction in soil-available Mn and Cu concentrations	[201]
corn stover	500	25.0 $Mg \cdot ha^{-1}$ soil	acidic soil	bean (*Phaseolus vulgaris*)	reducing soil acidity, decreasing bioavailability of Ni	[317]

One of the most important environmental problems of the 21st century is global warming, caused by the rising emissions of greenhouse gases (GHGs), and the carbon cycle plays a key role in both causing and mitigating it [318]. The main long-lived GHGs are, in descending order, atmospheric carbon dioxide (CO_2), methane (CH_4), and nitrous oxide (N_2O) [319]. Therefore, biochar represents huge potential in terms of mitigating climate change, primarily through carbon sequestration and reducing greenhouse gas emissions, as well as reducing waste biomass combustion and producing clean bioenergy [320].

According to Woolf et al. [321], the global use of biochar would offset 12% of current anthropogenic CO_2-C-equivalent greenhouse gas emissions. In addition, Jia et al. [322]

reported that adding biochar to the soil reduces emissions not only of CO_2, but also of other potent greenhouse gases, especially nitrous oxide and methane (Table 12). The reduction in CO_2 emissions is due to the close binding of biochar to soil particles [323]. For example, peanut shell biochar reduced the total CO_2 emissions of soil by 33%, while biochar from wheat straw reduced CO_2 emissions by 90.25% [324].

Moreover, biochar can reduce emissions of both carbon dioxide and nitrous oxide [325]. Zhang et al. [326] showed that long-term biochar (2 or 7 years) reduced N_2O emissions by 48% in acidic soils and by 22% in alkaline soils compared to control soils.

For instance, Rondon et al. [327] observed almost complete inhibition of CH_4 emissions from a grassland (*Brachiaria humidicola*) and soybean crop, with the application of biochar at 20 $g \cdot kg^{-1}$ soil. These authors also reported a decrease in N_2O emissions from the same crops, by 50 and 80%, respectively.

Table 12. Impact of biochar on reducing greenhouse gas emissions.

Type of Biochar	Pyrolysis Temperature [°C]	Dose of Biochar	Type of Soil	Effect on Reducing GHG	Reference
wood shavings	650	12% (by volume) of dairy cattle slurry	sandy loam	77% NH_3 emission reduction, 63% N_2O emission reduction, 84% CO_2 emission reduction	[328]
pinewood	495–505	1.14 and 4.56 $kg \cdot m^{-2}$ soil	dairy cattle slurry	23% NH_3 emission reduction, 30% H_2S emission reduction, 26% indole emission reduction	[329]
pine chips	400	5.0 and 20.0% of mass of poultry litter	poultry litter	NH_3 emission reduction by 58%, H_2S emission reduction by 71%	[330]
poultry litter	550	10.0% of dry mass of poultry manure-wheat straw	mixture (poultry manure wheat straw)	N_2O emission reduction of up to 75%, up to 40% of total N retained in the composting mixture	[331]
willow woodchips	350	5.0 and 10.0% mass of poultry manure–wheat straw	mixture (poultry manure–wheat straw)	44% NH_3 emission reduction	[332]

9. Threats Related of Application Biochar in Crop Production and the Environment

The use of biochar in the cultivation of crops on agricultural soils has excellent potential. Although the literature reports positive effects of biochar on plants, such as improving plant growth and reducing the accumulation of harmful substances in plants, the toxic effects of biochar on plants cannot be ignored (Table 13). Much depends on the composition of the biomass used to produce biochar and the conditions prevailing during the pyrolysis process [333].

Potential disadvantages related to the soil application of biochar include binding and deactivation of agrochemicals, mainly pesticides [334], influence on seed germination [335], processes, and potential release of toxic substances such as heavy metals [336] and PAHs [337,338] that may be present in biochar.

The use of biochar, derived from raw materials containing heavy metals, significantly increased the Cd(II) content in plants, even in minimally contaminated areas [339]. Similar results were obtained by Zhao et al. [340], who found that the addition of pig manure and sewage may release heavy metals—Cu(II) and Zn(II)—into soils.

Using biochar to retain organic and inorganic substances in the soil may contribute to disturbances in the nutrient uptake of plants and thus the quality of the final yield [341,342]. Especially in the case of food, where high quality and compliance with specific standards are required, it has not been clearly stated whether the changes occurring due to biochar affect its quality parameters [343].

Biochar can also have a negative effect on plant growth and development. Huang et al. [344] indeed revealed that the addition of this substance reduces the germination rate and root length of Chinese cabbage (*Brassica juncea* L.).

Additionally, the presence of biochar in the soil can reduce the biomass of microbes evenly distributed across different functional groups. As a result of changes in the soil environment, some phylotypes will become competitive or limit certain groups of microbes, which in turn leads to changes in the composition of the community and the diversity of populations of bacteria and fungi [345,346].

It is extremely difficult to interpret the results confirming the risk of using biochar in crop production, but in the future, the focus should be on studying the toxic effects of biochar at the cellular and molecular level to reveal the underlying mechanisms [347].

It is, therefore, crucial to implement mitigation strategies to address these potential risks and minimize their impact. These strategies should include testing for toxic content in biochar, assessing its quality, and considering the specific feedstock–application combinations and geographical contexts in which it is applied [348].

Furthermore, integrating biochar deployment scenarios with climate models can help quantify its climate interactions and develop simplified metrics for individual studies [349]. Overall, thorough investigation and careful consideration of potential risks are necessary when applying biochar to soil to ensure sustainable and responsible agricultural practices [350].

Table 13. Threats related to the application of biochar in crop production and the environment.

Type of Biochar	Pyrolysis Temperature [°C]	Dose of Biochar	Type of Soil	Threads to Crops and the Environment	Reference
beech wood	550	72.0 Mg·ha^{-1} soil	chernozem, Cambisol	reducing nitrogen absorption of corn by 44%	[351]
greenhouse vegetable waste	500	85.0 Mg·ha^{-1} soil	paddy soil	inhibiting the scale of the soil microbial community	[352]
paper fiber sludge, grain husks	550	20.0 Mg·ha^{-1} soil	Haplic Luvisol	decreasing soil water content, due to increasing soil salt content and blocking pores	[353]
poultry manure	400	15.0, 20.0, 25.0, and 30.0 Mg·ha^{-1} soil	Red Yellow Latosol	increasing soil salinity, amplifying the sodium adsorption ratio and exchangeable sodium percentages	[354]
apple branch	550	1.0, 3.0, 5.0, and 7.0% soil volume	silt loam	increasing total soil erosion	[355]
eggplant shoots	<500	12.0 Mg·ha^{-1} soil	clay loam soil	increasing soil pH and conductivity, which translated into a 6% reduction in potato yield	[356]

10. Conclusions and Future Perspectives

Using biochar in crops as a soil additive is an ideal alternative to other admixtures with high carbon content. There is a noticeable growing interest in this subject, as more and more scientific papers describing the advantages and disadvantages of its use are being published. Currently, the obtained results confirm the efficiency of its use, but it depends mainly on the conditions of pyrolysis, the substrate used, and the properties of the soil. Biochar may modify soil properties; however, due to its poor fertilization values, it is best used as an addition to fertilizers already in use.

The influence of biochar on the soil structure remains a big mystery. Numerous studies have shown soil aggregation, while others have shown no effect. However, its sorption abilities have found application in cleaning soils from residues of plant protection products, heavy metals, and polycyclic aromatic hydrocarbons.

Furthermore, its influence on the colonies of soil microorganisms has not been fully understood. Based on the currently obtained research results, biochar can successfully be an additive to composts because it is an excellent structure-forming material, binds excess water, and increases the activity of microorganisms and retention of nutrients.

The long-term impact of application of biochar in crop production shows a lot of promise in improving soil quality and increasing yields, but the effectiveness of its use in the long term is still a multifaceted problem. When evaluating the effectiveness of application of biochar, it is necessary to consider its stability and aging in the soil due to a number of mechanisms (biological degradation by soil organisms, abiotic oxidation, and changes in temperature and moisture).

To obtain the best and most reliable results in determining the appropriate method of biochar production and use in the future, it is necessary to carefully consider and characterize the climatic and soil conditions of the study area, considering the long period of research, and determine the best physicochemical parameters for the produced biochar. It will allow us to create a certified, high-quality product that is safe and positively impacts the natural environment in the future.

Author Contributions: Conceptualization, A.K. and D.P.; methodology, D.P. and A.K.; validation, G.M.; formal analysis, G.M. and N.B.; investigation, D.P.; resources, A.K.; data curation, A.K., D.P. and N.B.; writing—original draft preparation, A.K., D.P. and G.M.; writing—review and editing, G.M., D.P. and A.K.; visualization, N.B.; supervision, G.M. All authors have read and agreed to the published version of the manuscript.

Funding: This research received no external funding.

Data Availability Statement: No new data were created or analyzed in this study. Data sharing is not applicable to this article.

Conflicts of Interest: The authors declare no conflicts of interest.

References

1. Intergovernmental Panel on Climate Change (IPCC). *Climate Change and Land. An IPCC Special Report on Climate Change, Desertification, Land Degradation, Sustainable Land Management, Food Security, and Greenhouse Gas Fluxes in Terrestrial Ecosystems*; Shukla, P.R., Skea, J., Calvo Buendia, E., Masson-Delmotte, V., Pörtner, H.-O., Roberts, D.C., Zhai, P., Slade, R., Connors, S., van Diemen, R., et al., Eds.; Cambridge University Press: Cambridge, UK; New York, NY, USA, 2019.
2. Loboguerrero, A.M.; Campbell, B.M.; Cooper, P.J.M.; Hansen, J.W.; Rosenstock, T.; Wollenberg, E. Food and earth systems: Priorities for climate change adaptation and mitigation for agriculture and food systems. *Sustainability* **2019**, *11*, 1372. [CrossRef]
3. Wiegmann, M.; Maurer, A.; Pham, A.; March, T.J.; Al-Abdallat, A.; Thomas, W.T.; Bull, H.J.; Shahid, M.; Eglinton, J.; Baum, M.; et al. Barley yield formation under abiotic stress depends on the interplay between flowering time genes and environmental cues. *Sci. Rep.* **2019**, *9*, 6397. [CrossRef] [PubMed]
4. Aiqing, S.; Somayanda, I.; Sebastian, S.V.; Singh, K.; Gill, K.; Prasad, P.V.V.; Jagadish, S.K. Heat stress during flowering affects time of day of flowering, seed set, and grain quality in spring wheat. *Crop Sci.* **2018**, *58*, 380–392. [CrossRef]
5. Carr, T.; Balkovič, J.; Dodds, P.; Folberth, C.; Skalský, R. The impact of water erosion on global maize and wheat productivity. *Agr. Ecosyst. Environ.* **2021**, *322*, 107655. [CrossRef]
6. Bashagaluke, J.B.; Logah, V.; Opoku, A.; Sarkodie-Addo, J.; Quansah, C. Soil nutrient loss through erosion: Impact of different cropping systems and soil amendments in Ghana. *PLoS ONE* **2018**, *13*, e0208250. [CrossRef]
7. Lu, L.; Yu, W.; Wang, Y.; Zhang, K.; Zhu, X.; Zhang, Y.; Wu, Y.; Ullah, H.; Xiao, X.; Chen, B. Application of biochar-based materials in environmental remediation: From multi-level structures to specific devices. *Biochar* **2020**, *2*, 1–31. [CrossRef]
8. Jalal, F.; Arif, M.; Akhtar, K.; Khan, A.; Naz, M.; Said, F.; Zaheer, S.; Hussain, S.; Imtiaz, M.; Khan, M.A.; et al. Biochar integration with legume crops in summer gape synergizes nitrogen use efficiency and enhance maize yield. *Agronomy* **2020**, *10*, 58. [CrossRef]
9. Zulfiqar, F.; Wei, X.; Shaukat, N.; Chen, J.; Raza, A.; Younis, A.; Nafees, M.; Abideen, Z.; Zaid, A.; Latif, N.; et al. Effects of biochar and biochar–compost mix on growth, performance and physiological responses of potted Alpinia zerumbet. *Sustainability* **2021**, *13*, 11226. [CrossRef]
10. Qambrani, N.A.; Rahman, M.M.; Won, S.; Shim, S.; Ra, C. Biochar properties and eco-friendly applications for climate change mitigation, waste management, and wastewater treatment: A review. *Renew. Sustain. Energy Rev.* **2017**, *79*, 255–273. [CrossRef]
11. Li, Z.; Song, Z.; Singh, B.P.; Wang, H. The impact of crop residue biochars on silicon and nutrient cycles in croplands. *Sci. Total Environ.* **2019**, *659*, 673–680. [CrossRef]
12. Mandal, S.; Sarkar, B.; Bolan, N.; Novak, J.; Ok, Y.S.; Van Zwieten, L.; Singh, B.P.; Kirkham, M.B.; Choppala, G.; Spokas, K.; et al. Designing advanced biochar products for maximizing greenhouse gas mitigation potential. *Crit. Rev. Environ. Sci. Technol.* **2016**, *46*, 1367–1401. [CrossRef]
13. Agbna, G.H.; Dongli, S.; Zhipeng, L.; Elshaikh, N.A.; Guangcheng, S.; Timm, L.C. Effects of deficit irrigation and biochar addition on the growth, yield, and quality of tomato. *Sci. Hortic.* **2017**, *222*, 90–101. [CrossRef]

14. El-Naggar, A.; Lee, S.S.; Awad, Y.M.; Yang, X.; Ryu, C.; Rizwan, M.; Rinklebe, J.; Tsang, D.C.W.; Ok, Y.S. Influence of soil properties and feedstocks on biochar potential for carbon mineralization and improvement of infertile soils. *Geoderma* **2018**, *332*, 100–108. [CrossRef]
15. Kätterer, T.; Roobroeck, D.; Andrén, O.; Kimutai, G.; Karltun, E.; Kirchmann, H.; Nyberg, G.; Vanlauwe, B.; de Nowina, K.R. Biochar addition persistently increased soil fertility and yields in maize-soybean rotations over 10 years in sub-humid regions of Kenya. *Field Crops Res.* **2019**, *235*, 18–26. [CrossRef]
16. Melo, T.M.; Bottlinger, M.; Schulz, E.; Leandro, W.M.; de Oliveira, S.B.; de Aguiar Filho, A.M.; El-Naggar, A.; Bolan, N.; Wang, H.; Ok, S.Y.; et al. Management of biosolids-derived hydrochar (Sewchar): Effect on plant germination, and farmers acceptance. *J. Environ. Manag.* **2019**, *237*, 200–214. [CrossRef]
17. Yu, L.; Lu, X.; He, Y.; Brookes, P.C.; Liao, H.; Xu, J. Combined biochar and nitrogen fertilizer reduces soil acidity and promotes nutrient use efficiency by soybean crop. *J. Soils Sediment.* **2017**, *17*, 599–610. [CrossRef]
18. Dong, D.; Li, J.; Ying, S.; Wu, J.; Han, X.; Teng, Y.; Zhou, M.; Ren, Y.; Jiang, P. Mitigation of methane emission in a rice paddy field amended with biochar-based slow-release fertilizer. *Sci. Total Environ.* **2021**, *792*, 148460. [CrossRef]
19. Wang, C.; Wang, H.; Cao, Y. Pb (II) sorption by biochar derived from Cinnamomum camphora and its improvement with ultrasound-assisted alkali activation. *Colloids Surf. A Physicochem. Eng.* **2018**, *556*, 177–184. [CrossRef]
20. Speratti, A.B.; Johnson, M.S.; Sousa, H.M.; Dalmagro, H.J.; Couto, E.G. Biochars from local agricultural waste residues contribute to soil quality and plant growth in a Cerrado region (Brazil) Arenosol. *Glob. Change Biol. Bioenergy* **2018**, *10*, 272–286. [CrossRef]
21. Sarma, B.; Borkotoki, B.; Narzari, R.; Kataki, R.; Gogoi, N. Organic amendments: Effect on Carbon mineralization and crop productivity in acidic soil. *J. Clean. Prod.* **2017**, *152*, 157–166. [CrossRef]
22. Ahmad Bhat, S.; Kuriqi, A.; Dar, M.U.D.; Bhat, O.; Sammen, S.S.; Towfiqul Islam, A.R.M.; Elbeltagi, A.; Shah, O.; AI-Ansari, N.; Ali, R.; et al. Application of biochar for improving physical, chemical, and hydrological soil properties: A systematic review. *Sustainability* **2022**, *14*, 11104. [CrossRef]
23. Xu, X.; Cheng, K.; Wu, H.; Sun, J.; Yue, Q.; Pan, G. Greenhouse gas mitigation potential in crop production with biochar soil amendment—A carbon footprint assessment for cross-Site field experiments from China. *Glob. Chang. Biol. Bioenergy* **2019**, *11*, 592–605. [CrossRef]
24. Usowicz, B.; Lipiec, J.; Łukowski, M.; Bis, Z.; Usowicz, J.; Latawiec, A.E. Impact of biochar addition on soil thermal properties: Modelling approach. *Geoderma* **2020**, *376*, 114574. [CrossRef]
25. Xiong, J.; Yu, R.; Islam, E.; Zhu, F.; Sohail, M.I. Effect of biochar on soil temperature under high soil surface temperature in coal mined arid and semiarid regions. *Sustainability* **2020**, *12*, 8238. [CrossRef]
26. Ghodake, G.S.; Shinde, S.K.; Kadam, A.A.; Saratale, R.G.; Saratale, G.D.; Kumar, M.; Palem, R.R.; Al-Shwaiman, H.A.; Elgorban, A.M.; Syed, A.; et al. Review on biomass feedstocks, pyrolysis mechanism and physicochemical properties of biochar: State-of-the-art framework to speed up vision of circular bioeconomy. *J. Clean. Prod.* **2021**, *297*, 126645. [CrossRef]
27. Bezerra, J.; Turnhout, E.; Vasquez, I.M.; Rittl, T.F.; Arts, B.; Kuyper, T.W. The promises of the Amazonian soil: Shifts in discourses of Terra Preta and biochar. *J. Environ. Pol. Plan.* **2019**, *21*, 623–635. [CrossRef]
28. Han, C.; Chen, F.; Lian, C.; Liang, R.; Liang, W.; Chen, M.; Luo, A.; Gao, T. Development of preparation method and application of biochar. *Int. J. Emerging Technol. Adv. Eng.* **2020**, *10*, 6–13. [CrossRef]
29. Kamarudin, N.S.; Dahalan, F.A.; Hasan, M.; An, O.S.; Parmin, N.A.; Ibrahim, N.; Hamdzah, M.; Mohd Zain, N.A.; Muda, K.; Wikurendra, E.A. Biochar: A review of its history, characteristics, factors that influence its yield, methods of production, application in wastewater treatment and recent development. *Biointerface Res. Appl. Chem.* **2022**, *12*, 7914–7926. [CrossRef]
30. Chen, W.; Meng, J.; Han, X.; Lan, Y.; Zhang, W. Past, present, and future of biochar. *Biochar* **2019**, *1*, 75–87. [CrossRef]
31. De Oliveira, E.A.; Marimon-Junior, B.H.; Marimon, B.S.; Iriarte, J.; Morandi, P.S.; Maezumi, S.Y.; Nogueira, D.S.; Aragão, L.E.O.C.; da Silva, I.B.; Feldpausch, T.R. Legacy of Amazonian Dark Earth soils on forest structure and species composition. *Glob. Ecol. Biogeogr.* **2020**, *29*, 1458–1473. [CrossRef]
32. El-Azazy, M.; El-Shafie, A.S.; Fawzy, S.; Rooney, D.W.; Osman, A.I. Competitive adsorptive removal of promazine and promethazine from wastewater using olive tree pruning biochar: Operational parameters, kinetics, and equilibrium investigations. *Environ. Sci. Pollut. Res.* **2023**, *30*, 82387–82405. [CrossRef]
33. El-Shafie, A.S.; Mahmoud, R.; Ahsan, I.; Hamdi, H.; Shibl, M.F.; El-Azazy, M. Removal of antibiotics from aqueous solutions: Insights of competitive adsorption onto Ni-impregnated biochar of spent coffee grounds. *Appl. Water Sci.* **2024**, *14*, 209. [CrossRef]
34. Kumari, K.; Kumar, R.; Bordoloi, N.; Minkina, T.; Keswani, C.; Bauddh, K. Unravelling the recent developments in the production technology and efficient applications of biochar for agro-ecosystems. *Agriculture* **2023**, *13*, 512. [CrossRef]
35. Lehmann, J.; Joseph, S. *Biochar for Environmental Management: Science, Technology and Implementation*; Routledge; Taylor & Francis: New York, NY, USA, 2024; pp. 277–312. [CrossRef]
36. Rombel, A.; Krasucka, P.; Oleszczuk, P. Sustainable biochar-based soil fertilizers and amendments as a new trend in biochar research. *Sci. Total Environ.* **2020**, *816*, 151588. [CrossRef]
37. Feliz Florian, G.; Ragoubi, M.; Leblanc, N.; Taouk, B.; Abdelouahed, L. Biochar production and its potential application for biocomposite materials: A comprehensive review. *J. Compos. Sci.* **2024**, *8*, 220. [CrossRef]
38. Gabhane, J.W.; Bhange, V.P.; Patil, P.D.; Bankar, S.T.; Kumar, S. Recent trends in biochar production methods and its application as a soil health conditioner: A review. *SN Appl. Sci.* **2020**, *2*, 1–21. [CrossRef]

39. Mohanty, A.K.; Vivekanandhan, S.; Das, O.; Romero Millán, L.M.; Klinghoffer, N.B.; Nzihou, A.; Misra, M. Biocarbon materials. *Nat. Rev. Methods Primers* **2024**, *4*, 19. [CrossRef]
40. Azargohar, R.; Nanda, S.; Dalai, A.K.; Koziński, J.A. Physico-chemistry of biochars produced through steam gasification and hydro-thermal gasification of canola hull and canola meal pellets. *Biomass Bioenerg.* **2019**, *120*, 458–470. [CrossRef]
41. Amin, F.R.; Huang, Y.; He, Y.; Zhang, R.; Liu, G.; Chen, C. Biochar applications and modern techniques for characterization. *Clean Technol. Environ.* **2016**, *18*, 1457–1473. [CrossRef]
42. Kazimierski, P.; Januszewicz, K.; Godlewski, W.; Fijuk, A.; Suchocki, T.; Chaja, P.; Barczak, B.; Kardaś, D. The course and the effects of agricultural biomass pyrolysis in the production of high-calorific biochar. *Materials* **2022**, *15*, 1038. [CrossRef]
43. Hossain, N.; Nizamuddin, S.; Griffin, G.; Selvakannan, P.; Mubarak, N.M.; Mahlia, T.M.I. Synthesis, and characterization of rice husk biochar via hydrothermal carbonization for wastewater treatment and biofuel production. *Sci. Rep.* **2020**, *10*, 18851. [CrossRef]
44. Tratsch, M.V.M.; Ceretta, C.A.; Silva, L.S.D.; Ferreira, P.A.A.; Brunetto, G. Composition and mineralization of organic compost derived from composting of fruit and vegetable waste. *Rev. Ceres* **2019**, *66*, 307–315. [CrossRef]
45. Mahdi, Z.; El Hanandeh, A.; Yu, Q.J. Preparation, characterization, and application of surface modified biochar from date seed for improved lead, copper, and nickel removal from aqueous solutions. *J. Environ. Chem. Eng.* **2019**, *7*, 103379. [CrossRef]
46. Das, S.K.; Ghosh, G.K.; Avasthe, R.K.; Sinha, K. Compositional heterogeneity of different biochar: Effect of pyrolysis temperature and feedstocks. *J. Environ. Manag.* **2021**, *278*, 111501. [CrossRef]
47. Palniandy, L.K.; Yoon, L.W.; Wong, W.Y.; Yong, S.T.; Pang, M.M. Application of biochar derived from different types of biomass and treatment methods as a fuel source for direct carbon fuel cells. *Energies* **2019**, *12*, 2477. [CrossRef]
48. Tomczyk, A.; Sokołowska, Z.; Boguta, P. Biochar physicochemical properties: Pyrolysis temperature and feedstock kind effects. *Rev. Environ. Sci. Bio/Technol.* **2020**, *19*, 191–215. [CrossRef]
49. Břendová, K.; Száková, J.; Lhotka, M.; Krulikovská, T.; Punčochář, M.; Tlustoš, P. Biochar physicochemical parameters as a result of feedstock material and pyrolysis temperature: Predictable for the fate of biochar in soil? *Environ. Geochem. Health* **2017**, *39*, 1381–1395. [CrossRef]
50. Murtaza, G.; Ahmed, Z.; Usman, M. Feedstock type, pyrolysis temperature and acid modification effects on physiochemical attributes of biochar and soil quality. *Arab. J. Geosci.* **2022**, *15*, 305. [CrossRef]
51. Antonangelo, J.A.; Zhang, H.; Sun, X.; Kumar, A. Physicochemical properties and morphology of biochars as affected by feedstock sources and pyrolysis temperatures. *Biochar* **2019**, *1*, 325–336. [CrossRef]
52. Zolfi-Bavariani, M.; Ronaghi, A.; Ghasemi-Fasaei, R.; Yasrebi, J. Influence of poultry manure–derived biochars on nutrients bioavailability and chemical properties of a calcareous soil. *Arch. Agron. Soil Sci.* **2016**, *62*, 1578–1591. [CrossRef]
53. Cely, P.; Gascó, G.; Paz-Ferreiro, J.; Méndez, A. Agronomic properties of biochars from different manure wastes. *J. Anal. Appl. Pyrolysis* **2015**, *111*, 173–182. [CrossRef]
54. Qin, J.; Qian, S.; Chen, Q.; Chen, L.; Yan, L.; Shen, G. Cow manure-derived biochar: Its catalytic properties and influential factors. *J. Hazard. Mater.* **2019**, *371*, 381–388. [CrossRef]
55. Chen, T.; Luo, L.; Deng, S.; Shi, G.; Zhang, S.; Zhang, Y.; Deng, O.; Wang, L.; Zhang, J.; Wei, L. Sorption of tetracycline on H_3PO_4 modified biochar derived from rice straw and swine manure. *Bioresour. Technol.* **2018**, *267*, 431–437. [CrossRef]
56. Meng, J.; Wang, L.; Liu, X.; Wu, J.; Brookes, P.C.; Xu, J. Physicochemical properties of biochar produced from aerobically composted swine manure and its potential use as an environmental amendment. *Bioresour. Technol.* **2013**, *142*, 641–646. [CrossRef]
57. Suman, S.; Gautam, S. Pyrolysis of coconut husk biomass: Analysis of its biochar properties. *Energy Sources Part A Recovery Util. Environ. Eff.* **2017**, *39*, 761–767. [CrossRef]
58. Pratiwi, E.P.A.; Shinogi, Y. Rice husk biochar application to paddy soil and its effects on soil physical properties, plant growth, and methane emission. *Paddy Water Environ.* **2016**, *14*, 521–533. [CrossRef]
59. Dittakit, P.; Singkham, J.; Viriyasuthee, W.; Sangmanee, K. Properties of biochar from coconut waste and application in agriculture. *ASEAN J. Sci. Tech. Rep.* **2023**, *26*, 52–58. [CrossRef]
60. Ye, L.; Zhang, J.; Zhao, J.; Luo, Z.; Tu, S.; Yin, Y. Properties of biochar obtained from pyrolysis of bamboo shoot shell. *J. Anal. Appl. Pyrolysis* **2015**, *114*, 172–178. [CrossRef]
61. Angın, D. Effect of pyrolysis temperature and heating rate on biochar obtained from pyrolysis of safflower seed press cake. *Bioresour. Technol.* **2013**, *128*, 593–597. [CrossRef]
62. Karunanithi, R.; Sik, O.Y.; Dharmarajan, R.; Ahmad, M.; Seshadri, B.; Bolan, N.; Naidu, R. Sorption, kinetics and thermodynamics of phosphate sorption onto soybean stover derived biochar. *Environ. Technol. Innov.* **2017**, *8*, 113–125. [CrossRef]
63. Aon, M.; Aslam, Z.; Hussain, S.; Bashir, M.A.; Shaaban, M.; Masood, S.; Iqbal, S.; Khalid, M.; Rehim, A.; Mosa, W.F.A.; et al. Wheat straw biochar produced at a low temperature enhanced maize growth and yield by influencing soil properties of typic calciargid. *Sustainability* **2023**, *15*, 9488. [CrossRef]
64. Chandra, S.; Bhattacharya, J. Influence of temperature and duration of pyrolysis on the property heterogeneity of rice straw biochar and optimization of pyrolysis conditions for its application in soils. *J. Clean Prod.* **2019**, *215*, 1123–1139. [CrossRef]
65. Kongnine, D.M.; Kpelou, P.; Sodoga, K.; Napo, K. Evaluation of some combustion characteristics of biochar produced from coconut husks, corn cobs and palm kernel shells. *Innt. J. Innov. Stud.* **2018**, *24*, 1124–1130.
66. Yang, Z.; Liu, X.; Zhang, M.; Liu, L.; Xu, X.; Xian, J.; Cheng, Z. Effect of temperature and duration of pyrolysis on spent tea leaves biochar: Physiochemical properties and Cd(II) adsorption capacity. *Water Sci. Technol.* **2020**, *81*, 2533–2544. [CrossRef]

67. Razzaq, S.; Zhou, B.; Zia-ur-Rehman, M.; Aamer Maqsood, M.; Hussain, S.; Bakhsh, G.; Zhang, Z.; Yang, Q.; Altaf, A.R. Cadmium stabilization and redox transformation mechanism in maize using nanoscale zerovalent-iron-enriched biochar in cadmium-contaminated soil. *Plants* **2022**, *11*, 1074. [CrossRef]
68. Junior, A.; Guo, M. Efficacy of sewage sludge derived biochar on enhancing soil health and crop productivity in strongly acidic soil. *Front. Soil Sci.* **2023**, *3*, 1066547. [CrossRef]
69. Chen, T.; Zhang, Y.; Wang, H.; Lu, W.; Zhou, Z.; Zhang, Y.; Ren, L. Influence of pyrolysis temperature on characteristic and heavy metal adsorptive performance of biochar derived from municipal sewage sludge. *Bioresour. Technol.* **2015**, *164*, 47–54. [CrossRef]
70. Jin, H.; Capareda, S.; Chang, Z.; Gao, J.; Xu, Y.; Zhang, J. Biochar pyrolytically produced from municipal solid wastes for aqueous As (V) removal: Adsorption property and its improvement with KOH activation. *Bioresour. Technol.* **2014**, *169*, 622–629. [CrossRef]
71. Blanco-Vargas, A.; Chacón-Buitrago, M.A.; Quintero-Duque, M.C.; Poutou-Piñales, R.A.; Díaz-Ariza, L.A.; Devia-Castillo, C.A.; Castillo-Carvajal, L.C.; Toledo-Aranda, D.; da Conceição de Matos, C.; Olaya-González, W.; et al. Production of pine sawdust biochar supporting phosphate-solubilizing bacteria as an alternative bioinoculant in *Allium cepa* L., culture. *Sci. Rep.* **2022**, *12*, 12815. [CrossRef] [PubMed]
72. Novak, J.; Sigua, G.; Watts, D.; Cantrell, K.; Shumaker, P.; Szogi, A.; Johnson, M.G.; Spokas, K. Biochars impact on water infiltration and water quality through a compacted subsoil layer. *Chemosphere* **2016**, *142*, 160–167. [CrossRef] [PubMed]
73. Wijitkosum, S.; Sriburi, T. Applying *Cassava* stems biochar produced from agronomical waste to enhance the yield and productivity of maize in unfertile soil. *Fermentation* **2021**, *7*, 277. [CrossRef]
74. Zhao, S.-X.; Ta, N.; Wang, X.-D. Effect of temperature on the structural and physicochemical properties of biochar with apple tree branches as feedstock material. *Energies* **2017**, *10*, 1293. [CrossRef]
75. Choi, Y.K.; Kan, E. Effects of pyrolysis temperature on the physicochemical properties of alfalfa-derived biochar for the adsorption of bisphenol A and sulfamethoxazole in water. *Chemosphere* **2019**, *218*, 741–748. [CrossRef] [PubMed]
76. Moradi-Choghamarani, F.; Moosavi, A.A.; Baghernejad, M. Determining organo-chemical composition of sugarcane bagasse-derived biochar as a function of pyrolysis temperature using proximate and Fourier transform infrared analyses. *J. Therm. Anal. Calorim.* **2019**, *138*, 331–342. [CrossRef]
77. Ding, W.; Dong, X.; Ime, I.M.; Gao, B.; Ma, L.Q. Pyrolytic temperatures impact lead sorption mechanisms by bagasse biochars. *Chemosphere* **2014**, *105*, 68–74. [CrossRef]
78. Nagel, K.; Hoilett, N.O.; Mottaleb, M.A.; Meziani, M.J.; Wistrom, J.; Bellamy, M. Physicochemical characteristics of biochars derived from corn, hardwood, miscanthus, and horse manure biomasses. *Commun. Soil Sci. Plant Anal.* **2019**, *50*, 987–1002. [CrossRef]
79. Hernández-Mena, L.; Pecora, A.; Beraldo, A. Slow pyrolysis of bamboo biomass: Analysis of biochar properties. *Chem. Eng. Trans.* **2014**, *37*, 115–120. [CrossRef]
80. Yao, Y.; Gao, B.; Zhang, M.; Inyang, M.; Zimmerman, A.R. Effect of biochar amendment on sorption and leaching of nitrate, ammonium and phosphate in a sandy soil. *Chemosphere* **2012**, *89*, 1467–1471. [CrossRef]
81. Gezahegn, A.; Selassie, Y.G.; Agegnehu, G.; Addisu, S.; Mihretie, F.A.; Kohira, Y.; Lewoyehu, M.; Sato, S. The impact of water hyacinth biochar on maize growth and soil properties: The influence of pyrolysis temperature. *J. Sustain. Agric. Environ.* **2024**, *3*, e12117. [CrossRef]
82. Adekiya, A.O.; Agbede, T.M.; Olayanju, A.; Ejue, W.S.; Adekanye, T.A.; Adenusi, T.T.; Ayeni, J.F. Effect of biochar on soil properties, soil loss, and cocoyam yield on a tropical sandy loam Alfisol. *Sci. World J.* **2020**, 1–9. [CrossRef]
83. Liu, Z.; Dugan, B.; Masiello, C.A.; Gonnermann, H.M. Biochar particle size, shape, and porosity act together to influence soil water properties. *PLoS ONE* **2017**, *12*, e0179079. [CrossRef]
84. Rehman, A.; Nawaz, S.; Alghamdi, H.A.; Alrumman, S.; Yan, W.; Zohaib Nawaz, M. Effects of manure-based biochar on uptake of nutrients and water holding capacity of different types of soils. *Case Stud. Chem. Environ. Eng.* **2020**, *2*, 100036. [CrossRef]
85. Prasad, M.; Chrysargyris, A.; McDaniel, N.; Kavanagh, A.; Gruda, N.S.; Tzortzakis, N. Plant nutrient availability and pH of biochars and their fractions, with the possible use as a component in a growing media. *Agronomy* **2019**, *10*, 10. [CrossRef]
86. Sandhu, S.S.; Ussiri, D.A.; Kumar, S.; Chintala, R.; Papiernik, S.K.; Malo, D.D.; Schumacher, T.E. Analysing the impacts of three types of biochar on soil carbon fractions and physiochemical properties in a corn-soybean rotation. *Chemosphere* **2017**, *184*, 473–481. [CrossRef]
87. Han, L.; Sun, K.; Yang, Y.; Xia, X.; Li, F.; Yang, Z.; Xing, B. Biochar's stability and effect on the content, composition, and turnover of soil organic carbon. *Geoderma* **2020**, *364*, 114184. [CrossRef]
88. Kizito, S.; Luo, H.; Lu, J.; Bah, H.; Dong, R.; Wu, S. Role of nutrient-enriched biochar as a soil amendment during maize growth: Exploring practical alternatives to recycle agricultural residuals and to reduce chemical fertilizer demand. *Sustainability* **2019**, *11*, 3211. [CrossRef]
89. Ippolito, J.A.; Spokas, K.A.; Novak, J.; Lentz, R.D.; Cantrell, K.B. Biochar elemental composition and factors influencing nutrient retention. In *Biochar for Environmental Management: Science, Technology and Implementation*; Routledge: Abingdon, UK, 2015.
90. Wijitkosum, S.; Sriburi, T. Aromaticity, polarity, and longevity of biochar derived from disposable bamboo chopsticks waste for environmental application. *Heliyon* **2023**, *9*, e19831. [CrossRef]
91. Domingues, R.R.; Trugilho, P.F.; Silva, C.A.; Melo, I.C.N.D.; Melo, L.C.; Magriotis, Z.M.; Sanchez-Monedero, M.A. Properties of biochar derived from wood and high-nutrient biomasses with the aim of agronomic and environmental benefits. *PLoS ONE* **2017**, *12*, e0176884. [CrossRef]

92. Yang, K.; Yang, J.; Jiang, Y.; Wu, W.; Lin, D. Correlations and adsorption mechanisms of aromatic compounds on a high heat temperature treated bamboo biochar. *Environ. Pollut.* **2016**, *210*, 57–64. [CrossRef]
93. Dróżdż, D.; Malińska, K.; Wystalska, K.; Meers, E.; Robles-Aguilar, A. The Influence of poultry manure-derived biochar and compost on soil properties and plant biomass growth. *Materials* **2023**, *16*, 6314. [CrossRef]
94. Clough, T.J.; Condron, L.M.; Kammann, C.; Müller, C. A review of biochar and soil nitrogen dynamics. *Agronomy* **2013**, *3*, 275–293. [CrossRef]
95. Laird, D.; Fleming, P.; Wang, B.; Horton, R.; Karlen, D. Biochar impact on nutrient leaching from a Midwestern agricultural soil. *Geoderma* **2010**, *158*, 436–442. [CrossRef]
96. Manolikaki, I.I.; Mangolis, A.; Diamadopoulos, E. The impact of biochars prepared from agricultural residues on phosphorus release and availability in two fertile soils. *J. Environ. Manag.* **2016**, *181*, 536–543. [CrossRef] [PubMed]
97. Farrell, M.; Macdonald, L.M.; Butler, G.; Chirino-Valle, I.; Condron, L.M. Biochar and fertiliser applications influence phosphorus fractionation and wheat yield. *Biol. Fertil. Soils* **2014**, *50*, 169–178. [CrossRef]
98. Lehmann, J.; Rilig, M.C.; Thies, J.; Masiello, C.A.; Hockaday, W.C.; Crowley, D. Biochar affects on soil biota—A review. *Soil Biol. Biochem.* **2011**, *43*, 1812–1836. [CrossRef]
99. Tag, A.T.; Duman, G.; Ucar, S.; Yanik, J. Effects of feedstock type and pyrolysis temperature on potential applications of biochar. *J. Anal. Appl.* **2016**, *120*, 200–206. [CrossRef]
100. Kajina, W.; Rousset, P. Coupled effect of feedstock and pyrolysis temperature on biochar as soil amendment. In Proceedings of the IC-STAR 4.0: Interdisciplinary Research Enhancement for Industrial Revolution 4.0, Belintung, Indonesia, 29–30 August 2018.
101. Weber, K.; Quicker, P. Properties of biochar. *Fuel* **2018**, *217*, 240–261. [CrossRef]
102. Zhao, B.; O'Connor, D.; Zhang, J.; Peng, T.; Shen, Z.; Tsang, D.C.; Hou, D. Effect of pyrolysis temperature, heating rate, and residence time on rapeseed stem derived biochar. *J. Clean. Prod.* **2018**, *174*, 977–987. [CrossRef]
103. Wyn, H.K.; Zárate, S.; Carrascal, J.; Yermán, L. A novel approach to the production of biochar with improved fuel characteristics from biomass waste. *Waste Biomass Valori.* **2020**, *11*, 6467–6481. [CrossRef]
104. Pariyar, P.; Kumari, K.; Jain, M.K.; Jadhao, P.S. Evaluation of change in biochar properties derived from different feedstock and pyrolysis temperature for environmental and agricultural application. *Sci. Total Environ.* **2020**, *713*, 136433. [CrossRef]
105. Domingues, R.R.; Sánchez-Monedero, M.A.; Spokas, K.A.; Melo, L.C.A.; Trugilho, P.F.; Valenciano, M.N.; Silva, C.A. Enhancing cation exchange capacity of weathered soils using biochar: Feedstock, pyrolysis conditions and addition rate. *Agronomy* **2020**, *10*, 824. [CrossRef]
106. Khaledian, Y.; Kiani, F.; Ebrahimi, S.; Brevik, E.C.; Aitkenhead-Peterson, J. Assessment and monitoring of soil degradation during land use change using multivariate analysis. *Land Degrad. Dev.* **2017**, *28*, 128–141. [CrossRef]
107. Bordoloi, S.; Kumar, H.; Hussain, R.; Karangat, R.; Lin, P.; Sreedeep, S.; Zhu, H.H. Assessment of hydro-mechanical properties of biochar-amended soil sourced from two contrasting feedstock. *Biomass Convers. Bior.* **2020**, *14*, 5803–5818. [CrossRef]
108. Adhikari, S.; Timms, W.; Mahmud, M.P. Optimising water holding capacity and hydrophobicity of biochar for soil amendment–a review. *Sci. Total Environ.* **2022**, *851*, 158043. [CrossRef]
109. Wong, J.T.F.; Chen, Z.; Ng, C.W.W.; Wong, M.H. Gas permeability of biochar-amended clay: Potential alternative landfill final cover material. *Environ. Sci. Pollut. Res. Int.* **2016**, *23*, 7126–7131. [CrossRef]
110. Suliman, W.; Harsh, J.B.; Abu-Lail, N.I.; Fortuna, A.M.; Dallmeyer, I.; Garcia-Pérez, M. The role of biochar porosity and surface functionality in augmenting hydrologic properties of a sandy soil. *Sci. Total Environ.* **2017**, *574*, 139–147. [CrossRef]
111. Randolph, P.; Bansode, R.R.; Hassan, O.A.; Rehrah, D.J.; Ravella, R.; Reddy, M.R.; Watts, D.W.; Novak, J.M.; Ahmedna, M. Effect of biochars produced from solid organic municipal waste on soil quality parameters. *J. Environ. Manag.* **2017**, *192*, 271–280. [CrossRef]
112. Khater, E.S.; Bahnasawy, A.; Hamouda, R.; Sabahy, A.; Abbas, W.; Morsy, O.M. Biochar production under different pyrolysis temperatures with different types of agricultural wastes. *Sci. Rep.* **2024**, *14*, 2625. [CrossRef]
113. Banitalebi, G.; Mosaddeghi, M.R.; Shariatmadari, H. Evaluation of physico-chemical properties of biochar-based mixtures for soilless growth media. *J. Mater. Cycles Waste Manag.* **2021**, *23*, 950–964. [CrossRef]
114. Verheijen, F.G.; Zhuravel, A.; Silva, F.C.; Amaro, A.; Ben-Hur, M.; Keizer, J.J. The influence of biochar particle size and concentration on bulk density and maximum water holding capacity of sandy vs. sandy loam soil in a column experiment. *Geoderma* **2019**, *347*, 194–202. [CrossRef]
115. Mankasingh, U.; Choi, P.C.; Ragnarsdottir, V. Biochar application in a tropical, agricultural region: A plot scale study in Tamil Nadu, India. *J. Appl. Geochem.* **2011**, *26*, S218–S221. [CrossRef]
116. Toková, L.; Igaz, D.; Horák, J.; Aydin, E. Effect of biochar application and re-application on soil bulk density, porosity, saturated hydraulic conductivity, water content and soil water availability in a silty loam haplic luvisol. *Agronomy* **2020**, *10*, 1005. [CrossRef]
117. Yang, C.; Liu, J.; Lu, S. Pyrolysis temperature affects pore characteristics of rice straw and canola stalk biochars and biochar-amended soils. *Geoderma* **2021**, *397*, 115097. [CrossRef]
118. Rasse, D.P.; Budai, A.; O'Toole, A.; Ma, X.; Rumpel, C.; Abiven, S. Persistence in soil of Miscanthus biochar in laboratory and field conditions. *PLoS ONE* **2017**, *12*, e0184383. [CrossRef]
119. Adhikari, S.; Moon, E.; Paz-Ferreiro, J.; Timms, W. Comparative analysis of biochar carbon stability methods and implications for carbon credits. *Sci. Total Environ.* **2024**, *914*, 169607. [CrossRef]

120. Xu, Z.; He, M.; Xu, X.; Cao, X.; Tsang, D.C.W. Impacts of different activation processes on the carbon stability of biochar for oxidation resistance. *Bioresour. Technol.* **2021**, *338*, 125555. [CrossRef]
121. Kuzyakov, Y.; Bogomolova, I.; Glaser, B. Biochar stability in soil: Decomposition during eight years and transformation as assessed by compound-specific 14C analysis. *Soil Biol. Biochem.* **2014**, *70*, 229–236. [CrossRef]
122. Sheng, M.H.; Ai, X.Y.; Huang, B.C.; Zhu, M.K.; Liu, Z.Y.; Ai, Y.W. Effects of biochar additions on the mechanical stability of soil aggregates and their role in the dynamic renewal of aggregates in slope ecological restoration. *Sci. Total Environ.* **2023**, *898*, 165478. [CrossRef]
123. Khadem, A.; Raiesi, F.; Besharati, H.; Khalaj, M.A. The effects of biochar on soil nutrients status, microbial activity, and carbon sequestration potential in two calcareous soils. *Biochar* **2021**, *3*, 105–116. [CrossRef]
124. Ling, L.; Luo, Y.; Jiang, B.; Lv, J.; Meng, C.; Liao, Y.; Reid, B.J.; Ding, F.; Lu, Z.; Kuzyakov, Y.; et al. Biochar induces mineralization of soil recalcitrant components by activation of biochar responsive bacteria groups. *Soil Biol. Biochem.* **2022**, *172*, 108778. [CrossRef]
125. Chen, Z.; Kumar, A.; Fu, Y.; Singh, B.P.; Ge, T.; Tu, H.; Luo, Y.; Xu, J. Biochar decreased rhizodeposits stabilization via opposite effects on bacteria and fungi: Diminished fungi-promoted aggregation and enhanced bacterial mineralization. *Biol. Fertil. Soils* **2021**, *57*, 533–546. [CrossRef]
126. Sarfaraz, Q.; da Silva, L.S.; Drescher, G.L.; Zafar, M.; Severo, F.F.; Kokkonen, A.; Dal Molin, G.; Shafi, M.I.; Shafique, Q.; Solaiman, Z.M. Characterization and carbon mineralization of biochars produced from different animal manures and plant residues. *Sci. Rep.* **2020**, *10*, 955. [CrossRef] [PubMed]
127. Saffari, N.; Hajabbasi, M.A.; Shirani, H.; Mosaddeghi, M.R.; Owens, G. Influence of corn residue biochar on water retention and penetration resistance in a calcareous sandy loam soil. *Geoderma* **2021**, *38*, 114734. [CrossRef]
128. Garbuz, S.; Mackay, A.; Camps-Arbestain, M.; DeVantier, B.; Minor, M. Biochar amendment improves soil physico-chemical properties and alters root biomass and the soil food web in grazed pastures. *Agric. Ecosyst. Environ.* **2021**, *319*, 107517. [CrossRef]
129. Izilan, N.I.S.; Sari, N.A.; Othman, N.M.I.; Mustaffha, S. The effects of biochar-compost on soil properties and plant growth performance grown in a sandy-loam soil. *IOP Conf. Ser. Earth Environ. Sci.* **2022**, *10591*, 012021. [CrossRef]
130. Prakasha, H.C.; Gayathri, B.; Reddy, K.M.; Desai, N.; Mamatha, B.; Rohitha, D.S. Effect of coconut shell biochar on physical, chemical properties and available major nutrient status of acidic soil. *Int. J. Plant Sci.* **2022**, *34*, 1147–1153. [CrossRef]
131. Safari, S.; Nazari, F.; Vafaee, Y.; Teixeira da Silva, J.A. Impact of rice husk biochar on drought stress tolerance in perennial ryegrass (*Lolium perenne* L.). *J. Plant Growth Regul.* **2023**, *42*, 810–826. [CrossRef]
132. El-Naggar, A.; Awad, Y.M.; Tang, X.Y.; Liu, C.; Niazi, N.K.; Jien, S.H.; Tsang, D.C.W.; Song, H.; Ok, Y.S.; Lee, S.S. Biochar influences soil carbon pools and facilitates interactions with soil: A field investigation. *Land Degrad. Dev.* **2018**, *29*, 2162–2171. [CrossRef]
133. Pandian, K.; Subramaniayan, P.; Gnasekaran, P.; Chitraputhirapillai, S. Effect of biochar amendment on soil physical, chemical and biological properties and groundnut yield in rainfed Alfisol of semi-arid tropics. *Arch. Agron. Soil Sci.* **2016**, *62*, 1293–1310. [CrossRef]
134. Simarani, K.; Azlan Halmi, M.F.; Abdullah, R. Short-term effects of biochar amendment on soil microbial community in humid tropics. *Arch. Agron. Soil Sci.* **2018**, *64*, 1847–1860. [CrossRef]
135. Oladele, S.O. Changes in physicochemical properties and quality index of an Alfisol after three years of rice husk biochar amendment in rainfed rice—Maize cropping sequence. *Geoderma* **2019**, *353*, 359–371. [CrossRef]
136. Mclennon, E.; Solomon, J.K.; Neupane, D.; Davison, J. Biochar and nitrogen application rates effect on phosphorus removal from a mixed grass sward irrigated with reclaimed wastewater. *Sci. Total Environ.* **2020**, *715*, 137012. [CrossRef] [PubMed]
137. He, Y.; Zhou, X.; Jiang, L.; Li, M.; Du, Z.; Zhou, G.; Shao, J.; Wang, X.; Xu, Z.; Bai, S.H.; et al. Effects of biochar application on soil greenhouse gas fluxes: A meta-analysis. *Glob. Chang. Biol. Bioenergy* **2017**, *9*, 743–755. [CrossRef]
138. Ye, J.; Joseph, S.D.; Ji, M.; Nielsen, S.; Mitchell, D.R.; Donne, S.; Horvat, J.; Wang, J.; Munroe, P.; Thomas, T. Chemolithotrophic processes in the bacterial communities on the surface of mineral-enriched biochars. *ISME J.* **2017**, *11*, 1087–1101. [CrossRef]
139. Lu, H.; Yan, M.; Wong, M.H.; Mo, W.Y.; Wang, Y.; Chen, X.W.; Wang, J.J. Effects of biochar on soil microbial community and functional genes of a landfill cover three years after ecological restoration. *Sci. Total Environ.* **2020**, *717*, 137133. [CrossRef]
140. Araujo, J.; Díaz-Alcántara, C.A.; Urbano, B.; González-Andrés, F. Inoculation with native Bradyrhizobium strains formulated with biochar as carrier improves the performance of pigeonpea (*Cajanus cajan* L.). *Eur. J. Agron.* **2020**, *113*, 125985. [CrossRef]
141. Hale, L.; Luth, M.; Crowley, D. Biochar characteristics relate to its utility as an alternative soil inoculum carrier to peat and vermiculite. *Soil Biol. Biochem.* **2015**, *81*, 228–235. [CrossRef]
142. Głuszek, S.; Sas-Paszt, L.; Sumorok, B.; Kozera, R. Biochar-Rhizosphere Interactions—A Review. *Pol. J. Microbiol.* **2017**, *66*, 151–161. [CrossRef]
143. Abujabhah, I.S.; Doyle, R.B.; Bound, S.A.; Bowman, J.P. Assessment of bacterial community composition, methanotrophic and nitrogen-cycling bacteria in three soils with different biochar application rates. *J. Soils Sediments* **2018**, *18*, 148–158. [CrossRef]
144. Lilleskov, E.A.; Kuyper, T.W.; Bidartondo, M.I.; Hobbie, E.A. Atmospheric nitrogen deposition impacts on the structure and function of forest mycorrhizal communities: A review. *Environ. Pollut.* **2019**, *246*, 148–162. [CrossRef]
145. Lenoir, I.; Fontaine, J.; Sahraoui, A.L.H. Arbuscular mycorrhizal fungal responses to abiotic stresses: A review. *Phytochemistry* **2016**, *123*, 4–15. [CrossRef] [PubMed]
146. Lone, A.H.; Najar, G.R.; Ganie, M.A.; Sofi, J.A.; Ali, T. Biochar for sustainable soil health: A review of prospects and concerns. *Pedosphere* **2015**, *25*, 639–653. [CrossRef]

147. Yu, H.W.; Zou, W.X.; Chen, J.J.; Chen, H.; Yu, Z.B.; Huang, J.; Tang, H.R.; Wei, X.Y.; Gao, B. Biochar amendment improves crop production in problem soils: A review. *J. Environ. Manag.* **2019**, *232*, 8–21. [CrossRef] [PubMed]
148. Warnock, D.D.; Lehmann, J.; Kuyper, T.W.; Rillig, M.C. Mycorrhizal responses to biochar in soil–concepts and mechanisms. *Plant Soil* **2007**, *300*, 9–20. [CrossRef]
149. Solaiman, Z.M.; Abbott, L.K.; Murphy, D.V. Biochar phosphorus concentration dictates mycorrhizal colonisation, plant growth and soil phosphorus cycling. *Sci. Rep.* **2019**, *9*, 5062. [CrossRef]
150. Solaiman, Z.M.; Blackwell, P.; Abbott, L.K.; Storer, P. Direct and residual effect of biochar application on mycorrhizal root colonisation, growth and nutrition of wheat. *Soil Res.* **2010**, *48*, 546–554. [CrossRef]
151. Anderson, C.R.; Condron, L.M.; Clough, T.J.; Fiers, M.; Stewart, A.; Hill, R.A.; Sherlock, R.R. Biochar induced soil microbial community change: Implications for biogeochemical cycling of carbon, nitrogen and phosphorus. *Pedobiologia* **2011**, *54*, 309–320. [CrossRef]
152. Song, D.; Xi, X.; Zheng, Q.; Liang, G.; Zhou, W.; Wang, X. Soil nutrient and microbial activity responses to two years after maize straw biochar application in a calcareous soil. *Ecotoxicol. Environ. Saf.* **2019**, *180*, 348–356. [CrossRef]
153. Oladele, S.O.; Adeyemo, A.J.; Awodun, M.A. Influence of rice husk biochar and inorganic fertilizer on soil nutrients availability and rain-fed rice yield in two contrasting soils. *Geoderma* **2019**, *336*, 1–11. [CrossRef]
154. El-Naggar, A.; Lee, M.H.; Hur, J.; Lee, Y.H.; Igalavithana, A.D.; Shaheen, S.M.; Ryu, C.; Rinklebe, J.; Tsang, D.C.W.; Ok, Y.S. Biochar-induced metal immobilization and soil biogeochemical process: An integrated mechanistic approach. *Sci. Total Environ.* **2020**, *698*, 134112. [CrossRef]
155. Tripti; Kumar, A.; Usmani, Z.; Kumar, V.; Anshumali. Biochar and flyash inoculated with plant growth promoting rhizobacteria act as potential biofertilizer for luxuriant growth and yield of tomato plant. *J. Environ. Manag.* **2017**, *190*, 20–27. [CrossRef] [PubMed]
156. Medyńska-Juraszek, A.; Latawiec, A.; Królczyk, J.; Bogacz, A.; Kawałko, D.; Bednik, M.; Dudek, M. Biochar improves maize growth but has a limited effect on soil properties: Evidence from a three-year field experiment. *Sustainability* **2021**, *13*, 3617. [CrossRef]
157. Mollick, M.O.A.; Paul, A.K.; Alam, I.; Sumon, M.M. Effect of biochar on yield and quality of potato (*Solanum tuberosum*) tuber. *Int. J. Stress Manag.* **2020**, *11*, 445–450. [CrossRef]
158. Thomas, S.C.; Frye, S.; Gale, N.; Garmon, M.; Launchbury, R.; Machado, N.; Melamed, S.; Murray, J.; Petroff, A.; Winsborough, C. Biochar mitigates negative effects of salt additions on two herbaceous plant species. *J. Environ. Manag.* **2013**, *129*, 62–68. [CrossRef]
159. Chen, Z.; Liu, J.; Sun, H.; Xing, J.; Zhang, Z.; Jiang, J. Effects of biochar applied in either rice or wheat seasons on the production and quality of wheat and nutrient status in paddy profiles. *Plants* **2023**, *12*, 4131. [CrossRef]
160. Yi, Z.; Jeyakumar, P.; Yin, C.; Sun, H. Effects of biochar in combination with varied N inputs on grain yield, N uptake, NH_3 volatilization, and N_2O emission in paddy soil. *Front. Microbiol.* **2023**, *14*, 1174805. [CrossRef]
161. Aneseyee, A.B.; Wolde, T. Effect of biochar and inorganic fertilizer on the soil properties and growth and yield of onion (*Allium cepa*) in tropical Ethiopia. *Sci. World J.* **2021**, *1*, 5582697. [CrossRef]
162. Carpenter, B.H.; Nair, A. *Effect of Biochar on Carrot Production*; Iowa State University: Iowa, IA, USA, 2014; pp. 24–26.
163. Obia, A.; Mulder, J.; Hale, S.E.; Nurida, N.L.; Cornelissen, G. The potential of biochar in improving drainage, aeration and maize yields in heavy clay soils. *PLoS ONE* **2018**, *13*, e0196794. [CrossRef]
164. Latini, A.; Fiorani, F.; Galeffi, P.; Cantale, C.; Bevivino, A.; Jablonowski, N.D. Phenotyping of different Italian durum wheat varieties in early growth stage with the addition of pure or digestate-activated biochars. *Front. Plant Sci.* **2021**, *12*, 782072. [CrossRef]
165. Liu, C.; Sun, B.; Zhang, X.; Liu, X.; Drosos, M.; Li, L.; Pan, G. The water-soluble pool in biochar dominates maize plant growth promotion under biochar amendment. *J. Plant Growth Regul.* **2021**, *40*, 1466–1476. [CrossRef]
166. Petrova, T.; Naydenova, I.; Ribau, J.; Ferreira, A.F. Biochar from agro-forest residue: Application perspective based on decision support analysis. *Appl. Sci.* **2023**, *13*, 3240. [CrossRef]
167. Noguera, D.; Rondón, M.; Laossi, K.R.; Hoyos, V.; Lavelle, P.; de Carvalho, M.H.C.; Barot, S. Contrasted effect of biochar and earthworms on rice growth and resource allocation in different soils. *Soil Biol. Biochem.* **2010**, *42*, 1017–1027. [CrossRef]
168. Alburquerque, J.A.; Calero, J.M.; Barrón, V.; Torrent, J.; del Campillo, M.C.; Gallardo, A.; Villar, R. Effects of biochars produced from different feedstocks on soil properties and sunflower growth. *J. Plant Nutr. Soil Sci.* **2014**, *177*, 16–25. [CrossRef]
169. Hien, T.T.T.; Shinogi, Y.; Taniguchi, T. The different expressions of draft cherry tomato growth, yield, quality under bamboo and rice husk biochars application to clay loamy soil. *Agric. Sci.* **2017**, *8*, 934–948. [CrossRef]
170. Li, S.; Shangguan, Z. Positive effects of apple branch biochar on wheat yield only appear at a low application rate, regardless of nitrogen and water conditions. *J. Soils Sediments* **2018**, *18*, 3235–3243. [CrossRef]
171. Tan, G.; Wang, H.; Xu, N.; Liu, H.; Zhai, L. Biochar amendment with fertilizers increases peanut N uptake, alleviates soil N2O emissions without affecting NH_3 volatilization in field experiments. *Environ. Sci. Pollut. Res.* **2018**, *25*, 8817–8826. [CrossRef]
172. Hafeez, Y.; Iqbal, S.; Jabeen, K.; Shahzad, S.; Jahan, S.; Rasul, F. Effect of biochar application on seed germination and seedling growth of *Glycine max* (L.) Merr. under drought stress. *Pak. J. Bot.* **2017**, *49*, 7–13.
173. Uslu, O.S.; Babur, E.; Alma, M.H.; Solaiman, Z.M. Walnut shell biochar increases seed germination and early growth of seedlings of fodder crops. *Agriculture* **2020**, *10*, 427. [CrossRef]

174. Zhang, Z.; Dong, X.; Wang, S.; Pu, X. Benefits of organic manure combined with biochar amendments to cotton root growth and yield under continuous cropping systems in Xinjiang, China. *Sci. Rep.* **2020**, *10*, 4718. [CrossRef]
175. Chang, Y.; Rossi, L.; Zotarelli, L.; Gao, B.; Shahid, M.A.; Sarkhosh, A. Biochar improves soil physical characteristics and strengthens root architecture in Muscadine grape (*Vitis rotundifolia* L.). *Chem. Biol. Technol. Agric.* **2021**, *8*, 1–11. [CrossRef]
176. Żebrowska, J.; Michalek, W. Chlorophyll fluorescence in two strawberry (*Fragaria x ananassa* Duch.) cultivars. *J. Cent. Eur. Agric.* **2014**, *15*, 12–21. [CrossRef]
177. Agegnehu, G.; Bass, A.M.; Nelson, P.N.; Muirhead, B.; Wright, G.; Bird, M.I. Biochar and biochar-compost as soil amendments: Effects on peanut yield, soil properties and greenhouse gas emissions in tropical North Queensland, Australia. *Agric. Ecosyst. Environ.* **2015**, *213*, 72–85. [CrossRef]
178. Petruccelli, R.; Bonetti, A.; Traversi, M.L.; Faraloni, C.; Valagussa, M.; Pozz, A. Influence of biochar application on nutritional quality of tomato (*Lycopersicon esculentum*). *Crop. Pasture Sci.* **2015**, *66*, 747–755. [CrossRef]
179. Yang, W.; Zhang, L. Biochar and cow manure organic fertilizer amendments improve the quality of composted green waste as a growth medium for the ornamental plant *Centaurea cyanus* L. *Environ. Sci. Pollut. Res. Int.* **2022**, *29*, 45474–45486. [CrossRef]
180. Abid, M.; Danish, S.; Zafar-ul-Hye, M.; Shaaban, M.; Iqbal, M.M.; Rehim, A.; Qayyum, M.F.; Naqqash, M.N. Biochar increased photosynthetic and accessory pigments in tomato (*Solanum lycopersicum* L.) plants by reducing cadmium concentration under various irrigation waters. *Environ. Sci. Pollut. Res. Int.* **2017**, *24*, 22111–22118. [CrossRef]
181. Jabborova, D.; Ma, H.; Bellingrath-Kimura, S.D.; Wirth, S. Impacts of biochar on basil (*Ocimum basilicum*) growth, root morphological traits, plant biochemical and physiological properties and soil enzymatic activities. *Sci. Hortic.* **2021**, *290*, 110518. [CrossRef]
182. Hafez, E.M.; Kheir, A.M.; Badawy, S.A.; Rashwan, E.; Farig, M.; Osman, H.S. Differences in physiological and biochemical attributes of wheat in response to single and combined salicylic acid and biochar subjected to limited water irrigation in saline sodic soil. *Plants* **2020**, *9*, 1346. [CrossRef]
183. Lévesque, V.; Rochette, P.; Hogue, R.; Jeanne, T.; Ziadi, N.; Chantigny, M.H.; Dorais, M.; Antoun, H. Greenhouse gas emissions and soil bacterial community as affected by biochar amendments after periodic mineral fertilizer applications. *Biol. Fertil. Soils.* **2020**, *56*, 907–925. [CrossRef]
184. Batool, A.; Taj, S.; Rashid, A.; Khalid, A.; Qadeer, S.; Saleem, A.R.; Ghufran, M.A. Potential of soil amendments (biochar and gypsum) in increasing water use efficiency of *Abelmoschus esculentus* L. Moench. *Front. Plant Sci.* **2015**, *6*. [CrossRef]
185. Kannan, V.; Srinivasan, G.; Babu, R.; Thiyageshwari, S.; Sivakumar, T. Effect of biochar, mulch and PPFM spray on leaf relative water content, leaf proline, chlorophyll stability index and yield of cotton under moisture stress condition. *Int. J. Curr. Microbiol. Appl. Sci.* **2017**, *6*, 604–611. [CrossRef]
186. Qian, Z.H.U.; Kong, L.J.; Shan, Y.Z.; Yao, X.D.; Zhang, H.J.; Xie, F.T.; Xue, A.O. Effect of biochar on grain yield and leaf photosynthetic physiology of soybean cultivars with different phosphorus efficiencies. *J. Integr. Agric.* **2019**, *18*, 2242–2254. [CrossRef]
187. Jabborova, D.; Wirth, S.; Halwani, M.; Ibrahim, M.F.M.; Azab, I.H.E.; El-Mogy, M.M.; Elkelish, A. Growth Response of Ginger (*Zingiber officinale*), its physiological properties and soil enzyme activities after biochar application under greenhouse conditions. *Horticulturae* **2021**, *7*, 250. [CrossRef]
188. De la Rosa, J.M.; Campos, P.; Diaz-Espejo, A. Soil biochar application: Assessment of the effects on soil water properties, plant physiological status, and yield of super-intensive olive groves under controlled irrigation conditions. *Agronomy* **2022**, *12*, 2321. [CrossRef]
189. Arshad, U.; Raheel, M.; Ashraf, W.; Ur Rehman, A.; Zahid, M.S.; Moustafa, M.; Ali, M.A. Influence of biochar application on morpho-physiological attributes of tomato (*Lycopersicon esculentum* Mill.) and soil properties. *Commun. Soil Sci. Plant Anal.* **2023**, *54*, 515–525. [CrossRef]
190. Hagner, M.; Kemppainen, R.; Jauhiainen, L.; Tiilikkala, K.; Setälä, H. The effects of birch (*Betula* spp.) biochar and pyrolysis temperature on soil properties and plant growth. *Soil Tillage Res.* **2016**, *163*, 224–234. [CrossRef]
191. Cao, Y.; Gao, Y.; Qi, Y.; Li, J. Biochar-enhanced composts reduce the potential leaching of nutrients and heavy metals and suppress plant-parasitic nematodes in excessively fertilized cucumber soils. *Environ. Sci. Pollut. Res.* **2018**, *25*, 7589–7599. [CrossRef]
192. Kamau, S.; Karanja, N.K.; Ayuke, F.O.; Lehmann, J. Short-term influence of biochar and fertilizer-biochar blends on soil nutrients, fauna and maize growth. *Biol. Fertil. Soils* **2019**, *55*, 661–673. [CrossRef]
193. Xu, M.; Gao, P.; Yang, Z.; Su, L.; Wu, J.; Yang, G.; Zhang, X.; Ma, J.; Peng, H.; Xiao, Y. Biochar impacts on phosphorus cycling in rice ecosystem. *Chemosphere* **2019**, *225*, 311–319. [CrossRef]
194. Tisserant, A.; Cherubini, F. Potentials, limitations, Co-benefits, and trade-offs of biochar applications to soils for climate change mitigation. *Land* **2019**, *8*, 179. [CrossRef]
195. Gaskin, J.W.; Speir, R.A.; Harris, K.; Das, K.C.; Lee, R.D.; Morris, L.A.; Fisher, D.S. Effect of peanut hull and pine chip biochar on soil nutrients, corn nutrient status, and yield. *Agronomy* **2010**, *102*, 623–633. [CrossRef]
196. Hossain, M.K.; Strezov, V.; Chan, K.Y.; Nelson, P.F. Agronomic properties of wastewater sludge biochar and bioavailability of metals in production of cherry tomato (*Lycopersicon esculentum*). *Chemosphere* **2010**, *78*, 1167–1171. [CrossRef]
197. Ch'ng, H.Y.; Ahmed, O.H.; Majid, N.M.A. Improving phosphorus availability, nutrient uptake and dry matter production of *Zea mays* L. on a tropical acid soil using poultry manure biochar and pineapple leaves compost. *Exp. Agric.* **2015**, *52*, 447–465. [CrossRef]

198. Sigua, G.C.; Novak, J.M.; Watts, D.W.; Johnson, M.G.; Spokas, K. Efficacies of designer biochars in improving biomass and nutrient uptake of winter wheat grown in a hard setting subsoil layer. *Chemosphere* **2016**, *142*, 176–183. [CrossRef]
199. Sorrenti, G.; Masiello, C.A.; Toselli, M. Biochar interferes with kiwifruit Fe-nutrition in calcareous soil. *Geoderma* **2016**, *272*, 10–19. [CrossRef]
200. Guo, L.; Yu, H.; Kharbach, M.; Wang, J. The response of nutrient uptake, photosynthesis and yield of tomato to biochar addition under reduced nitrogen application. *Agronomy* **2021**, *11*, 1598. [CrossRef]
201. Yan, P.; Shen, C.; Zou, Z.; Fu, J.; Li, X.; Zhang, L.; Zhang, L.; Han, W.; Fan, L. Biochar stimulates tea growth by improving nutrients in acidic soil. *Sci. Hortic.* **2021**, *283*, 110078. [CrossRef]
202. Chew, J.; Joseph, S.; Chen, G.; Zhang, Y.; Zhu, L.; Liu, M.; Taherymoosavi, S.; Munroe, P.; Mitchell, D.R.G.; Pan, G.; et al. Biochar-based fertiliser enhances nutrient uptake and transport in rice seedlings. *Sci. Total Environ.* **2022**, *826*, 154174. [CrossRef]
203. Gao, T.; Bian, R.; Joseph, S.; Taherymoosavi, S.; Mitchell, D.R.; Munroe, P.; Xu, J.; Shi, J. Wheat straw vinegar: A more cost-effective solution than chemical fungicides for sustainable wheat plant protection. *Sci. Total Environ.* **2020**, *725*, 138359. [CrossRef]
204. Mansoor, S.; Kour, N.; Manhas, S.; Zahid, S.; Wani, O.A.; Sharma, V.; Wijaya, L.; Alyemeni, M.N.; Alsahli, A.A.; El-Serehhy, H.A.; et al. Biochar as a tool for effective management of drought and heavy metal toxicity. *Chemosphere* **2021**, *271*, 129458. [CrossRef]
205. Tang, H.; Shubin, W.; Ying, L.; Hassan, M.U.; Ying, S.; Huang, G.; Hasem, M.; Alamri, S.; Mostafa, Y.S. Biochar: A promising soil amendment to mitigate heavy metals toxicity in plants. *Not. Bot. Hortic. Agrobot. Cluj-Napoca* **2022**, *50*, 12778. [CrossRef]
206. Moragues Saitua, L.; Arias-González, A.; Blanco, F.; Benito-Carnero, G.; Gartzia-Bengoetxea, N. Effects of biochar and wood ash amendments in the soil-water-plant environment of two temperate forest plantations. *Front. For. Glob. Change* **2023**, *5*, 290. [CrossRef]
207. Iacomino, G.; Idbella, M.; Laudonia, S.; Vinale, F.; Bonanomi, G. The suppressive effects of biochar on above-and belowground plant pathogens and pests: A review. *Plants* **2022**, *11*, 3144. [CrossRef] [PubMed]
208. Lou, Y.; Joseph, S.; Li, L.; Graber, E.R.; Liu, X.; Pan, G. Water extract from straw biochar used for plant growth promotion: An initial test. *Bioresources* **2016**, *11*, 249–266. [CrossRef]
209. Gu, Y.; Hou, Y.; Huang, D.; Hao, Z.; Wang, X.; Wei, Z.; Jousset, A.; Tan, S.; Xu, D.; Shen, O.; et al. Application of biochar reduces *Ralstonia solanacearum* infection via effects on pathogen chemotaxis, swarming motility, and root exudate adsorption. *Plant Soil* **2017**, *415*, 269–281. [CrossRef]
210. Akanmu, A.O.; Sobowale, A.A.; Abiala, M.A.; Olawuyi, O.J.; Odebode, A.C. Efficacy of biochar in the management of *Fusarium verticillioides* Sacc. causing ear rot in *Zea mays* L. *Biotechnol. Rep.* **2020**, *26*, e00474. [CrossRef]
211. Mehari, Z.H.; Elad, Y.; Rav-David, D.; Graber, E.R.; Meller Harel, Y. Induced systemic resistance in tomato (*Solanum lycopersicum*) against *Botrytis cinerea* by biochar amendment involves jasmonic acid signaling. *Plant Soil* **2015**, *395*, 31–44. [CrossRef]
212. Bass, A.M.; Bird, M.I.; Kay, G.; Muirhead, B. Soil properties, greenhouse gas emissions and crop yield under compost, biochar and co-composted biochar in two tropical agronomic systems. *Sci. Total Environ.* **2016**, *550*, 459–470. [CrossRef]
213. Singh, B.P.; Hatton, B.J.; Singh, B.; Cowie, A.L.; Kathuria, A. Influence of biochars on nitrous oxide emission and nitrogen leaching from two contrasting soils. *J. Environ. Qual.* **2010**, *39*, 1224–1235. [CrossRef]
214. Xiang, L.; Harindintwali, J.D.; Wang, F.; Redmile-Gordon, M.; Chang, S.X.; Fu, Y.; He, C.; Muhoza, B.; Brahushi, F.; Bolan, N.; et al. Integrating biochar, bacteria, and plants for sustainable remediation of soils contaminated with organic pollutants. *Environ. Sci. Technol.* **2022**, *56*, 16546–16566. [CrossRef]
215. De Medeiros, E.V.; dos Santos, M.D.C.H.; da Costa, D.P.; Duda, G.P.; de Oliveira, J.B.; da Silva, J.A.; de Sousa Lima, J.R.; Hammecker, C. Effect of biochar and inoculation with *Trichoderma aureoviride* on melon growth and sandy entisol quality. *Aust. J. Crop Sci.* **2020**, *14*, 971–977. [CrossRef]
216. Meller Harel, Y.; Elad, Y.; Rav-David, D.; Borenstein, M.; Shulchani, R.; Lew, B.; Graber, E.R. Biochar mediates systemic response of strawberry to foliar fungal pathogens. *Plant Soil* **2012**, *357*, 245–257. [CrossRef]
217. Zwart, D.C.; Kim, S.H. Biochar amendment increases resistance to stem lesions caused by *Phytophthora* spp. in tree seedlings. *Res. HortScience* **2012**, *47*, 1736–1740. [CrossRef]
218. Jaiswal, A.K.; Elad, Y.; Graber, E.R.; Frenkel, O. Rhizoctonia solani suppression and plant growth promotion in cucumber as affected by biochar pyrolysis temperature, feedstock and concentration. *Soil Biol. Biochem.* **2014**, *69*, 110–118. [CrossRef]
219. Jaiswal, A.K.; Alkan, N.; Elad, Y.; Sela, N.; Philosoph, A.M.; Graber, E.R.; Frenkel, O. Molecular insights into biochar-mediated plant growth promotion and systemic resistance in tomato against *Fusarium* crown and root rot disease. *Sci. Rep.* **2020**, *10*, 13934. [CrossRef]
220. Singh, J.; Thakur, J.K. Photosynthesis and abiotic stress in plants. In *Biotic and Abiotic Stress Tolerance in Plants*; Vats, S., Ed.; Springer: Berlin/Heidelberg, Germany, 2018; pp. 27–46. [CrossRef]
221. Ali, E.F.; Al-Yasi, H.M.; Kheir, A.M.; Eissa, M.A. Effect of biochar on CO_2 sequestration and productivity of pearl millet plants grown in saline sodic soils. *J. Soil Sci. Plant Nutr.* **2021**, *21*, 897–907. [CrossRef]
222. Trupiano, D.; Cocozza, C.; Baronti, S.; Amendola, C.; Vaccari, F.P.; Lustrato, G.; Di Lonardo, S.; Fantasma, F.; Tognetti, R.; Scippa, G.S. The effects of biochar and its combination with compost on lettuce (*Lactuca sativa* L.) growth, soil properties, and soil microbial activity and abundance. *J. Agron.* **2017**, *1*, 3158207. [CrossRef]
223. Parihar, P.; Singh, S.; Singh, R.; Singh, V.P.; Prasad, S.M. Effect of salinity stress on plants and its tolerance strategies: A review. *Environ. Sci. Pollut. Res.* **2015**, *22*, 4056–4075. [CrossRef]

224. Azeem, M.; Iqbal, N.; Kausar, S.; Javed, M.T.; Akram, M.S.; Sajid, M.A. Efficacy of silicon priming and fertigation to modulate seedlings vigor and ion homeostasis of wheat (*Triticum aestivum* L.) under saline environment. *Environ. Sci. Pollut. Res.* **2015**, *22*, 14367–14371. [CrossRef]
225. Hussain, R.A.; Ahmad, R.; Waraich, E.A.; Nawaz, F. Nutrient uptake, water relations, and yield performance of different wheat cultivars (*Triticum aestivum* L.) under salinity stress. *J. Plant Nutr.* **2015**, *38*, 2139–2149. [CrossRef]
226. Mohamed, A.K.; Qayyum, M.F.; Abdel-Hadi, A.M.; Rehman, R.A.; Ali, S.; Rizwan, M. Interactive effect of salinity and silver nanoparticles on photosynthetic and biochemical parameters of wheat. *Arch. Agron. Soil Sci.* **2017**, *63*, 1736–1747. [CrossRef]
227. Hussien Ibrahim, M.E.; Adam Ali, A.Y.; Zhou, G.; Ibrahim Elsiddig, A.M.; Zhu, G.; Ahmed Nimir, N.E.; Ahmad, I. Biochar application affects forage sorghum under salinity stress. *Chil. J. Agric. Res.* **2020**, *80*, 317–325. [CrossRef]
228. Parkash, V.; Singh, S. Potential of biochar application to mitigate salinity stress in eggplant. *HortScience* **2020**, *55*, 1946–1955. [CrossRef]
229. Kerbab, S.; Silini, A.; Chenari, B.A.; Cherif-Silini, H.; Eshelli, M.; El Houda, R.N.; Belbahri, L. Mitigation of NaCl stress in wheat by rhizosphere engineering using salt habitat adapted PGPR halotolerant bacteria. *Appl. Sci.* **2021**, *11*, 1034. [CrossRef]
230. Abo-Elyousr, K.A.; Mousa, M.A.; Ibrahim, O.H.; Alshareef, N.O.; Eissa, M.A. Calcium-rich biochar stimulates salt resistance in pearl millet (*Pennisetum glaucum* L.) plants by improving soil quality and enhancing the antioxidant defense. *Plants* **2022**, *11*, 1301. [CrossRef]
231. Akhtar, S.S.; Andersen, M.N.; Naveed, M.; Zahir, Z.A.; Liu, F. Interactive effect of biochar and plant growth-promoting bacterial endophytes on ameliorating salinity stress in maize. *Funct. Plant Biol.* **2015**, *42*, 770–781. [CrossRef]
232. Akhtar, S.S.; Andersen, M.N.; Liu, F. Residual effects of biochar on improving growth, physiology and yield of wheat under salt stress. *Agric. Water Manag.* **2015**, *158*, 61–68. [CrossRef]
233. Liang, G.; Liu, J.; Zhang, J.; Guo, J. Effects of drought stress on photosynthetic and physiological parameters of tomato. *J. Am. Soc. Hortic. Sci.* **2020**, *145*, 12–17. [CrossRef]
234. Latif, M.; Bukhari, S.A.H.; Alrajhi, A.A.; Alotaibi, F.S.; Ahmad, M.; Shahzad, A.N.; Dewidar, A.Z.; Mattar, M.A. Inducing drought tolerance in wheat through exopolysaccharide-producing rhizobacteria. *Agronomy* **2022**, *12*, 1140. [CrossRef]
235. AlKahtani, M.D.; Hafez, Y.M.; Attia, K.; Rashwan, E.; Husnain, L.A.; AlGwaiz, H.I.; Abdelaal, K.A. Evaluation of silicon and proline application on the oxidative machinery in drought-stressed sugar beet. *Antioxidants* **2021**, *10*, 398. [CrossRef]
236. Sahin, U.; Ekinci, M.; Ors, S.; Turan, M.; Yildiz, S.; Yildirim, E. Effects of individual and combined effects of salinity and drought on physiological, nutritional and biochemical properties of cabbage (*Brassica oleracea* var. capitata). *Sci. Hortic.* **2018**, *240*, 196–204. [CrossRef]
237. Razavi, F.; Pollet, B.; Steppe, K.; Van Labeke, M.C. Chlorophyll fluorescence as a tool for evaluation of drought stress in strawberry. *Photosynthetica* **2008**, *46*, 631–633. [CrossRef]
238. Boyer, J. Leaf enlargement and metabolic rates in corn, soybean, and sunflower at various leaf water potentials. *Plant Physiol.* **1970**, *46*, 233–235. [CrossRef] [PubMed]
239. Singer, S.M.; Helmy, Y.I.; Karas, A.N.; Abou-Hadid, A.F. Influences of different water-stress treatments on growth, development and production of snap bean (*Phaseolus vulgaris* L.). *Acta Hortic.* **2003**, *614*, 605–611. [CrossRef]
240. Ali, S.; Rizwan, M.; Qayyum, M.F.; Ok, Y.S.; Ibrahim, M.; Riaz, M.; Arif, M.S.; Hafeez, F.; Al-Wabel, M.I.; Shahzad, A.N. Biochar soil amendment on alleviation of drought and salt stress in plants: A critical review. *Environ. Sci. Pollut. Res. Int.* **2017**, *24*, 12700–12712. [CrossRef] [PubMed]
241. Langeroodi, A.R.S.; Campiglia, E.; Mancinelli, R.; Radicetti, E. Can biochar improve pumpkin productivity and its physiological characteristics under reduced irrigation regimes? *Sci. Hortic.* **2019**, *247*, 195–204. [CrossRef]
242. Gharred, J.; Derbali, W.; Derbali, I.; Badri, M.; Abdelly, C.; Slama, I.; Koyro, H.W. Impact of biochar application at water shortage on biochemical and physiological processes in *Medicago ciliaris*. *Plants* **2022**, *11*, 2411. [CrossRef]
243. Keabetswe, L.; Shao, G.C.; Cui, J.; Lu, J.; Stimela, T. A combination of biochar and regulated deficit irrigation improves tomato fruit quality: A comprehensive quality analysis. *Folia Hortic.* **2019**, *31*, 181–193. [CrossRef]
244. Poormansour, S.; Razzaghi, F.; Sepaskhah, A.R. Wheat straw biochar increases potassium concentration, root density, and yield of faba bean in a sandy loam soil. *Commun. Soil Sci. Plant Anal.* **2019**, *50*, 1799–1810. [CrossRef]
245. Huang, M.; Yin, X.; Chen, J.; Cao, F. Biochar application mitigates the effect of heat stress on rice (*Oryza sativa* L.) by regulating the root-zone environment. *Front. Plant Sci.* **2021**, *12*, 711725. [CrossRef]
246. Kanwal, S.; Ilyas, N.; Shabir, S.; Saeed, M.; Gul, R.; Zahoor, M.; Batool, N.; Mazhar, R. Application of biochar in mitigation of negative effects of salinity stress in wheat (*Triticum aestivum* L.). *J. Plant Nutr.* **2018**, *41*, 526–538. [CrossRef]
247. Castañeda, W.; Toro, M.; Solorzano, A.; Zúñiga-Dávila, D. Production and nutritional quality of tomatoes (*Solanum lycopersicum* var. Cerasiforme) are improved in the presence of biochar and inoculation with arbuscular mycorrhizae. *Am. J. Plant Sci.* [CrossRef]
248. Soothar, M.K.; Mounkaila Hamani, A.K.; Kumar Sootahar, M.; Sun, J.; Yang, G.; Bhatti, S.M.; Traore, A. Assessment of acidic biochar on the growth, physiology and nutrients uptake of maize (*Zea mays* L.) seedlings under salinity stress. *Sustainability* **2021**, *13*, 3150. [CrossRef]
249. Rizwan, M.; Ali, S.; Abbas, T.; Adrees, M.; Zia-ur-Rehman, M.; Ibrahim, M.; Abbas, F.; Qayyum, M.F.; Nawaz, R. Residual effects of biochar on growth, photosynthesis and cadmium uptake in rice (*Oryza sativa* L.) under Cd stress with different water conditions. *J. Environ. Manag.* **2018**, *206*, 676–683. [CrossRef] [PubMed]

250. Hashem, A.; Kumar, A.; Al-Dbass, A.M.; Alqarawi, A.A.; Al-Arjani, A.B.F.; Singh, G.; Farooq, M.; Abd_Allah, E.F. Arbuscular mycorrhizal fungi and biochar improves drought tolerance in chickpea. *Saudi J. Biol. Sci.* **2019**, *26*, 614–624. [CrossRef] [PubMed]
251. Zoghi, Z.; Hosseini, S.M.; Kouchaksaraei, M.T.; Kooch, Y.; Guidi, L. The effect of biochar amendment on the growth, morphology and physiology of *Quercus castaneifolia* seedlings under water-deficit stress. *Eur. J. Forest Res.* **2019**, *138*, 967–979. [CrossRef]
252. Durukan, H.; Demirbas, A.; Turkekul, I. Effects of biochar rates on yield and nutrient uptake of sugar beet plants grown under drought stress. *Commun. Soil Sci. Plant Anal.* **2020**, *51*, 2735–2745. [CrossRef]
253. Haider, I.; Raza, M.A.S.; Iqbal, R.; Aslam, M.U.; Habib-ur-Rahman, M.; Raja, S.; Khan, M.T.; Aslam, M.M.; Waqas, M.; Ahmad, S. Potential effects of biochar application on mitigating the drought stress implications on wheat (*Triticum aestivum* L.) under various growth stages. *J. Saudi Chem. Soc.* **2020**, *24*, 974–981. [CrossRef]
254. Khan, Z.; Khan, M.N.; Zhang, K.; Luo, T.; Zhu, K.; Hu, L. The application of biochar alleviated the adverse effects of drought on the growth, physiology, yield and quality of rapeseed through regulation of soil status and nutrients availability. *Ind. Crops Prod.* **2021**, *171*, 113878. [CrossRef]
255. Li, Y.; Feng, H.; Chen, J.; Lu, J.; Wu, W.; Liu, X.; Li, C.; ge Dong, Q.; Siddique, K.H. Biochar incorporation increases winter wheat (*Triticum aestivum* L.) production with significantly improving soil enzyme activities at jointing stage. *Catena* **2022**, *211*, 105979. [CrossRef]
256. Borchard, N.; Schirrmann, M.; Cayuela, M.L.; Kammann, C.; Wrage-Mönnig, N.; Estavillo, J.M.; Fuertes-Mendizabal, T.; Sigua, G.; Spokas, K.; Ippolito, J.A.; et al. Biochar, soil and land-use interactions that reduce nitrate leaching and N_2O emissions: A meta-analysis. *Sci. Total Environ.* **2019**, *651*, 2354–2364. [CrossRef]
257. Hüppi, R.; Neftel, A.; Lehmann, M.F.; Krauss, M.; Six, J.; Leifeld, J. N use efficiencies and N_2O emissions in two contrasting, biochar amended soils under winter wheat—Cover crop—Sorghum rotation. *Environ. Res. Lett.* **2016**, *11*, 084013. [CrossRef]
258. Lin, X.W.; Xie, Z.B.; Zheng, J.Y.; Liu, Q.; Bei, Q.C.; Zhu, J.G. Effects of biochar application on greenhouse gas emissions, carbon sequestration and crop growth in coastal saline soil. *Eur. J. Soil Sci.* **2015**, *66*, 329–338. [CrossRef]
259. Rodríguez-Eugenio, N.; McLaughlin, M.; Pennock, D. *Soil Pollution: A Hidden Reality*; FAO: Rome, Italy, 2018; p. 142.
260. Abdel-Shafy, H.I.; Mansour, M.S. A review on polycyclic aromatic hydrocarbons: Source, environmental impact, effect on human health and remediation. *Egypt. J. Pet.* **2016**, *25*, 107–123. [CrossRef]
261. Alengebawy, A.; Abdelkhalek, S.T.; Qureshi, S.R.; Wang, M.Q. Heavy metals and pesticides toxicity in agricultural soil and plants: Ecological risks and human health implications. *Toxics* **2021**, *9*, 42. [CrossRef]
262. Tóth, G.; Hermann, T.; Da Silva, M.R.; Montanarella, L.J.E.I. Heavy metals in agricultural soils of the European Union with implications for food safety. *Environ. Int.* **2016**, *88*, 299–309. [CrossRef] [PubMed]
263. Javed, M.T.; Tanwir, K.; Akram, M.S.; Shahid, M.; Niazi, N.K.; Lindberg, S. Phytoremediation of cadmium-polluted water/sediment by aquatic macrophytes: Role of plant-induced pH changes. In *Cadmium Toxicity and Tolerance in Plants from Physiology to Remediation*; Hasanuzzaman, M., Prasad, M.N.V., Fujita, M., Eds.; Elsevier: Amsterdam, The Netherlands, 2019; pp. 495–529. [CrossRef]
264. Hu, B.; Xue, J.; Zhou, Y.; Shao, S.; Fu, Z.; Li, Y.; Chen, S.; Qi, L.; Shi, Z. Modelling bioaccumulation of heavy metals in soil-crop ecosystems and identifying its controlling factors using machine learning. *Environ. Pollut.* **2020**, *262*, 114308. [CrossRef]
265. Ahmad, S.M.; Yusafzai, F.; Bari, T. Assessment of heavy metals in surface water of River Panjkora Dir Lower, KPK Pakistan. *Biomed. Res. Environ. Sci.* **2014**, *5*, 144–152.
266. Gunatilake, S.K. Methods of removing heavy metals from industrial wastewater. *J. Multidiscip. Eng. Sci. Technol.* **2015**, *1*, 12–18.
267. Uchimiya, M.; Lima, I.M.; Klasson, K.T.; Wartelle, L.H. Contaminant immobilization and nutrient release by biochar soil amendment: Roles of natural organic matter. *Chemosphere* **2010**, *80*, 935–940. [CrossRef]
268. Dong, X.; Li, G.; Lin, Q.; Zhao, X. Quantity and quality changes of biochar aged for 5 years in soil under field conditions. *Catena* **2017**, *159*, 136–143. [CrossRef]
269. Guo, X.; Li, H. Effects of iron-modified biochar and AMF inoculation on the growth and heavy metal uptake of senna occidentalis in heavy metal-contaminated soil. *Pol. J. Environ. Stud.* **2019**, *28*, 2611–2621. [CrossRef]
270. Rathika, R.; Srinivasan, P.; Alkahtani, J.; Al-Humaid, L.A.; Alwahibi, M.S.; Mythili, R.; Selvankumar, T. Influence of biochar and EDTA on enhanced phytoremediation of lead contaminated soil by *Brassica juncea*. *Chemosphere* **2021**, *271*, 129513. [CrossRef] [PubMed]
271. Leng, L.; Leng, S.; Chen, J.; Yuan, X.; Li, J.; Li, K.; Wang, Y.; Zhou, W. The migration and transformation behavior of heavy metals during co-liquefaction of municipal sewage sludge and lignocellulosic biomass. *Bioresour. Technol.* **2018**, *259*, 156–163. [CrossRef] [PubMed]
272. Fahmi, A.H.; Samsuri, A.W.; Jol, H.; Singh, D. Bioavailability and leaching of Cd and Pb from contaminated soil amended with different sizes of biochars. *R. Soc. Open Sci.* **2018**, *5*, 181328. [CrossRef] [PubMed]
273. Gong, X.; Huang, D.; Liu, Y.; Zeng, G.; Wang, R.; Wei, J.; Huang, C.; Xu, P.; Wan, J.; Zhang, C. Pyrolysis and reutilization of plant residues after phytoremediation of heavy metals contaminated sediments: For heavy metals stabilization and dye adsorption. *Bioresour. Technol.* **2018**, *253*, 64–71. [CrossRef]
274. Koh, I.H.; Kim, J.; Kim, G.S.; Park, M.S.; Kang, D.M.; Ji, W.H. Stabilization of agricultural soil contaminated by arsenic and heavy metals using biochar derived from Buffalo weed. *J. Soil Groundwater Environ.* **2016**, *21*, 87–100. [CrossRef]

275. Neeraj, G.; Krishnan, S.; Kumar, P.S.; Shriaishvarya, K.R.; Kumar, V.V. Performance study on sequestration of copper ions from contaminated water using newly synthesized high effective chitosan coated magnetic nanoparticles. *J. Mol. Liq.* **2016**, *214*, 335–346. [CrossRef]
276. Jiang, T.Y.; Xu, R.K.; Gu, T.X.; Jiang, J. Effect of crop-straw derived biochars on Pb (II) adsorption in two variable charge soils. *J. Integr. Agric.* **2014**, *13*, 507–516. [CrossRef]
277. Ren, X.; Sun, H.; Wang, F.; Cao, F. The changes in biochar properties and sorption capacities after being cultured with wheat for 3 months. *Chemosphere* **2016**, *144*, 2257–2263. [CrossRef]
278. Lebrun, M.; Macri, C.; Miard, F.; Hattab-Hambli, N.; Motelica-Heino, M.; Morabito, D.; Bourgerie, S. Effect of biochar amendments on As and Pb mobility and phytoavailability in contaminated mine technosols phytoremediated by Salix. *J. Geochem. Explor.* **2017**, *182*, 149–156. [CrossRef]
279. Dai, J.; Meng, X.; Zhang, Y.; Huang, Y. Effects of modification and magnetization of rice straw derived biochar on adsorption of tetracycline from water. *Bioresour. Technol.* **2020**, *311*, 123455. [CrossRef]
280. Kończak, M.; Oleszczuk, P. Co-pyrolysis of sewage sludge and biomass in carbon dioxide as a carrier gas affects the total and leachable metals in biochars. *J. Hazard. Mater.* **2020**, *400*, 123144. [CrossRef] [PubMed]
281. Yaashikaa, P.R.; Kumar, P.S.; Saravanan, A.; Vo, D.V.N. Advances in biosorbents for removal of environmental pollutants: A review on pretreatment, removal mechanism and future outlook. *J. Hazard. Mater.* **2021**, *420*, 126596. [CrossRef] [PubMed]
282. Li, Z.; Su, M.; Duan, X.; Tian, D.; Yang, M.; Guo, J.; Wang, S.; Hu, S. Induced biotransformation of lead (II) by Enterobacter sp. in SO_4-PO_4-Cl solution. *J. Hazard. Mater.* **2018**, *357*, 491–497. [CrossRef] [PubMed]
283. Forján, R.; Rodríguez-Vila, A.; Cerqueira, B.; Covelo, E.F. Comparison of the effects of compost versus compost and biochar on the recovery of a mine soil by improving the nutrient content. *J. Geochem. Explor.* **2017**, *183*, 46–57. [CrossRef]
284. Bashir, S.; Shaaban, M.; Mehmood, S.; Zhu, J.; Fu, Q.; Hu, H. Efficiency of C3 and C4 plant derived-biochar for cd mobility, nutrient cycling and microbial biomass in contaminated soil. *Bull. Environ. Contam. Toxicol.* **2018**, *100*, 834–838. [CrossRef] [PubMed]
285. Igalavithana, A.D.; Kwon, E.E.; Vithanage, M.; Rinklebe, J.; Moon, D.H.; Meers, E.; Tsang, D.C.; Ok, Y.S. Soil lead immobilization by biochars in short-term laboratory incubation studies. *Environ. Int.* **2019**, *127*, 190. [CrossRef]
286. Krzyszczak, A.; Dybowski, M.P.; Czech, B. Formation of polycyclic aromatic hydrocarbons and their derivatives in biochars: The effect of feedstock and pyrolysis conditions. *J. Anal. Appl. Pyrol.* **2021**, *160*, 105339. [CrossRef]
287. Acosta-Luque, M.P.; López, J.E.; Henao, N.; Zapata, D.; Giraldo, J.C.; Saldarriaga, J.F. Remediation of Pb-contaminated soil using biochar-based slow-release P fertilizer and biomonitoring employing bioindicators. *Sci. Rep.* **2023**, *13*, 1657. [CrossRef]
288. Bousdra, T.; Papadimou, S.G.; Golia, E.E. The use of biochar in the remediation of Pb, Cd, and Cu-contaminated soils. The impact of biochar feedstock and preparation conditions on its remediation capacity. *Land* **2023**, *12*, 383. [CrossRef]
289. Younis, U.; Malik, S.A.; Rizwan, M.; Qayyum, M.F.; Ok, Y.S.; Shah, M.H.R.; Rehman, R.A.; Ahmad, N. Biochar enhances the cadmium tolerance in spinach (*Spinacia oleracea*) through modification of Cd uptake and physiological and biochemical attributes. *Environ. Sci. Pollut. Res.* **2016**, *23*, 21385–21394. [CrossRef]
290. Houben, D.; Sonnet, P. Impact of biochar and root-induced changes on metal dynamics in the rhizosphere of *Agrostis capillaris* and *Lupinus albus*. *Chemosphere* **2015**, *139*, 644–651. [CrossRef] [PubMed]
291. Sharma, A.; Kumar, V.; Shahzad, B.; Ramakrishnan, M.; Singh Sidhu, G.P.; Bali, A.S.; Handa, N.; Kapoor, D.; Yadav, P.; Khanna, K.; et al. Photosynthetic response of plants under different abiotic stresses: A review. *J. Plant Growth Regul.* **2020**, *39*, 509–531. [CrossRef]
292. Nabi, R.B.S.; Tayade, R.; Hussain, A.; Kulkarni, K.P.; Imran, Q.M.; Mun, B.G.; Yun, B.W. Nitric oxide regulates plant responses to drought, salinity, and heavy metal stress. *Environ. Exp. Bot.* **2019**, *161*, 120–133. [CrossRef]
293. Nowicka, B. Heavy metal–induced stress in eukaryotic algae—Mechanisms of heavy metal toxicity and tolerance with particular emphasis on oxidative stress in exposed cells and the role of antioxidant response. *Environ. Sci. Pollut. Res.* **2022**, *29*, 16860–16911. [CrossRef]
294. Rizvi, A.; Khan, M.S. Heavy metal induced oxidative damage and root morphology alterations of maize (*Zea mays* L.) plants and stress mitigation by metal tolerant nitrogen fixing *Azotobacter chroococcum*. *Ecotoxicol. Environ. Saf.* **2018**, *157*, 9–20. [CrossRef]
295. Ahmad, M.; Rajapaksha, A.U.; Lim, J.E.; Zhang, M.; Bolan, N.; Mohan, D.; Vithanage, M.; Lee, S.S.; Ok, Y.S. Biochar as a sorbent for contaminant management in soil and water: A review. *Chemosphere* **2014**, *99*, 19–33. [CrossRef]
296. Ali, N.; Khan, S.; Yao, H.; Wang, J. Biochars reduced the bioaccessibility and (bio)uptake of organochlorine pesticides and changed the microbial community dynamics in agricultural soils. *Chemosphere* **2019**, *224*, 805–815. [CrossRef]
297. Rehman, M.Z.; Zafar, M.; Waris, A.A.; Rizwan, M.; Ali, S.; Sabir, M.; Usman, M.; Ayub, M.A.; Ahmad, Z. Residual effects of frequently available organic amendments on cadmium bioavailability and accumulation in wheat. *Chemosphere* **2020**, *244*, 125548. [CrossRef]
298. Quartacci, M.I.K.E. Biochar amendment reduces oxidative stress in lettuce grown under copper excess. *Agrochimica* **2015**, *59*, 188–202. [CrossRef]
299. Zhang, F.; Liu, M.; Li, Y.; Che, Y.; Xiao, Y. Effects of arbuscular mycorrhizal fungi, biochar and cadmium on the yield and element uptake of *Medicago sativa*. *Sci. Total Environ.* **2019**, *655*, 1150–1158. [CrossRef]
300. Lu, K.; Yang, X.; Shen, J.; Robinson, B.; Huang, H.; Liu, D.; Bolan, N.; Pei, J.; Wang, H. Effect of bamboo and rice straw biochars on the bioavailability of Cd, Cu, Pb and Zn to *Sedum plumbizincicola*. *Agric. Ecosyst. Environ.* **2014**, *191*, 124–132. [CrossRef]

301. Alaboudi, K.A.; Ahmed, B.; Brodie, G. Effect of biochar on Pb, Cd and Cr availability and maize growth in artificial contaminated soil. *Ann. Agric. Sci.* **2019**, *64*, 95–102. [CrossRef]
302. Shetty, R.; Prakash, N.B. Effect of different biochars on acid soil and growth parameters of rice plants under aluminium toxicity. *Sci. Rep.* **2020**, *10*, 12249. [CrossRef] [PubMed]
303. Zhu, Y.; Wang, H.; Lv, X.; Zhang, Y.; Wang, W. Effects of biochar and biofertilizer on cadmium-contaminated cotton growth and the antioxidative defense system. *Sci. Rep.* **2020**, *10*, 20112. [CrossRef]
304. Stanek-Tarkowska, J.; Czyż, E.A. Cultivation systems for winter wheat (*Triticum aestivum* L.) and soil susceptibility to erosion. *J. Water Land Dev.* **2024**, *62*, 29–36. [CrossRef]
305. Gunarathne, V.; Senadeera, A.; Gunarathne, U.; Biswas, J.K.; Almaroai, Y.A.; Vithanage, M. Potential of biochar and organic amendments for reclamation of coastal acidic-salt affected soil. *Biochar* **2020**, *2*, 107–120. [CrossRef]
306. Xia, H.; Riaz, M.; Zhang, M.; Liu, B.; El-Desouki, Z.; Jiang, C. Biochar increases nitrogen use efficiency of maize by relieving aluminum toxicity and improving soil quality in acidic soil. *Ecotoxicol. Environ. Saf.* **2020**, *196*, 110531. [CrossRef]
307. Bakari, R.; Mungai, N.; Thuita, M.; Masso, C. Impact of soil acidity and liming on soybean (*Glycine max*) nodulation and nitrogen fixation in Kenyan soils. *Acta Agric. Scand.* **2020**, *70*, 667–678. [CrossRef]
308. Zhang, Y.; Chen, H.; Xiang, J.; Xiong, J.; Wang, Y.; Wang, Z.; Zhang, Y. Effect of rice-straw biochar application on the acquisition of rhizosphere phosphorus in acidified paddy soil. *Agronomy* **2022**, *12*, 1556. [CrossRef]
309. He, L.; Zhao, J.; Wang, M.; Liu, Y.; Wang, Y.; Yang, S.; Wang, S.; Zhao, X.; Lyu, H. Long-term successive seasonal application of rice straw-derived biochar improves the acidity and fertility of red soil in Southern China. *Agronomy* **2023**, *13*, 505. [CrossRef]
310. Zhao, R.; Jiang, D.; Coles, N.; Wu, J. Effects of biochar on the acidity of a loamy clay soil under different incubation conditions. *J. Soils. Sedi.* **2015**, *15*, 1919–1926. [CrossRef]
311. Phares, C.A.; Amoakwah, E.; Danquah, A.; Akaba, S.; Frimpong, K.A.; Mensah, T.A. Improved soil physicochemical, biological properties and net income following the application of inorganic NPK fertilizer and biochar for maize production. *Acta Ecol. Sin.* **2022**, *42*, 289–295. [CrossRef]
312. Novak, J.M.; Busscher, W.J.; Laird, D.L.; Ahmedna, M.; Watts, D.W.; Niandou, M.A.S. Impact of biochar amendment on fertility of a Southeastern Coastal Plain soil. *Soil Sci.* **2009**, *174*, 105–112. [CrossRef]
313. Das, S.K.; Ghosh, G.K.; Avasthe, R. Ecotoxicological responses of weed biochar on seed germination and seedling growth in acidic soil. *Environ. Technol. Innov.* **2020**, *20*, 101074. [CrossRef]
314. Nguyen, B.T.; Le, L.B.; Pham, L.P.; Nguyen, H.T.; Tran, T.D.; Van Thai, N. The effects of biochar on the biomass yield of elephant grass (*Pennisetum purpureum* Schumach) and properties of acidic soils. *Ind. Crops Prod.* **2021**, *161*, 113224. [CrossRef]
315. Lin, Y.C.; Ho, S.H.; Zhou, Y.; Ren, N.Q. Highly efficient adsorption of dyes by biochar derived from pigments-extracted macroalgae pyrolyzed at different temperature. *Bioresour. Technol.* **2018**, *259*, 104–110. [CrossRef]
316. Masud, M.M.; Abdulaha-Al Baquy, M.; Akhter, S.; Sen, R.; Barman, A.; Khatun, M.R. Liming effects of poultry litter derived biochar on soil acidity amelioration and maize growth. *Ecotoxicol. Environ. Saf.* **2020**, *202*, 110865. [CrossRef]
317. Becerra-Agudelo, E.; López, J.E.; Betancur-García, H.; Carbal-Guerra, J.; Torres-Hernández, M.; Saldarriaga, J.F. Assessment of the application of two amendments (lime and biochar) on the acidification and bioavailability of Ni in a Ni-contaminated ag-ricultural soils of northern Colombia. *Heliyon* **2022**, *8*, e10221. [CrossRef]
318. Bolan, N.; Hoang, S.A.; Beiyuan, J.; Gupta, S.; Hou, D.; Karakoti, A.; Joseph, S.; Jung, S.; Kim, K.H.; Kirkham, M.B.; et al. Multifunctional applications of biochar beyond carbon storage. *Int. Mater. Rev.* **2022**, *67*, 150–200. [CrossRef]
319. WMO. *The Role of Climatological Normals in a Changing Climate. WMO/TD-No. 1377*; WMO: Geneva, Switzerland, 2007.
320. Castaldi, S.; Riondino, M.; Baronti, S.; Esposito, F.R.; Marzaioli, R.; Rutigliano, F.A.; Vaccari, F.P.; Miglietta, F. Impact of biochar application to a Mediterranean wheat crop on soil microbial activity and greenhouse gas fluxes. *Chemosphere* **2011**, *85*, 1464–1471. [CrossRef]
321. Woolf, D.; Amonette, J.E.; Street-Perrott, F.A.; Lehmann, J.; Joseph, S. Sustainable biochar to mitigate global climate change. *Nat. Commun.* **2010**, *1*, 56. [CrossRef] [PubMed]
322. Jia, J.; Li, B.; Chen, Z.; Xie, Z.; Xiong, Z. Effects of biochar application on vegetable production and emissions of N_2O and CH_4. *Soil Sci. Plant Nutr.* **2012**, *58*, 503–509. [CrossRef]
323. Gupta, D.K.; Gupta, C.K.; Dubey, R.; Fagodiya, R.K.; Sharma, G.; Noor Mohamed, M.B.; Dev, R.; Shukla, A.K. Role of biochar in carbon sequestration and greenhouse gas mitigation. In *Biochar Applications in Agriculture and Environment Management*; Sing, J.S., Singh, C., Eds.; Springer Nature: Berlin/Heidelberg, Germany, 2020; pp. 141–165. [CrossRef]
324. Wang, H.; Ren, T.; Yang, H.; Feng, Y.; Feng, H.; Liu, G.; Yin, Q.; Shi, H. Research and application of biochar in soil CO_2 emission, fertility, and microorganisms: A sustainable solution to solve China's agricultural straw burning problem. *Sustainability* **2020**, *12*, 1922. [CrossRef]
325. Jacobs, J.; Work, T.; Paré, D.; Bergeron, Y. Paludification of boreal soils reduces wood decomposition rates and increases wood-based carbon storage. *Ecosphere* **2015**, *6*, 1–20. [CrossRef]
326. Zhang, Q.; Zhang, X.; Duan, P.; Jiang, X.; Shen, H.; Yan, X.; Xiong, X. The effect of long-term biochar amendment on N_2O Emissions: Experiments with N15–O18 isotopes combined with specific inhibition approaches. *Sci. Total Environ.* **2021**, *769*, 144533. [CrossRef]

327. Rondon, M.; Ramirez, J.A.; Lehmann, J. Greenhouse gas emissions decrease with charcoal additions to tropical soils. In Proceedings of the 3rd USDA Symposium on Greenhouse Gases and Carbon Sequestration, Maryland, MD, USA, 21–24 March 2005; pp. 21–24.
328. Steiner, C.; Das, K.C.; Melear, N.; Lakly, D. Reducing nitrogen loss during poultry litter composting using biochar. *J. Environ. Qual.* **2010**, *39*, 1236–1242. [CrossRef]
329. Brennan, R.B.; Healy, M.G.; Fenton, O.; Lanigan, G.J. The effect of chemical amendments used for phosphorus abatement on greenhouse gas and ammonia emissions from dairy cattle slurry: Synergies and pollution swapping. *PLoS ONE* **2015**, *10*, e0111965. [CrossRef]
330. Maurer, D.L.; Koziel, J.A.; Kalus, K.; Andersen, D.S.; Opalinski, S. Pilot-scale testing of non-activated biochar for swine manure treatment and mitigation of ammonia, hydrogen sulfide, odorous Volatile Organic Compounds (VOCs), and Greenhouse Gas Emissions. *Sustainability* **2017**, *9*, 929. [CrossRef]
331. Agyarko-Mintah, E.; Cowie, C.; Singh, P.S.; Joseph, S.; van Zwieten, L.; Cowie, A.; Harden, S.; Smillie, R. Biochar increases nitrogen retention and lowers greenhouse gas emissions when added to composting poultry litter. *Waste Manag.* **2017**, *61*, 138–149. [CrossRef]
332. Janczak, D.; Malińska, K.; Czekała, W.; Cáceres, R.; Lewicki, A.; Dach, J. Biochar to reduce ammonia emissions in gaseous and liquid phase during composting of poultry manure with wheat straw. *Waste Manag.* **2017**, *66*, 36–45. [CrossRef]
333. Shaheen, S.M.; Mosa, A.; Natasha; Arockiam Jeyasundar, P.G.S.; Hassan, N.E.; Yang, X.; Antoniadis, V.; Li, R.; Wang, J.; Zhang, T.; et al. Pros and cons of biochar to soil potentially toxic element mobilization and phytoavailability: Environmental implications. *Earth Syst. Environ.* **2023**, *7*, 321–345. [CrossRef]
334. Liu, Y.; Lonappan, L.; Brar, S.K.; Yang, S. Impact of biochar amendment in agricultural soils on the sorption, desorption, and degradation of pesticides: A review. *Sci. Total Environ.* **2018**, *645*, 60–70. [CrossRef] [PubMed]
335. Liao, S.; Pan, B.; Li, H.; Zhang, D.; Xing, B. Detecting free radicals in biochars and determining their ability to inhibit the germination and growth of corn, wheat and rice seedlings. *Environ. Sci. Technol.* **2014**, *48*, 8581–8587. [CrossRef] [PubMed]
336. Cui, H.; Li, D.; Liu, X.; Fan, Y.; Zhang, X.; Zhang, S.; Zhou, J.; Fang, G.; Zhou, J. Dry-wet and freeze-thaw aging activate endogenous copper and cadmium in biochar. *J. Clean. Prod.* **2021**, *288*, 125605. [CrossRef]
337. Dutta, T.; Kwon, E.; Bhattacharya, S.S.; Jeon, B.H.; Deep, A.; Uchimiya, M.; Kim, K.H. Polycyclic aromatic hydrocarbons and volatile organic compounds in biochar and biochar-amended soil: A review. *Glob. Change Biol. Bioenergy* **2017**, *9*, 990–1004. [CrossRef]
338. Qiu, M.; Sun, K.; Jin, J.; Han, L.; Sun, H.; Zhao, Y.; Xia, X.; Wu, F.; Xing, B. Metal/metalloid elements and polycyclic aromatic hydrocarbon in various biochars: The effect of feedstock, temperature, minerals, and properties. *Environ. Pollut.* **2015**, *206*, 298–305. [CrossRef]
339. Shen, X.; Huang, D.Y.; Ren, X.F.; Zhu, H.H.; Wang, S.; Xu, C.; He, Y.B.; Luo, Z.C.; Zhu, Q.H. Phytoavailability of Cd and Pb in crop straw biochar-amended soil is related to the heavy metal content of both biochar and soil. *J. Environ. Manag.* **2016**, *168*, 245–251. [CrossRef]
340. Zhao, Y.; Zhao, L.; Mei, Y.; Li, F.; Cao, X. Release of nutrients and heavy metals from biochar-amended soil under environmentally relevant conditions. *Environ. Sci. Pollut. Res.* **2018**, *25*, 2517–2527. [CrossRef]
341. Kim, H.S.; Kim, K.R.; Yang, J.E.; Ok, Y.S.; Owens, G.; Nehls, T.; Wessolek, G.; Kim, K.H. Effect of biochar on reclaimed tidal land soil properties and maize (*Zea mays* L.) response. *Chemosphere* **2016**, *142*, 153–159. [CrossRef]
342. Joseph, S.; Kammann, C.I.; Shepherd, J.G.; Conte, P.; Schmidt, H.P.; Hagemann, N.; Rich, A.M.; Marjo, C.E.; Allen, J.; Munroe, P.; et al. Microstructural and associated chemical changes during the composting of a high temperature biochar: Mechanisms for nitrate, phosphate and other nutrient retention and release. *Sci. Total Environ.* **2018**, *618*, 1210–1223. [CrossRef]
343. Kuppusamy, S.; Thavamani, P.; Megharaj, M.; Venkateswarlu, K.; Naidu, R. Agronomic and remedial benefits and risks of applying biochar to soil: Current knowledge and future research directions. *Environ. Int.* **2016**, *87*, 1–12. [CrossRef] [PubMed]
344. Huang, H.; Yao, W.; Li, R.; Ali, A.; Du, J.; Guo, D.; Xiao, R.; Guo, Z.; Zhang, Z.; Awasthi, M.K. Effect of pyrolysis temperature on chemical form, behavior and environmental risk of Zn, Pb and Cd in biochar produced from phytoremediation residue. *Bioresour. Technol.* **2018**, *249*, 487–493. [CrossRef] [PubMed]
345. Palansooriya, K.N.; Wong, J.T.F.; Hashimoto, Y.; Huang, L.; Rinklebe, J.; Chang, S.X.; Bolan, N.; Wang, H.; Ok, Y.S. Response of microbial communities to biochar-amended soils: A critical review. *Biochar* **2019**, *1*, 3–22. [CrossRef]
346. Li, X.; Wang, T.; Chang, S.X.; Jiang, X.; Song, Y. Biochar increases soil microbial biomass but has variable effects on microbial diversity: A meta-analysis. *Sci. Total Environ.* **2020**, *749*, 141593. [CrossRef]
347. Jeffery, S.; Bezemer, T.M.; Cornelissen, G.; Kuyper, T.W.; Lehmann, J.; Mommer, L.; Sohi, S.P.; van de Voorde, T.F.J.; Wardle, D.A.; van Groenigen, J.W. The way forward in biochar research: Targeting trade-offs between the potential wins. *Glob. Change Biol. Bioenergy* **2015**, *7*, 1–13. [CrossRef]
348. Oleszczuk, P.; Jośko, I.; Kuśmierz, M. Biochar properties regarding to contaminants content and ecotoxicological assessment. *J. Hazard. Mater.* **2013**, *260*, 375–382. [CrossRef]
349. Lychuk, T.E.; Izaurralde, R.C.; Hill, R.L.; McGill, W.B.; Williams, J.R. Biochar as a global change adaptation: Predicting biochar impacts on crop productivity and soil quality for a tropical soil with the Environmental Policy Integrated Climate (EPIC) model. *Mitig. Adapt. Strateg. Glob. Chang.* **2015**, *20*, 1437–1458. [CrossRef]

350. Jatav, H.S.; Rajput, V.D.; Minkina, T.; Singh, S.K.; Chejara, S.; Gorovtsov, A.; Barakov, A.; Bauer, T.; Sushkova, S.; Mandzhieva, S.; et al. Sustainable approach and safe use of biochar and its possible consequences. *Sustainability* **2021**, *13*, 10362. [CrossRef]
351. Karer, J.; Wimmer, B.; Zehetner, F.; Kloss, S.; Soja, G. Biochar application to temperate soils: Effects on nutrient uptake and crop yield under field conditions. *Agric. Food Sci.* **2013**, *22*, 390–403. [CrossRef]
352. Igalavithana, A.D.; Le, S.E.; Lee, Y.H.; Tsang, D.C.W.; Rinklebe, J.; Kwon, E.E.; Ok, Y.S. Heavy metal immobilization and microbial community abundance by vegetable waste and pine cone biochar of agricultural soils. *Chemosphere* **2017**, *174*, 593–603. [CrossRef]
353. Vitkova, J.; Kondrlova, E.; Rodny, M.; Surda, P.; Horak, J. Analysis of soil water content and crop yield after biochar application in field conditions. *Plant Soil Environ.* **2017**, *63*, 569–573. [CrossRef]
354. Fernandes, J.D.; Chaves, L.H.; Mendes, J.S.; Chaves, I.B.; Tito, G.A. Alterations in soil salinity with the use of different biochar doses. *Rev. Ciências Agrárias* **2019**, *42*, 89–98. [CrossRef]
355. Li, Y.; Zhang, F.; Yang, M.; Zhang, J.; Xie, Y. Impacts of biochar application rates and particle sizes on runoff and soil loss in small cultivated loess plots under simulated rainfall. *Sci. Total Environ.* **2019**, *649*, 1403–1413. [CrossRef] [PubMed]
356. Salem, T.M.; Refaie, K.M.; Sherif, A.E.-H.E.-G.A.E.-L.; Eid, M.A.M. Biochar application in alkaline soil and its effect on soil and plant. *Acta Agric. Slov.* **2019**, *114*, 85–96. [CrossRef]

agriculture

MDPI

Review

Climate Change as an Existential Threat to Tropical Fruit Crop Production—A Review

Chinnu Raju [1], Sellaperumal Pazhanivelan [2,*], Irene Vethamoni Perianadar [3], Ragunath Kaliaperumal [2], N. K. Sathyamoorthy [1] and Vaithiyanathan Sendhilvel [1]

1 Agro Climate Research Centre, Tamil Nadu Agricultural University, Coimbatore 641003, Tamil Nadu, India; chinnuraju2015@gmail.com (C.R.); nks6@tnau.ac.in (N.K.S.); patsendhil@gmail.com (V.S.)
2 Centre for Water and Geospatial Studies, Tamil Nadu Agricultural University, Coimbatore 641003, Tamil Nadu, India; ragunathkp@tnau.ac.in
3 Horticultural College & Research Institute, Tamil Nadu Agricultural University, Coimbatore 641003, Tamil Nadu, India; irenevetha17@gmail.com
* Correspondence: pazhanivelans@tnau.ac.in

Abstract: Climate change is an emerging threat to global food and nutritional security. The tropical fruits such as mango, bananas, passionfruit, custard apples, and papaya are highly sensitive to weather changes especially; changes of monsoon onset and elevated temperature are influencing crop growth and production. There is a need for more specific studies concerning individual crops and regional variations. Long-term effects and interactions of weather parameters and increased concentration of greenhouse gases, especially carbon dioxide, with phenological stages of the plant, pests, and diseases remain understudied, while adaptation strategies require further exploration for comprehensive understanding and effective mitigation. Few researchers have addressed the issues on the effect of climate change on tropical fruits. This paper focuses on the impact of abiotic (temperature, rainfall, humidity, wind speed, evaporation, carbon dioxide concentration) and biotic (pest and pathogens dynamics) factors affecting the fruit crop ecosystem. These factors influence flowering, pollination, fruit set, fruit yield and quality. This review paper will help develop adaptive strategies, policy interventions and technological innovations aimed at mitigating the adverse effects of climate change on tropical fruit production and safeguarding global food and nutritional security.

Keywords: fruit quality; pest and pathogen dynamics; phenology; pollination; weather; yield

1. Introduction

Climate change represents one of the most pressing challenges of our time, with profound implications for multiple facets of human life, particularly agriculture. Tropical regions are more vulnerable to climate change than other parts of the world because of their relatively small temperature ranges, high reliance on ecosystem services and natural resources and the poor socioeconomic status of the people [1]. Recently, rainfall patterns in the tropics have become highly irregular with increased frequency and intensity leading to flash floods, while at the same time there have been absences of rainfall for prolonged periods causing droughts [2]. According to the Intergovernmental Panel on Climate Change (IPCC) sixth assessment report, the global surface temperature is expected to rise by 2.5 °C to 4 °C in 2100, with the most likely estimate being around 3 °C [3]. The frequency and intensity of heat waves have increased in recent decades, with severe impacts from multi-day heat waves that are accompanied by warm nighttime temperatures and high relative humidity [4,5]. Rise in temperature and heat waves cause increased thermally induced morbidity and mortality of species [3]. Carbon dioxide (CO_2) concentration in the atmosphere is expected to reach 550–800 ppm by 2100, which may accelerate global warming and the rise in temperature across the globe [6]. Fruit production, a significant activity in tropical areas, has not received much attention in the comprehensive assessment of climate change impacts and adaptation research [7].

Tropical fruit crops are grown in the region between the Tropic of Cancer (23° (27) N latitude) and the Tropic of Capricorn (23° (27) S latitude). Unlike subtropical crops, tropical crops are extremely sensitive to frost. The average mean temperature of the coldest month should be higher than 50 °F (10 °C) for the growth and development of tropical fruits [8]. Tropical crops include 2700 species of fruits belonging to families like *Anacardiaceae* (mango), *Musaceae* (banana), *Sapindaceae* (rambutan), *Passifloraceae* (passion fruit), *Bromeliaceae* (pineapple), and *Annonaceae* (custard apple, soursop, sugar apple) [9]. They are vital components of both local diets and international markets, providing essential nutrients and economic opportunities for millions of people in tropical regions. Nearly 99% of tropical fruit production is concentrated in Asia and Latin America and a smaller share in African countries. Tropical fruits are the third most important fruit group, contributing to 3% of world agricultural food exports with an average export unit value of more than USD 1000 per tonne. It is estimated that 11 million tonnes of tropical fruits were exported in 2023 and the export volume of tropical fruits shows the fastest average annual growth among other internationally traded food commodities. European countries are the major importers of tropical fruits, consuming 50% of the world's exports [10]. However, their cultivation is intricately linked to climatic conditions, making them highly susceptible to the effects of a changing climate. The production and quality can be significantly affected by any departure from the ideal conditions. The interplay of changed temperatures, different fruit growth stages, suitability of various species and cultivars, increased CO_2 levels, limited water availability, pollinators, pests, diseases and management practices has shown to have a significant influence and will heavily influence tropical fruit production [7].

In this review, we delve into the ecological impacts of climate change on tropical fruit crop ecosystems, exploring how shifts in temperature, precipitation, CO_2 concentration and weather patterns influence crop phenology, water availability, pollination dynamics, pest and disease outbreaks, crop yields, fruit quality and supply chain networks. Ultimately, this comprehensive review seeks to inform policymakers, agricultural practitioners, researchers and stakeholders about the urgent need for adaptation and mitigation strategies to safeguard tropical fruit crop ecosystems against the existential threat of climate change.

2. Methodology

Databases like Scopus, ScienceDirect, Springer, Centre for Agriculture and Bioscience International (CABI), and Google Scholar were used for collection of literature (Table 1). The PRISMA diagram (Figure 1) depicts the systematic review process followed while preparing this review article. BioRender (Science Suite Inc., Toronto, Canada) and Canva (Canva Pty Ltd., Sydney, Australia) softwares were used to prepare figures and illustrations in this review.

Table 1. Inclusion and exclusion criteria.

Sl. No.	Article Characteristics	Included	Excluded
1	Study area	Articles from agricultural and biological sciences	Articles from study areas other than agricultural and biological sciences
2	Study population	Tropical fruit crops	Crops other than tropical fruits
3	Document type	Research articles, review papers, and book chapters	Conference papers, preprints, note, letter, short survey
4	Language	English	Languages other than English
5	Year of publishing	2000 to 2024	Article older than 2000

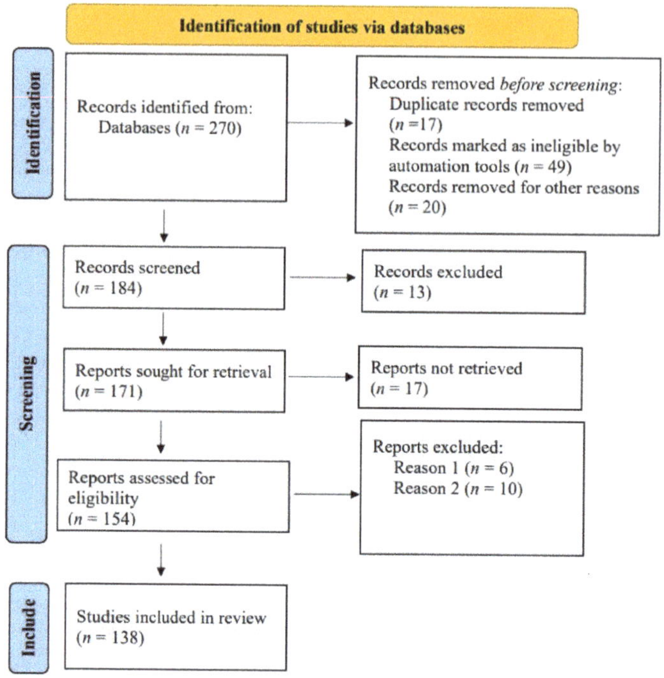

Figure 1. PRISMA flow diagram.

3. Abiotic Factors Affecting Tropical Fruit Ecosystem

Abiotic factors are the non-living components of an ecosystem that influence the living organisms within it. In a tropical fruit ecosystem, several abiotic factors such as temperature, rainfall, humidity, wind speed, evaporation and changes in atmospheric CO_2 level play a crucial role in determining the growth, geographical distribution and health of fruit-bearing plants.

3.1. Impact of Temperature Variations on Tropical Fruit Crops

Temperature plays an important role in tropical fruit crop growth, development, and yield [11]. Plants experience two types of temperature stress: high-temperature stress and low-temperature stress [12]. The cardinal temperatures of some tropical fruits are depicted in Table 2, which includes the minimum, maximum, and optimum temperature requirements of each species. Generally, tropical fruit crops thrive best in temperatures ranging from 24 °C to 30 °C, but their growth is adversely affected by temperatures below 10 °C or above 42 °C [7].

Table 2. Cardinal temperature requirements of tropical fruits.

Fruits	Minimum Threshold Temperature (°C)	Optimum Temperature Range (°C)	Maximum Threshold Temperature (°C)	References
Mango (*Mangifera indica*)	10	24–30	42	[7,13]
Banana (*Musa* sp.)	10	25–35	38	[14,15]
Guava (*Psidium guajava*)	10	23–28	51.2	[16,17]

Table 2. *Cont.*

Fruits	Minimum Threshold Temperature (°C)	Optimum Temperature Range (°C)	Maximum Threshold Temperature (°C)	References
Papaya (*Carica papaya*)	20	21–33	35	[18,19]
Custard apple (*Annona squamosa*)	11	21–30	38	[20,21]
Watermelon (*Citrullus lanatus*)	18	22–30	35	[22,23]
Jack fruit (*Artocarpus heterophyllus*)	5	22–35	35	[24,25]
Sapota (*Manikara zapota*)	10	16–38	43	[26,27]
Pineapple (*Ananas comosus*)	10	20–30	35	[28,29]
Aonla (*Emblica officinalis*)	4	20–34	43	[30,31]
Carambola (*Averrhoa carambola*)	15	21–32	30	[7,32]
Passion fruit (*Passiflora edulis*)	1–2	20–30	30	[33,34]
Mangosteen (*Garcinia mangostana*)	20	25–35	35	[7,35]
Rambutan (*Nephelium lappaceum*)	10	22–30	40	[36,37]

3.1.1. High Temperature Stress

High temperature stress is said to occur when the ambient temperature rises above a threshold limit for a sufficient period such that it causes irreversible damage to plant growth and development [38]. The response of plants to high temperature stress depends on the degree of rise in temperature, duration and crop characteristics. Plants undergo morphological, anatomical, phenological and physiological changes when exposed to high temperature stress [39]. The morphological changes are initially manifested as scorching and sunburn on all plant parts followed by senescence and abscission of leaves, stunted growth and damage and discolouration of fruit [40]. The anatomical changes include reduction in cell size, bigger xylem vessels in both the root and the shoot, as well as higher stomatal and trichomatous densities [41]. High temperature damages thylakoid membranes, decreases the number of photosynthetic pigments, affects the performance of photosystem II and alters the carbon metabolism of stroma and photochemical reactions of thylakoid lamellae, in turn, resulting in reduced photosynthetic rate in plants [42, 43]. High temperature affects all the stages of the plant right from germination, growth and development, reproduction, yield and quality of the fruits as depicted in Figure 2. Seed germination is inhibited due to abscisic acid (ABA) production in heat-stressed plants. The reproductive stage is significantly impacted by heat stress, leading to reduced flower bud formation, sterility of male and female organs, reduced pollen germination and pollen tube growth, alterations in stigma and style positions, disturbed fertilization and embryo abortion [44]. The physiological response of plants to high temperature stress involves overproduction of reactive oxygen species (ROS) such as singlet oxygen ($^1O_2$), superoxide (O_2^-), hydrogen peroxide (H_2O_2) and hydroxyl radicals (OH^-) resulting in oxidative stress in plants causing damage to cell membrane, protein denaturation and DNA fragmentation [45]. During certain plant stages, especially seed germination, embryogenesis, microsporogenesis and fruit maturity, heat shock proteins (HSPs) are expressed when a plant is subjected to high temperatures. This serves to preserve the stability of the cell membrane and shields the plant against protein denaturation [39].

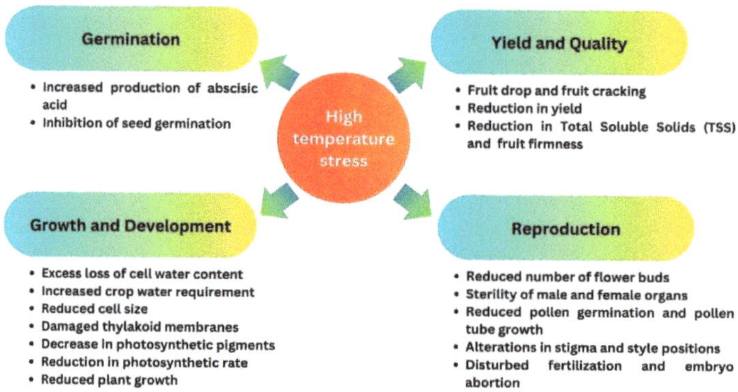

Figure 2. Effect of high temperature stress on different phenological stages of fruit crops.

In recent decades, an increase in the frequency and intensity of heat waves and droughts was reported in the tropical region. Nath et al., opined that the rise in temperature by 1 °C may alter the geographical distribution and potential area of cultivation of tropical fruits. For example, an increase in temperature by 1 °C reduced the area suitable for the *Dussaheri* variety of mango, while *Alphonso* became confined to the *Ratnagiri* area of Maharashtra state in India [7]. However, if the temperature increases by 10 °C, the crop water requirement rises by 50% [12]. High temperature delays flowering, resulting in a pseudo-setting called clustering in mango. Heat stress during the panicle development stage enhances rapid growth, resulting in a reduced number of effective days for pollination, pollen desiccation and poor pollinator activities, leading to poor fruit set. Warm night conditions result in the conversion of flower buds into vegetative buds [14,46]. In traditional guava-growing regions, higher summer temperatures result in suppressed blossom bud differentiation and a prolonged growth period [14]. If the temperature goes below 23 °C and above 27 °C during the flowering stage of guava, it reduces the fruit set [16]. A temperature above 38 °C causes growth cessation and leaf burn in bananas [47]. Leaf emergence in bananas depends on the health of the crop and climate. In summer, 4–5 leaves are produced, whereas in winter only half a leaf is produced.

Temperature also affects the orientation of leaves in bananas; low temperature results in upright leaves, whereas high temperature results in horizontal leaves [48]. When the temperature increases by 1 °C, there will be a shift in the area of cultivation of bananas by 200–300 km in mid-latitudes [14]. In papaya, at higher temperatures, there is a tendency for bisexual cultivars to produce male flowers with poorly developed female parts; in addition, the net photosynthetic rate also decreases above 30 °C [47]. In annona, poor pollination occurs under high temperatures (30 °C) and low humidity (30%) with effective pollination at 25 °C temperature and 80% relative humidity. Pollen germination and pollen tube growth are also reduced under high temperatures [47,49]. Rise in leaf temperature, reduced stomatal conductance and leaf vapor pressure deficit occur in annona due to high temperature and irradiance. Hence, photosynthesis and shoot growth are affected [50]. Figure 3 represents the effect of high temperatures on major tropical fruits such as mango, guava, banana, annona and papaya.

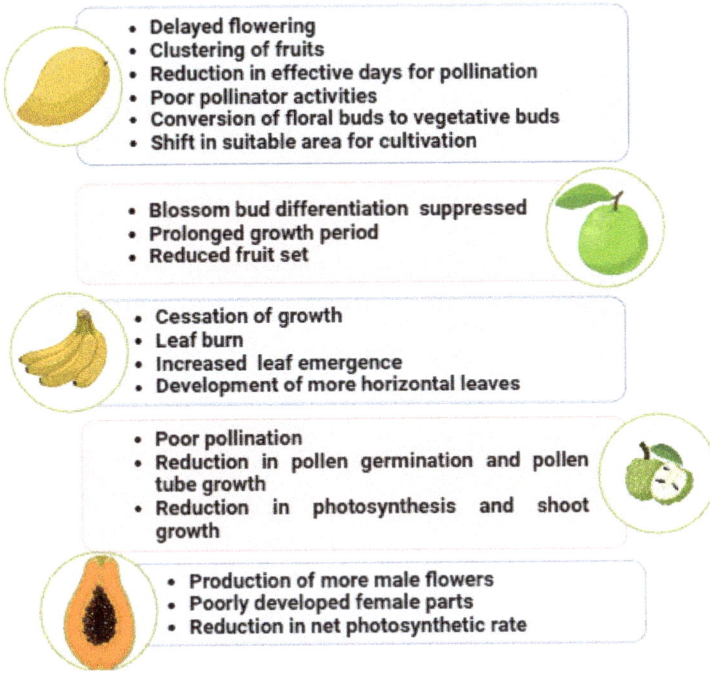

Figure 3. Effects of high-temperature stress on major tropical fruits.

3.1.2. Low Temperature Stress

Injuries due to low temperature or cold stress in plants are of two types: chilling injury and freezing injury. Chilling injury occurs at temperature range of 0–15 °C, it will not cause ice crystal formation in plant tissues. The chilling injury cause damage to cell membrane, reduces uptake function of roots leading to water balance disorders, lowers photosynthetic rate and retards plant growth. Freezing injury occurs at a temperature below 0 °C, with ice crystal formation in plant tissues. The ice crystal formation can be intercellular (when temperature falls gradually) or intracellular (when there is a sudden fall in temperature), latter one directly affects the cell metabolism and hence more dangerous. These injuries further lead to suffocation and heaving, ultimately affecting yield of the crop. Tropical and subtropical plants are more susceptible to low temperatures than temperate plants due to the lack of cold acclimatization [12,51]. The impact of low temperatures on tropical fruit crops, in general, is illustrated in Figure 4. Dinesh and Reddy reported that temperature less than 20 °C for mangosteen trees leads to slow growth and lower numbers of flowers [46]. When the temperature drops below 10 °C, bananas may develop the 'choke throat' disorder, which adversely affects the emergence of leaves and inflorescence, leading to poor fruit development [47]. In annona, anthesis time is delayed by one hour when the temperature decreases from 25 °C to 15 °C [49]. Low temperatures in the range of 4–11 °C combined with high humidity and cloudy weather delay panicle emergence in mango. Early flowering and poor fruit set in mangoes occur as a result of low night temperatures, whereas low day temperatures result in a greater proportion of male flowers and poor pollinator activities, which causes poor fruit set [14]. In addition, low temperatures can lead to flower deformation, loss of pollen viability, reduced number of perfect flowers, and ovule abortion, leading to the development of parthenocarpic fruit and malformation of fruits in mango [52,53].

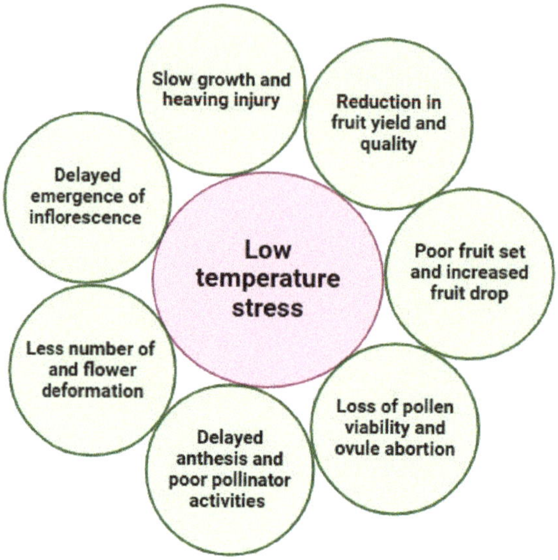

Figure 4. Effects of low-temperature stress on major tropical fruits.

3.2. Impact of Erratic Rainfall Pattern on Growth and Development of Tropical Fruit Crops

Rainfall variability due to climate change can lead to water shortages, decreased crop yields, increased pest pressure and even the total failure of crop stands [54]. Table 3 represents the optimum rainfall requirement and maximum altitude level up to which some of the important tropical crops could be grown without severe impacts on yield.

Table 3. Optimum rainfall and maximum altitude for the growth of important tropical fruits.

Fruits	Optimum Rainfall (mm)	Maximum Elevation from Mean Sea Level (m)	References
Mango	400–3600	1200	[52,55]
Banana	1200	2000	[56]
Guava	1000–2000	1500	[14,16]
Papaya	1200	2395	[57,58]
Annona	1500	1500	[59]
Passion fruit	1000–1500	2000	[60]
Pineapple	760–1000	1000	[28]
Aonla	630–800	1800	[31]
Carambola	1800	1200	[61]
Rambutan	2000–5000	700	[36]

Mango is basically drought tolerant and can withstand occasional flooding. The crop prefers even distribution of rainfall for flowering and fruit set. Heavy rain accompanied by high humidity and temperature variations induces more vegetative flushes, leading to alteration in flowering and delay in panicle emergence and fruit set [52]. In India, heavy rains during the flowering time of mango washed out pollen grains accompanied by poor pollinator activities, leading to poor pollination and fruit set [62]. Heavy rainfall causes the blackening of mango fruits [46]. Water stress during the first 4-6 weeks will affect fruit development and also cause fruit drop in mango. The summer rains during the pre-monsoon period will lead to early flowering in guava when grown as a winter crop [14].

In papaya, drought leads to rapid shedding of leaves and poor fruit set. Flooding causes chlorosis and abscission of leaves, followed by death of the leaves due to root rot [58].

Hail is a form of precipitation of small ice pieces with size ranging from 5 to 50 mm usually. Hailstorms lead to heavy defoliation, peeling of bark, flower and fruit drop, fruit cracking and in severe cases lodging or uprooting of trees may also occur. Among tropical fruit crops, banana and papaya are severely affected, nearly 50% damage to plants were reported for these fruit crops in India [63].

3.3. Effect of Relative Humidity, Wind Speed, and Evaporation on the Growth and Yield of Tropical Fruits

Meteorological parameters like relative humidity (RH), wind speed, and evaporation play crucial roles in the growth and yield of tropical fruits. Research indicates that different humidity levels have a substantial effect on the growth of plants, with increased humidity levels positively impacting both photosynthesis and water regulation [64]. But low temperature combined with high humidity favors the occurrence of plant diseases. Germination rates were low under high-temperature conditions (30 °C) and high humidity (90% RH), as well as under low-temperature conditions (15 °C) and low humidity (40% RH) in annonas [49]. During guava flowering, low relative humidity results in the abortion of flowers, decreases fruit set and raises fruit drop rates. Additionally, it diminishes fruit size and quality [16].

Wind leads to increased stomatal transpiration, speeds up the drying of leaf surfaces, making them vulnerable to pest and disease infestations [65]. The accelerated transpiration rate results in wilting and stunted growth. Mango yields decrease if strong winds occur during the flowering and early fruiting stages. Papaya trees are fragile and require shelter from strong winds, as strong winds can lead to uprooting. Even if uprooting does not happen, strong winds can still cause significant damage to the large leaves, resulting in the shedding of flowers and fruits. Banana plants are uprooted when the wind speed exceeds 80 km/h. In such areas, people prefer the *Dwarf Cavendish* types. Wind speeds ranging from 18 to 30 km/h cause the splitting of leaf lamina and this reduces bunch yield. High-speed hot winds during the summer months shred and desiccate the leaves, significantly reducing photosynthetic efficiency when the leaves are torn into 10 cm wide strips [66]. The wind, in combination with rain, assists in spreading primary inoculums, leading to the increased spread of plant diseases.

Papaya growth and yield are positively influenced by higher evaporation-replenishment rates, resulting in improved relative water content, transpiration rate, plant height, stem girth, fruit number and yield when using drip irrigation [67]. Furthermore, tropical fruits like bananas and avocados have high water needs because of their increased evapotranspiration rates. This highlights the significance of employing efficient irrigation methods to maximize water usage and improve crop yield [68].

3.4. Influence of Rising Carbon Dioxide (CO_2) Levels on Tropical Fruit Production

The primary cause of global warming is the rise in carbon dioxide levels and it is projected that by the year 2100, the concentration of CO_2 will reach 600–700 parts per million (ppm). Studies conducted in Columbia proved that elevated CO_2 has a positive impact on fruit trees like citrus, papaya, dragon fruit, grapevine and strawberry, causing increased photosynthesis, efficient use of water, growth and greater biomass. Under elevated levels of CO_2, C_3 plants exhibit an increase in growth rate by 30% [6]. Sugar accumulation in fruits is enhanced by higher levels of CO_2, leading to improved weight and size as well as biochemical aspects of the fruit [69].

The reaction of plant species to increased CO_2 levels might not only depend on their photosynthetic capacity and ability to distribute nutrients to different parts of the plant, but also on how effectively they can transport assimilates to the areas where they are needed. This suggests that certain species, varieties or rootstocks with narrow vessels (such as many dwarf fruit tree rootstocks) may struggle to move the excess carbohydrates produced under elevated CO_2 to the appropriate areas, making them susceptible to feedback inhibition of photosynthesis [70]. Researchers have identified positive [71–73] and negative effects [74–76] of elevated CO_2 levels on fruit crops as illustrated in Figure 5. The higher photosynthetic rates are primarily responsible for increased availability of carbohydrates in the apical meristem region leading to increased cell division. Higher CO_2 levels cause the cell wall to loosen up and become more extensible, which eventually results in bigger cells. The increased internodal length, high root and shoot growth, increased tree height and fruit biomass as mentioned in Figure 5. would all result from high cell division and larger cell size. It is anticipated that higher CO_2 levels will cause leaf area of tree species to increase by 14%, producing larger leaves [77,78]. The stomatal density and stomatal conductance are expected to decrease leading to reduced transpiration rates [79]. Reduced transpiration would affect the mass flow mechanism resulting in decreased nutrient absorption and transportation in plants [74,80].

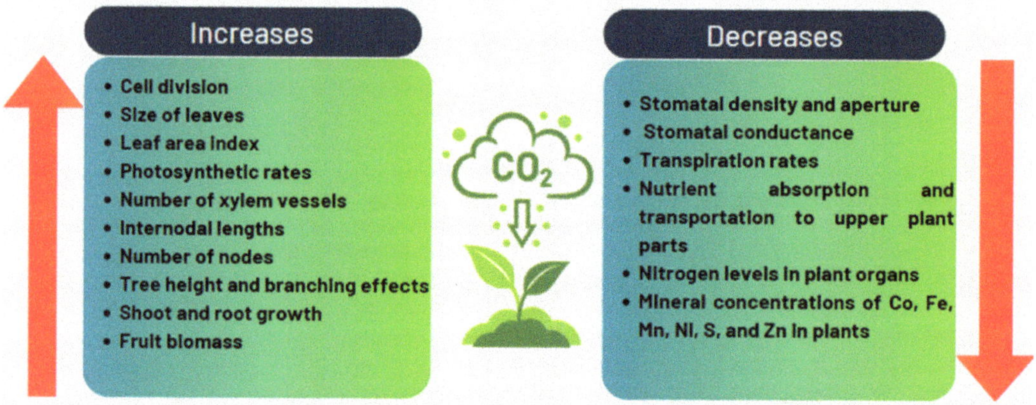

Figure 5. Responses of fruit crops to elevated CO_2 levels in the atmosphere.

3.5. Impact of Environmental Factors on the Behavior of Pollinators and Pollination of Tropical Fruit Crops

Crop production in the tropics will be significantly affected with a reduction in the population of pollinating insects due to climate change [11]. There will be a reduction in pollinator population by 61% with an increase in temperature. The potential threat to crop yield in 2050 due to declines in insect pollinator numbers, relative to overall production in a specific area, is most pronounced in the tropical areas of Sub-Saharan Africa, South America and Southeast Asia. Cocoa is considered to be at the greatest risk among crops, especially in Africa, followed by mango (especially in India) and watermelon (particularly in China) [81]. In mango temperatures, more than 32 °C caused a reduction in the activity of the honey bee pollinators (*Apis florea*) while dipteran pollinators were not affected (*Chrysomya megacephala*) [82]. Nitidulid beetles are the pollinators of annona flowers; in the absence of these pollinators, a 25% reduction in fruit set was observed. The natural occurrence of nitidulid beetles was favored by high soil temperatures (range: 20–30 °C), but was adversely affected by rainfall greater than 5 mm/day [83]. Climate change not only affects pollination activities but also pollen dehiscence, pollen tube growth and synchrony of male and female organs. In annonas, pollen is discharged in the afternoon between 3 and 6 pm, the temperature at that time should be greater than22 °C and relative humidity above 80%; if the temperature is less than 22 °C, pollen dehiscence will not occur [83].

3.6. Climate Change Impact on Biochemical Aspects and Quality of Tropical Fruit Crops

High quality standards are essential for securing a competitive price in the export market [84]. The changing climate leads to changes in the total soluble solids (TSSs), titrable acidity and content of ascorbic acid in the fruit, as well as fruit's color, flavor and firmness. Exposure of mango to high temperatures along with intense solar radiation results in scalds, which makes the fruits unmarketable [14]. In addition, exposure to high UV-B radiation reduces yield and quality of mangoes [85].

The total soluble solids (TSSs), fruit firmness, and percentage dry mass in guava all show a negative correlation with temperature [86]. The development of anthocyanin pigments, which give a red color to the pericarp of Guava, requires a night temperature of 8–10 °C or even lower. Better-quality guava fruits will come from cooler nights rather than warmer ones [14]. Rise in temperature above 26.7 °C reduces pulp firmness and changes the color of the skin in banana [87]. Guava fruit that has experienced heavy rainfall becomes watery, loses firmness, TSS, titrable acidity, ascorbic acid content and may also crack and drop [16,88]. In annona, low humidity (<60% RH) and low temperature (<13 °C) in the months prior to fruit maturity can exacerbate fruit-skin russeting and postpone fruit maturity [89].

4. Biotic Factors Affecting Tropical Fruit Ecosystem

4.1. Impact of Climate Change on Pest Population Dynamics of Tropical Crops

A changing climate directly affects pest reproduction, development, survival and dispersal and indirectly affects natural enemies, predators, competitors, vectors and mutualists [90]. Temperature is the most important parameter affecting insect physiology since their metabolism doubles when temperatures increase by 10 °C [91]. In tropical regions, however, a decrease in growth rate will occur because temperatures in this region have already reached the optimum level for pest development. If the temperature rises further, then growth and development will decrease [92]. Climate change results in a rise in temperature, uneven precipitation and an increase in CO_2 concentration, affecting insect population dynamics, geographical distribution, growth and development, abundance, etc., as illustrated in Figure 6 [93].

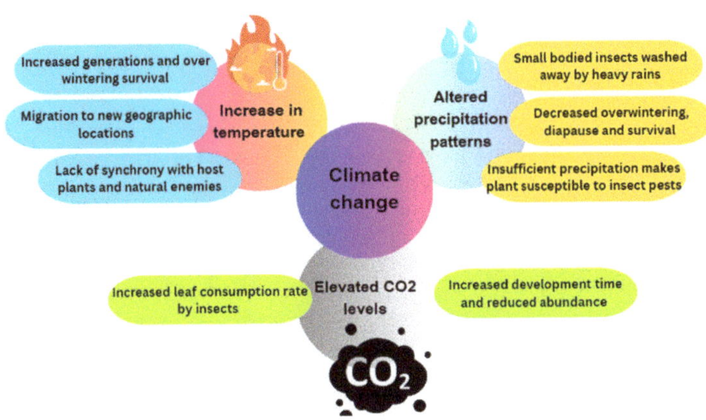

Figure 6. Responses of insects under climate change scenario.

In the tropics, rainfall frequency has decreased but the intensity has increased, affecting overwintering, diapause and insect survival. Heavy rainfall also affects the flying and reproduction of insects [94]. In addition, insect eggs, larvae and small-bodied insects like aphids, mites, jassids, whiteflies, etc., will be washed away due to heavy rain [95]. Reduced rainfall will lead to drought, making the plant susceptible to insect pests [96].

Mango hoppers cause damage to leaves, inflorescence and fruits by sucking sap from plant tissues. Leaf hoppers transmit viruses while they are sucking sap from the plants. The optimum temperature for leaf hopper growth and development is 15–35 °C and a decline in hopper population was observed when the temperature goes above 35 °C [97]. High humidity and unseasonal rains increase the attack of mango hoppers, leading to heavy flower and fruit shedding [98]. Cloudy weather with high relative humidity favors the multiplication of hoppers [99]. The population of hoppers decrease with an increase in wind speed [100].

In mango, thrips attack is seen on leaves, flowers and fruits. Female thrips lay eggs on any non-woody structures like stems, leaves, flowers and fruits. The optimum temperature for the multiplication of eggs by female thrips is 25–30 °C [101]. Fruit fly attack occurs during the fruit-ripening stage of mango. The rate of development of fruit flies increases when the temperature rises from 20 to 35 °C [102]. Mango fruit fly attacks increase with an increase in temperature up to 28 °C but after that, they decrease. Heavy rainfall along with fruit fly attacks will accelerate the yield loss [103].

The aphid (*Pentalonia nigronervosa*) transmits Banana Bunchy Top Virus (BBTV) more efficiently at 25 °C and 30 °C as compared to 20 °C [104]. It is reported that aphids are not able to recognize alarm pheromones properly under high-temperature conditions, which makes them more susceptible to predators and parasitoids [105]. High temperatures along with high humidity favor whitefly population buildup [106]. Based on a study conducted on 1100 insect species, the rise in temperature would lead to the extinction of 15–37% of species by 2050 [107].

4.2. *Diseases of Tropical Fruit Crops in the Context of Climate Change*

A plant disease is a dynamic process in which a host and a pathogen intimately related to the environment are mutually influenced, resulting in morphological and physiological changes [108]. In phytopathology, the environment encompasses temperature, humidity, precipitation and ultraviolet (UV) radiation, along with water, air and soil [109]. Among the environmental parameters, temperature and humidity are important parameters which favour the development of plant disease. Climate change can lead to the emergence of a pre-existing pathogen as a major disease, allow the introduction of new pathogens, alter the stage and rate of development of the prevailing disease, cause changes in the physiology of host–pathogen interactions and even affect host resistance, making the plant more susceptible to the pathogen [110].

4.2.1. Fungal Diseases

Fungal diseases like downy mildews, powdery mildews, anthracnose, phytophthora rot and leaf blight that affect horticultural crops are weather-driven [14]. Rise in temperatures, especially night temperatures and milder winters, have increased the winter survival of pathogens, shortened pathogen life cycles and increased sporulation and infectiousness of foliar fungi [111]. In general, during warmer and drier summers fungal infection tends to decrease, but some pathogens like *Podosphaera*, *Sphaerotheca*, *Uncinula* and *Ustilago*. thrive in such conditions. High temperature and drought stress will increase vascular wilt diseases caused by *Fusarium*, *Verticillium* and *Ganoderma* in annual and perennial crops [112]. Severe summers will cause early infection and longer epidemics in the case of air and seed-borne pathogens like *Pernospora* and *Puccinia* but the survival of soil-borne fungi will be impacted by reduced soil moisture [14].

The prevalence of weather parameters like temperature in the range of 25–30 °C, a greater number of rainy days, low sunshine hours and a relative humidity of over 90% for a prolonged period leads to the development of *Phytopthora* infections in crops. On the other hand, increased temperatures, less precipitation and extended drought periods can reduce the incidence of *Phytophthora* infections. Temperature in the range of 10–22 °C along with high relative humidity favors the occurrence of powdery mildew disease. However, the development of pathogen-causing powdery mildew disease will be inhibited by temperatures below 9–10 °C and above 34–35 °C [14].

A major fungal disease of tropical fruit crops is anthracnose, which affects all the phenological stages of the crop. This disease is aggravated under high relative humidity growth rate and tends to increase with humidity and temperature conditions [113–115]. Rain splashes disperse the spores of anthracnose disease [115]. In mango, anthracnose disease affects leaves, inflorescence and fruits. This disease will lead to twig dieback, defoliation and dropping of flowers and fruit in soursop. Heavy rain causes leaching of the fruit's protective layer, making it susceptible to blackening disorder and post-harvest diseases in mango [46]. Climate change affects the disease dynamics of two major diseases, banana black sigatoka (*Mycosphaerella fijiensis*) and Panama wilt or Fusarium wilt (*Fusarium oxysporum* f. sp. *cubense*). The changes in relative humidity may reduce the spatial distribution of black sigatoka because the pathogen survival is negatively affected below 70% relative humidity [116,117]. Panama wilt is positively correlated with increased temperature and water stress [118]. Pineapple fusariosis (*Fusarium subglutinans* f. sp. *ananas*) disease incidence decreases when temperature rises above 35 °C [119]. The temperature in the range of 23–27 °C and heavy rainfall increases the severity of black spot disease caused by *Asperisporium caricae* in papaya [120].

4.2.2. Bacterial Diseases

Bacterial diseases generally affect the yield and quality of tropical fruits. Common bacterial diseases in tropical fruits are bacterial wilt and bacterial canker of stone fruits. The bacterium *Pseudomonas syringe* pv. *morsprunorum* causes bacterial canker and gummosis of stone fruits. This bacterium attacks trunks, branches, shoots, spurs, blossoms, dormant buds, leaves and even fruits. The dispersal and infection of pathogens are facilitated by frequent rainfall, high humidity, cool temperatures and wind. Rainfall in the orchard during the growing season ensures the spread of disease. Typically, autumn and winter are when the trunk and branches are impacted [14].

Bacterial blight of pomegranate is caused by *Xanthomonas axonopodis* pv. *punicae*, which attacks all the above-ground parts, including fruits, causing the splitting of fruits. High temperature, low humidity and rainfall favour disease development. The increase in day temperature (38.6 °C) and afternoon RH of 30.4% along with cloudy weather and intermittent rainfall favour the initiation and further spread [121]. Wind-splashed rains increase the disease spread from healthy plants to unhealthy plants.

4.2.3. Viral Diseases

Pineapple mealybug wilt-associated virus (PMWaV-1, PMWaV-2 and PMWaV-3) causes mealybug wilt disease, prevalent in all pineapple-growing areas. The vectors of the disease are *Dysmicoccus brevipes* and *Dysmicoccus neobrevipes*, commonly known as pineapple mealy bugs. *Dysmicoccus brevipes* also acts as a vector of banana streak virus (BSV) in Uganda, which causes banana streak disease [122]. Studies suggest that the vector population increases with the temperature rise; hence, the incidence of pineapple mealybug wilt and banana streak disease may increase in the future [25,123]. The incidence of papaya ring spot virus transmitted by aphids is more severe when the temperature is in the range of 26–31 °C, increasing temperature favors the spread of the disease [27].

5. Prospects: Adaptation and Mitigation Strategies in Tropical Fruit Production

A combination of adaptation and mitigation strategies is necessary for better outcomes. Even if emissions were completely halted, the adverse effects of climate change would persist for years [124].

Orchard establishment: Due to changing climate scenarios, tropical regions may experience a northward shift in agroclimatic zones in the northern hemisphere and a southward shift in the southern hemisphere in the future [125]. Hence, identification of potential areas for orchard establishment could be done using climate suitability models. The suitable areas for establishment of mango and guava orchards in India were mapped using models developed using algorithms like rule set prediction (GARP), maximum entropy (MAXENT) and bio-climate (BIOCLIM) [126].

Variety development: Breeding techniques could be adopted to develop varieties that can endure both biotic and abiotic stresses like high temperature, drought, salinity, pest and diseases. For example, in regions with limited water resources, varieties like Arka sahan (Annona) and ruby (pomegranate), that can withstand droughts show great potential. Mango cultivars that are monoembryonic are best suited for areas with well-defined winter season, while coastal regions benefit from polyembryonic varieties [7]. In vegetatively propagated plants virus infection is a major issue. Certified virus-free plants could be used to avoid the spread of graft-transmissible diseases [127,128].

Management practices: Pruning for canopy management shapes young plants and increases leaf and fruit exposure, leading to enhanced fruit yield and quality. Innovative techniques like fruit bagging could be adopted to prevent fruit drop at early stages and avoid physiological disorders. Bagging of mango fruits at the marble stage using brown paper or newspaper improved fruit retention and reduced the occurrence of spongy tissue [129]. The use of prgmen bags to cover pomegranate fruits helped decrease fruit cracking and sunburn issues [130]. Nets could be used to provide shade and protect fruits from birds, insects and hail. The abiotic stress tolerance could be imparted by the exogeneous application of plant growth regulators like cytokinins, abscisic acid, salicylic acid, jasmonic acid and proline [131,132].

Soil and water conservation: Mulching and microsite modification using fillers improves the water-holding capacity of soils. Reuse of wastewater and solid waste in agriculture, along with implementing water-harvesting technologies, is advisable. Conservation techniques such as levelling, constructing bunds, bench terracing, etc., need to be implemented, as well as effective management of irrigation water through the use of drip systems in growing bananas, papayas, pomegranates, mangoes and sapotas [65]. Transformation of degraded lands and barren areas into tropical fruit orchards can serve as a significant carbon-sequestration tool. Over a year, perennial crops can store 320 to 1100 kg of soil carbon per hectare, while annual crops can store 0 to 450 kg; additionally, perennial crops tend to yield better than annual crops in hotter conditions [133].

Microclimate modifications: Modifications to the microclimate can be utilized to mitigate extreme weather events. Overcoming heat and cold stress is possible by utilizing methods such as overhead irrigation, sprinklers and shade nets. The use of mulch enhances soil microclimate, microbial activity and soil health. Plastic mulch resulted in higher yields for papaya (64.24%), mango (45.23%), banana (33.95%), ber (27.06%), guava (25.93%), pineapple (14.63%), and litchi (12.61%) compared to no mulch [134]. Antitranspirants can be used to reduce water loss through transpiration and temperature on leaf and fruit surfaces. The use of antitranspirant chitosan at a concentration of 2% resulted in significantly higher average finger weight, average hand weight and bunch weight in bananas compared to the other treatments [135]. Using terra alba in treatment improves the quality of pomegranate fruits by lowering fruit and leaf temperature compared to the control, as reported by [136]. Kaolin is also key in preventing sunburn in pomegranate fruits [137]. Windbreaks or shelter belts alter the microclimate and soil of orchards, offering shelter for pollinating insects and guarding against wind erosion and natural disasters. Fruit plant mortality due to frost is lower in orchards with wind breaks (2.97 to 30.81%) compared to those without (up to 91.43%) [138].

Pest and disease control: Integrated pest and disease management along with choice of tolerant cultivars will help to ensure good yield and avoid economic losses.

6. Conclusions

Tropical fruit ecosystems are threatened by rising temperatures, shifting precipitation and unprecedented extreme weather events. The changing climate plays a crucial role right from seed germination, growth and development, flowering, fruiting, occurrence of pests and diseases, production, productivity and nutritional quality in tropical fruit crops. Adoption of proper adaptation and mitigation measures is the only possible way to overcome these challenges. The existing strategies include proper site selection for orchard establishment, development of tolerant varieties using breeding techniques, proper tree/canopy management, adoption of microclimate modifications and pest and disease control measures. Though the existing technologies are capable of minimizing short-term negative effects, thorough research is required in this field to overcome long-term impacts of climate change on tropical fruit production. Hence, in future, research and development should focus in developing advanced technologies that are economically viable and user-friendly for stakeholders.

Author Contributions: Conceptualization and the overall framework of research, S.P. and C.R.; Literature review, Data collection, and Writing—original draft C.R.; Supervision; S.P.; Data visualization; V.S. and Writing—review and editing; S.P., I.V.P., N.K.S., R.K. and V.S. All authors have read and agreed to the published version of the manuscript.

Funding: This research received external funding from Hindustan Coca-cola Beverages Pvt. Ltd. Creating spatial database on mango growing areas with varietal information using remote sensing, drone and mobile mapping techniques project (Coca-Cola-F-37 APT).

Data Availability Statement: All relevant data are included in the manuscript.

Acknowledgments: This paper was supported by Tamil Nadu Agricultural University, Coimbatore, Tamil Nadu, India.

Conflicts of Interest: The authors declare no conflicts of interest related to this article.

References

1. Martin, R. Climate change: Why the tropical poor will suffer most. *Technol. Rev.* **2015**.
2. Haque, S.; Akbar, D.; Kinnear, S. Identifying impacts & adaptation strategies for tropical fruit farms affected by extreme weather events in sub-tropical Australia: Stakeholders' insights. *Heliyon* **2024**, *10*, e26097. [CrossRef] [PubMed]
3. Lee, H.; Calvin, K.; Dasgupta, D.; Krinner, G.; Mukherji, A.; Thorne, P.; Trisos, C.; Romero, J.; Aldunce, P.; Barret, K.; et al. *IPCC, 2023: Climate Change 2023: Synthesis Report, Summary for Policymakers. Contribution of Working Groups I, II and III to the Sixth Assessment Report of the Intergovernmental Panel on Climate Change*; Core Writing Team, Lee, H., Romero, J., Eds.; IPCC: Geneva, Switzerland, 2023.
4. Fischer, E.M.; Schär, C. Consistent geographical patterns of changes in high-impact European heatwaves. *Nat. Geosci.* **2010**, *3*, 398–403. [CrossRef]
5. Liu, X.; He, B.; Guo, L.; Huang, L.; Chen, D. Similarities and Differences in the Mechanisms Causing the European Summer Heatwaves in 2003, 2010, and 2018. *Earth's Future* **2020**, *8*, e2019EF001386. [CrossRef]
6. Fischer, G.; Melgarejo, L.M.; Balaguera-López, H.E. Review on the impact of elevated CO_2 concentrations on fruit species in the face of climate change. *Cienc. Tecnol. Agropecu.* **2022**, *23*, e2475.
7. Nath, V.; Kumar, G.; Pandey, S.; Pandey, S. Impact of climate change on tropical fruit production systems and its mitigation strategies. In *Climate Change and Agriculture in India: Impact and Adaptation*; Springer: Berlin/Heidelberg, Germany, 2019; pp. 129–146.
8. Watson, B.J.; Moncur, M. *Criteria for Determining Survival: Commercial and Best Minimum July Temperatures for Various Tropical Fruits in Australia (S. Hemisphere)*; West Tropical Regional Publication: Cape York, QLD, Australia, 1985.
9. Mansyah, E.; Budiyanti, T.; Hadiati, S.; Riska Indriyani, N.L. Biodiversity of Fruit Crops and Utilization in Food and Nutritional Security. In *Sustainable Utilization and Conservation of Plant Genetic Diversity*; Springer Nature: Singapore, 2024; pp. 127–170.
10. Mohamed, Z.; AbdLatif, I.; Abdullah, A.M. Economic importance of tropical and subtropical fruits. In *Postharvest Biology and Technology of Tropical and Subtropical Fruits*; Elsevier: Amsterdam, The Netherlands, 2011; pp. 1–20.
11. Haokip, S.W.; Shankar, K.; Lalrinngheta, J. Climate change and its impact on fruit crops. *J. Pharmacogn. Phytochem.* **2020**, *9*, 435–438.
12. Tewari, G.S.; Pareek, N.; Pandey, S.; Nath, V. Impact of temperature stress on growth and development of fruit crops and its mitigation strategies. *Int. J. Innov. Hortic.* **2018**, *7*, 1–9.
13. National Horticultural Board (NHB). Mango. 2024. Available online: https://www.nhb.gov.in/report_files/mango/MANGO.htm (accessed on 6 February 2024).
14. Dhillon, W.S.; Aulakh, P.S. *Impact of Climate Change on Fruit Crops*; Narendra Publishing House: Delhi, India, 2011.
15. Joshi, R.U.; Singh, A.K.; Singh, V.P.; Rai, R.; Joshi, P. A review on adaptation of banana (*Musa* spp.) to cold in subtropics. *Plant Breed.* **2023**, *142*, 269–283. [CrossRef]
16. Fischer, G.; Melgarejo, L.M. Ecophysiological aspects of guava (*Psidium guajava* L.). A review. *Rev. Colomb. Cienc. Hortíc.* **2021**, *15*, e12355. [CrossRef]
17. Gill. *Encyclopedia of Food and Health*; Academic Press: Cambridge, MA, USA, 2016; pp. 270–277. [CrossRef]

18. Salinas, I.; Hueso, J.J.; Cuevas, J. Active Control of Greenhouse Climate Enhances Papaya Growth and Yield at an Affordable Cost. *Agronomy* **2021**, *11*, 378. [CrossRef]
19. Lange, A.H. The Effect of Temperature and Photoperiod on the Growth of Carica Papaya. *Ecology* **1961**, *42*, 481–486. [CrossRef]
20. Kome, G.K.; Enang, R.K.; Silatsa, F.B.; Yerima, B.P.K.; Van Ranst, E. Baseline edaphic requirements of soursop (*Annona muricata* L.). *Trop. Plants* **2024**, *3*, 1–9. [CrossRef]
21. Alves Gomes, A.; Marangon Debastiani, M.; Pizzatto, M.; Carvalho Da Silva, A.; Pacheco De Souza, A. Thermal requirements and estimates of number of leaves of *Annona squamosa* L. grown under different shading conditions. *Comun. Sci.* **2023**, *14*, e3973.
22. Jang, Y.; Moon, J.-H.; Kim, S.-G.; Kim, T.; Lee, O.-J.; Lee, H.-J.; Wi, S.-H. Effect of Low-Temperature Tolerant Rootstocks on the Growth and Fruit Quality of Watermelon in Semi-Forcing and Retarding Culture. *Agronomy* **2022**, *13*, 67. [CrossRef]
23. Melo, T.K.D.; Espínola, J.; Medeiros, J.F.D.; Figueiredo, V.B.; Silva, J.S.D.; Sá, F.V.D.S. Impacts of climate change scenarios in the Brazilian semiarid region on watermelon cultivars. *Rev. Caatinga* **2020**, *33*, 794–802. [CrossRef]
24. Devi, S.P.; Talaulikar, S.; Gupta, M.J.; Thangam, M.; Singh, N.P. *A Guide on Jack Fruit: Cultivation and Value Addition*; Indian Council of Agricultural Research: Hyderabad, India, 2014.
25. Sether, D.; Melzer, M.; Busto, J.; Zee, F.; Hu, J. Diversity and mealybug transmissibility of ampeloviruses in pineapple. *Plant Dis.* **2005**, *89*, 450–456. [CrossRef]
26. Baskar, M.; Hemalatha, G.; Muneeshwari, P. Traditional and medicinal importance of sapota–Review. *Int. J. Curr. Microbiol. Appl. Sci.* **2020**, *9*, 1711–1717. [CrossRef]
27. Mangrauthia, S.K.; Singh Shakya, V.P.; Jain, R.; Praveen, S. Ambient temperature perception in papaya for papaya ringspot virus interaction. *Virus Genes* **2009**, *38*, 429–434. [CrossRef]
28. Hossain, M. World pineapple production: An overview. *Afr. J. Food Agric. Nutr. Dev.* **2016**, *16*, 11443–11456. [CrossRef]
29. Williams, P.A.; Crespo, O.; Atkinson, C.J.; Essegbey, G.O. Impact of climate variability on pineapple production in Ghana. *Agric. Food Secur.* **2017**, *6*, 1–14. [CrossRef]
30. Sarkar, T.; Roy, A.; Choudhary, S.M.; Sarkar, S.K. Impact of climate change and adaptation strategies for fruit crops. In *India: Climate Change Impacts, Mitigation and Adaptation in Developing Countries*; Springer International Publishing: Cham, Switzerland, 2021; pp. 79–98.
31. Kishore, K. Phenological growth stages of Indian gooseberry (*Phyllanthus emblica* L.) according to the extended BBCH scale. *Sci. Hortic.* **2017**, *225*, 607–614. [CrossRef]
32. Danjuma, S.; Thaochan, N.; Permkam, S.; Satasook, C. Effect of temperature on the development and survival of immature stages of the carambola fruit fly, *Bactrocera carambolae*, and the Asian papaya fruit fly, *Bactrocera papayae*, reared on guava diet. *J. Insect Sci.* **2014**, *14*, 126. [CrossRef] [PubMed]
33. Utsunomiya, N. Effect of temperature on shoot growth, flowering and fruit growth of purple passionfruit (*Passiflora edulis Sims* var. *edulis*). *Sci. Hortic.* **1992**, *52*, 63–68. [CrossRef]
34. de Souto, A.G.L.; da Costa, J.C.F.; Campos, N.L.F.; de Azevedo, J.L.F.; dos Santos, C.E.M. Effect of temperature on passion fruit emergence and seedling vigor. *J. Seed Sci.* **2017**, *39*, 50–57. [CrossRef]
35. Jaroensutasinee, K.; Jaroensutasinee, M.; Boonsanong, P. Climatic factor differences and mangosteen fruit quality between on-and off-season productions. *Emerg. Sci. J.* **2023**, *7*, 578–588. [CrossRef]
36. Tripathi, P.; Karunakaran, G. Rambutan Cultivation. In *Fruit Production in India*; Narendra Publishing House: Delhi, India, 2013.
37. De Andrade, R.A.; Bagatim, A.G.; Nacata, G. Rambutan seed germination: Temperature and storage. *Comun. Sci.* **2017**, *8*, 383–388. [CrossRef]
38. Wahid, A. Physiological implications of metabolite biosynthesis for net assimilation and heat-stress tolerance of sugarcane (Saccharum officinarum) sprouts. *J. Plant Res.* **2007**, *120*, 219–228. [CrossRef]
39. Hasanuzzaman, M.; Nahar, K.; Alam, M.M.; Roychowdhury, R.; Fujita, M. Physiological, biochemical, and molecular mechanisms of heat stress tolerance in plants. *Int. J. Mol. Sci.* **2013**, *14*, 9643–9684. [CrossRef]
40. Rodríguez, M.; Canales, E.; Borrás-Hidalgo, O. Molecular aspects of abiotic stress in plants. *Biotechnol. Appl.* **2005**, *22*, 1–10.
41. Wahid, A.; Gelani, S.; Ashraf, M.; Foolad, M.R. Heat tolerance in plants: An overview. *Environ. Exp. Bot.* **2007**, *61*, 199–223. [CrossRef]
42. Marchand, F.L.; Mertens, S.; Kockelbergh, F.; Beyens, L.; Nijs, I. Performance of High Arctic tundra plants improved during but deteriorated after exposure to a simulated extreme temperature event. *Glob. Chang. Biol.* **2005**, *11*, 2078–2089. [CrossRef] [PubMed]
43. Wang, J.Z.; Cui, L.J.; Wang, Y.; Li, J.L. Growth, lipid peroxidation and photosynthesis in two tall fescue cultivars differing in heat tolerance. *Biol. Plant.* **2009**, *53*, 237–242. [CrossRef]
44. Cao, Y.; Duan, H.; Yang, L.; Wang, Z.; Zhou, S.; Yang, J. Effect of heat-stress during meiosis on grain yield of rice cultivars differing in heat-tolerance and its physiological mechanism. *Acta Agron. Sin.* **2008**, *34*, 2134–2142. [CrossRef]
45. Awasthi, R.; Bhandari, K.; Nayyar, H. Temperature stress and redox homeostasis in agricultural crops. *Front. Environ. Sci.* **2015**, *3*, 11. [CrossRef]
46. Dinesh, M.; Reddy, B. Physiological basis of growth and fruit yield characteristics of tropical and sub-tropical fruits to temperature. In *Tropical Fruit Tree Species and Climate Change*; Bioversity International: New Delhi, India, 2012; pp. 45–65.
47. Paull, R.E.; Duarte, O. *Tropical Fruits*; CABI: Wallingford, UK, 2010.
48. Ravi, I.; Mustaffa, M.M. Impact, adaptation and mitigation strategies for climate resilient banana production. In *Climate-Resilient Horticulture: Adaptation and Mitigation Strategies*; Springer: Berlin/Heidelberg, Germany, 2013; pp. 45–52.

49. Matsuda, H.; Higuchi, H. Effects of temperature and humidity conditions on anthesis and pollen germinability of cherimoya (*Annona cherimola* Mill.). *Trop. Agric. Dev.* **2015**, *59*, 57–62.
50. Higuchi, H.; Utsunomiya, N.; Sakuratani, T. Effects of temperature on growth, dry matter production and CO_2 assimilation in cherimoya (Annona cherimola Mill.) and sugar apple (*Annona squamosa* L.) seedlings. *Sci. Hortic.* **1998**, *73*, 89–97. [CrossRef]
51. Wang, Y.; Wang, J.; Sarwar, R.; Zhang, W.; Geng, R.; Zhu, K.M.; Tan, X.L. Research progress on the physiological response and molecular mechanism of cold response in plants. *Front. Plant Sci.* **2024**, *15*, 1334913. [CrossRef]
52. Chadha, K. Global climate change and Indian horticulture. *Author Copy* **2015**, *1*, 26.
53. Datta, M.; Singh, N.P.; Daschaudhuri, D. *Climate Change & Food Security*; New India Publishing: Delhi, India, 2008.
54. Makhmale, S.; Bhutada, P.; Yadav, L.; Yadav, B. Impact of climate change on phenology of mango-the case study. *Ecol. Environ. Conserv.* **2016**, *22*, S127–S132.
55. Bally, I.S. Mangifera indica (mango). *Species Profiles Pac. Isl. Agrofor.* **2006**, *6*, 1–25.
56. Ranjitkar, S.; Sujakhu, N.; Budhamagar, K.; Rimal, S.; Xu, J.; Merz, J.; Zomer, R.J. *Projected Climate Change Impacts on Climatic Suitability and Geographical Distribution of Banana and Coffee Plantations in Nepal*; World Agroforestry Center (ICRAF): Nairobi, Kenya, 2015.
57. Hernández-Salinas, G.; Luna-Cavazos, M.; Soto-Estrada, A.; García-Pérez, E.; Pérez-Vázquez, A.; Córdova-Téllez, L. Distribution and eco-geographic characterization of *Carica papaya* L. native to Mexico. *Genet. Resour. Crop. Evol.* **2022**, *69*, 99–116. [CrossRef]
58. Salinas, I.; Hueso, J.J.; Cuevas, J. Fruit growth model, thermal requirements and fruit size determinants in papaya cultivars grown under subtropical conditions. *Sci. Hortic.* **2019**, *246*, 1022–1027. [CrossRef]
59. de Pinto, A.Q.; Cordeiro, M.; De Andrade, S.; Ferreira, F.; de Filgueiras, H.C.; Alves, R.; Kinpara, S.I. *Annona Species*; DFID: London, UK, 2005.
60. Deshmukh, N.; Patel, R.; Okram, S.; Rymbai, H.; Roy, S.; Jha, A. Passion fruit (*Passiflora* spp.). *Magnes. Mglitre* **2017**, *100*, 200.
61. Manda, H.; Vyas, K.; Pandya, A.; Singhal, G. A complete review on: Averrhoa carambola. *World J. Pharm. Pharm. Sci.* **2012**, *1*, 17–33.
62. Rajan, S. Phenological responses to temperature and rainfall: A case study of mango. In *Tropical Fruit Tree Species and Climate Change*; Bioversity International: New Delhi, India, 2012; pp. 71–96.
63. Jaybhaye, P. Mitigation and Adaptation Strategies of Plants against Hailstorm under Changing Climate. In *Plant Defense Mechanisms*; BoD—Books on Demand: Norderstedt, Germany, 2019; pp. 123–142.
64. Lorenz, D.J.; DeWeaver, E.T.; Vimont, D.J. Evaporation change and global warming: The role of net radiation and relative humidity. *J. Geophys. Res. Atmos.* **2010**, *115*, D20118. [CrossRef]
65. Gao, Y.; Ren, C.; Liu, Y.; Zhu, J.; Li, B.; Mu, W.; Liu, F. Pepper-maize intercropping affects the occurrence of anthracnose in hot pepper. *Crop Prot.* **2021**, *148*, 105750. [CrossRef]
66. Balraj, G.; Preethi, P.G.M.; Selvaraj, R. Banana. In *Tropical Fruit Crop: Theory to Practical*; Jaya Publishing House: Delhi, India, 2021; pp. 43–114.
67. Srinivas, K. Plant water relations, yield, and water use of papaya (*Carica papaya* L.) at different evaporation-replenishment rates under drip irrigation. *Trop. Agric.* **1996**, *73*, 264–269.
68. Arslan, F.; Kartal, S. Water management effect on tropical fruits: Case study of Alanya, Turkey. In Proceedings of the 22nd International Scientific Conference "Engineering for Rural Development": Proceedings, Jelgava, Latvia, 24–26 May 2023; pp. 533–538.
69. Ma, Y.; Xie, Y.; Ha, R.; Cao, B.; Song, L. Effects of elevated CO_2 on photosynthetic accumulation, sucrose metabolism-related enzymes, and genes identification in goji berry (*Lycium barbarum* L.). *Front. Plant Sci.* **2021**, *12*, 643555. [CrossRef]
70. Pritchard, S.G.; Rogers, H.H.; Prior, S.A.; Peterson, C.M. Elevated CO_2 and plant structure: A review. *Glob. Chang. Biol.* **1999**, *5*, 807–837. [CrossRef]
71. Lauriks, F.; Salomón, R.L.; Steppe, K. Temporal variability in tree responses to elevated atmospheric CO_2. *Plant Cell Environ.* **2021**, *44*, 1292–1310. [CrossRef]
72. Leibar-Porcel, E.; Dodd, I.C. Role of Plant Hormones in Plant Response to Elevated CO_2 Concentrations: Above-and Below-ground Interactions. In *Plant Hormones and Climate Change*; Springer: Berlin/Heidelberg, Germany, 2023; pp. 55–74.
73. Yong, J.W.; Wong, S.C.; Letham, D.S.; Hocart, C.H.; Farquhar, G.D. Effects of elevated [CO_2] and nitrogen nutrition on cytokinins in the xylem sap and leaves of cotton. *Plant Physiol.* **2000**, *124*, 767–780. [CrossRef] [PubMed]
74. Cruz, J.L.; Alves, A.A.; LeCain, D.R.; Ellis, D.D.; Morgan, J.A. Interactive effects between nitrogen fertilization and elevated CO_2 on growth and gas exchange of papaya seedlings. *Sci. Hortic.* **2016**, *202*, 32–40. [CrossRef]
75. Elbasiouny, H.; El-Ramady, H.; Elbehiry, F.; Rajput, V.D.; Minkina, T.; Mandzhieva, S. Plant nutrition under climate change and soil carbon sequestration. *Sustainability* **2022**, *14*, 914. [CrossRef]
76. Kiba, T.; Takebayashi, Y.; Kojima, M.; Sakakibara, H. Sugar-induced de novo cytokinin biosynthesis contributes to Arabidopsis growth under elevated CO_2. *Sci. Rep.* **2019**, *9*, 7765. [CrossRef]
77. Roussos, P.A. Climate Change Challenges in Temperate and Sub-Tropical Fruit Tree Cultivation. *Encyclopedia* **2024**, *4*, 558–582. [CrossRef]
78. Ramírez, F.; Kallarackal, J. Response of trees to CO_2 Increase. In *Responses of Fruit Trees to Global Climate Change*; Springer International Publishing: Cham, Switzerland, 2015; pp. 3–7.

79. Hao, L.; Chang, Z.; Lu, Y.; Tian, Y.; Zhou, H.; Wang, Y.; Liu, L.; Wang, P.; Zheng, Y.; Wu, J. Drought dampens the positive acclimation responses of leaf photosynthesis to elevated CO2 by altering stomatal traits, leaf anatomy, and Rubisco gene expression in Pyrus. *Environ. Exp. Bot.* **2023**, *211*, 105375. [CrossRef]
80. Mcgrath, J.M.; Lobell, D.B. Reduction of transpiration and altered nutrient allocation contribute to nutrient decline of crops grown in elevated CO_2 concentrations. *Plant Cell Environ.* **2013**, *36*, 697–705. [CrossRef]
81. Millard, J.; Outhwaite, C.L.; Ceaușu, S.; Carvalheiro, L.G.; Silva, F.D.d.S.e.; Dicks, L.V.; Ollerton, J.; Newbold, T. Key tropical crops at risk from pollinator loss due to climate change and land use. *Sci. Adv.* **2023**, *9*, eadh0756. [CrossRef]
82. Reddy, P.R.; Kavitha, S.J. Thermal sensitivity of major pollinators of mango: Dipterans score high in climate resilience. *Pest Manag. Hortic. Ecosyst.* **2023**, *29*, 161–165.
83. George, A.; Nissen, R.; Ironside, D.; Anderson, P. Effects of nitidulid beetles on pollination and fruit set of *Annona* spp. hybrids. *Sci. Hortic.* **1989**, *39*, 289–299. [CrossRef]
84. Jones, G.V.; Davis, R.E. Climate influences on grapevine phenology, grape composition, and wine production and quality for Bordeaux, France. *Am. J. Enol. Vitic.* **2000**, *51*, 249–261. [CrossRef]
85. Wang, H.; Guo, Y.; Zhu, J.; Yue, K.; Zhou, K. Characteristics of mango leaf photosynthetic inhibition by enhanced UV-B radiation. *Horticulturae* **2021**, *7*, 557. [CrossRef]
86. Rajan, S. Implications of climate change in mango. In *Impact Assessment of Climate Change for Research Priority Planning in Horticultural Crops*; Central Potato Research Institute: Shimla, India, 2008; pp. 36–42.
87. Peacock, B. Banana ripening-effect of temperature of fruit quality. *Qld. J. Agric. Anim. Sci.* **1980**, *37*, 39–45.
88. Fischer, G.; Parra-Coronado, A. Influence of some environmental factors on the feijoa (Acca sellowiana [Berg] Burret): A review. *Agron. Colomb.* **2020**, *38*, 388–397. [CrossRef]
89. Jiménez-Zurita, J.O.; Alia-Tejacal, I.; Balois-Morales, R.; Villarreal-Fuentes, J.M.; Núñez-Colín, C.A.; Berumen-Varela, G. Phenological growth stages of soursop trees (Annona muricata L.) based on the extended BBCH-scale. *Rev. Chapingo Ser. Hortic.* **2023**, *29*, 5–18. [CrossRef]
90. Prakash, A.; Rao, J.; Mukherjee, A.K.; Berliner, J.; Pokhare, S.S.; Adak, T.; Munda, S.; Shashank, P.R. *Climate Change: Impact on Crop Pests*; Applied Zoologists Research Association (AZRA), Central Rice Research: Cuttack, India, 2014.
91. Dukes, J.S.; Pontius, J.; Orwig, D.; Garnas, J.R.; Rodgers, V.L.; Brazee, N.; Cooke, K.; Theoharides, K.A.; Stange, E.E.; Harrington, R.; et al. Responses of insect pests, pathogens, and invasive plant species to climate change in the forests of northeastern North America: What can we predict? *Can. J. For. Res.* **2009**, *39*, 231–248. [CrossRef]
92. Deutsch, C.A.; Tewksbury, J.J.; Tigchelaar, M.; Battisti, D.S.; Merrill, S.C.; Huey, R.B.; Naylor, R.L. Increase in crop losses to insect pests in a warming climate. *Science* **2018**, *361*, 916–919. [CrossRef]
93. Skendžić, S.; Zovko, M.; Živković, I.P.; Lešić, V.; Lemić, D. The impact of climate change on agricultural insect pests. *Insects* **2021**, *12*, 440. [CrossRef]
94. Cilas, C.; Goebel, F.R.; Babin, R.; Avelino, J. Tropical crop pests and diseases in a climate change setting—A few examples. In *Climate Change and Agriculture Worldwide*; SpringerNature: Berlin/Heidelberg, Germany, 2016; pp. 73–82.
95. Pathak, H.; Aggarwal, P.K.; Singh, S. *Climate Change Impact, Adaptation and Mitigation in Agriculture: Methodology for Assessment and Applications*; Indian Agricultural Research Institute: New Delhi, India, 2012; Volume 302, p. 19.
96. Yihdego, Y.; Salem, H.S.; Muhammed, H.H. Agricultural pest management policies during drought: Case studies in Australia and the state of Palestine. *Nat. Hazards Rev.* **2019**, *20*, 05018010. [CrossRef]
97. Du, F.; Deng, W.; Yang, M.; Wang, H.; Mao, R.; Shao, J.; Fan, J.; Chen, Y.; Fu, Y.; Li, C.; et al. Protecting grapevines from rainfall in rainy conditions reduces disease severity and enhances profitability. *Crop Prot.* **2015**, *67*, 261–268. [CrossRef]
98. Bhriguvanshi, S. Impact of climate change on mango and tropical fruits. In *Challenges of Climate Change-Indian Horticulture*; Westville Publishing House: New Delhi, India, 2010; pp. 1–224.
99. Chowdhury, S.K. Diversity and nature of damage of mango insect pests at Kaliachak-II Block of Malda, West Bengal, India. *J. Entomol. Zool. Stud.* **2015**, *3*, 307–311.
100. Raut, P.; Desai, V.; Narangalkar, A.; Naik, K.V.; Mehendale, S.; Karmarkar, M. Effect of weather parameters on mango hoppers population. *J. Entomol. Zool. Stud.* **2018**, *6*, 112–114.
101. Stuart, R.R.; Gao, Y.; Lei, Z. Thrips: Pests of concern to China and the United States. *Agric. Sci. China* **2011**, *10*, 867–892. [CrossRef]
102. Tora, T.I.; Rahman, M.M.; Afroz, M.; Miah, R.U.; Amin, M.R.; Kabir, M.H. Effect of different temperatures on developmental periods and reproduction of fruit fly (*Bactrocera cucurbitae*). *Serangga* **2023**, *27*, 110–123.
103. Patel, K.; Saxena, S.; Patel, K. Fluctuation of fruit fly oriented damage in mango in relation to major abiotic factors. *HortFlora Res. Spectr.* **2013**, *2*, 197–201.
104. Anhalt, M.; Almeida, R. Effect of temperature, vector life stage, and plant access period on transmission of Banana bunchy top virus to banana. *Phytopathology* **2008**, *98*, 743–748. [CrossRef]
105. Awmack, C.; Woodcock, C.; Harrington, R. Climate change may increase vulnerability of aphids to natural enemies. *Ecol. Entomol.* **1997**, *22*, 366–368. [CrossRef]

106. Pathania, M.; Verma, A.; Singh, M.; Arora, P.K.; Kaur, N. Influence of abiotic factors on the infestation dynamics of whitefly, Bemisia tabaci (Gennadius 1889) in cotton and its management strategies in North-Western India. *Int. J. Trop. Insect. Sci.* **2020**, *40*, 969–981. [CrossRef]
107. Thomas, C.D.; Cameron, A.; Green, R.E.; Bakkenes, M.; Beaumont, L.J.; Collingham, Y.C.; Erasmus, B.F.N.; de Siqueira, M.F.; Grainger, A.; Hannah, L.; et al. Extinction risk from climate change. *Nature* **2004**, *427*, 145–148. [CrossRef]
108. De Cal, A.; Gell, I.; Usall, J.; Viñas, I.; Melgarejo, P. First report of brown rot caused by Monilinia fructicola in peach orchards in Ebro Valley, Spain. *Plant Dis.* **2009**, *93*, 763. [CrossRef]
109. Cayan, D.R.; Maurer, E.P.; Dettinger, M.D.; Tyree, M.; Hayhoe, K. Climate change scenarios for the California region. *Clim. Chang.* **2008**, *87*, 21–42. [CrossRef]
110. Garrett, K.A.; Dendy, S.P.; Frank, E.E.; Rouse, M.N.; Travers, S.E. Climate change effects on plant disease: Genomes to ecosystems. *Annu. Rev. Phytopathol.* **2006**, *44*, 489–509. [CrossRef] [PubMed]
111. Harvell, C.D.; Mitchell, C.E.; Ward, J.R.; Altizer, S.; Dobson, A.P.; Ostfeld, R.S.; Samuel, M.D. Climate warming and disease risks for terrestrial and marine biota. *Science* **2002**, *296*, 2158–2162. [CrossRef] [PubMed]
112. Ownley, B.H.; Trigiano, R.N. *Plant Pathology Concepts and Laboratory Exercises*; CRC Press: Boca Raton, FL, USA, 2016.
113. Benatar, G.V.; Wibowo, A. First report of Colletotrichum asianum associated with mango fruit anthracnose in Indonesia. *Crop Prot.* **2021**, *141*, 105432. [CrossRef]
114. Muimba-Kankolongo, A. Root and Tuber Crops. In *Food Crop Production by Smallholder Farmers in Southern Africa*; Academic Press: Cambridge, MA, USA, 2018; pp. 1–382.
115. Uddin, M.; Shefat, S.; Afroz, M.; Moon, N. Management of anthracnose disease of mango caused by Colletotrichum gloeosporioides: A review. *Acta Sci. Agric.* **2018**, *2*, 169–177.
116. Ghini, R.; Hamada, E.; Gonçalves, R.R.; Gasparotto, L.; Pereira, J.C.R. Análise de risco das mudanças climáticas globais sobre a sigatoka-negra da bananeira no Brasil. *Fitopatol. Bras.* **2007**, *32*, 197–204. [CrossRef]
117. Ploetz, R.C.; Thomas, J.E.; Slabaugh, W.R. Diseases of banana and plantain. In *Diseases of Tropical Fruits*; CABI Publishing, CABI International: Wallingford, UK, 2003; pp. 73–134.
118. Ghini, R.; Bettiol, W.; Hamada, E. Diseases in tropical and plantation crops as affected by climate changes: Current knowledge and perspectives. *Plant Pathol.* **2011**, *60*, 122–132. [CrossRef]
119. Matos, A.; Cabral, J.; Sanches, N.; Caldas, R. Effect of temperature and rainfall on the incidence of Fusarium subglutinans on pineapple fruits. *Acta Hortic.* **1998**, *529*, 265–272. [CrossRef]
120. Shetty, S.; Kumar, V.S.; Yogananda, S.; Kumar, N.K. A Review of Papaya Black Spot-A Fungal Disease (Asperisporium caricae). *Mysore J. Agric. Sci.* **2021**, *55*, 1–11.
121. Kumar, R.; Shamarao Jahagirdar, M.; Yenjerappa, S.; Patil, H. Epidemiology and management of bacterial blight of pomegranate caused by Xanthomonas axonopodis pv. punicae. *Acta Hortic.* **2006**, *818*, 291–296. [CrossRef]
122. Kubiriba, J.; Legg, J.; Tushemereirwe, W.; Adipala, E. Vector transmission of Banana streak virus in the screenhouse in Uganda. *Ann. Appl. Biol.* **2001**, *139*, 37–43. [CrossRef]
123. Sether, D.; Karasev, A.; Okumura, C.; Arakawa, C.; Zee, F.; Kislan, M.; Busto, J.L.; Hu, J.S. Differentiation, distribution, and elimination of two different pineapple mealybug wilt-associated viruses found in pineapple. *Plant Dis.* **2001**, *85*, 856–864. [CrossRef] [PubMed]
124. Sthapit, B.; Ramanatha Rao, V.; Sthapit, S. *Tropical Fruit Tree Species and Climate Change*; CGSpace: Milano, Italy, 2012.
125. Akritidis, D.; Georgoulias, A.K.; Lorilla, R.S.; Kontoes, C.; Ceglar, A.; Toreti, A.; Kalisoras, A.; Zanis, P. On the Northward Shift of Agro-Climatic Zones in Europe under Different Climate Change Scenarios. *Environ. Sci. Proc.* **2023**, *26*, 20. [CrossRef]
126. Singh, K.; Rani, D. Climate change impacts on orchard management: A review. In *PLANTA-Research Book Series*; Association of Plant Science Researchers (APSR): Dehradun, India, 2020; pp. 1428–1446.
127. Hull, R. Ecology, Epidemiology, and Control of Plant Viruses. In *Plant Virology*, 5th ed.; Academic Press: Cambridge, MA, USA, 2014; pp. 809–876.
128. Kumar, B.N.; Kumar, P.V.; Naidu, L.N.; Rajasekhar, M. Studies on fruit development in different cultivars of sapota. *Andhra Pradesh J. Agric. Sci.* **2015**, *1*, 24–28.
129. Haldankar, P.; Parulekar, Y.; Kireeti, A.; Kad, M.; Shinde, S.; Lawande, K. Studies on influence of bagging of fruits at marble stage on quality of mango cv. *J. Plant Stud.* **2015**, *4*, 12. [CrossRef]
130. Abou El-Wafa, M. Effect of bagging type on reducing pomegranate fruit disorders and quality improvement. *Egypt. J. Hortic.* **2014**, *41*, 263–278.
131. Abobatta, W.F. Managing citrus orchards under climate change. *MOJ Eco Environ. Sci.* **2021**, *6*, 43–44. [CrossRef]
132. Kang, G.Z.; Wang, Z.X.; Xia, K.F.; Sun, G.C. Protection of ultrastructure in chilling-stressed banana leaves by salicylic acid. *J. Zhejiang Univ. Sci. B* **2007**, *8*, 277–282. [CrossRef]
133. Glover, J.D.; Cox, C.M.; Reganold, J.P. Future farming: A return to roots? *Sci. Am.* **2007**, *297*, 82–89. [CrossRef]
134. Patil Shirish, S.; Kelkar Tushar, S.; Bhalerao Satish, A. Mulching: A soil and water conservation practice. *Res. J. Agric. Sci.* **2013**, *2320*, 6063.
135. Ahmed, Y.; Ahmed, M. Impact of spraying some antitranspirants on fruiting of williams bananas grown under Aswan region conditions. *Stem Cell* **2014**, *5*, 34–39.
136. Parashar, A.; Ansari, A. A therapy to protect pomegranate (*Punica granatum* L.) from sunburn. *Int. J. Compr. Pharm.* **2012**, *3*, 1–3.

137. Ehteshami, S.; Sarikhani, H.; Ershadi, A. Effect of kaolin and gibberellic acid application on some qualitative characteristics and reducing the sunburn in pomegranate fruits (*Punica granatum* L. cv. *Hicaznar*). *Acta Hortic.* **2011**, *818*, 167–174.
138. Rathore, A.; Raizada, A.; Prakash, J.J.; Sharda, V. Impact of chilling injury on common fruit plants in the Doon Valley. *Curr. Sci.* **2012**, *102*, 1107–1111.

agriculture

MDPI

Review

Conception, Consequences and Design of Cool Climate Viticulture Training Systems

Richard Danko [1,*], Pavel Pavloušek [1], Magdalena Kapłan [2] and Kamila E. Klimek [3]

[1] Department of Viticulture and Enology, Faculty of Horticulture, Mendel University in Brno, 69144 Lednice, Czech Republic; pavel.pavlousek@mendelu.cz

[2] Institute of Horticulture Production, University of Life Sciences in Lublin, 20-612 Lublin, Poland; magdalena.kaplan@up.lublin.pl

[3] Department of Applied Mathematics and Computer Science, University of Life Sciences in Lublin, 20-612 Lublin, Poland; kamila.klimek@up.lublin.pl

* Correspondence: richard.danko@mendelu.cz; Tel.: +420-722920884

Abstract: In this review, the problems, challenges and opportunities of trellis design are dealt with in the conditions of cool climate viticulture influenced by climate changes. Viticulture in so-called cool climate regions faces a number of weather and climatic extremes that directly or indirectly damage the grapes and so the wine. A suitable option is to use the structural and technical implementation of vine trellises—training systems, canopy management, and pruning methods which can help the plant withstand various extremes. At the same time, it is essential to choose trellis design training systems that growers can maintain and that support the appropriate quality of the grapes. Viticultural regions of warmer climate are strengthening the shading potential of training systems. Even so, the central viticultural areas withstand highly variable extremes of previous vintages with numerous shortcomings in the shading potential of trellis design. Meanwhile, the cool climate regions tend to use a trellis design with a simple canopy and easy sunlight exposition to reach the maximum solar contribution.

Keywords: VSP; canopy; SHW; shoot; pruning; grapevine; viticulture; Guyot; cordon; climate change

1. Introduction

The craft of viticulture is an ever-evolving field that must continuously adapt to diverse climatic conditions, soils, terrain, and geographical features across the globe. Historically, grapevines have thrived as heliophilous plants, spreading out to capture maximum sunlight. This natural adaptation is still reflected in the plant's structure, making its spreading habit an unnecessary feature used for grape-growing.

However, the factors of climate change are dramatically reshaping the landscape of grape production. Most of the growing areas are annually posing the increased frequency and intensity of extreme weather challenges. Spring frosts, heavy rainfall, hail storms, or longer drought periods are occurring with a higher prevalence. In fact, most of these circumstances occur in historically planted areas. As a response to these changes, there are new grape-growing regions arising in cool climate fields, facing extreme conditions and changes year-by-year [1]. Viticulture faces climate adaptations. These changes have a massive impact on ripening, harvest dates, the spreading of new diseases, and water stress. One of the common interceptions is that the grapes are facing a lot of stress factors; the wines are weak in aromatic compounds and there is a lack of freshness with an excessive alcohol content [2]. There is a significant threat of spring frost damage in growing regions with mild and inconsistent winter periods. Massive hail storms and heavy rainfall are actually common extremes that damage the crop, canopy, and the plant's structure. Also, more of the risks associated with new diseases are being detected, particularly with wood-decay fungi, rots, and phytoplasma [3].

The emergence of complex conditions caused by various wood-decay fungi has become a major concern. These diseases spread well with increased rainfall and the humidity driven by extreme weather conditions. Also, the use of improper pruning techniques that leave large wounds on the wood is facilitating infection to enter. To address these challenges, innovative solutions are essential [4]. It is necessary to modify pruning practices with maximal effort for sustainability and compact design. It is really important to reduce the concentration of pruning wounds, to support the natural sap flow and structural archetype with minimized infection risks and wood tissue damage. As a result of these challenges, it will be quite complicated to face all of the specific extremes with unsuitable vine-training styles. The goal is to develop plants of suitable trellis and training, with a significant advantage in evading the stress factors introduced by climate change. It would be appropriate to use the common features offered by the habitus of the plant. With the development of new pruning methods, a new point of view has emerged about how canopy architecture, clear sap-flow, and the structure of the plant can be considered as a game-changing trait of protecting the yield and the plant itself.

As viticulture expands into diverse cool regions, adapting vineyard management systems has become essential. The vertical shoot positioning (VSP) has been widely used to support upward vine growth and to control shading, crucial in a cool climate. However, in warmer regions, or even in the regions of Central Europe, we can observe serious complications with the traditional simple vertical canopy, as the sunlight exposure of previous vintages was really problematic. Therefore, if the climate conditions will continue to change this way for the upcoming years, it will be a serious challenge to face exposure with the low-shading capacity of a canopy. VSP's limitations are prompting a shift toward alternative systems like the single-high wire (SHW). The SHW system already enhances natural shading with improved air circulation and sufficient light distribution. These conditions can be helpful during the hot daily period, but also to face the overexposure to intensive solar radiation that can lead to overheating, premature ripening, and to decreased grape quality. These challenges are driving the development and adoption of different trellis systems, new grape varieties, and innovative cultivation techniques. The evolution of cultivation is inevitable for sustaining high-quality wine production [5]. The integration of biological measurement and disease-resistant varieties further supports the resilience of vineyards. These adaptive strategies can ensure us to thrive across various regions, maintaining exceptional wine quality amidst the shifting climate landscape using a proper growing technique.

2. Grapevine, the Modifications of the Liana in Modern Viticulture

Until now, the original non-domesticated *Vitis vinifera* spp. *silvestris* was still located in the range of the Middle East and in the large woods of the Mediterranean basin. However, the origin of grapevine domestication starts in the regions of the Lesser Caucasus. As is known, the first cultivation of the vine was without any harassment or canopy modification. A really important step in vine modification was the natural selection of plants bearing the largest crop load [6,7]. Another important step in the domestication timeline was when pruning was implemented as a means of obtaining a stable and loaded yield, and to manage and facilitate the vegetative growth [8].

Grapevines, *Vitis vinifera* L., have the characteristics of perennial lianas that combine the morphology of herbaceous and woody plants. Molecular biology with use of specific SSR Markers with geographical location has proven the evolution of a growing repertoire of grapevine genes [9]. However, not every gene donor tends to clear clustering. According to the gene analysis included in this study, most of the taxonomy of the grapevine is described as perennial liana with characteristics of both herbaceous and woody plants [10]. Moreover, several biological and morphological mechanisms are responsible for the sustainability and perennial plant cycle, supported by lignified tissues, bark cover protection, and the cambial development of the vascular system [11].

Vitis vinifera L., the Eurasian grapevine, remains as one of the main fruit crops worldwide. The determination of phylloxera (*Daktulosphaira vitifoliae*) in European vineyards during the 19th century forced the producers to use the American *Vitis* species with habitats similar to the wild grapevine as rootstocks. On the other hand, it was essential to not interrupt the natural tension of bud differentiation and primary meristematic activities [12].

Modern Limitations

The viticulture sector faces a large scale of challenges—strategies of sustainable production, water availability, or weather extremes. Many of the viticulturists have set a goal of balanced berry and wine composition, the yield optimization, or the collection of functional rootstocks. However, the impact of climate change is expected to transform the grapevine breeding market over Europe, as all of the regions face the increased intensity of extreme conditions. According to the study of [13], one of the suitable strategies would be to use different varieties to adapt in warmer climates, than to lose the suitability over the traditional regions. The distribution of suitable rootstocks and varieties is crucial for climate variability and the bioclimatic overlap. According to the study of [14], there are various consequences of changing the vegetation, significantly disrupting the dormant period and advanced phenological events [14]. Therefore, the berry composition is changing, twisting the wine parameters. When the overall impact of conditions becomes unbearable, berries ripen unproperly and wines tend to become unbalanced with higher alcohol and poor flavors [15]. According to the study of [16,17], one of the worst scenarios is that former traditional warm viticulture regions may become unsuitable for grapevines, as there are more vulnerable areas over the world, such as the Mediterranean area.

As the scenarios find out, wild grapevines provide a valuable gene pool for adapting viticulture to the improved management of the natural grapevine habitat. There is also a variability to counteract the relatives of future climate limitations, as it stands in countries like Poland, Latvia, or Sweden. In particular, there is a specific strategy that makes grapevine propagation and reproduction a different level of breeding—the management of biomass production and regulation of shoot system development [18,19]. Canopy management has been an underestimated aspect of viticulture. However, in recent years, it has become evident how much working with the plant's physiology can help to facilitate grape cultivation.

3. Grapevine Wood Morphology

3.1. Green Shoots

As a perennial higher plant, the grapevine tends to grow primarily from winter buds, well-protected vegetative postpone organs. The bud includes a primordial apical meristem, the green shoot. Organogenesis appears after budburst with new growing units for the primary stem. The buds remain unbalanced in production, vigor, and shape, as its fertility depends on the position and character of the wood from which they originally burst [20,21].

Grapevines feature a specific structure of primary and secondary growing tissues, the specific meristems. All of them are considered as a narrative of specific cell pools, the morphogenetic meristems with particular behaviors [22].

Many of those attributes result in specific vigor and require cultivation management to control the vegetative overgrowth. Due to this and many other reasons, the grapevine became one of the most suspicious plants when it comes to controlling the quantitative and qualitative development by pruning and thinning [23,24].

The grapevine structure is a precisely organized system with several stages of plant tissues ranging from cell development to branching. With years of progress, the shape of the shoot system has diverged, evolved, and now it leads to adapt for the mechanization of pruning, shoot positioning, spraying, and harvest [25].

As we observe a lignified cane, which was a green shoot even before, there is a morphologic signature typical of lianas—nodes and internodes. Described in an article as phytomers and dedicated to metamers, they grow in length when the apical dominance is present. After decapitation, the shoots and nodes of the main axis establish in length, growing in thickness, and giving a signal to burst the side shoots, also known as lateral or sylleptic shoots. The regulation of shoot development is useful technique for yield control and limiting phytosanitary inputs [26–28].

The vegetative architecture of grapevine plants needs to be modified periodically. Possibly, the most useful of all interruptions is to modify the plant vigor and overgrowth. In particular, the parallel is to balance the trophic competition between vegetative axes, source-sink (leaf-cluster) balance, and maintain the vigor of buds and shoots [29].

As they grow, the shoots bear the fundamental sources of the next-gen axis, the fruitful winter buds. These vegetative organs obtain the quantity of yield per vegetative axis in a close correlation with variety dispositions. It is necessary to select the correct level of buds during the pruning season, as there numerous types can occur with various prepositions [30]. Grapevine growing management is estimated by the strength and vigor of the shoots. The proportion of dry matter within the shoots is generally constant, i.e., 50%, and the rest of the tissues and components are related as annual accumulated biomass.

Ravaz Index and the Plant Vitality

The strength and vigor of the shoots is related to the plant's resource management ability during the vegetation period. A wider scale of extreme conditions impact the plant's ability to develop more or less vital shoots. As Ref. [31] mentioned in their study, the plant's vitality should be related to drought, but also to the mechanic tissue damage (hail storms) and larger yield reduction (cluster or berry thinning) from previous vegetation. Most of this matter comes out during lignification, harvest, and pruning, making a footprint in the plant's vitality. The Ravaz index is capable of bringing a closer reference to the complexity of the issue with plant vitality, to understand the plant's reactions in different situations [31]. The value of plant balance is related to the fresh pruning wood/yield ratio. This index tends to express the quality of the canopy during vegetation with the correlation of lignified shoots. The index can provide a useful interpretation of the plant's reaction to extremes. It should be a useful tool, giving the viticulturist an accurate chance to obtain more or less weak plants with vigorous and less vigorous shoots, or with lower biomass production [32].

3.2. Other Morphological Factors of Inner Tissues

The grapevine is a liana, so the inner tissues have a special composition and its morphology should be a useful tool for building the functional structure of vessels and tissues. The internal structure of grapevine wood is morphologically segmented. Starting with the internodes, their diameter after lignification points out all of the precursors for good or weak sap-drawing features—intensity of lignification and the proportion of secondary tissues and flesh [33].

Grapevines, in line with other perennial plants, develop many secondary tissues located within the cortex of structures. Many of these consist of less-lignified cells, including vascular cambium, the phelloderm, and phellogen tissues (suberin, cork) [32,34]. The phelloderm is suberic cambium and plays a role in shoot development and canopy architecture. Sideways, the vascular cambium has an essential influence on grapevine morphology. This tissue is primarily involved in the production of the xylem, the main tissue of the stem, and is responsible for the process of structural ramification [35,36]. Some diversity in the length and diameter of primary axis shoots and internodes was observed. During the process of lignification, the primary structure of the trunk is developed year by year until it reaches the demanded shape and height. It consists of less supporting tissues with a larger collenchyma in the cortex. This ability is useful for arranging the ideal anatomical flexibility of the apex with plastic-like growing direction, composed of three main directions—gravitropism (root–soil relationship), trimorphism (mechanosensory movement), and phototropism (canopy–sun relationship) [37].

All of these factors combined have a significant role in shoot morphology, affinity of scion and rootstock, growing potential, and vigor [38]. Following the composition and continuity of tissues, a complex structure is formed of physiological and morphological traits, forming the system of living nutrition channels. This sap-flow path remains unchanged throughout the plant's lifespan and any intervention—dried wounds, scars—disrupts it. Pruning without considering the natural sap-flow conducts a large decrease in the trunk, canopy, and production [39].

3.2.1. The Dormancy

In areas where the annual winter temperatures are under 10 °C, the grapevine tends to perform one vegetative cycle per year. Considered as an end of the green period, facing a few days of lesser temperatures than 10 °C, dormancy is a subordinate process to the traits of various climate influences. It is supposed to end one growth cycle and to culminate nutrition and set up the winter buds for the next vegetation; therefore, dormancy is an essential period for the plant's rest [40–42].

Three successive stages are typical for dormancy. It starts with pre-dormancy, where the inhibition process is set to prevent the winter buds from breaking with influence of external conditions and limiting the internal resources to plant level [43]. Dormancy (endo-dormancy) is where all the internal processes are regulated by physiological factors, mainly the cold temperatures of abiotic stress [44]. Eco-dormancy is not affected by internal factors, but only by the environment, particularly the temperature initiation. After this stage, budburst is determined by the rate of nutrient resources and temperatures. As is known, pruning during eco-dormancy has a lower effect on budburst [45,46].

During the dormant stages, pruning is performed to adjust the fruitful potential for the next vegetation period, but also to limit the shoot development. Late pruning after the budburst limits the flowering and the ripening process. This factor is a useful tool for facing the spring frosts [47–49].

3.2.2. The Consequences of Pruning

The pruning process can be considered as the top of a glacier through the dormant period of plant. Pruners are dismantling the canopy from unnecessary shoots and living wood, giving an advantage to the residual fruitful wood in terms of nutrition and spacing, focusing the production of the future canopy directly to the reserved canes or spurs. On the other hand, during the pruning season, a large scale of damage and scars occurs, that will never be completely healed and which are sensitive to the trunk pathogens [50,51].

The pruning period is considered to be the most important and longest work undertaken in a vineyard during one plant cycle. This operation is necessary to maintain the health and production of the plants on a long-term basis. Moreover, every pruning system is dedicated to the number of fertile dormant buds per vine. This feature will determine the yield per plant, the number of primary shoots, or the number of buds per spur/cane [52]. It ends/starts the annual cycle, as we tend to prune after leaf fall and end it before or during the bud burst. It is necessary to obey the dormancy, to avoid abiotic disorders, such as spring frost or the lethal effect of susceptibility to pruning wound consequences [53].

As plants are facing the increased temperatures and unsuitable climate changes, several varieties and regions are affected by unnecessary metabolite accumulation, color intensity, and polyphenol concentration. One of the possible traits of seeking balance to these changes is to adapt the start of the pruning period. Later shoot pruning has a significant effect through the blue varieties, enhancing the canopy architecture and the photosynthesis efficiency [54]. Standard winter pruning is taking place in February. Late-performed pruning has an impact on the grapes, accelerating the ripening, especially the veraison of blue varieties, as late-pruning takes a one-week pace from mid-May until the end of June, synchronizing all the threats during veraison. This treatment may give an advantage to the blue varieties, syncing the phenolic compound of grape, must, and wine [55].

Pruning Wounds

With the conditions of pruning, there comes a significant obstacle of high humidity during the fresh cuts. Infections of living wood are caused by saprophytic parasitic fungi, which invade the pruning wounds and are associated with the complex of grapevine trunk diseases (GTDs) [56,57]. All of the spores and infections are located on the fresh pruning wounds within the dormant season, creating multiple entrance portals for pathogens [58].

Decay, death arms, and diebacks might be susceptible for the trunk establishment mistakes. The plant structure gains a large surface of open scars each year without capability of prompt regeneration. Older permanent structures are stacked by numerous pruning injuries and this study observation proved a higher loss of yield and canopy capacity of older, damaged plants [59]. The perennial wood structures, such as trunk and branches, contribute more of the fungal diseases within the structure, as there is an unbalanced texture of living wood and desiccated rotten wounds. Mostly, the struggling of vigor and fall of growing intensity includes concrete symptoms such as cordon strangulation, diebacks, and commonly the foliar symptoms of canopy with low vitality. The quantification of increased symptoms might be boosted by several surrounding factors, mostly by the extremes of climatic conditions—drying winds, frosts, and high humidity [60].

During the dormant period, the wood of grapevines is prone to the agents of infection, mainly the large pruning wounds, rainfall, and high humidity. Infectious spores are spread to the fresh wounds of woody tissues following rainfall. On the surface of pruning wounds, there was detected the presence of wound-colonizing microbes. What is notable for the practices of pruning, following the studies, is that the number of microbes found within rainfall-exposed pruning wounds was much higher than in those protected from rainfall [61].

Pruning is an essential practice dictated by the morphology of grapevine plants. To optimize yield and plant productivity, growers manage the bud load on the youngest, productive wood—the lignified shoots, also known as canes or spurs. This process creates wounds on the living wood, triggering physical and biochemical defense mechanisms, with different pruning techniques eliciting distinct defensive responses [62,63].

The pruning of shoots and old wood is causing long-range damage to the vine structure, equal to the range of the opened surface. This leads to the development of tyloses in damaged xylem vessels, disrupting the transport of water and minerals. There is a correlation between the gap of nodes and the distance from the permanent structure. Nodes closer to the trunk developed tylosis faster than distal ones. The development of tyloses might block up to 40% of vessels. However, the water transportation system may be restored [64,65]. The selection and character of 1-year-old wood, spurs, or canes, the fruitful vectors of future yield, are the main part of any pruning style. With new trends, there are many experiments with the length, density, and position of fruitful wood [66]. In 3-year-old grapevine wood, about 2 months after pruning, there is an increased amount of lignin development in the cortical parenchyma. Below the opened wound, many hyphae of fungi were detected [67].

Unlike in other fruitful trees, in the 3-year-old pruning wounds of grapevines, no suberin or calloses were detected. Pre-observation of desiccated pruning wounds verified that the preserved basal buds protected the lower vascular system from desiccation. The cones appeared greater in the 3-year-old wood than the 1-year-old of the same pruning style and the same age [68].

As it stands, with damaged wood, there is a higher chance of fungi infection within specific condition, penetrating into the living wood and stem tissues. Except for the roots, all the types of living tissues are prone to the spreading of trunk diseases. The highest infection rates were detected within the trunks and cane ends, proving that GTDs are primarily shoot- and trunk-focused pathogens, spreading through the open wounds [69].

There are many techniques of grapevine plant establishment in terms of morphology and production. However, there is not any significant proof that different plant establishment can interrupt the grape yield. Mostly, the habitus of grafted plants can withstand more injuries and damage. On the other hand, the own-rooted plants tend to increase the trunk weight [70].

Grafted plants have a strong influence against the leaf symptoms of GTDs. Different grafting techniques have an effect on plant xylem development within the grafting point. An important factor, besides the correct technique, is the choice of grape cultivar and rootstock [71].

3.2.3. GTDs, the Main Threat to Wooden Structures

Heat and water stress may affect the plant's physiology, production, and development. Also, the pathogen's life cycle is affected by individual or combined stress factors. The response of grapevines against stress may interfere with the recovery of the plant, such as the intensity, duration, and timing. On different occasions, most of the grapevine pathogens have a negative effect on the canopy architecture [72]. Local shoot necrosis and deformed vigor is similar for most of the diseases or specific fungi. All of the biotic stressors tend to produce an irregular plant morphology, limiting the lignification and causing specific damage to stem tissues [73].

Vineyards all over the world are facing issues with production and vitality loses subject through the vegetation stages. Many of the losses are caused by different fungi species. With the changes in vineyard cultivation, we have to face specific symptoms of fungi with a common injurious range in the old wood. Those fungi are a wide-ranging species, so-called grapevine trunk diseases (GTDs). According to the study of [74], the vascular system of the grapevine produces morphological and physiological mechanisms as a response to pathogen activities. GTD-fungi cause expressed symptoms relating to the formation of tyloses inside the xylem vessels. The complex also causes extracellular enzymatic ligninolytic activities of laccase, manganese-peroxidase, or lignin-peroxidase [74].

According to the authors of [75,76], GTDs consist of up to 135 fungal species of 35 genera and are considered the largest group of grapevine pathogens. Most of the GTDs are fungi ascomycetes; however, some basidiomycetous species also occur in the system. Stress and GTD infections, the main vascular pathogen disrupting vineyard sustainability in the last 30 years, are closely linked, influenced by factors such as the host–pathogen nature, rootstock, clone, and cultivar. While studies have suggested that water and thermal stress may trigger these diseases, the research on the combined effects of abiotic stress and GTD infections is lacking. Additionally, as per [77], endophytic fungi can remain latent in vines for years, complicating detection and the use of prophylaxis methods. Grapevine trunk diseases (GTDs), such as Eutypa, ESCA, Botryosphaeria dieback, or Petri disease, are related to the wilt of the vascular system. They are caused by set of taxonomically unrelated ascomycetes with different techniques of colonization, and host limitations. The plant objective is limited by the loss of hydraulic conductivity and the xylem transport capability. Many *V. vinifera* cultivars reflect different stages of vascular disease tolerance [39,78].

As per the studies of [77,79], the main part of the plant defense mechanism has not been completely evaluated. There is a significant evidence that commercial *V. vinifera* cultivars are ESCA-prone by the correlation with the open vessel diameter. There are many organism taxons which have a direct or indirect impact on plant biomass production and so on the strength and vigor of shoots. A significant example is the use of green manure of weeds as a cover-crop on the field. Green cultivation has a massive effect on plant development, and nutrient or water transport is boosted or reduced by its presence. Inside these tissues, there was detected minimal indication of spreading fungi. A similar relationship is seen with leaf fungal pathogens, as the competition for carbon and mineral resources, reducing the biomass assimilation by limiting the sources. And so, many of these pathogens are affecting the root system.

The main result of the study [80] is that there is significant proof of GTD infection caused by the removal of grapevine suckers and green shoots. The infection caused by GTD is on the verge if trunk the fungi penetrates the vascular system through the gap caused by pruning wounds. Another possible entrance for infection are the wounds caused by removal of green shoots and suckers. According to the results of the mentioned study conducted in South Africa, 62% of isolated sucker wounds were infected by GTD pathogens. Most common of the fungi species was *Diaporthe ampelina*, followed by *Eutypa lata, Diplodia seriata, Cryptovalsa ampelina*, and many others like *Phaeomoniella chlamydospora, Phaeoacremonium minimum*, or *Eutypella microtheca*.

4. Features of Shoot Architecture Occupying Different Trellis Systems

Different types of trellising are used around the world to suit their style of production. In choosing different trellis/pruning intensity, it is important to suit different cultivars, growing habits, and the vigor dispositions unique to the region. Growers around the globe are focusing on two different treatments—the vertical shoot position (VSP) trellis, including such strong advocates Guyot and spurred cordon and on the other hand, there are hanging high-wire trellises [81].

Selection of Canopy Architecture and Training System

A combination of pruning, plant management, and training systems results in a grapevine architecture model. It has to be calculated by using the plant spacing in the row (the distance of each plant represents density), and so the geometry of the trellis [82]. According to the variety, soil, and conditions for cultivation and the maintenance of buds per plant, it is really important to choose the right strategy of production using the training system. Choosing the spur pruning or cane pruning is a main task for the correct establishment of the training system and architectural model. For example, Guyot is a main cane-pruning system of the VSP (vertical shoot position) architectural group. The bud fertility is related to the position of buds on a cane and top–bottom sap-flow gradient. This training system is well-known to be suitable for many varieties and climate conditions. On the other hand, cordon pruning is based on the spurs of two visible buds and the number of spurs fixes the production of plant, but also stimulates the plant production [83].

A canopy is adjusted by green shoots with considered position on 1-year-old-wood of the plant structure. Those fruitful shoots are determined by interactions with biological agents, environmental factors, sinks, and reproductive organs. These interactions are used to optimize the mineral and water consumption of vegetative organs, but also to moderate the canopy microclimate. Grapevine growing dispositions are nearly unlimited in the natural habitat [84,85].

The training system is a vector of crop placement to ease the position for laboring, protection, and harvesting features. However, there is the density of the canopy biomass, which causes humidity inside the canopy and so supports the spreading of fungal infections [86,87].

Modern viticulture faces a huge challenge to pose a water limitation in many viticultural areas. In many production countries, water supplements are needed to preserve the economical yield and quality. The strategy is focused on improvement of water used by crops per dry matter produced, to consider the goals of sustainability and functionality of training [88,89].

Viticulturists tend to manage the canopy for sunlight exposition, yield control, and microclimate conditions [90]. Despite the yield demands, they have to consider total underwater availability, potential water consumption, and nutrition requirements. There is also the task connected with the prompt management of solar radiation inside the canopy, which is allied with cultivar dispositions, design of canopy, and trellis [91].

A large scale of trellis and training configurations is available at the moment. Therefore, many factors need to be considered—local climate, the official regulations, terrain and soil type, variety and clone, mechanization, management, human work resources, and so on [92].

5. The Selection of Suitable Training System—Effectiveness of Vertical Shoot Positioned (VSP) Training Systems in Comparison to Hanging Systems (SHWs)

The production of grapes is highly connected with shoot distribution and canopy architecture, as it features the potential vegetation, crop production, canopy gas exchanges, and thus the potential light interception [93]. In any event, the model of the shoot system is a matter of complex growing, design, and architectural attributes which have to work at many levels—shoot, plant, and site [47,94]. All of those integrations were approached many times with issues of creating variation in the plant model with local proposals and well-integrated sink–source interactions with the needs of plant, canopy, vineyard management, and climate properties [95]. As it stands, one of the biggest traits is shoot branching and control of the main growing axis [96,97]. As to control, there are many genetic factors of vegetative development that need to be included while choosing suitable viticultural model—the length of metameter, primary growth rate, above-ground biomass, tendril differentiation, or para-dormancy of winter buds [98,99].

- Facing the spring frosts in cool regions

The viticulture sector faces climate change adaptations since those charges have a massive influence on ripening, harvest dates, spreading of new diseases, and water stress. One of the common interceptions is that the wines have a lack of aromatic compounds, a lack of freshness with excessive alcohol content. There is a significant threat from spring frost damage, in growing regions with mild and inconsistent winter periods. As noted by [100,101], spring frosts are pushing the phenolic stages of plants. The worst scenario is if spring frosts disrupt the most sensitive parts of plant—the apex and inflorescences. A damaged apex does not tend to grow upwards in a natural position anymore, seeking new ways of branching out. On the other hand, disrupted inflorescences may cost us the crucial part of yield content. As mentioned in the previous studies, grapevines always burst new shoots, even if they are just suckers. But the crucial stage is the damage to the yield. Southern regions face a larger scale of traits, especially increased dryness with reduced yields, while the actual CO_2 accumulation may compensate for the dryness effects in Northern Europe, with increase in the maximal yield. According to the narrative of this article, one of the possible solutions for facing the spring frost is the trait of plant's growing nature. Suitable training and a higher altitude of the trellis should help to evade these dangerous spring frosts up to the delicate zone of burst buds. Actual problems also impact the production of a chained market and the potential changes in the suitability of viticulture [102,103].

5.1. Modern Threats of Selecting the Suitable Training System

The grapevine is a widely grown plant with a perennial life span. Viticulture and winemaking share an important cultural and agricultural legacy with social and economic features [104]. Furthermore, viticulture and winemaking share an environmental impact, with many pesticides used on small areas of cultivated land. For instance, in 2011 in France, there was a commission of 20% of pesticides used only for vineyards, leaving environmental impacts such as groundwater pollution and potential ecotoxicity of air, water, and soil [105,106].

Training systems are chosen to support our demands for grapes—the fruit composition, sugar content, volatile and aromatic compounds, and as a result, the phenolic content of grapes and must. Also, if we check different studies, changes might occur in the parameters, such as the acidity and sugar content, pH, or YAN, but also, many of the authors have suggested that different training systems have some impact on the wine sensory [107–109]. Training grapevines with appropriate canopy development may stimulate the vegetation period. This attribute will be hugely important in the future through the process of climate change adaptation. All the regions of Northern Europe start the ripening period much earlier than in recent years. This fact may cause the unwanted pre-maturing process or overripening of the early varieties and possibly a higher number of pests. One of the existing critical points, according to the sources, is when the variety cannot suit the climatic dispositions. Therefore, it needs to be grubbed up or regrafted and to be replaced by a different variety/crop [110,111]. On the other hand, training systems such as minimal pruning are suitable for the effect of delayed ripening. This is because of the higher leaf/fruit ratio in such a massive clutch of biomass; therefore, it is able to mitigate the negative effects of climate change and high temperatures [112,113].

5.1.1. Considering Vertical Shoot Position (VSP)

According to [114], there will always be competition between training with vertical shoot position and the free growing canopy. Both of those groups are locally considered as the category of hedgerow training systems, which account for 70% of total vineyard plantings worldwide, and still counting in a sense of smart mechanization. The essential feature of the VSP training model is the support of natural liana spreading on the trellis [114,115]. Moreover, the sprawled canopy is a revolutionary type of canopy habitus and the importance of having an upward or downward canopy is perceived. VSP obeys the factor of the narrow, vertical extension of the canopy with limitless vigor and ripening, gaining optimal leaf expression to the sunlight, giving this training an advantage by equalizing the ratio between the maximum canopy height and row-spacing to 1:1. Many viticulturists insist that various types of training demand different pruning and correct support trellis systems [116,117].

Many factors need to be considered when selecting the combination of trellis and training system. One of the huge and therefore very important challenges for the future is work with the sunlight exposure and shading dispositions. Really good examples are open-shaped trellises, such as Y-shape and Lyra, where the sunlight exposure of vigorous varieties is increased, pulling back the canopy density, making this system useful for improving the health condition and grape quality of humid and cool climate regions. However, higher canopy exposition has a threat of increased water consumption and may lead to different plant stress [32,118,119]. Summer periods in various regions became the hottest in the history of measurement and many vineyards over the world were affected—sunburned leaves and clusters are the first stage. After reaching 35 °C, the leaf is forced to enclose the stomata, cutting off the photosynthesis. Damaged foliar tissues reduce the photosynthesis and sugars gained, interrupting the C6 cycles and reducing phenolic content—the aroma of white varieties and pigments of blue ones. Reduced absorption of foliar and radical nutrients is exhibited in the plant's health by reduction in the trunk-accumulated reserves [120].

While choosing the VSP, to gain the best canopy exposure to sunlight, it is usually required to consider the orientation of the rows with the maximum of day light interception and daily photosynthesis rate of the whole canopy and display the effectiveness of CO_2 exchange in relation to the amount of intercepted light and the part of the best-lit part of the day [121,122].

By these inductions, the VSP-type management of the canopy is the most widespread system, uniformly exhibiting both sides of the canopy to the sun and contributing maximum sunlight [123]. However, the effectiveness of VSP fades in warmer areas. This can happen because the conditions of total sunlight in the canopy management interfere with grape ripening, exerting with berry composition, must content, and exact wine quality [124,125]. By the theory, in relation to controlling the canopy, there might be a highest effectiveness for the climate change, planting vineyards in an East-to-West orientation with sunlight exposition taking aim to the North–South, trying out the reduction in solar radiation interception in the afternoon. Under these conditions, this scenario would give more freshness and typicity to wines produced in Mediterranean areas with issues of using VSP or SHW [126,127].

5.1.2. Considering Hanging Canopy (SHW)

According to the study of [113], different training systems may change the wine sensory. To choose the criterion to consider while selecting the trellis system, there is the main trait of shoot posing—the upright shoot systems tent vegetation to face to the vertical trellis design and the hanging shoots are managed as descending vegetation models or minimal pruning. However, the genus of *Vitis* needs to be characterized more by the phenotypic and genotypic plasticity. It gives a perimeter for grapevine breeders to manage the shoot architecture traits, via the selection of rootstock and scion for better management, and biotic or abiotic stress adaptation [113].

It is necessary to consider that diverse trellis types have different reactions to main traits, significantly related to the climate. With different canopy treatments, there is a different valuation of vigor and yield control, pathogen-related sensitivity, or sustainability for charges. The final results are different in cool and warm climate areas [128]. Recent diversification of grapevine pruning system was based on low-level specification of canopy management, split into three main groups—minimally pruned canopy, vertical shoot position (VSP), and hanging (sprawl) shoot canopy, whereas all of them can be encompassed with the metabolic signature in final wines. A higher level of training system is an option of compositional preferences [129].

The SHW training model tends to become an extremely important growing system for the next generation of viticulturists. There is a massive advantage of cost due to minimal trellis structure, cultivation needs, and mechanization used. Also, in a long-term comparison of different training systems, as previously mentioned, with the correct establishment, management, and support, there is also similar grape and yield composition as in the case of VSP. As confirmed in [130], by the use of different SHW trellis systems on representative warm-climate varieties, we can promote the accumulation of total soluble solids (TSSs) and more stable anthocyanins and flavonols in berry skins. Nevertheless, the main factor of success remains in the interaction of the canopy with cultivation issues, such as the factors of vine growing— the association of scion with rootstock, that develops the appropriate canopy with both good ripening and wood maturation potential [130,131].

5.2. Different Groups of Trellis Systems

Trellis systems can be divided into various groups, as the architecture of the canopy is followed by vigor of the trunk, wires, and poll system. All of the trellis types have different features and attributes, such as susceptibility to fungal diseases, sustainability, wine style, mechanization model, plant vigor dispositions, yield control, and the climate-related features in general.

According to [132], we can divide pruning styles into two main categories—cane- or spur-pruned types, as they are followed by the nature of fruitful shoots. By the evaluation of canopy management, pruning attitude, and wooden structure, we can divide the styles as follows:

- Classic VSP (vertical shoot position): the majority of plantings all around the world—spurred cordons, Guyot, and more. They are suitable for the majority of viticulture regions.
- Spreading and hanging trellis types: (pergolas, chamber system, Raggi-Bellussi, Tendone) are sustainable, easy to maintain, and have high yield potential. We can find them more in the cool and humid areas of Northern Italy or Northern China.
- Single-high-wire training (SHW, Curtain, Sylvoz) are suitable for higher yield varieties. However, they are considered to be the future, because of the shading potential and easy costs.
- Split canopy systems (GDC, Scott Henry, Vertiko, Kniffin system) are considered as really high-yield trellis systems. With mid-easy management of the canopy, they are commonly used in warmer climates such as New Zealand or Mediterranean areas.
- Closing shape (Y-shape, SAYM, Lyra) is still in development, but obtains serious potential for VSP variations in regions with higher shading demands.
- Single-pole systems (Trier Slamka wheel, Mosel stake training, pole cultivation...) are low demand trellis models with higher participation of hand work. The design is planned for the conditions of steep riverbank slopes and narrow terraces.

- Traditional spatial systems (Gobelet, Bush vine. . .) are the remains of elder wide-space trainings. The design is planned for semi-arid areas with high sun exposure such as regions of Southern Spain, Southern France, Portugal, California, Middle East, and so on.
- Special pruning styles with local traditions (Chablis, Vallé de la Marne, Cordon Royat, Nest/Basket/Kouloura) are designed to face specific conditions of the regions from which they come. What is typical for these trellises is that only a few pruners are authorized to take care of them.

In a different way of diversity, we can divide the training systems by the shape/orientation of the canopy—vertical, hanging, or minimal pruning canopy.

5.2.1. VSP—Vertical Shoot Position Variations

Vertical position is suitable for low to medium vigor cultivars. First, the base wire is installed less than 3 feet above the ground with multiple pairs of movable catching wires above the base cordon wire. The training of shoots is vertical from the base to reach and tuck the flexible catch wires. Those trainings initiate a well-organized canopy with hanging clusters and easier access inside. However, it is more expensive and requires more maintenance [86,133].

As you can see in the Figure 1, here are various pruning styles and trellis systems with a controlled vertical canopy fixed more or less at trellis construction. Many of them are on stakes, such as (A) arm stake and (B) Mosela heart stake. This type of trellis management is related to very specific conditions of steep slopes with harder mechanic management. Also, there are trainings represented by canes with a horizontal wire trellis, such as (C) Guyot and (D) Sylvoz. These trainings are versatile and fit to various conditions over the whole winemaking map. Following the production goals, there exist more cordon training variations with numerous multiplications of cordon arms. A few of them are regional, for example, (E) Lyra or (F) three-tier top cordon, but there also exist versatile variations for the general part of the winemaking world, such as (G) spurred cordon or (H) four-armed cordon. The potential of these trellis systems is connected with specific regions all over the world. Different traditions, varieties, production goals, and terrain exposition give growers more or less developed opportunities for vineyard management.

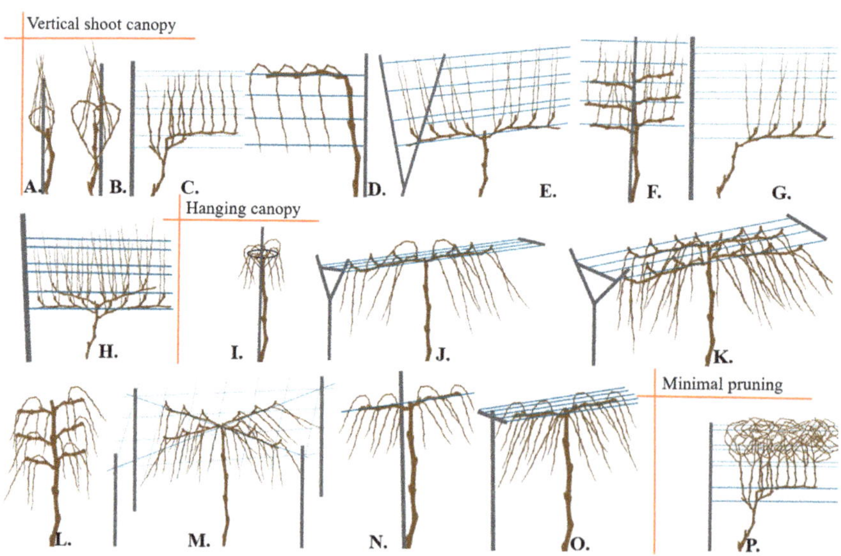

Figure 1. Different types of canopy, sorted by trellis model and dispositions of shoots—vertical shoot canopy (**A–H**); hanging canopy (**I–O**); and minimal pruning (**P**).

5.2.2. Hanging Shoot System Variations

Moving to the hanging trainings, many of those systems were designed to be easily cultivated by mechanization such as the (I) Slamka system on the steep slopes of the Mosel river, or the (J) cordon curtain. There are numerous regions over the world, such as Veneto or Trentino, where producers aim for a quantity of grapes and where the conditions of cultivation are suitable for high-yield hanging trellises, such as (K) GDC, (L) GDC duplex, (M) pergola, (N) Vertiko, (O) curtain, or (P) tied curtain training.

Single high wire system (SHW) is suitable for cool climate viticulture, suiting well cold-hardy varieties and hybrids with higher vigor through vegetation. SHW requires less labor, mechanization, and investment, moreover demands less pruning practice and shoot positioning than more complex training systems. All of the trainings consist of 1–3 wires 1.2–1.8 m above the ground. Two cordon arms should be trained along the top wire. The high position allows good sun exposure for the fruit and the "renewal zone", establishing fruiting buds for next season [134,135].

5.2.3. Minimal Pruning System

This (no)pruning system (P) takes place if there is minimal effort given to the canopy. It is based on the natural habit of lianas for acrotony. Yield and quality is consistent and provides a serious impact of minimal growing costs. Agronomists are satisfied with the minimal human work resources and maximum of work obtained by tractor. It is important to mention that the trellis needs to be strengthened by adding more wires. The biomass of living tissue grows annually. The basis of this training is the classic cane, with controlled growth upwards, then twisted into the wires [136].

5.3. *The Well-Suited Training Systems—Guyot and Spurred Cordon*

According to [82], two of the most effective VSP training systems are Guyot and spurred cordon. As we can observe in the Figure 2, both of these models are considered to be part a training and part a pruning system, according to many styles and variations in them. Both of them can be used for the majority of grape cultivars. Guyot is determined by the cane, obtaining as many buds as needed by its length and production rate. This cane needs to be bent in a horizontal position, avoiding acrotony. The density of the vegetation depends on the length of internodes. Spurs are providing the supporting role, giving this training a supporting basement of stronger shoots, obtaining more possibilities of selection of future canes for the next season. Guyot training is also well suited to cool climate regions. The fruitful cane is better preserved and in case of frosts that bring different dangers. On the other hand, cordons are determined by the number of spurs, the tertiary structure, located on the secondary structure, the cordon arm. The production capacity and density of the vegetation will depend on the number and distance of spurs. The selection of a cordon trellis is notable for higher quality production goals. Various countries around the globe use cordons in different shapes, as mentioned in the Section 5.2. What is also really important to mention is that historically cordons were more suitable for warmer climate regions, such as South Africa, California, Southern Italy, and Spain. Nowadays, the popularity of cordon training is also increasing in cool climate areas. The main reason is to minimalize hand work by machinery [137].

The modern threats of viticulture have made growers reflect on all of the factors that bring functional and sustainable training systems for both of the high-quality pruning styles. Research studies from Italy have developed sustainable and functional method of pruning, using Guyot and cordon with a special setup. These rules give the wooden structure a beneficial opportunity to gain as much compact wooden structure as possible, using a special technique of old-wood ramification, as displayed at Figure 3. This method is called the Simonit & Sirch pruning technique and its role in modern viticulture is becoming crucial. Grape-growers are facing the modern threats of climate change, diebacks, and new diseases. Plants are hardly withstanding all of the pressures of climate change, desiccation, and spreading infections. Compact pruning with structure work is obtaining large-scale popularity across the field of growers. One of the benefits is to properly preserve the old vineyards, as they produce a different quality of grapes [138].

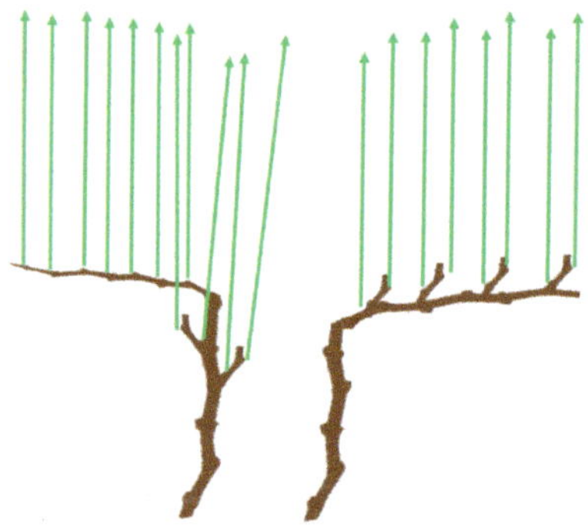

Figure 2. Comparison of canopy of Guyot and spurred cordon.

Figure 3. Comparison of structure, growth, and canopy of two classic Simonit & Sirch trainings—bilateral Guyot and double-armed spurred cordon.

5.4. Choice of Suitable Trellis and Training System

If it comes to the comparison of cool and warm climate training models, lateral cordons are the most common training at all. Its establishment is lower in cost and easier for pruning, harvesting, and manual work. As we can observe on warm climate training representatives, such as gobelet of Bush wine, the canopy of the warm area is configured to prevent the excessive sunlight exposure of the grapes. Highly vigorous vines tend to require the shading and higher humidity inside the canopy. However, as mentioned in the article [139], popularity is also increasing when used as a combination of upward and downward shoot positioning, making this system a hybrid of a vertical and hanging canopy. Generally, it is used in coastal regions. The lower fruiting zone always becomes weak over time. A cane pruning alternative with a combination of up and down canopy is also desired to be suitable for vigorous vineyards in warmer climates [139].

To consider the trellis system, there is the main trait of shoot posing—the upright shoot systems tent vegetation to face the vertical trellis design and the hanging shoots are managed as descending vegetation models or minimal pruning. In general, all the sorts of *Vitis* genus need to be characterized more by the phenotypic and genotypic plasticity. These charges should be a perimeter for grapevine breeding, to face the shoot architecture traits. Either way, this should be considered as an opportunity for proper selection of rootstocks and scions for better biotic or abiotic stress adaptation [116,140].

The liana-like nature of the grapevine makes the reaction of plants to the pruning more sensible. Moreover, vertical shoot position (VSP)-based training systems, such as cordon and Guyot, are generously adapted worldwide. However, the presence of longer than one-year-old wood makes cane-pruned training systems suitable for the pruning style transformations with a lower limitation of plants. Global climate changes have created a lot of new challenges for grapevine development. The task is concerned with tackling features such as the length of growing season, phenology formation, ripening, overheating and sunburnt stress, and extreme weather events. However, is there any coincidence that makes the training system an adaptation tool of viticulture against climate changes [141,142]?

5.5. Future of Training Systems and Canopy Modifications

The future of training systems, regardless of the mechanization demands, has various needs depending on the climate conditions of the relevant areas. Therefore, we can divide into two main climate models—cool and warm climate areas. According to [143], interest in the heresy of pruning and trellis models broke out as average temperatures rose through vine-growing regions worldwide. The laboring conditions became inconsistent in various areas [143]. Vintners understand that under these changes, they have to utilize the physiological tools of the vine. Nowadays, they are working with different varieties and rootstock selections, genotypes, and training systems to accelerate the ripening process through the controlled yield and vigor of plants. In line with the study of [144], vineyards are receiving ideal shade content, better bunch exposure, aromatic and phenolic expression, and leaf-to-fruit nutrient ratios. Meanwhile, in the scenario of [145], current warm areas head to becoming sub-tropical and the cool climate areas would rise in quality. Traditional SHW, VSP, or gobelets are limited by the number of canopy manipulations, as the trend tends to increase the number of manipulations, building more from fresh leaves and shade. Despite the selection of suitable training models, it will be essential to change our minds about plant demands, especially cluster shading, delayed or slowed ripening, drought, and frost damage. It will be necessary to start choosing natural-shading systems such as curtains of different SHW representatives that are suitable for mechanization [145].

In humid and relatively cool areas, such as Poland, where hybrid varieties are preferred and where white wines represent most of the production, a downward canopy should be preferred for it is easier to maintain by mechanical trimming close-to-the-ground. Meanwhile, in Mediterranean areas, the SHW cordon trellis is also developing, which has an ability to produce more shade for clusters inside the canopy. The effect of shading is one of the present precipices of viticulture [86,146]. The trimming of VSP tends to promote the habitus of canopy and is very useful to control the vigor of reluctant varieties, such as Pinots. In fact, there are also varieties which tend to develop a straight and modified canopy, such as Cabernets [147].

Moreover, recent publications have compared the characteristics based on the development and quality of gas and solar exchange in the canopy. Many 3D-spaced models of canopy have been integrated, which are declared as the non-spreading ones. These are enabled to intercept higher light ratio into the best favorable cluster microclimate with lesser requirements for cultivation works as the classic VSP demands. The research into which deals with efficient sun lit perceptive by canopy—leaf area index (LAI)—showed that the VSP system lacks more than 25–30% of this index in comparison to the bilateral hang-cordon [148].

However, in different validity, there is a study [149] investigating the measurement of CO_2 rate (NCER) in an enclosed vine system. However, this study had to deal with uncertain factors of compact leaf area, row-spacing, different light exposure, or aging. The vines were manipulated in two different forms, simulating the VSP canopy conditions and free-form sprawl canopy. Through additional studies by the team, the hanging sprawl canopy received a 26% higher photosynthetic rate due to better sunlight penetration inside the layers of canopy. In addition, the VSP model achieved less midday light capture. According to these results, the sprawling canopy has a lot of potential. Even if we consider it formally as a shading variant, there is a possibility for this training to reach better physiologic content than VSP [149].

There has been recent research into grapevine training systems and related pruning consequences, vine spacing, precise viticulture, and structure development. As a topic of research, it attracts less funding than any other area of research (e.g., smart viticulture, mechanized viticulture) [150,151]. Nevertheless, global climate changes obtain more of the focus, so there are explored areas such as New World (South Africa, Argentina, Chile) or Mediterranean lands (Balkan, Italy), where a training model gains more relevance as a remedy of viticultural plasticity. Adaptation to climate change is the main part of modulation of a regional grape-growing business by matching trunk geometry and canopy shape. As the conditions are changing year by year, it will soon become a relevant field to study the potential of trellis adaptation in growing areas as a trait of sustainable viticulture [152].

6. Conclusions

Due to the impacts of climate change, it is possible that the typical conditions for grapevine cultivation will need to be adjusted. These changes in average temperatures or sunlight distribution primarily concern the training systems that help plants withstand extreme weather conditions. However, high average temperatures, sunlight exposure, and strong UV have a serious impact on the balanced quality of grapes—high alcohol content, lack of freshness, lower acidity, and subtle aromatic compounds. This review examined both well-known and traditional, as well as unconventional, vine training systems typical of the so-called cool climate zone of viticulture. The canopy of these cool-climate available training vine systems, is set up on trellises of various shapes, sizes, and spatial arrangements. Use of various trellis designs depends on the terrain or site conditions with the canopy opened for sunlight. Grapes located within these training systems are exposed to the sunlight but also well-protected by a vigorous canopy. The most common vine training system in the Central European area is the VSP type, or vertical shoot positioning. Any training within this shape of canopy involves pruning to canes or spurs and this style is nearly universal for the majority of grape-growing regions. The main reason for this preference is the comfortable position of the grapes and so their exposure to sunlight. However, in recent years, even vine-growing regions in cool climate zone have experienced conditions typical for warmer, southern wine regions, such as long periods of unusual higher temperature and intense sunlight. Therefore, it is essential to assess the efficiency of the training systems with trellis models. Growers will have to evaluate how well the chosen trellis system suits in the upcoming extremes. It would become an essential aspect of viticulture to consider the technical aspects of the chosen trellis design. Selecting the suitable training system should be a visionary ace for growers, as vineyards are established for several generations and must adapt to changing conditions over time. With prompt conditions for cultivation, there will be possibilities of stabilizing the shape of the most vulnerable varieties, such as Chardonnay, Sauvignon, Viognier, Gamay, or Merlot, in various regions. Recent analysis has proven numerous problems for grapes caused by excessive sunlight exposure. With upgrades to the shading capability of the trellis, there would be better freshness and even improvements to the aromatic and sensory profile of wines.

To assess the appropriate growing conditions in cool regions requires consideration of the total sunlight capability and the annual temperatures of vegetation period for different varieties. Using classic VSP in northern cool climate viticulture is probably the most suitable option for plenty of cultivars. The winemaker can profit from simple vertical exposition of grapes to the sun. Viticulturists are available to easily maintain the canopy if they need to obtain more the sun exposure, to empower the aromatic and sugar content. A simply designed trellis for a canopy of Guyot or spurred cordon training turns the main part of the grapes to the sunlight. With the combination of sloping terrain and east-to-west canopy orientation, it is much easier to contribute the benefits of terroir. However, the further south we go on the map of wine regions, the more areas face challenges with excessive grape exposure and overheating. Southern wine-growing states are gradually moving away from simple vertical shoot positioning (VSP) and instead favoring either spatial or hanging trellis systems. Somewhere between the northern and southern wine regions lies the challenging zone where vintage variability is highly unpredictable. Countries like Germany, Austria, Slovenia, Northern Italy, Czechia or Georgia are increasingly struggling with the classic, single VSP training system. For these regions, it is advisable to weigh the pros and cons of VSP and potentially start exploring shading solutions. Suitable systems for areas like Germany, Austria, as well as the Czech Republic and Slovenia, could include Lyre, Y-type, or SAYM vertical variations. However, these systems come with high costs due to trellis installation, along with significantly higher yields, which is less typical for central regions that usually avoid high-yielding production.

In contrast, regions like Northern Italy, specifically Franciacorta, traditionally used vertical Sylvoz training before it was largely replaced by cordon and Guyot systems. However, local growers now lean towards hanging systems for the future. In recent vintages, vintners of several wine regions have begun experimenting with variations in cordon curtains. Producers have to seek a balance between vigor and yield to maintain wine quality.

Author Contributions: Conceptualization, R.D.; methodology, R.D.; software, R.D.; validation, P.P.; formal analysis, R.D.; investigation, R.D.; resources, R.D.; data collection, R.D. and P.P.; writing—original draft preparation, R.D.; writing—review and editing, R.D.; visualization, M.K.; supervision, P.P.; project administration, K.E.K.; funding acquisition, R.D. All authors have read and agreed to the published version of the manuscript.

Funding: This article was performed within the project of the Internal Grant Agency of Mendel University in Brno, Faculty of Horticulture in Lednice, Czech Republic—Possibilities of transforming grapevine training systems and pruning according to the Simonit&Sirch method, ZF/2023-SI2-003. This study was conducted as part of the project financed by the Mendel University in Brno, Czech Republic—Implementation of ecosystem services with a focus on water balance in viticultural practice, QK21010189.

Institutional Review Board Statement: Not applicable.

Data Availability Statement: The data presented in this study are available on request from the corresponding author.

Conflicts of Interest: The authors declare no conflicts of interest.

Glossary

Branches—portion of living wood from previous vegetations, connected with structure of trunk and secondary structures.
Budburst—the annual onset of grapevine's growing season. Buds swell and burst due to the activity of inner tissues.
Cane—term used to determine one-year-old wood considered as the bearer of yield after pruning. It has to be longer than three fertile buds.
Canopy—the controlled complex of green shoots developed by plant.
Decay, death arm, dieback—different terms for large, decreased pruning wounds with various overlap into the inner tissues.
Desiccation (cone)—process, in which pruning wound dries out into the inner tissues of plant, marking a specific shape of cone/arrow.
Dormancy—physiological stage of grapevine. It is considered as the period of plant's rest during winter. Its length is tied to regional climatic conditions.
GTDs—complex of grapevine trunk diseases, dangerous group of wood-spreading fungi of vine tissues. Well-known symptoms are those of Eutypa, ESCA, Botryosphaeria dieback or Petri disease.
Permanent structures—that part of plant, which is designed and trained to stay and develop in time with no needs for reposition. Simonit&Sirch divided permanent structures into several parts:
primary—the trunk—vertical position, based since the start
secondary—the arms—horizontal position, ever-evolving with each year
tertiary—one-year-old wood, cane or spurs.
Phloem—specific inner tissue of woody plants, bast texture.
Pruning (vine)—specific method of selection of one-year-old fruitful, lignified shoots, that will bear the crop and canopy production for the next vegetation.
Regrafting—special technique in tree and vine production, that involves replacing the existing scion with new, more desirable variety while retaining the established root system. This technique is used to improve crop quality or plant adaptability without replanting.

Ramification—specific technique of building structural head grapevine plant. This technique uses portion of living wood from each of previous growing cycles, branches. It was promptly summarized by Simonit&Sirch pruning method.
Ravaz index—is used for evaluation of vine's vigor and vitality compared with yield and canopy distribution.
Sap-flow—is a term commonly used for the full size of inner vessels, to point out the direction, health and capacity of plant's structures.
Spur—term used to determine short (less than three fertile buds) one-year-old wood of cordon training or the whole ramified part of cordon arm, as known as tertiary structure.
Sucker (water-shoot)—green shoots bursting randomly anywhere on plant, not included cane/spur location. So-called water-shoots, they are not bearers of regular inflorescences.
Vegetation (stage)—in terms of viticulture, it is used to summarize the whole period of intense green biomass growth of the plant.
Veraison—a part of the grape's growing cycle, during which the skins undergo coloration.
Terroir (of vineyard)—refers to the unique signature of climate, topography, soil and human work that influence the character and quality of the grapes grown there.
Training system (grapevine)—is a vector of plant development in the trellis space. It also indicates the crop dispositions for laboring, protection and harvesting features.
Trellis model—is a specific construction (wires, poles, stakes) suitable for specific grapevine training system. Its role is to bear vegetation and to ease the sunlight exposure for shoots.
Xylem—specific tissue of woody plants—wood texture.

References

1. Bakos, J.L.; Ladányi, M.; Szalay, L. Frost Hardiness of Flower Buds of 16 Apricot Cultivars during Dormancy. *Folia Hortic.* **2024**, *36*, 81–93. [CrossRef]
2. Kobayashi, Y.; Yamamoto, T.; Ikeda, H.; Sugihara, R.; Kaihori, H.; Kawabata, M.; Suzuki, S. Effects of Constantly High Soil Water Content on Vegetative Growth and Grape Quality in Japan with High Rainfall during Grapevine Growing Season. *Folia Hortic.* **2020**, *32*, 135–145. [CrossRef]
3. Grabowski, M.; Pawlikowski, M. Biomineralogical Investigation of Late-Harvest Grapes Colonised by Pers. *Folia Hortic.* **2020**, *32*, 171–178. [CrossRef]
4. Živković, S.P.; Vasić, T.P.; Marković, J.P.; Jevremović, D.R. Susceptibility of Grapevine Cultivars to Eutypa Lata in Serbia. *Acta Sci. Pol. Hortorum Cultus* **2023**, *22*, 105–116. [CrossRef]
5. Zohary, D.; Spiegel-Roy, P. Beginnings of Fruit Growing in the Old World. *Science* **1975**, *187*, 319–327. [CrossRef]
6. Kopta, T.; Ragasová, L.N.; Sotolář, R.; Sedláček, J.; Ferby, V.; Hurajová, E.; Winkler, J. The Influence of Different Methods of Under-Vine Management on the Structure of Vegetation and the Qualitative Parameters of the Grapes in the Moravian Wine Region. *Folia Hortic.* **2024**, *36*, 235–257. [CrossRef]
7. Vezzulli, S.; Doligez, A.; Bellin, D. Molecular Mapping of Grapevine Genes. In *The Grape Genome*; Cantu, D., Walker, M.A., Eds.; Compendium of Plant Genomes; Springer International Publishing: Cham, Switzerland, 2019; pp. 103–136, ISBN 978-3-030-18600-5.
8. Smart, R.; Robinson, M. *Sunlight into Wine: A Handbook for Winegrape Canopy Management*; Winetitles: Adelaide, Australia, 1991; ISBN 978-1-875130-10-8.
9. Baránková, K.; Sotolář, R.; Baránek, M. Identification of Rare Traditional Grapevine Cultivars Using SSR Markers and Their Geographical Location within the Czech Republic. *Czech J. Genet. Plant Breed.* **2020**, *56*, 71–78. [CrossRef]
10. Shangguan, L.; Wang, C.; Kayesh, E.; Zhang, Y.P.; Korir, N.; Han, J.; Fang, J.G. Review and Structural Analysis of the Evolution of Grapevine (*Vitis vinifera* L.) Genes Involved in Flower and Fruit Development. *J. Hortic. Sci. Biotechnol.* **2012**, *87*, 243–249. [CrossRef]
11. Palonen, P.; Buszard, D. Current State of Cold Hardiness Research on Fruit Crops. Available online: https://cdnsciencepub.com/doi/abs/10.4141/P96-013 (accessed on 10 September 2024).
12. Petitpierre, B.; Arnold, C.; Phelps, L.; Guisan, A. A Tale of Three Vines: Current and Future Threats to Wild Eurasian Grapevine by Vineyards and Invasive Rootstocks. *Divers. Distrib.* **2023**, *29*, 1594–1608. [CrossRef]
13. Sgubin, G.; Swingedouw, D.; Mignot, J.; Gambetta, G.A.; Bois, B.; Loukos, H.; Noël, T.; Pieri, P.; García De Cortázar-Atauri, I.; Ollat, N.; et al. Non-linear Loss of Suitable Wine Regions over Europe in Response to Increasing Global Warming. *Glob. Chang. Biol.* **2023**, *29*, 808–826. [CrossRef]
14. Naulleau, A.; Gary, C.; Prévot, L.; Hossard, L. Evaluating Strategies for Adaptation to Climate Change in Grapevine Production—A Systematic Review. *Front. Plant Sci.* **2021**, *11*, 607859. [CrossRef] [PubMed]
15. Xynas, B.; Barnes, C. Yeast or Water: Producing Wine with Lower Alcohol Levels in a Warming Climate: A Review. *J. Sci. Food Agric.* **2023**, *103*, 3249–3260. [CrossRef] [PubMed]
16. Droulia, F.; Charalampopoulos, I. Future Climate Change Impacts on European Viticulture: A Review on Recent Scientific Advances. *Atmosphere* **2021**, *12*, 495. [CrossRef]

17. Luković, J.; Burić, D.; Mihajlović, J.; Pejović, M. Spatial and Temporal Variations of Aridity-Humidity Indices in Montenegro. *Theor. Appl. Climatol.* **2024**, *155*, 4553–4566. [CrossRef]
18. Albani, M.C.; Coupland, G. Chapter Eleven—Comparative Analysis of Flowering in Annual and Perennial Plants. In *Current Topics in Developmental Biology*; Timmermans, M.C.P., Ed.; Plant Development; Academic Press: Cambridge, MA, USA, 2010; Volume 91, pp. 323–348.
19. Dinu, D.-G.; Ricciardi, V.; Demarco, C.; Zingarofalo, G.; De Lorenzis, G.; Buccolieri, R.; Cola, G.; Rustioni, L. Climate Change Impacts on Plant Phenology: Grapevine (*Vitis vinifera*) Bud Break in Wintertime in Southern Italy. *Foods* **2021**, *10*, 2769. [CrossRef]
20. Greb, T.; Lohmann, J.U. Plant Stem Cells. *Curr. Biol.* **2016**, *26*, R816–R821. [CrossRef]
21. Keller, M. *The Science of Grapevines*; Academic Press: Cambridge, MA, USA, 2020; ISBN 978-0-12-816702-1.
22. Costes, E. Physiology and Genetics of Plant Architecture. In *Annual Plant Reviews Online*; John Wiley & Sons: Hoboken, NJ, USA, 2019; pp. 1031–1068, ISBN 978-1-119-31299-4.
23. Naor, A.; Gal, Y.; Bravdo, B. Shoot and Cluster Thinning Influence Vegetative Growth, Fruit Yield, and Wine Quality of "Sauvignon Blanc" Grapevines. *J. Am. Soc. Hortic. Sci.* **2002**, *127*, 628–634. [CrossRef]
24. Lecomte, P.; Cholet, C.; Bruez, E.; Martignon, T.; Giudici, M.; Simonit, M.; Alonso Ugaglia, A.; Forget, D.; Miramon, J.; Arroyo, M.; et al. Recovery after Curettage of Grapevines with Esca Leaf Symptoms. *Phytopathol. Mediterr.* **2022**, *61*, 473–489. [CrossRef]
25. Barthélémy, D.; Caraglio, Y. Plant Architecture: A Dynamic, Multilevel and Comprehensive Approach to Plant Form, Structure and Ontogeny. *Ann. Bot.* **2007**, *99*, 375–407. [CrossRef]
26. Zimmermann, M.H.; Milburn, J.A. Transport and Storage of Water. In *Physiological Plant Ecology II: Water Relations and Carbon Assimilation*; Lange, O.L., Nobel, P.S., Osmond, C.B., Ziegler, H., Eds.; Springer: Berlin/Heidelberg, Germany, 1982; pp. 135–151, ISBN 978-3-642-68150-9.
27. Louarn, G. Analyse et Modélisation de l'Organogenèse et de l'Architecture du Rameau de Vigne (*Vitis vinifera* L.). Ph.D. Thesis, École Nationale Supérieure Agronomique de Montpellier, Montpellier, France, 2005.
28. Costes, E.; Lauri, P.-E.; Simon, S.; Andrieu, B. Plant Architecture, Its Diversity and Manipulation in Agronomic Conditions, in Relation with Pest and Pathogen Attacks. *Eur. J. Plant Pathol.* **2013**, *135*, 455–470. [CrossRef]
29. Carbonneau, A.; Torregrosa, L. *Traité de la Vigne*, 3rd ed.; Dunod: Malakoff, France, 2020; ISBN 978-2-10-079857-5.
30. Bruez, E.; Cholet, C.; Giudici, M.; Simonit, M.; Martignon, T.; Boisseau, M.; Weingartner, S.; Poitou, X.; Rey, P.; Geny-Denis, L. Pruning Quality Effects on Desiccation Cone Installation and Wood Necrotization in Three Grapevine Cultivars in France. *Horticulturae* **2022**, *8*, 681. [CrossRef]
31. De Barros, M.I.L.F.; Frölech, D.B.; De Mello, L.L.; Manica-Berto, R.; Malgarim, M.B.; Costa, V.B.; Mello-Farias, P. Impact of Cluster Thinning on Quality of "Malbec" Grapes in Encruzilhada Do Sul-RS. *Am. J. Plant Sci.* **2018**, *9*, 495–506. [CrossRef]
32. Ravaz, M.K. Influence Spécifique Réciproque Du Greffon Et Du Sujet Chez La Vigne. *Bull. Société Bot. Fr.* **1903**, *50*, 87–100. [CrossRef]
33. Torregrosa, L.; Carbonneau, A.; Kelner, J.-J. The Shoot System Architecture of *Vitis vinifera* ssp. *Sativa. Sci. Hortic.* **2021**, *288*, 110404. [CrossRef]
34. Scorza, L.C.T.; Dornelas, M.C. Plants on the Move: Towards Common Mechanisms Governing Mechanically-Induced Plant Movements. *Plant Signal. Behav.* **2011**, *6*, 1979–1986. [CrossRef]
35. Torregrosa, L.J.-M.; Rienth, M.; Romieu, C.; Pellegrino, A. The Microvine, a Model for Studies in Grapevine Physiology and Genetics. *OENO One* **2019**, *53*, 373–391. [CrossRef]
36. Pratt, C. Vegetative Anatomy of Cultivated Grapes—A Review. *Am. J. Enol. Vitic.* **1974**, *25*, 131–150. [CrossRef]
37. Pavloušek, P. 10—Grapevine Breeding in Central and Eastern Europe. In *Grapevine Breeding Programs for the Wine Industry*; Reynolds, A., Ed.; Woodhead Publishing Series in Food Science, Technology and Nutrition; Woodhead Publishing: Oxford, UK, 2015; pp. 211–244, ISBN 978-1-78242-075-0.
38. Galet, P. *Cépages et Vignobles de France*, 2nd ed.; Entièrement Refondue; Dehan: Montpellier, France, 1990; ISBN 978-2-902771-04-2.
39. Adão, F.; Santos, J.A.; Fraga, H.; Malheiro, A.C. Assessment of Grapevine Sap Flow and Trunk Diameter Variations in Mediterranean Climate Using Time Series Decomposition. *VITIS-J. Grapevine Res.* **2023**, *62*, 97–105. [CrossRef]
40. Claverie, M.; Lecomte, P.; Delorme, G.; Dumot, V.; Jacquet, O.; Cochard, H. Xylem Water Transport Is Influenced by Age and Winter Pruning Characteristics in Grapevine (*Vitis vinifera*). *OENO One* **2023**, *57*, 53–68. [CrossRef]
41. Nigond, J. Quelques Aspects de La Dormance Des Bourgeons de La Vigne Sous Le Climat Du Languedoc. *Bull. Société Bot. Fr.* **1966**, *113*, 85–99. [CrossRef]
42. Gu, S.; McCarthy, B.; Gohil, H. Fruit Quality and Vine Vigor of Cabernet Sauvignon Grapevines Under Crop Forcing in a Warm Region to Produce Cool Climate Quality Fruit. In Proceedings of the 2011 ASHS Annual Conference, Waikoloa, HI, USA, 23–30 September 2011.
43. Chervin, C.; Fennell, A. Ethanol Sprays to Release Grapevine Bud Dormancy: A Potential Alternative to Cyanamides. *OENO One* **2019**, *53*, 661–666. [CrossRef]
44. Sudawan, B.; Chang, C.-S.; Chao, H.; Ku, M.S.B.; Yen, Y. Hydrogen Cyanamide Breaks Grapevine Bud Dormancy in the Summer through Transient Activation of Gene Expression and Accumulation of Reactive Oxygen and Nitrogen Species. *BMC Plant Biol* **2016**, *16*, 202. [CrossRef] [PubMed]
45. Pou, A.; Balda, P.; Albacete, A.; Martínez De Toda, F. Forcing Vine Regrowth to Delay Ripening and Its Association to Changes in the Hormonal Balance. *VITIS-J. Grapevine Res.* **2019**, *58*, 95–101. [CrossRef]

46. Camargo-A, H.A.; Salazar-G., M.R.; Zapata, D.M.; Hoogenboom, G. Predicting the Dormancy and Bud Break Dates for Grapevines. *Acta Hortic.* **2017**, *1182*, 153–160. [CrossRef]
47. Ribeiro, T.H.C.; Fernandes-Brum, C.N.; De Souza, C.R.; Dias, F.A.N.; Almeida-Junior, O.D.; Regina, M.D.A.; De Oliveira, K.K.P.; Dos Reis, G.L.; Oliveira, L.M.; Fernandes, F.D.P.; et al. Transcriptome Analyses Suggest That Changes in Fungal Endophyte Lifestyle Could Be Involved in Grapevine Bud Necrosis. *Sci. Rep.* **2020**, *10*, 9514. [CrossRef]
48. Lebon, E. Shoot Development in Grapevine (*Vitis vinifera*) Is Affected by the Modular Branching Pattern of the Stem and Intra- and Inter-Shoot Trophic Competition. *Ann. Bot.* **2004**, *93*, 263–274. [CrossRef]
49. Gatti, M.; Pirez, F.J.; Chiari, G.; Tombesi, S.; Palliotti, A.; Merli, M.C.; Poni, S. Phenology, Canopy Aging and Seasonal Carbon Balance as Related to Delayed Winter Pruning of *Vitis vinifera* L. Cv. Sangiovese Grapevines. *Front. Plant Sci.* **2016**, *7*, 659. [CrossRef]
50. Zheng, W.; García, J.; Balda, P.; Martínez De Toda, F. Effects of Late Winter Pruning at Different Phenological Stages on Vine Yield Components and Berry Composition in La Rioja, North-Central Spain. *OENO One* **2017**, *51*, 363. [CrossRef]
51. Rosace, M.C.; Legler, S.E.; Salotti, I.; Rossi, V. Susceptibility of Pruning Wounds to Grapevine Trunk Diseases: A Quantitative Analysis of Literature Data. *Front. Plant Sci.* **2023**, *14*, 1063932. [CrossRef]
52. De Almeida Junior, O.; De Souza, C.R.; Alcantara Novelli Dias, F.; De Paula Fernandes, F.; Torregrosa, L.; Noronha Fernandes-Brum, C.; Chalfun, A., Jr.; Vieira Da Mota, R.; Peregrino, I.; De Albuquerque Regina, M. Effect of Pruning Strategy on "Syrah" Bud Necrosis and Fruitfulness in Brazilian Subtropical Southeast. *VITIS-J. Grapevine Res.* **2019**, *58*, 87–94. [CrossRef]
53. Pellegrino, A.; Rogiers, S.; Deloire, A. Grapevine Latent Bud Dormancy and Shoot Development: Original Language of the Article: English. *IVES Tech. Rev. Vine Wine* **2020**. [CrossRef]
54. Meza, L.; Deyett, E.; Vallance, J.; Gendre, L.; Garcia, J.F.; Cantu, D.; Rey, P.; Lecomte, P.; Rolshausen, P.E. Grapevine Pruning Strategy Affects Trunk Disease Symptoms, Wood Pathobiome and Mycobiome. *Phytopathol. Mediterr.* **2024**, *63*, 91–102. [CrossRef]
55. Perin, C.; Verma, P.K.; Harari, G.; Suued, Y.; Harel, M.; Ferman-Mintz, D.; Drori, E.; Netzer, Y.; Fait, A. Influence of Late Pruning Practice on Two Red Skin Grapevine Cultivars in a Semi-Desert Climate. *Front. Plant Sci.* **2023**, *14*, 1114696. [CrossRef] [PubMed]
56. Netzer, Y.; Suued, Y.; Harel, M.; Ferman-Mintz, D.; Drori, E.; Munitz, S.; Stanevsky, M.; Grünzweig, J.M.; Fait, A.; Ohana-Levi, N.; et al. Forever Young? Late Shoot Pruning Affects Phenological Development, Physiology, Yield and Wine Quality of *Vitis vinifera* Cv. Malbec. *Agriculture* **2022**, *12*, 605. [CrossRef]
57. Luque, J.; Elena, G.; Garcia-Figueres, F.; Reyes, J.; Barrios, G.; Legorburu, F.J. Natural Infections of Pruning Wounds by Fungal Trunk Pathogens in Mature Grapevines in Catalonia (Northeast Spain): Natural Fungal Infections of Pruning Wounds. *Aust. J. Grape Wine Res.* **2014**, *20*, 134–143. [CrossRef]
58. Lawrence, D.P.; Nouri, M.T.; Trouillas, F.P. Taxonomy and Multi-Locus Phylogeny of Cylindrocarpon-like Species Associated with Diseased Roots of Grapevine and Other Fruit and Nut Crops in California. *Fungal Syst. Evol.* **2019**, *4*, 59–75. [CrossRef]
59. Gramaje, D.; Úrbez-Torres, J.R.; Sosnowski, M. Managing Grapevine Trunk Diseases with Respect to Etiology and Epidemiology: Current Strategies and Future Prospects. *Plant Dis.* **2018**, *102*, 12–39. [CrossRef]
60. Sosnowski, M.R.; Ayres, M.R.; Billones-Baaijens, R.; Savocchia, S.; Scott, E.S. Susceptibility of Pruning Wounds to Grapevine Trunk Disease Pathogens *Eutypa lata* and *Diplodia seriata* in Three Climatic Conditions in Australia. *Fungal Ecol.* **2023**, *64*, 101260. [CrossRef]
61. O'Brien, P.; Bei, R.; Collins, C. Assessing the Relationship between Cordon Strangulation, Dieback, and Fungal Trunk Disease Symptom Expression in Grapevine. *OENO One* **2023**, *57*, 151–160. [CrossRef]
62. Makatini, G.J.; Halleen, F.; Mutawila, C.; Moyo, P.; Mostert, L. Susceptibility of Grapevine Sucker and Green Shoot Wounds to Trunk Disease Pathogens. *S. Afr. J. Enol. Vitic.* **2023**, *44*, 55–63. [CrossRef]
63. Smart, R.E. Principles of Grapevine Canopy Microclimate Manipulation with Implications for Yield and Quality. A Review. *Am. J. Enol. Vitic.* **1985**, *36*, 230–239. [CrossRef]
64. Bates, T. Pruning Level Affects Growth and Yield of New York Concord on Two Training Systems. *Am. J. Enol. Vitic.* **2008**, *59*, 276–286. [CrossRef]
65. Schultz, H.R.; Stoll, M. Some Critical Issues in Environmental Physiology of Grapevines: Future Challenges and Current Limitations. *Aust. J. Grape Wine Res.* **2010**, *16*, 4–24. [CrossRef]
66. Zhao, X.H.; Liu, L.Y.; Nan, L.J.; Wang, H.; Li, H. Development of Tyloses in the Xylem Vessels of Meili Grapevine and Their Effect on Water Transportation. *Russ. J. Plant Physiol.* **2014**, *61*, 194–203. [CrossRef]
67. Todaro, T.; Dami, I. Cane Morphology and Anatomy Influence Freezing Tolerance in *Vitis vinifera* Cabernet Franc. *Int. J. Fruit Sci.* **2017**, *17*, 391–406. [CrossRef]
68. De Souza, C.R.; Vieira Da Mota, R.; Novelli Dias, F.A.; De Melo, E.T.; De Souza, L.C.; De Souza, A.L.; Meirelles De Azevedo Pimentel, R.; De Albuquerque Regina, M. Starch Accumulation and Agronomical Performance of Syrah under Winter Cycle: Responses to Pruning and Ethephon Management. *VITIS-J. Grapevine Res.* **2015**, *54*, 195–201. [CrossRef]
69. Falsini, S.; Moretti, S.; Battiston, E.; Tani, C.; Papini, A.; Carella, G.; Nocentini, M.; Mugnai, L.; Schiff, S. Phytopathologia Mediterranea Grapevine Histological Responses to Pruning: The Influence of Basal Buds on Tissue Defence Reactions. *Phytopathol. Mediterr.* **2023**, *62*, 321–332. [CrossRef]
70. Amponsah, N.T.; Jones, E.E.; Ridgway, H.J.; Jaspers, M.V. Susceptibility of Grapevine Tissues to *Neofusicoccum luteum* Conidial Infection. *Plant Pathol.* **2012**, *61*, 719–729. [CrossRef]

71. Thomas, A.L.; Harris, J.L.; Bergmeier, E.A.; Striegler, R.K. Establishment Technique and Rootstock Impact 'Chambourcin' Grapevine Morphology and Production in Missouri. *HortTechnology* **2017**, *27*, 248–256. [CrossRef]
72. Pinto, C.; Custódio, V.; Songy, A.; Rabenoelina, F.; Courteaux, B.; Clément, C.; Gomes, A.; Florence, F. Understand the Potential Role of *Aureobasidium pullulans*, a Resident Microorganism From Grapevine, to Prevent the Infection Caused by *Diplodia seriata*. *Front. Microbiol.* **2018**, *9*, 3047. [CrossRef]
73. Songy, A.; Fernandez, O.; Clément, C.; Larignon, P.; Fontaine, F. Grapevine Trunk Diseases under Thermal and Water Stresses. *Planta* **2019**, *249*, 1655–1679. [CrossRef] [PubMed]
74. Wilcox, W.F.; Gubler, W.D.; Uyemoto, J.K. (Eds.) *Compendium of Grape Diseases, Disorders, and Pests, Second Edition*; The American Phytopathological Society: St. Paul, MN, USA, 2015; ISBN 978-0-89054-481-5.
75. Gómez, P.; Báidez, A.G.; Ortuño, A.; Del Río, J.A. Grapevine Xylem Response to Fungi Involved in Trunk Diseases. *Ann. Appl. Biol.* **2016**, *169*, 116–124. [CrossRef]
76. Aigoun-Mouhous, W.; Elena, G.; Cabral, A.; León, M.; Sabaou, N.; Armengol, J.; Chaouia, C.; Mahamedi, A.E.; Berraf-Tebbal, A. Characterization and Pathogenicity of Cylindrocarpon-like Asexual Morphs Associated with Black Foot Disease in Algerian Grapevine Nurseries, with the Description of *Pleiocarpon algeriense* sp. Nov. *Eur. J. Plant Pathol.* **2019**, *154*, 887–901. [CrossRef]
77. Berlanas, C.; Ojeda, S.; López-Manzanares, B.; Andrés-Sodupe, M.; Bujanda, R.; Del Pilar Martínez-Diz, M.; Díaz-Losada, E.; Gramaje, D. Occurrence and Diversity of Black-Foot Disease Fungi in Symptomless Grapevine Nursery Stock in Spain. *Plant Dis.* **2020**, *104*, 94–104. [CrossRef] [PubMed]
78. Pouzoulet, J.; Pivovaroff, A.; Santiago, L.; Rolshausen, P. Can Vessel Dimension Explain Tolerance toward Fungal Vascular Wilt Diseases in Woody Plants? Lessons from Dutch Elm Disease and Esca Disease in Grapevine. *Front. Plant Sci.* **2014**, *5*, 253. [CrossRef]
79. Claverie, M.; Notaro, M.; Fontaine, F.; Wery, J. Current Knowledge on Grapevine Trunk Diseases with Complex Etiology: A Systemic Approach. *Phytopathol. Mediterr.* **2020**, *59*, 29–53. [CrossRef]
80. Celette, F.; Gary, C. Dynamics of Water and Nitrogen Stress along the Grapevine Cycle as Affected by Cover Cropping. *Eur. J. Agron.* **2013**, *45*, 142–152. [CrossRef]
81. Battiston, E.; Falsini, S.; Giovannelli, A.; Schiff, S.; Tani, C.; Panaiia, R.; Papini, A.; Di Marco, S.; Mugnai, L. Xylem Anatomy and Hydraulic Traits in *Vitis* Grafted Cuttings in View of Their Impact on the Young Grapevine Decline. *Front. Plant Sci.* **2022**, *13*, 1006835. [CrossRef]
82. Martinson, T.; Mansfield, A.K.; Luby, J.; Gartner, W.; Dharmadhikari, M.; Domoto, P. The Northern Grapes Project: Integrating Viticulture, Enology, and Marketing of New Cold-Hardy Wine Grape Cultivars in the Midwest and Northeast United States. *Acta Hortic.* **2016**, *1115*, 3–12. [CrossRef]
83. Deloire, A. A Few Thoughts on Grapevine Training Systems. *Wineland Mag.* **2012**, *6*, 82–86.
84. Xyrafis, E.; Gambetta, G.; Biniari, K. A Comparative Study on Training Systems and Vine Density in Santorini Island: Physiological, Microclimate, Yield and Quality Attributes. *OENO One* **2023**, *57*, 141–152. [CrossRef]
85. Lebon, E.; Pellegrino, A.; Louarn, G.; Lecoeur, J. Branch Development Controls Leaf Area Dynamics in Grapevine (*Vitis vinifera*) Growing in Drying Soil. *Ann. Bot.* **2006**, *98*, 175–185. [CrossRef] [PubMed]
86. Louarn, G.; Guedon, Y.; Lecoeur, J.; Lebon, E. Quantitative Analysis of the Phenotypic Variability of Shoot Architecture in Two Grapevine (*Vitis vinifera*) Cultivars. *Ann. Bot.* **2007**, *99*, 425–437. [CrossRef] [PubMed]
87. Reynolds, A.G.; Heuvel, J.E.V. Influence of Grapevine Training Systems on Vine Growth and Fruit Composition: A Review. *Am. J. Enol. Vitic.* **2009**, *60*, 251–268. [CrossRef]
88. Coupel-Ledru, A.; Lebon, E.; Christophe, A.; Gallo, A.; Gago, P.; Pantin, F.; Doligez, A.; Simonneau, T. Reduced Nighttime Transpiration Is a Relevant Breeding Target for High Water-Use Efficiency in Grapevine. *Proc. Natl. Acad. Sci. USA* **2016**, *113*, 8963–8968. [CrossRef]
89. Schwabe, K.; Albiac, J.; Connor, J.; Hassan, R.; Meza Gonzalez, L. *Drought in Arid and Semi-Arid Regions: A Multi-Disciplinary and Cross-Country Perspective*; Schwabe, K., Ed.; Springer: Dordrecht, The Netherlands; New York, NY, USA, 2013; ISBN 978-94-007-6635-8.
90. Viveros Santos, I.; Renaud-Gentié, C.; Roux, P.; Levasseur, A.; Bulle, C.; Deschênes, L.; Boulay, A.-M. Prospective Life Cycle Assessment of Viticulture under Climate Change Scenarios, Application on Two Case Studies in France. *Sci. Total Environ.* **2023**, *880*, 163288. [CrossRef]
91. Van Leeuwen, C.; Destrac-Irvine, A. Modified Grape Composition under Climate Change Conditions Requires Adaptations in the Vineyard. *OENO One* **2017**, *51*, 147–154. [CrossRef]
92. Medrano, H.; Pou, A.; Tomás, M.; Martorell, S.; Gulias, J.; Flexas, J.; Escalona, J.M. Average Daily Light Interception Determines Leaf Water Use Efficiency among Different Canopy Locations in Grapevine. *Agric. Water Manag.* **2012**, *114*, 4–10. [CrossRef]
93. Bem, B.P.D.; Bogo, A.; Everhart, S.; Casa, R.T.; Gonçalves, M.J.; Filho, J.L.M.; Cunha, I.C.D. Effect of Y-Trellis and Vertical Shoot Positioning Training Systems on Downy Mildew and Botrytis Bunch Rot of Grape in Highlands of Southern Brazil. *Sci. Hortic.* **2015**, *185*, 162–166. [CrossRef]
94. Prieto, J.A.; Louarn, G.; Perez Peña, J.; Ojeda, H.; Simonneau, T.; Lebon, E. A Functional–Structural Plant Model That Simulates Whole- Canopy Gas Exchange of Grapevine Plants (*Vitis vinifera* L.) under Different Training Systems. *Ann. Bot.* **2020**, *126*, 647–660. [CrossRef]

95. Tandonnet, J.-P.; Marguerit, E.; Cookson, S.J.; Ollat, N. Genetic Architecture of Aerial and Root Traits in Field-Grown Grafted Grapevines Is Largely Independent. *Theor. Appl. Genet.* **2018**, *131*, 903–915. [CrossRef]
96. Louarn, G.; Lecoeur, J.; Lebon, E. A Three-Dimensional Statistical Reconstruction Model of Grapevine (*Vitis vinifera*) Simulating Canopy Structure Variability within and between Cultivar/Training System Pairs. *Ann. Bot.* **2007**, *101*, 1167–1184. [CrossRef] [PubMed]
97. Pallas, B.; Louarn, G.; Christophe, A.; Lebon, E.; Lecoeur, J. Influence of Intra-Shoot Trophic Competition on Shoot Development in Two Grapevine Cultivars (*Vitis vinifera*). *Physiol. Plant.* **2008**, *134*, 49–63. [CrossRef] [PubMed]
98. Houel, C.; Chatbanyong, R.; Doligez, A.; Rienth, M.; Foria, S.; Luchaire, N.; Roux, C.; Adivèze, A.; Lopez, G.; Farnos, M.; et al. Identification of Stable QTLs for Vegetative and Reproductive Traits in the Microvine (*Vitis vinifera* L.) Using the 18 K Infinium Chip. *BMC Plant Biol.* **2015**, *15*, 205. [CrossRef] [PubMed]
99. Coupel-Ledru, A.; Lebon, É.; Christophe, A.; Doligez, A.; Cabrera-Bosquet, L.; Péchier, P.; Hamard, P.; This, P.; Simonneau, T. Genetic Variation in a Grapevine Progeny (*Vitis vinifera* L. Cvs Grenache×Syrah) Reveals Inconsistencies between Maintenance of Daytime Leaf Water Potential and Response of Transpiration Rate under Drought. *J. Exp. Bot.* **2014**, *65*, 6205–6218. [CrossRef]
100. Zurowietz, A.; Lehr, P.P.; Kleb, M.; Merkt, N.; Gödde, V.; Bednarz, H.; Niehaus, K.; Zörb, C. Training Grapevines Generates a Metabolomic Signature of Wine. *Food Chem.* **2022**, *368*, 130665. [CrossRef]
101. Leolini, L.; Moriondo, M.; Fila, G.; Costafreda-Aumedes, S.; Ferrise, R.; Bindi, M. Late Spring Frost Impacts on Future Grapevine Distribution in Europe. *Field Crop. Res.* **2018**, *222*, 197–208. [CrossRef]
102. Van Leeuwen, C.; Destrac-Irvine, A.; Dubernet, M.; Duchêne, E.; Gowdy, M.; Marguerit, E.; Pieri, P.; Parker, A.; De Rességuier, L.; Ollat, N. An Update on the Impact of Climate Change in Viticulture and Potential Adaptations. *Agronomy* **2019**, *9*, 514. [CrossRef]
103. Victorino, G.F.; Braga, R.; Santos-Victor, J.; Lopes, C.M. Yield Components Detection and Image-Based Indicators for Non-Invasive Grapevine Yield Prediction at Different Phenological Phases. *OENO One* **2020**, *54*, 833–848. [CrossRef]
104. Williams, L.E.; Ayars, J.E. Grapevine Water Use and the Crop Coefficient Are Linear Functions of the Shaded Area Measured beneath the Canopy. *Agric. For. Meteorol.* **2005**, *132*, 201–211. [CrossRef]
105. Santillán, D.; Iglesias, A.; La Jeunesse, I.; Garrote, L.; Sotes, V. Vineyards in Transition: A Global Assessment of the Adaptation Needs of Grape Producing Regions under Climate Change. *Sci. Total Environ.* **2019**, *657*, 839–852. [CrossRef] [PubMed]
106. Viveros Santos, I.; Bulle, C.; Levasseur, A.; Deschênes, L. Regionalized Terrestrial Ecotoxicity Assessment of Copper-Based Fungicides Applied in Viticulture. *Sustainability* **2018**, *10*, 2522. [CrossRef]
107. Ugalde, D.; Renaud-Gentié, C.; Symoneaux, R. Perception of French Wine Buyers Regarding Environmental Issues in Wine Production. *J. Wine Res.* **2021**, *32*, 77–102. [CrossRef]
108. Favero, A.C.; Amorim, D.A.D.; Mota, R.V.D.; Souza, C.R.D.; Regina, M.D.A. Physiological Responses and Production of "Syrah" Vines as a Function of Training Systems. *Sci. Agric.* **2010**, *67*, 267–273. [CrossRef]
109. Vanden Heuvel, J.E.; Lerch, S.D.; Lenerz, C.C.; Meyers, J.M.; Mansfield, A.K. Training System and Vine Spacing Impact Vine Growth, Yield, and Fruit Composition in a Vigorous Young 'Noiret' Vineyard. *HortTechnology* **2013**, *23*, 505–510. [CrossRef]
110. Xu, X.-Q.; Cheng, G.; Duan, L.-L.; Jiang, R.; Pan, Q.-H.; Duan, C.-Q.; Wang, J. Effect of Training Systems on Fatty Acids and Their Derived Volatiles in Cabernet Sauvignon Grapes and Wines of the North Foot of Mt. Tianshan. *Food Chem.* **2015**, *181*, 198–206. [CrossRef]
111. Wolf, T.K.; Dry, P.R.; Iland, P.G.; Botting, D.; Dick, J.; Kennedy, U.; Ristic, R. Response of Shiraz Grapevines to Five Different Training Systems in the Barossa Valley, Australia. *Aust. J. Grape Wine Res.* **2003**, *9*, 82–95. [CrossRef]
112. Klimek, K.; Postawa, K.; Kapłan, M.; Kułażyński, M. Evaluation of the Influence of Rootstock Type on the Yield Parameters of Vines Using a Mathematical Model in Nontraditional Wine-Growing Conditions. *Appl. Sci.* **2022**, *12*, 7293. [CrossRef]
113. Kyraleou, M.; Kallithraka, S.; Koundouras, S.; Chira, K.; Haroutounian, S.; Spinthiropoulou, H.; Kotseridis, Y. Effect of Vine Training System on the Phenolic Composition of Red Grapes (*Vitis vinifera* L. Cv. Xinomavro). *OENO One* **2015**, *49*, 71. [CrossRef]
114. Pappaccogli, G.; Carlomagno, A.; De Simei, G.; Confalonieri, M.; Buccolieri, R.; Montanaro, G.; Nuzzo, V.; Rustioni, L. Effects of the Training System on Water Productivity and Water Footprint in Mediterranean Vineyards. *Arch. Agron. Soil Sci.* **2024**, *70*, 1–12. [CrossRef]
115. Ryugo, K. *Fruit Culture: Its Science and Art*; Wiley: New York, NY, USA, 1988; ISBN 978-0-471-89191-8.
116. Greer, D.H.; Weston, C.; Weedon, M. Shoot Architecture, Growth and Development Dynamics of *Vitis vinifera* Cv. Semillon Vines Grown in an Irrigated Vineyard with and without Shade Covering. *Funct. Plant Biol.* **2010**, *37*, 1061–1070. [CrossRef]
117. Valín, M.I.; Araújo-Paredes, C.; Mendes, S.; Dafonte, J.; Alonso, J.; Rodrigues, A.S.; Cancela, J.J. Training Systems Evaluation of *Vitis vinifera* L. 'Alvarinho' (Vinhos Verdes PDO Region) to Physiological and Productive Parameters. *Acta Hortic.* **2021**, *1314*, 375–382. [CrossRef]
118. Marín, D.; Armengol, J.; Carbonell-Bejerano, P.; Escalona, J.M.; Gramaje, D.; Hernández-Montes, E.; Intrigliolo, D.S.; Martínez-Zapater, J.M.; Medrano, H.; Mirás-Avalos, J.M.; et al. Challenges of Viticulture Adaptation to Global Change: Tackling the Issue from the Roots. *Aust. J. Grape Wine Res.* **2021**, *27*, 8–25. [CrossRef]
119. Deloire, A.; Ojeda, H.; Zebic, O.; Bernard, N.; Hunter, J.J.; Carbonneau, A. Influence de l'Etat hydrique de la Vigne sur le Style de Vin. 2006. Available online: https://www.researchgate.net/publication/267421113 (accessed on 10 September 2024).
120. Palliotti, A. A New Closing Y-Shaped Training System for Grapevines: A New Mechanised Closing Y-Shape Training System. *Aust. J. Grape Wine Res.* **2012**, *18*, 57–63. [CrossRef]

121. Parry, C.; Blonquist, J.M.; Bugbee, B. In Situ Measurement of Leaf Chlorophyll Concentration: Analysis of the Optical/Absolute Relationship. *Plant Cell Environ.* **2014**, *37*, 2508–2520. [CrossRef]
122. Intrigliolo, D.S.; Lakso, A.N. Effects of light interception and canopy orientation on grapevine water status and canopy gas exchange. *Acta Hortic.* **2011**, *889*, 99–104. [CrossRef]
123. Poni, S.; Magnanini, E.; Bernizzoni, F. Degree of Correlation between Total Light Interception and Whole-Canopy Net CO_2 Exchange Rate in Two Grapevine Growth Systems. *Aust. J. Grape Wine Res.* **2003**, *9*, 2–11. [CrossRef]
124. Baeza, P.; Ruiz, C.; Cuevas, E.; Sotés, V.; Lissarrague, J.R. Ecophysiological and Agronomic Response of Tempranillo Grapevines to Four Training Systems. *Am. J. Enol. Vitic.* **2005**, *56*, 129–138. [CrossRef]
125. Poni, S.; Gatti, M.; Palliotti, A.; Dai, Z.; Duchêne, E.; Truong, T.-T.; Ferrara, G.; Matarrese, A.M.S.; Gallotta, A.; Bellincontro, A.; et al. Grapevine Quality: A Multiple Choice Issue. *Sci. Hortic.* **2018**, *234*, 445–462. [CrossRef]
126. Buesa, I.; Caccavello, G.; Basile, B.; Merli, M.C.; Poni, S.; Chirivella, C.; Intrigliolo, D.S. Delaying Berry Ripening of Bobal and Tempranillo Grapevines by Late Leaf Removal in a Semi-Arid and Temperate-Warm Climate under Different Water Regimes: Late Leaf Removal Effects in Bobal and Tempranillo. *Aust. J. Grape Wine Res.* **2019**, *25*, 70–82. [CrossRef]
127. Fraga, H.; García De Cortázar Atauri, I.; Malheiro, A.C.; Santos, J.A. Modelling Climate Change Impacts on Viticultural Yield, Phenology and Stress Conditions in Europe. *Glob. Chang. Biol.* **2016**, *22*, 3774–3788. [CrossRef] [PubMed]
128. Lorenzo, M.N.; Ramos, A.M.; Brands, S. Present and Future Climate Conditions for Winegrowing in Spain. *Reg. Environ. Chang.* **2016**, *16*, 617–627. [CrossRef]
129. Fernandes, A.; Kovač, N.; Fraga, H.; Fonseca, A.; Radonjić, S.; Simeunović, M.; Ratkovic, K.; Menz, C.; Costafreda-Aumedes, S.; Santos, J. Challenges to Viticulture in Montenegro under Climate Change. *ISPRS Int. J. Geo-Inf.* **2024**, *13*, 270. [CrossRef]
130. Ferree, D.; Steiner, T.; Gallander, J.; Scurlock, D.; Johns, G.; Riesen, R. Performance of "Seyval Blanc" Grape in Four Training Systems Over Five Years. *HortScience A Publ. Am. Soc. Hortic. Sci.* **2002**, *37*, 1023–1027. [CrossRef]
131. Wimmer, M.; Workmaster, B.A.; Atucha, A. Training Systems for Cold Climate Interspecific Hybrid Grape Cultivars in Northern Climate Regions. *HortTechnology* **2018**, *28*, 202–211. [CrossRef]
132. O'Brien, P.; De Bei, R.; Sosnowski, M.; Collins, C. A Review of Factors to Consider for Permanent Cordon Establishment and Maintenance. *Agronomy* **2021**, *11*, 1811. [CrossRef]
133. Dufourcq, T.; Gassiolle, E.; Lopez, F.; Gontier, L.; Gaviglio, C. Behaviour of Two Training Systems for Mechanical Pruning Combined with Different Nitrogen Fertilizations on cv. *Colombard*. In Proceedings of the 21th International GiESCO Symposium, Thessaloniki, Greece, 23–28 June 2019. [CrossRef]
134. Rojo, F.; Zaccaria, D.; Gonçalves-Voloua, R.; Del Rio, R.; Pérez, F.; Lagos, L.O.; Snyder, R.L. Evapotranspiration and Water Productivity of Microirrigated Wine Grape Vineyards Grown with Different Trellis Systems in the Central Valley of Chile. *J. Irrig. Drain Eng.* **2023**, *149*, 04023005. [CrossRef]
135. Liu, M.-Y.; Chi, M.; Tang, Y.-H.; Song, C.-Z.; Xi, Z.-M.; Zhang, Z.-W. Effect of Three Training Systems on Grapes in a Wet Region of China: Yield, Incidence of Disease and Anthocyanin Compositions of *Vitis vinifera* Cv. Cabernet Sauvignon. *Molecules* **2015**, *20*, 18967–18987. [CrossRef]
136. Valentini, G.; Pastore, C.; Allegro, G.; Mazzoleni, R.; Chinnici, F.; Filippetti, I. Vine Physiology, Yield Parameters and Berry Composition of Sangiovese Grape under Two Different Canopy Shapes and Irrigation Regimes. *Agronomy* **2022**, *12*, 1967. [CrossRef]
137. Pennington, T.; Kraus, C.; Alakina, E.; Entling, M.; Hoffmann, C. Minimal Pruning and Reduced Plant Protection Promote Predatory Mites in Grapevine. *Insects* **2017**, *8*, 86. [CrossRef]
138. Faúndez-López, P.; Delorenzo-Arancibia, J.; Gutiérrez-Gamboa, G.I.; Moreno-Simunovic, Y. Pruning Cuts Affect Wood Necrosis but Not the Percentage of Budburst or Shoot Development on Spur Pruned Vines for Different Grapevine Varieties. *VITIS-J. Grapevine Res.* **2021**, *60*, 137–141. [CrossRef]
139. Cholet, C.; Bruez, É.; Lecomte, P.; Barsacq, A.; Martignon, T.; Giudici, M.; Simonit, M.; Dubourdieu, D.; Gény, L. Plant Resilience and Physiological Modifications Induced by Curettage of Esca-Diseased Grapevines. *OENO One* **2021**, *55*, 153–169. [CrossRef]
140. Williams, L.E.; Levin, A.D.; Fidelibus, M.W. Crop Coefficients (Kc) Developed from Canopy Shaded Area in California Vineyards. *Agric. Water Manag.* **2022**, *271*, 107771. [CrossRef]
141. Rogiers, S.Y.; Greer, D.H.; Liu, Y.; Baby, T.; Xiao, Z. Impact of Climate Change on Grape Berry Ripening: An Assessment of Adaptation Strategies for the Australian Vineyard. *Front. Plant Sci.* **2022**, *13*, 1094633. [CrossRef]
142. Peterlunger, E.; Celotti, E.; Dalt, G.D.; Stefanelli, S.; Gollino, G.; Zironi, R. Effect of Training System on Pinot Noir Grape and Wine Composition. *Am. J. Enol. Vitic.* **2002**, *53*, 14–18. [CrossRef]
143. Bernizzoni, F.; Gatti, M.; Civardi, S.; Poni, S. Long-Term Performance of Barbera Grown under Different Training Systems and Within-Row Vine Spacings. *Am. J. Enol. Vitic.* **2009**, *60*, 339–348. [CrossRef]
144. Del Zozzo, F.; Poni, S. Climate Change Affects Choice and Management of Training Systems in the Grapevine. *Aust. J. Grape Wine Res.* **2024**, *2024*, 7834357. [CrossRef]
145. Maj, G.; Klimek, K.; Kapłan, M.; Wrzesińska-Jędrusiak, E. Using Wood-Based Waste from Grapevine Cultivation for Energy Purposes. *Energies* **2022**, *15*, 890. [CrossRef]
146. Intrieri, C.; Poni, S. Integrated Evolution of Trellis Training Systems and Machines to Improve Grape Quality and Vintage Quality of Mechanized Italian Vineyards. *Am. J. Enol. Vitic.* **1995**, *46*, 116–127. [CrossRef]

147. Intrieri, C.; Poni, S.; Rebucci, B.; Magnini, E. Effects of Canopy Manipulations on Whole-Vine Photosynthesis: Results from Pot and Field Experiments. *VITIS-J. Grapevine Res.* **1997**, *36*, 167. [CrossRef]
148. Yu, R.; Torres, N.; Tanner, J.D.; Kacur, S.M.; Marigliano, L.E.; Zumkeller, M.; Gilmer, J.C.; Gambetta, G.A.; Kurtural, S.K. Adapting Wine Grape Production to Climate Change through Canopy Architecture Manipulation and Irrigation in Warm Climates. *Front. Plant Sci.* **2022**, *13*, 1015574. [CrossRef] [PubMed]
149. Smith, J.P.; Edwards, E.J.; Walker, A.R.; Gouot, J.C.; Barril, C.; Holzapfel, B.P. A Whole Canopy Gas Exchange System for the Targeted Manipulation of Grapevine Source-Sink Relations Using Sub-Ambient CO_2. *BMC Plant Biol* **2019**, *19*, 535. [CrossRef] [PubMed]
150. Bavougian, C.M.; Read, P.E.; Walter-Shea, E. Training System Effects on Sunlight Penetration, Canopy Structure, Yield, and Fruit Characteristics of 'Frontenac' Grapevine (*Vitis* spp.). *Int. J. Fruit Sci.* **2012**, *12*, 402–409. [CrossRef]
151. Cavallo, P.; Poni, S.; Rotundo, A. Ecophysiology and Vine Performance of Cv. "Aglianico" under Various Training Systems. *Sci. Hortic.* **2001**, *87*, 21–32. [CrossRef]
152. Previtali, P.; Dokoozlian, N.K.; Pan, B.S.; Wilkinson, K.L.; Ford, C.M. Crop Load and Plant Water Status Influence the Ripening Rate and Aroma Development in Berries of Grapevine (*Vitis vinifera* L.) Cv. Cabernet Sauvignon. *J. Agric. Food Chem.* **2021**, *69*, 7709–7724. [CrossRef]

agriculture

MDPI

Review

Review of the Report on the Nutritional and Health-Promoting Values of Species of the *Rubus* L. Genus

Kamil Buczyński *, Magdalena Kapłan and Zbigniew Jarosz

Institute of Horticulture Production, University of Life Sciences in Lublin, 28 Gleboka Str., 20-612 Lublin, Poland; magdalena.kaplan@up.lublin.pl (M.K.); zbigniew.jarosz@up.lublin.pl (Z.J.)

* Correspondence: kamil.buczynski@up.lublin.pl

Abstract: The genus *Rubus* encompasses over 1000 species, including raspberries and blackberries, known for their rich nutritional and health-promoting properties. This review aims to provide a comprehensive overview of the nutritional values, health benefits, and potential medical and pharmaceutical applications of *Rubus* species. The fruits, roots, shoots, and leaves of these plants are distinguished by their high content of polyphenols, vitamins, and minerals, which contribute to their potent anti-oxidant, anti-inflammatory, anti-neurodegenerative, and anti-cancer effects. The diverse phytochemical profiles of the *Rubus* species support their use in the prevention and treatment of chronic diseases such as cardiovascular diseases, diabetes, and certain cancers. Additionally, the *Rubus* species are valuable as pharmaceutical raw materials due to their bioactive compounds. Despite the focus on a few widely cultivated species, numerous wild and lesser-known *Rubus* species offer significant untapped potential as functional foods, nutraceuticals, and pharmaceuticals. Future research should explore the detailed mechanisms of their bioactivities, develop effective extraction and formulation techniques, and integrate these findings into public health strategies. The genus *Rubus* represents a promising resource for enhancing human health and nutrition, as well as for pharmaceutical and medical applications, justifying increased cultivation and utilization of species from this genus.

Keywords: raspberry; blackberry; black raspberry; polyphenols; health; yellow raspberry

1. Introduction

Defining what constitutes a healthy diet has been the subject of countless discussions over decades among the food-related community. As a result, numerous definitions of healthy food have emerged [1]. One partial cause of this problem is the wide variety of health issues for which different diets are used. A healthy diet should be free from harmful substances, without excess of any nutrients and health-promoting substances, and, above all, should result in high health outcomes and reduce the likelihood of diseases [2]. A diet rich in highly processed foods is characterized by low nutrient content, which results in a higher risk of chronic diseases [3]. Diseases caused by improper diets are currently a major cause of decreasing life expectancy and reduction in the quality of life [4]. Properly designed food systems can have a significant impact on the health of the entire human population; however, current food systems can pose a threat to both humans and the environment. The current food production system faces many problems, one of which is ensuring sufficient access to food for the entire world's population [5].

Plant-based foods, including fresh fruits, are a major component of a proper diet and can potentially have a significant impact on reducing the incidence of many diseases [6]. The benefits of consuming fruits and vegetables are now widely confirmed. The World Health Organization (WHO) recommends consuming at least five servings of fruits and vegetables daily [7]. The consumption of berry fruits is associated with lowering blood pressure and reducing the risk of diabetes and cardiovascular diseases. It has also been confirmed that berry fruits enhance resistance to viral diseases and play a role in their treatment.

Consumption of berry fruits rich in phytochemicals increases neuroplasticity, neurotransmission, and calcium buffering. Additionally, they possess anti-cancer, anti-oxidant, and anti-inflammatory properties, and they reduce age-related neurodegeneration [8,9].

To fully utilize the benefits of fruit consumption, changes in the food system are necessary [10]. Increasing fruit consumption and reducing the disparity in access to healthy food between adolescents from wealthier and less affluent families are crucial for public health [11].

The genus *Rubus* L. comprises over 1000 species [12]. Due to the large number of species, which are also difficult to identify, it is considered a very challenging genus for taxonomic classification [13]. This genus includes *Rubus idaeus* L., *R. strigosus* Michx., *R. occidentalis* L., *R. neglectus*, *R. xanthocarpus* Bur. et Fr. [14], *R. ursinus* Cham. et Schlecht., *R. candicans* Weihe, *R. alleghanensis* Porter, *R. plicatus* W. et N., and *R. lacinigtus* Willd. [15]. The distribution areas of species of the genus *Rubus* L. span six continents and include raspberries, blackberries, and their hybrids [16]. As early as ancient times, species belonging to the genus *Rubus* were associated with medicinal properties. They were used, among other things, in the treatment of colicky pain, inflammations, diarrhea, wounds, diabetes, and as anti-microbial agents [17]. Today, many species of the genus *Rubus* are used in the treatment of various diseases [18].

Due to their high nutritional and bioactive compound values, there has been an increased interest in the consumption of blackberries in recent years [19]. This growing interest also extends to red raspberries [8], whose fruits are noted for their appealing appearance, high content of beneficial compounds for human health, and delicious taste [20,21]. A key feature of consumable products in the context of health benefits is their anti-oxidant potential [22]. Species of the genus *Rubus* L. are considered valuable sources of polyphenols and bioactive compounds, fitting within the framework of functional foods and pharmaceutical raw materials. Their popularity among consumers is increasing, leading to greater market demand. This situation is mainly due to the nutritional and industrial significance of the fruits derived from these species. Knowledge regarding the biosynthesis pathways and mechanisms controlling the accumulation of valuable metabolites in raspberry fruits, as well as the role of individual phytochemicals, is rapidly advancing. Simultaneously, knowledge related to raspberry genetics is expanding, enabling the selection of varieties not only with high fruit flavor qualities but also with significant values that provide more effective protection against degenerative diseases [23].

The content of nutrients and bioactive compounds is influenced not only by the species or variety itself but also by the agricultural practices used in cultivating plants from the genus *Rubus*, which often play a fundamental role in this aspect. Similarly, the microclimate, despite the use of identical agricultural practices on closely located plantations, will lead to varied development at different sites, which will also determine the diverse content of nutrients and bioactive compounds.

This article primarily aims to review current knowledge and showcase the species diversity of the *Rubus* genus in the context of nutritional values, health benefits, and potential medical applications of the fruits, leaves, shoots and roots of these plants. This diversity is due to the large number of species within the genus, which are highly diverse in morphological, biological, and ecological characteristics, mainly resulting from their vast range and frequent interspecies hybridization. This diversity translates into numerous possibilities for utilizing these species not only for consumption but also for improving human health worldwide, with a very wide spectrum of potential effects.

2. Methodology

To prepare this review on the nutritional and health-promoting values of species from the genus *Rubus*, several databases were searched for relevant publications. The search was conducted using PubMed, Scopus, Web of Science, Google Scholar, and SpringerLink. To cover the topic as comprehensively as possible and minimize the chance of missing information on lesser-known species discussed in older publications, the search included publications from the year 2000 onwards. The search encompassed articles, review papers, and books. The keywords used for the search included the following phrases: phenols, polyphenols, *Rubus*, blackberry, raspberry, red raspberry, black raspberry, yellow raspberry, Korean raspberry. The use of five databases combined with ten keywords allowed for a comprehensive study of this topic, focusing primarily on the nutritional and health-promoting properties of the fruit of these plants, but also considering roots, shoots, and leaves. The review discusses the most widespread and cultivated species of this genus worldwide, while also considering lesser-known species that often occur only locally.

3. Nutritional Values

Species of the genus *Rubus* are characterized by high nutritional values, particularly for dietary fiber and vitamin C content, while being low in calories [24]. The fruits of *Rubus* species vary in their content of specific components, including fructose, glucose, minerals, protein, carotenoids, and ascorbic acid. Differences are especially noticeable in the group of phenolic compounds, particularly phenolic acids, anthocyanins, and other flavonoids. Generally, the most nutrient-dense and phenolic-rich species are *Rubus ulmifolius*, *R. idaeus*, and *R. fruticosus* [25]. Black raspberry fruits are notable for their high content of phosphorus, potassium, iron, organic acids, and vitamin C [26,27]. Red raspberry fruits are considered a rich source of pectin, fiber, calcium, potassium, iron, as well as vitamins C, E, B1, and B2 [28]. Similar to other berry species, raspberry fruits have a low glycemic index [29].

The glycemic index (GI) of carbohydrate-containing foods is used to determine the rate at which these carbohydrates are broken down during digestion and the speed of their absorption into the blood [30]. Carbohydrate-rich foods that break down quickly and are rapidly absorbed into the blood have a high GI. Foods with a high GI lead to rapid spikes in blood glucose and insulin levels after consumption. In contrast, low-GI foods, which are digested more slowly, have a smaller impact on postprandial blood glucose levels and the insulin response [31].

Table 1 presents the nutritional composition of raspberries and blackberries, gathered from the database FoodData of the Central United States Department of Agriculture. Both raspberries and blackberries are rich sources of energy, calcium, iron, magnesium, and vitamins, such as vitamin C, vitamin K, and B-group vitamins. Comparing the two species, raspberries have a slightly higher calorie and carbohydrate content than blackberries, which are richer in protein. Blackberries have higher levels of vitamin E, and raspberries have higher levels of fatty acids, lutein and zeaxanthin. Both fruits are also good sources of dietary fiber, which may have a beneficial impact on gastrointestinal health. It is also worth noting that blackberries and raspberries, according to the presented data, contain high levels of vitamins C, A, E and B, making them a valuable source of food that has a positive effect on the immune system.

Table 1. Raspberry and blackberry fruit nutrients [32,33].

Common Name	Energy (kcal)	Protein (g)	Total Lipid (g)	Carbohy-Drate (g)	Fiber (g)	Sugars (g)	Ca (mg)	Fe (mg)	Mg (mg)
Blackberries	43	1.39	0.49	9.61	5.3	4.88	29	0.62	20
Raspberries	52	1.2	0.65	11.9	6.5	4.42	25	0.69	22
Common name	**P (mg)**	**K (mg)**	**Na (mg)**	**Zn (mg)**	**Cu (mg)**	**Se (µg)**	**Lutein + zeaxanthin (µg)**	**Thiamin (mg)**	**Fatty acids (g)**
Blackberries	22	162	1	0.53	0.165	0.4	118	0.02	0.014
Raspberries	29	151	1	0.42	0.09	0.2	136	0.032	0.019
Common name	**Riboflavin (mg)**	**Niacin (mg)**	**Vit B-6 (mg)**	**Folate (µg)**	**Choline (mg)**	**Vit A (µg)**	**Vit E (mg)**	**Vit K (µg)**	**Vit C (mg)**
Blackberries	0.026	0.646	0.03	25	8.5	11	1.17	19.8	21
Raspberries	0.038	0.598	0.055	21	12.3	2	0.87	7.8	26.2

4. Health-Promoting Properties and Use in Medicine

Species of the genus *Rubus* have provided food and sources of medicinal raw materials for indigenous communities since the post-glacial period [16]. Raspberry fruits are widely regarded as an excellent source of bioactive compounds, including phenolic compounds, which are natural anti-oxidants with positive health effects [34]. Various species belonging to the genus *Rubus* possess significant application potential due to their high content of polyphenols, ellagitannins, and anthocyanins. These substances are valuable sources of natural plant-based anti-oxidants. Polyphenols are one of the most abundant chemical compounds in the *Rubus* genus, characterized by significant health-promoting properties [24].

Polyphenols are chemical compounds found in plants, characterized by the presence of one or more aromatic rings and at least one hydroxyl group [35]. Polyphenols can be broadly divided into two main types, flavonoids and non-flavonoids, primarily include phenolic acids, or are classified into many subcategories based on the number of phenolic units in their molecular structure, types of substituent groups, and types of bonds between these units. These compounds are commonly found in plant tissues, where they typically exist as glycosides or aglycones. The structural diversity of flavonoids results from differences in hydroxylation and oxidation levels, leading to various compound types, such as flavanols, anthocyanidins, anthocyanins, isoflavones, flavones, flavonols, flavanones, and flavanonols [36]. Flavonoids are among the most common dietary anti-oxidants and are widely present in plants and plant products, including fruits. In recent decades, there has been increasing scientific interest in their various health benefits, as polyphenols exhibit diverse bioactivities. In vitro and in vivo studies conducted over the past few decades have demonstrated that a diet rich in polyphenols or polyphenol-containing products contributes to improved human health and is associated with a reduced risk of chronic diseases. Furthermore, the development of new technologies, such as multi-omics, bioinformatics, and artificial intelligence, opens new possibilities for elucidating the molecular mechanisms by which polyphenols can support the treatment of chronic diseases [37]. Additionally, polyphenols are used in medicine to treat various ailments due to their numerous health properties. These compounds can mitigate oxidative effects in the human body, prevent damage to organs and cellular structures, and protect their functionality. Their health-promoting abilities stem from intense bioactivity, translating into strong anti-oxidant, anti-hypertensive, immunomodulatory, anti-bacterial, anti-viral, and anti-cancer properties [38].

Due to the medicinal properties of the fruits and leaves of *Rubus idaeus*, interest in this species is growing among dietitians, pharmacognosists, and phytochemists. Raspberry seed oil is also gaining popularity among consumers as a valuable source of essential unsaturated fatty acids, vitamins, and anti-oxidants [28]. Eleven anthocyanins have been identified in red raspberry fruits, including cyanidin-3-sophoroside, cyanidin-3-(2G-glucosylrutinoside), cyanidin-3-glucoside, cyanidin-3-rutinoside,pelargonidin-3-sophoroside, pelargonidin-3-(2G-glucosylrutinoside), and pelargonidin-3-glucoside. These fruits also contain ellagic acid, ellagitannin, lambertianin C, and sanguiin H-6 [39]. Among the phenolic compounds in *Rubus idaeus* fruits, anthocyanins, particularly cyanidin-3-sophoroside and cyanidin-3-glucoside, are noteworthy [21]. These compounds have chemo-preventive potential and anti-oxidant properties [40]. Additionally, they can be used to extract natural dyes efficiently and cost-effectively via ultrasound, offering an alternative to synthetic counterparts [41]. Red raspberries owe their anti-oxidant properties primarily to high levels of anthocyanins, vitamin C, and ellagic acid derivatives [42]. Yellow and red raspberry varieties contain carotenoids, mainly lutein esters, which in red raspberries are masked by anthocyanins [43]. *Rubus idaeus* leaf infusions are used to facilitate childbirth, cleanse the blood, and treat intestinal issues [44]. Extracts from *Rubus rugosus* leaves have astringent and emmenagogic effects, while those from *Rubus idaeus* show astringent, spasmolytic, and carminative properties. Extracts from the leaves and fruits of wild raspberry *Rubus idaeus* contain numerous secondary metabolites with anti-cancer, anti-oxidant, and anti-microbial activities [45]. Studies on *R. idaeus* shoots revealed that they are rich in compounds, such as ellagic acid and sanguiin H-6. The extracts from these shoots exhibit strong anti-oxidant, anti-microbial (especially against *Corynebacterium diphtheriae*), and cytotoxic properties, particularly effective against HL-60 cancer cells. The use of comprehensive LCxLC methods has allowed for effective separation and identification of present polyphenolic compounds, revealing their diversity and biological significance [46]. The shoots of this species are also used to treat fever, colds, and flu-like infections [16]. Studies on the properties of wild raspberry (*R. idaeus*) fruits from the Kelkit Valley in Turkey revealed their high content of active phenolic compounds and flavonoids, along with strong anti-bacterial properties, particularly against *Pseudomonas aeruginosa* bacteria, suggesting their potential application in cultivation and as raw material for pharmaceutical research [47]. Freeze-dried *R. idaeus* fruits are also an exceptional source of nutrients, including vitamin C, potassium, and magnesium, as well as bioactive compounds, such as proanthocyanidins, ellagitannins, and anthocyanins, making them comparable to fresh raspberries in terms of nutritional value. The main bioactive compounds among the ellagitannins are isomers of sanguiin H-6, lambertianin C, and sanguiin H-10, while the dominant anthocyanins include cyanidin-3-sophoroside, cyanidin-3-glucoside, and cyanidin-3-sambubioside [48]. Studies also indicate that daily consumption of raspberries significantly reduces serum glucose levels and postprandial inflammatory biomarkers, such as interleukin-6 and tumor necrosis factor-alpha. These results suggest that raspberries may be beneficial in alleviating hyperglycemia and inflammation in individuals with diabetes, warranting further investigation [49]. Research confirms the potential use of *Rubus hirsutus* berries, characterized by high anti-oxidant activity, as functional food ingredients and sources of phytochemicals for nutraceutical production [50]. Extracts from the fruits and leaves of *Rubus chingii* are notable for regulating the central nervous system, and their hypoglycemic, anti-cancer, anti-thrombotic, anti-osteoporotic, anti-oxidant, anti-fungal, and anti-inflammatory properties [51].

Drawing upon the data presented in the accompanying Table 2, this discussion elaborates on the multifaceted therapeutic potentials of *Rubus idaeus* (red raspberry) fruit extracts and derivatives. The extracts from *Rubus idaeus* have demonstrated a wide range of health benefits, spanning from cardiovascular and liver health to anti-inflammatory and anti-microbial effects. Drinking raspberry juice has been linked to the prevention of early atherosclerosis, primarily through improvements in anti-oxidant status and serum lipid profiles. This suggests that regular consumption could offer a natural strategy for managing and preventing this cardiovascular condition. In the realm of liver health, a natural phenolic compound extracted from *Rubus idaeus*, optimized through ethanol extraction and ultrasound, has shown efficacy in reducing inflammation and fatty liver disease caused by non-alcoholic fatty liver disease (NAFLD). This indicates its potential as a therapeutic agent in managing liver health. Moreover, red raspberry extracts have displayed significant synergistic anti-microbial activity, particularly when combined with conventional antibiotics, such as penicillin, cefoxitin, tetracycline, ciprofloxacin, and fusidic acid. This synergy enhances the effectiveness of these antibiotics, suggesting a role for raspberry flavonoids in anti-microbial treatments. The extracts are also notably effective against bacteria like *Pseudomonas aeruginosa*, *Bacillus cereus*, *Staphylococcus aureus*, and *S. saprophyticus*, underscoring their potential as natural anti-bacterial agents. Moreover, studies have shown that consumption of raspberries offers protective benefits against oxidative stress induced by diabetes, thanks to their polyphenol content and fiber, enhancing anti-oxidant defenses in a diabetic obesity model. In terms of inflammatory conditions, *Rubus idaeus* fruit extracts enriched with polyphenols have been shown to exhibit anti-arthritis effects and provide additional cartilage protection. These extracts inhibit key inflammatory pathways, such as NF-κB and MAPK, reducing joint inflammation, inhibiting osteophyte formation, and limiting joint destruction. They also decrease bone resorption and soft tissue swelling associated with joint diseases. Furthermore, raspberry ketone has been highlighted for its role in increasing norepinephrine-induced lipolysis in white adipocytes, which may help prevent obesity and fatty liver disease. This underlines the potential of raspberry components in weight management and metabolic health. Additionally, consumption of raspberries has been suggested to enhance detoxifying cellular defenses, offering protection against oxidative stress caused by conditions like diabetes. Lastly, extracts from *Rubus idaeus* leaves have also been utilized to relieve pain associated with osteoarthritis of the knee, presenting a safe and effective method for alleviating symptoms of this debilitating condition.

Based on the comprehensive data presented in the accompanying Table 3, red raspberries, where the species is not specified, have demonstrated potential health benefits in weight management and cardiovascular health. Research has shown that a concentrate obtained from red raspberry can limit weight gain induced by a high-fat and high-calorie diet. This highlights its potential utility in dietary interventions aimed at obesity prevention. Furthermore, red raspberry extracts have been noted for their anti-hypertensive effects, which are mediated by increasing the activation of nitric oxide and improving vascular endothelial function through their anti-oxidant actions. Additionally, raspberry ketone, a compound found in these berries, has shown the ability to reduce obesity caused by a high-fat diet by altering metabolism in adipocytes, indicating its promise for metabolic health enhancements. Moreover, red raspberry fruit extracts have also demonstrated protective properties against DNA damage, specifically limiting DNA cleavage induced by peroxinitrite and protecting against the formation of hydroxyl radicals. This protective effect against oxidative stress suggests potential for therapeutic use in preventing cellular damage. Specifically focusing on *Rubus chingii*, research has uncovered its notable anti-osteoporotic effects. Compounds such as kaempferol, quercetin, rubusin A, and B found in *Rubus chingii* fruits not only increase the activity of alkaline phosphatase but also inhibit the resorption of osteoclastic cells and stimulate osteoblastic activity, thereby potentially aiding in the treatment and prevention of osteoporosis. Furthermore, *Rubus chingii* fruit extract has shown promising anti-oxidant capabilities, specifically in scavenging DPPH and hydroxyl radicals and exhibiting a strong ability to reduce iron ions. This anti-oxidant capacity is paired with significant anti-cancer potential, particularly in preventing and treating bladder cancer by inhibiting cell proliferation and inducing apoptosis in cancer cells.

Table 2. Health benefits and applications of red raspberry (*Rubus idaeus*).

Common Name (Species)	Possibility of Use	Reference
Red raspberry (*R. ideaus*)	Extracts obtained from the fruits of *Rubus idaeus* have a strong anti-bacterial effect against *Pseudomonas aeruginosa*, *Bacillus cereu*, *Staphylococcus aureus* and *S. saprophyticus*	[52]
Red raspberry (*R. ideaus*)	Research results suggest that *Rubus idaeus* fruit extract enriched with polyphenols has an anti-arthritis effect and also provides additional cartilage protection.	[53]
Red raspberry (*R. ideaus*)	Red raspberry fruit extract has anti-inflammatory effects on joints, additionally inhibiting the formation of osteophytes and further joint destruction, and also reduces the degree of bone resorption and soft tissue swelling.	[54]
Red raspberry (*R. ideaus*)	Red raspberry extract with a high content of anthocyanin-rich fractions showed anti-inflammatory effects both in vitro and in vivo, primarily due to the inhibition of NF-κB and MAPK activation.	[55]
Red raspberry (*R. ideaus*)	Raspberry ketone (4-(4-hydroxyphenyl) butan-2-one), by increasing norepinephrine-induced lipolysis in white adipocytes, may prevent obesity and fatty liver disease	[56]
Red raspberry (*R. ideaus*)	The research showed the occurrence of synergism between fel compounds derived from red raspberry fruit concentrate and penicillin, cefoxitin, tetracycline, ciprofloxacin and fusidic acid. In turn, raspberry flavonoids themselves have shown potential for anti-microbial activity and may enhance the effect of conventional antibiotics.	[57]
Red raspberry (*R. ideaus*)	Based on the results, the authors of the study suggest the possibility of obtaining enhanced detoxifying cellular defense as a result of consuming raspberries, which leads to protection against oxidative stress caused by diabetes	[58]
Red raspberry (*R. ideaus*)	A study examining the impact of raspberry consumption on oxidative stress in obese mice with type 2 diabetes demonstrated that raspberries enhance anti-oxidant defense by reducing interleukin-6 levels and increasing glutathione peroxidase activity in the liver and blood. Although other disease risk biomarkers remained at similar levels, increases in total cholesterol and LDL cholesterol were observed. This first in vivo study using a rodent model suggests that regular consumption of raspberries may protect against oxidative stress induced by diabetes, attributed to the polyphenol and fiber content in the fruit.	[58]
Red raspberry (*R. ideaus*)	*Rubus idaeus* leaves can be used to produce an extract intended to relieve pain caused by osteoarthritis of the knee. Researchers also emphasize that it is a safe and, above all, effective method of alleviating the symptoms associated with this disease.	[59]
Red raspberry (*R. ideaus*)	Extracted and purified using ethanol extraction additionally supported by ultrasound, a natural phenolic compound (ketone) from *Rubus ideaus* fruit can reduce fatty liver disease and inflammation caused by non-alcoholic fatty liver disease	[60]

Table 3. Health benefits and applications of red raspberry (various *Rubus* species).

Common Name (Species)	Possibility of Use	Reference
Red raspberry (no data)	Raspberry ketone, due to its biological activity, can reduce obesity caused by a high-fat diet by changing the metabolism in 3T3-L1 adipocytes	[61,62]
Red raspberry (no data)	Red raspberry extracts, by increasing the activation of nitric oxide and improving vascular endothelial dysfunction through anti-oxidant action, have an anti-hypertensive effect	[63]
Red raspberry (no data)	It has been shown that red raspberry fruit extract limits DNA cleavage induced by peroxynitrite, while protecting against DNA damage and the formation of hydroxyl radicals	[64]
Red raspberry (*R. chingii*)	Research results suggest that kaempferol, quecentin, rubusin A and B found in *Rubus chingii* fruits have anti-osteoporotic effects, increase the activity of alkaline phosphatase, inhibit the resorption of osteoclastic cells and bones, and stimulate osteoblastic activity.	[65]

Table 3. *Cont.*

Common Name (Species)	Possibility of Use	Reference
Red raspberry (no data)	Limiting weight gain caused by a high-fat and high-calorie diet by consuming a concentrate obtained from red raspberry	[29]
Red raspberry (*R. chingii*)	*Rubus chingii* fruit extract has the ability to scavenge DPPH and OH- radicals and is also characterized by a strong ability to reduce iron ions. By inhibiting cell proliferation and cell apoptosis, this extract may help prevent and treat bladder cancer.	[66]

Blackberries are characterized by high biological potential, primarily due to their high flavonoid and polyphenol content [67]. They are a valuable source of bioactive compounds that can be utilized in both the prevention and treatment of various diseases, especially those related to oxidative stress [68]. These fruits contain significant amounts of anthocyanins, such as cyanidin-3-O-glucoside and cyanidin-3-O-arabinoside [69], as well as cyanidin-3-O-malonyl-glucoside, cyanidin-3-O-dioxalylglucoside, and cyanidin derivatives [70]. In the case of *Rubus laciniatus*, studies have also shown the presence of cyanidin-3-O-glucoside and cyanidin-3-O-rutinoside in the fruits [71]. For each species, *R. glaucus*, *R. alpinus*, and *R. alutaceus*, research has identified nine phenolic compounds: two catechins, pinocembrin, quercetin, and different concentrations of cyanidin 3-rutinoside, quercetin 3-glucoside, pelargonidin 3-glucoside, cyanidin, and kaempferol-3-glucoside. In the case of *R. alutaceus*, a phenolic marker was described for the first time, including kaempferol-3-glucoside as a potential chemotaxonomic marker for this species [72]. Polyphenol extracts from blackberries, containing proanthocyanidins, exhibit anti-inflammatory properties, suggesting that anthocyanins from these species could aid in wound healing [73]. Blackberries can be classified as functional foods due to their anti-diabetic and anti-oxidant properties [74]. Consuming their juice may reduce the risk of kidney disease, and some studies suggest it may also protect against urinary tract infections [75]. Due to the high phenol content and anti-oxidant, cytotoxic, and anti-bacterial activity against Gram-positive bacteria of the roots of *Rubus hyrcanus*, commonly known as Caspian blackberry, researchers highlight the potential of this raw material for producing plant products with bioactive properties [76]. Extracts from *Rubus fruticosus* plants have anthelmintic, relaxant, anti-diarrheal, and anti-inflammatory effects, and compresses made from the leaves are used on abscesses and skin ulcers [77–79]. Studies indicate that *Rubus fruticosus* fruit extract contains 42 compounds, with oleic acid being the most abundant. Among the 29 polyphenols in the extract, quercetin-3-O-glucoside is the most prevalent. Researchers highlight the enormous potential of this species' fruits as a source of natural anti-oxidants for nutraceutical production and as a component of functional foods [80]. Spreads made from blackberry pulp containing cream or cream cheese show increased anti-oxidant activity with higher blackberry pulp content, with equivalent anti-oxidant capacity increasing from 43.51% to 44.50%. These results suggest that spreads with higher blackberry pulp content can be considered functional and healthy foods due to their beneficial anti-oxidant and rheological properties [81]. Research on the anti-oxidant properties and bioactive components of blackberries from *R. fruticosus* and *R. laciniatus* species demonstrated that these fruits are rich in anthocyanins, primarily cyanidin-3-glucoside, and other phenolics. In particular, the "Loch Ness" and "Thorn Free" varieties, cultivated in northwestern Romania, stood out for their highest polyphenol content and greatest anti-oxidant capacity. The differences in anti-oxidant capacities among the seven studied varieties were significant, as confirmed by multivariate analysis. These findings underscore the potential of blackberries for use as functional foods or in the production of commercial products rich in bioactive compounds [82]. A study focused on analyzing the impact of different extraction solvents on the polyphenol, flavonoid, and tannin content in dried *Rubus armeniacus* leaves and their anti-oxidant activity. It was found that the 80% methanol extract showed the highest concentrations of TPC, TFC, and CTC and the most effective anti-oxidant activity according to DPPH, reducing power, and β-carotene bleaching assays. These results suggest the potential application of *R. armeniacus* extracts in the food and pharmaceutical industries as natural anti-oxidants [83].

The extensive data from the Table 4 showcase the diverse health benefits and therapeutic applications associated with various species of blackberries, including studies on multiple species, such as *Rubus fruticosus, Rubus liebmannii, Rubus palmeri, Rubus discolor, Rubus apetalus, Rubus imperialis,* and the hybrid *Rubus loganobaccus* × *baileyanus* (Boysenberry). Insights also include findings on unspecified blackberries and a combination study involving blackberries, Korean raspberries, and black raspberries. Ethanol extracts and anthocyanin fractions from these *Rubus* fruits have demonstrated protective effects against H_2O_2-induced cytotoxicity by reducing intracellular reactive oxygen species levels and enhancing anti-oxidant enzyme activity, while also suppressing the activation of inflammatory pathways, highlighting their potential as natural anti-inflammatory and anti-oxidant agents. Numerous studies have shown that blackberry extracts can increase glucose metabolism, prevent oxidative stress from exogenous pollutants, improve memory through anti-oxidant and anti-cholinesterase properties, regulate vascular functions, and inhibit telomerase in colorectal cancer cells, suggesting their utility in managing diabetes, protecting against neurodegenerative diseases, reducing thrombosis, and offering anti-cancer benefits. Additionally, recent studies on *Rubus erlanrige* leaf extract have confirmed its anti-diabetic properties, significantly reducing blood glucose levels in diabetic models without inducing hypoglycemia in normal subjects, suggesting a safe and effective profile for managing diabetes. Specifically, *Rubus fruticosus* extracts reduce cell damage by activating anti-oxidant enzymes and removing reactive oxygen species in HepG2 cells. Anthocyanin-enriched fractions from *Rubus liebmannii* and *Rubus palmeri* induce apoptotic responses in glioma cells, suggesting potential as a therapeutic raw material. *Rubus discolor* extracts have been used in the green synthesis of silver nanoparticles, showing anti-bacterial activity against multidrug-resistant bacteria and cytotoxic effects on cancer cell lines. *Rubus apetalus* extracts improve fertility parameters and reduce oxidative stress in rats with cryptorchidism, indicating potential for treating male infertility. *Rubus imperialis* extracts exhibit protective effects against stomach ulcers due to their anti-secretory properties. Boysenberry fruit extracts have shown in vivo anti-oxidant properties, enhancing serum anti-oxidant status and reducing oxidative damage markers, although these effects were influenced by diet. The broad scope of research across these diverse blackberry species highlights their significant potential in pharmaceutical and nutraceutical applications, addressing a wide array of health issues from oxidative stress and inflammation to cancer and fertility, supporting further exploration and development of these berries in health-related fields.

Table 4. Health benefits and applications of various *Rubus* species.

Common Name (Species)	Possibility of Use	Reference
Boysenberry (*R. loganobaccus* × *baileyanus*)	The research findings indicated that the extract from Boysenberry fruits (*Rubus loganobaccus* × *baileyanus* Britt) acted in vivo as an anti-oxidant, elevating serum anti-oxidant status. Concurrently, a reduction in certain oxidative damage markers was observed; however, this effect was significantly modified by the basal diet.	[84]
Blackberry (*R. imperilis*)	The extract from the aerial parts of *Rubus imperialis* and the isolated niga-ichigoside F1 and 2β,3β-19-α-trihydroxyursolic acid have protective properties on the stomach against damage caused by ulcers. Researchers suggest that this ability is related to its anti-secretory effect	[85]

Table 4. *Cont.*

Common Name (Species)	Possibility of Use	Reference
Blackberry, Korean raspberry, Black raspberry (no data)	The studies demonstrated that ethanol extracts (E) and anthocyanin fractions (AF) from *Rubus* fruits, including blackberries, Korean raspberries, and black raspberries, protect RAW 264.7 cells from H_2O_2-induced cytotoxicity by significantly reducing intracellular ROS levels and enhancing the activity of anti-oxidant enzymes, such as SOD and GPx. Additionally, AF markedly reduced DNA oxidation and inhibited the production of inflammatory mediators, such as NO and PGE2, decreasing the expression of pro-inflammatory proteins iNOS and COX-2 in LPS-stimulated cells. Furthermore, anthocyanin fractions effectively suppressed the activation of the NF-κB pathway, indicating their potential as natural anti-inflammatory and anti-oxidant agents.	[86]
Blackberry (*R. fruticosus*)	Blackberry fruit extract reduces H_2O_2-induced cell damage by activating the anti-oxidant enzyme SOD in HepG2 cells and removing reactive oxygen species	[87]
Blackberry (no data)	Blackberry fruit extract, by regulating the active substance of the vascular endothelium, as well as its anti-thrombotic effect and activation of blood flow, can reduce the formation of thrombi	[88]
Blackberry (no data)	It has been shown that blackberry extract can be used to prevent oxidative stress caused by exogenous pollutants. Study results also suggest improved effectiveness in mitigating oxidative damage when blackberries undergo gastrointestinal digestion.	[89]
Blackberry (no data)	Blackberry fruit extract, by reducing the expression of hTERT and methylating its promoter in colorectal cancer cell lines, has a telomerase inhibitory effect, thus having anti-cancer effects in CRC cells	[90]
Blackberry (no data)	Research results suggest that consuming blackberries through their neuroprotective, anti-oxidant and anti-inflammatory properties may reduce or help prevent the occurrence of manic episodes	[91]
Blackberry (*R. liebmannii* and *R. palmeri*)	Fractions enriched in anthocyanins derived from the fruits of wild blackberry species from Mexico have the ability to induce an apoptotic response in the cells of the most lethal primary brain tumor, glioma. The authors of the study suggest the possibility of using anthocyanins present in these fruits as a valuable raw material for the development of alternative therapies to increase the survival of patients suffering from this disease.	[92]
Blackberry (*R. apetalus*)	The research conducted on rats demonstrated that cryptorchidism (CPT) significantly affects male fertility parameters, such as testicular mass, sperm density, and levels of sex hormones, while also inducing oxidative stress. However, extracts from *Rubus apetalus* counteract these effects by improving sperm parameters, reducing oxidative stress, and increasing pregnancy and fertility rates in rats with CPT, suggesting the potential application of this plant as an alternative treatment for male infertility associated with cryptorchidism.	[93]
Blackberry (*R. erlanrige)*	Studies on the extract from *Rubus erlanrige* leaves have demonstrated its anti-diabetic potential in both in vitro and in vivo experiments. In vitro, the extract exhibited α-amylase inhibitory activity and anti-oxidant properties while, in vivo, it significantly reduced blood glucose levels in mice with streptozotocin-induced diabetes, without causing hypoglycemia in healthy mice. These results indicate that the extract, likely through the synergy of its secondary metabolites, offers a promising anti-diabetic effect with minimal risk of hypoglycemia.	[94]
Blackberry (no data)	Blackberry fruit extract has anti-oxidant properties and prevents the decrease in the activity of ketalase in brain structures and superoxide dismutase in the striatum. These results suggest the possibility of using this extract as an agent to prevent damage caused by inflammation of the nervous system	[95]
Blackberry (no data)	Research results suggest that consuming blackberry juice increases glucose metabolism by increasing insulin levels and improving the activity of enzymes responsible for glucose metabolism, which may translate into improved health of people with diabetes.	[96]
Blackberry (no data)	Research shows that blackberry fruit extract, thanks to its anti-oxidant and anti-cholinesterase properties, has a memory-improving effect and can be used as an agent in the treatment of neurodegenerative diseases	[97]

Table 4. *Cont.*

Common Name (Species)	Possibility of Use	Reference
Blackberry (*R. discolor*)	The study focused on the green synthesis of silver nanoparticles (AgNPs) using an extract from the leaves of *Rubus discolor*, resulting in the formation of stable nanoparticles with an approximate diameter of 37 nm. The AgNPs demonstrated effective anti-bacterial activity against multidrug-resistant strains of Escherichia coli and *Pseudomonas aeruginosa*, as well as significant cytotoxic effects on selected cancer cell lines, including cytotoxic activity on A431, MCF-7, and HepG2 cells.	[98]

In the case of *Rubus occidentalis* fruits, five anthocyanins are notable: pelargonidin-3-rutinoside, cyanidin-3-sambubioside, cyanidin-3-glucoside, with cyanidin-3-rutinoside and cyanidin-3-xylosylrutinoside being the most abundant and potent phenolic anti-oxidants. Due to the biological activity of these two anthocyanins, there is potential for their use in cancer treatment [99]. Fermented juice from *Rubus occidentalis* fruits using the lactic acid method with the Lactobacillus plantarum GBL17 strain shows potential as a functional beverage [100]. Among extracts from blackberry, strawberry, blueberry, cranberry, and black raspberry fruits, those from black raspberries and strawberries exhibited the strongest inhibitory properties against cancer cell lines of the colon, prostate, and oral cavity [9]. The extract obtained from flour made from black raspberry seeds has anti-proliferative effects, confirmed by the inhibition of HT 29 colon cancer cell proliferation. Researchers also suggest the possibility of using this flour as a natural source of anti-oxidants and anti-proliferative compounds, as an element of diets that ensure optimal human health [101]. Phytochemicals derived from the consumption of raspberry fruits, upon reaching the colon, can inhibit colon carcinogenesis in vitro [102].

The data from Table 5 present an extensive evaluation of the potential uses of black raspberry (*Rubus occidentalis*) extracts across a range of medical and health-related applications. Black raspberry extracts have been studied for their effects on various diseases and conditions. Not only promising for chemoprevention, they also show potential in the management of colorectal cancer, chronic muscle pain, benign prostatic hyperplasia, possibly even esophageal cancer, and in supporting diabetes management, as suggested by the beneficial effects of *R. occidentalis* extract on glucose control in pre-diabetic patients. This broad therapeutic potential makes black raspberry extracts a significant subject of interest for further pharmacological studies. The table highlights multiple mechanisms through which black raspberry extracts exert their effects, including modulation of gene expression, immunomodulation, apoptosis induction, tumor angiogenesis inhibition, DNA demethylation, and effects on the microbiome. These diverse mechanisms suggest that black raspberry extracts could interact with biological systems at multiple levels, which could be harnessed to develop multifaceted treatment strategies. The table also references several innovative delivery methods for black raspberry extracts, such as PLA ilicylindrical implants and gel preparations for local delivery, underlining the importance of pharmaceutical formulation in enhancing the efficacy and application of phytochemicals in clinical settings. Beyond their therapeutic effects, black raspberry extracts show promise in preventive health, with potential benefits in lowering cholesterol, preventing atherosclerosis, and even reducing DNA methylation in cancer suppressor genes. This preventive aspect could be pivotal in public health strategies, especially in dietary recommendations or supplemental formulations. The data suggest numerous avenues for future research, including further exploring the immunological effects, optimizing extract formulations for better bioavailability and efficacy, and conducting long-term clinical trials to validate the preventive and therapeutic benefits of black raspberry extracts.

Table 5. Health benefits and applications of black raspberry (*Rubus occidentalis*).

Common Name (Species)	Possibility of Use	Reference
Black raspberry (*R. occidentalis*)	Black raspberry extract has an inhibitory effect on the growth of pre-cancerous and malignant cells of the oral cavity. Researchers suggest that this action is related to specific compounds that target aberrant signaling pathways that regulate cell cycle progression	[103]
Black raspberry (*R. occidentalis*)	The chempreventive properties of black raspberry fruits and the obtained research results indicate the possibility of using them as an alternative to chemopreventive drugs in the treatment of esophageal cancer in humans.	[104]
Black raspberry (*R. occidentalis*)	Possibility of developing gel preparations containing dried black raspberries with chemopreventive effects for local delivery to the tissues of the oral mucosa with good absorption	[105]
Black raspberry (*R. occidentalis*)	Research indicates that the optimal dosage form for prolonged and continuous delivery of FBR chemopreventive anthocyanins to the body for a period of one month is the method of injecting PLA ilicylindrical implants containing RE:TA-β-CD IC	[106]
Black raspberry (*R. occidentalis*)	Taking black raspberry powder resulted in a reduction in profiling, as well as an increase in apoptosis of colorectal cancer cells and inhibition of tumor angiogenesis. Treatment with this method also reduced the expression of β-catenin and increased the expression of E-cadherin in colorectal cancer cells. DNA methylation studies indicate that ingredients derived from black raspberry fruits are able to demethylate tumor suppressor genes related to the Wnt pathway, partially inhibiting the expression of DNMT1 and DNA methyltransferase, which is overexpressed in colorectal cancer cells.	[107]
Black raspberry (*R.occidentalis*)	Black raspberry extracts and their metabolites contain phytochemicals that influence immune processes related to carcinogenesis and immunotherapy. Additionally, specific ingredients and their metabolites derived from black raspberries may in the future be used as a key source of raw material for the development of drugs with immunological effects and inhibition of specific STAT-regulated signaling pathways	[108]
Black raspberry (*R. occidentalis*)	Research results suggest that the extract of unripe *Rubus occidentalis* fruit may serve as an agent in the treatment of benign prostatic hyperplasia by inhibiting it.	[109]
Black raspberry (*R. occidentalis*)	Research suggests the possibility of using *Rubus occidentalis* fruit extract to reduce induced chronic muscle pain thanks to its anti-nociceptin effect	[110]
Black raspberry (*R. occidentalis*)	In a 12-week, randomized, double-blind, placebo-controlled trial, the impact of *Rubus occidentalis* (RO) extract on metabolic parameters in pre-diabetic patients was examined. It was found that high doses of the extract (1800 mg per day) significantly improved glucose control and inflammatory markers compared to the placebo group. These results suggest that RO extract may be beneficial in modulating glucose and lipid levels, which is promising for individuals at risk of developing type 2 diabetes.	[111]
Black raspberry (*R. occidentalis*)	Consuming raspberry juice may prevent the formation of fatty streaks in the aorta	[112]
Black raspberry (*R. occidentalis*)	Black raspberry fruit extract may influence cholesterol efflux and clearance by increasing hippuric acid levels in the cecum and modifying the intestinal microbiome, which then results in effects on microbial metabolites and cholesterol metabolizm	[113]
Black raspberry (*R. occidentalis*)	The results indicate the potential use of black raspberry extract in changing intestinal bacteria, lowering cholesterol levels by regulating gene expression and modulating bile acids.	[114]

For the species *Rubus fraxinifolius*, the leaf extract exhibited the highest anti-oxidant activity among extracts from the fruits, leaves, and shoots. The researchers identified it as a potential natural anti-oxidant for health improvement [115]. In China, *Rubus alceaefolius* is used as an herb in the treatment of osteoma, nasopharyngeal cancer, liver cancer, and lung cancer [116]. The aerial parts of *Rubus ulmifolius* contain flavonoids. The crude methanol extract and the flavonoid-rich fraction from this species exhibit anti-oxidant and anti-pyretic effects. This suggests the potential future use of this plant as a source of raw material for the production of anti-pyretic and anti-oxidant drugs, provided that pharmacologically active substances are isolated from this species [117]. Moreover, the fruits of *R. ulmifolius* are rich in nutrients and bioactive compounds, primarily cyanidin-3-O-glucoside and pentoside, and glucuronide of ellagic acid. They exhibit anti-microbial activity, making them a potentially valuable addition to the daily diet for its diversification and enrichment. Research findings suggest that these wild fruits can contribute to human health due to their high content of various health-promoting compounds, including phenols and anthocyanins [118]. The fruits of *Rubus coreanus* are widely used in East Asia, especially in Korea, China, and Japan, as a medicinal ingredient in traditional medicine for treating conditions such as cancer, diarrhea, and asthma (Seo et al. 2019). The leaves of *Rubus corchorifolius* are used as raw material for producing leaf tea, which provides hypoglycemic material [119].

Drawing on data presented in the accompanying Table 6, this discussion elaborates on the multifaceted therapeutic potentials of *Rubus coreanus* fruit extracts. The fruit extracts of *Rubus coreanus* are noted for their capacity to inhibit the growth of specific cancers, including hepatocellular and gastric cancers. These findings suggest a potential for these extracts in both therapeutic and preventive oncology settings, leveraging their natural compounds to target and inhibit cancer cell proliferation. Extracts from *Rubus coreanus* have shown strong anti-inflammatory effects, particularly effective in treating colon damage induced by DSS (Dextran Sulfate Sodium) and reducing inflammation triggered by LPS (Lipopolysaccharides). The isolated triterpenoid-rich fractions from the plant have demonstrated significant efficacy in alleviating symptoms and possibly providing therapeutic benefits in the treatment of inflammatory bowel diseases, such as colitis. Clinical studies have highlighted the effectiveness of freeze-dried *Rubus coreanus* fruit extract in capsule form for significantly reducing high blood cholesterol levels and managing lipid disorders. This suggests its potential as a natural alternative or complement to traditional lipid-lowering medications, offering a plant-based approach to managing cardiovascular risk factors. The extracts from both ripe and unripe fruits, as well as the leaves of *Rubus coreanus*, exhibit anti-bacterial activities, particularly against Gram-positive rod-shaped bacteria like *Bacillus cereus*. This anti-bacterial property underlines the potential of *Rubus coreanus* in developing natural anti-bacterial agents or supplements that could support the treatment or prevention of bacterial infections. The compounds in *Rubus coreanus* have been implicated in physiological changes in retinal pigment epithelium cells that may help counteract age-related macular degeneration. The reduction in vascular angiogenesis in these cells, alongside hypocholesterolemic effects that lower cholesterol levels, points to the neuroprotective and vision-supporting potential of these extracts. Other studies on black raspberry where the species is not specified suggest additional health benefits. Extracts have shown potential in improving intestinal health through probiotic strains and in modulating harmful effects of sunlight on the skin, acting as chemopreventive agents against non-melanoma skin cancer. Moreover, anthocyanins derived from these black raspberries are noted for their role in the cessation of the colorectal cancer cell cycle, reducing inflammation and apoptosis.

Table 6. Health benefits and applications of black raspberry (Various *Rubus* species).

Common Name (Species)	Possibility of Use	Reference
Black raspberry (*R. coreanus*)	Niga-ichigoside F1 (1,23-hydroxytormentic acid 28-O-glc) obtained from the fruit of *Rubus coreanus* and aglyconic-23-hydroxytormentic acid (1a) produced by hydrolysis in NaOH solution from the first compound have anti-nociceptive and anti-inflammatory effects	[120]
Black raspberry (no data)	Possibility of using black raspberries in modulating the harmful effects of sunlight on the skin as a chemopreventive/chemotherapeutic agent against non-melanoma skin cancer	[121]

Table 6. *Cont.*

Common Name (Species)	Possibility of Use	Reference
Black raspberry (*R. coreanus*)	Soluble compounds contained in ethanol may induce changes at the physiological level in retinal pigment epithelium cells and also counteract age-related macular degeneration. The mechanisms responsible for these effects are: decreased vascular angiogenesis in RPE cells, hypocholesterolemic effects caused by lowering cholesterol levels, and cellular cholesterol biosynthesis and lipoprotein uptake	[122]
Black raspberry (*R. coreanus*)	The isolated triterpenoid-rich fraction from *Rubus coreanus* has a strong anti-inflammatory effect on colon damage due to DSS, as well as macrophage activation due to LPS. Research also suggests therapeutic and preventive properties in the treatment of colitis disease	[123]
Black raspberry (no data)	The obtained results suggest the possibility of using *Lactobacillus plantarum* strains GBL 16 and 17 isolated from black raspberry in the future as probiotics improving human intestinal health	[124]
Black raspberry (*R. coreanus*)	Research has confirmed the potential use of freeze-dried *Rubus coreanus* fruit extract in the form of capsules as a potential therapeutic agent for people with extremely high blood cholesterol levels and lipid disorders.	[125]
Black raspberry (no data)	The obtained results indicate that protein acetylation may play a key role in the development of colorectal cancer regulated by anthocyanins derived from black raspberry fruit, which affect the cessation of the cancer cell cycle, alleviating inflammation and apoptosis.	[126]

Extract from the roots of *Rubus fairholmianus* can be used in the synthesis of ZnO nanoparticles. The results suggest that this extract could serve as a raw material for the large-scale synthesis of metal nanoparticles with high purity and crystalline nature. This process is also considered environmentally friendly and without negative environmental impacts. The synthesized RE-ZnO nanoparticles exhibit anti-microbial activity [127]. Extracts from the fruits of the commonly found *Rubus ellipticus* species in the Himalayas have anti-microbial and anti-oxidant properties. Studies indicate greater effectiveness of polar extracts compared to non-polar extracts [128]. *R. ellipticus* is not only a valuable natural anti-oxidant but may also play an important role in treating skin diseases, wounds, and cancers [129], and is a source of nephroprotective, anti-inflammatory, analgesic, anti-pyretic, and cytotoxic compounds. Due to its properties, it is suggested to use this species as a raw material for the production of nutraceuticals [130–132]. The fruits, leaves, shoots, and roots of *R. ellipticus* are used as raw materials in traditional remedies to treat fever, ischemic heart disease, sore throat, abdominal pain, cough, wounds, fractures, cancers, and bacterial infections [133]. *Rubus imperialis*, found in Brazil, is used in the folk medicine of local populations to treat viral infections, diabetes, pain, and burns [18,85]. Studies also indicate its anti-inflammatory effects (Tonin et al. 2016). In traditional medicine in Cameroon, the aerial parts of *Rubus rigidus* are used to treat respiratory and cardiovascular diseases. The ethanol extract from this plant shows strong anti-oxidant properties [134]. Studies on the composition of *Rubus chamaemorus* leaves revealed the presence of tannins (4-O-α-L-arabinofuranosylolagowy acid, 2 [(–)-epicatechin]) and glucuronide flavonoids (quercetin 3-O-β-D-glucuronide (3), quercetin 3-O-β-D-2″-galloglucuronide (4)), as well as kaempferol 3-O-β-D-glucuronide (5)). Due to the wide range of biological activities of these five compounds, researchers suggest their broad potential use as pharmacological ingredients, particularly for producing anti-cancer and anti-diabetic agents [135]. Research findings indicate an increased anti-bacterial, anti-viral, and anti-oxidant potential of honey enriched with ingredients derived from raspberry and blackberry leaves and fruits. The obtained honey could be a component of functional foods with high health-promoting value [136].

Based on the data extracted from the provided Table 7, this discussion elucidates the health benefits and therapeutic applications associated with three distinct raspberry species: *Rubus parvifolius* (Japanese Bramble), *Rubus ellipticus* (Yellow Raspberry), and *Rubus chamaemorus* (Cloudberry). *Rubus parvifolius* fruit extract has demonstrated significant anti-proliferative activity against myeloid leukemia cells both in vitro and in vivo, suggesting its potential as a treatment for leukemia. Further research supports the use of total saponins extracted from the roots of this plant for developing drugs aimed at combating chronic myeloid leukemia. These saponins are characterized by their ability to limit tumor growth and induce apoptosis, contributing to their therapeutic potential. Additionally, *Rubus parvifolius* extracts, rich in phenolic compounds such as caffeic acid conjugates, ellagic acid glycosides, and flannel glycosides, have shown hepatoprotective and anti-oxidant properties. These extracts mitigate hepatotoxicity induced by CCl4 primarily through their anti-oxidant capacity, which involves scavenging free radicals, alleviating oxidative stress, and reducing lipid peroxidation. Research on *Rubus ellipticus* has led to the development of copper oxide nanoparticles synthesized using aqueous fruit extracts, which exhibited potent anti-microbial activity against pathogens, like *Bacillus subtilis*, *Rosellinia necatrix*, *Staphylococcus aureus*, and *Escherichia coli*, while maintaining low toxicity towards healthy cells. These nanoparticles also showed anti-cancer activity against colorectal cancer cell lines, indicating their potential applications in the pharmaceutical and medical industries. Additionally, acetone and methanol extracts from the fruits have strong anti-oxidant properties and the ability to neutralize various reactive radicals. These extracts have shown potential in treating cervical cancer due to their anti-filtering properties. Studies on *Rubus chamaemorus* extract have confirmed its robust anti-oxidant effect in bacterial cultures, facilitated through multiple pathways, such as the direct inhibition of reactive oxygen species, chelation of iron ions, and induction of anti-oxidant genes. Notably, these extracts have been shown to activate the OxyR-regulon genes and influence the activity of the general oxidative stress response system controlled by RpoS, illustrating their capacity to modulate oxidative stress at a genetic level.

Table 7. Health benefits and applications of yellow raspberry, japanese bramsle and cloudberry (*Rubus ellipticus*, *R. parvifolius*, *R. chamaemorus*).

Common Name (Species)	Possibility of Use	Reference
Yellow raspberry (*R. ellipticus*)	The obtained research results suggest the possibility of using acetone and methanol extracts obtained from *Rubus ellipticus* fruits in the treatment of cervical cancer due to their anti-filtering properties. Extracts from these fruits have very strong anti-oxidant properties and the ability to capture all reactive radicals, including DPPH, ABTS, superoxide radicals, linoleate peroxide free radicals	[137]
Japanese bramble (*R. parvifolius*)	*Rubus parvifolius* extract contains many phenolic compounds, including caffeic acid conjugates, ellagic acid glycosides and flannel glycosides, and research results suggest that they may have hepatoprotective and anti-oxidant properties. Protection against hepatotoxicity due to CCl4 is due in part to anti-oxidant capacity, and more specifically to the scavenging of free radicals, which alleviates oxidative stress and reduces lipid peroxidation.	[138]
Cloudberry (*R. chamaemorus*)	Studies have confirmed the simultaneous anti-oxidant effect of *Rubus chamaemorus* extract in bacterial cultures through several different pathways. Direct inhibition of reactive oxygen species, chelation of iron ions and induction of anti-oxidant genes have also been demonstrated. Researchers managed to prove for the first time that the *R. chamaemorus* extract has the ability to activate the OxyR-regulon genes and influence the activity of the general oxidative stress response system controlled by RpoS	[139]
Japanese bramsle (*R. parvifolius*)	*Rubus parvifolius* fruit extract showed anti-proliferative activity against myeloid leukemia cells (K562) in vitro and in vivo	[140]
Japanese bramsle (*R. parvifolius*)	Research suggests the possibility of using total saponins from the roots of *Rubus parvifolius* as a raw material for the production of drugs against chronic myeloid leukemia. Total saponins from *Rubus parvifolius* are characterized by limiting the rate of tumor growth and inducing cell death through apoptosis	[141]
Yellow raspberry (*R. ellipticus*)	Synthesized copper oxide nanoparticles (Ru-Cu2O-NP) using aqueous extract of *Rubus ellipticus* fruits exhibited enhanced anti-microbial activity against various pathogens, such as *Bacillus subtilis*, *Rosellinia necatrix*, *Staphylococcus aureus*, and *Escherichia coli*, while showing no toxicity towards healthy cells at tested concentrations. Moreover, the nanoparticles demonstrated anti-cancer activity against the colorectal cancer cell line, indicating their potential application in the pharmaceutical and medical industries with maintained low toxicity.	[142]

The cultivation of the genus *Rubus* is primarily limited to a few species, overlooking a large number of other non-conventional and wild-growing representatives of this genus, resulting in a significant loss of their potential utility. Due to the exceptional characteristics of *Rubus* species, they can be termed "superfoods". It is crucial to expand cultivation to include new species while also achieving the full potential of both the already popular and lesser-known representatives, which are mainly appreciated by specific communities [143]. The primary step for wild species to become food plants is their popularization [144]. Currently, wild berries in Europe are gaining increasing significance in the context of the cultural role of ecosystems. An example of a locally used species that gained nationwide popularity is *Rubus caesius* (European dewberry, salmbär 'Solomon berries'), which has become an ingredient in many dishes, especially desserts. One such dish was even served at the Nobel Prize banquet in December 2014 [145].

5. Conclusions

In summary, species of the genus *Rubus* exhibit a wide range of nutritional and health-promoting properties, making them valuable components of a healthy diet. These plants, particularly raspberries and blackberries, are rich in polyphenols, flavonoids, vitamins, and minerals, contributing to their anti-oxidant, anti-inflammatory, anti-neurodegenerative, and anti-cancer effects. The diverse phytochemical composition of *Rubus* species supports their use in the prevention and treatment of various chronic diseases, including cardiovascular diseases, diabetes, and certain cancers. Research highlights the significant potential of lesser-known and wild-growing *Rubus* species, which are often overlooked in favor of more commonly cultivated varieties. These wild species, with their unique phytochemical profiles and health benefits, present untapped opportunities for expanding the range of functional foods and nutraceuticals available to consumers. The extensive bioactivity of *Rubus* species underscores the need for continued exploration and utilization of these plants in both dietary and medicinal contexts. Future studies should focus on the detailed mechanisms underlying their health benefits, the development of effective extraction and formulation techniques, and the integration of these findings into public health strategies and clinical practices. The genus *Rubus* represents a valuable resource for improving human health, and its increased cultivation and use could significantly contribute to global health and nutrition.

Author Contributions: Conceptualization, K.B. and M.K.; methodology, K.B.; software, K.B.; validation, K.B., M.K. and Z.J.; formal analysis, K.B.; investigation, K.B.; resources, K.B.; data curation, K.B.; writing—original draft preparation, K.B.; writing—review and editing, K.B. and M.K.; visualization, K.B.; supervision, M.K. and Z.J. All authors have read and agreed to the published version of the manuscript.

Funding: This research received no external funding.

Data Availability Statement: Data sharing not applicable.

Conflicts of Interest: The authors declare no conflicts of interest.

References

1. Cena, H.; Calder, P.C. Defining a Healthy Diet: Evidence for the Role of Contemporary Dietary Patterns in Health and Disease. *Nutrients* **2020**, *12*, 334. [CrossRef] [PubMed]
2. Neufeld, L.M.; Hendriks, S.; Hugas, M. Healthy Diet: A Definition for the United Nations Food Systems Summit 2021. In *Science and Innovations for Food Systems Transformation*; Von Braun, J., Afsana, K., Fresco, L.O., Hassan, M.H.A., Eds.; Springer International Publishing: Cham, 2023; pp. 21–30. ISBN 978-3-031-15702-8.
3. Seligman, H.K.; Schillinger, D. Hunger and Socioeconomic Disparities in Chronic Disease. *N. Engl. J. Med.* **2010**, *363*, 6–9. [CrossRef] [PubMed]
4. Gordon, L.J.; Bignet, V.; Crona, B.; Henriksson, P.J.G.; Van Holt, T.; Jonell, M.; Lindahl, T.; Troell, M.; Barthel, S.; Deutsch, L.; et al. Rewiring Food Systems to Enhance Human Health and Biosphere Stewardship. *Environ. Res. Lett.* **2017**, *12*, 100201. [CrossRef]

5. Willett, W.; Rockström, J.; Loken, B.; Springmann, M.; Lang, T.; Vermeulen, S.; Garnett, T.; Tilman, D.; DeClerck, F.; Wood, A.; et al. Food in the Anthropocene: The EAT–Lancet Commission on Healthy Diets from Sustainable Food Systems. *Lancet* **2019**, *393*, 447–492. [CrossRef] [PubMed]
6. Slimko, M.L.; Mensah, G.A. The Role of Diets, Food, and Nutrients in the Prevention and Control of Hypertension and Prehypertension. *Cardiol. Clin.* **2010**, *28*, 665–674. [CrossRef] [PubMed]
7. World Health Organization. Available online: https://www.who.int/News-Room/Fact-Sheets/Detail/Healthy-Diet (accessed on 8 May 2024).
8. Miller, M.G.; Shukitt-Hale, B. Berry Fruit Enhances Beneficial Signaling in the Brain. *J. Agric. Food Chem.* **2012**, *60*, 5709–5715. [CrossRef] [PubMed]
9. Seeram, N.P.; Adams, L.S.; Zhang, Y.; Lee, R.; Sand, D.; Scheuller, H.S.; Heber, D. Blackberry, Black Raspberry, Blueberry, Cranberry, Red Raspberry, and Strawberry Extracts Inhibit Growth and Stimulate Apoptosis of Human Cancer Cells In Vitro. *J. Agric. Food Chem.* **2006**, *54*, 9329–9339. [CrossRef] [PubMed]
10. Lovell, S.T.; Krishnaswamy, K.; Lin, C.-H.; Meier, N.; Revord, R.S.; Thomas, A.L. Nuts and Berries from Agroforestry Systems in Temperate Regions Can Form the Foundation for a Healthier Human Diet and Improved Outcomes from Diet-Related Diseases. *Agrofor. Syst.* **2023**, *97*, 1347–1360. [CrossRef] [PubMed]
11. Nicolas, C.; Rouche, M.; Dierckens, M.; Kelly, C.; Fismen, A.-S.; Nardone, P.; Castetbon, K.; Chatelan, A. Sixteen-Year Trends in Fruit Consumption and Related Socioeconomic Inequalities among Adolescents in Western European Countries. *Eur. J. Nutr.* **2023**, *62*, 3287–3296. [CrossRef]
12. Kurtto, A.; Lampinen, R.; Junikka, L.E. Atlas Florae Europaeae. Distribution of Vascular Plants in Europe. 13. Rosaceae (Spiraea to Fragaria, Excl. Rubus). In *Atlas Florae Europaeae: Distribution of Vascular Plants in Europe*; Committee for Mapping the Flora of Europe: Helsinki, Finland, 2004.
13. Zieliński, J.; Rejman, A. The Genus Rubus (Rosaceae) in Poland. *Pol. Bot. Stud.* **2004**, *16*, 1–300.
14. Smolarz, K. *Pomologia*; Rejman, A., Ed.; PWRiL: Warszawa, Polska, 1994; pp. 521–535.
15. Żurawicz, E. *Pomologia. Aneks*; Żurawicz, E., Ed.; PWRiL: Warszawa, Polska, 2003; pp. 234–242.
16. Hummer, K.E. Rubus Pharmacology: Antiquity to the Present. *HortScience* **2010**, *45*, 1587–1591. [CrossRef]
17. Rocabado, G.O.; Bedoya, L.M.; Abad, M.J.; Bermejo, P. *Rubus*—A Review of Its Phytochemical and Pharmacological Profile. *Nat. Prod. Commun.* **2008**, *3*, 423–436. [CrossRef]
18. Niero, R.; Filho, V.C. Therapeutic Potential and Chemical Composition of Plants from the Genus *Rubus*: A Mini Review of the Last 10 Years. *Nat. Prod. Commun.* **2008**, *3*, 437–444. [CrossRef]
19. Du, X.; Finn, C.E.; Qian, M.C. Volatile Composition and Odour-Activity Value of Thornless 'Black Diamond' and 'Marion' Blackberries. *Food Chem.* **2010**, *119*, 1127–1134. [CrossRef]
20. Dragišić Maksimović, J.J.; Milivojević, J.M.; Poledica, M.M.; Nikolić, M.D.; Maksimović, V.M. Profiling Anti-oxidant Activity of Two Primocane Fruiting Red Raspberry Cultivars (Autumn Bliss and Polka). *J. Food Compos. Anal.* **2013**, *31*, 173–179. [CrossRef]
21. Chen, J.; Du, J.; Li, M.; Li, C. Degradation Kinetics and Pathways of Red Raspberry Anthocyanins in Model and Juice Systems and Their Correlation with Color and Anti-oxidant Changes during Storage. *LWT* **2020**, *128*, 109448. [CrossRef]
22. McCarty, M.F. Proposal for a Dietary "Phytochemical Index". *Med. Hypotheses* **2004**, *63*, 813–817. [CrossRef] [PubMed]
23. Hancock, R.D.; Petridis, A.; McDougall, G.J. Raspberry Fruit Chemistry in Relation to Fruit Quality and Human Nutrition. In *Raspberry*; Graham, J., Brennan, R., Eds.; Springer International Publishing: Cham, Switzerland, 2018; pp. 89–119. ISBN 978-3-319-99030-9.
24. Moreno-Medina, B.L.; Casierra-Posada, F.; Cutler, J. Phytochemical Composition and Potential Use of Rubus Species. *Gesunde Pflanz.* **2018**, *70*, 65–74. [CrossRef]
25. Schulz, M.; Chim, J.F. Nutritional and Bioactive Value of Rubus Berries. *Food Biosci.* **2019**, *31*, 100438. [CrossRef]
26. Lee, Y.; Kim, J.C.; Hwang, K.T.; Kim, D.-H.; Jung, C.M. Quality Characteristics of Black Raspberry Wine Fermented with Different Yeasts. *J. Korean Soc. Food Sci. Nutr.* **2013**, *42*, 784–791. [CrossRef]
27. Nile, S.H.; Park, S.W. Edible Berries: Bioactive Components and Their Effect on Human Health. *Nutrition* **2014**, *30*, 134–144. [CrossRef] [PubMed]
28. Baranowska, A.; Radwańska, K.; Zarzecka, K.; Gugała, M.; Mystkowska, I. Health Benefits of Red Raspberry Fruit (*Rubus Idaeus* L.). *Probl. Hig. Epidemiol.* **2015**, *96*, 406–409.
29. Luo, T.; Miranda-Garcia, O.; Adamson, A.; Sasaki, G.; Shay, N.F. Development of Obesity Is Reduced in High-Fat Fed Mice Fed Whole Raspberries, Raspberry Juice Concentrate, and a Combination of the Raspberry Phytochemicals Ellagic Acid and Raspberry Ketone. *JBR* **2016**, *6*, 213–223. [CrossRef]
30. Esfahani, A.; Wong, J.M.W.; Mirrahimi, A.; Srichaikul, K.; Jenkins, D.J.A.; Kendall, C.W.C. The Glycemic Index: Physiological Significance. *J. Am. Coll. Nutr.* **2009**, *28*, 439S–445S. [CrossRef] [PubMed]
31. Arvidsson-Lenner, R.; Asp, N.-G.; Axelsen, M.; Bryngelsson, S.; Haapa, E.; Järvi, A.; Karlström, B.; Raben, A.; Sohlström, A.; Thorsdottir, I.; et al. Glycaemic Index. *Scand. J. Nutr.* **2004**, *48*, 84–94. [CrossRef]
32. FoodData Central United States Department of Agriculture. Available online: https://fdc.nal.usda.gov/fdc-app.html#/Food-Details/167755/Nutrients (accessed on 1 May 2024).
33. FoodData Central United States Department of Agriculture. Available online: https://fdc.nal.usda.gov/fdc-app.html#/Food-Details/173946/Nutrients (accessed on 1 May 2024).

34. Das, P.R.; Eun, J.-B. A Comparative Study of Ultra-Sonication and Agitation Extraction Techniques on Bioactive Metabolites of Green Tea Extract. *Food Chem.* **2018**, *253*, 22–29. [CrossRef] [PubMed]
35. De Paulo Farias, D.; De Araújo, F.F.; Neri-Numa, I.A.; Pastore, G.M. Antidiabetic Potential of Dietary Polyphenols: A Mechanistic Review. *Food Res. Int.* **2021**, *145*, 110383. [CrossRef] [PubMed]
36. Singla, R.K.; Dubey, A.K.; Garg, A.; Sharma, R.K.; Fiorino, M.; Ameen, S.M.; Haddad, M.A.; Al-Hiary, M. Natural Polyphenols: Chemical Classification, Definition of Classes, Subcategories, and Structures. *J. AOAC Int.* **2019**, *102*, 1397–1400. [CrossRef]
37. Li, W.; Chen, H.; Xu, B.; Wang, Y.; Zhang, C.; Cao, Y.; Xing, X. Research Progress on Classification, Sources and Functions of Dietary Polyphenols for Prevention and Treatment of Chronic Diseases. *J. Future Foods* **2023**, *3*, 289–305. [CrossRef]
38. Rathod, N.B.; Elabed, N.; Punia, S.; Ozogul, F.; Kim, S.-K.; Rocha, J.M. Recent Developments in Polyphenol Applications on Human Health: A Review with Current Knowledge. *Plants* **2023**, *12*, 1217. [CrossRef]
39. Mullen, W.; Lean, M.E.J.; Crozier, A. Rapid Characterization of Anthocyanins in Red Raspberry Fruit by High-Performance Liquid Chromatography Coupled to Single Quadrupole Mass Spectrometry. *J. Chromatogr. A* **2002**, *966*, 63–70. [CrossRef] [PubMed]
40. Teng, H.; Fang, T.; Lin, Q.; Song, H.; Liu, B.; Chen, L. Red Raspberry and Its Anthocyanins: Bioactivity beyond Anti-oxidant Capacity. *Trends Food Sci. Technol.* **2017**, *66*, 153–165. [CrossRef]
41. Teixeira, B.A.; Vidigal, M.C.T.R.; De Castro Leite Júnior, B.; Vieira, É.N.R.; Martins, E.M.F.; Stringheta, P.C. Optimization, Kinetic and Phenomenological Modeling of Ultrasound-Assisted Extraction Process of Bioactive Compounds from Raspberries (*Rubus Idaeus* L.). *Food Anal. Methods* **2023**, *16*, 759–770. [CrossRef]
42. Mullen, W.; McGinn, J.; Lean, M.E.J.; MacLean, M.R.; Gardner, P.; Duthie, G.G.; Yokota, T.; Crozier, A. Ellagitannins, Flavonoids, and Other Phenolics in Red Raspberries and Their Contribution to Anti-oxidant Capacity and Vasorelaxation Properties. *J. Agric. Food Chem.* **2002**, *50*, 5191–5196. [CrossRef]
43. Carvalho, E.; Fraser, P.D.; Martens, S. Carotenoids and Tocopherols in Yellow and Red Raspberries. *Food Chem.* **2013**, *139*, 744–752. [CrossRef] [PubMed]
44. Khare, C.P. Rubus Ellipticus Sm. In *Indian Medicinal Plants*; Khare, C.P., Ed.; Springer: New York, NY, USA, 2007; p. 1. ISBN 978-0-387-70637-5.
45. Veljkovic, B.; Djordjevic, N.; Dolicanin, Z.; Licina, B.; Topuzovic, M.; Stankovic, M.; Zlatic, N.; Dajic-Stevanovic, Z. Anti-oxidant and Anticancer Properties of Leaf and Fruit Extracts of the Wild Raspberry (*Rubus Idaeus* L.). *Not. Bot. Horti. Agrobot.* **2018**, *47*, 359–367. [CrossRef]
46. Krauze-Baranowska, M.; Głód, D.; Kula, M.; Majdan, M.; Hałasa, R.; Matkowski, A.; Kozłowska, W.; Kawiak, A. Chemical Composition and Biological Activity of Rubus Idaeus Shoots—A Traditional Herbal Remedy of Eastern Europe. *BMC Complement. Altern. Med.* **2014**, *14*, 480. [CrossRef]
47. Aglar, E.; Sumbul, A.; Karakaya, O.; Erturk, O.; Ozturk, B. Phytochemical and Antimicrobial Characteristics of Raspberry Fruit Growing Naturally in Kelkit Valley, Turkey. *Erwerbs-Obstbau* **2023**, *65*, 65–70. [CrossRef]
48. Marino, M.; Gardana, C.; Rendine, M.; Klimis-Zacas, D.; Riso, P.; Porrini, M.; Del Bo', C. Nutritional and Phytochemical Characterization of Freeze-Dried Raspberry (Rubus Idaeus): A Comprehensive Analysis. *Foods* **2024**, *13*, 1051. [CrossRef]
49. Schell, J.; Betts, N.M.; Lyons, T.J.; Basu, A. Raspberries Improve Postprandial Glucose and Acute and Chronic Inflammation in Adults with Type 2 Diabetes. *Ann. Nutr. Metab.* **2019**, *74*, 165–174. [CrossRef]
50. Fu, Y.; Zhou, X.; Chen, S.; Sun, Y.; Shen, Y.; Ye, X. Chemical Composition and Anti-oxidant Activity of Chinese Wild Raspberry (*Rubus Hirsutus* Thunb.). *LWT-Food Sci. Technol.* **2015**, *60*, 1262–1268. [CrossRef]
51. Sheng, J.-Y.; Wang, S.-Q.; Liu, K.-H.; Zhu, B.; Zhang, Q.-Y.; Qin, L.-P.; Wu, J.-J. Rubus Chingii Hu: An Overview of Botany, Traditional Uses, Phytochemistry, and Pharmacology. *Chin. J. Nat. Med.* **2020**, *18*, 401–416. [CrossRef] [PubMed]
52. Velicanski, A.; Cvetkovic, D.; Markov, S. Screening of Antibacterial Activity of Raspberry (*Rubus Idaeus* L.) Fruit and Pomace Extracts. *Acta Period. Technol.* **2012**, *43*, 305–313. [CrossRef]
53. Jean-Gilles, D.; Li, L.; Ma, H.; Yuan, T.; Chichester, C.O.; Seeram, N.P. Anti-Inflammatory Effects of Polyphenolic-Enriched Red Raspberry Extract in an Antigen-Induced Arthritis Rat Model. *J. Agric. Food Chem.* **2012**, *60*, 5755–5762. [CrossRef]
54. Figueira, M.E.; Câmara, M.B.; Direito, R.; Rocha, J.; Serra, A.T.; Duarte, C.M.M.; Fernandes, A.; Freitas, M.; Fernandes, E.; Marques, M.C.; et al. Chemical Characterization of a Red Raspberry Fruit Extract and Evaluation of Its Pharmacological Effects in Experimental Models of Acute Inflammation and Collagen-Induced Arthritis. *Food Funct.* **2014**, *5*, 3241–3251. [CrossRef]
55. Li, L.; Wang, L.; Wu, Z.; Yao, L.; Wu, Y.; Huang, L.; Liu, K.; Zhou, X.; Gou, D. Anthocyanin-Rich Fractions from Red Raspberries Attenuate Inflammation in Both RAW264.7 Macrophages and a Mouse Model of Colitis. *Sci. Rep.* **2014**, *4*, 6234. [CrossRef] [PubMed]
56. Morimoto, C.; Satoh, Y.; Hara, M.; Inoue, S.; Tsujita, T.; Okuda, H. Anti-Obese Action of Raspberry Ketone. *Life Sci.* **2005**, *77*, 194–204. [CrossRef] [PubMed]
57. Bokan, A.; Vukmirović, S.; Medić, D. The Antibacterial Effect of the Combination of Red Raspberry Concentrate and Antibiotics on Staphylococcus Aureus. *BMC Proc.* **2015**, *9*, A1. [CrossRef]
58. Noratto, G.D.; Chew, B.P.; Atienza, L.M. Red Raspberry (*Rubus Idaeus* L.) Intake Decreases Oxidative Stress in Obese Diabetic (Db/Db) Mice. *Food Chem.* **2017**, *227*, 305–314. [CrossRef]
59. Henrotin, Y.; Cozannet, R.L.; Fança-Berthon, P.; Truillet, R.; Cohen-Solhal, M.; DunnGalvin, G.; Grouin, J.-M.; Doolan, A. Rubus Idaeus Extract Improves Symptoms in Knee Osteoarthritis Patients: Results from a Phase II Double-Blind Randomized Controlled Trial. *BMC Musculoskelet. Disord.* **2022**, *23*, 650. [CrossRef]

60. Ma, Y.; Xiu, W.; Wang, X.; Yang, Q. Extraction of Raspberry Ketone from Red Raspberry and Its Intervention in the Non-Alcoholic Fatty Liver Disease. *Appl. Biol. Chem.* **2022**, *65*, 78. [CrossRef]
61. Park, K. Raspberry Ketone Increases Both Lipolysis and Fatty Acid Oxidation in 3T3-L1 Adipocytes. *Planta Med.* **2010**, *76*, 1654–1658. [CrossRef]
62. Park, K.S. Raspberry Ketone, a Naturally Occurring Phenolic Compound, Inhibits Adipogenic and Lipogenic Gene Expression in 3T3-L1 Adipocytes. *Pharm. Biol.* **2015**, *53*, 870–875. [CrossRef] [PubMed]
63. Jia, H.; Liu, J.; Ufur, H.; He, G.; Liqian, H.; Chen, P. The Antihypertensive Effect of Ethyl Acetate Extract from Red Raspberry Fruit in Hypertensive Rats. *Pharmacogn. Mag.* **2011**, *7*, 19. [CrossRef] [PubMed]
64. Chen, W.; Su, H.; Huang, Z.; Feng, L.; Nie, H. Neuroprotective Effect of Raspberry Extract by Inhibiting Peroxynitrite-Induced DNA Damage and Hydroxyl Radical Formation. *Food Res. Int.* **2012**, *49*, 22–26. [CrossRef]
65. Liang, W.-Q.; Xu, G.-J.; Weng, D.; Gao, B.; Zheng, X.-F.; Qian, Y. Anti-Osteoporotic Components of Rubus Chingii. *Chem. Nat. Compd.* **2015**, *51*, 47–49. [CrossRef]
66. Li, K.; Zeng, M.; Li, Q.; Zhou, B. Identification of Polyphenolic Composition in the Fruits of Rubus Chingii Hu and Its Anti-oxidant and Antiproliferative Activity on Human Bladder Cancer T24 Cells. *J. Food Meas. Charact.* **2019**, *13*, 51–60. [CrossRef]
67. Jazić, M.; Kukrić, Z.; Vulić, J.; Četojević-Simin, D. Polyphenolic Composition, Anti-oxidant and Antiproliferative Effects of Wild and Cultivated Blackberries (*Rubus Fruticosus* L.) Pomace. *Int. J. Food Sci. Technol.* **2019**, *54*, 194–201. [CrossRef]
68. Gil-Martínez, L.; Mut-Salud, N.; Ruiz-García, J.A.; Falcón-Piñeiro, A.; Maijó-Ferré, M.; Baños, A.; De La Torre-Ramírez, J.M.; Guillamón, E.; Verardo, V.; Gómez-Caravaca, A.M. Phytochemicals Determination, and Anti-oxidant, Antimicrobial, Anti-Inflammatory and Anticancer Activities of Blackberry Fruits. *Foods* **2023**, *12*, 1505. [CrossRef]
69. Moyer, R.A.; Hummer, K.E.; Finn, C.E.; Frei, B.; Wrolstad, R.E. Anthocyanins, Phenolics, and Anti-oxidant Capacity in Diverse Small Fruits: *Vaccinium*, *Rubus*, and *Ribes*. *J. Agric. Food Chem.* **2002**, *50*, 519–525. [CrossRef]
70. Zhang, L.; Zhou, J.; Liu, H.; Khan, M.A.; Huang, K.; Gu, Z. Compositions of Anthocyanins in Blackberry Juice and Their Thermal Degradation in Relation to Anti-oxidant Activity. *Eur. Food Res. Technol.* **2012**, *235*, 637–645. [CrossRef]
71. Stintzing, F.C.; Stintzing, A.S.; Carle, R.; Wrolstad, R.E. A Novel Zwitterionic Anthocyanin from Evergreen Blackberry (*Rubus Laciniatus* Willd.). *J. Agric. Food Chem.* **2002**, *50*, 396–399. [CrossRef]
72. Moreno-Medina, B.L.; Casierra-Posada, F.; Medina-Vargas, O.J. Phenolic Profile and Anti-oxidant Capacity of Blackberry Fruits (*Rubus* Spp.) Grown in Colombia. *Erwerbs-Obstbau* **2023**, *65*, 1047–1056. [CrossRef]
73. Van De Velde, F.; Esposito, D.; Grace, M.H.; Pirovani, M.E.; Lila, M.A. Anti-Inflammatory and Wound Healing Properties of Polyphenolic Extracts from Strawberry and Blackberry Fruits. *Food Res. Int.* **2019**, *121*, 453–462. [CrossRef]
74. Gowd, V.; Bao, T.; Wang, L.; Huang, Y.; Chen, S.; Zheng, X.; Cui, S.; Chen, W. Anti-oxidant and Antidiabetic Activity of Blackberry after Gastrointestinal Digestion and Human Gut Microbiota Fermentation. *Food Chem.* **2018**, *269*, 618–627. [CrossRef] [PubMed]
75. De Gomes, M.G.; Del Fabbro, L.; Rossito Goes, A.T.; Souza, L.C.; Donato, F.; Boeira, S.P.; Prigol, M.; Jesse, C.R. Blackberry Juice Anthocyanidins Limit Cisplatin-Induced Renal Pathophysiology in Mice. *Pathophysiology* **2019**, *26*, 137–143. [CrossRef] [PubMed]
76. Yousefbeyk, F.; Ghasemi, S.; Evazalipour, M.; Dabirian, S.; Schubert, C.; Hekmatnia, S.; Habibi, Y.; Eghbali Koohi, D.; Böhm, V. Phytochemical Analysis, Anti-oxidant, Antibacterial, and Cytotoxic Activities of Leaves and Roots of Rubus Hyrcanus Juz. *Eur. Food Res. Technol.* **2022**, *248*, 141–152. [CrossRef]
77. Hassan, S. Ethnoecological Evaluation of Some Medicinal and Aromatic Plants of Kot Malakand Agency, Pakistan. *Sci. Res. Essays* **2011**, *6*, 2164–2173. [CrossRef]
78. Riaz, M.; Ahmad, M.; Rahman, N. Antimicrobial Screening of Fruit, Leaves, Root and Stem of Rubus Fruticosus. *J. Med. Plants Res.* **2011**, *5* (Suppl. 24), 5920–5924.
79. Ali, N.; Aleem, U.; Ali Shah, S.W.; Shah, I.; Junaid, M.; Ahmed, G.; Ali, W.; Ghias, M. Acute Toxicity, Brine Shrimp Cytotoxicity, Anthelmintic and Relaxant Potentials of Fruits of Rubus Fruticosus Agg. *BMC Complement. Altern. Med.* **2013**, *13*, 138. [CrossRef]
80. Cadi, H.E.; Bouzidi, H.E.; Selama, G.; Ramdan, B.; Majdoub, Y.O.E.; Alibrando, F.; Brigui, J.; Altemimi, A.B.; Dugo, P.; Mondello, L.; et al. Characterization of *Rubus Fruticosus* L. Berries Growing Wild in Morocco: Phytochemical Screening, Anti-oxidant Activity and Chromatography Analysis. *Eur. Food Res. Technol.* **2021**, *247*, 1689–1699. [CrossRef]
81. Salazar-Montoya, J.A.; Hereira-Pacheco, S.; Cruz-Orea, A.; Ramos-Ramírez, E.G. Composition, Anti-oxidant Activity and Rheological Characteristics of Spreadable Pastes with Blackberry Pulp (Rubus Fruticosus). *Food Meas.* **2022**, *16*, 1459–1471. [CrossRef]
82. Memete, A.R.; Sărac, I.; Teusdea, A.C.; Budău, R.; Bei, M.; Vicas, S.I. Bioactive Compounds and Anti-oxidant Capacity of Several Blackberry (*Rubus* Spp.) Fruits Cultivars Grown in Romania. *Horticulturae* **2023**, *9*, 556. [CrossRef]
83. Karimkhani, M.M.; Jamshidi, A.; Zeinali, T.; Fathi, B. Effect of Extraction Solvents on Total Phenolic Compound, Lipid Peroxidation, Anti-oxidant and Cytotoxic Activity of Leaves of Rubus Armeniacus (Himalayan Blackberry). *J. Food Meas. Charact.* **2021**, *15*, 5725–5734. [CrossRef]
84. Barnett, L.E.; Broomfield, A.M.; Hendriks, W.H.; Hunt, M.B.; McGhie, T.K. The In Vivo Anti-oxidant Action and the Reduction of Oxidative Stress by Boysenberry Extract Is Dependent on Base Diet Constituents in Rats. *J. Med. Food* **2007**, *10*, 281–289. [CrossRef] [PubMed]
85. Berté, P.E.; Da Silva Lopes, J.; Comandulli, N.G.; Rangel, D.W.; Monache, F.D.; Filho, V.C.; Niero, R.; De Andrade, S.F. Evaluation of the Gastroprotective Activity of the Extracts, Fractions, and Pure Compounds Obtained from Aerial Parts of Rubus Imperialis in Different Experimental Models. *Naunyn-Schmiedebergs Arch. Pharmacol.* **2014**, *387*, 313–319. [CrossRef] [PubMed]

86. Jung, H.; Lee, H.J.; Cho, H.; Lee, K.; Kwak, H.-K.; Hwang, K.T. Anthocyanins in Rubus Fruits and Anti-oxidant and Anti-Inflammatory Activities in RAW 264.7 Cells. *Food Sci. Biotechnol.* **2015**, *24*, 1879–1886. [CrossRef]
87. Cho, B.-O.; Ryu, H.-W.; Lee, C.-W.; Jin, C.-H.; Seo, W.-D.; Ryu, J.; Kim, D.-S.; Kang, S.Y.; Yook, H.-S.; Jeong, I.-Y. Protective Effects of New Blackberry Cultivar MNU-32 Extracts against H_2O_2-Induced Oxidative Stress in HepG2 Cells. *Food Sci. Biotechnol.* **2015**, *24*, 643–650. [CrossRef]
88. Xie, P.; Zhang, Y.; Wang, X.; Wei, J.; Kang, W. Antithrombotic Effect and Mechanism of *Rubus* Spp. Blackberry. *Food Funct.* **2017**, *8*, 2000–2012. [CrossRef]
89. Chen, W.; Su, H.; Xu, Y.; Jin, C. In Vitro Gastrointestinal Digestion Promotes the Protective Effect of Blackberry Extract against Acrylamide-Induced Oxidative Stress. *Sci. Rep.* **2017**, *7*, 40514. [CrossRef]
90. Tatar, M.; Bagheri, Z.; Varedi, M.; Naghibalhossaini, F. Blackberry Extract Inhibits Telomerase Activity in Human Colorectal Cancer Cells. *Nutr. Cancer* **2019**, *71*, 461–471. [CrossRef] [PubMed]
91. Chaves, V.C.; Soares, M.S.P.; Spohr, L.; Teixeira, F.; Vieira, A.; Constantino, L.S.; Pizzol, F.D.; Lencina, C.L.; Spanevello, R.M.; Freitas, M.P.; et al. Blackberry Extract Improves Behavioral and Neurochemical Dysfunctions in a Ketamine-Induced Rat Model of Mania. *Neurosci. Lett.* **2020**, *714*, 134566. [CrossRef] [PubMed]
92. Sánchez-Velázquez, O.A.; Cortés-Rodríguez, M.; Milán-Carrillo, J.; Montes-Ávila, J.; Robles-Bañuelos, B.; Del Ángel, A.S.; Cuevas-Rodríguez, E.O.; Rangel-López, E. Anti-Oxidant and Anti-Proliferative Effect of Anthocyanin Enriched Fractions from Two Mexican Wild Blackberries (*Rubus* Spp.) on HepG2 and Glioma Cell Lines. *JBR* **2020**, *10*, 513–529. [CrossRef]
93. Alumeti Munyali, D.; Tetsatsi Momo, A.C.; Bonsou Fozin, G.R.; Deeh Defo, P.B.; Petnga Tchatat, Y.; Lieunang, B.; Watcho, P. Rubus Apetalus (Rosaceae) Improves Spermatozoa Characteristics, Anti-oxidant Enzymes and Fertility Potential in Unilateral Cryptorchid Rats. *Basic Clin. Androl.* **2020**, *30*, 8. [CrossRef] [PubMed]
94. Ayele, A.G.; Kumar, P.; Engidawork, E. Antihyperglycemic and Hypoglycemic Activities of the Aqueous Leaf Extract of Rubus Erlangeri Engl (Rosacea) in Mice. *Metab. Open* **2021**, *11*, 100118. [CrossRef] [PubMed]
95. Soares, M.S.P.; Luduvico, K.P.; Chaves, V.C.; Spohr, L.; Meine, B.D.M.; Lencina, C.L.; Reginatto, F.H.; Spanevello, R.M.; Simões, C.M.O.; Stefanello, F.M. The Protective Action of Rubus Sp. Fruit Extract Against Oxidative Damage in Mice Exposed to Lipopolysaccharide. *Neurochem. Res.* **2021**, *46*, 1129–1140. [CrossRef] [PubMed]
96. Tony, S.K.; Hassan, M.S.; Ismail, H.A.; El-Naem, G.F.A.; Gazwi, H.S.S. Effect of Anthocyanin-Rich Blackberry Juice on Endoplasmic Reticulum Stress in Streptozotocin-Induced Diabetic Rats. *Environ. Sci. Pollut. Res.* **2023**, *30*, 79067–79081. [CrossRef]
97. De Mello, J.E.; Luduvico, K.P.; Dos Santos, A.; Teixeira, F.C.; De Souza Cardoso, J.; De Aguiar, M.S.S.; Cunico, W.; Vizzotto, M.; Stefanello, F.; Spanevello, R. Therapeutic Potential of Blackberry Extract in the Preventing Memory Deficits and Neurochemical Alterations in the Cerebral Cortex, Hippocampus and Cerebellum of a Rat Model with Amnesia. *Metab. Brain Dis.* **2023**, *38*, 1261–1272. [CrossRef] [PubMed]
98. Ghasemi, S.; Dabirian, S.; Kariminejad, F.; Koohi, D.E.; Nemattalab, M.; Majidimoghadam, S.; Zamani, E.; Yousefbeyk, F. Process Optimization for Green Synthesis of Silver Nanoparticles Using Rubus Discolor Leaves Extract and Its Biological Activities against Multi-Drug Resistant Bacteria and Cancer Cells. *Sci. Rep.* **2024**, *14*, 4130. [CrossRef]
99. Tulio, A.Z.; Reese, R.N.; Wyzgoski, F.J.; Rinaldi, P.L.; Fu, R.; Scheerens, J.C.; Miller, A.R. Cyanidin 3-Rutinoside and Cyanidin 3-Xylosylrutinoside as Primary Phenolic Anti-oxidant in Black Raspberry. *J. Agric. Food Chem.* **2008**, *56*, 1880–1888. [CrossRef]
100. Ryu, E.-H.; Yoon, H.-H.; Jung, J.-H. Characteristics of Lactic Acid Fermentation of Black Raspberry Juice Using the Lactobacillus Plantarum GBL17 Strain. *Korean J. Food Cook. Sci.* **2015**, *31*, 773–780. [CrossRef]
101. Parry, J.; Su, L.; Moore, J.; Cheng, Z.; Luther, M.; Rao, J.N.; Wang, J.-Y.; Yu, L.L. Chemical Compositions, Anti-oxidant Capacities, and Antiproliferative Activities of Selected Fruit Seed Flours. *J. Agric. Food Chem.* **2006**, *54*, 3773–3778. [CrossRef] [PubMed]
102. Coates, E.M.; Popa, G.; Gill, C.I.; McCann, M.J.; McDougall, G.J.; Stewart, D.; Rowland, I. Colon-Available Raspberry Polyphenols Exhibit Anti-Cancer Effects on in Vitro Models of Colon Cancer. *J. Carcinog.* **2007**, *6*, 4. [CrossRef] [PubMed]
103. Han, C.; Ding, H.; Casto, B.; Stoner, G.D.; D'Ambrosio, S.M. Inhibition of the Growth of Premalignant and Malignant Human Oral Cell Lines by Extracts and Components of Black Raspberries. *Nutr. Cancer* **2005**, *51*, 207–217. [CrossRef]
104. Chen, T.; Hwang, H.; Rose, M.E.; Nines, R.G.; Stoner, G.D. Chemopreventive Properties of Black Raspberries in *N* - Nitrosomethylbenzylamine-Induced Rat Esophageal Tumorigenesis: Down-Regulation of Cyclooxygenase-2, Inducible Nitric Oxide Synthase, and c- *Jun*. *Cancer Res.* **2006**, *66*, 2853–2859. [CrossRef]
105. Mallery, S.R.; Stoner, G.D.; Larsen, P.E.; Fields, H.W.; Rodrigo, K.A.; Schwartz, S.J.; Tian, Q.; Dai, J.; Mumper, R.J. Formulation and In-Vitro and In-Vivo Evaluation of a Mucoadhesive Gel Containing Freeze Dried Black Raspberries: Implications for Oral Cancer Chemoprevention. *Pharm. Res.* **2007**, *24*, 728–737. [CrossRef] [PubMed]
106. Desai, K.G.H.; Olsen, K.F.; Mallery, S.R.; Stoner, G.D.; Schwendeman, S.P. Formulation and In Vitro-In Vivo Evaluation of Black Raspberry Extract-Loaded PLGA/PLA Injectable Millicylindrical Implants for Sustained Delivery of Chemopreventive Anthocyanins. *Pharm. Res.* **2010**, *27*, 628–643. [CrossRef] [PubMed]
107. Stoner, G.D.; Wang, L.-S.; Sardo, C.; Arnold, M.; Martin, E.; Frankel, W.; Pearl, D. Effects of Black Raspberries on Cellular and Epigenetic Biomarkers of Colon Cancer Development in Humans. In *Berries and Cancer Prevention*; Seeram, N.P., Stoner, G.D., Eds.; Springer: New York, NY, USA, 2011; pp. 281–303. ISBN 978-1-4419-7553-9.
108. Mace, T.A.; King, S.A.; Ameen, Z.; Elnaggar, O.; Young, G.; Riedl, K.M.; Schwartz, S.J.; Clinton, S.K.; Knobloch, T.J.; Weghorst, C.M.; et al. Bioactive Compounds or Metabolites from Black Raspberries Modulate T Lymphocyte Proliferation, Myeloid Cell Differentiation and Jak/STAT Signaling. *Cancer Immunol. Immunother.* **2014**, *63*, 889–900. [CrossRef]

109. Lee, S.J.; Choi, H.R.; Lee, J.-H.; Kwon, J.W.; Lee, H.K.; Jeong, J.T.; Lee, T.-B. Effects of Unripe Black Raspberry Extracts on Prostate Cancer Cell Line and Rat Model of Benign Prostatic Hyperplasia. *J. Korean Soc. Food Sci. Nutr.* **2014**, *43*, 507–515. [CrossRef]
110. Choi, G.J.; Kang, H.; Kim, W.J.; Baek, C.W.; Jung, Y.H.; Woo, Y.C.; Kwon, J.W. Rubus Occidentalis Alleviates Hyperalgesia Induced by Repeated Intramuscular Injection of Acidic Saline in Rats. *BMC Complement. Altern. Med.* **2016**, *16*, 202. [CrossRef]
111. An, J.H.; Kim, D.-L.; Lee, T.-B.; Kim, K.J.; Kim, S.H.; Kim, N.H.; Kim, H.Y.; Choi, D.S.; Kim, S.G. Effect of *Rubus Occidentalis* Extract on Metabolic Parameters in Subjects with Prediabetes: A Proof-of-Concept, Randomized, Double-Blind, Placebo-Controlled Clinical Trial: Effects of *Rubus Occidentalis* in Prediabetes. *Phytother. Res.* **2016**, *30*, 1634–1640. [CrossRef] [PubMed]
112. Bolling, B.W.; DiMarco, D.M.; Lainas, K.; Kranz, S. Raspberry Juice. In *Handbook of Functional Beverages and Human Health*; CRC Press: Boca Raton, FL, USA, 2016; pp. 527–540.
113. Lim, T.; Lee, K.; Kim, R.H.; Cha, K.H.; Koo, S.Y.; Moon, E.C.; Hwang, K.T. Black Raspberry Extract Can Lower Serum LDL Cholesterol via Modulation of Gut Microbial Composition and Serum Bile Acid Profile in Rats Fed Trimethylamine-N-Oxide with a High-Fat Diet. *Food Sci. Biotechnol.* **2022**, *31*, 1041–1051. [CrossRef] [PubMed]
114. Lim, T.; Lee, K.; Kim, R.H.; Ryu, J.; Cha, K.H.; Park, S.Y.; Koo, S.Y.; Hwang, K.T. Effects of Black Raspberry Extract on Gut Microbiota, Microbial Metabolites, and Expressions of the Genes Involved in Cholesterol and Bile Acid Metabolisms in Rats Fed Excessive Choline with a High-Fat Diet. *Food Sci. Biotechnol.* **2023**, *32*, 577–587. [CrossRef] [PubMed]
115. Shamsudin, N.A.; Matawali, A.; Gansau, J.A. Comparison of Anti-oxidant Activity and Phytochemical Content of Borneo Wild Berry, Rubus Fraxinifolius (Rogimot). *Trans. Sci. Technol.* **2019**, *6*, 36–41.
116. Chen, P.; Yan, H.; Mei, Q.; Chen, Y.; Wang, X.; Ming, L.; Fang, Z. Triterpenoids from the Roots and Stems of Rubus Alceaefolius. *Chem. Nat. Compd.* **2016**, *52*, 248–251. [CrossRef]
117. Ali, N.; Shaoib, M.; Shah, S.W.A.; Shah, I.; Shuaib, M. Pharmacological Profile of the Aerial Parts of Rubus Ulmifolius Schott. *BMC Complement. Altern. Med.* **2017**, *17*, 59. [CrossRef]
118. Da Silva, L.P.; Pereira, E.; Pires, T.C.S.P.; Alves, M.J.; Pereira, O.R.; Barros, L.; Ferreira, I.C.F.R. Rubus Ulmifolius Schott Fruits: A Detailed Study of Its Nutritional, Chemical and Bioactive Properties. *Food Res. Int.* **2019**, *119*, 34–43. [CrossRef] [PubMed]
119. Li, F.; Luo, T.; Hou, J.; Fei, T.; Zhang, J.; Wang, L. Natural α-Glucosidase and α-Amylase Inhibitors from Raspberry (*Rubus Corchorifolius* L.) Leaf-Tea: Screening, Identification and Molecular Docking Analysis. *LWT* **2023**, *181*, 114763. [CrossRef]
120. Choi, J.; Lee, K.-T.; Ha, J.; Yun, S.-Y.; Ko, C.-D.; Jung, H.-J.; Park, H.-J. Antinociceptive and Antiinflammatory Effects of Niga-Ichigoside F1 and 23-Hydroxytormentic Acid Obtained from Rubus Coreanus. *Biol. Pharm. Bull.* **2003**, *26*, 1436–1441. [CrossRef]
121. Oberyszyn, T. Effects of Black Raspberries on UV-Induced Cutaneous Inflammation and Tumor Development. In *Berries and Cancer Prevention*; Seeram, N.P., Stoner, G.D., Eds.; Springer: New York, NY, USA, 2011; pp. 131–142. ISBN 978-1-4419-7553-9.
122. Jun, H.; Wen, Q.; Lee, J.H.; Jeun, J.; Lee, H.-J.; Lee, K.-G.; Oh, S.-K.; Lee, S.-J. Effects of Korean Black Raspberry Wines on Hepatic Cholesterol Metabolism and Retinal Vascular Formation in Vitro. *J. Korean Soc. Appl. Biol. Chem.* **2012**, *55*, 249–257. [CrossRef]
123. Shin, J.-S.; Cho, E.-J.; Choi, H.-E.; Seo, J.-H.; An, H.-J.; Park, H.-J.; Cho, Y.-W.; Lee, K.-T. Anti-Inflammatory Effect of a Standardized Triterpenoid-Rich Fraction Isolated from Rubus Coreanus on Dextran Sodium Sulfate-Induced Acute Colitis in Mice and LPS-Induced Macrophages. *J. Ethnopharmacol.* **2014**, *158*, 291–300. [CrossRef] [PubMed]
124. Choi, H.R.; Chung, Y.H.; Yuk, H.-G.; Lee, H.; Jang, H.S.; Kim, Y.; Shin, D. Characterization of Lactobacillus Plantarum Strains Isolated from Black Raspberry and Their Effect on BALB/c Mice Gut Microbiota. *Food Sci. Biotechnol.* **2018**, *27*, 1747–1754. [CrossRef] [PubMed]
125. Cho, J.M.; Chae, J.; Jeong, S.R.; Moon, M.J.; Ha, K.-C.; Kim, S.; Lee, J.H. The Cholesterol-Lowering Effect of Unripe Rubus Coreanus Is Associated with Decreased Oxidized LDL and Apolipoprotein B Levels in Subjects with Borderline-High Cholesterol Levels: A Randomized Controlled Trial. *Lipids Health Dis.* **2020**, *19*, 166. [CrossRef] [PubMed]
126. Chen, L.; Li, M.; Zhou, H.; Liu, Y.; Pang, W.; Ma, T.; Niu, C.; Yang, Z.; Chang, A.K.; Li, X.; et al. Sirtuin1 (SIRT1) Is Involved in the Anticancer Effect of Black Raspberry Anthocyanins in Colorectal Cancer. *Eur. J. Nutr.* **2023**, *62*, 395–406. [CrossRef] [PubMed]
127. Rajendran, N.K.; George, B.P.; Houreld, N.N.; Abrahamse, H. Synthesis of Zinc Oxide Nanoparticles Using Rubus Fairholmianus Root Extract and Their Activity against Pathogenic Bacteria. *Molecules* **2021**, *26*, 3029. [CrossRef] [PubMed]
128. Kumar, T.; Pande, K.; Mathur, A.; Koranga, M.; Pande, B.; Sharma, H. Antimicrobial and Anti-oxidant Profile of Solvent Extracts of Fruits of Rubus Ellipticus. *J. Adv. Sci. Res.* **2020**, *11*, 100–105. Available online: https://www.sciensage.info/index.php/JASR/article/view/562 (accessed on 10 May 2024).
129. George, B.P.; Parimelazhagan, T.; Kumar, Y.T.; Sajeesh, T. Antitumor and Wound Healing Properties of Rubus Ellipticus Smith. *J. Acupunct. Meridian Stud.* **2015**, *8*, 134–141. [CrossRef] [PubMed]
130. Ahmad, M.; Masood, S.; Sultana, S.; Hadda, T.B.; Bader, A.; Zafar, M. Report: Anti-oxidant and Nutraceutical Value of Wild Medicinal Rubus Berries. *Pak. J. Pharm. Sci.* **2015**, *28*, 241–247.
131. Karn, A.; Quasim, M.; Hmar, E.B.; Paul, S.; Sharma, H. An Updated Review of Rubus Ellipticus (an Edible Shrub), Its Bioactive Constituents and Functional Properties. *Sci. Phytochem.* **2022**, *1*, 22–33. [CrossRef]
132. Kewlani, P.; Tiwari, D.; Rawat, S.; Bhatt, I.D. Pharmacological and Phytochemical Potential of *Rubus Ellipticus*: A Wild Edible with Multiple Health Benefits. *J. Pharm. Pharmacol.* **2023**, *75*, 143–161. [CrossRef]
133. Sharmaa, M.; Kaura, J.; Kumar, V.; Sharmaa, K. Nutraceutical Potential of Rubus Ellipticus: A Critical Review on Phytochemical Potential, Health Benefits, and Utilization. *Think India J.* **2019**, *22*, 878–898.

134. Nguelefack, T.B.; Mbakam, F.H.K.; Tapondjou, L.A.; Watcho, P.; Nguelefack-Mbuyo, E.P.; Ponou, B.K.; Kamanyi, A.; Park, H.-J. A Dimeric Triterpenoid Glycoside and Flavonoid Glycosides with Free Radical-Scavenging Activity Isolated from Rubus Rigidus Var. Camerunensis. *Arch. Pharm. Res.* **2011**, *34*, 543–550. [CrossRef] [PubMed]
135. Whaley, A.K.; Ponkratova, A.O.; Orlova, A.A.; Serebryakov, E.B.; Smirnov, S.N.; Proksh, P.; Ionov, N.S.; Poroikov, V.V.; Luzhanin, V.G. Phytochemical Analysis of Polyphenol Secondary Metabolites in Cloudberry (*Rubus Chamaemorus* L.) Leaves. *Pharm. Chem. J.* **2021**, *55*, 253–258. [CrossRef]
136. Grabek-Lejko, D.; Miłek, M.; Sidor, E.; Puchalski, C.; Dżugan, M. Antiviral and Antibacterial Effect of Honey Enriched with Rubus Spp. as a Functional Food with Enhanced Anti-oxidant Properties. *Molecules* **2022**, *27*, 4859. [CrossRef] [PubMed]
137. Saini, R.; Dangwal, K.; Singh, H.; Garg, V. Anti-oxidant and Antiproliferative Activities of Phenolics Isolated from Fruits of Himalayan Yellow Raspberry (Rubus Ellipticus). *J. Food Sci. Technol.* **2014**, *51*, 3369–3375. [CrossRef]
138. Gao, J.; Sun, C.; Yang, J.; Shi, J.; Du, Y.; Zhang, Y.; Li, J.; Wan, H. Evaluation of the Hepatoprotective and Anti-oxidant Activities of *Rubus Parvifolius* L. *J. Zhejiang Univ. Sci. B* **2011**, *12*, 135–142. [CrossRef]
139. Bezmaternykh, K.V.; Shirshova, T.I.; Beshlei, I.V.; Matistov, N.V.; Smirnova, G.V.; Oktyabr'skii, O.N.; Volodin, V.V. Anti-oxidant Activity of Extracts from *Allium Schoenoprasum* L. and *Rubus Chamaemorus* L. Growing in the Komi Republic. *Pharm. Chem. J.* **2014**, *48*, 116–120. [CrossRef]
140. Zhang, X.; Xu, X.; Gao, R.; Xu, J. *Rubus Parvifolius* L. Inhibited the Growth of Leukemia K562 Cells in Vitro and in Vivo. *Chin. J. Integr. Med.* **2014**, *20*, 36–42. [CrossRef] [PubMed]
141. Xu, X.; Cheng, R.; Zhang, X.; Gao, R. Total Saponins of *Rubus Parvifolius* L. Exhibited Anti-Leukemia Effect in Vivo through STAT3 and eIF4E Signaling Pathways. *Chin. J. Integr. Med.* **2018**, *24*, 920–924. [CrossRef]
142. Dhatwalia, J.; Kumari, A.; Chauhan, A.; Batoo, K.M.; Banerjee, A.; Radhakrishnan, A.; Thakur, S.; Guleria, I.; Lal, S.; Ghotekar, S.; et al. Rubus Ellipticus Fruits Extract-Mediated Cuprous Oxide Nanoparticles: In Vitro Anti-oxidant, Anti-oxidant, Antimicrobial, and Toxicity Study. *Chem. Pap.* **2023**, *77*, 1377–1393. [CrossRef]
143. Bhatt, S.C.; Naik, B.; Kumar, V.; Gupta, A.K.; Kumar, S.; Preet, M.S.; Sharma, N.; Rustagi, S. Untapped Potential of Non-Conventional Rubus Species: Bioactivity, Nutrition, and Livelihood Opportunities. *Plant Methods* **2023**, *19*, 114. [CrossRef] [PubMed]
144. Łuczaj, Ł.; Pieroni, A.; Tardío, J.; Pardo-de-Santayana, M.; Sõukand, R.; Svanberg, I.; Kalle, R. Wild Food Plant Use in 21st Century Europe: The Disappearance of Old Traditions and the Search for New Cuisines Involving Wild Edibles. *Acta Soc. Bot. Pol.* **2012**, *81*, 359–370. [CrossRef]
145. Svanberg, I.; Ståhlberg, S. Wild European Dewberry, *Rubus Caesius* L. (Fam. Rosaceae), in Sweden: From Traditional Regional Consumption to Exotic Dessert at the Nobel Prize Banquet. *J. Ethn. Food* **2021**, *8*, 37. [CrossRef]

MDPI AG
Grosspeteranlage 5
4052 Basel
Switzerland
Tel.: +41 61 683 77 34

Agriculture Editorial Office
E-mail: agriculture@mdpi.com
www.mdpi.com/journal/agriculture

www.ingramcontent.com/pod-product-compliance
Lightning Source LLC
Chambersburg PA
CBHW061559220926
51850CB00005B/374
* 9 7 8 3 7 2 5 8 5 3 1 5 1 *